AF684781

33033
TEXAS A & M
BOOKSTORE
10$44.95

FILTER THEORY AND DESIGN: ACTIVE AND PASSIVE

Circuit Analysis with Computer Application to Problem Solving
Gupta, Bayless, and Peikari

Dynamics of Physical Circuits and Systems
Lindsay and Katz

Filter Theory and Design: Active and Passive
Sedra and Brackett

Introduction to Linear Systems Analysis
Swisher

Linear Dynamic Systems
Lewis

Linear Systems Control
Sage

MATRIX SERIES
IN CIRCUITS AND SYSTEMS

Andrew P. Sage, *Editor*

FILTER THEORY AND DESIGN: ACTIVE AND PASSIVE

ADEL S. SEDRA

Department of Electrical Engineering
University of Toronto

PETER O. BRACKETT

Department of Electrical Engineering
Queen's University

MATRIX PUBLISHERS, INC., Beaverton, Oregon

To Doris and Mary

© COPYRIGHT, MATRIX PUBLISHERS, INC. 1978

All rights reserved. No part of this book may be reproduced or utilized in any form or by any means, electronic or mechanical, including photocopying, recording or by any information storage and retrieval system, without written permission from the publisher.

10 9 8 7 6 5 4 3

Library of Congress catalog card number: 76-39745

Matrix Publishers, Inc.
Beaverton, OR 97005

ISBN: 0-916469-14-2

This book was prepared by typewriter composition.
Illustrations by Scientific Illustrators.
In-house Editor was Merl K. Miller.
Printing and Binding by Pantagraph Printing, Bloomington Illinois.

CONTENTS

4 THE EQUIRIPPLE PASSBAND

5 ITERATIVE STOPBAND APPROXIMATION

6 FREQUENCY AND NETWORK TRANSFORMATIONS

7 THE DESIGN OF REACTIVE LADDER FILTERS

8 ACTIVE FILTERS–FUNDAMENTALS

9 SECOND ORDER SECTIONS–THE BIQUADS

10 CASCADE AND COUPLED BIQUAD STRUCTURES

11 COMPONENT SIMULATION OF LC LADDERS

12 OPERATIONAL SIMULATION OF LC LADDERS

PREFACE

The purpose of this book is to present a unified and combined treatment of the two phases of filter design; namely, Approximation and Realization. It is written both for the student and the design engineer. Although a prior course in network analysis and synthesis, of the classical type, would be beneficial it is not essential.

Filter design is one of these rare areas that require a deep understanding of topics that range from the highly theoretical to the extremely practical. Although there are texts on various aspects of the filter design problem such as approximation theory, network synthesis, active networks etc. we felt that the need existed for a book that "tells the complete story". Our objective therefore was to take the reader through the entire process starting from the Specification Phase, where practical system design considerations play a major role, through the Approximation Phase, which is almost entirely a mathematical exercise, and finally to the Realization Phase, which relies both on highly developed theories of network synthesis as well as on practical knowledge of "real life" components and their imperfections.

In addition to the obvious usefulness of this approach to the design engineer we have found it of equal benefit to the student. At present the book serves as a text for a sequence of two one-term courses at the University of Toronto. The first course is offered to both senior-year and graduate students while the second course is taken primarily by graduate students. Graduate students in these courses come from a variety of research areas such as circuit theory, communications, biomedical engineering, etc., as well as engineers from industry.

In order to achieve our goal, as stated in the above, and still keep the book a reasonable size, a number of decisions were made. The treatment of the approximation problem is quite general, however attention is focussed on the approximation for filters realized using passive LC and active RC technologies. A short section outlines the application of approximation methods to discrete-time filters. Also, although the formulation is general, the approximation problem is solved only for filters which are specified to meet attenuation requirements.

Concerning the realization technologies, it was decided to concentrate on Active-RC, especially those circuits that use the Integrated Circuit (IC) Operational Amplifier (Op Amp). However, a concise and practical treatment of LC design methods was deemed beneficial since many LC filters continue to be designed and manufactured every day. Also, two of the major methods for active filter design start from a passive LC prototype.

In the active filter area, we consider in detail the design of high-order (greater than two) filters. However, since biquadratic sections are useful in their own right as well as building blocks in higher order filter structures, a thorough treatment of biquads is presented. In all cases, only circuits that have proven practical are considered. This eliminated all methods that do not use op amps.

Throughout the text we make frequent reference to the FILTOR 2 package of filter design programs. This is a set of over 100 Fortran IV routines that have been developed, at the University of Toronto, based on the material presented here, and will shortly be made generally available. Although the book can be used for teaching without the programs, we have found it highly beneficial to assign problems and design projects that require the use of the programs. The programs are described in a separate manual which will become available in the near future.

We now present an overview of the material contained in this book. Chapter 1 introduces the reader to the filter design problem. The problem is defined and background information is presented. This chapter also serves to establish terminology and notation. Chapter 2 is an introduction to the approximation part of the book. Chapters 3 to 5 are concerned with the solution of the approximation problem: Chapter 3 deals with filters having a maximally-flat passband attenuation, while Chapter 4 deals with the equiripple passband case. Chapter 5 completes the approximation phase for attenuation specifications by presenting an iterative stopband approximation method.

Chapter 6 presents various frequency and network transformations. The material presented enables the design of classical filters using closed-form expressions and/or filter tables. This chapter concludes with an important section dealing with the bilinear z-transform. There we show how the material on approximation may be directly applied to the design of discrete-time filters of the recursive type. These include both digital filters and filters realized using Charge Transfer Devices (CTD's).

The design of passive LC ladder filters, as well as a study of their sensitivity properties, is the subject of Chapter 7.

In Chapter 8 we present the basic concepts and necessary background knowledge for the design of active filters. This chapter also serves as a concise overview of the active filter part. Chapter 9 deals with the theory and design of second-order active filters, or biquads. In an introductory filters course, Chapters 8 and 9 contain sufficient information on the active filters area.

Each of the three remaining chapters deals with a particular method for the design of high-order active filters. Chapter 10 is concerned with the cascade method of design. A particular coupled-biquad structure is also presented. Design methods based on the simulation of LC ladders are studied in Chapters 11 and 12. Methods which essentially replace the ladder inductors by simulated inductances are considered in Chapter 11. The other approach to ladder simulation, namely, simulating the internal working or the operation of the ladder network is the subject of Chapter 12.

There are many ways for teaching from this book or for using it for self study. In fact we have attempted to present the material in a manner that allows the reader to study the part that interests him without having to read the entire book. For instance, most of the material on approximation could be skipped by a reader specifically interested in realization. Also, it is not essential to study the details of passive design in order to study active filters.

A specific package which we have found quite successful for a one-term (40 lecture hours) course at the senior/first year graduate level is as follows: Chapters 1, 2, 3, 4 (excluding Sections 4.3 and 4.4), 5 (a light treatment), 6 (excluding Sections 6.7 and 6.8), 8 and 9. The students in this course are assigned problems selected from those given in the book in addition to few design problems requiring the use of the programs. A follow-up course intended for graduate students covers the rest of the book and is highly design oriented. Here again extensive use is made of the programs in design projects.

We would like to express our sincere thanks and gratitude to a number of people who contributed in a variety of ways to the development of this work. Professor K. C. Smith provided encouragement throughout the course of this undertaking. Martin Snelgrove has been the principal contributor to the development of the filter design programs. He has been assisted in this task by Moez Dharssi. Ken Martin's research strongly influenced the material of the last two chapters. Les Brown's work provided some of the material on two-amplifier biquads in Chapter 9. Needless to say, the research of almost all of the graduate students who worked with the first author during the last few years helped in the development of the material on active filters. The second author is grateful to the management of ESE Ltd., Rexdale, Ont. for providing him with the opportunity to work on filter design during his tenure as chief development engineer. Elisa Krissilas, Maggie Byrne and Sarah Cherian participated in typing the various drafts. The final copy was most ably typed by Joan Lidkie who deserves a special note of thanks. Our research in the filter design area has been supported by the National Research Council of Canada through research grants (A.S.) and fellowships (P.B.). Last, but not least, special thanks go to our families for their patience and understanding.

A. S. Sedra
P. O. Brackett

Toronto, Canada
April, 1978

CHAPTER 1

INTRODUCTION

This book is mainly concerned with the design of a certain class of linear systems which are usually referred to as Electric Wave Filters. In the broadest sense, a filter may be considered to be a signal processing device which operates on an input signal to produce an output signal bearing a prescribed relationship to the input signal. Design of a linear system to satisfy such an objective is not, generally, an easy task. This book is devoted to the design of a special class of such devices, for which the input-output relationship may be described in terms of the ability of the system to "pass" a certain class of signals having a specified spectrum, and to "stop" a certain class of signals having a different (and usually disjoint) spectrum. Such filters are frequently called "frequency selective networks". We shall make these ideas more precise in the following sections.

Few proofs are included in this introductory chapter, and the interested reader is referred to the selected references listed at the end of the chapter.

1.1 LINEAR SYSTEMS

It is assumed that the reader has a background in network analysis and a working knowledge of elementary linear system theory. The following material is presented as a brief review, and for the purpose of establishing notation. Fig. 1.1 illustrates the "black box" concept of a linear system. The system is characterized in the time domain by its impulse response g(t), which is the output signal y(t) produced in response to a unit impulse $\delta(t)$, applied at the input. For an arbitrary input signal u(t), the output signal y(t) is given by the well known convolution integral

$$y(t) = \int_{-\infty}^{\infty} g(t - \tau)\, u(\tau)\, d\tau\ . \tag{1.1}$$

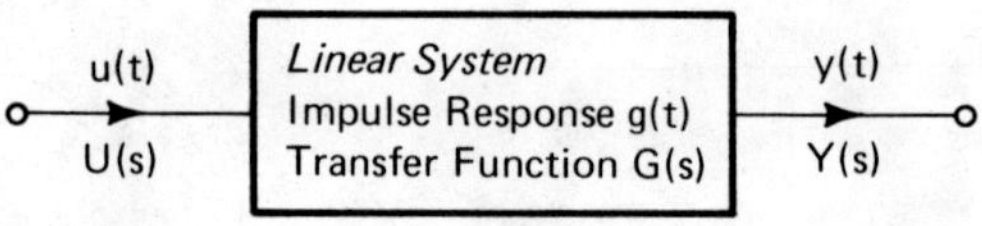

Figure 1.1 Black box representation of a linear system.

In the usual manner, we denote the Laplace transforms of the time signals by the use of upper case letters. Thus we have the system transfer function G(s), which is the transform of g(t), defined by

$$G(s) = \int_{0}^{\infty} g(t)\, e^{-st}\, dt,$$

with the inverse transform given by

$$g(t) = \frac{1}{2\pi j} \int_{-\infty}^{\infty} G(s)\, e^{st}\, ds$$

where $s = \sigma + j\omega$ is the complex frequency variable. Under the Laplace transformation (1.1) becomes simply

$$Y(s) = G(s)\ U(s)$$

so that the transfer function G(s) is the ratio of the output variable to the input variable,

$$G(s) = \frac{Y(s)}{U(s)} = \frac{\text{output variable}}{\text{input variable}} \tag{1.2}$$

The most general types of linear systems we shall consider are those consisting of a finite number of lumped, linear and time-invariant elements. The system is characterized by an Nth order ordinary linear differential equation which results, in the most general case, in a transfer function G(s) which is a real rational polynomial function of the complex variable s. Thus we may write G(s) in general as

$$G(s) = \frac{P(s)}{E(s)}$$

$$= \frac{p_M s^M + p_{M-1} s^{M-1} + \ldots + p_o}{e_N s^N + e_{N-1} s^{N-1} + \ldots + e_o} = \frac{\sum_{k=o}^{M} p_k s^k}{\sum_{k=o}^{N} e_k s^k} \tag{1.3}$$

where p_k and e_k are real numbers so that G(s) is real for real s, and the roots of the polynomials P(s) and E(s) must either be real or occur in conjugate pairs.

In general, for the kinds of systems we shall consider, the degree of the numerator (deg. [P(s)] = M) is less than or equal to the degree of the denominator (deg. [E(s)] = N), and E(s) is a strictly Hurwitz polynomial (all roots in the open left-half s-plane)†. The denominator polynomial E(s) is variously known as

†A "closed" half of the s-plane includes the $j\omega$-axis while an "open" half excludes the $j\omega$-axis.

the characteristic polynomial or natural mode polynomial of the linear system, and the degree of E(s), that is N, is the order or degree of the system.

The transfer function G(s) may be split into even and odd parts as

$$G(s) = G_{ev}(s) + G_{od}(s)$$

where

$$G_{ev}(s) = \frac{1}{2}\,[G(s) + G(-s)]$$

is the even part of G(s), so that $G_{ev}(s) = G_{ev}(-s)$, while

$$G_{od}(s) = \frac{1}{2}\,[G(s) - G(-s)]$$

is the odd part of G(s), so that $G_{od}(s) = -G_{od}(-s)$.

We also note that

$$G(s)\,G(-s) = [G_{ev}^2(s) - G_{od}^2(s)]$$

The function G(s) G(-s) is usually called the "squared" transfer function.

We may obtain the physical frequency behaviour of G(s) by letting $s = j\omega$ and evaluating G(s) along the imaginary axis. Thus,

$$G(s)\Big|_{s=j\omega} = G(j\omega) = \frac{P(j\omega)}{E(j\omega)}$$

In general $G(j\omega)$ is a complex function and we see that,

$$R(\omega) = \frac{1}{2}\,[G(j\omega) + G(-j\omega)] = G_{ev}(j\omega)$$

is the real part of $G(j\omega)$ while

$$X(\omega) = \frac{1}{2j}\,[G(j\omega) - G(-j\omega)] = \frac{1}{j}G_{od}(j\omega)$$

is the imaginary part of $G(j\omega)$. Thus we may write $G(j\omega)$ in terms of its real and imaginary parts as

$$G(j\omega) = R(\omega) + j\ X(\omega).$$

We note that $R(\omega)$ is an even function of ω while $X(\omega)$ is an odd function of ω.

Alternatively, $G(j\omega)$ may be expressed in polar form as

$$G(j\omega) = |G(j\omega)|\ e^{-j\phi(\omega)} \tag{1.4}$$

where $|G(j\omega)|$ is the magnitude of $G(j\omega)$ with the magnitude squared given by

$$|G(j\omega)|^2 = R^2(\omega) + X^2(\omega)$$

or

$$|G(j\omega)|^2 = G(j\omega)\ G(-j\omega) = G(s)\ G(-s)\ |_{s=j\omega} \tag{1.5}$$

which is even function of ω, and $\phi(\omega)$ is the phase lag of $G(j\omega)$

$$\phi(\omega) = \tan^{-1}\left[\frac{X(\omega)}{R(\omega)}\right] \tag{1.6}$$

which is an odd function of ω.

Considered as a mathematical function, the roots of the numerator polynomial of $G(s)$, that is the roots of $P(s)$, are the zeros of $G(s)$, while the roots of the denominator polynomial $E(s)$ are the poles of $G(s)$. In system terminology (as distinct from filter terminology, to be introduced in the following sections), the roots of $P(s)$ are the transmission zeros of the system while the roots of $E(s)$ are the natural modes or transmission poles of the system.

As the reader will find, the single most important mathematical operation used in filter design is the factoring of higher order polynomials. Indeed it was not until the advent of the modern digital computer, and hence the ability to rapidly and accurately find the roots of high degree polynomials, that modern filter design became a reality. By factoring P(s) and E(s), we may express G(s) as a product of factors containing the transmission zeros s_{zk} and the natural modes s_{nk} as

$$G(s) = C \frac{\prod_{k=1}^{M} (s - s_{zk})}{\prod_{k=1}^{N} (s - s_{nk})} = \frac{P(s)}{E(s)} \qquad (1.7)$$

where C is a constant extracted from G(s) such that E(s) and P(s) become monic polynomials (leading coefficients equal to unity). We shall adopt the usual convention here of saying that there are n_i = N-M transmission zeros at infinity.

There exist several well known numerical techniques for factoring high order polynomials. Among these methods, Mueller's method or the "method of quadratic interpolation"[1] is perhaps the most reliable general purpose technique. A FORTRAN subroutine called MULLER, which implements this root-finding algorithm, is one of the basic routines of the FILTOR-2 package[2].

1.2 FILTERS AS A SPECIAL CLASS OF LINEAR SYSTEMS

The reader is probably aware that analysis of linear systems is an exact process, which may be accomplished mathematically by solving the equations that describe the network. In contrast, the design of linear systems is generally an inexact process. The design problem may be generally stated as follows. Given an input signal u(t) and a desired output signal y(t), find first the impulse response g(t), such that (1.1) is satisfied, and second, a physical system, that is a real network, which has an impulse response identical to g(t). Stated in this way, the network design problem is one of time-domain synthesis.

We note at the outset that there may not be a lumped linear time-invariant system which has the required impulse response g(t). Thus we must seek a function, say g*(t), which corresponds to a class of linear system that we know exists, such that $g^*(t)$ approximates $g(t)$ in some way. This illustrates the inexactness of the system design process. If we are satisfied that $g^*(t)$ is close enough to $g(t)$ we must then synthesize an electrical network which realizes $g^*(t)$.

From the above, it follows that the system design process can be split into two parts or phases. Phase (1), called the approximation phase, consists of finding a "realizable" impulse response which approximates the required response, while phase (2), the realization phase, deals with the synthesis of a physical network which realizes the obtained impulse response. Accordingly, this book is divided into two parts. The first part develops the necessary theory and presents methods for solving the approximation problems, while the second part provides techniques for designing practical network realizations. The material presented applies only to a restricted (yet still quite extensive) class of systems, and although we have described the system design problem in the time-domain, this book deals exclusively with the design of filters specified in terms of their frequency domain characteristics. (Those interested in the time-domain synthesis problem are referred to[3]). The subsequent sections of this chapter will clearly delineate the classes of linear systems with which we will be concerned.

At this juncture, it should be mentioned that it is theoretically possible to solve the linear system design problem by assuming a network structure and adjusting (mathematically) the element values of the network to minimize some previously defined error measure[4]. Such optimization techniques are generally much more difficult and expensive to perform than the more direct methods described in this book.

Most of this book is concerned with a special category of the general system design problem, termed "filter design". In particular, for these networks, both approximation and realization can be accomplished in the frequency domain. We should like to point out the special nature of the filter design problem by introducing the dichotomy of Fig. 1.2. Here we separate linear systems into those whose purpose is to pass, undistorted, signals with a spectrum which lies only in a designated region (the passband) of the physical frequency ($j\omega$) axis while attenuating all others, and those whose purpose is to provide a specified output for a given input. The former, henceforth simply referred to as filters, are frequency selective networks, while the latter are spectral shaping networks. Although one might argue that filters are a special class of spectral shaping networks, their separation from the latter is quite profitable. Indeed, it is this separability that makes possible many of the relatively simple design techniques to be presented.

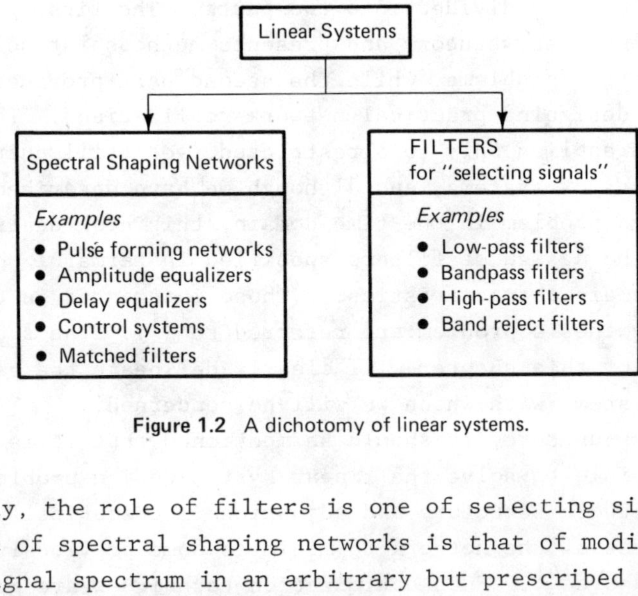

Figure 1.2 A dichotomy of linear systems.

Clearly, the role of filters is one of selecting signals while the role of spectral shaping networks is that of modifying the input signal spectrum in an arbitrary but prescribed manner. Specifically, we desire that a filter should do as little shaping as possible on signals in its passband; any shaping is considered a distortion of the signal. On the other hand networks which perform pulse forming fall within the spectral shaping category.

Another notable example of spectral shaping networks is that of equalizers. Fig. 1.3 illustrates the concept of a special kind of equalizer; namely, amplitude equalizers. Curve (a) is the measured transfer function magnitude, expressed in decibels (dB), of a transmission system. We wish to have the transmission characteristic of this system modified to that indicated by Curve (c). It follows that a network having the transfer function magnitude of Curve (b), when connected in tandom with the transmission system, will result in an overall magnitude function identical to that of Curve (c). Such a network is an amplitude equalizer. In general, the design of filters is much simpler than the design of equalizers which usually entails an optimization process.

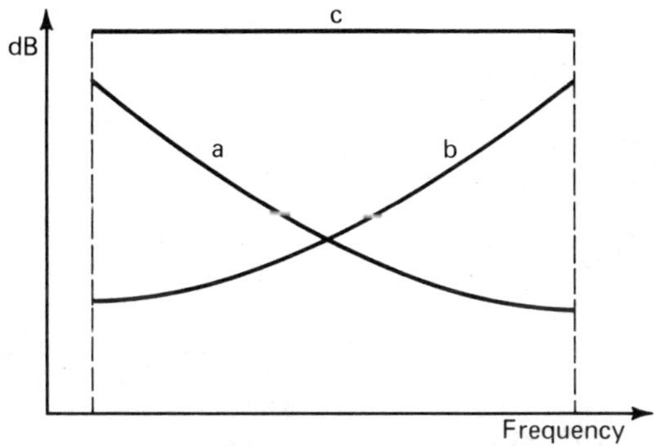

Figure 1.3 Illustrating the concept of amplitude equalization.

1.3 ATTENUATION, PHASE AND DELAY FUNCTIONS

A considerable amount of terminology and jargon has grown around the area of filter design and we now introduce the reader to some of these special terms and definitions. One of the most confusing features of the filter literature is the use of the term "transfer function", or some similar term, for the inverse of the system function G(s) introduced in Section 1.1. The transfer function which we designate by H(s) where

$$H(s) = \frac{1}{G(s)} = \frac{E(s)}{P(s)} = \frac{\text{input variable}}{\text{output variable}} \tag{1.8}$$

is used almost universally by those who specify and design filters. The result of this is that the roles of the poles and zeros of G(s) become reversed for H(s). Note that while G(s) is a measure of the "transmission" through the system, H(s) is a measure of the "loss" through the system. For this reason the roots of P(s) or "transmission zeros" are also known as the "loss poles". However, the term "natural modes" is usually kept for the roots of E(s), although some authors refer to these as simply "the poles".

The "effective transmission constant" Θ is defined by

$$H(s) = e^{\Theta} \tag{1.9}$$

so that

$$\Theta = \ln H(s) \tag{1.10}$$

For $s = j\omega$, we may write

$$\Theta(j\omega) = \alpha(\omega) + j\phi(\omega) \tag{1.11}$$

whereby using (1.4) and (1.8),

$$\alpha(\omega) = \ln \mid H(j\omega) \mid = - \ln \mid G(j\omega) \mid \tag{1.12}$$

is the attenuation function in Nepers, while $\phi(\omega)$ is the phase lag function already defined in (1.4). Equation (1.12) suggests that the transmission zeros may also be called "attenuation poles". It is common practice to express the filter attenuation in decibels through the use of the function

$$A(\omega) = 20 \log \mid H(j\omega) \mid \simeq 8.686\, \alpha(\omega) \quad \text{dB}$$

As will be shown in Section (1.6), if one is to completely specify a distortionless passband for a filter, then both the magnitude $\mid H(j\omega) \mid$ and phase $\phi(\omega)$ of the transfer function have to be simultaneously specified. We note here that the phase lag of a linear system is given by a trancendental function; namely, $\tan^{-1}$, and is thus more difficult to evaluate and manipulate analytically than $\mid H(j\omega) \mid^2$, which is a rational polynomial function.

In addition, the measurement of phase lag is a difficult and not very accurate process. For these reasons, filter designers do not work with the phase lag function $\phi(\omega)$, but with the derivative of the phase lag. This function, designated by $\tau(\omega)$, is defined by

$$\tau(\omega) = \frac{d\phi(\omega)}{d\omega} \tag{1.13}$$

and is termed the group delay or envelope delay or just simply delay, and has units of seconds. It is easily shown that $\tau(\omega)$ is a rational polynomial function since

$$\tau(\omega) = \frac{d}{d\omega} \tan^{-1} \left[\frac{-X(\omega)}{R(\omega)}\right]$$

$$= \frac{-R(\omega)\ (dX(\omega)/d\omega) + X(\omega)\ (dR(\omega)/d\omega)}{R^2(\omega) + X^2(\omega)}$$

is free of trancendental functions. Note also that the denominator of $\tau(\omega)$ is identical to $|G(j\omega)|^2 = 1/|H(j\omega)|^2$, and that since both numerator and denominator are even functions, then $\tau(\omega)$ is an even function of ω. Using analytic continuation, that is, substituting $\omega = \frac{s}{j}$ in the $\tau(\omega)$ expression, we obtain $\tau(s)$, which is a real rational function in s^2.

As mentioned in the above, it is rather difficult to accurately measure $\phi(\omega)$. On the other hand, there exists a number of delay measuring sets which accurately determine, by differential means, the quantity $(\frac{\Delta\phi}{\Delta\omega})$. This provides a close approximation to $\tau(\omega)$. The conveniences intrinsic in $\tau(\omega)$ make this function a natural substitute for the phase function $\phi(\omega)$ in specifying filters.

The phase function $\phi(\omega)$ may be expressed in terms of $\tau(\omega)$ as

$$\phi(\omega) = \int_0^{\omega} \tau(x)\ dx + \phi(0) \tag{1.14}$$

Evaluating (1.14) as $\omega \to \infty$ results in

$$\phi(\infty) = \int_0^\infty \tau(x)\,dx + \phi(0).$$

Thus, we have

$$\int_0^\infty \tau(x)\,dx = [\phi(\infty) - \phi(0)], \tag{1.15}$$

indicating that the area under the delay function, corresponding to a given transfer function (1.3), is a fixed constant dependent only upon the phase asymptotes and not on the particular values of the transfer function coefficients. These concepts will be further illustrated by the following example.

Example 1.1
Consider a second order transfer function

$$H(s) = s^2 + e_1 s + e_0$$

It follows that the loss-poles are at infinity and that $N = 2$, $M = 0$. The natural modes are the roots of H(s), which is a polynomial here. It is convenient, for reasons that will become clear at a later stage, to express H(s) in the form

$$H(s) = s^2 + s\left(\frac{\omega_o}{Q}\right) + \omega_o^2 \tag{1.16}$$

where ω_o is called the natural-mode frequency and Q is the natural-mode Q-factor. As mentioned before, the natural modes are sometimes referred to as the poles and correspondingly ω_o and Q are called the pole-frequency and the pole-Q, respectively. For the natural modes in (1.16) to be complex, Q has to be greater than 0.5. Fig. 1.4 shows the possible location of the pair of complex conjugate natural modes in the s-plane. Note that ω_o determines the radial distance of the natural modes from the origin, while Q determines their distance to the jω-axis; the higher the Q, the closer the natural modes are to the jω-axis. As $Q \to \infty$, the natural modes lie on the $j\omega$-axis. For physical frequencies, the transfer function is given by

$$H(j\omega) = (\omega_o^2 - \omega^2) + j\frac{\omega\,\omega_o}{Q} \qquad (1.17)$$

Thus,

$$|\,H(j\omega)\,|^2 = (\omega_o^2 - \omega^2)^2 + \frac{\omega^2\,\omega_o^2}{Q^2}\ .$$

It is convenient here to normalize the magnitude squared function such that $|H(j0)| = 1$.

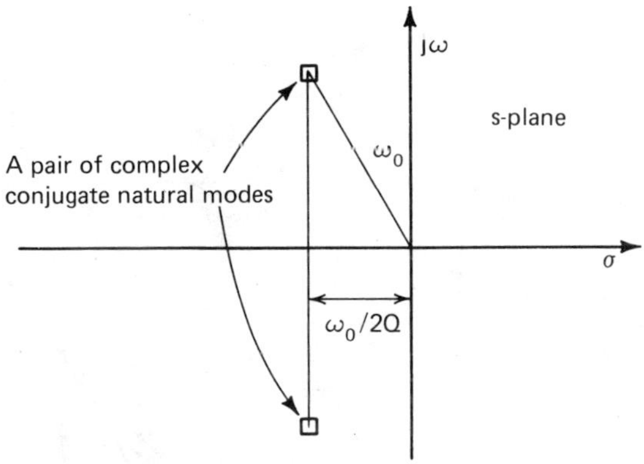

Figure 1.4 The natural modes of a second-order filter.

This results in the modified function

$$|\,H(j\omega)\,|^2 = [1 - (\frac{\omega}{\omega_o})^2]^2 + (\frac{\omega}{\omega_o})^2\ \frac{1}{Q^2}$$

Normalizing the frequency variable relative to ω_o, the magnitude-squared function can be expressed in terms of $\Omega = \frac{\omega}{\omega_o}$ as,

$$|\,H(j\Omega)\,|^2 = (1 - \Omega^2)^2 + \frac{\Omega^2}{Q^2}$$

The attenuation function A(ω), or A(Ω) in terms of the normalized frequency, is given by

$$A(\Omega) = 10 \log_{10} [(1 - \Omega^2)^2 + \frac{\Omega^2}{Q^2}]$$

Fig. 1.5 shows plots for the attenuation A versus the normalized frequency Ω for various values of Q. Note that for Q values less than $1/\sqrt{2}$, the attenuation has no peak. The case of $Q = 1/\sqrt{2}$ is therefore referred to as the "maximally flat magnitude" case. For large values of Q, the peak occurs at $\Omega \simeq 1$.

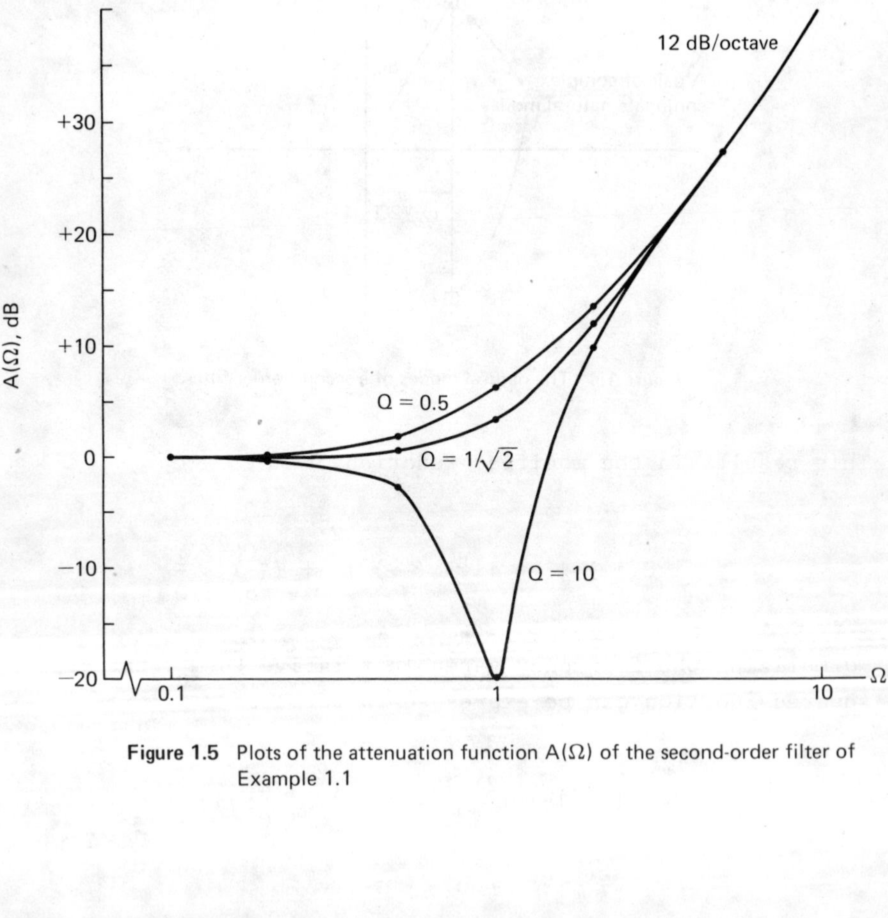

Figure 1.5 Plots of the attenuation function A(Ω) of the second-order filter of Example 1.1

The phase function $\phi(\omega)$ is obtained from (1.17)

$$\phi(\omega) = \tan^{-1} \frac{\omega\,\omega_o}{Q(\omega_o^2 - \omega^2)} \tag{1.18}$$

Thus,

$$\phi(0) = 0 \text{ and } \phi(\infty) = \pi$$

The delay function $\tau(\omega)$ is obtained as

$$\tau(\omega) = \frac{d\phi(\omega)}{d\omega}$$

$$\tau(\omega) = \left(\frac{\omega_o}{Q}\right) \frac{\omega^2 + \omega_o^2}{(\omega_o^2 - \omega^2)^2 + \dfrac{\omega^2\,\omega_o^2}{Q^2}} \tag{1.19}$$

The delay function $\tau(\omega)$ in (1.19) may be multiplied by the natural-mode frequency ω_o to obtain a normalized delay function $\tau_n(\omega)$, which is dimensionless and a function of the normalized frequency Ω; that is,

$$\tau_n(\omega) \equiv \omega_o\,\tau(\omega)$$

$$\tau_n(\Omega) = \frac{1}{Q} \frac{\Omega^2 + 1}{(1-\Omega^2)^2 + \dfrac{\Omega^2}{Q^2}} \tag{1.20}$$

This relation enables us to obtain universal curves for the delay associated with a pair of natural modes. Such plots are shown in Fig. 1.6 for various values of Q. Note that as the Q is increased, the delay becomes more "peaky". Thus a highly selective attenuation characteristic has an associated "peaky" delay response.

The area under the delay response curve must, however, remain constant, equal to π, for varying values of Q.

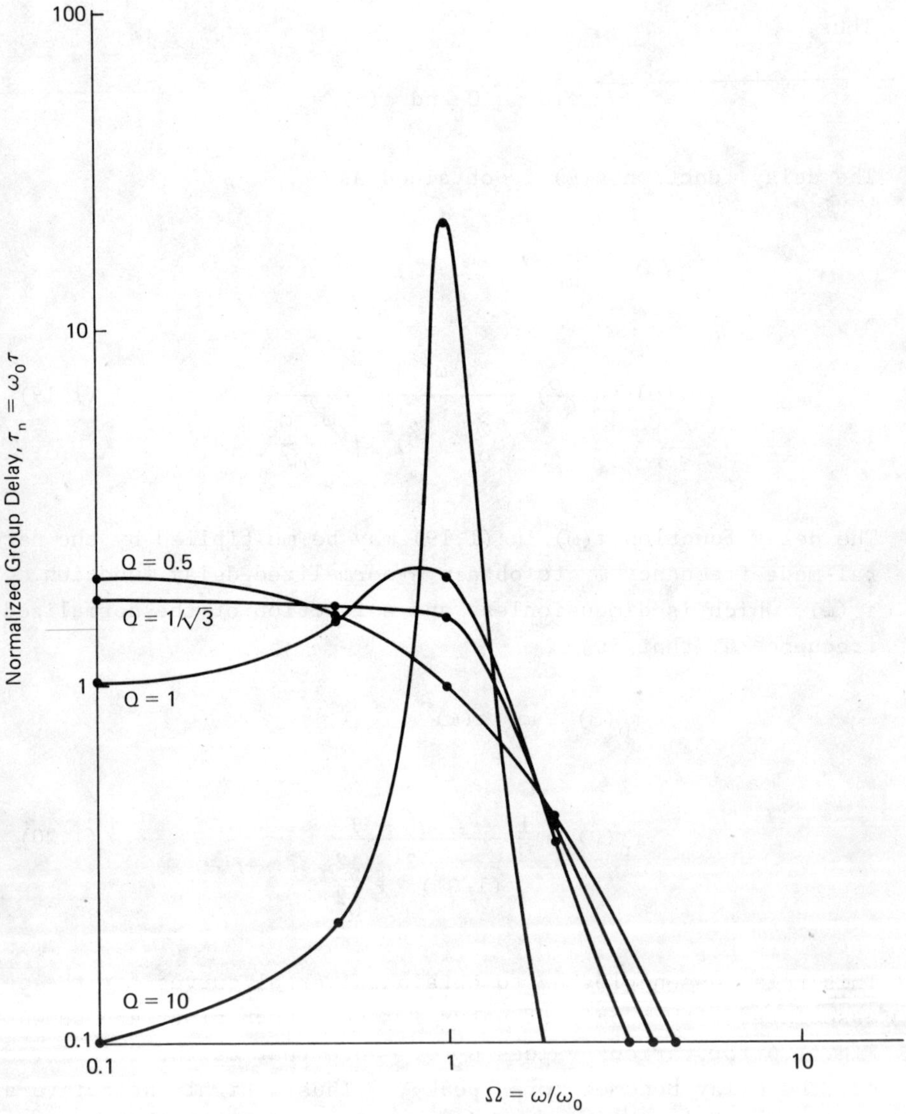

Figure 1.6 Group delay associated with a pair of natural modes.

From equation (1.20), the frequency at which the delay peaks is obtained as

$$\Omega = \sqrt{-1 + \sqrt{4 - \frac{1}{Q^2}}}$$

Thus the lowest Q for which a peak occurs is

$$Q = \frac{1}{\sqrt{3}}$$

The delay characteristic corresponding to this value of Q is included in Fig. 1.6. The Maximally-flat delay characteristic is therefore obtained for $Q = \frac{1}{\sqrt{3}}$.

Before leaving this example, we note the general resemblance between the delay characteristics and attenuation characteristics associated with a pair of natural modes. If the natural modes lie on the jω-axis, $Q \rightarrow \infty$, the attenuation reaches $-\infty$ while the group delay becomes infinite, at $\omega = \omega_o$. Thus, for a natural mode on the jω-axis, the phase angle ϕ changes by step π at $\omega = \omega_o$ leading to the delay characteristic having an impulse at that frequency. Although we will normally have no natural modes on the jω-axis, the above comments are important as they apply equally well to a pair of transmission zeros (loss poles) located on the jω-axis. In fact, all the results of this example may be applied to the case of a pair of loss poles. This will be further illustrated in Section 1.5, where we consider biquadratic transfer functions in detail.

1.4 TYPES OF FILTERING FUNCTIONS

The first type of functions we should like to consider are those H(s) for which all of the loss poles lie in the closed left-half s-plane. These functions are termed minimum-phase transfer functions. For minimum-phase functions the real and imaginary parts, or what is equivalent the magnitude and phase, are not independent, but are related by the so-called Hilbert transform. Thus if one of the two parts of the transfer function is prescribed, the other part can be calculated from it via the Hilbert transform relationship. As an example, the group delay $\tau(\omega)$ and the attenuation $\alpha(\omega)$ are related by the contour integral,

$$\tau(\omega) = -\frac{1}{\pi}\int_{-\infty}^{\infty}\left[\frac{d\alpha(x)}{dx}\right]\frac{dx}{\omega - x} \tag{1.21}$$

Thus, for minimum phase transfer functions, one cannot specify $\alpha(\omega)$ and $\tau(\omega)$ independently. Note also that (1.21) implies that $\tau(\omega)$ is large where $\alpha(\omega)$ is steep.

Non-minimum-phase transfer functions are those H(s) for which some or all of the loss poles lie in the open right-half s-plane. For such functions, the attenuation and group delay functions are independent of one another. Thus, filters which are designed to meet simultaneous attenuation and group delay requirements usually make use of non-minimum-phase transfer functions.

A particular class of non-minimum-phase transfer functions are the all-pass functions for which the loss poles (in the open right-half plane) are mirror images of the natural modes (in the open left-half plane). Thus if $H_A(s)$ is an all-pass transfer function, then P(s) = E(-s), and we may express $H_A(s)$ as

$$H_A(s) = \frac{E(s)}{E(-s)} = \frac{\prod_{k=1}^{N}(s - s_{nk})}{\prod_{k=1}^{N}(-s - s_{nk})} \tag{1.22}$$

For $s = j\omega$, $H_A(j\omega)$ may be written in terms of the real and imaginary parts of $E(j\omega)$ as

$$H_A(j\omega) = \frac{E_r(\omega) + j\,E_i(\omega)}{E_r(\omega) - j\,E_i(\omega)} = e^{j\phi(\omega)} \tag{1.23}$$

where

$$E(j\omega) = E_r(\omega) + j\,E_i(\omega) \quad , \text{ and}$$

$$\phi(\omega) = 2\tan^{-1}\left[\frac{E_i(\omega)}{E_r(\omega)}\right] \qquad (1.24)$$

We note from (1.23) that

$$|H_A(j\omega)| = 1$$

Thus, the attenuation function of an all-pass transfer function does not exhibit any frequency selectivity. The phase lag function and the group delay function, however, have frequency selective properties. This fact enables one to design an all-pass network which has zero attenuation and a prescribed delay function.

All-pass networks find application in the design of "delay equalizers". In Section 1.2, we described the function of amplitude equalizers. Delay equalizers perform a similar task as far as the group-delay of a transmission system is concerned.

The reader should note from (1.24) that the phase associated with an all-pass function is twice that obtained from the related function H(s) = E(s). Consequently, the delay associated with $H_A(s)$ is twice that obtained from E(s). Adding mirror image loss poles in the right-half plane doubles the delay obtained from the natural modes alone.

Example 1.2

In example 1.1 we studied the attenuation and delay characteristics of the quadratic transfer function

$$H(s) = s^2 + s\left(\frac{\omega_o}{Q}\right) + \omega_o^2$$

We now consider the related all-pass function

$$H_A(s) = \frac{s^2 + s\left(\frac{\omega_o}{Q}\right) + \omega_o^2}{s^2 - s\left(\frac{\omega_o}{Q}\right) + \omega_o^2} \tag{1.25}$$

Fig. 1.7 shows possible locations of the natural modes and loss poles of $H_A(s)$ in the s-plane. The attenuation associated with $H_A(s)$ is constant at 0 dB from $\omega = 0$ to $\omega = \infty$. On the other hand, the phase and delay are twice those given in equations (1.18) and (1.19), respectively. Thus, the normalized group delay curves given in Fig. 1.6 can be used to determine the delay of the biquadratic $H_A(s)$ in (1.25) by simply multiplying the delay scale by a factor of 2.

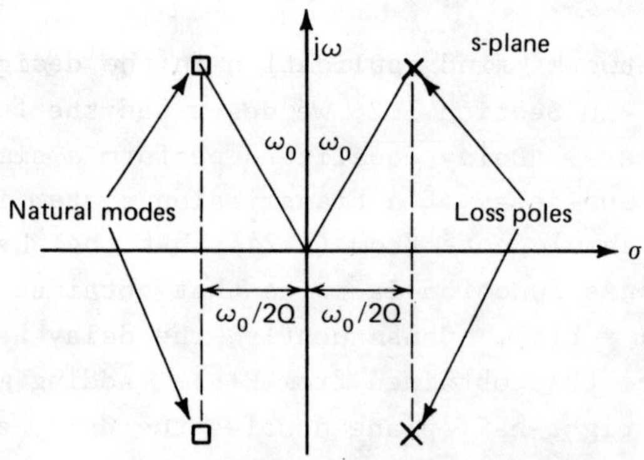

Figure 1.7 Singularities of a second-order all-pass function.

Making use of the properties of all-pass functions, it is easily shown that any arbitrary non-minimum-phase transfer function H(s) may be written as a product of a minimum-phase function $H_m(s)$ and an all-pass function $H_A(s)$,

$$H(s) = H_m(s)\, H_A(s) \tag{1.26}$$

This suggests that, if a filter is required to meet simultaneous attenuation and delay specifications, one may split the design problem into two parts. First, find a minimum-phase transfer function $H_m(s)$ which meets the given attenuation requirements, while disregarding the group delay requirements. Subsequently, find an all-pass function $H_A(s)$ whose delay function, when added to that of $H_m(s)$, should meet the given group delay specifications. Connecting the networks realizing $H_m(s)$ and $H_A(s)$ in cascade, provided they do not interact, will realize the function $H(s)$, whose attenuation and delay now meet the specifications.

At this point, we present a broad classification of minimum-phase transfer functions, and of the corresponding network realizations, in terms of their attenuation characteristics over the frequency axis. The four most common types of filter functions are:

(1) Low-Pass (LP)
(2) High-Pass (HP)
(3) Band-Pass (BP)
(4) Band-Reject (BR)

Sketches of attenuation versus frequency for these four filter types are shown in Fig. 1.8. As shown, each filter has a well defined "passband" over which the attenuation is zero, and a well defined "stopband" in which the attenuation is large.

In some realizations, the filter may exhibit constant but non zero attenuation (positive or negative) over the passband. This is usually of no major concern. Variation in attenuation over the passband is, however, an important consideration. Such variation, which is inevitable in a real filter, represents distortion in the sense that frequencies which we wish to pass through the filter will suffer different amounts of attenuation. The system designer, therefore, has to specify the maximum allowed variation in the filter attenuation over the passband.

In the sketches of Fig. 1.8, the stopband attenuation is shown to be large but finite. This is again a property of physically realizable filters. Although the attenuation might be made infinite at one or more frequencies, by placing transmission zeros or loss poles at such frequencies, it cannot be infinite over the entire stopband. The stopband attenuation in Fig. 1.8 is shown to be constant. A real filter will clearly have variable stopband attenuation, but the designer can specify a minimum amount of required attenuation in the stopband. All these points will be elaborated on in Section 1.6, which deals with filter specifications.

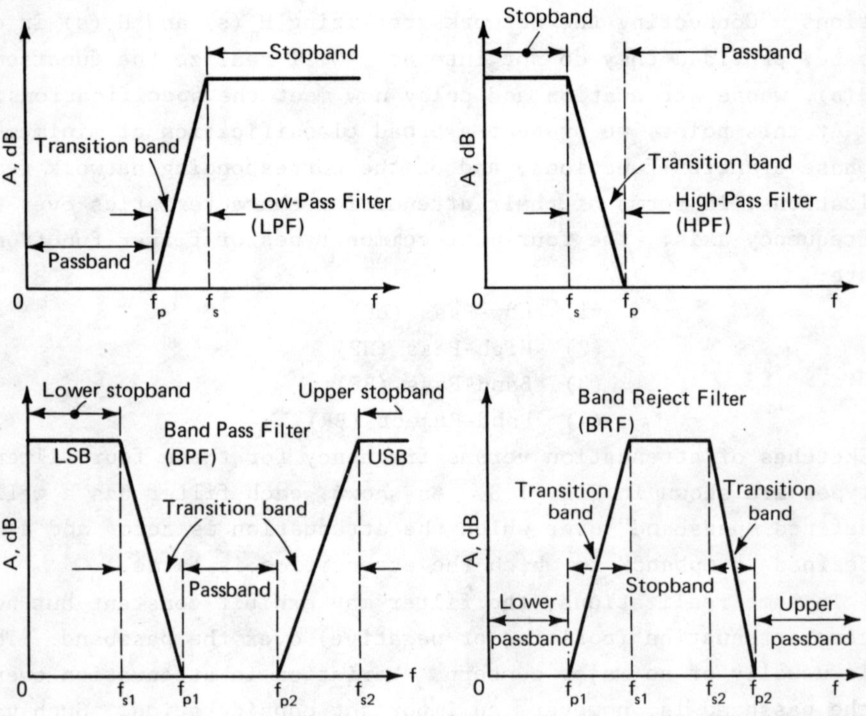

Figure 1.8 Attenuation characteristics of the four common types of filters.

The frequency range separating the stopband from the passband is referred to as the transition band. Although for an ideal filter one desires the transition band to be zero in width, this is impossible to obtain in reality. In fact, the narrower the transition band, i.e. the more selective the filter is, the more complex its realization will be; this naturally means an increase in cost.

1.5 BEQUADRATIC TRANSFER FUNCTIONS

The filter transfer function H(s), expressed in a polynomial quotient form, is given by

$$H(s) = \frac{E(s)}{P(s)}$$

The natural mode polynomial E(s) may be factored and expressed as

$$E(s) = (s + \sigma_{no}) \prod_{k=1}^{(N-1)/2} (s^2 + 2\,\sigma_{nk}\,s + \sigma_{nk}^2 + \omega_{nk}^2) \qquad (1.27)$$

for N odd, or as

$$E(s) = \prod_{k=1}^{N/2} (s^2 + 2\,\sigma_{nk}\,s + \sigma_{nk}^2 + \omega_{nk}^2) \qquad (1.28)$$

for N even. Since all the natural modes lie in the open left half plane $\sigma_{nk} > 0$. Usually, with the exception of a possible real root as in (1.27), all the natural modes occur in complex conjugate pairs. Rather than expressing each pair in terms of σ_{nk} and ω_{nk}, where n denotes natural mode, we may use the ω_o and Q notation already introduced in example 1.1. For instance, for N even we may write E(s) as

$$E(s) = \prod_{k=1}^{N/2} (s^2 + \frac{\omega_{ok}}{Q_k}\,s + \omega_{ok}^2) \qquad (1.29)$$

where ω_{ok} is the frequency and Q_k is the Q factor of the kth natural mode or pole. Fig. 1.9 shows possible locations of the natural modes in the s-plane for the case N = 5.

The loss-pole polynomial can also be factored and expressed as

$$P(s) = s^{n_o}\,(s + \sigma_{\ell o}) \prod_{k=1}^{L} (s^2 + 2\,\sigma_{\ell k}\,s + \sigma_{\ell k}^2 + \omega_{\ell k}^2) \qquad (1.30)$$

where ℓ indicates loss (for loss-pole polynomial roots). In (1.30) it is arbitrarily assumed that the filter has n_o loss-poles at the origin, one real loss-pole at $s = -\sigma_{\ell o}$, and L complex conjugate loss-pole pairs at $s = -\sigma_{\ell k} \pm j\omega_{\ell k}$. Thus, the degree of P(s) is obtained as

$$M = n_o + 1 + 2L$$

Since there are generally no restrictions on the location of the loss-poles, there may be more than one real loss-pole and $\sigma_{\ell k}$ can be positive, negative or zero. We may also wish to use the ω_o and Q notation to express the quadratic terms in the P(s) factorization.

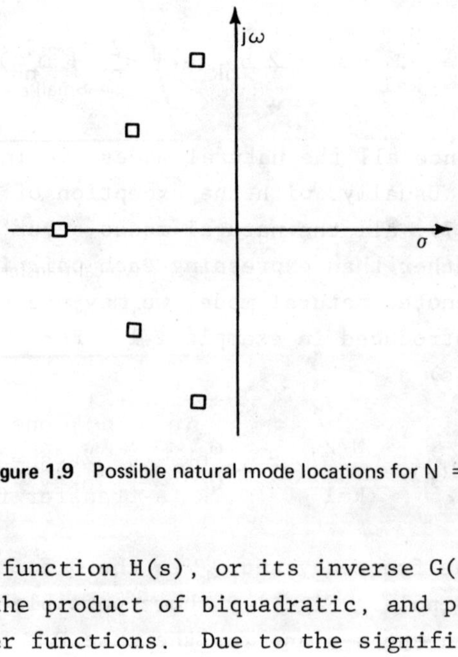

Figure 1.9 Possible natural mode locations for N = 5

The transfer function H(s), or its inverse G(s), can thus be factored into the product of biquadratic, and possibly one bilinear, transfer functions. Due to the significance of this factorization and its application in the design of active filters, we will consider biquadratic transfer functions in some detail. Bilinear factors are straightforward and it is assumed that the reader is familiar with such functions.

Since biquadratic functions play an important role in active filter design, we consider such functions in their G(s) or output/input form, as has been customary in the active filter literature.[†] A general second-order transfer function or "biquad" function may be written as

$$G(s) = \frac{p_2 s^2 + p_1 s + p_o}{s^2 + e_1 s + e_o} = \frac{P(s)}{E(s)} \tag{1.31}$$

where P(s) is the loss-pole or, more appropriately here, the transmission-zero polynomial, and E(s) is the natural mode or biquad pole polynomial. It is a usual practice to express the denominator polynomial in terms of ω_o and Q, referred to here as the pole-ω_o and pole-Q. Thus (1.31) becomes

$$G(s) = \frac{p_2 s^2 + p_1 s + p_o}{s^2 + \frac{\omega_o}{Q}s + \omega_o^2} \tag{1.32}$$

The numerator coefficients determine the location of the transmission zeros and hence, the type of filter function the biquad provides. Special cases of interest are:

(1) <u>Low-Pass (LP)</u> for which $p_1 = p_2 = 0$, thus the two transmission zeros are at infinity, and

$$G(s) = \frac{p_o}{s^2 + \frac{\omega_o}{Q} s + \omega_o^2} \tag{1.33}$$

[†]In the Chapters dealing with active filters, we will use T(s) rather than G(s) to denote the output/input transfer function. This is done to conform with the active filter literature.

(2) <u>High-Pass (HP)</u> for which $p_o = p_1 = 0$, thus the two transmission zeros are at $s = 0$, and

$$G(s) = \frac{p_2 \, s^2}{s^2 + \frac{\omega_o}{Q} s + \omega_o^2} \tag{1.34}$$

(3) <u>Band-Pass (BP)</u> for which $p_o = p_2 = 0$, thus one transmission zero is at infinity while the other is at the origin, and

$$G(s) = \frac{p_1 s}{s^2 + \frac{\omega_o}{Q} s + \omega_o^2} \tag{1.35}$$

(4) <u>Band-Reject (BR)</u> for which $p_1 = 0$ and the two transmission zeros are at $s = \pm j\omega_o$, leading to

$$G(s) = p_2 \frac{s^2 + \omega_o^2}{s^2 + (\frac{\omega_o}{Q}) s + \omega_o^2} \tag{1.36}$$

(5) <u>Low-Pass Notch (LPN)</u> for which $p_1 = 0$ and the two transmission zeros are at $s = \pm j\omega_n$, $\omega_n > \omega_o$, leading to[†]

$$G(s) = p_2 \frac{s^2 + \omega_n^2}{s^2 + (\frac{\omega_o}{Q}) s + \omega_o^2} \tag{1.37}$$

[†] The subscript n in ω_n denotes notch. We will use ω_z, denoting zero, and ω_n interchangeably.

(6) High-Pass Notch (HPN) for which $p_1 = 0$ and the two transmission zeros are at $s = \pm j\omega_n$, $\omega_n < \omega_o$, leading to

$$G(s) = p_2 \frac{s^2 + \omega_n^2}{s^2 + (\frac{\omega_o}{Q}) s + \omega_o^2} \tag{1.38}$$

(7) All-Pass (AP) for which the pair of zeros are at the mirror image locations of the pair of poles, that is

$$G(s) = p_2 \frac{s^2 - (\frac{\omega_o}{Q}) s + \omega_o^2}{s^2 + (\frac{\omega_o}{Q}) s + \omega_o^2} \tag{1.39}$$

Fig. 1.10 shows sketches for the $|G(j\omega)|$ in decibels versus frequency. Also, Table 1.1 presents some important parameters of biquad functions [7]. Note that in Example 1.1, we have shown in detail how to evaluate the attenuation and group delay functions of the low-pass biquad. Other biquads can be treated similarly. An alternative approach is given in the following.

The transfer function for a general biquad is expressed as

$$G(j\omega) = p_2 \frac{P(j\omega)}{E(j\omega)}$$

where the constant p_2 is extracted to make P and E monic polynomials. Thus $A(\omega)$ is given by

$$A(\omega) = 20 \log_{10} p_2 + 20 \log_{10} |P(j\omega)| - 20 \log_{10} |E(j\omega)|$$

which suggests that the attenuation function of the biquad may be evaluated using the attenuation function of a quadratic polynomial. The latter attenuation functions have been given in normalized form in connection with example 1.1.

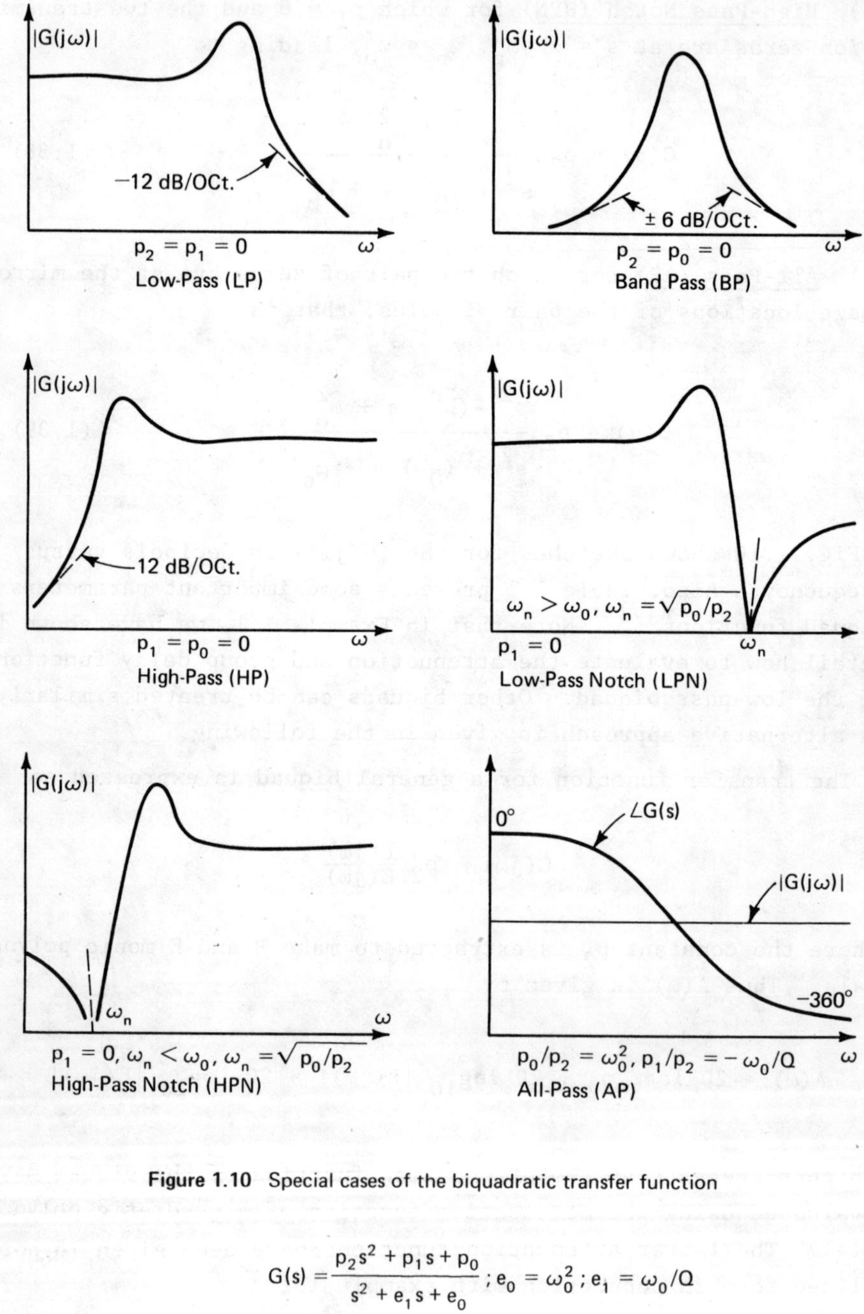

Figure 1.10 Special cases of the biquadratic transfer function

$$G(s) = \frac{p_2 s^2 + p_1 s + p_0}{s^2 + e_1 s + e_0} \,;\, e_0 = \omega_0^2 \,;\, e_1 = \omega_0/Q$$

Table 1.1 Some important parameters of biquadratic transfer functions. Note that ω_m denotes the frequency at which $|G(j\omega)|$ has a peak while $F(\omega) = |G(j\omega)/K|^2$.

Type	G(s)	ω_m^2	$F(\omega_m)$	ω_m real if	F(0)	F(∞)
LP	$\dfrac{K}{s^2 + s\dfrac{\omega_o}{Q} + \omega_o^2}$	$\omega_o^2\left(1 - \dfrac{1}{2Q^2}\right)$	$\dfrac{Q^2}{\omega_o^4\left(1 - \dfrac{1}{4Q^2}\right)}$	$Q^2 \geq \frac{1}{2}$	$\dfrac{1}{\omega_o^4}$	0
HP	$\dfrac{K s^2}{s^2 + s\dfrac{\omega_o}{Q} + \omega_o^2}$	$\dfrac{\omega_o^2}{1 - \dfrac{1}{2Q^2}}$	$\dfrac{Q^2}{1 - \dfrac{1}{4Q^2}}$	$Q^2 \geq \frac{1}{2}$	0	1
BP	$\dfrac{K s}{s^2 + s\dfrac{\omega_o}{Q} + \omega_o^2}$	ω_o^2	Q^2 / ω_o^2	always real	0	0
Notch	$K\dfrac{s^2 + \omega_n^2}{s^2 + s\dfrac{\omega_o}{Q} + \omega_o^2}$	$\dfrac{\left(\dfrac{\omega_n}{\omega_o}\right)^2\left(1 - \dfrac{1}{2Q^2}\right) - 1}{\left(\dfrac{\omega_n}{\omega_o}\right)^2 + \dfrac{1}{2Q^2} - 1}\,\omega_o^2$	$\dfrac{(\omega_n^2 - \omega_m^2)^2}{(\omega_o^2 - \omega_m^2)^2 + \left(\dfrac{\omega_c}{Q}\right)^2\omega_m^2}$	$\dfrac{\omega_n}{\omega_o} > 1 : Q^2 \geq \dfrac{\left(\dfrac{\omega_n}{\omega_o}\right)^2}{2\left[\left(\dfrac{\omega_n}{\omega_o}\right)^2 - 1\right]}$ $\dfrac{\omega_n}{\omega_o} < 1 : Q^2 \geq \dfrac{1}{2\left[1 - \left(\dfrac{\omega_n}{\omega_o}\right)^2\right]}$	$\left(\dfrac{\omega_n}{\omega_o}\right)^4$	1

Similarly for the delay function we have

$$\tau(\omega) = \tau_E(\omega) - \tau_P(\omega)$$

where τ_E and τ_P are the group delay functions associated with the E and P polynomials, respectively. Again equation (1.20) provides a universal expression for the normalized delay function associated with a quadratic polynomial. Thus, the universal delay curves in Fig. 1.6 can be used to evaluate the delay function associated with a general biquad transfer function.

It can be shown that the phase function exhibits step discontinuities if the transfer function has $j\omega$-axis loss poles. Correspondingly, the delay function contains impulses at these loss-pole locations.

Before leaving this section, we show in Fig. 1.11 some passive RLC realizations of the special biquad functions presented. The reader is urged to verify that the circuits given indeed realize the corresponding transfer functions.

1.6 FILTER SPECIFICATIONS

There are usually many facets to a filter specification, including such things as: physical dimensions, power requirements, operating temperature range, types of components, input and output impedance levels, etc. In this section, we address ourselves to the most fundamental specification for a filter; namely, its required frequency-domain transfer characteristics.

As already pointed out, the primary function of a filter is to pass, undistorted, a certain band of frequencies, while highly attenuating frequencies in other bands. Thus, over the passband, the filter should present zero, or small but constant, attenuation, while the attenuation should be infinite or "very high" over the stop band. Constant attenuation over the passband, however, does not alone guarantee distortionless transmission of signals. To illustrate this point, consider an input signal u(t) whose frequency spectrum $U(j\omega)$ lies entirely within the filter passband. We therefore desire the output signal y(t) to have the same wave-

(1) Low-Pass

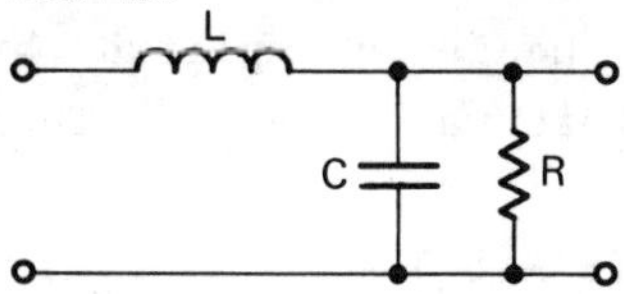

$p_0 = (1/LC)$
$\omega_0 = 1/\sqrt{LC}$
$Q = R\sqrt{C/L}$

(2) High-Pass

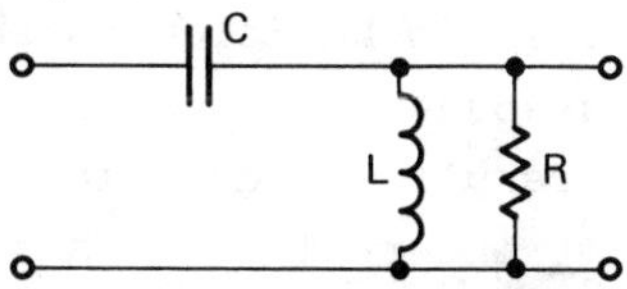

$p_2 = 1$
$\omega_0 = 1/\sqrt{LC}$
$Q = R\sqrt{C/L}$

(3) Bandpass

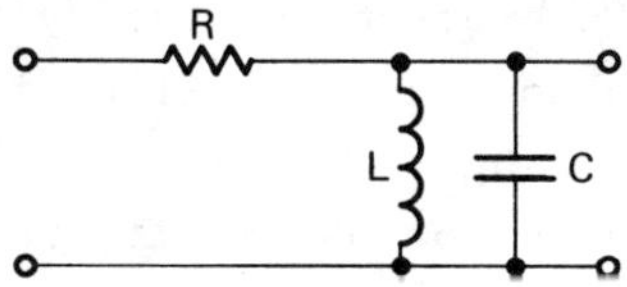

$p_1 = (1/CR)$
$\omega_0 = 1/\sqrt{LC}$
$Q = R\sqrt{C/L}$

(4) Band reject

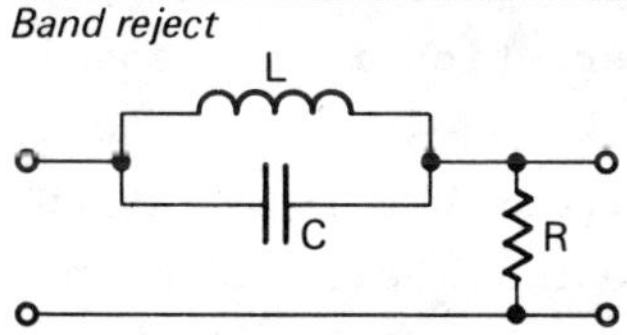

$p_2 = 1$
$\omega_0 = 1/\sqrt{LC}$
$Q = R\sqrt{C/L}$

(5) Low-Pass Notch

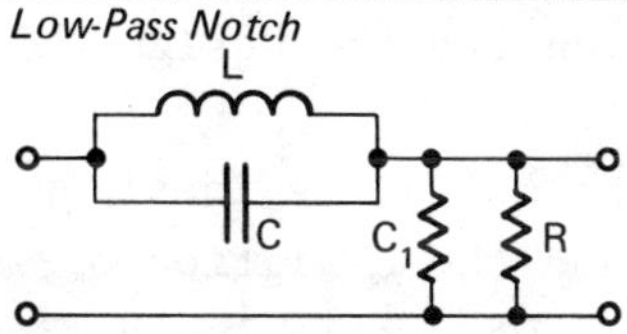

$p_2 = C/(C + C_1)$
$\omega_n = 1/\sqrt{LC}$
$\omega_0 = 1/\sqrt{L(C + C_1)}$
$Q = R\sqrt{(C + C_1)/L}$

(6) High-Pass Notch

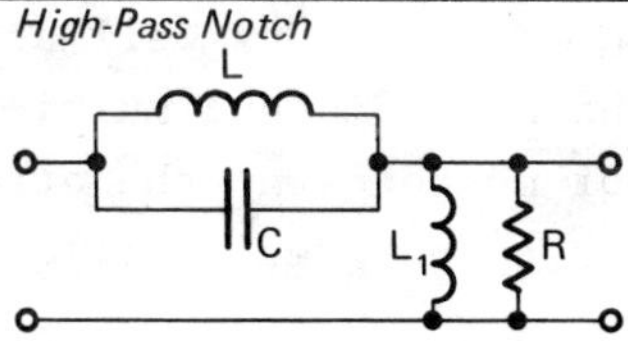

$p_2 = 1$
$\omega_n = 1/\sqrt{LC}$
$\omega_0 = \sqrt{1 + (L/L_1)}\,/\sqrt{LC}$
$Q = \sqrt{1 + (L/L_1)}\,R\sqrt{C/L}$

(7) All-Pass

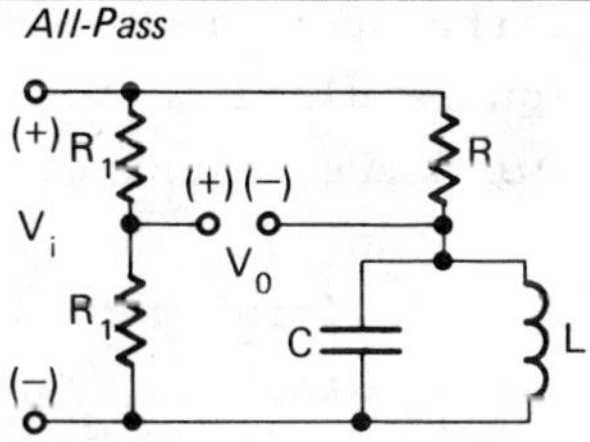

$p_2 = 0.5$
$\omega_0 = 1/\sqrt{LC}$
$Q = R\sqrt{C/L}$

Figure 1.11 Some passive realizations of the special biquad functions.

shape as u(t). We can tolerate a change in amplitude by a factor, say k, and a delay in time by an interval, say t_o. Thus, ideally,

$$y(t) = k\,u(t - t_o)$$

In the frequency domain, this entails, in addition to the constant transmission over the filter passband, a linear phase function $\phi(\omega)$. The latter requirement results in delaying each frequency component of u(t) by the same interval t_o. This linear-phase requirement can be restated as a constant group delay over the passband. Note that there is no reason to specify the phase or delay performance over the stopband, as the filter is supposed to block signals in that frequency range.

The ideal filter specified above is not physically realizable. Thus, some compromise, tempered by engineering judgement, must be substituted for the ideal characteristics in the specifications for a filter. The actual specification should allow some deviation from ideality so that the designer has some latitude, or margin, for realization. It should be remembered when specifying a filter that, as with all design problems, the closer one wishes to approach ideality, the more expensive the design solution will become.

Let us now consider methods of relaxing the ideal specifications in such a way that we can bound the deviation of filter performance from the ideal behaviour. We must allow a finite nonzero amount of passband distortion to occur, as well as a finite amount of attenuation in the stopbands. Since a multiplicative scale factor on the magnitude is of no concern, the attenuation specification is usually referenced to an ideal zero decibel (dB) passband attenuation.

Keeping in mind the cost involved, the specifications are drawn up such that the amount of attenuation required in the stopbands is as small as possible, consistent with the amount of signal suppression desired over various frequency ranges. This will entail judgement on the part of the engineer specifying the filter, since only he knows the minimum amount of stopband suppression his system will require. Similarly, the maximum amount of distortion that can be tolerated in the passband must be tempered by the knowledge that decreasing this tolerance will lead to an increase in the cost of the final network.

Usual practice is to give the filter loss specifications in the form of idealized tolerance contours, which are almost always in the form of straight line segments. Fig. 1.12 is an example of a typical set of loss versus frequency specifications for a bandpass filter. The dotted line illustrates the character of a typical rational polynomial approximation to these specifications. Note that the relative attenuation in the passband can be no greater than the specified maximum value designated by the tolerance contour marked A_p. Thus, A_p bounds the maximum deviation of the filter attenuation from ideality in the passband. Similarly, the tolerance boundaries marked A_s bound the minimum allowable attenuation in the stopbands. Thus, the tolerance contours A_p and A_s specify the absolute worst case deviation in magnitude response of the filter from the ideal filter performance.

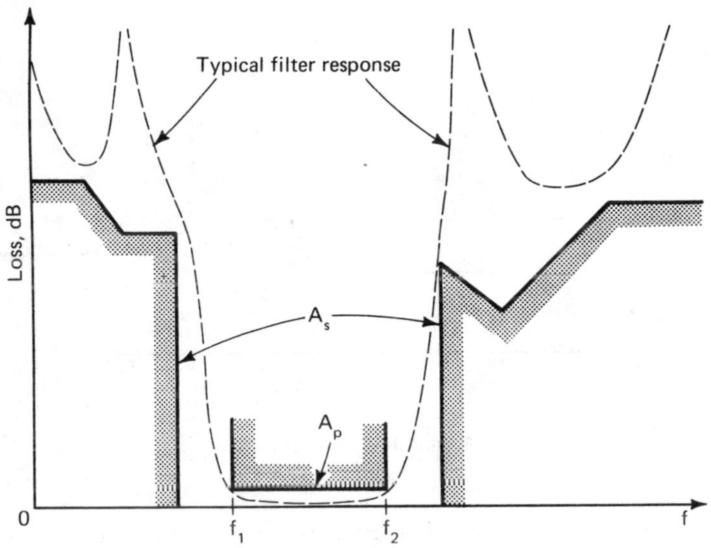

Figure 1.12 A typical loss versus frequency specification for a bandpass filter.

In order to bound the linearity of the phase response, usual practice is to place limiting tolerance contours on the group-delay. Here, as with magnitude specifications, it is usually the relative group delay that is of importance, and a constant group delay in the passband is not considered to be a distortion. The specifications for group delay usually take the form of straight line segment tolerance boundaries within which the desired delay function must lie. Engineering judgement must be used in order to obtain the maximum allowable deviation from "flatness" in the passband delay. The reader will note here the great similarity between the passband attenuation specifications in dB's, and the specifications for delay. It must be remembered in drawing up the specifications on delay, that the smaller the delay deviation allowed, the more expensive the resulting network is likely to become. For purposes of illustration Fig. 1.13 shows a typical delay specification for a bandpass filter.

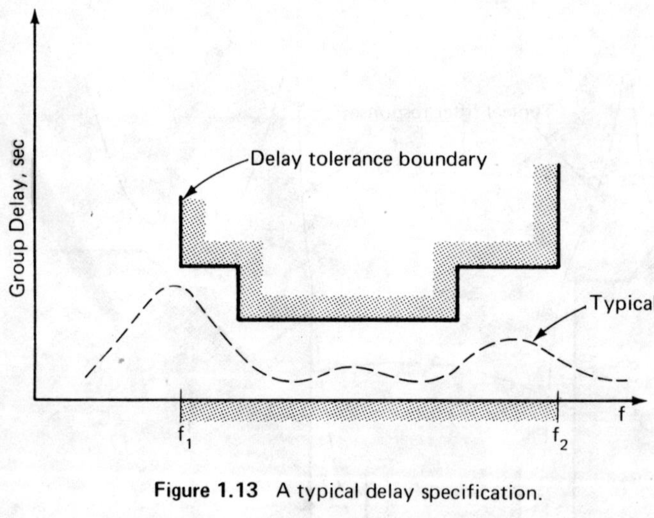

Figure 1.13 A typical delay specification.

As already pointed out, for minimum phase networks, the group delay and attenuation are not independent of each other. However, in many cases, the latitude allowed by the magnitude tolerance boundaries may enable the designer to find several alternative solutions to the attenuation requirements, one of which might also meet the delay requirements. In general, the simultaneous satisfaction of both attenuation and delay requirements is very difficult and, of necessity, involves the use of non-minimum phase networks. In these cases one usually obtains a solution to the attenuation requirements and subsequently designs a delay equalizer to be placed in cascade with the filter. It should be noted, however, that in many cases only the attenuation response of the filter is of interest. In this book we will only be concerned with the design of filters which meet given attenuation specifications.

In summary, we have illustrated the typical methods by which filter transfer functions are specified. These methods consist of a set of idealized tolerance contours, which bound the performance of the desired filter transfer function.

1.7 THE FILTER DESIGN PROCESS

The filter design process can be broken down into two fundamental phases; namely,

(1) the approximation phase, and

(2) the realization phase.

In the approximation phase, the filter designer finds a real rational polynomial transfer function $H(s)$, whose associated attenuation function $A(\omega)$ and delay function $\tau(\omega)$ "fit" within the given tolerance boundaries in some (hopefully) optimum manner. The realization phase starts with the function $H(s)$, obtained as a result of solving the approximation problem, and finds a two-port network, whose transfer function (voltage or current) is identical to $H(s)$, to within a multiplicative real constant.

There are various technologies for realizing a given transfer function $H(s)$. Currently available technologies include passive LC, active-RC, crystal and mechanical filters[5,6]. Due to topological considerations, each one of these technologies poses restrictions on the class of transfer functions which it is capable of realizing. These restrictions are in terms of the permissible locations of the loss-poles in the s-plane. Consequently, in seeking a solution to the approximation problem, the filter designer has to consider the type of technology intended to realize the required filter.

In spite of the constraints imposed by the realization technology on the solution to the approximation problem, in this book the two phases of filter design are treated separately. In dealing with realization, we will be concerned with two technologies; namely, passive LC and active-RC. Consequently, our treatment of the approximation phase will be highly influenced by these two technologies. Nevertheless, the presentation of approximation methods is of a general nature and many of the results given are applicable to the design of other linear filter networks, such as crystal and mechanical filters. In addition, with suitable transformation, given in Chapter 6, the approximation techniques can be applied to the design of discrete-time filters. Special classes of the latter are digital filters, and continuous-amplitude discrete-time filters, such as those employing charge coupled devices.

We conclude this section by presenting the filter design flow chart shown in Fig. 1.14, which illustrates the relationships between the specification, approximation, and realization phases of filter design.

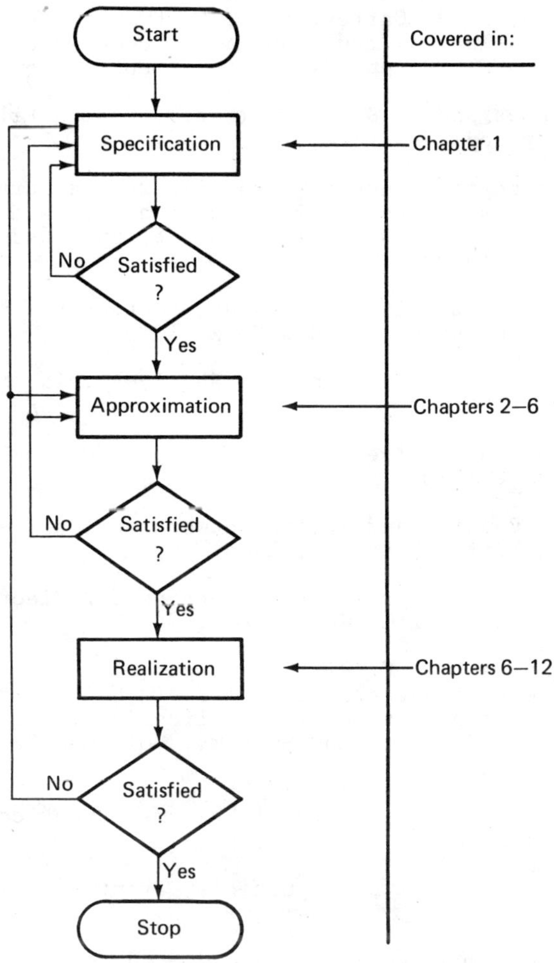

Figure 1.14 Flow chart for filter design.

1.8 REFERENCES AND SOURCES FOR FURTHER READING

(1) S. D. Conte and G. de Boor, Elementary Numerical Analysis, 2nd ed., McGraw-Hill, New York, 1972.

(2) W. M. Snelgrove and A. S. Sedra, FILTOR 2--A Computer Aided Filter Design Package, Matrix Publishers, Champaign, Ill.

(3) K. Su, Time Domain Synthesis, Prentice-Hall, Englewood Cliffs, N.J., 1971.

(4) H. B. Lee, P. P. Carrey and D. S. Evans, "MATCH, a new, generally available circuit optimization program," Frequency Technol., Vol. 8, pp. 24-29, Aug./Sept. 1976.

(5) G. C. Temes and S. K. Mitra (Eds.), Modern Filter Theory and Design, Wiley, New York, 1973.

(6) G. S. Moschytz, "Inductorless filters: a survey, I. Electromechanical filters. II. Linear active and digital filters", IEEE Spectrum, Vol. 7, No. 8, pp. 30-36 and No. 9, pp. 63-75, 1970.

(7) G. S. Moschytz, Linear Integrated Networks: Fundamentals, Van Nostrand Reinhold, New York, 1975.

(8) L. Weinberg, Network Analysis and Synthesis, McGraw-Hill, New York, 1962.

(9) S. S. Haykin, Active Network Theory, Addison-Wesley, Reading, Mass., 1970.

(10) A Budak, Passive and Active Network Analysis and Synthesis, Houghton Mifflin, Boston, 1974.

(11) R. W. Daniels, Approximation Methods for Electronic Filter Design, McGraw-Hill, New York, 1974.

(12) M. L. Liou and C. F. Kurth, "Computation of group delay from attenuation characteristics via Hilbert transformation and spline function and its application to filter design," IEEE Trans. Circuits and Systems, Vol. CAS-22, No. 9, pp. 729-734, Sept. 1975.

(13) B. C. Kuo, Automatic Control Systems, Prentice-Hall, Englewood Cliffs, N.J., 1967.

(14) N. Balabanian, Network Synthesis, Prentice-Hall, Englewood Cliffs, N.J., 1958.

(15) H. W. Bode, Network Analysis and Feedback Amplifiers, Van Nostrand Reinhold, New York, 1945.

(16) D. S. Humphreys, The Analysis, Design and Synthesis of Electrical Filters, Prentice-Hall, Englewood Cliffs, N.J., 1970.

(17) J. Vlach, Computerized Approximation and Synthesis of Linear Networks, Wiley, New York, 1969.

(18) M. E. Van Valkenburg, Introduction To Modern Network Synthesis, Wiley, New York, 1960.

(19) A. I. Zverev, Handbook of Filter Synthesis, Wiley, New York, 1967.

PROBLEMS

P1.1 Sketch the s-plane natural modes and loss poles associated with the following H(s) functions:

(a) $H(s) = (s+1)\ (s^2+s+1)$

(b) $H(s) = \dfrac{s^2+\sqrt{2}\ s+1}{s^2+2}$

(c) $H(s) = \dfrac{s^2+0.1\ s+1}{s^2-0.1\ s+1}$

P1.2 Give the transfer function H(s) for each of the filters whose loss poles and natural modes are shown in Fig. P1.2.

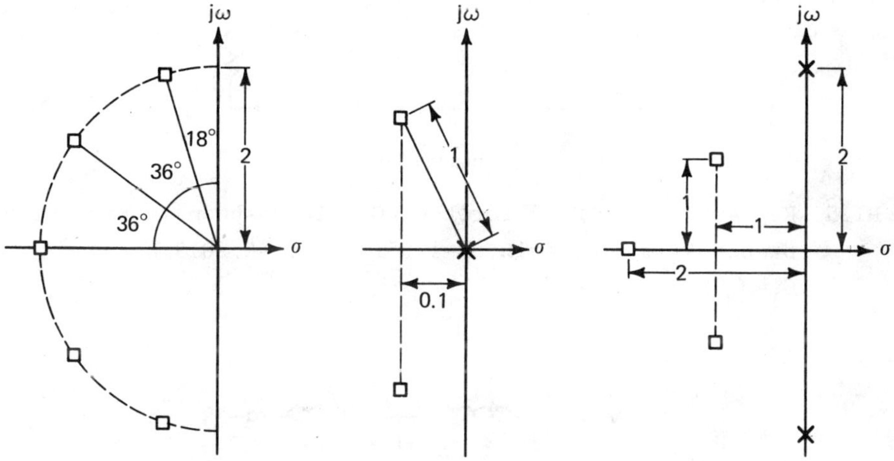

(a) All loss poles at ∞ (b) One loss pole at ∞ (c) One loss pole at ∞

Figure P1.2

P1.3 Find the s-plane locations of the loss poles and natural modes of the RC bridged-T network of Fig. P1.3. (Use the ω_o and Q notation for complex singularities and denote CR=T)

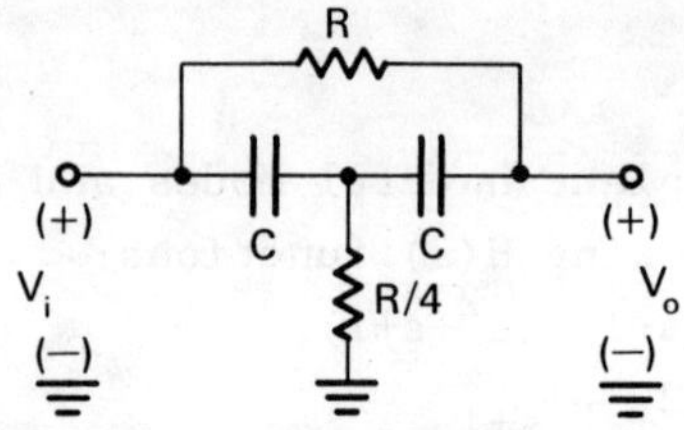

Figure P1.3

P1.4 Find the transfer function $H(s)=(V_i/V_o)(s)$ for the LC filter shown in Fig. P1.4. Give the location of the loss poles and natural modes (using the ω_o and Q notation). For R=1Ω, L=1H and C=2F sketch $A(j\omega)$. What type of filter is this?

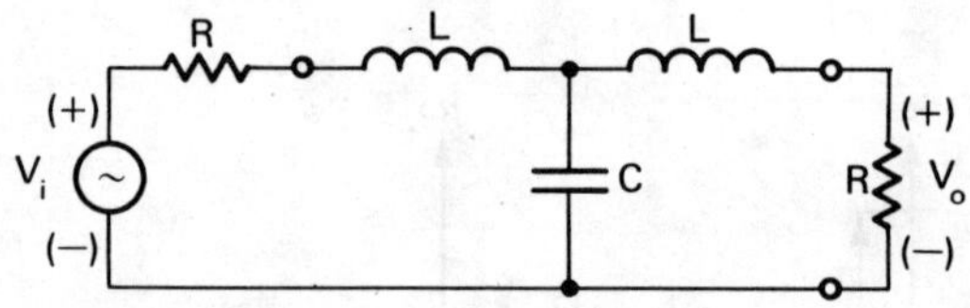

Figure P1.4

P1.5 Repeat problem P1.4 for the LC filter shown in Fig. P1.5. Sketch $A(j\omega)$ for R=1Ω, L=1.0512 H, L_1=0.2019 H and C_1=0.9612 F.

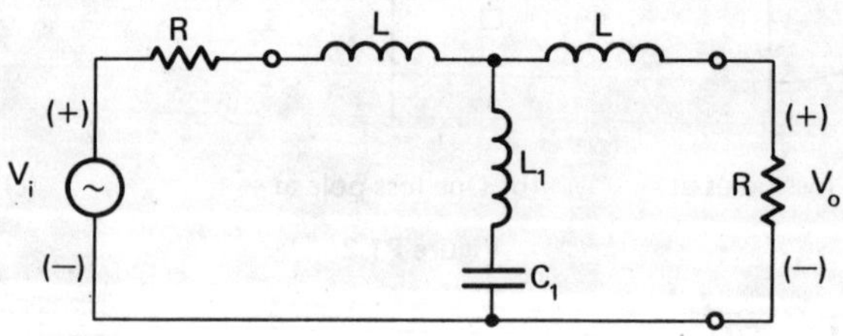

Figure P1.5

P1.6 Find an expression for the delay of a first-order transfer function $G(s) = K\ (s+z_1)/(s+p_1)$.

P1.7 (a) Find $\tau(\omega)$ of a first-order all-pass function $G(s) = \frac{s-a}{s+a}$

(b) Consider a second-order low-pass filter with $\omega_o = 1$ rad/sec and Q=2. Plot its $\tau(\omega)$ and find its maximum variation over the frequency range $o \le \omega \le 1$. In order to reduce this variation we wish to connect a first-order all-pass section in cascade. Find the pole location (i.e. the value of "a") for this all-pass section such that for the overall filter $\tau(o)=\tau(j1)$. Plot $\tau(\omega)$ for the overall filter and find its maximum variation over the band $o \le \omega \le 1$.

P1.8 Show that a pair of complex natural modes characterized by (ω_o, Q) gives rise to a group delay function $\tau(\omega)$ characterized by: $\tau(o)=1/(\omega_o Q)$ and $\tau_{max} \simeq 2Q/\omega_o$ for Q>1.

P1.9 A low-pass filter which has a maximally flat group delay function is called a Bessel filter. For such a filter the maximum group delay occurs at $\omega=0$. A fourth order Bessel filter, which is normalized such that the magnitude response drops by 3 dB at $\omega=1$ rad/sec., has the following natural modes:

$- 0.9952088 \pm j\, 1.2571058$

$- 1.3700679 \pm j\, 0.4102497$

Find the group delay at $\omega=o$.

P1.10 Expressing the biquadratic transfer function G(s) in the form:

$$G(s) = K \frac{s^2+s(\frac{\omega_z}{Q_z})+\omega_z^2}{s^2+s(\frac{\omega_o}{Q})+\omega_o^2},$$

identify the parameters K, ω_z, Q_z, ω_o and Q in the following G(s) functions:

(a) $5\ \dfrac{s^2+4s+16}{s^2+2s+25}$

(b) $\dfrac{6s^2+24}{2s^2+0.6s+18}$

(c) $10\ \dfrac{s^2-0.4s+4}{s^2+0.4s+4}$

P1.11 Determine the biquadratic transfer functions G(s) whose associated magnitude functions are shown in Fig. P1.11.

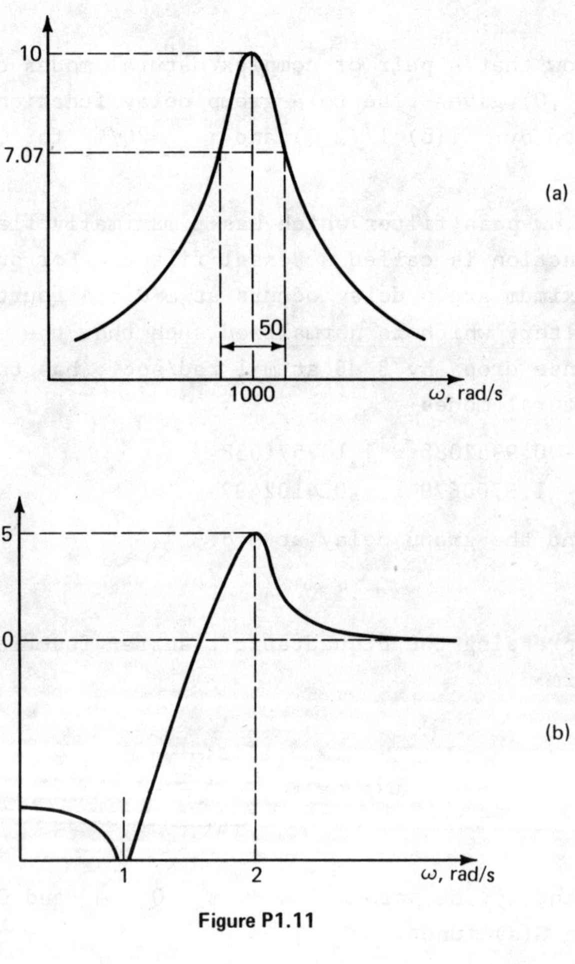

Figure P1.11

P1.12 Give the transfer function G(s) for a second-order low-pass filter having:

DC gain = 10

Maximum gain = 20 at $\omega = 1$ rad/sec

Specify the value of ω_o and Q.

P1.13 Give the transfer function and plot the attenuation $A(\omega)$, the phase $\phi(\omega)$ and the delay $\tau(\omega)$ for a second-order low-pass notch (LPN) filter having ω_o=1 rad/sec, ω_n=2 rad/sec, Q=1 and DC gain = 1.

P1.14 A second-order amplitude equalizer has the transfer function

$$G(s) = \frac{s^2+(\frac{\omega_o}{Q_z})s+\omega_o^2}{s^2+(\frac{\omega_o}{Q_p})s+\omega_o^2}$$

Sketch $|G(j\omega)|$ for (a) $Q_z=Q_p/2$, and (b) $Q_z=2Q_p$.

P1.15 Derive all the results given in Table 1.1

P1.16 Derive all the results given in Fig.1.11

P1.17 Identify the filter types (e.g. LP, HP, ...) corresponding to the loss poles - natural modes patterns shown in Fig. P1.17. Also give the filter order N.

P1.18 A LP filter with the following attenuation specifications:

$A \le 0.1$ dB for $o \le \omega \le 2.5$ rad/sec

$A \ge 80$ dB for $\omega \ge 3$ rad/sec

is connected in cascade with a HP filter whose attenuation specifications are

$A \le 0.1$ dB for $1.5 \le \omega \le \infty$ rad/sec

$A \ge 50$ dB for $\omega \le 1$ rad/sec

Sketch the contour specifications for the attenuation of the overall filter and state its type.

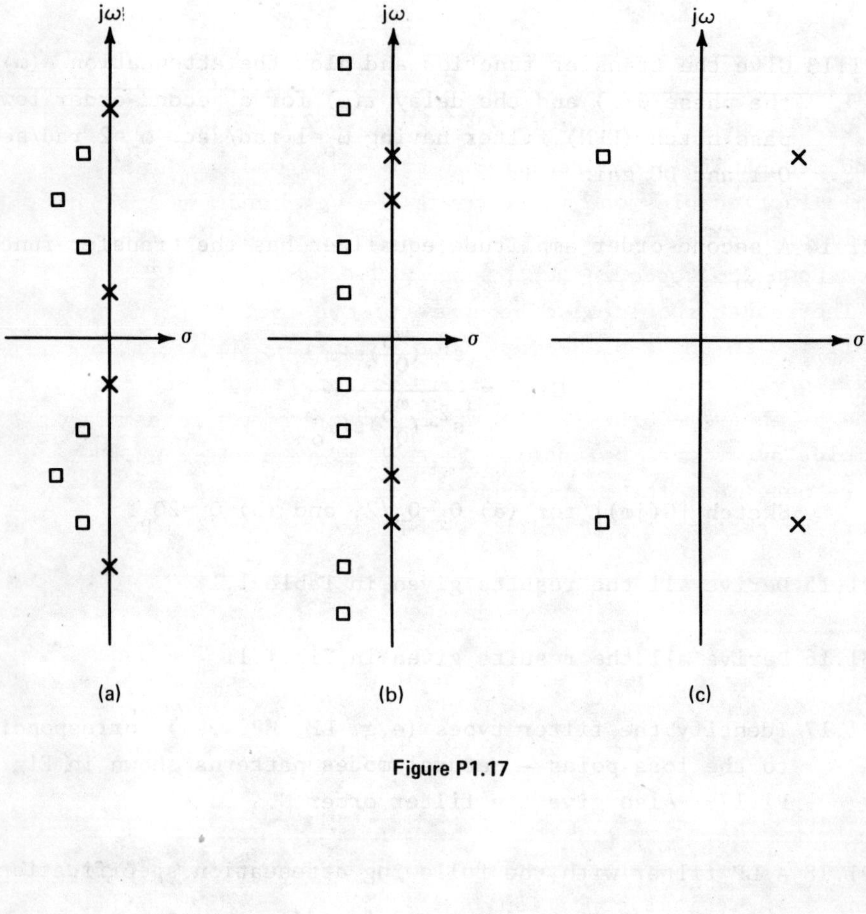

Figure P1.17

CHAPTER 2

THE FILTER APPROXIMATION PROBLEM

In this chapter, we begin our study of the first phase of filter design; namely, the solution of the approximation problem. Briefly, this may be stated as the problem of determining the order and parameters of a rational polynomial transfer function, which corresponds to a given realization method, such that the function "fits" within a set of prescribed frequency domain tolerance boundaries in some "hopefully" optimum manner. This statement is purposely rather vague and will be made more precise in the sections to follow.

To proceed with the solution of this problem, we first consider the various classes of transfer function and their relationship to the available realization or synthesis methods, thus putting the roles of the loss poles and natural modes in a proper perspective with respect to the approximation problem. The first section of this chapter presents just such an overview. In following sections, we introduce several new functions, related to the transfer function, which are useful in the approximation problem and in the description of filter characteristics. We then discuss the idea of optimality in approximation and introduce the two most widely used criteria for optimum filter approximation, as well as a strategy for obtaining such approximations. Finally we discuss the transformed variable and product method for solving the numerical accuracy problem in filter design, and present an algorithm for solving the central design equation of filter approximation theory.

2.1 CLASSES OF TRANSFER FUNCTIONS

The reader will recall that various types of networks, classified according to their element types and topological structure, give rise to certain classes of transfer functions. A filter designer should have the details of such classifications at his disposal during the solution of a filter design problem. For this reason, we shall review this material and place it in a proper perspective with respect to the filter approximation theory.

The transfer function we are considering here is the rational polynomial

$$H(s) = \frac{E(s)}{P(s)} , \tag{2.1}$$

as described in Chapter 1. The roots of E(s) are the natural modes, and the roots of P(s) are the loss poles. It is well known that the natural modes must always lie in the open left-half plane. For certain classes of networks (RC and RL networks) the natural modes must also be simple and lie on the negative real axis. We shall present further details on a certain class of RC network in Chapter 8. For the remainder of our discussion, however, we shall allow the natural modes to lie anywhere in the open left-half-plane.

We shall now classify transfer functions in correspondence with the permissible locations of the loss poles of certain classes of networks. Table 2.1 provides a classification of loss poles into seven groups. These seven groups have been arranged in an order which corresponds to the generality of the kind of network which can produce them. Recalling the broad transfer function classification of Chapter 1 into minimum-phase and non-minimum phase types, we recognize that the first three loss-pole groups in Table 2.1 belong to minimum phase functions, while loss pole groups four to seven belong to non-minimum phase transfer functions.

The first class (the least general in our classification) of networks to be considered is the lossless (LC) ladder network with no mutual inductance. A qualification equivalent to requiring no mutual inductance is the requirement that all of the L and C elements be positive, since negative elements may be "absorbed" by structures having mutual inductance[1]. In ladder networks, the loss poles are produced by the impedance poles of the series arms, and by the impedance zeros of the shunt arms. If the arms consist exclusively of +L and +C elements, then loss poles can only occur in conjugate pairs at points along the imaginary axis, in addition to possible loss poles at zero and infinity.

The first three groups of loss poles in Table 2.1 correspond to loss poles on the imaginary axis at: infinity, zero and finite non-zero conjugate locations. These three groups can be produced by a +L +C ladder structure, provided certain conditions on the loss pole locations, relative to driving point impedance singularities, are satisfied. These conditions[2] are seldom violated for "reasonable" filter specifications and will not be considered here. If we denote the number of loss poles at zero by n_o, and the number of pairs of finite non-zero complex conjugate loss poles by n_f, then a typical loss pole polynomial for this class of network has the form

$$P(s) = s^{n_o} \prod_{k=1}^{n_f} (s^2 + \omega_k^2) \tag{2.2}$$

It should be noted that, for a ladder structure, there must be at

Table 2.1 Loss pole (roots of P(s)) classification.

Group	Typical Factor	s-plane	Applicable Networks
1 (n_i)	Loss Poles at Infinity ($s = \infty$) No Factors Present in P(s)	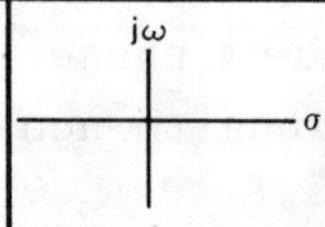	+L+C Ladder Networks (Groups 1–3); M +L+C or ±L±C Ladder Networks or +L+C Non Ladder Networks (Groups 1–5); General RLCM or R±L±C or Active RC (Groups 1–7)
2 (n_o)	Loss Poles at Zero ($s = o$) Factor Is Of Form s	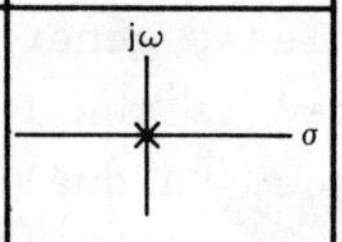	
3 (n_f)	Finite Non-Zero Conjugate Imaginary Axis Loss Poles. Factor Is Of Form $(s^2+\omega_o^2)$	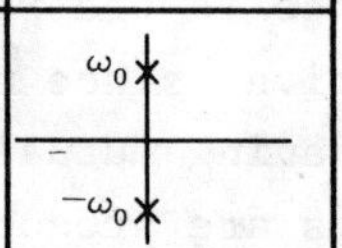	
4 (n_p)	Finite Non-Zero Real Axis Mirror Image Loss Poles. Factor Is Of Form $(s^2-\sigma_o^2)$	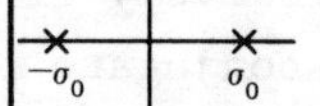	
5 (n_q)	Finite Non-Zero Complex Conjugate Loss Poles Having Quadrantal Symmetry $[s^4+2(\omega_o^2-\sigma_o^2)s^2+(\sigma_o^2+\omega_o^2)^2]$	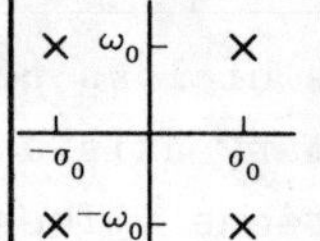	
6 (n_r)	Finite Non-Zero Real Axis Loss Poles $(s-\sigma_o)$ σ_o is a real number $-\infty<\sigma_o<\infty$	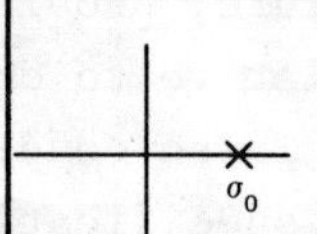	
7 (n_c)	Finite Non-Zero Complex Conjugate Loss Poles $(s^2-2\sigma_o s+\sigma_o^2+\omega_o^2)$ $-\infty<\sigma_o<\infty$ $o<\omega_o<\infty$	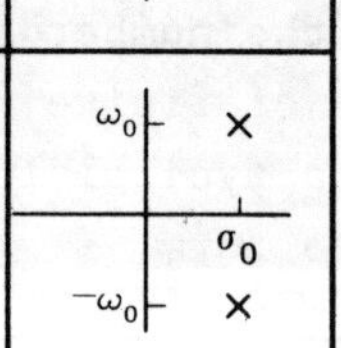	(Non-Ladder)

R - resistance C - capacitance

L - inductance M - mutual inductance

least one loss pole at infinity or zero.

We turn now to consider the loss pole locations designated as groups four and five in Table 2.1. These groups are: loss poles which occur in symmetrically displaced pairs on the real axis, and pairs of complex conjugate loss poles which have quadrantal symmetry. These two groups of loss poles can be realized by lossless ladder structures containing positive and negative elements. If the element values are restricted to be positive, then mutual inductance must be allowed. Darlington's[3,9] celebrated C and D sections are examples of networks having mutual inductance which can realize loss poles of groups four and five. If one desires a network that has positive LC elements with no mutual inductance, then the structure has to be a non-ladder one.†

Denoting by n_p the number of real axis pairs, and by n_q the number of quadrantally symmetric quads, a general form for P(s) for a lossless LC ladder network, which contains both positive and negative elements is

$$P(s) = s^{n_o} \prod_{k=1}^{n_f} (s^2 + \omega_k^2) \prod_{\ell=1}^{n_p} (s^2 - \sigma_\ell^2)$$

$$\prod_{m=1}^{n_q} [s^4 + 2(\omega_m^2 - \sigma_m^2)s^2 + (\sigma_m^2 + \omega_m^2)^2] \qquad (2.3)$$

Loss pole groups six and seven of Table 2.1 remove us completely from the realm of LC networks and in general will require either RLCM or R ± L ± C networks for realization. Also, ladder structures are not possible if these loss pole groups are to be included. Of course, active-RC networks are capable of realizing these loss poles.

Group six includes real axis loss poles, which do not occur in

---

†Note that from a practical point of view, we are specifically interested in ladder structures. Also, ladder structures that contain negative elements are still of interest, since they can be simulated by active networks.

symmetrical pairs, and group seven includes complex conjugate loss poles, which do not occur with quadrantal symmetry. If we denote by n_r the number of loss poles of group six, and by n_c the number of pairs of complex conjugate loss poles of group seven, then the most general form of loss pole polynomial may be written as

$$P(s) = s^{n_o} \prod_{k=1}^{n_f} (s^2 + \omega_k^2) \prod_{\ell=1}^{n_p} (s^2 - \sigma_\ell^2)$$

$$\prod_{m=1}^{n_q} [s^4 + 2(\omega_m^2 - \sigma_m^2)s^2 + (\sigma_m^2 + \omega_m^2)^2] \prod_{n=1}^{n_r} (s + \sigma_n)$$

$$\prod_{i=1}^{n_c} [s^2 + 2\sigma_i s + (\sigma_i^2 + \omega_i^2)] \qquad (2.4)$$

If the transfer function $H(s) = E(s)/P(s)$ corresponding to this loss pole polynomial has n_i loss poles at infinity, then the order of the resulting network may be computed from

$$N = n_i + n_o + 2n_f + 2n_p + 4n_q + n_r + 2n_c \qquad (2.5)$$

The most general transfer function can always be realized, within a multiplicative constant, by a passive RLCM network or an active-RC network. The latter consists of RC elements and gain blocks (amplifiers). Our treatment of the approximation problem will be quite general and the results will be applicable to the design of general active-RC filters; the design of passive LC filters will make use of special cases of the general theory.

In conclusion, before embarking on a solution to the approximation problem, the filter designer must decide on the type of network he wishes to use for realization. This in turn will determine, with reference to Table 2.1, the form of loss pole polynomial to be used in carrying out the approximation process.

2.2 THE CHARACTERISTIC FUNCTION AND REFLECTION ZEROS

Since an arbitrary multiplicative constant may be associated with a transfer function, the usual practice is to obtain a transfer function H(s) whose magnitude approximates unity in the passband. In particular the magnitude squared of H(s) along the j-axis, $|H(j\omega)|^2$, is required to approximate unity in some manner. Note that since the attenuation function $A(\omega)$ is given by

$$A(\omega) = 10 \log_{10} |H(j\omega)|^2 \tag{2.6}$$

then $A(\omega)$ approximates zero dB in the passband.

It is usually a simpler task to obtain rational functions which approximate zero rather than unity. Towards this end, we introduce the function K(s), which is called the characteristic function, as it "characterizes" the manner in which $|H(j\omega)|$ approximates unity in the passband. The function K(s) is defined as follows

$$|H(j\omega)|^2 = 1 + |K(j\omega)|^2 \tag{2.7}$$

Thus, $|K(j\omega)|$ approximates zero where $|H(j\omega)|$ approximates unity. Using analytic continuation (simply replacing ω by $\frac{s}{j}$) in (2.7), we obtain

$$H(s)\ H(-s) = 1 + K(s)\ K(-s) \tag{2.8}$$

This equation is central to the design of filter networks and is known in the literature as Feldtkeller's equation[4]. It should be noted that K(s) plays an important role in passive filter synthesis, where it has a physical interpretation in terms of the power reflected from the filter. This will be shown in Chapter 7. At present, however, we are only interested in K(s) as a mathematical function useful in solving the approximation problem.

Fig. 2.1 shows a typical bandpass $|H(j\omega)|^2$ and its associated $|K(j\omega)|^2$; both related by (2.7). We observe that $|H(j\omega)|^2$ approximates unity over the passband, extending from ω_1 to ω_2, with a maximum deviation of ε^2. Correspondingly, $|K(j\omega)|^2$ approximates zero over the passband with a maximum deviation of ε^2. Thus, the parameter ε represents the approximation error in the passband or the passband distortion. Since some approximations provide a function $|K(j\omega)|^2$ that oscillates over the passband, the parameter ε is often referred to as the "passband ripple factor". At this stage, we use ε to indicate the maximum passband magnitude distortion. The passband distortion may be also expressed in decibels, using the attenuation function $A(\omega)$ in (2.6). First, we may express $A(\omega)$ as

$$A(\omega) = 10 \log_{10} (1 + |K(j\omega)|^2) \tag{2.9}$$

Then, we denote the maximum deviation of $A(\omega)$ in the passband from the ideal zero dB by A_p and obtain,

$$A_p = 10 \log_{10} (1 + \varepsilon^2) \tag{2.10}$$

It is this parameter A_p that is normally given in the filter specifications. Since A_p denotes maximum passband deviation, some authors refer to it as A_{max}. Given A_p, the parameter ε may be determined using (2.10) as

$$\varepsilon = \sqrt{10^{A_p/10} - 1} \tag{2.11}$$

Fig. 2.1 also shows the attenuation function $A(\omega)$ corresponding to the bandpass $|H|^2$ and $|K|^2$.

As can be seen from (2.8), the characteristic function $K(s)$ is also a real rational function in s, having the same order N as $H(s)$. It is convenient to extract a real constant equal to ε from $K(s)$ and express the squared characteristic function as

$$K(s)\ K(-s) = \varepsilon^2 \frac{F(s)\ F(-s)}{P(s)\ P(-s)} \tag{2.12}$$

In writing (2.12), we have used the fact that $K(s)\ K(-s)$ and $H(s)\ H(-s)$ have the same denominator; namely, the squared loss pole polynomial, $P(s)\ P(-s)$. This is evident from (2.8). We have also introduced a new polynomial $F(s)$. Since $|K|^2$ is zero, and the filter has perfect transmission (zero attenuation) at the zeros of F(s), the latter polynomial is called the "reflection zero" or

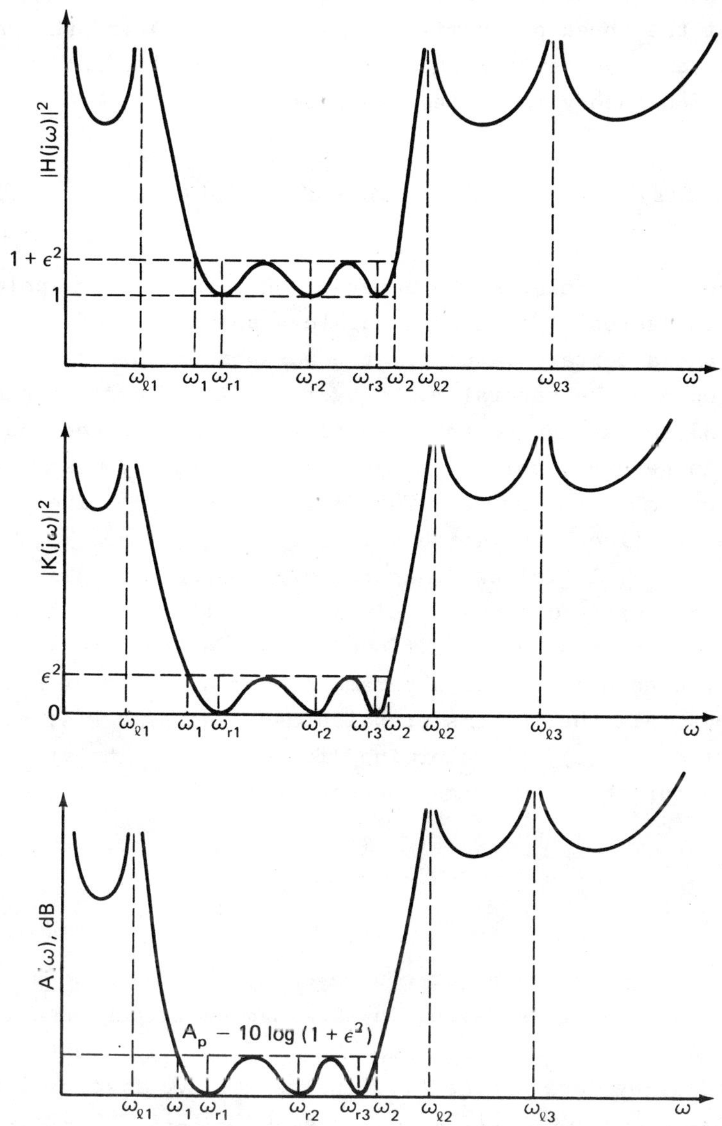

Figure 2.1 $|H|^2$, $|K|^2$ and A for a bandpass filter.

"attenuation zero" polynomial. Note that for the filter function shown in Fig. 2.1, F(s) will necessarily have zeros at $\pm j\omega_{r_1}$, $\pm j\omega_{r_2}$ and $\pm j\omega_{r_3}$.

Returning now to Feldtkeller's equation and rewriting it in terms of the three polynomials E(s), P(s), F(s) and the parameter ε, we obtain the central equation of approximation and passive realization theory for filter networks,

$$E(s)\,E(-s) = P(s)\,P(-s) + \varepsilon^2\,F(s)\,F(-s) \qquad (2.13)$$

Equation (2.13) relates the natural modes to the loss poles and reflection zeros. Given a set of loss poles and reflection zeros, and a certain value for the passband ripple parameter ε, we can obtain the natural modes by factoring the right-hand-side of (2.13) and assigning the resulting left-half-plane roots to E(s). In general, this task must be performed by means of an accurate computer routine. The FILTOR 2 package[5] includes a computer routine FELDT which solves this equation.

From (2.13), it follows that for a given set of loss poles and reflection zeros, different values of ε will result in different natural modes, and hence, different transfer functions H(s). The value of ε then serves as a parameter for a family of filter functions, all the members of which have the same P(s) and F(s) but different E(s), as determined from (2.13). The transfer function for this family may be expressed as

$$H(s,\varepsilon) = \frac{E(s,\varepsilon)}{P(s)}$$

This concept of using ε, or equivalently the passband distortion A_p, as a parameter is useful in cataloguing filter functions in tables, as will be seen in Chapter 6. Fig. 2.2 illustrates the above ideas by showing the $A(\omega)$ curves for two filters having the same loss poles and reflection zeros, but different passband distortion.

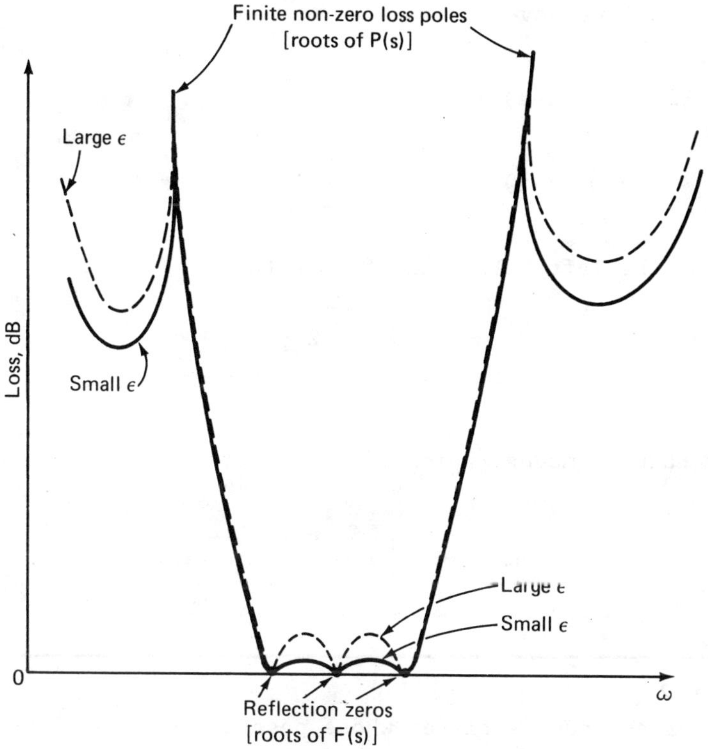

Figure 2.2 Illustrating the effect of ϵ upon a typical transfer function.

Example 2.1

Given one loss pole at infinity ($n_i = 1$), one loss pole at zero ($n_o = 1$), and a pair of reflection zeros at $s = \pm j\omega_r$ with $\varepsilon = 1$, find the order of the network and its transfer function.

Solution

From (2.5) we have

$$N = n_o + n_i = 2$$

Here, $P(s) = s$ and $P(s)\,P(-s) = -s^2$ with $F(s) = (s^2 + \omega_r^2)$ and $F(s)\,F(-s) = (s^2 + \omega_r^2)^2$.

From (2.13) we have

$$E(s)\ E(-s) = -s^2 + (s^2 + \omega_r^2)^2 \qquad \text{or,}$$

$$E(s)\ E(-s) = (s^2 + s + \omega_r^2)\ (s^2 - s + \omega_r^2)$$

Selecting the left-half-plane factor for E(s) yields

$$H(s) = \frac{E(s)}{P(s)} = \frac{s^2 + s + \omega_r^2}{s}$$

and the natural modes, which are the roots of E(s), are given by

$$s_{1,2} = -\frac{1}{2} \pm j\frac{1}{2}\sqrt{4\omega_r^2 - 1}$$

Example 2.2

Consider a bandpass filter whose passband extends from f_1 = 1 KHz to f_2 = 1.4 KHz. If the frequency axis is normalized with respect to the upper edge of the passband, the filter is characterized by:

Loss-poles: two at infinity ($n_i = 2$), two at zero ($n_o = 2$), a pair at $s = \pm j1.2501$ and a pair at $s = \pm j0.3996$ ($n_f = 2$)

Passband: equiripple with A_p = 0.4 dB ($\varepsilon = 0.31061$)

Reflection zeros: one pair at each of $s = \pm j0.98919$, $s = \pm j0.90953$, $s = \pm j0.79696$ and $s = \pm j0.72391$. Also, the leading coefficient of F(s) is 733.3508.

It is required to find the natural modes via Feldtkeller's equation. To illustrate the concept of a family of filters parameterized by A_p (or ε), we will also find the natural modes for the cases A_p = 0.8 and 1.6 dB.

Solution

The filter order is obtained from (2.5) as

$$N = n_o + n_i + 2n_f = 8$$

The reflection zeros, the loss-poles and the given value of ε can be used in Feldtkeller's equation to determine E(s) E(-s) and hence E(s). However, since the order of the network is relatively high it is quite difficult to do this manually. A computer subroutine such as FELDT of FILTOR 2[5] may be used to solve the Feldtkeller equation. For the case A_p = 0.4 dB the natural modes are found to be

$$\begin{aligned} \omega_{o1} &= 0.7106 & Q_1 &= 15.256 \\ \omega_{o2} &= 0.7911 & Q_2 &= 6.192 \\ \omega_{o3} &= 0.9188 & Q_3 &= 6.743 \\ \omega_{o4} &= 1.0048 & Q_4 &= 19.113 \end{aligned}$$

where the values of ω_o are normalized with respect to 2π x 1400 rad/sec.

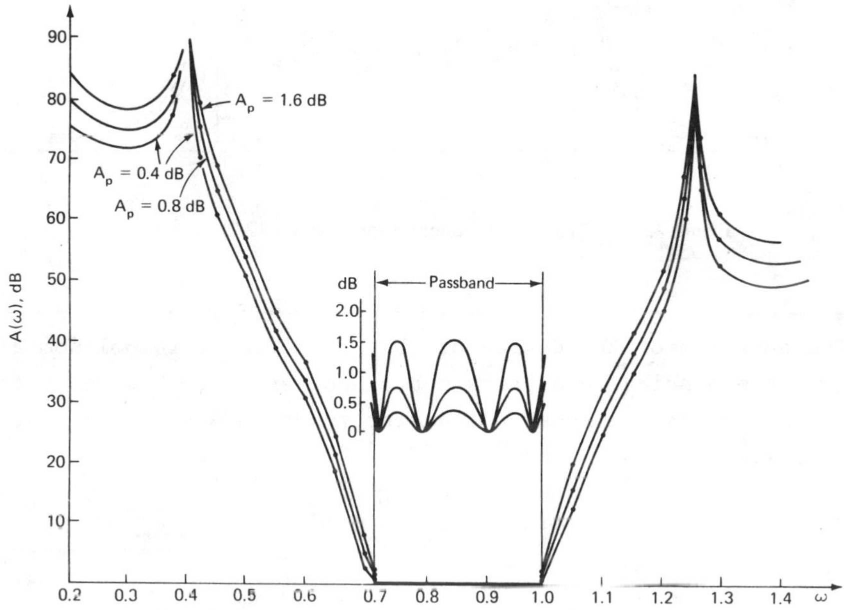

Figure 2.3(a) Attentuation function for Example (2.2).

The natural modes for A_p = 0.8 and 1.6 dBs can be similarly obtained. Fig. 2.3 shows plots of the attenuation and group-delay functions for the three cases. These plots were obtained using an evaluating subroutine of FILTOR 2. The attenuation plots clearly show the trade-off between passband ripple and stopband attenuation. The reader should also note that the group delay is far from the ideal constant, and that its distortion becomes worse as the maximum passband attenuation distortion A_p is increased. This is because the attenuation characteristics become steeper; an observation we have predicted by examining (1.21) of Chapter 1. This is further illustrated by the fact that $\tau(\omega)$ peaks where $A(\omega)$ is steepest, at the passband edges.

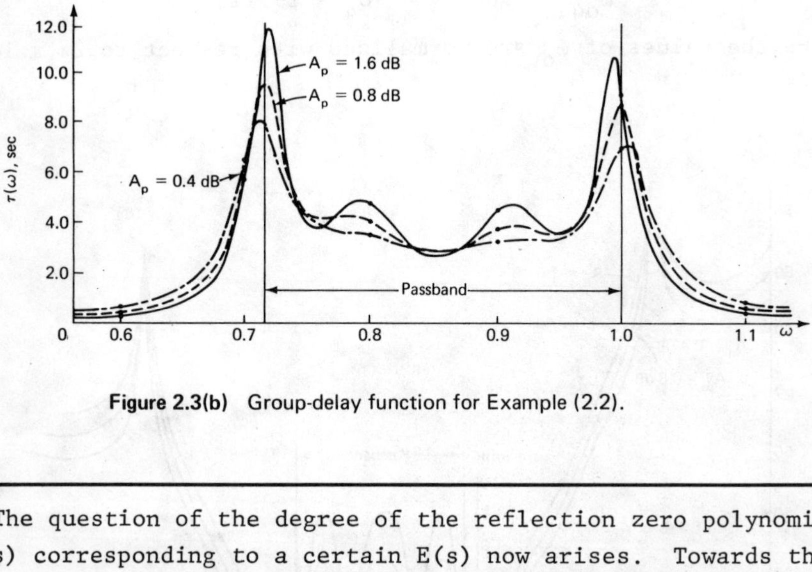

Figure 2.3(b) Group-delay function for Example (2.2).

The question of the degree of the reflection zero polynomial F(s) corresponding to a certain E(s) now arises. Towards this end, we express the squared transfer function H(s) H(-s) in terms of its singularities; that is

$$H(s)\,H(-s) = (-1)^{n_i}\, h^2 \frac{\prod_{n=1}^{N} (s^2 - s_n^2)}{\prod_{p=1}^{M} (s^2 - s_p^2)} \tag{2.14}$$

where h is a multiplicative constant and n_i = N - M is the number of loss poles at infinity. Similarly, the squared characteristic function may be expressed as

$$K(s)\,K(-s) = (-1)^{R-M} k^2 \frac{\prod_{r=1}^{R} (s^2 - s_r^2)}{\prod_{p=1}^{M} (s^2 - s_p^2)} \tag{2.15}$$

where k is a multiplicative constant (in general not equal to ε), R is the number of finite reflection zeros, and (M-R) is the number of reflection zeros at infinity. Now, rewriting Feldtkeller's equation in the above terms, we obtain an explicit relationship between natural modes, loss poles and reflection zeros as

$$(-1)^{n_i} h^2 \prod_{n=1}^{N} (s^2 - s_n^2) = \prod_{p=1}^{M} (s^2 - s_p^2) + (-1)^{R-M} k^2 \prod_{r=1}^{R} (s^2 - s_r^2) \tag{2.16}$$

Examining (2.16) and recalling that $M \le N$ it follows that

(a) If $M < N$ then R must equal N, that is, the order of F(s) is equal to the filter order.

(b) If $M = N$ then R may be less than N, and there may be reflection zeros at infinity.

Thus, if there are no reflection zeros at infinity (the usual case for low-pass and bandpass filters), the order of F(s) equals that of E(s), which is the filter order N. For high-pass filters we may well have reflection zeros at infinity, in which case the order of P(s) equals that of E(s) which is again equal to the filter order.

2.3 THE TRANSMISSION COEFFICIENT, REFLECTION COEFFICIENT AND ECHO (RETURN) LOSS

In the previous sections, we have introduced the transfer function H(s), the characteristic function K(s), and the three polynomials E(s), P(s), and F(s), which are related by the Feldtkeller equation. Some authors (in particular North American authors) do not use the two rational functions H and K, but rather use the so-called transmission coefficient t(s) and the reflection coefficient $\rho(s)$ in the characterization of filters. All four of these rational functions may be related to the physics of passive filter realizations; however, we are considering them here only as mathematical functions useful in the development of approximation methods.

First, we note that the transmission coefficient t(s) is either equal to the output/input transfer function G(s), defined in Chapter 1, or related to it by a multiplicative constant, depending upon how the input and output variables for the particular network are defined. The only reason for using the symbol t(s) rather than G(s) in the following is that the former has been prevalent in the literature. We will now show how t(s) and $\rho(s)$ are related to the three polynomials, E, P and F.

Dividing both sides of the Feldtkeller equation in (2.13) by the squared natural mode polynomial E(s) E(-s) results in

$$\frac{P(s)\ P(-s)}{E(s)\ E(-s)} + \varepsilon^2 \frac{F(s)\ F(-s)}{E(s)\ E(-s)} = 1 \tag{2.17}$$

an equation which contains two new rational functions, $t(s) = \frac{P(s)}{E(s)}$ and $\rho(s) = \varepsilon\frac{F(s)}{E(s)}$. Thus (2.17) may be written as

$$t(s)\ t(-s) + \rho(s)\ \rho(-s) = 1 \tag{2.18}$$

As will be shown in Chapter 7 the reflection coefficient can be assigned a physical meaning in the context of passive filter designs.

For $s = j\omega$, (2.18) yields

$$|t(j\omega)|^2 + |\rho(j\omega)|^2 = 1 \tag{2.19}$$

Thus, over the passband where $|t(j\omega)|^2$ approximates unity, $|\rho(j\omega)|^2$ approximates zero. Since the maximum deviation of

$|t(j\omega)|^2$ from unity is $[1 - \frac{1}{1+\varepsilon^2}]$, then the maximum deviation of $|\rho(j\omega)|$ from zero, denoted by $|\rho|_{max}$, is obtained from (2.19)

$$|\rho|_{max} = \varepsilon / \sqrt{1+\varepsilon^2} \tag{2.20}$$

Thus, in addition to ε and A_p, we have a third parameter which may be used to specify the filter passband distortion; namely, $|\rho|_{max}$. It is a usual practice to multiply $|\rho|_{max}$ by 100 and specify the passband distortion in terms of the maximum reflection coefficient in per cent. The relation between A_p and $|\rho|_{max}$ can be easily shown to be

$$A_p = -10 \log (1 - |\rho|^2_{max}) \tag{2.21}$$

To illustrate the above concepts, we present in Fig. 2.4 sketches of $|t(j\omega)|^2$ and $|\rho(j\omega)|^2$ for a bandpass filter.

A number of authors define another function termed the "return loss" or "echo loss", which we shall denote by $A_e(\omega)$ and which is a logarithmic measure of $|\rho(j\omega)|$. Thus the echo loss in dB is defined by

$$A_e(\omega) = 10 \log \frac{1}{|\rho(j\omega)|^2} \tag{2.22}$$

$$\text{i.e.} \quad A_e(\omega) = -20 \log |\rho(j\omega)| \tag{2.23}$$

We note that the echo loss is large in the passband where $|\rho(j\omega)|$ approximates zero, and small in the stopband where $|\rho(j\omega)|$ becomes large. Thus, rather than bounding the passband distortion in terms of a maximum value for the reflection coefficient $|\rho|_{max}$, one can alternatively specify a minimum value for the return loss in the passband, A_{emin}. The two quantities are related by

$$A_{emin} = -20 \log |\rho|_{max} \tag{2.24}$$

which may be also expressed in terms of ε as

$$A_{emin} = -20 \log \frac{\varepsilon}{\sqrt{1 + \varepsilon^2}} \tag{2.25}$$

It is easily established that A_{emin} and A_p are related by the symmetric formulae

$$A_{emin} = -10 \log (1 - 10^{-A_p/10}) \tag{2.26}$$

$$A_p = -10 \log (1 - 10^{-A_{emin}/10}) \tag{2.27}$$

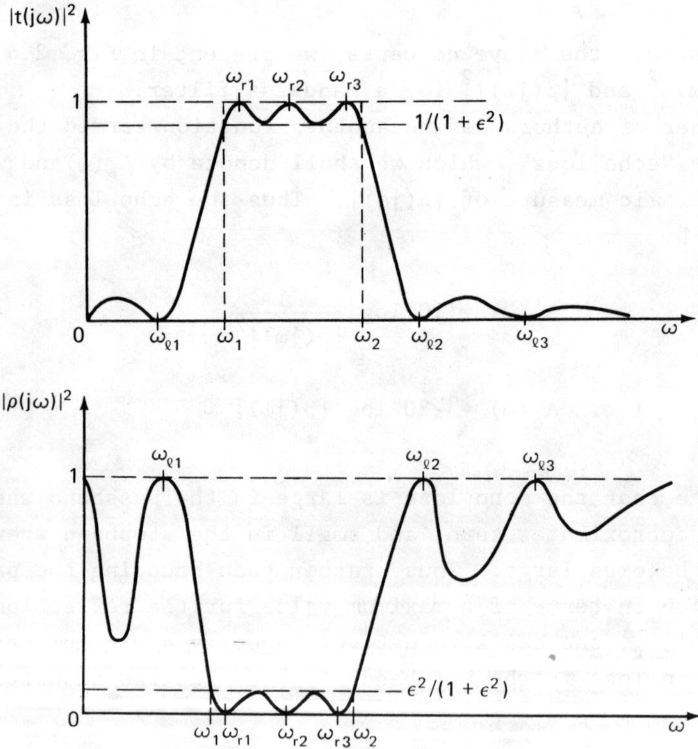

Figure 2.4 Typical transmission and reflection coefficient magnitude squared for a bandpass filter.

Fig. 2.5 illustrates typical plots of A and A_e for a bandpass filter, across its passband. Note here that we are using A_e merely as a mathematical function useful in describing approximations. In the special case of passive filters, A_e may be assigned a physical meaning.

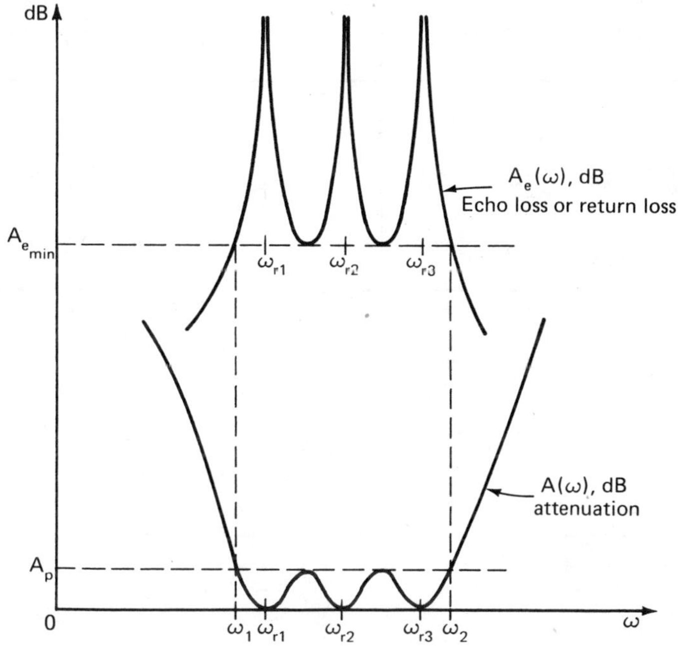

Figure 2.5 Typical plots for $A(\omega)$ and $A_e(\omega)$ for a bandpass filter.

Thus, we find that there are four parameters in common use for specifying the passband performance of filters. These are ε, A_p, $|\rho|_{max}$ and A_{emin}. Table 2.2 provides a useful collection of formulae relating these four parameters.

Of all the four parameters, the minimum echo loss is probably the most useful from the designer's point of view, since there exists a direct approximate relationship between A_{emin}, which specifies passband distortion, and the stopband attenuation. As will be shown, this relationship enables the designer to directly trade off passband distortion for stopband attenuation and vice-versa.

Table 2.2 Formulae relating the four parameters used for specifying passband distortion.

$A_p = 10 \log (1+\varepsilon^2)$ dB	$A_e = -10 \log [\varepsilon^2/(1+\varepsilon^2)]$ dB
$A_p = -10 \log (1-\lvert\rho\rvert^2)$ dB	$A_e = -10 \log \lvert\rho\rvert^2$ dB
$A_p = -10 \log (1-10^{-A_e/10})$ dB	$A_e = -10 \log (1-10^{-A_p/10})$ dB
$\varepsilon = \sqrt{10^{A_p/10} - 1}$	$\lvert\rho\rvert = \varepsilon/\sqrt{1+\varepsilon^2}$
$\varepsilon = \sqrt{10^{-A_e/10}/(1-10^{-A_e/10})}$	$\lvert\rho\rvert = \sqrt{1 - 10^{-A_p/10}}$
$\varepsilon = \lvert\rho\rvert / \sqrt{(1-\lvert\rho\rvert^2)}$	$\lvert\rho\rvert = \sqrt{10^{-A_e/10}}$

Notes:

A_p = maximum passband attenuation in dB

A_e = minimum passband echo loss in dB

$|\rho|$ = maximum magnitude of reflection coefficient in passband

Consider the attenuation function

$$A(\omega) = 10 \log |H|^2 = 10 \log (1 + |K|^2)$$

evaluated in the stopband region where $|K|^2 >> 1$ (say for $A(\omega) > 20$ dB). Here we have the approximate relationship

$$A(\omega) \simeq 10 \log |K|^2 = 10 \log \varepsilon^2 \left|\frac{F}{P}\right|^2$$

which may be rewritten as

$$A(\omega) = 10 \log \left|\frac{F}{P}\right|^2 + 10 \log \varepsilon^2 \tag{2.28}$$

We recognize F and P as the reflection zero and loss pole polynomials which characterize a family of transfer functions as explained before. Thus, equation (2.28) tells us that, in the stopband, the sum of 10 log $|F/P|^2$ and 10 log ε^2 equals the attenuation.

Now making the approximation

$$\frac{\varepsilon^2}{1 + \varepsilon^2} \simeq \varepsilon^2$$

which holds for small A_p (say $A_p < 1$ dB), we recognize that the minimum echo loss approximates the second term of (2.28); that is

$$A_{emin} = -10 \log \frac{\varepsilon^2}{1 + \varepsilon^2} \simeq -10 \log \varepsilon^2$$

Substituting in (2.28) results in

$$A(\omega) + A_{emin} \simeq \text{Constant} \tag{2.29}$$

where the constant is a function only of the particular family. This relation holds for A > 20 dB and $A_p < 1$ dB, which is true for most practical filters. Using (2.29), the filter designer can tell immediately that he must trade x dB of passband performance A_{emin} in order to gain x dB of stopband attenuation. This is further illustrated in Example 2.3 below.

Example 2.3

The minimum attenuation of the first filter function of Example 2.2, corresponding to $\varepsilon = 0.31061$ ($A_p = 0.4$ dB), is 50.6 dB in the upper stopband. Let us say that we wish to increase this attenuation by 5 dB to 55.6 dB. By how much should A_p change?

From equation (2.29) we note that A_{emin} must decrease by 5 dB. Using the equations of Table 2.2, we find that $A_p = 0.4$ dB corresponds to $A_{emin} = 10.65$ dB. The new A_{emin} will thus be 5.65 dB, which corresponds to $A_p = 1.35$ dB. Thus, an increase of 5 dB in stopband attenuation is obtained at the price of an increase of about 1 dB in passband distortion, for this particular filter.

2.4 APPROXIMATION CRITERIA

We are now in a position to give a more precise statement of the approximation problem for ideal attenuation characteristics. Recalling the factored form of the "squared" characteristic function in equation (2.15), where the s_r are the reflection zeros, and the s_p are the loss poles, we have the following problem.

Given a set of attenuation tolerance boundaries designated A_p and A_s, as described in Chapter 1, find the locations of loss poles s_p and reflection zeros s_r, and the value of the constant k (related to ε), such that the attenuation function

$$A(\omega) = 10 \log [\, 1 + |K|^2] \qquad (2.30)$$

"fits" within the tolerance contours in an "optimum" manner.

Still, there remains the ambiguity of the words "fit" and "optimum". In order to make these notions more precise, we now discuss certain approximation criteria. Generally, the optimality of an approximation lies in the "eye of the beholder", since a "good fit" is a subjective measure, and in the absence of a well defined measure of optimality, there is no criteria on which to base an objective judgement. In fact, in many cases of design, there may be certain constraints, not explicitly given by the attenuation boundaries, which the designer might like to consider. For instance, the designer usually has some tacit notions about the "worst" delay distortion he will tolerate in his design, and he might well subjectively include this in his criteria of optimality. A further example might be a constraint on the maximum Q of the natural modes which can be tolerated. In general, it is very difficult to include such criteria in an approximation method without introducing some form of numerical optimization, which departs from the spirit of the more direct methods which we are exposing here.

Clearly, one good measure of optimality is to satisfy the specifications with a function having the lowest possible order. This leads to the conclusion that we should place the loss poles so that we get the maximum possible stopband attenuation effect from them. Similarly, we should place the reflection zeros in a way that provides the "best" approximation to zero dB in the passband. It follows that the loss poles will be most effective if they lie on the $j\omega$-axis in the stopband, while the reflection zeros should lie on the $j\omega$-axis in the passband. Thus, we are led to consider only those transfer functions which have imaginary axis loss poles and reflection zeros.

The above conclusion might not be valid if we wish to satisfy both attenuation and group delay requirements. Clearly, the imaginary axis loss poles in the stopbands have no effect on the passband delay, other than that produced by the relationship of the natural modes to them through Feldtkeller's equation. Thus, for those kinds of problems, where both an attenuation and delay tolerance are prescribed, we should not rule out the use of non-imaginary axis loss poles. In addition to this, there exist certain problems for which we will find it desirable to have non-imaginary axis reflection zeros. In the remainder of this section, however, we are concerned only with the case of imaginary axis loss poles and reflection zeros.

Let us now consider the problem of defining an optimum solution for a typical set of stopband attenuation specifications by some placement of loss poles along the imaginary axis. The form of squared loss pole polynomial we are considering here is

$$P(s)\,P(-s) = (-1)^{n_o}\, s^{2n_o} \prod_{k=1}^{n_f} (s^2 + \omega_k^2)^2 \tag{2.31}$$

and the order of the corresponding filter function is

$$N = n_o + n_i + 2n_f \tag{2.32}$$

The stopband attenuation of this class of filter will have the general character of that shown in Fig. 2.6. The attenuation function will exhibit peaks at the corresponding loss poles, and it will arc or loop downward between these peaks. We must find a placement of the loss pole locations (loss peaks), such that the stopband attenuation everywhere exceeds the stopband loss specification contour, and we require that the order of the function needed to do this be as small as possible. In some cases, there may be the additional restriction that some of the loss poles are prescribed. For instance, if a +L +C ladder realization is required, we must ensure that we have at least one loss pole at infinity for the low pass case, and at least one loss pole each at zero and infinity for the bandpass case. Another example of this is the case of a fixed loss peak which is required to produce a very large attenuation at a specific frequency, such as may be required for the rejection of a particular frequency or unwanted tone in the stopband.

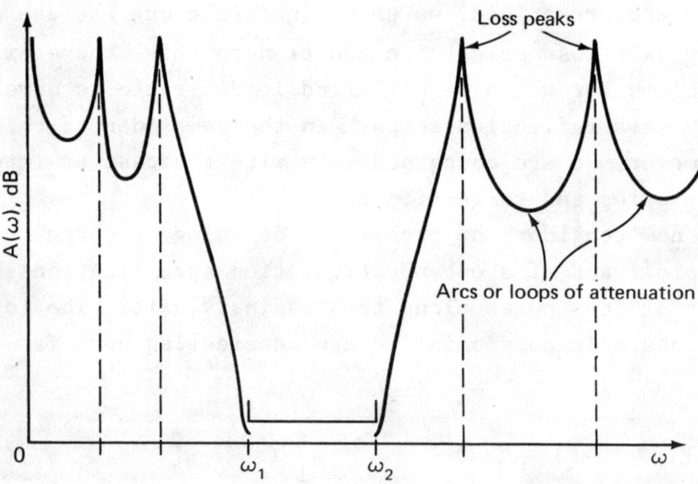

Figure 2.6 Illustrating stopband loss peaks and attenuation arcs.

Consider now the arcs or loops where the attenuation function dips downward between loss peaks. We define an error function or "margin" in dB, which is the difference between the specification contour $A_s(\omega)$ and the actual attenuation of the filter response

$A(\omega)$. Thus, the margin can be expressed as

$$E(\omega) = A(\omega) - A_s(\omega) \tag{2.34}$$

We note that the margin is positive when $A(\omega)$ exceeds $A_s(\omega)$, and is negative when $A(\omega)$ is less than $A_s(\omega)$. For typical specification contours $A_s(\omega)$, which are usually straight line segments, there will be one or more points in each arc where the margin reaches a minimum. A typical set of minima are illustrated in Fig. 2.7.

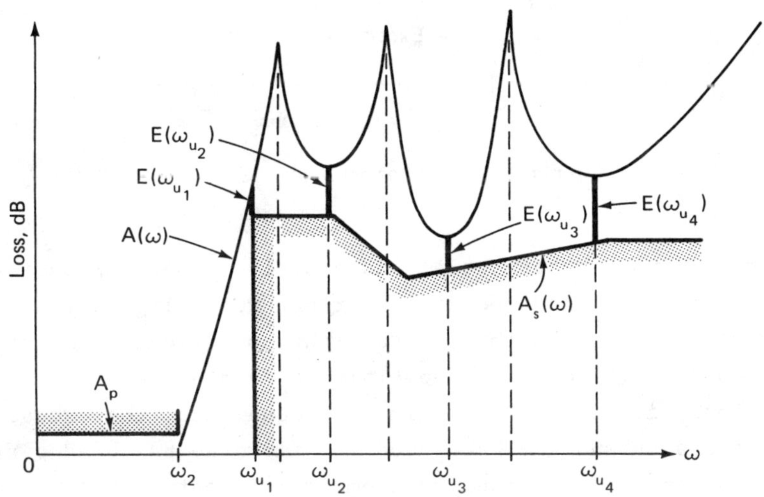

Figure 2.7 Illustrating stopband minima $E(\omega_u) = A(\omega_u) - A_s(\omega_u)$ dB

Obviously, to satisfy our requirements, all of these minima should be positive. The reader should now recall from section 2.3 that the minima will be a function of ε (and the maximum passband loss $A_p = 10 \log (1 + \varepsilon^2)$), since increasing ε increases $A(\omega)$. Thus, if all of the minima are positive, we may decrease ε and achieve a lower passband distortion. In fact, we may decrease A_p until the smallest positive minimum just vanishes. This is illustrated in Fig. 2.8. Clearly, this situation is not optimum since we now have excess margin at the remaining minima.

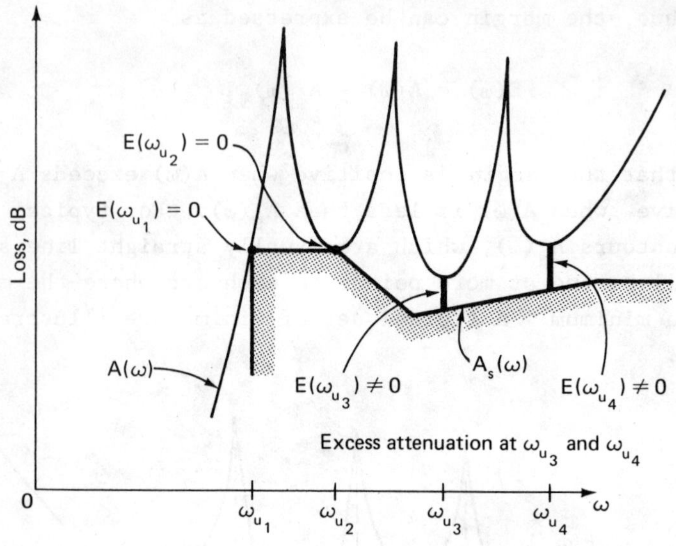

Figure 2.8 Illustrating the case of non-equal minima, when A_p is reduced.

If the minima had been equal and positive, then the passband ripple A_p might be reduced still further resulting in less passband distortion. This implies that the specification might be satisfied with the original specified A_p and a lower order function. Thus, if all of the minima are equalized, the excess margin in the stopband may be traded for improved passband performance (smaller A_p). If the improvement is great enough, then a function of lower order will satisfy the specification. The reader should realize that it is not always possible to reduce the order in this manner, since order changes occur in discrete integer steps, and the excess margin may not be large enough to allow the decrease in order. In these cases, we must be satisfied with a response that slightly exceeds the specifications.

We now have our objective clearly before us. We must place the loss poles such that the minima of the margin are forced to become equal. The procedure to be followed is to estimate the order required, then estimate an initial set of loss poles and adjust these loss poles until the margins are equalized. If the resulting margins are negative, we have failed, and we must use a higher order function. If the resulting margins are positive, then we have succeeded, and we may either reduce A_p or attempt a design with a lower order function.

In the above discussion, we have not explicitly described what is to be done about the reflection zeros. The ideal attenuation in the passband is, of course, constant (equal to zero) and we seek a placement of the reflection zeros, such that zero attenuation is approximated in an optimum manner. By arguments similar to those given for the stopband loss, we should place the reflection zeros along the imaginary axis in the passband region, such that the maximum deviation of attenuation from zero is minimized. This again occurs when all of the maxima are forced to become equal, which results in a mini-max (commonly called Chebyshev) or equiripple approximation, as illustrated in Fig. 2.9. Note that any other distribution would result in a greater passband distortion and hence a smaller stopband attenuation for the given order.

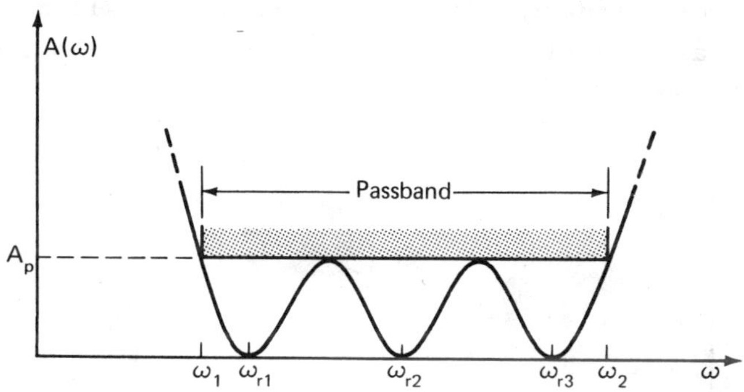

Figure 2.9 The reflection zeros, ω_r's are placed in a way to produce an equiripple passband attenuation.

A possible (but inefficient) procedure for obtaining numerically this optimum solution to attenuation specifications has been described[6] and consists of a modification of Remez method[7] for the gradual equalization of extrema. Briefly, this consists of guessing an initial set of loss poles and reflection zeros, searching all passband and stopband attenuation "arcs" for their extrema, and then adjusting all loss poles and reflection zeros to force the stopband margins to become equal while simultaneously forcing the passband margins to become equal. This procedure is performed iteratively until the desired equalization is obtained. However, as will be shown subsequently, there is a simpler and more efficient method, which allows the separation of the stopband and passband approximation problems by taking advantage of the generally simpler passband tolerance boundary.

Before proceeding to an outline of these methods, we should like to discuss another commonly encountered passband approximation. The so-called maximally flat (also called Taylor) approximation of zero is obtained by choosing the reflection zeros such that, if the characteristic function $|K|$ is expanded in Taylor's series about the bandcenter (zero for a low-pass), then as many terms as possible of the series are forced to be zero. Such a function is termed maximally flat at the bandcenter, since the function and as many of its derivatives as possible are forced to zero at that point. It is left as an exercise for the reader to show that this situation is obtained only when all of the reflection zeros are coincident. Figure 2.10 illustrates a maximally flat passband.

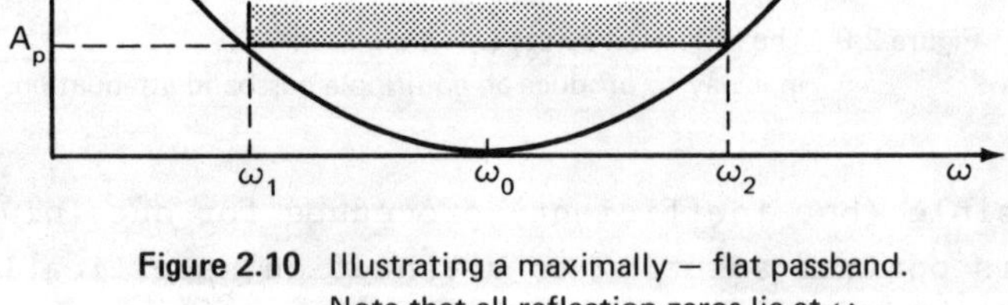

Figure 2.10 Illustrating a maximally — flat passband. Note that all reflection zeros lie at ω_0.

Comparing the maximally flat and equi-ripple passband approximations, we note that the former approximates zero only at a "point", while the latter provides an approximation of zero over a "band". Of course, by using arguments similar to those above, we can show that the equi-ripple approximation in the passband is more efficient than the maximally flat in that a larger stopband attenuation may be obtained when the reflection zeros are distributed such that the attenuation maxima in the passband are all equal (the maximally flat has the two extreme maxima equal while all of the remaining possible maxima are zero). It is thus clear that, for an optimum attenuation approximation, the equi-ripple passband is superior, and the maximally flat passband finds its usefulness in those areas where other more subjective criteria such as transient response, delay distortion, or natural mode-Q constraints are involved.

2.5 TRANSFORMED VARIABLE

In this section, we introduce a complex-variable transformation which proves to be very useful in filter design. The use of this transformation has come about for two reasons. The first is that many of the functions involved in approximation for filters are very complex when expressed in the Laplace frequency variable s. The introduction of a new variable simplifies these expressions and allows a greater insight to be obtained. The second reason is that the filter design problem is notorious for its numerical ill-conditioning. This is caused by the "clustering" of the roots of the polynomials near the band edges in the s-plane. Introducing a new variable maps the s-plane in such a way that the clustered roots are "spread out" in the new plane, thus alleviating the difficult numerical situations which arise due to the clustering.

Consider the bandpass filter problem where ω_1 is the lower edge of the passband and ω_2 is the upper edge. Without loss of generality, we introduce the frequency normalization (ω/ω_2) so that the lower edge of the passband becomes $a = \omega_1/\omega_2$ and the upper edge of the passband becomes unity. Fig. 2.11 illustrates this normalized s-plane. Note that a low-pass situation is obtained if ω_1/ω_2 is set equal to zero.

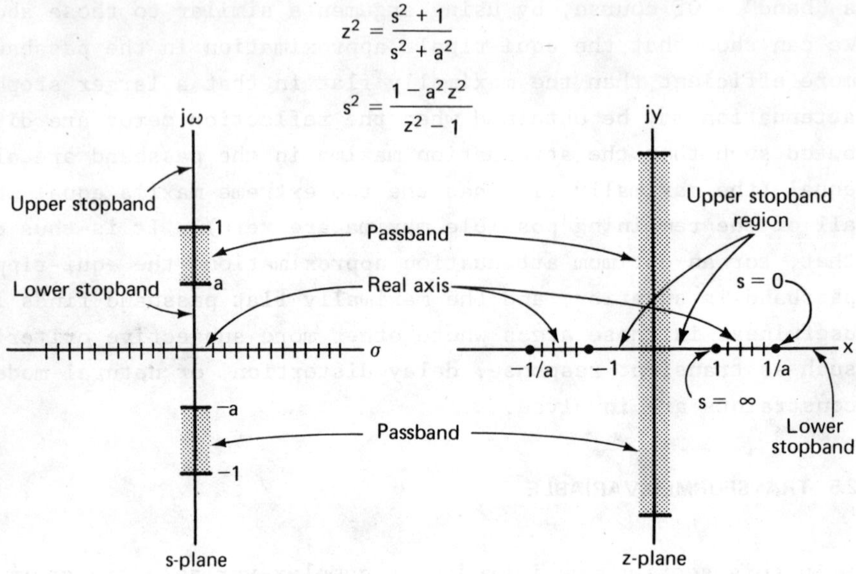

Figure 2.11 Illustrating the mapping between the s-plane and z-plane.

There are several closely related transformations of the s-plane which map the passband region $[a \leq |\omega| \leq 1]$ onto the entire imaginary axis of the new plane. The most widely used transformation is the bilinear mapping between the squared variables,

$$z^2 = \frac{s^2 + 1}{s^2 + a^2} \tag{2.35}$$

and its inverse mapping,

$$s^2 = \frac{1 - a^2 z^2}{z^2 - 1} \tag{2.36}$$

The reader will note that the two variables $s = \sigma + j\omega$ and $z = x + jy$ are related through their squares. This will not cause difficulties in approximation, since we shall be dealing exclusively with transfer functions in their squared form.

Fig. 2.11 illustrates the mapping between the s and z planes.

Note that the passband $[a \leq |\omega| \leq 1]$ of the s-plane is mapped onto the entire imaginary axis of the z-plane. The real axis of the s-plane is mapped onto the regions $[1 \leq |x| \leq 1/a]$ of the real axis in the z-plane. The stopband and transition band regions on either side of the passband in the s-plane are mapped to the real axis of the z-plane, on either side of the mapped real axis of the s-plane. Specifically, the upper stopband region lies in $[-1 < x < 1]$ and the lower stopband region lies in $[1/a \leq |x| \leq \infty]$

We see that in the z-plane, the passband (along imaginary axis) and stopbands (along real axis) are orthogonal to each other. This fact enables one to separate the passband and stopband approximation problems and express one in terms of the other, as will be shown in subsequent chapters.

The reader should also be aware of the fact that the point at infinity in the s-plane maps to the finite points ± 1 on the real axis of the z-plane. This is due to the rational character of the bilinear mapping (2.35) which, when applied to a polynomial, yields a rational polynomial in the new variable.

We note that the transformation is applicable for low-pass filters by setting $a = 0$ which yields

$$z^2 = 1 + \frac{1}{s^2} \tag{2.37}$$

and

$$s^2 = \frac{1}{z^2 - 1} \tag{2.38}$$

Here also, the transformation serves the dual purpose of enhancing the numerical conditioning by "spreading" the passband, and of placing the passband and stopband on orthogonal axes, thus enabling the passband and stopband approximations to be separated.

As will be seen, a major part of the solution to the filter approximation problem will be performed in the z-plane. Therefore, we need to find expressions for the characteristic function and loss-pole polynomial in terms of the z variable. In the following, we do this for the various loss-pole groupings of Table 2.1. To avoid introducing new symbols, we simply use the same symbol, with an added prime, to denote the mapped z-plane function. For instance, the z-transformation of the function $G(s^2)$ will be denoted by $G'(z^2)$ where

$$G'(z^2) = G(s^2) \Big|_{s^2 = \frac{1 - a^2 z^2}{z^2 - 1}}$$

Table 2.3 shows the transformations of the various groups of loss poles. For instance, loss poles at $s = \infty$ transform to $z = \pm 1$, and hence result in the factor $(-1)^{n_i} (z^2 - 1)^{n_i}$ in the z-plane squared loss-pole polynomial $P'(z) P'(-z)$. The reader is urged to verify the transformations of Table 2.3 and to derive expressions for the locations in the z-plane as a function of the corresponding s-plane locations. As an example, it can be shown, using the transformation equation (2.35), that a pair of conjugate $j\omega$-axis loss poles at $s = \pm j\omega_o$ transform into a pair of real axis poles at $z = \pm x_o$ where,

$$x_o = \sqrt{\frac{1 - \omega_o^2}{a^2 - \omega_o^2}}$$

We are now in a position to write down the most general form of loss pole polynomial in the z-plane. Referring to Table 2.3 and equation (2.4), we find that for a filter having n_o loss poles at zero, n_i loss poles at infinity, n_f pairs of finite non-zero conjugate imaginary-axis loss poles, n_p real-axis mirror-image pairs, n_q quads, n_r real-axis poles, and n_c complex-conjugate pairs, the most general form of squared z-plane loss pole polynomial is,

$$P'(z)P'(-z) = (-1)^{n_o + n_i + n_r} (z^2 - 1/a^2)^{n_o} (z^2 - 1)^{n_i} \prod_{k=1}^{n_f} (z^2 - x_k^2)^2$$

$$\prod_{\ell=1}^{n_p} (z^2 - x_\ell^2)^2 \prod_{m=1}^{n_q} [z^4 + 2(y_m^2 - x_m^2) z^2 + (x_m^2 + y_m^2)^2]^2$$

$$\prod_{n=1}^{n_r} (z^2 - x_n^2) \prod_{i=1}^{n_c} [z^4 + 2(y_i^2 - x_i^2)\, z^2 + (x_i^2 + y_i^2)^2] \qquad (2.39)$$

Note that equation (2.39) is written in terms of the loss pole locations in the z-plane. It is, however, not a direct mapping of P(s)P(-s); the direct mapping of the latter differs from (2.39) by a real constant, as well as by certain denominator factors which are not used. This constant is unimportant since we will carry out the entire process in the z-plane. (as outlined in the next section dealing with approximation strategy.)

To convert polynomials from the s-plane to the z-plane, and the reverse, a filter design program requires two routines that implement equations (2.35) and (2.36). In FILTOR 2[5] these routines are named STOZ and ZTOS.

2.6 APPROXIMATION STRATEGY

We are now in a position to outline the strategy used in this book to obtain approximations to the ideal attenuation characteristics. In this context, we will also outline the contents of the following three chapters.

The z-mapping introduced in section 2.5 is very useful in solving the approximation problem. We have already seen that such a mapping allows the separation of the passband and stopband, which share the imaginary axis of the s-plane, into separate and orthogonal axes in the z-plane. This, in turn, will enable us to completely separate the passband approximation problem from the stopband approximation problem.

The passband approximation problem is one of finding the squared reflection zero polynomial F'(z)F'(-z), such that for a given squared loss pole polynomial P'(z)P'(-z) the characteristic function K'(z)K'(-z) approximates zero in the passband (extending over the entire jy axis) in a prescribed optimum manner. The latter refers to either a maximally flat passband or an equiripple passband. In Chapter 3, we solve the passband approximation problem for a maximally flat passband, while in Chapter 4, we do the same for an equiripple passband. In both cases, the

Table 2.3 Loss-pole polynomials in the z-plane.

	Loss-pole group	$P'(z)\,P'(-z)$	z-plane locations
(1)	n_i	$(-1)^{n_i}(z^2-1)^{n_i}$	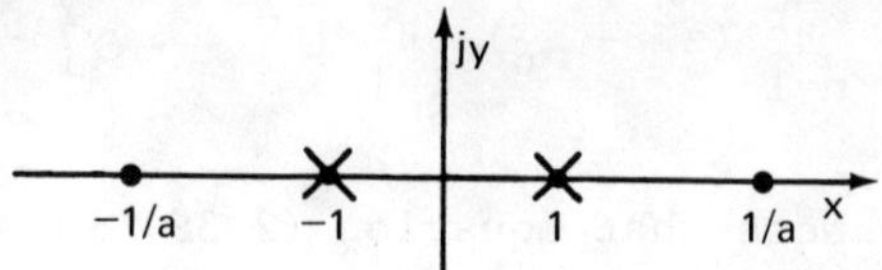
(2)	n_0	$(-1)^{n_0}(z^2-1/a^2)^{n_0}$	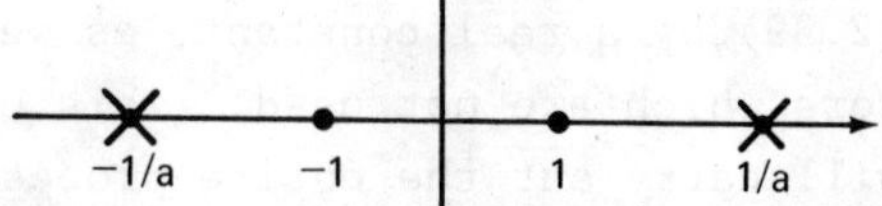
(3)	n_f	$\prod_{k=1}^{n_f}(z^2-x_k^2)^2$	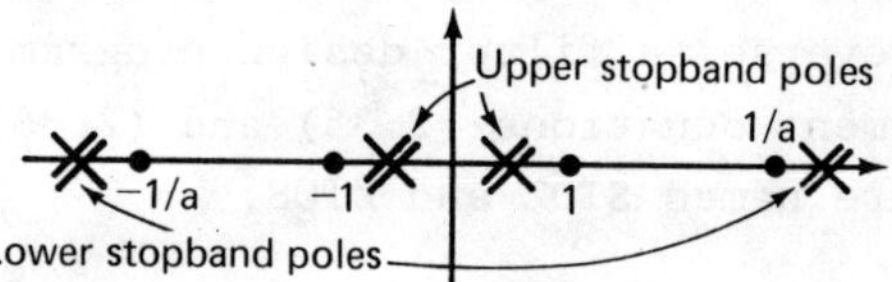
(4)	n_p	$\prod_{\ell=1}^{n_p}(z^2-x_\ell^2)^2$	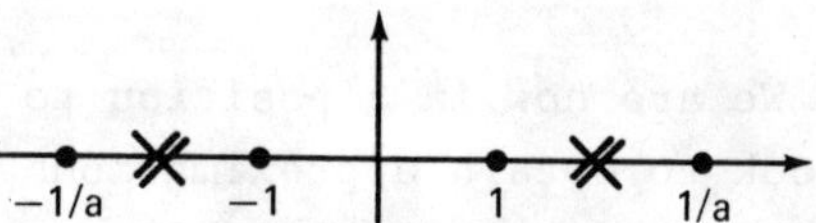
(5)	n_q	$\prod_{m=1}^{n_q}[z^4+2(y_m^2-x_m^2)z^2+(x_m^2+y_m^2)^2]^2$	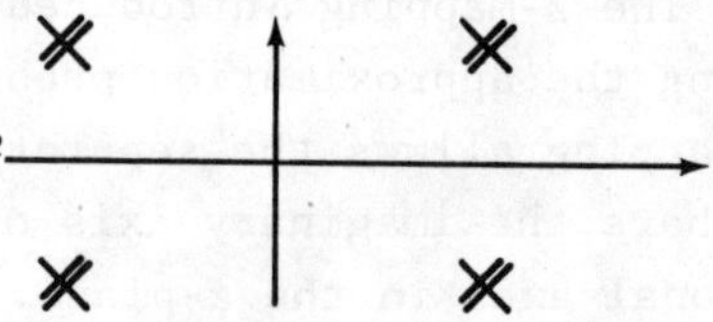
(6)	n_r	$(-1)^{n_r}\prod_{n=1}^{n_r}(z^2-x_n^2)$	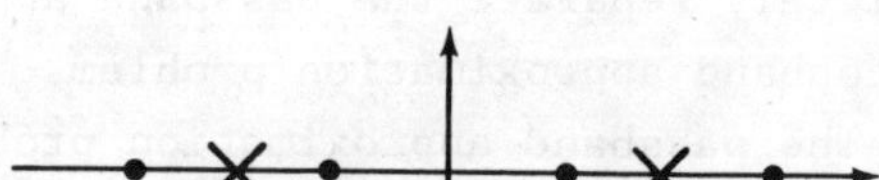
(7)	n_c	$\prod_{i=1}^{n_c}[z^4+2(y_i^2-x_i^2)z^2+(x_i^2+y_i^2)^2]$	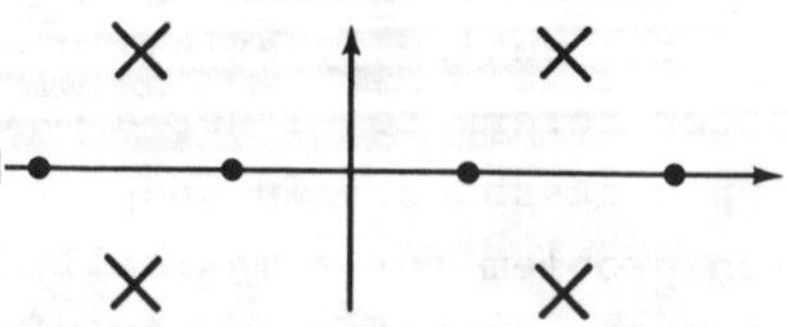

starting point is a given squared loss pole polynomial $P'(z)P'(-z)$ which, in the most general case, takes the form (2.39). The result in each case is a closed form expression for the squared reflection zero polynomial $F'(z)F'(-z)$ and hence the characteristic function,

$$K'(z)K'(-z) = \varepsilon^2 \frac{F'(z)\ F'(-z)}{P'(z)\ P'(-z)} \tag{2.40}$$

Note that, as already mentioned in section 2.5, the constant multiplier in the expression for $P'(z)P'(-z)$, that is equation (2.39), is ignored, since we will force $K'(z)K'(-z)$ of (2.40) to equal ε^2 at the passband edges, $z = 0$ and $z = \infty$. This will ensure that we end up with the squared characteristic function correctly scaled. In each of Chapters 3 and 4, an algorithm is given for finding the squared reflection zero polynomial.

Having determined the characteristic function that provides the required passband performance for the given loss poles, we will then proceed to determine the natural modes, using the z-plane form of Feldtkeller's equation,

$$E'(z)E'(-z) = P'(z)P'(-z) + \varepsilon^2\ F'(z)\ F'(-z) \tag{2.41}$$

Solving (2.41) involves a root finding process, as outlined in Section 2.8. Finally, the squared natural mode polynomial and the squared loss pole polynomial are transformed back to the s-plane to obtain the transfer function $H(s)$.

The starting point for solving the passband approximation problem is a squared loss pole polynomial. The latter is obtained by solving the stopband approximation problem. Such a solution uses iterative schemes and is given in Chapter 5 It should be noted that, in solving the stopband approximation problem, use is made of the closed form expressions for the characteristic function, as obtained in Chapter 3 and Chapter 4 This will ensure that the set of loss poles obtained will always provide the prescribed passband performance, when used in the appropriate characteristic function expression.

The flow chart of Fig. 2.12 clearly outlines the strategy used in solving the approximation problem. Also indicated is the particular chapter of the book dealing with each step and the name of the FILTOR 2 routine that implements that step. For simplicity, the chart shows the process in the s-plane. In practice, however, the entire procedure is carried out in the z-plane with the conversions performed at the beginning and the end.

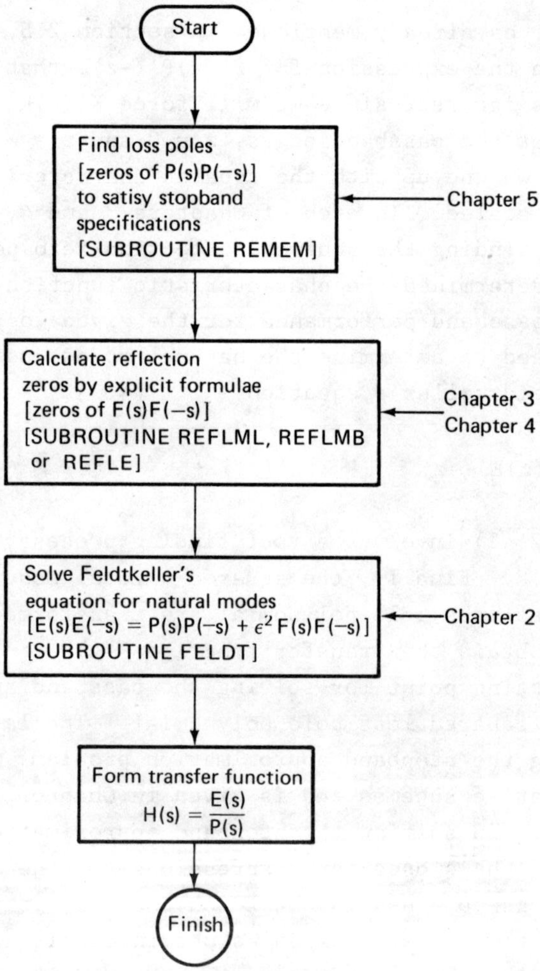

Figure 2.12 Flow chart of filter approximation process.

Both Chapters 3 and 4 are concerned with the passband approximation problem for bandpass and low-pass filters. The high-pass case is easily obtained by using the frequency-band transformation presented in Chapter 6.

2.7 CLASSICAL FILTER APPROXIMATIONS

The iterative procedure given in Chapter 5 for solving the stopband approximation problem accommodates frequency-dependent stopband attenuation specifications. It places the loss poles so that these specifications are satisfied in an optimum manner; the criteria for optimality have been given in section 2.4. This stopband approximation procedure requires a computer routine for its implementation. If one does not have access to a computer, or if the stopband specifications are simple, one reverts to one of the "classical approximations", whose parameters have been extensively tabulated.

Classical approximations to the ideal attenuation characteristics can accommodate only constant stopband attenuation. Tabulations of such approximations are usually available only for the low-pass case shown in Fig. 2.13. The results can be applied, with some limitations, to the design of other filter types, that is BP, HP and BR filters, through the use of the frequency band transformations presented in Chapter 6.

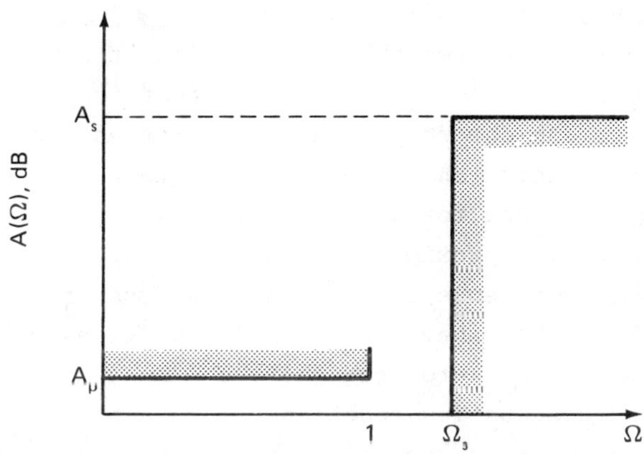

Figure 2.13 Normalized tolerance plot of LP filters for which classical approximations exist.

There are four main types of classical approximations to the LP attenuation characteristics of Fig. 2.13. Fig. 2.14 illustrates typical attenuation functions for these four types. They are classified according to the manner in which the function approximates the constant attenuation specifications in the two bands of interest. We consider maximally flat and equiripple approximations to the passband attenuation, and monotonic and equiminima, or simply equiripple, approximations to the stopband attenuation. The four approximation types follow.

(1) The Butterworth approximation provides a maximally flat passband (around zero frequency) and a monotonically increasing attenuation function in the stopband. This approximation follows as a special case of the general maximally flat passband approximation of Chapter 3, by letting all the loss poles lie at infinite frequency. The details of the Butterworth approximation are given in Chapter 3. Note that since all the loss poles are at infinity, H(s) for a Butterworth filter is a polynomial and hence the name "polynomial filter" is usually used.

(2) The Chebyshev approximation provides an equiripple passband and a monotonically increasing attenuation function in the stopband. This approximation follows as a special case of the general equiripple passband approximation of Chapter 4 by letting all the loss poles lie at infinity. The details of the Chebyshev approximation are given in Chapter 4. Note that the Chebyshev filter is also a polynomial filter.

Compared to the Butterworth approximation, the Chebyshev approximation is more efficient in the sense that it provides an approximation over the entire passband rather than around a single frequency, the origin, as in the Butterworth case. Consequently, for the same order and for an equal amount of passband distortion the Chebyshev approximation exhibits a higher cutoff rate, that is a steeper attenuation function, at the passband edge.

(3) The Inverse Chebyshev approximation provides a maximally flat passband (around zero frequency) and an equiminima (equiripple) stopband attenuation function. Although the Inverse Chebyshev approximation is a special case of the general maximally-flat-passband approximation of Chapter 3, where the stopband att-

enuation approximates a constant, it is most easily derived by "inverting" the Chebyshev approximation. This will be demonstrated in Chapter 4. Note that the Inverse Chebyshev filter has finite loss poles.

(4) The Elliptic approximation provides an equiripple passband and an equiminima, or equiripple, stopband attenuation function. This approximation follows as a special case of the general equiripple-passband approximation of Chapter 4, where the loss poles are placed in a way which yields an equiminima stopband attenuation. Of the four classical attenuation approximations, the Elliptic is the most efficient. In other words, for a given A_p, A_s and Ω_s, the Elliptic approximation will have the lowest order N. Details of the Elliptic approximation are given in Chapter 4.

The Butterworth approximation is named after the English engineer who first proposed this type of attenuation characteristic. The Chebyshev approximation is named after the Russian mathematician who first used the polynomials involved in this approximation. The elliptic approximation is given this name as it uses elliptic functions. It is sometimes referred to as the Cauer or Darlington approximation, since Cauer and Darlington independently suggested this form of filter approximation functions. It should be noted that each of the classical attenuation functions is optimum in its own way[8].

2.8 THE PRODUCT METHOD IN FILTER APPROXIMATIONS

The essentials of modern filter design (as opposed to the older image parameter methods) have been in existence since the now classic work of Darlington[9] and Cauer[10]. However, due to the mathematical complexity of the modern theory, little practical filter design using the modern methods was undertaken until the use of high speed digital computers became widespread. Early work in computer-aided filter design, using the modern methods implemented in a straightforward manner, was disappointing. This was due to the inherent ill-conditioning of the numerical operations involved. In fact, early workers found that, when working directly in the s plane and using the modern methods and compu-

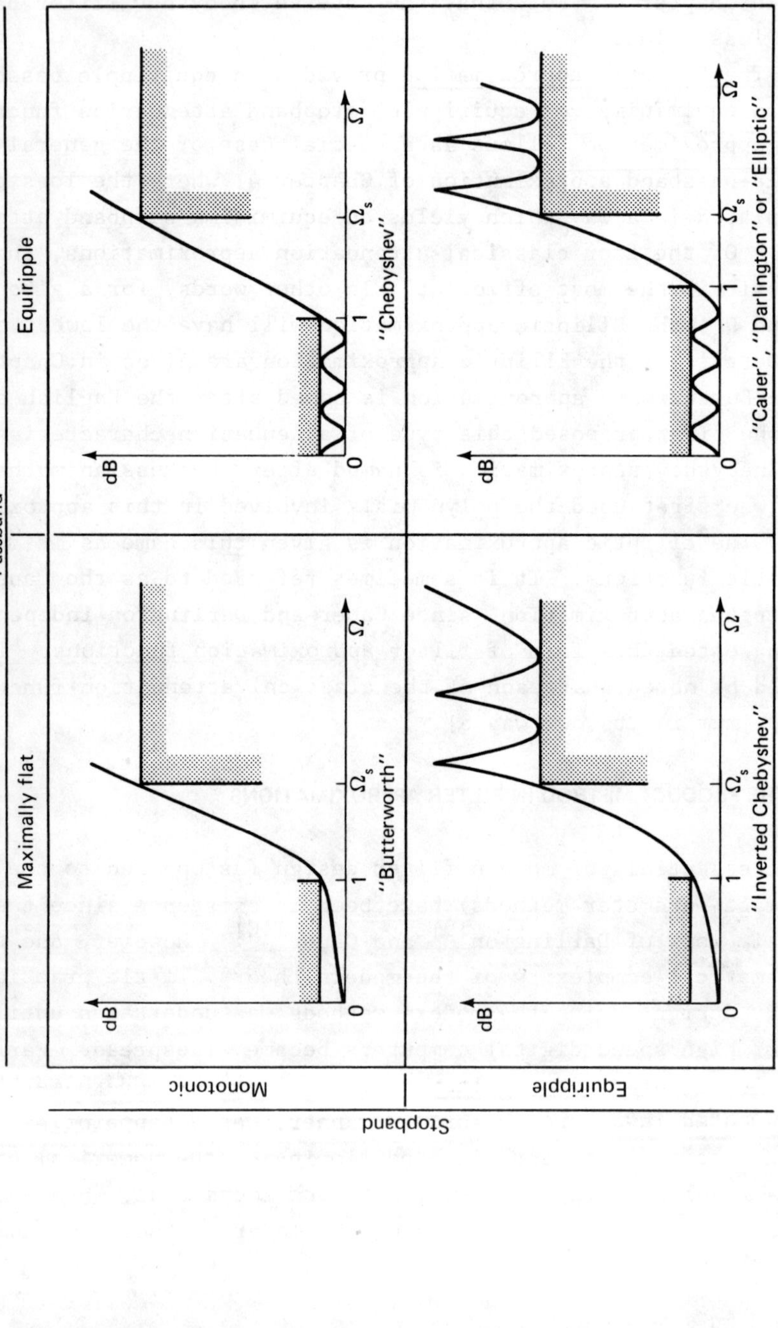

Figure 2.14 The four main classes of tabulated classical approximations.

ters with N decimal digits of precision, a moderately stringent design could be carried out, with sufficient accuracy, only for filter orders less than about N. (Most modern computers have 6-8 decimal digits in single precision.)

These early results were quite disappointing, since much higher order designs can be easily undertaken using the older image parameter methods. This numerical accuracy problem in filter design then became the object of much research and, after recognition of the cause of the numerical difficulties (polynomial root clustering in the s-plane), led to the development of two separate and quite successful approaches to solving the problem.

The two competing methods, which were apparently developed simultaneously[11] are the "Transformed Variable"[12,13] method, and the "Product Method"[14,15]. Both of these methods take cognizance of the fact that the determination of the roots of a polynomial, when they are clustered, is a numerically ill-conditioned problem. This is because most of the information concerning the relative spacing of closely located roots lies in the last few significant digits of the polynomial coefficients. As a simple example of this effect, consider the following polynomial having the closely spaced roots $s_1 = -1.001$ and $s_2 = -0.999$,

$$(s + 1.001)\ (s + 0.999) = s^2 + 2.000s + 0.999999.$$

On a 4 digit floating point machine, the polynomial in coefficient form might be

$$s^2 + 2.000s + 1.000 = (s + 1.000)^2 \quad ,$$

in which case the roots cannot be separated. We note, in this example, that all of the information about the separation of the two roots lies in the last 3 digits of the constant term, which are truncated or rounded off by the machine. This effect occurs to varying degrees in all of the filter polynomials, in the regions near the band edges.

The Transformed Variable Method makes use of a mapping (discussed in Section 2.5), which expands the clustering region (the passband) in the s-plane over an infinite axis in the new plane. In the new plane (z-plane), the filter polynomials are therefore much better conditioned and, providing that the complete design is carried out in the z-plane before finally transforming to the s-plane, one finds that on a machine using N decimal digits of precision, filters up to an order of about 3N can be successfully designed without taking undue precautions.

The Product Method does not revert to a transformed plane, but carries out all calculations in the s-plane in a "product form". That is, polynomial roots are never multiplied together to obtain polynomials in coefficient form, which would lead to the coefficient truncation effect. This means that simple operations, such as polynomial addition and subtraction, must necessarily involve a root finding process, and the product form of a sum is found directly by factoring (root finding) the sum expressed as a sum of factored forms. Thus, one treats the sum of two polynomials as a function whose roots are to be found. Again, this method has led to comparable results in accuracy enhancement.

It can now be seen that a still better method is a combination of the two approaches. In the combined method used in the computer programs of FILTOR 2[5], all calculations are performed in the transformed plane using the product method. Thus, all of the z-plane polynomials are kept in factored form and the product method is used for all polynomial operations.

As an example, consider the problem of solving Feldtkeller's equation. The FILTOR 2 routine used is called FELDT. A complete listing of FELDT can be found in[5]. The basic function of FELDT is to take a set of loss poles and reflection zeros together with the constant ε, and calculate the corresponding natural modes via the Feldtkeller's equation (2.41). In this routine, the two polynomials on the right hand side of Feldtkeller's equation are evaluated separately in product form, and the results added to form a function which is used by a root finding routine called MULLER. The roots of this function are of course the "squared" natural modes, and are thus found directly from the roots of the original two polynomials, without ever resorting to the actual multiplication of roots to obtain polynomial coefficients with the attendant loss of accuracy due to coefficient rounding.

In summary, subroutine FELDT accepts P'(z) and F'(z), each represented in terms of the leading coefficient and the roots. Another input parameter to FELDT is the passband distortion factor ε. FELDT then uses the root finding routine MULLER to calculate the roots of the function $[P'(z)P'(-z) + \varepsilon^2 F'(z)F'(-z)]$ which are also the roots of E'(z)E'(-z). Using ZTOS, these roots can be transformed to the s-plane and those in the left-half-plane assigned to E(s).

2.9 REFERENCES AND SOURCES FOR FURTHER READING

(1) L. Weinberg, Network Analysis and Synthesis, McGraw-Hill, New York, 1962, Chapter 10.

(2) T. Fujisawa, "Realizability theorem for mid-series or mid-shunt low-pass ladders without mutual induction", IRE Trans. Circuit Theory, Vol. CT-2, No.4, pp. 320-325, Dec. 1955.

(3) N. Balabanian, Network Synthesis, Prentice-Hall, Englewood Cliffs, N.J., 1958, Chapter 6.

(4) G. C. Temes and S. K. Mitra (Eds.), Modern Filter Theory and Design, Wiley, New York, 1973, Chapters 1 and 2.

(5) W. M. Snelgrove and A. S. Sedra, FILTOR 2 - A Computer Aided Filter Design Package, Matrix Publishers, Champaign, Ill.

(6) W. F. McGee, "Numerical approximation technique for filter characteristic functions", IEEE Trans. on Circuit Theory, Vol. CT-14, pp. 92-94, Mar. 1967. Reprinted in (16).

(7) E. Remez, "Sur le calcul effectif des polynomes de Tchebicheff", Compt. Rend. Acad. Sci. (France), Vol. 199, pp. 337-340, July 1934.

(8) Reference (1), Chapter 11.

(9) S. Darlington, "Synthesis of reactance four-poles", J. Math. Phys., Vol. 18, pp. 257-353, Sept. 1939.

(10) W. Cauer, Synthesis of Linear Communication Networks, McGraw Hill, New York, 1958.

(11) G. Szentirmai, Introduction to Reference (16), pp. 1-4.

(12) G. Szentirmai, "Theoretical basis of a digital computer program for filter synthesis", Proc. 1st Allerton Conf. Circuit and System Theory, pp. 37-49, 1963.

(13) H. J. Orchard and G. C. Temes, "Filter design using transformed variables", IEEE Trans. Circuit Theory, Vol. CT-15, pp. 385-408, 1968.

(14) C. Norek, "Product method for the calculation of the effective loss in LC filters" in Proc. Int. Symp. Network Theory (Belgrade Yugoslavia), pp. 353-365, 1968.

(15) J. K. Skwirzynski, "On synthesis of filters", IEEE Trans. Circuit Theory, Vol. CT-18, No. 1, pp. 152-163, Jan. 1971.

(16) G. Szentirmai (Ed), Computer Aided Filter Design, IEEE Press New York, 1973.

(17) D. S. Humphreys, The Analysis, Design and Synthesis of Electrical Filters, Prentice-Hall, Englewood Cliffs, N.J., 1970.

(18) H. Watanabe, "Approximation theory for filter networks", IRE Trans. Circuit Theory, Vol. CT-8, No. 3, pp. 341-356, Sept. 1961.

(19) A. I. Zverev, Handbook of Filter Synthesis, Wiley, New York, 1967.

(20) R. W. Daniels, Approximation Methods for Electronic Filter Design, McGraw-Hill, New York, 1974.

(21) A. Papoulis, "On the approximation problem in filter design", IRE Conv. Rec., 5, Pt. 2, pp. 175-185, 1975.

PROBLEMS

P2.1 A filter has the following loss poles: $n_o=2$, $n_i=2$, one pair at $s=\pm j0.5$ and a pair at $s=\pm j2$. Give an expression for P(s). What is the order of this filter? Is the filter realizable as +L+C ladder network? What is the filter type?

P2.2 A low-pass filter has three loss poles at ∞, one reflection zero at zero and a pair of reflection zeros at $s=\pm j$ 0.866025. Find the squared characteristic function $K(s)\,K(-s)$. The passband extends from $\omega=o$ to $\omega=1$ rad/sec with $\varepsilon=1$. Sketch $A(j\omega)$. Use Feldtkeller's equation to determine the transfer function H(s) and hence the natural modes.

P2.3 (a) A bandpass filter has two loss poles at zero and two loss poles at ∞. The reflection zeros are all at $s=\pm j\omega_o$ where ω_o is a frequency in the passband which extends from $\omega=\omega_1$ to $\omega=\omega_2$ rad/sec with $\varepsilon=1$. Find $K(s)\,K(-s)$ and show that ω_o lies at the geometric center of the passband (i.e. $\omega_o^2=\omega_1\omega_2$).

(b) Show that this bandpass filter may be derived from a second-order maximally flat low-pass filter whose pole frequency is 1 rad/sec using the transformation:

$$s \to \frac{s^2+\omega_o^2}{s(\omega_2-\omega_1)}$$

(c) Use the transformation of (b) above to find the natural modes of the bandpass filter given that $\omega_o=10(\omega_2-\omega_1)$.

P2.4 For the low-pass filter of Problem P2.2 calculate $|\rho|_{max}$ and A_{emin}. Derive expressions for $\rho(s)$ and $t(s)$ and check that $\rho(s)\,\rho(-s) + t(s)\,t(-s)=1$. Sketch $|\rho(j\omega)|$ and $A_e(\omega)$.

P2.5 For a doubly terminated LC ladder filter the reflection coefficient function $\rho(s)$ is identical to the usual physical definition of the reflection coefficient that is

$$\rho(s) = \frac{Z_1-R}{Z_1+R} \quad \text{(see Fig. P2.5)}$$

Use this definition to derive an expression for $\rho(s)$ for the filter in Fig. P2.5. For $R=1\Omega$, $L=1H$ and $C=2F$ find the location of the reflection zeros, natural modes and loss poles. Hence, give H(s). Sketch $A(\omega)$. What type of filter is this? What is the type of classical approximation used?

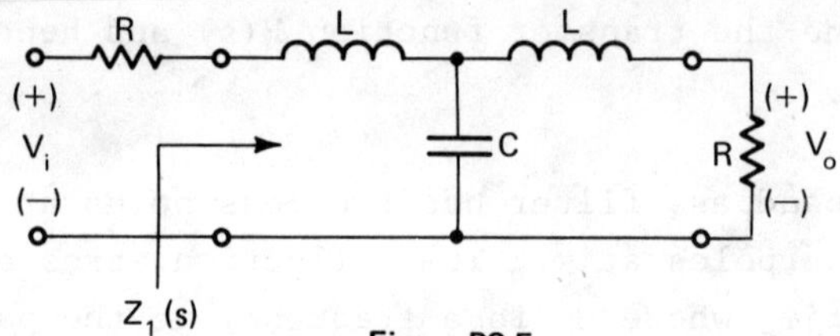

Figure P2.5

P2.6 The LC ladder network in Fig. P2.6 realizes a low-pass filter:

(a) Find the loss poles by inspection

(b) Find the reflection zeros which are the values of s at which $Z_1(s)=R$.

(c) If the passband edge is at $\omega=1$ rad/sec and if $R=1\Omega$, $L=1.0512$ H, $L_1=0.2019$ H and $C_1 = 0.9612$F find A_p and ε.

(d) Sketch $A(\Omega)$ and find the minimum attenuation in the stopband (A_s).

(e) Use Feldtkeller's equation to find the natural modes.

(f) If the loss poles and reflection zeros are held constant but it is desired to increase A_s by 5dB, what is the new value of A_p?

(g) What is the type of classical approximation used in this filter?

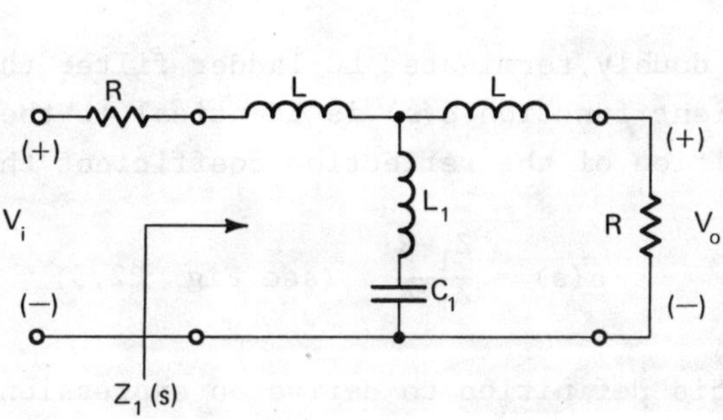

Figure P2.6

P2.7 A filter has 3 reflection zeros at $s=\infty$ and three loss poles at $s=o$. The attenuation A=3dB at $\omega=1$ rad/sec. Sketch $A(\Omega)$ and show that this is a high pass filter.

Find the transfer function H(s) and show that it may be obtained from the transfer function of a low pass filter having 3 reflection zeros at zero and three loss poles at ∞ (i.e. a third order Butterworth) by applying the transformation $s \to (1/s)$.

P2.8 Consider a low-pass filter approximation that meets the normalized specifications of Fig. 2.13 where $\varepsilon=1 (A_p \simeq 3\text{dB})$. Let the filter have three reflection zeros at zero, one loss pole at ∞ and a pair of loss poles at $\Omega=\Omega_\ell$. For $\Omega_s=\sqrt{3}$, find Ω_ℓ and A_s.

P2.9 A bandpass filter has A_p=1dB and A_s=50dB. It is required to reduce A_p to 0.5dB while keeping the reflection zeros and loss poles unchanged. Find the new value of A_s.

P2.10 A third-order low-pass Butterworth filter has the transfer function

$$H(s) = (s+1)(s^2+s+1)$$

Give the expression for the squared transfer function in the z-plane. Sketch the loss poles and natural modes in the z-plane.

P2.11 A bandpass filter has a passband extending from $\omega=1/\sqrt{2}$ to $\omega=1$ rad/sec with all the reflection zeros at a frequency ω_o in the passband. The value of $\varepsilon=1$ $(A_p \simeq 3\text{dB})$. The loss poles are:
one at zero, one at ∞, a pair at $\pm j0.5$ and a pair at $\pm j\sqrt{2}$.

(a) Find the filter order N
(b) Find $P'(z)\,P'(-z)$
(c) Find $K'(z)\,K'(-z)$
(d) Find the value of ω_o
(e) Find $K(s)\,K(-s)$
(f) Find $|K(j\omega)|^2$

(g) Sketch $A(\omega)$ and show that it is geometrically symmetric; that is, every two frequencies ω_1 and ω_2 for which $A(\omega_1)=A(\omega_2)$, are related by $\omega_1\omega_2=\omega_o^2$.

CHAPTER 3

THE MAXIMALLY FLAT PASSBAND

This chapter is concerned with the solution of the passband approximation problem for bandpass and low-pass filters having a maximally flat passband and arbitrary stopband attenuation characteristics. Closed form expressions are derived for the location of the reflection zeros, in terms of an arbitrary set of loss poles. The latter may be determined from a solution to the stopband approximation problem (Chapter 5), but in the following they are considered to be merely an arbitrary set, which may have been determined in any manner.

Section 3.1 presents the general formulation for the bandpass case, and the low pass case is treated in section 3.2. High-pass filters are not specifically considered, but their design is readily obtainable, using the low-pass formulae together with the low-pass to high-pass frequency transformation presented in Chapter 6.

The Butterworth approximation is studied in section 3.3 where it is derived as a special case of the formulation of section 3.2.

3.1 MAXIMALLY FLAT BANDPASS FILTERS

Objective and Strategy

The object of this section is to develop an expression for the characteristic function of a bandpass filter having a maximally flat passband attenuation characteristic, such as that shown in Fig. 3.1. The stopband attenuation characteristics are arbitrary and it may be assumed that the loss poles have already been determined, using the iterative procedure of Chapter 5. The starting point, therefore, is a given squared loss pole polynomial, $P'(z)P'(-z)$, as well as a specified maximum passband distortion, A_p dB. It is required to find the squared reflection zero polynomial $F'(z)F'(-z)$ such that the passband attenuation function $A(\Omega)$ is maximally flat around a "center" frequency Ω_o (yet to be determined), and that $A(\Omega)$ attains the specified maximum distortion at the passband edges, $\Omega=a$ and $\Omega=1$, as shown in Fig. 3.1.

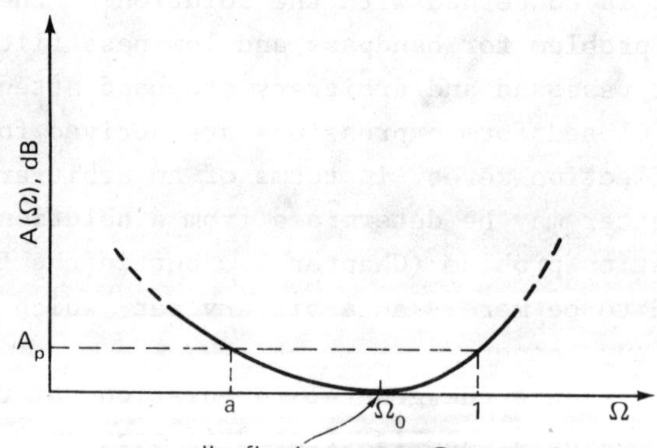

Figure 3.1 A bandpass maximally – flat attenuation function with arbitrary loss poles.

As already implied, we will carry out the process of developing the squared reflection zero polynomial in the z-plane. Once the reflection zeros are determined, they can be used together with the loss poles and the value of ε (determined from A_p), to obtain the natural modes using the Feldtkeller equation. All the singularities can then be transformed back to the s-plane to obtain K(s) and H(s).

Determining the Squared Reflection Zero Polynomial

Consider the attenuation function shown in Fig. 3.1, where the loss poles can be located anywhere except in the passband. Clearly, the characteristic function must have its zeros at some point in the passband, say $s = \pm j\Omega_o$. Thus, the reflection zero polynomial in the s-plane must be of the form

$$F(s) = (s^2 + \Omega_o^2)^{N/2} \qquad (3.1)$$

which we note requires an even order function; that is N is even.

For a given loss pole polynomial P(s), using F(s) from (3.1) results in the characteristic function

$$K(s) = \varepsilon \frac{(s^2 + \Omega_o^2)^{N/2}}{P(s)} \qquad (3.2)$$

which is the most general form of a characteristic function for a BPF with a maximally flat passband attenuation. Note that P(s) is generally of the form in (2.4), where $(n_i + n_o + n_r)$ must be an even number.

We note that the prescribed loss function must attain the value zero at $s = \pm j\Omega_o$ and that it must attain the value A_p dB at the band edges $s = ja$ and $s = j$. Thus, the loss poles and the reflection zeros, together with the parameter ε, must be forced into a relationship which satisfies this requirement. This implies that the reflection zero location Ω_o must be a function of the prescribed loss pole locations. Such a relationship is difficult to determine in the s-plane; however, it is quite simple to set up in the z-plane.

From (3.2), it follows that the squared characteristic function in the z-plane is given by

$$K'(z)\, K'(-z) = \frac{\varepsilon^2 (z^2 + y_o^2)^N}{P'(z)P'(-z)} \tag{3.3}$$

where y_o will be the z-transform of Ω_o; that is,

$$y_o^2 = \frac{1 - \Omega_o^2}{\Omega_o^2 - a^2}$$

and $P'(z)\, P'(-z)$ is of the form in (2.39) which with $(n_o + n_i + n_r)$ an even number, becomes

$$P'(z)P'(-z) = (z^2 - \frac{1}{a^2})^{n_o} (z^2 - 1)^{n_i} \prod_{k=1}^{n_f} (z^2 - x_k^2)^2$$

$$\prod_{\ell=1}^{n_p} (z^2 - x_\ell^2)^2 \prod_{m=1}^{n_q} [z^4 + 2(y_m^2 - x_m^2)\, z^2 + (x_m^2 + y_m^2)^2]^2$$

$$\prod_{n=1}^{n_r} (z^2 - x_n^2) \prod_{i=1}^{n_c} [z^4 + 2(y_i^2 - x_i^2)\, z^2 + (x_i^2 + y_i^2)^2] \tag{3.4}$$

The reader will recall that the band edges s = j (upper edge) and s = ja (lower edge) map to z = 0 and z = ∞, respectively. Thus, we must force y_o to take on a value such that $K'(z)K'(-z)$ equals ε^2 at both z = 0 and z = ∞, i.e.

$$K'(z)\, K'(-z)|_{z=0} = K'(z)\, K'(-z)|_{z=\infty} = \varepsilon^2 \tag{3.5}$$

Clearly, this is already satisfied by (3.3) at $z = \infty$, since both the numerator and denominator have equal coefficients for the highest degree terms. All that remains is to enforce condition (3.5) at $z = 0$; that is,

$$(z^2 + y_o^2)^N \big|_{z=0} = P'(z)P'(-z)\big|_{z=0}$$

Making use of (3.4) results in

$$y_o^{2N} = \left(\frac{1}{a^2}\right)^{n_o} \prod_{k=1}^{n_f} (x_k^4) \prod_{\ell=1}^{n_p} x_\ell^4 \prod_{m=1}^{n_q} (x_m^2 + y_m^2)^4 \prod_{n=1}^{n_r} x_n^2 \prod_{i=1}^{n_c} (x_i^2 + y_i^2)^2 \qquad (3.6)$$

where, as before, $(n_o + n_i + n_r)$ must be even. Condition (3.6) is the required result and shows that y_o is a function of the loss poles for a prescribed maximally flat passband. We note from (3.6) that, in fact, y_o is a function only of the magnitude squared of the loss pole locations in the z-plane.

In summary, given the loss poles, equation (3.6) is used to determine the z-plane location y_o of the reflection zeros. Then the squared reflection zero polynomial is simply given by

$$F'(z)\ F'(-z) = (z^2 + y_o^2)^N$$

and the squared characteristic function is given by (3.3). A computer routine, called REFLMB, is included in the FILTOR 2 package[1] for determining the value of y_o. If we are restricted to +L +C ladder realizations, then, from the discussion in Chapter 2, we can only allow loss poles on the $j\omega$-axis, including zero and infinity. The maximally flat condition in (3.6) then reduces to

$$y_o^{2N} = \left(\frac{1}{a^2}\right)^{n_o} \prod_{k=1}^{n_f} x_k^4 \qquad (3.7)$$

where $(n_o + n_i)$ is an even integer. Moreover, if a +L +C ladder realization is required, both n_o and n_i should have a minimum value of one.

Example 3.1

Find E(s), P(s), F(s) and all other pertinent parameters for a second-order bandpass filter having the following specifications:

Maximally flat passband, $\frac{1}{2} \leq \Omega \leq 1$ with $\varepsilon = 1$(Ap ≃ 3 dB)

Loss-Poles: one pair at $s = \pm j2$

Solution

$$n_f = 1 \quad N = 2 \quad a = \frac{1}{2}$$

The z-mapping and its inverse are given by:

$$z^2 = \frac{s^2 + 1}{s^2 + \frac{1}{4}}$$

$$s^2 = \frac{1 - \frac{1}{4}z^2}{z^2 - 1}$$

The loss poles at $\Omega = 2$ map to $x = \sqrt{0.8}$.

From (3.6),

$$y_o^4 = x^4$$

Thus, $y_o = \sqrt{0.8}$ and $\Omega_o = \sqrt{\frac{2}{3}} = 0.816497$

$$P'(z)P'(-z) = (z^2 - 0.8)^2$$

$$F'(z)F'(-z) = (z^2 + 0.8)^2$$

Since $\varepsilon = 1$ we have,

$$E'(z)E'(-z) = (z^2 - 0.8)^2 + (z^2 + 0.8)^2$$
$$= 2\,(z^4 + 0.64)$$

which results in the z-plane roots $z^2 = \pm j0.8$. Using the inverse z-mapping, the s-plane natural modes are obtained as

$$s = -0.210969 \pm j0.867079$$

The s-plane natural mode polynomial is thus given (to within a constant multiplier to be determined later) by

$$E(s) = s^2 + 0.421938\ s + 0.796333$$

Note that the natural modes are characterized by a frequency of 0.892375 and a Q-factor of 2.11494.
The transfer function H(s) is now obtained as

$$H(s) = 9.05539\ \frac{s^2 + 0.421938\ s + 0.796333}{s^2 + 4}$$

where the constant multiplier has been determined from $|H(j\Omega_o)| = 1$. The loss pole polynomial is given by

$$P(s) = s^2 + 4$$

and the reflection zero polynomial is

$$F(s) = f(s^2 + \Omega_o^{\ 2}) = f(s^2 + \tfrac{2}{3}) ,$$

where the value of the constant multiplier f may be determined by setting $|\frac{F}{P}| = 1$ at the passband edge, say $s = j1$. We thus obtain $f = 9$. The characteristic function K(s) is given by

$$K(s) = 9\ \frac{s^2 + \frac{2}{3}}{s^2 + 4}$$

The attenuation function $A(\Omega)$ is now given by,

$$A(\Omega) = 10 \log_{10} |H(j\Omega)|^2 = 10 \log_{10} [1 + |K(j\Omega)|^2] ,$$

which the reader is urged to sketch. From such a sketch it will be apparent that, although the filter considered has been obtained as a bandpass filter in this case, it may also be thought of as a low-pass notch (LPN) filter [See section 1.5].

3.2 MAXIMALLY FLAT LOW-PASS FILTERS

We now proceed to consider the low-pass maximally flat case, where $a = 0$ and there are no loss poles at $s = 0$. The usual situation is to force the maximum flatness at $\Omega = 0$; hence, we require all the reflection zeros to occur at $s = 0$, resulting in the s-plane characteristic function

$$K(s) = \varepsilon \frac{s^N}{P(s)} \tag{3.8}$$

where ε is determined from the passband distortion A_p [See Fig. 3.2] and $P(s)$ is the loss pole polynomial. Note that $P(s)$ can be quite general, with the obvious restriction that no loss poles should occur in the passband. It should also be noted that in the low-pass case N can be either odd or even.

Since $s = 0$ maps to $z = \infty$, the squared characteristic function in the z-plane will be given by

$$K'(z)K'(-z) = \varepsilon^2 \frac{C}{P'(z)P'(-z)} \tag{3.9}$$

where all the reflection zeros are placed at $z = \infty$, resulting in a constant numerator in (3.9). The squared loss pole polynomial $P'(z)P'(-z)$ is of the general form of (2.39). To determine the value of C, we impose the condition to obtain A_p dB at the band edge; namely,

$$K'(z)K'(-z)|_{z=0} = \varepsilon^2 \tag{3.10}$$

where z = 0 is the mapped location of the upper edge of the passband. This leads to

$$C = \prod_{k=1}^{n_f} x_k^4 \prod_{\ell=1}^{n_p} x_\ell^4 \prod_{m=1}^{n_q} (x_m^2 + y_m^2)^4 \prod_{n=1}^{n_r} x_n^2$$

$$\prod_{i=1}^{n_c} (x_i^2 + y_i^2)^2 \tag{3.11}$$

in which we have used the general form of P'(z)P'(-z) of (2.39)

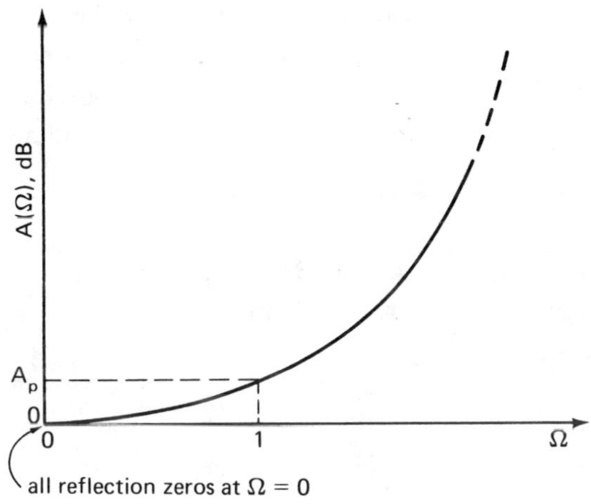

Figure 3.2 A low-pass maximally – flat attenuation function with arbitrary loss poles.

Thus, the value of the constant C is a function of the squared magnitude of the loss pole locations in the z-plane. A computer routine called REFLML is included in FILTOR 2[1] for determining the value of the constant C.

For a +L +C ladder realization of the low-pass filter, the loss poles are restricted to the jω-axis, including ∞, and (3.11) reduces to

$$C = \prod_{k=1}^{n_f} x_k^4 \tag{3.12}$$

Furthermore, to obtain a +L +C ladder realization, n_i has to be a minimum of one.

The low-pass formulae in the above apply only to the case where the attenuation is maximally flat at $\Omega=0$. This method, however, can be simply extended to cases in which maximal flatness is obtained at one or more other frequencies in the passband. In such a case, the attenuation may be nonzero at the origin. These kinds of low-pass (high-pass) filters may be thought of as bandpass filters with n_o (n_i) set to zero.

Table 3.1 provides a useful summary of the pertinent formulae for bandpass and low-pass filters with a maximally flat passband and arbitrary loss poles. The high-pass case can be simply obtained by using the low-pass formulation, together with the low-pass to high-pass transformation presented in Chapter 6.

Example 3.2

Find H(s) and K(s) for a low-pass filter which has maximally flat attenuation at $\Omega=0$, has $\varepsilon=1$ and has a pair of complex conjugate loss poles at $s = \pm j2$.

Solution

$$n_f = 1 \quad N = 2$$

$$P(s) = s^2 + 4$$

The z-mapping and its inverse are given by

$$z^2 = 1 + \frac{1}{s^2} \text{ and } s^2 = \frac{1}{z^2-1}$$

Thus, the mapped loss pole is at $x = \sqrt{\frac{3}{4}}$ and,

Table 3.1 Summary of maximally flat characteristic functions for the most general loss pole polynomial.

<table>
<tr><td colspan="2">

$$P^{`}(z)\,P^{`}(-z) = (-1)^{n_i+n_o+n_r}\left(z^2 - \frac{1}{a^2}\right)^{n_o}(z^2-1)^{n_i}\prod_{k=1}^{n_f}(z^2-x_k^2)^2\prod_{\ell=1}^{n_p}(z^2-x_\ell^2)^2$$

$$\prod_{m=1}^{n_q}\left[z^4 + 2(y_m^2 - x_m^2)\,z^2 + (x_m^2 + y_m^2)^2\right]^2\prod_{n=1}^{n_r}(z^2-x_n^2)$$

$$\prod_{i=1}^{n_c}\left[z^4 + 2(y_i^2 - x_i^2)\,z^2 + (x_i^2 + y_i^2)^2\right]$$

</td></tr>
<tr><td>

<u>Band Pass</u>

$(n_i + n_o + n_r)$ must be even

</td><td>

<u>Low Pass</u>

"a" is zero n_o must be zero

</td></tr>
<tr><td>

$$K^{`}(z)\,K^{`}(-z) = \varepsilon^2\,\frac{(z^2 + y_o^2)^N}{P^{`}(z)\,P^{`}(-z)}$$

where

$$y_o^{2N} = \left(\frac{1}{a^2}\right)^{n_o}\prod_{k=1}^{n_f}(x_k^4)\prod_{\ell=1}^{n_p}x_\ell^4\prod_{m=1}^{n_q}(x_m^2 + y_m^2)^4$$

$$\prod_{n=1}^{n_r}x_n^2\prod_{i=1}^{n_c}(x_i^2 + y_i^2)^2$$

</td><td>

$$K^{`}(z)\,K^{`}(-z) = \varepsilon^2\,\frac{C}{P^{`}(z)\,P^{`}(-z)}$$

where

$$C = \prod_{k=1}^{n_f}x_k^4\prod_{\ell=1}^{n_p}x_\ell^4\prod_{m=1}^{n_q}(x_m^2 + y_m^2)^4$$

$$\prod_{n=1}^{n_r}x_n^2\prod_{i=1}^{n_c}(x_i^2 + y_i^2)^2$$

</td></tr>
<tr><td>

For +L+C ladder realization:

$$y_o^{2N} = \left(\frac{1}{a^2}\right)^{n_o}\prod_{k=1}^{n_f}x_k^4\ ,$$

$(n_i + n_o)$ is an even integer,

$n_i \geq 1$ and $n_o \geq 1$

</td><td>

For a +L+C ladder realization:

$$C = \prod_{k=1}^{n_f}x_k^4\ ,$$

$n_i \geq 1$

</td></tr>
</table>

$$P'(z)P'(-z) = (z^2 - \frac{3}{4})^2 .$$

From (3.11), we obtain $C = \frac{9}{16}$ and hence, the squared z-plane characteristic function is given by

$$K'(z)K'(-z) = \frac{9}{16} \frac{1}{(z^2 - \frac{3}{4})^2}$$

Thus,

$$E'(z)E'(-z) = (z^2 - \frac{3}{4})^2 + \frac{9}{16} = z^4 - \frac{3}{2} z^2 + \frac{9}{8}$$

The z-plane roots of this equation are at

$$z = \pm \sqrt{\frac{3}{4} (1 \pm j)}$$

Using the inverse z-mapping results in the natural mode polynomial

$$E(s) = s^2 + 1.315227s + 1.264911$$

The transfer function is thus given by

$$H(s) = h \frac{s^2 + 1.315227s + 1.264911}{s^2 + 4}$$

where the value of h may be obtained from,

$$H(0) = 1, \text{ leading to}$$

$$h = 3.16228$$

As a check, the reader may verify that $|H(j1)|^2 = 2$

The characteristic function is given by

$$K(s) = \varepsilon \frac{F(s)}{P(s)} = f \frac{s^2}{s^2+4}$$

where the constant f is determined from

$$|K(j1)| = 1 \text{ leading to } f = 3. \quad \text{Thus,}$$

$$K(s) = \frac{3\,s^2}{s^2+4}$$

The reader is urged to sketch the attenuation function $A(\Omega) = 10 \log_{10} |H(j\Omega)|^2$. From such a sketch, it will become apparent that, according to the biquad classification in Chapter 1, this filter is a low-pass notch (LPN).

3.3 THE BUTTERWORTH FILTER APPROXIMATION

In Chapter 2, we mentioned four special approximations to the ideal low-pass attenuation characteristics. The simplest of these classical approximations is the Butterworth approximation, which exhibits a maximally-flat attenuation response at $\Omega=0$ and monotonically increasing attenuation, as shown in Fig. 3.3. The Butterworth approximation follows as a special case of the general maximally flat low-pass approximation derived in section 3.2, if all the loss poles are located at infinity; that is, $N = n_i$. Although there are a number of ways of deriving the Butterworth approximation[2,3], we will use the formulae developed in the previous section, with the specialization $N = n_i$.

Note that although the name "Butterworth approximation" is used here to imply a low-pass filter approximation, other filter types can be designed using the Butterworth approximation. This is accomplished through the use of the frequency band transformations presented in Chapter 6.

Since all the loss-poles for a Butterworth filter lie at ∞, the squared loss-pole polynomial is given by $P'(z)P'(-z) = (-1)^N (z^2 - 1)^N$. The constant C may be calculated as

$$C = P'(z')P'(-z)|_{z=0} = 1 .$$

The z-plane squared characteristic function is obtained by subs-

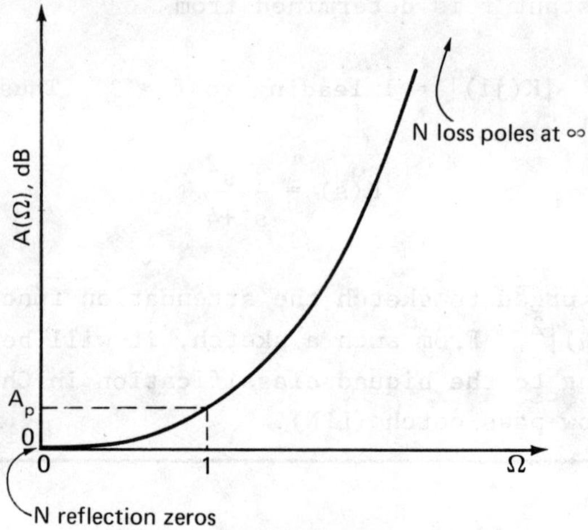

Figure 3.3 Attenuation function of an Nth order Butterworth filter approximation.

titution in (3.9),

$$K'(z)K'(-z) = (-1)^N \frac{\varepsilon^2}{(z^2 - 1)^N} \tag{3.13}$$

Substituting $z^2 = 1 + \frac{1}{s^2}$, which is the inverse z-mapping for the low-pass case, the squared characteristic function in the s-plane is obtained as

$$K(s)\ K(-s) = (-1)^N\ \varepsilon^2\ s^{2N}\ . \tag{3.14}$$

Thus,

$$K(s) = \varepsilon\ s^N\ .$$

The squared transfer function may be obtained as,

$$H(s)\ H(-s) = 1 + (-1)^N\ \varepsilon^2\ s^{2N}\ . \tag{3.15}$$

For $s = j\Omega$, equations (3.14) and (3.15) provide

$$|K(\Omega)|^2 = \varepsilon^2 \Omega^{2N} \tag{3.16}$$

$$|H(\Omega)|^2 = 1 + \varepsilon^2 \Omega^{2N} \tag{3.17}$$

The attenuation function is thus given by

$$A(\Omega) = 10 \log_{10} (1 + \varepsilon^2 \Omega^{2N}) \tag{3.18}$$

We next find an expression for the order N of the Butterworth filter, that satisfies the attenuation specifications given in the form of the tolerance plot of Fig. 3.4. Evaluating (3.18) at $\Omega = \Omega_s$ yields,

$$A_s = 10 \log_{10} (1 + \varepsilon^2 \Omega_s^{2N}) \tag{3.19}$$

The passband distortion A_p is related to the factor ε by the familiar relationship,

$$A_p = 10 \log_{10} (1 + \varepsilon^2) \tag{3.20}$$

Equations (3.19) and (3.20) are combined to provide,

$$N = \left[\log_{10} \sqrt{\frac{10^{A_s/10} - 1}{10^{A_p/10} - 1}}\right] \Bigg/ \left[\log_{10} \Omega_s\right] \tag{3.21}$$

Note that if (3.21) results in a noninteger value for N, the next higher integer value should be selected.

The natural modes for a Butterworth filter are the left-half-plane zeros of the polynomial in (3.15). These can be found as follows:

$$(-1)^N \varepsilon^2 s_k^{2N} = 1$$

$$s_k = \frac{1}{\varepsilon^{1/N}} e^{j\left(\frac{\pi}{2} + \frac{2k-1}{2N}\pi\right)}, \quad k = 1,2,\ldots,N \tag{3.22}$$

Thus, we may express H(s) in terms of its natural modes as,

$$H(s) = \varepsilon \prod_{k=1}^{N} (s - s_k) \tag{3.23}$$

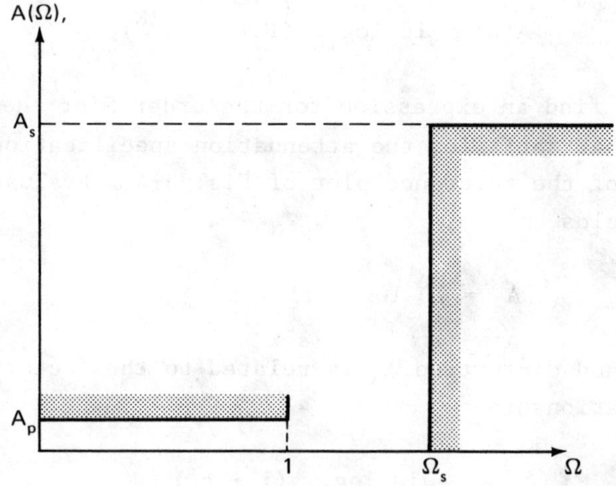

Figure 3.4 Normalized tolerance plot.

We note from (3.22) that the natural modes for a Butterworth filter lie on a circle of radius $(\frac{1}{\varepsilon})^{\frac{1}{N}}$, equally spaced by an angle of $\frac{\pi}{N}$, with the first mode at an angle of $(\frac{\pi}{2} + \frac{\pi}{2N})$. A graphical construction which determines the natural mode locations is illustrated in Fig. 3.5. In terms of ω_o and Q notation, it follows that all the natural modes of a Butterworth filter have the same frequency of $(\frac{1}{\varepsilon})^{\frac{1}{N}}$, while the Q-factor of the kth mode is given by

$$Q_k = \frac{1}{2 \sin (\frac{2k-1}{2N} \pi)}, \quad k = 1,2,\ldots,N \tag{3.24}$$

As seen from (3.23), the Butterworth filter transfer function is a polynomial, hence the name polynomial filter. The resulting polynomials are termed Butterworth polynomials.

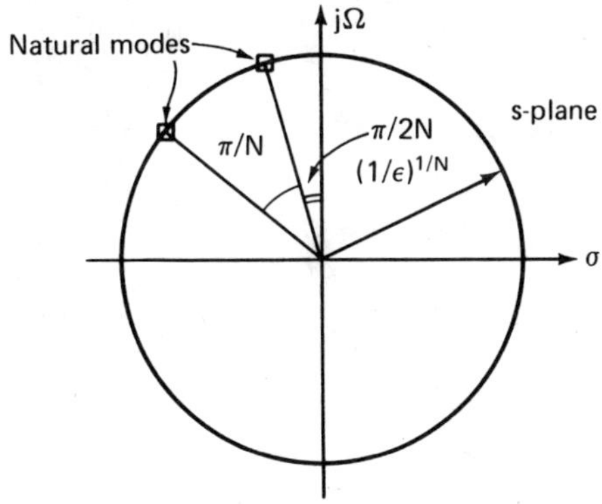

Figure 3.5 The natural modes of a Butterworth filter lie on a circle of radius $(1/\epsilon)^{1/N}$ as shown.

There exist extensive tabulations of the parameters of Butterworth filters[2]-[5]. However, the formulae given in this section, and the use of a calculator, should enable the designer to find the parameters and characteristics of any Butterworth approximation.

We conclude our discussion of Butterworth filters by noting that the Butterworth approximation is optimum in the sense that, of all the polynomial filter approximations which provide a monotonic attenuation function and a maximally flat passband, it has the highest rate of cut-off; that is, the steepest attenuation function at $\Omega = 1$.

Example 3.3

Find the pertinent parameters of the Butterworth approximation to the following low-pass filter specifications:

Passband extends from 0 to 10 KHz with $A_p \leq 0.5$ dB

Stopband extends from 20 KHz to ∞ with $A_s \geq 30$ dB.

Solution

From given data, $\Omega_s = \frac{20}{10} = 2$

The order N is determined from (3.21) as

$N = 6.49$

Thus, we select $N = 7$

From $A_p = 0.5$ dB, the value of ϵ is obtained,

$$\varepsilon = \sqrt{10^{A_p/10} - 1} = 0.3493$$

$$K(s) = \varepsilon\, s^N = 0.3493\, s^7$$

$$|K(\Omega)|^2 = 0.122\, \Omega^{14}$$

$$A(\Omega) = 10 \log_{10} (1 + 0.122\, \Omega^{14})$$

The realized A_s is

$$A_s = A(2) = 33 \text{ dB},$$

which is 3 dB in excess of the specifications.
If desired, one may use this excess margin to obtain a lower passband distortion. This can be accomplished by using the method outlined in Chapter 2. Alternatively, we can simply calculate the new value of ε as follows,

$$A(\Omega_s) = 10 \log_{10} (1 + \grave{\varepsilon}^2\, \Omega_s^{14}) = 30 \text{ dB}$$

Thus, $\grave{\varepsilon} = 0.2469$ and $A_p^{`} = 0.26$ dB.

To complete our example, we shall select the initial value of $\varepsilon = 0.3493$. The natural modes will be as shown in Fig. 3.6, where all have the same frequency $\Omega_o = 1.162132$, while the Q-factors of the complex poles are [using (3.24)],

$$Q_1 = 2.247 \quad Q_2 = 0.802 \quad Q_3 = 0.555$$

Using (3.25), the transfer function H(s) is obtained as,

$$\begin{aligned} H(s) = 0.3493\, &(s^2 + 0.517197\, s + 1.350550) \\ &(s^2 + 1.449154\, s + 1.350550) \\ &(s^2 + 2.09409\, s + 1.350550) \\ &(s + 1.162132) \end{aligned}$$

Note that the transfer function is normalized with respect to the passband edge; that is, 10 KHz. It can be denormalized by replacing s with $\frac{s}{2\pi x 10^4}$.

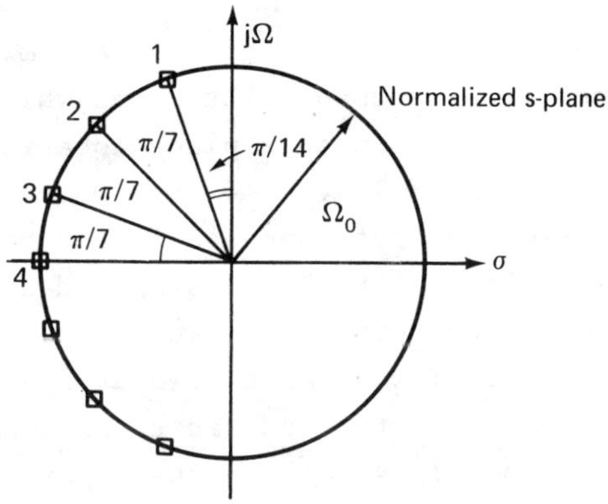

Figure 3.6 The natural modes of the seventh order Butterworth filter of Example (3.3).

3.4 REFERENCES AND SOURCES FOR FURTHER READING

(1) W. M. Snelgrove and A. S. Sedra, FILTOR 2 - A Computer Aided Filter Design Package, Matrix Publishers, Champaign, Ill.

(2) L. Weinberg, Network Analysis and Synthesis, McGraw-Hill, New York, 1962, Chapter 11.

(3) D. S. Humphreys, The Analysis, Design and Synthesis of Electrical Filters, Prentice-Hall, Englewood Cliffs, N.J., 1970, Chapter 6.

(4) A. I. Zverev, Handbook of Filter Synthesis, Wiley, New York, 1967.

(5) E. Christian and E. Eisenmann, Filter Design Tables and Graphs, Wiley, New York, 1966.

(6) G. C. Temes and S. K. Mitra (Eds.), Modern Filter Theory and Design, Wiley, New York, 1973, Chapter 2.

(7) R. W. Daniels, Approximation Methods for Electronic Filter Design, McGraw-Hill, New York, 1974.

PROBLEMS

P3.1 Find the order N and center frequency Ω_o of a maximally-flat bandpass filter which has a=0.8, n_o=1, n_i=2, one real axis loss pole at s=+1, a pair of loss poles at s=±j0.6 and a pair of loss poles at s=±j1.2.

P3.2 (a) Show that two frequencies ω_1 and ω_2 in the two stop-bands of a bandpass filter and which are related by $\omega_1\omega_2$=a transform to z-plane locations x_1 and x_2 related by x_1x_2=1/a.

(b) A maximally flat bandpass filter has n_o=n_i, (n_f/2) loss poles in the lower stopband and (n_f/2) loss poles in the upper stopband. Show that in order for the reflection zeros to lie at $\Omega_o=\sqrt{a}$, each loss pole in the lower stopband has to be geometrically symmetric with a loss pole in the upper stopband, with Ω_o as the center of symmetry.

(c) Show that the function $|K(j\Omega)|^2$ and hence $A(\Omega)$ are geometrically symmetric.
(Hint: work in the z-plane)

P3.3 In an LC ladder network the reflection coefficient is defined as

$$\rho(s) = \frac{Z_1(s)-R}{Z_1(s)+R}$$

(a) Show that for the network in Fig. P3.3 to realize a filter whose reflection zeros are all at zero we must have $L=(CR^2/2)$.

(b) For R=1Ω and assuming that we require ε=1 at ω=1 rad/sec find the value of L and C.

(c) From the expression for ρ(s) verify that the natural modes obtained are those of a third order Butterworth filter.

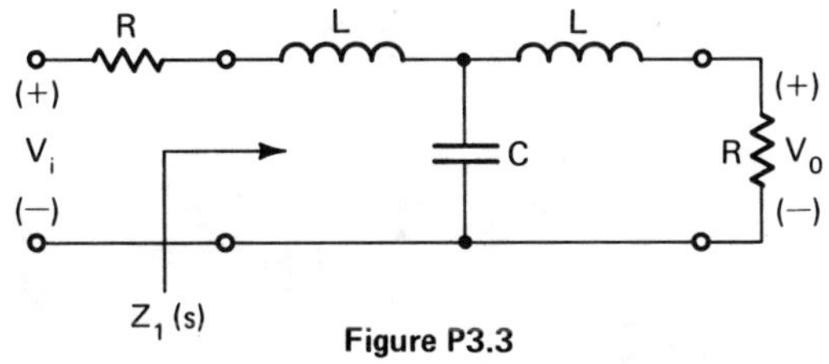

Figure P3.3

P3.4 Find the transfer function and other pertinent parameters for a low-pass filter that has a maximally flat attenuation characteristic around a frequency in the passband. The passband extends between Ω=o and Ω=1 with a ripple factor ε=1($A_p \approx$3dB). The filter has two loss poles at infinity. Sketch A(Ω).

P3.5 Consider a fifth order Butterworth filter with the passband edge at Ω=1 and with ε=1. Find K'(z) K'(-z), K(s) K(-s), H(s) and A(Ω). Give the natural modes in terms of the Ω_o and Q notation. Calculate the attenuation at Ω=2.

P3.6 Find the order N and the natural modes (in terms of ω_o and Q) of the Butterworth filter that meets the specifications

$A \leq 1$ dB	$f \leq 1$ KHz
$A \geq 40$ dB	$f \geq 2$ KHz

What is the excess attenuation (if any) at f=2KHz? Keeping the order constant, find the new value of the A_p and the new locations of the natural modes, if the excess stopband attenuation is reduced to zero.

P3.4 Find the transfer function and other pertinent parameters for a lowpass filter that has a maximally flat attenuation characteristic around a frequency in the passband. The passband extends between 0 [illegible] and [illegible] with [illegible] unit [illegible] factor [illegible] The filter has [illegible] zero [illegible] at infinity. See [illegible]

P3.5 Consider a [illegible] order Butterworth filter with the passband edge at [illegible] and [illegible] [illegible] K(ω), K [illegible] [illegible] K [illegible] and [illegible] Give the natural modes in terms of the [illegible] and [illegible]. Calculate the attenuation at [illegible]

P3.6 Find the order n and the natural modes (in terms of [illegible] and [illegible]) of the Butterworth filter that meets the specifications

[illegible] dB [illegible]

40 dB [illegible]

What is the [illegible] (if any) at [illegible] Keeping the order constant, find the new value of the [illegible] and the new locations of the natural modes if the [illegible] stop band [illegible] is reduced to [illegible]

CHAPTER 4

THE EQUIRIPPLE PASSBAND

A more efficient approximation to the ideal passband attenuation characteristics is the mini-max or equiripple approximation. In section 4.1, closed form expressions are developed for the characteristic function of filters having an equiripple passband attenuation response and arbitrary stopband attenuation characteristics. The formulation is in terms of an arbitrary set of loss poles, which may have been obtained as a result of solving the stopband approximation problem using the methods of Chapter 5.

In the remainder of this Chapter, a number of special cases are considered. Section 4.2 provides a detailed study of Chebyshev filters. These are low-pass filter approximations having equiripple passband with all loss poles at infinite frequency. Since even-order Chebyshev filters do not have reflection zeros at zero frequency, their LC ladder realizations cannot be equally terminated. A frequency transformation is presented in section 4.3 that modifies the even order Chebyshev functions, causing them to have zero-frequency reflection zeros.

Section 4.4 introduces another classical filter approximation which has a maximally flat passband and an equi-minima stopband attenuation. This is the inverse Chebyshev approximation and is derived from the Chebyshev approximation. If the stopband attenuation specification is constant and frequency independent, then an optimum approximation is one that is equiripple in the passband and has equi-minima in the stopband. Such an approximation is called "Elliptic", since elliptic functions may be used to describe its characteristic function. In section 4.5, we study the elliptic filter approximation without making use of these relatively complicated elliptic functions.

4.1 CHARACTERISTIC FUNCTIONS FOR FILTERS WITH EQUIRIPPLE PASSBAND

In this section, we derive formulae for the characteristic functions of filters having an equiripple passband and arbitrary stopband attenuation characteristics. We assume that the stopband approximation problem may have been solved using the methods of Chapter 5 and as a result, we have a prescribed set of loss poles given in terms of the squared loss pole polynomial in the z-plane; $P'(z)P'(-z)$. Thus, our task here is to find an explicit expression for the squared characteristic function

$$K'(z)K'(-z) = \varepsilon^2 \frac{F'(z)F'(-z)}{P'(z)P'(-z)} \tag{4.1}$$

given $P'(z)P'(-z)$, and for which the attenuation function oscillates throughout the passband between zero dB and A_p dB. Fig. 4.1 shows such an equiripple passband, defined in terms of its maximum distortion A_p, where

$$\varepsilon = \sqrt{10^{A_p/10} - 1}\ , \tag{4.2}$$

as well as the two passband edges, s = ja and s = j, on the normalized frequency axis. Note that the attenuation function attains the value A_p at both passband edges. Such a passband is clearly applicable for bandpass filters. The results derived below are also directly applicable to the low-pass case by setting a = 0. In this case, as will be shown, the attenuation at the lower passband edge, s = 0 will be either zero dB or A_p dB.

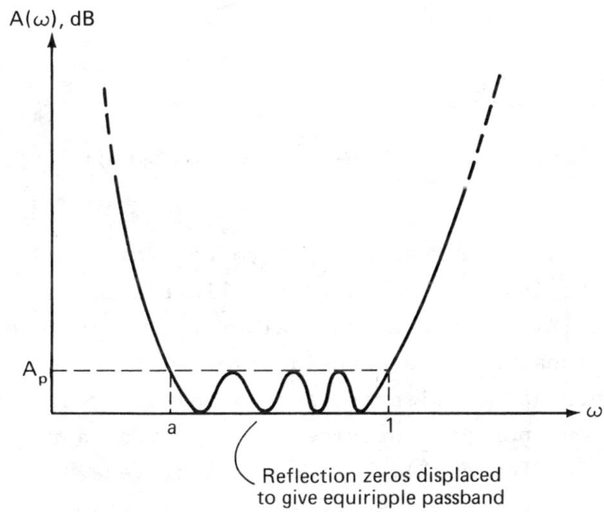

Figure 4.1 An equiripple bandpass attenuation function.

Although not specifically dealt with here, high-pass filters with equiripple passband can be obtained from the corresponding low-pass filters, through the application of the low-pass to high-pass transformation presented in Chapter 6.

Proceeding with the problem at hand, we note that we now must find an expression for the squared reflection zero polynomial, F'(z)F'(-z), such that for the given P'(z)P'(-z), the passband attenuation is equiripple, as shown in Fig. 4.1. We will carry out the entire solution in the z-plane and recall that the passband in the s-plane maps onto the entire jy axis in the z-plane with the upper passband edge at $z = 0$ and the lower passband edge at $z = \infty$. We therefore seek a "squared" characteristic function which oscillates between 0 and ε^2 when evaluated along the jy axis. The character of the equiripple response would suggest making use of trigonometric functions.

Consider a squared characteristic function of the form

$$K'(z)K'(-z)\Big|_{z=jy} = \varepsilon^2 \cos^2 \phi(y) \tag{4.3}$$

that is,

$$K'(z)K'(-z)\Big|_{z=jy} = \frac{1}{2}\varepsilon^2[1 + \cos 2\phi(y)] \tag{4.4}$$

where $\phi(y)$ is a phase function yet to be determined. The squared characteristic function of (4.4) is illustrated in Fig. 4.2. We observe that $|K|^2$ starts at the value of ε^2 near zero, and oscillates with equal ripples between zero and ε^2 as the phase function $\phi(y)$ increases. Clearly, such a characteristic function would serve our purpose, provided one can find a rational polynomial expression for $\cos^2\phi(y)$. We shall digress here to remind the reader of the properties of all-pass functions, which will be used to accomplish the above task.

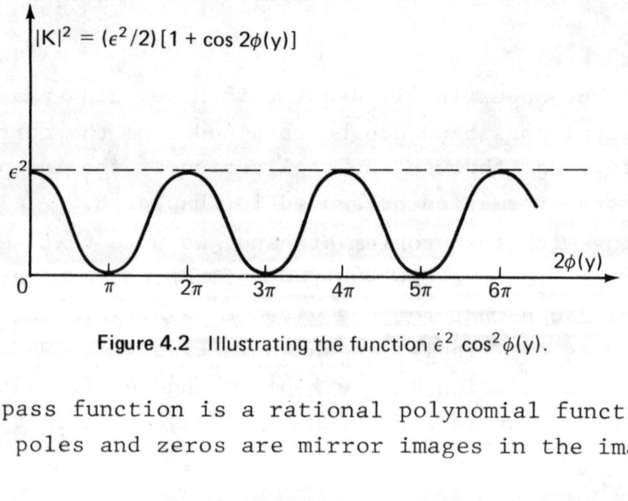

Figure 4.2 Illustrating the function $\epsilon^2 \cos^2 \phi(y)$.

An all-pass function is a rational polynomial function for which the poles and zeros are mirror images in the imaginary axis. A first-order all-pass function in the complex variable z may be expressed as

$$R_1(z) = \frac{-z + a}{z + a} \tag{4.5}$$

while a second-order all-pass function takes the form

$$R_2(z) = \frac{z^2 - 2x_o z + (x_o^2 + y_o^2)}{z^2 + 2x_o z + (x_o^2 + y_o^2)} \tag{4.6}$$

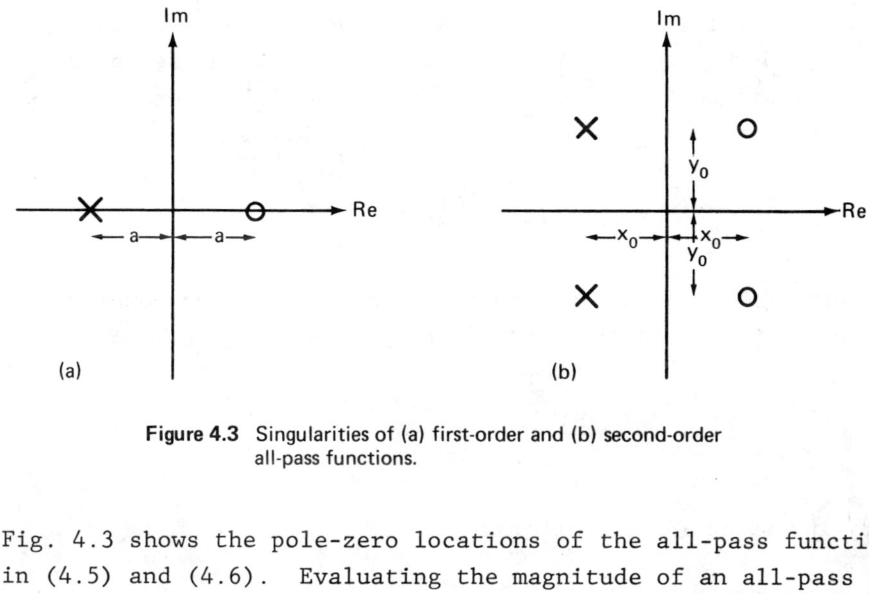

Figure 4.3 Singularities of (a) first-order and (b) second-order all-pass functions.

Fig. 4.3 shows the pole-zero locations of the all-pass functions in (4.5) and (4.6). Evaluating the magnitude of an all-pass function along the jy-axis results in a constant magnitude equal to unity, that is,

$$|R(jy)| = 1$$

and an associated phase function, say $\beta(y)$. An all-pass function may therefore be expressed in the form

$$R(jy) = e^{j\beta(y)} \tag{4.7}$$

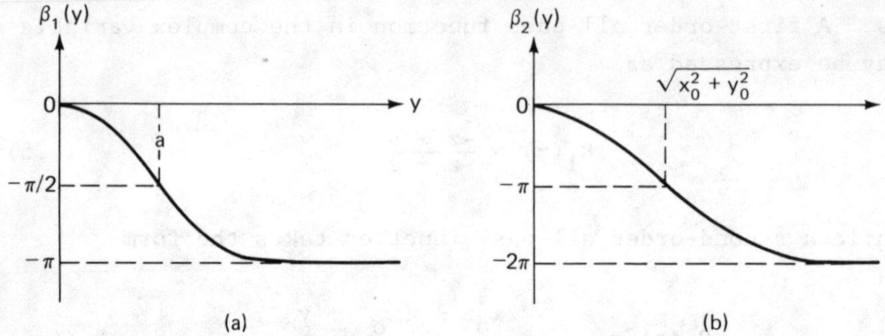

Figure 4.4 Illustrating the phase characteristics of: (a) first order and (b) second order all-pass functions.

Fig. 4.4 illustrates $\beta(y)$ for $R_1(jy)$ and $R_2(jy)$. Further, by analogy to the expressions for the first-order and second-order all-pass functions in (4.5) and (4.6), an Nth order all-pass function can be expressed as

$$R_N(z) = \frac{Q(-z)}{Q(z)} \tag{4.8}$$

where Q(z) is a real polynomial of order N in the complex variable z. Combining (4.7) and (4.8) yields

$$\left.\frac{Q(-z)}{Q(z)}\right|_{z = jy} = e^{j\beta(y)} \tag{4.9}$$

By inverting (4.9) we obtain

$$\left.\frac{Q(z)}{Q(-z)}\right|_{z = jy} = e^{-j\beta(y)} \tag{4.10}$$

Recalling the identity

$$\text{Cos } \beta(y) = \frac{1}{2} [e^{j\beta(y)} + e^{-j\beta(y)}], \tag{4.11}$$

we can use (4.4) and (4.10) to obtain

$$\text{Cos } \beta(y) = \frac{1}{2} \left[\frac{Q(z)}{Q(-z)} + \frac{Q(-z)}{Q(z)}\right]_{z=jy} \tag{4.12}$$

Equation (4.12) is significant, as it allows one to express a cosine function in terms of rational polynomial functions. We can use this result directly in (4.4), by substituting $\beta(y)$ for $2\phi(y)$ to obtain

$$K'(z)K'(-z)\Big|_{z=jy} = \frac{1}{4}\,\varepsilon^2 \left[2 + \frac{Q(z)}{Q(-z)} + \frac{Q(-z)}{Q(z)}\right]_{z=jy}$$

By analytic continuation we obtain

$$K'(z)K'(-z) = \frac{\varepsilon^2}{4} \left[2 + \frac{Q(z)}{Q(-z)} + \frac{Q(-z)}{Q(z)}\right] \tag{4.13}$$

where $Q(z)$ is an order N real polynomial in z, yet to be determined. For any such polynomial, however, we know that a characteristic function formed as in (4.13) will exhibit a perfectly equiripple passband. To determine $Q(z)$, we express (4.13) in the form

$$K'(z)K'(-z) = \frac{\varepsilon^2}{4} \frac{Q^2(z) + 2Q(z)\,Q(-z) + Q^2(-z)}{Q(z)\,Q(-z)} \tag{4.14}$$

Comparing (4.14) and (4.1) results in

$$Q(z)\,Q(-z) = P'(z)P'(-z) \tag{4.15}$$

Thus, (4.14) can now be written in terms of the given squared loss pole polynomial $P'(z)P'(-z)$ as

$$K'(z)K'(-z) = \frac{\varepsilon^2}{4} \frac{P'^2(z) + 2P'(z)\,P'(-z) + P'^2(-z)}{P'(z)\,P'(-z)} \tag{4.16}$$

Thus,

$$K'(z)K'(-z) = \frac{\varepsilon^2}{4} \frac{[P'(z) + P'(-z)]^2}{P'(z)\,P'(-z)} \tag{4.17}$$

We further recognize that

$$P'(z) + P'(-z) = 2\,P'_{ev}(z)$$

which when substituted in (4.17) yields

$$K'(z)K'(-z) = \varepsilon^2 \frac{[P'_{ev}(z)]^2}{P'(z)P'(-z)} \qquad (4.18)$$

Equation (4.18) is the required result.

In summary, the procedure for constructing characteristic functions for filters with equiripple passbands and arbitrary stopbands is as follows:

(1) Split the given squared loss pole polynomial $P'(z)P'(-z)$ into right-half-plane and left-half-plane factors and assign, say, the left-half-plane factors to $P'(z)$ and the rest to $P'(-z)$.

(2) Find the even part of $P'(z)$; that is $P'_{ev}(z)$.

(3) Determine ε from the specified passband ripple A_p dB using (4.2).

(4) Determine the squared characteristic function using (4.18).

We note by comparing (4.18) to (4.1) that the squared reflection zero polynomial is given by

$$F'(z)F'(-z) = [P'_{ev}(z)]^2 \qquad (4.19)$$

The natural mode locations in the z-plane can therefore be obtained from the Feldtkeller equation

$$E'(z)E'(-z) = P'(z)P'(-z) + \varepsilon^2\,[P'_{ev}(z)]^2 \qquad (4.20)$$

Subsequently, the s-plane transfer function H(s) can be obtained, in the usual manner, by transforming back to the s-plane.

Before considering an illustrative example, we wish to make the following points. The reflection zeros occur in conjugate pairs at jy axis locations obtained by equating $|(jy)|^2$ in (4.4) to zero. Thus, the reflection zeros are given by the solutions of

$$\text{Cos}\ 2\phi(y) = -1 \qquad (4.21)$$

Also, since $2\phi(y)$ is the phase angle of an all-pass function of order N then,

$$2\phi(y)\Big|_{y\to\infty} = -N\pi \tag{4.22}$$

Thus, evaluating at infinity

$$|K'(jy)|^2\Big|_{y=\infty} = \frac{\varepsilon^2}{2}(1 + \cos N\pi)$$

which implies that the magnitude of the characteristic function at the lower edge of the passband, $s = ja$, is given by

$$|K(ja)|^2 = \frac{\varepsilon^2}{2}[1 + \cos N\pi] \tag{4.23}$$

It follows from (4.23) that, for the $|K|^2$ to attain the value ε^2 at the lower passband edge in a bandpass filter, then N must be even. This constraint is identical to that imposed on bandpass filters with maximally flat passband in Chapter 3.

There is no reason to impose this constraint on low-pass filters. Thus, the above formulae apply for low-pass filters with N either odd or even. In fact, for an equiripple low-pass filter with an odd order, (4.23) indicates that $|K(0)|^2 = 0$; that is, the filter has perfect transmission at zero frequency. This latter property is desirable as it enables realization using equally terminated LC ladders, as will be shown in Chapter 6.

Example 4.1

Find the transfer function and other pertinent parameters of a BP filter having an equiripple passband extending from $\Omega = 0.5$ to $\Omega = 1$ with $\varepsilon = 1$. From a solution to the stopband approximation problem, it has been found that the filter requires two loss poles at zero and two loss poles at infinity. Sketch the attenuation function $A(\Omega)$.

Solution

From given data,

$$n_o = 2, \ n_i = 2 \text{ thus } N = 4.$$

$$a = 0.5$$

The z-mapping in this case is given by

$$z^2 = \frac{s^2 + 1}{s^2 + \frac{1}{4}}$$

and the inverse mapping is

$$s^2 = \frac{1 - \frac{1}{4} z^2}{z^2 - 1}$$

The squared z-plane loss pole polynomial is

$$P'(z)P'(-z) = (z^2 - 4)^2 \ (z^2 - 1)^2$$

Thus, $$P'(z) = (z + 2)^2 \ (z + 1)^2$$

$$Ev[P'(z)] = z^4 + 13 z^2 + 4$$

Thus from (4.19)

$$F'(z)F'(-z) = (z^4 + 13 z^2 + 4)^2$$

$$F'(z) = z^4 + 13 z^2 + 4$$

The z-plane locations of the reflection zeros are obtained from

$$z^4 + 13 z^2 + 4 = 0$$

leading to

$$z^2 = - 12.68466, \ - 0.315342$$

Using the inverse z-mapping, the s-plane reflection zeros are obtained as

$$\pm j\ 0.552092 \text{ and } \pm j\ 0.905646$$

The characteristic function is therefore given by

$$K(s) = 8\ \frac{[s^2 + (0.552092)^2]\ [s^2 + (0.905646)^2]}{s^2}$$

where the constant 8 is obtained by setting $|K| = 1$ at $s = j1$. The reader is urged to check that $|K(j\frac{1}{2})| = 1$, as it should be.

To determine the natural modes, it is preferable, from the point of view of numerical accuracy, to solve Feldtkeller's equation in the z-plane and then transform the roots to the s-plane. The Feldtkeller equation is given by

$$E'(z)E'(-z) = (z^2 - 4)^2\ (z^2 - 1)^2 + (z^4 + 13\ z^2 + 4)^2$$

After solving this equation and using the inverse z-mapping, we obtain

$$E(s) = (s^2 + 0.205533\ s + 0.863771)$$

$$(s^2 + 0.1189745\ s + 0.289429)$$

In terms of the ω_o and Q notation, the natural modes are characterized by

$$\Omega_{01} = 0.929393 \qquad Q_1 = 4.5218$$

$$\Omega_{02} = 0.537986 \qquad Q_2 = 4.5218$$

Thus, the transfer function is given by

$$H(s) = 63.3704\ \frac{E(s)}{s^2}$$

where the multiplicative constant has been determined by setting

$|H| = 1$ at one of the reflection zeros.

The attenuation function $A(\Omega)$ is plotted in Fig. 4.5. Note that this function is geometrically symmetric about a center frequency $\Omega_0 = \sqrt{0.5}$. The reader might verify that any two frequencies Ω_1 and Ω_2 with equal attenuation are related by

$$\Omega_1 \Omega_2 = \Omega_o^2$$

Thus, if we denote the reflection zeros frequencies by Ω_{r1} and Ω_{r2}, then

$$\Omega_{r1} \Omega_{r2} = \Omega_0^2 = 0.5$$

Also note that the natural mode frequencies are related by

$$\Omega_{0_1} \Omega_{0_2} = 0.5$$

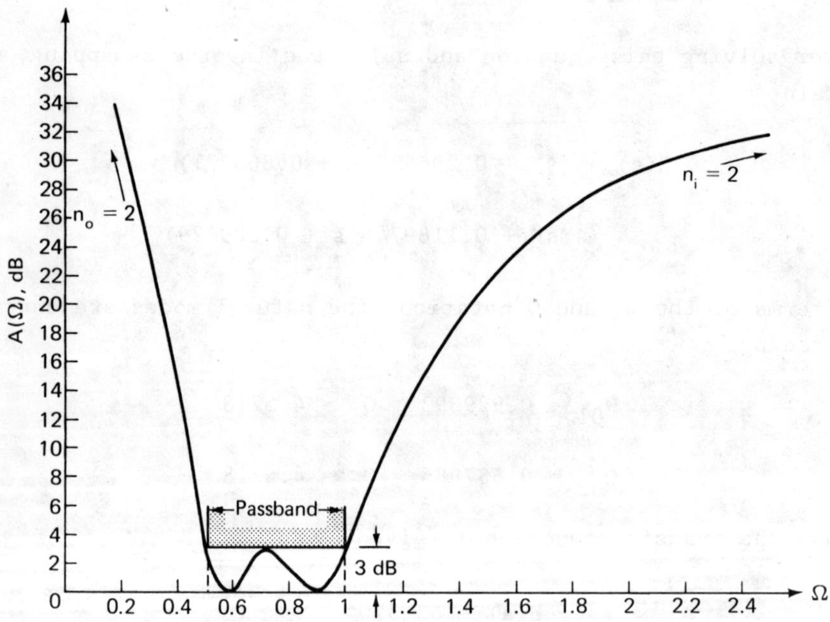

Figure 4.5 Attenuation characteristics of the bandpass filter of Example (4.1).

As will be shown in Chapter 6, the BP filter we have derived here can be obtained by applying the LP to BP transformation to a low pass prototype of the Chebyshev type. The latter is considered in the next section. All bandpass filters obtained via this transformation are geometrically symmetric.

The FILTOR 2 package includes a subroutine called REFLE which uses the formulae developed above, in conjunction with the product method, to compute the reflection zero locations for a given set of loss poles.

Even-order LP Filters with A(0) = 0

It was pointed out in the previous paragraphs that odd order equiripple low pass filters have zero loss at the origin (perfect transmission at DC), while the even order filters have a loss at the origin equal to the maximum passband distortion A_p. It is sometimes required that an even order low pass filter have zero loss at the origin. One example where this requirement arises is when a passive ladder terminated in equal resistances is desired. Such a characteristic function cannot be obtained explicitly by the methods previously described. It can be obtained, however, by a simple transformation applied to a conventional even order characteristic function, once the reflection zero locations are known explicitly. These may, of course, be obtained using the formulae developed above. The transformation given in the following obtains the desired result ($A(0) = 0$), at the expense of slightly shifting all of the loss poles. Unfortunately, the amount of this shift cannot be predicted in advance, and a couple of iterations of the design procedure may be required to meet exacting specifications near the transition band.

We require an order preserving (hence bilinear) complex variable transformation which will map the largest reflection zero in the z-plane, say $z = jy_r$, to infinity, while simultaneously preserving the value of the loss at the upper passband edge ($z = 0$) and at infinity ($z = \pm 1$). We do this to introduce a frequency axis warping in the s-plane, as illustrated in Fig. 4.6. Such a transformation is easily constructed and is given in terms of the largest reflection zero y_r as,

$$\xi^2 = \frac{(1 + y_r^2)\, z^2}{(z^2 + y_r^2)} \quad . \tag{4.24}$$

The reader will note, by substitution, the correspondence of $z = \pm 1$ ($\Omega = \pm\infty$) to $\xi = \pm 1$, $z = 0$ ($\Omega = 1$) to $\xi = 0$, and $z = \pm j y_r$ ($\Omega = \Omega_r$) to $\xi = \infty$. The formula (4.24) serves as the basis for a simple computer routine called SUBROUTINE WARP which can be found in[1], and which may be used in conjunction with the output of SUBROUTINE REFLE to obtain even order low pass filters having $A(0) = 0$, by simply transforming the poles and zeros of $K'(z)$ $K'(-z)$ through (4.24).

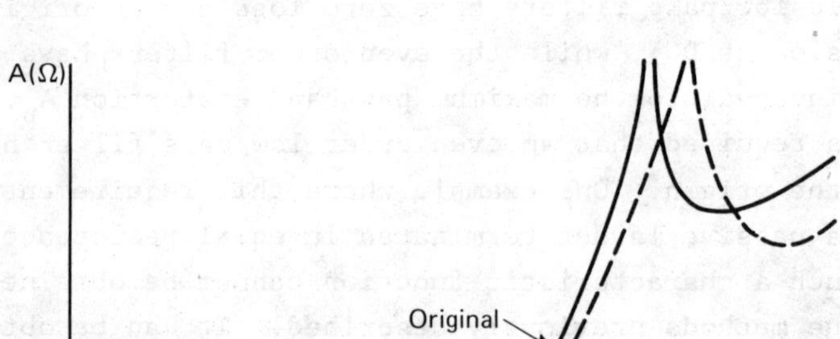

Figure 4.6 Illustrating the "warping" of the frequency axis as a result of transforming even-order equiripple low-pass filters to obtain reflection zeros at $\Omega = 0$.

4.2 THE CHEBYSHEV FILTER APPROXIMATION

The Chebyshev filter approximation is a low-pass filter approximation with equiripple passband and a monotonically increasing attenuation in the stopband. Fig. 4.7 shows the attenuation characteristics for a fifth order (N = 5) Chebyshev filter.

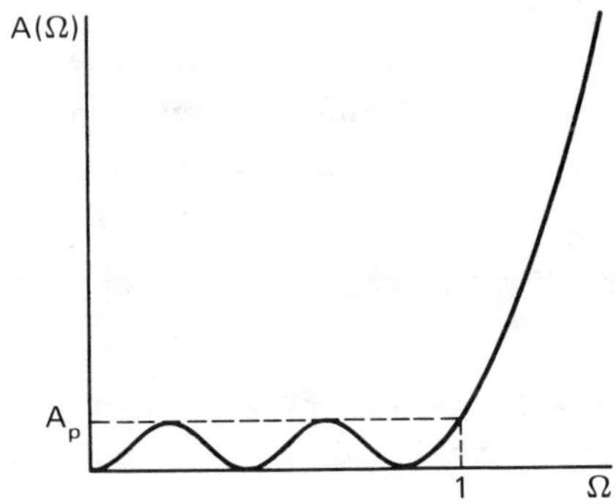

Figure 4.7 Attenuation function for a fifth-order Chebyshev filter.

Since the attenuation increases monotonically in the stopband, it follows that the loss poles all lie at infinity. Chebyshev filters are therefore a special case of the filters described in section 4.1, which have arbitrary stopband attenuation characteristics. Although Chebyshev filters do not provide an efficient approximation to general stopband attenuation requirements, they have closed form expressions and their parameters have been extensively tabulated. Therefore, they can be useful in simple filtering problems, as well as in cases where the designer has no access to a computer.

Compared to the Butterworth filter studied in section 3.2, the Chebyshev filter provides a more efficient approximation to the ideal passband attenuation characteristics. Chebyshev filters have a higher cut-off rate; that is, a larger attenuation slope at $\Omega = 1$, than a Butterworth filter having an equal order and equal passband distortion. Like Butterworth filters, Chebyshev filters have their loss poles at infinity and are hence described by a polynomial transfer function H(s). In the following, we study in detail the characteristics of Chebyshev filters. Although the name "Chebyshev approximation" is used here to imply a low-pass filter approximation, the frequency band transformations of Chapter 6 enable the design of BP, HP and BR filters, using the Chebyshev approximation.

Chebyshev Characteristic Functions

The squared characteristic function, in the z-plane, for a filter having an equiripple passband and arbitrary loss-pole distribution, has been derived in section 4.1 as

$$K'(z)\ K'(-z) = \frac{\varepsilon^2}{4}\left[\frac{P'(z)}{P'(-z)} + 2 + \frac{P'(-z)}{P'(z)}\right] \tag{4.25}$$

If we let all the loss poles lie at infinite frequency, the Chebyshev case results, with the squared loss-pole polynomial given by,

$$P'(z)\ P'(-z) = (-1)^N\ (z^2 - 1)^N = (1-z^2)^N \tag{4.26}$$

Thus, $P'(z) = (1+z)^N$ (4.27)

and, $P'(-z) = (1-z)^N$ (4.28)

Substituting in(4.25) obtains,

$$K'(z)\ K'(-z) = \frac{\varepsilon^2}{4}\left[\left(\frac{1+z}{1-z}\right)^N + \left(\frac{1-z}{1+z}\right)^N + 2\right] \tag{4.29}$$

Consider, for the moment, the passband $0 \le \Omega \le 1$, which maps onto $z = jy$. Substituting $z=jy$ in (4.29), the squared characteristic function in the passband is obtained. A more convenient expression is given in section 4.1 equation (4.3),

$$|K(jy)|^2 = \varepsilon^2 \cos^2 \phi(y) \tag{4.30}$$

where, $\phi(y) = -\frac{1}{2} \arg \left[\frac{P'(z)}{P'(-z)}\right]_{z=jy}$ (4.31)

Making use of (4.27) and (4.28),

$$\phi(y) = -\frac{1}{2} \arg \left(\frac{1+jy}{1-jy}\right)^N$$

$$= \frac{-N}{2} \cdot 2 \cdot \tan^{-1} y$$

$$\phi(y) = -N \tan^{-1} y \tag{4.32}$$

Recalling that for the low-pass case the z-mapping is given by

$$z^2 = 1 + \frac{1}{s^2} \tag{4.33}$$

then for $s = j\Omega$, $0 \le \Omega \le 1$, $z = jy$ where

$$y^2 = \frac{1}{\Omega^2} - 1$$

$$\text{or,} \quad \Omega^2 = \frac{1}{1+y^2} \tag{4.34}$$

From simple trigonometry it follows that,

$$\tan^{-1} y = \cos^{-1} \frac{1}{\sqrt{1+y^2}} = \cos^{-1} \Omega$$

Substituting in (4.32) obtains

$$\phi(\Omega) = -N \cos^{-1} \Omega$$

Finally, using (4.30), we obtain the squared characteristic function in the passband as,

$$|K(\Omega)|^2 = \varepsilon^2 \cos^2 (N \cos^{-1} \Omega) \tag{4.35}$$

At this point, we note that the function $\cos(N \cos^{-1} \Omega)$ is a polynomial in the variable Ω, and is termed the Chebyshev polynomial of order N;

$$C_N(\Omega) = \cos(N \cos^{-1} \Omega) \quad , \quad \Omega \le 1 \tag{4.36}$$

For real arguments, the above expression is valid only for $\Omega \le 1$.

It can be easily shown, however, that for $\Omega \geq 1$, (4.36) becomes,

$$C_N(\Omega) = \cosh(N \cosh^{-1}\Omega), \; \Omega \geq 1 \tag{4.37}$$

Equations (4.36) and (4.37) define the Chebyshev polynomial over the entire Ω axis. We digress briefly to consider some of the properties of these polynomials.

First, we use (4.36) and (4.37) to derive the first few Chebyshev polynomials,

$$C_0(\Omega) = \cos(0) = 1 \tag{4.38}$$

$$C_1(\Omega) = \cos(\cos^{-1}\Omega) = \Omega \tag{4.39}$$

$$C_2(\Omega) = \cos(2\cos^{-1}\Omega)$$

$$= 2\cos^2(\cos^{-1}\Omega) - 1$$

$$= 2\Omega^2 - 1 \tag{4.40}$$

$$C_3(\Omega) = \cos(3\cos^{-1}\Omega)$$

$$= 4\cos^3(\cos^{-1}\Omega) - 3\cos(\cos^{-1}\Omega)$$

$$= 4\Omega^3 - 3\Omega \tag{4.41}$$

To derive higher order polynomials, our work may be considerably facilitated by the use of the following recursive formula,

$$C_{N+1}(\Omega) = 2\Omega C_N(\Omega) - C_{N-1}(\Omega) \tag{4.42}$$

This formula can be easily derived using simple trigonometric identities, and the derivation is left as an exercise. As an example of the use of (4.42) we derive $C_4(\Omega)$,

$$C_4(\Omega) = 2\Omega C_3(\Omega) - C_2(\Omega)$$

$$= 2\Omega\ (4\Omega^3 - 3\Omega) - 2\Omega^2 + 1$$

$$\text{i.e. } C_4\ (\Omega) = 8\Omega^4 - 8\Omega^2 + 1 \tag{4.43}$$

Fig. 4.8 shows plots of $C_N\ (\Omega)$, for N = 1,2,....,5, over the passband $0 \leq \Omega \leq 1$. From equations (4.38)-(4.43), as well as from Fig. 4.8, we note the following properties of Chebyshev polynomials:

(1) The Chebyshev polynomial is odd for N odd and even for N even.

(2) The value of $C_N\ (\Omega)$ oscillates between +1 and -1 over the passband and equals +1 at the band edge, $\Omega = 1$. Furthermore $C_N\ (0) = 0$ for N odd and ± 1 for N even.

(3) The roots of $C_N\ (\Omega)$, which are the reflection zeros of the Chebyshev filter (see 4.35), are given by

$$\cos\ [N \cos^{-1} \Omega_r] = 0\ ,$$

$$N \cos^{-1} \Omega_{rk} = (2k+1)\ \frac{\pi}{2}\ ,$$

$$\text{Thus, } \Omega_{rk} = \cos\ [\frac{2k+1}{N}\ \frac{\pi}{2}]\ ,\quad k=0,1,2,\ldots,\ \frac{N-1}{2} \text{ for N odd}$$
$$\text{or } k=0,1,2,\ldots,\ \frac{N}{2} -1 \text{ for N even} \tag{4.44}$$

(4) The Chebyshev polynomial may be expressed in terms of its roots as

$$C_N(\Omega) = 2^{N-1} \prod_k\ (\Omega^2 - \Omega_{rk}^2) \tag{4.45}$$

where k takes on the values specified in (4.44)

From the above properties, it should be clear why the Chebyshev polynomials yield an equiripple passband. Also, since $C_N(\Omega)$ is bound by ± 1 in the passband, the value of ϵ determines the passband ripple. This is further illustrated by writing the various Chebyshev filter functions,

$$|K(\Omega)|^2 = \varepsilon^2 C_N^{\ 2}(\Omega) \tag{4.46}$$

$$|H(\Omega)|^2 = 1 + \varepsilon^2 C_N^{\ 2}(\Omega) \tag{4.47}$$

$$A(\Omega) = 10 \log_{10} [1 + \varepsilon^2 C_N^{\ 2}(\Omega)] \tag{4.48}$$

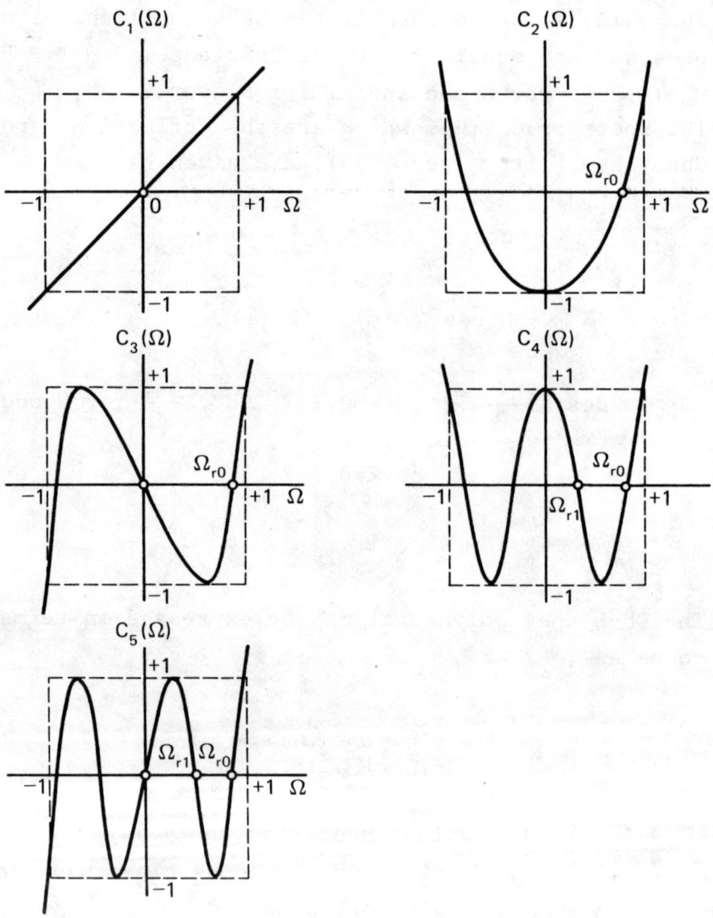

Figure 4.8 Chebyshev functions $C_N(\Omega)$. Note that the reflection zeros are denoted by ○.

Fig. 4.9 shows sketches of $|H(\Omega)|^2$ for Chebyshev filters of order N = 2,3,4 and 5 where we have numbered the reflection zeros in accordance with equation (4.44). Note that, as evident in Fig. 4.9 and from property (2) above, even order Chebyshev filters obtain their maximum passband ripple at $\Omega = 0$. The reader may also observe, in Fig. 4.9, that the total number of peaks and valleys in the passband is equal to the order N.

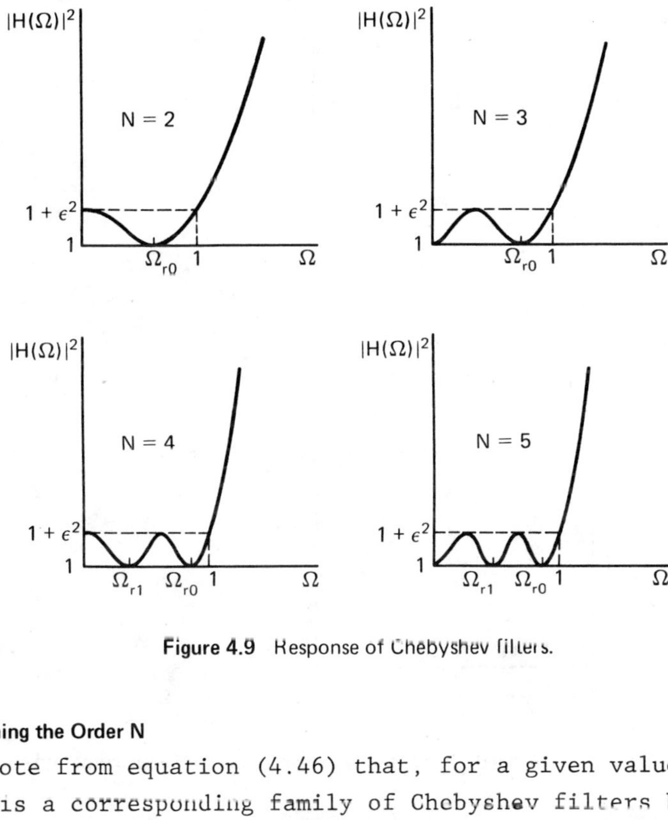

Figure 4.9 Response of Chebyshev filters.

Determining the Order N

We note from equation (4.46) that, for a given value of N, there is a corresponding family of Chebyshev filters having different passband ripple A_p (that is, different values of ε), and correspondingly different values for the stopband attenuation. All the members of the family will have identical reflection zeros, which are the zeros of the Chebyshev polynomial $C_N(\Omega)$, as given by (4.44).

The concept of a family of approximations parameterized by ε

has been introduced and illustrated in Chapter 2. It is useful to apply this concept to the Chebyshev case. Using equation (4.48) to evaluate the attenuation of an Nth order Chebyshev filter at the edge of the stopband Ω_s (See Fig. 4.10), we obtain

$$A_s = A(\Omega_s) = 10 \log_{10} [1 + \varepsilon^2 C_N^{\,2} (\Omega_s)]$$

For A_s greater than about 10 dB; that is, for $\varepsilon^2 C_N^{\,2}(\Omega_s) > 10$, we can make the approximation

$$A_s \simeq 10 \log_{10} \varepsilon^2 C_N^{\,2} (\Omega_s)$$

Thus,

$$A_s + 20 \log_{10}(\frac{1}{\varepsilon}) = 20 \log_{10} C_N(\Omega_s)$$

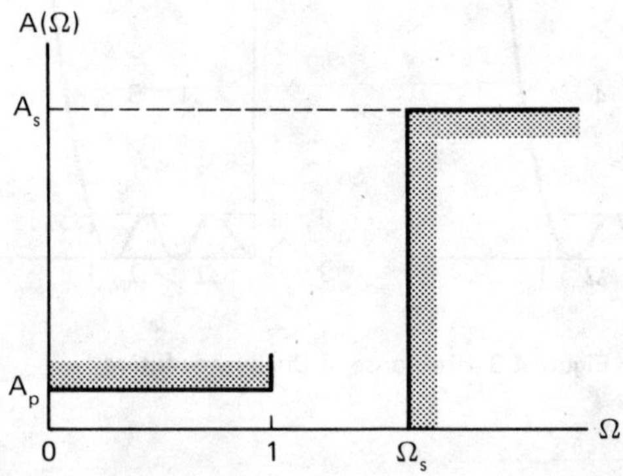

Figure 4.10 Normalized specifications for a classical LPF approximation.

Note that, in this relationship, the term on the right-side is a constant for a given order N and, of course, a given Ω_s. The amount of attenuation specified by this constant can be split between the two terms on the left-side according to the designer's wish. Increasing the stopband attenuation A_s will mean an increase in ε and hence in the passband distortion A_p. The reader should note that equation (2.29) of Chapter 2 is an approximation of (4.49), where 20 log $(\frac{1}{\varepsilon})$ is approximated by the minimum passband return loss A_{emin}. Such an approximation is valid for $A_p < 1$ dB.

Equation (4.49) can be used to determine the filter order N required to meet attenuation specifications in the form of the tolerance plot of Fig. 4.10. To illustrate this process, we present in Fig. 4.11 plots of the quantity 20 $\log_{10} C_N(\Omega_s)$ for $\Omega_s = 1$ to 2 and for N = 3 to 10. Thus given A_p, A_s and Ω_s, these graphs may be used to determine the required order N. The procedure is best illustrated by an example.

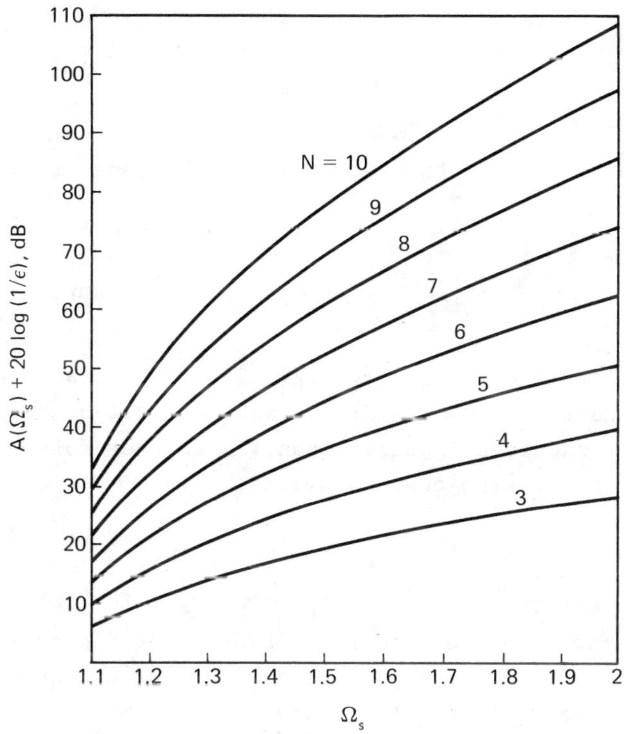

Figure 4.11 Chebyshev design chart.

Example 4.2

Find the order N and the characteristic function of the Chebyshev approximation which meets the following low-pass specifications:

$$f_p = 10 \text{ KHz} \quad f_s = 15 \text{ KHz} \quad A_p \le 0.5 \text{ dB}$$

$$A_s \ge 50 \text{ dB}$$

Solution

Normalizing with respect to the passband edge f_p, we have

$$\Omega_p = 1 \text{ rad/sec} \ , \ \Omega_s = 1.5 \text{ rad/sec}$$

From A_p, the parameter ε is obtained using (4.2) as

$$\varepsilon = 0.3493$$

We next find the sum of the two terms on the left-side of (4.49) to be

$$A_s + 20 \log_{10} \frac{1}{\varepsilon} = 50 + 9.13 = 59.13 \text{ dB}$$

The point $\Omega_s = 1.5$ and $(A_s + 20 \log_{10} \frac{1}{\varepsilon}) = 59.13$ is located on the design chart as shown in Fig. 4.11. It follows that the order N required should be greater than 7. We therefore use N = 8.

For N = 8, the design chart provides

$$A_s + 20 \log_{10} \frac{1}{\varepsilon} = 60.86 \text{ dB}$$

which is higher than the specified value. The designer now has two choices. The first is to keep $A_p = 0.5$ dB and accept a higher than required value for A_s; that is,

$$A_s = 60.86 - 9.13 = 51.73 \text{ dB}$$

The second choice is to keep A_s at the specified 50 dB value and utilize the design margin in obtaining a lower value for ε; that is, a smaller passband distortion A_p'. This is obtained as follows,

$$20 \log \frac{1}{\varepsilon'} = 60.86 - 50 = 10.86 \text{ dB}$$

Thus, $\varepsilon' = 0.2864$ and $A_p' = 0.34$ dB

Both choices, of course, have identical reflection zeros. Using (4.45), we obtain

$$|K(\Omega)| = \varepsilon\, 2^{N-1} (\Omega^2 - \Omega_{r0}^2)(\Omega^2 - \Omega_{r1}^2)(\Omega^2 - \Omega_{r2}^2)(\Omega^2 - \Omega_{r3}^2)$$

where $\varepsilon = 0.3493$ or 0.2864, depending on our choice of option, and Ω_{r_k} are given by (4.44). Choosing $\varepsilon = 0.3493$ and substituting $\Omega = \frac{s}{j}$, the characteristic function K(s) is obtained as

$$K(s) = 44.7104\,(s^2 + 0.961940)(s^2 + 0.691342)(s^2 + 0.308658)(s^2 + 0.0380602)$$

The design chart of Fig. 4.11 is useful not only in determining the required order N, but also to exercise the tradeoff between the passband and stopband performance, provided the freedom exists to do so.

The order N can be alternatively determined from the following approximate expression:

$$N \geq \frac{A_s + 20 \log \frac{1}{\varepsilon} + 6}{8.68 \cosh^{-1}(\Omega_s)} \tag{4.50}$$

The proof of this formula is left to the reader as an exercise.

In all of our presentation, the frequency is normalized to yield a ripple bandwidth $\Omega_p = 1$. It is useful, however, to find an expression for the 3-dB bandwidth Ω_{3dB}. This is obtained by setting $|H(\Omega)|^2 = 2$, resulting in

$$\Omega_{3dB} = \cosh\left[\frac{1}{N}\cosh^{-1}\frac{1}{\varepsilon}\right] \tag{4.51}$$

Natural Modes of Chebyshev Filters

We now proceed to find the natural modes of Chebyshev filters and consequently, the transfer function H(s). Toward this end, we make use of (4.35), substituting $\Omega = \frac{s}{j}$, to obtain the squared characteristic function

$$K(s)\,K(-s) = \varepsilon^2 \cos^2\left[N \cos^{-1}\left(\frac{s}{j}\right)\right] \tag{4.52}$$

Introducing the new complex variable W = u + jv, defined as

$$W = \cos^{-1}\left(\frac{s}{j}\right), \tag{4.53}$$

equation (4.52) becomes,

$$K^2 = \varepsilon^2 \cos^2 (NW), \tag{4.54}$$

where we use K^2 to denote the squared characteristic function. From (4.53),

$$s = j \cos W$$

$$= j \cos (u + jv)$$

$$s = \sin u \sinh v + j \cos u \cosh v$$

$$\text{Thus, } \sigma = \sin u \sinh v, \tag{4.55}$$

$$\Omega = \cos u \cosh v \tag{4.56}$$

Using the Feldtkeller equation, the squared transfer function is given by

$$H^2 = 1 + \varepsilon^2 \cos^2 (NW) \tag{4.57}$$

The natural modes can now be found by setting $H^2 = 0$; that is,

$$\cos NW = \pm \left(\frac{j}{\varepsilon}\right) \qquad \cos N(u + jv) = \pm \frac{j}{\varepsilon}$$

Thus, $$\cos Nu \cosh Nv - j \sin Nu \sinh Nv = \pm \frac{j}{\varepsilon} \tag{4.58}$$

Equating the real terms on both sides of (4.58) yields,

$$\cos Nu \cosh Nv = 0$$

Thus, $$u = \left(\frac{2k+1}{N}\right)\frac{\pi}{2} \qquad k = 0,1,2,\ldots,2N-1 \tag{4.59}$$

Similarly, equating the imaginary terms on both sides of (4.58) results in

$$-\sin Nu \sinh Nv = \pm \frac{1}{\varepsilon}$$

Substituting for u from (4.59) obtains,

$$\sinh Nv = \pm \frac{1}{\varepsilon} \tag{4.60}$$

We may now ignore the negative sign in (4.60) and take it into account in the final result; that is

$$\sinh Nv = \frac{1}{\varepsilon} \qquad \text{or } v = \frac{1}{N}\sinh^{-1}\left(\frac{1}{\varepsilon}\right) \tag{4.61}$$

The left-half-plane roots of $H^2 = 0$ can be found from (4.55), (4.56), (4.59) and (4.61) as,

$$s_k = -\sin\left(\frac{2k+1}{N}\right)\frac{\pi}{2}\sinh\left(\frac{1}{N}\sinh^{-1}\frac{1}{\varepsilon}\right)$$

$$+ j\cos\left(\frac{2k+1}{N}\right)\frac{\pi}{2}\cosh\left(\frac{1}{N}\sinh^{-1}\frac{1}{\varepsilon}\right)$$

$$k = 0,1,2,\ldots,N-1 \tag{4.62}$$

In terms of the ω_o and Q notation the natural modes for a Chebyshev filter are characterized by

$$\Omega_{ok} = \sqrt{\sin^2 \left(\frac{2k+1}{N}\right) \frac{\pi}{2} \sinh^2 \left(\frac{1}{N} \sinh^{-1} \frac{1}{\varepsilon}\right) + \cos^2 \left(\frac{2k+1}{N}\right) \frac{\pi}{2} \cosh^2 \left(\frac{1}{N} \sinh^{-1} \frac{1}{\varepsilon}\right)}$$

and, $$Q_k = 1 \Big/ 2 \operatorname{Cos} \left\{ \tan^{-1} \left[\cot \left(\frac{2k+1}{N}\right) \frac{\pi}{2} \coth \left(\frac{1}{N} \sinh^{-1} \frac{1}{\varepsilon}\right)\right] \right\}$$

$$k = 0, 1, 2, \ldots, N-1$$

Finally, the transfer function H(s) may be obtained as,

$$H(s) = h \prod_{k=0}^{N-1} (s-s_k) \tag{4.63}$$

where h is a multiplicative constant determined from,

$$H(0) = h \prod_{k=0}^{N-1} (-s_k) \tag{4.64}$$

We now summarize the procedure for determining the transfer function H(s) for a Chebyshev filter satisfying a given tolerance plot, such as that in Fig. 4.10:

(1) From the passband ripple A_p, determine ε, using (4.2).

(2) From the given Ω_s and A_s, determine the order N, using the chart of Fig. 4.11 or equation (4.50). If desired, the pass band distortion may be reduced to take advantage of any excess stopband attenuation; (see example 4.2).

(3) Calculate the natural modes s_k from (4.62).

(4) Determine H(s) from (4.63) and (4.64).

In Chapter 3, we have shown that the natural modes of a Butterworth filter lie on a circle in the s-plane. For Chebyshev filters, the locus is an ellipse, as can be derived from (4.62). Expressing $s_k = \sigma_k + j\Omega_k$, it follows that

$$\frac{\sigma_k^2}{\sinh^2(\frac{1}{N}\sinh^{-1}\frac{1}{\varepsilon})} + \frac{\Omega_k^2}{\cosh^2(\frac{1}{N}\sinh^{-1}\frac{1}{\varepsilon})} = 1 \tag{4.65}$$

which is the equation of an ellipse with its major axis along the $j\Omega$-axis and its minor axis along the σ-axis.

Comparing the Butterworth natural modes given by (3.21) with those for the Chebyshev filter in equation (4.62), one senses that they are somehow related. In fact it can be shown that the Chebyshev natural modes reduce to those of the Butterworth, as the ripple ε is reduced to zero.

Example 4.3

Find the natural modes and the transfer function H(s) for the Chebyshev filter of example 4.2.

Solution

The natural modes are obtained using (4.62), with N = 8 and ε = 0.3493.

$$S_{0,7} = -0.0436208 \pm j\,1.005003$$

$$S_{1,6} = -0.1242217 \pm j\,0.0852000$$

$$S_{2,5} = -0.1859109 \pm j\,0.569288$$

$$S_{3,4} = -0.219297 \pm j\,0.1999075$$

The natural modes' frequencies and Q factors are given by:

$S_{0,7}$: $\Omega_o = 1.005949$ Q = 11.53

$S_{1,6}$: $\Omega_o = 0.861008$ Q = 3.47

$S_{2,5}$: $\Omega_o = 0.5988767$ Q = 1.61

$$S_{3,4} : \Omega_o = 0.296739 \quad Q = 0.68$$

Thus, H(s) is given by

$$H(s) = 44.7106\,(s^2 + 0.0872417\,s + 1.01193)(s^2 + 0.248443\,s + 0.741336)(s^2 + 0.371822\,s + 0.358652)(s^2 + 0.438594\,s + 0.0880541)$$

where the multiplicative constant has been determined from $|H(0)|^2 = 1 + \varepsilon^2$. Note that the above H(s) is normalized with respect to the passband edge (10 KHz). If desired, the transfer function can be denormalized by replacing s by $\frac{s}{2\pi \times 10^4}$. Similarly, the denormalized natural modes are characterized by $\omega_{ok} = \Omega_{ok} \times 2\pi \times 10^4$ and the same Q factors.

4.3 MODIFIED CHEBYSHEV FILTERS

Even order Chebyshev filters have their maximum passband distortion at $\Omega = 0$, as has already been seen. This does not allow their LC ladder realization to be equally terminated. We therefore seek a modification of the even order Chebyshev filter function to provide a reflection zero at $\Omega = 0$. A procedure for accomplishing this in the z-plane was given at the end of Section 4.1 Consider Fig. 4.12a, which shows the squared characteristic function $|K|^2$ versus Ω for a sixth order Chebyshev filter. We wish to modify the characteristic function to yield the plot shown in Fig. 4.12b. Note that the reflection zero at Ω_{r3} has been moved to the origin.

In order to find the modified Chebyshev function, we first recall the result regarding the zeros of Chebyshev polynomials. Equation (4.44) provides those zeros on the positive Ω-axis. Recalling that even order Chebyshev polynomials are even, their singularities will be symmetric. It follows that one can express an even order Chebyshev polynomial in terms of its zeros, as

$$C_N(\Omega) = C \prod_{k=1}^{N/2} (\Omega^2 - \Omega_{rk}^2) \tag{4.66}$$

$$\text{where} \quad \Omega_{rk} = \cos\left(\frac{2k-1}{2N}\pi\right) \tag{4.67}$$

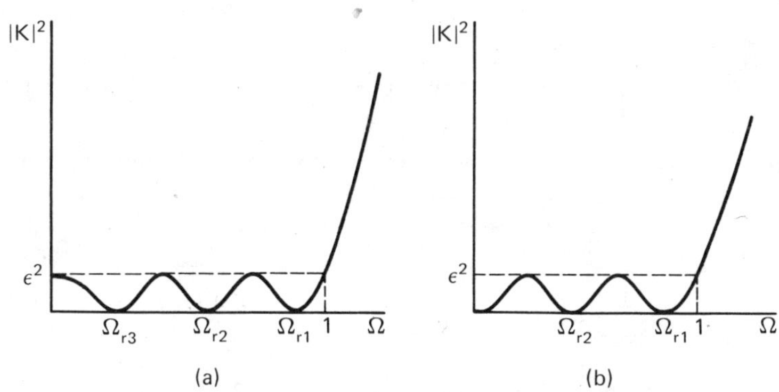

Figure 4.12 (a) 6th order Chebyshev characteristic function; (b) The characteristic function in (a) modified by moving Ω_{r3} to the origin.

Note that we have changed the index k of (4.44) to (k-1) to make the counting, in (4.66) and (4.67), start at k = 1. The constant C in (4.66) can be determined from $C_N(0) = 1$ for $\frac{N}{2}$ even or $C_N(0) = -1$ for $\frac{N}{2}$ odd. It can be shown that $C = 2^{N-1}$. Let the lowest reflection zero be denoted by Ω_x; that is,

$$\Omega_x = \cos\left(\frac{N-1}{2N}\pi\right) \tag{4.68}$$

We now desire to change the variable Ω to a new frequency variable v, such that the modified Chebyshev polynomial $\overline{C}_N(v)$ has a pair of zeros at the origin and no zeros at Ω_x. Simple reasoning suggests the change of variable

$$\Omega^2 - \Omega_x^2 = v^2 \tag{4.69}$$

This, indeed, will accomplish the desired goal, but with the disadvantage that the ripple bandwidth is not unity any longer. In fact, for $\Omega = 1$,

$$v^2 = 1 - \Omega_x^2 \tag{4.70}$$

All that needs to be done, however, is to renormalize v in (4.69), using the factor in (4.70). Thus, the required change of variables becomes

$$\Omega^2 - \Omega_x^2 = v^2 \; (1-\Omega_x^2) \tag{4.71}$$

Notice now, that at $\Omega = 1$, $v = 1$ as desired. Substituting for Ω^2 from (4.71) into (4.66) obtains

$$\overline{C}_N \; (v) = C \; (1-\Omega_x^2) \; v^2 \prod_{k=1}^{(N/2)-1} [v^2 \; (1-\Omega_x^2) + \Omega_x^2 - \Omega_{rk}^2]$$

Thus,

$$\overline{C}_N \; (v) = C \; (1-\Omega_x^2)^{\frac{N}{2}} \; v^2 \prod_{k=1}^{(N-2)/2} [v^2 + \frac{\Omega_x^2 - \Omega_{rk}^2}{1 - \Omega_x^2}] \tag{4.72}$$

Equation (4.72) together with (4.67) and (4.68) provide the modified Chebyshev function. Of course we may rewrite (4.72) with v replaced by a frequency variable Ω, with the resultant squared characteristic function for the modified Chebyshev filters,

$$|K(\Omega)|^2 = \varepsilon^2 \; \overline{C}_N^{\;2} \; (\Omega)$$

The reflection zeros of the modified Chebyshev filters are readily available from (4.72). The natural modes can be calculated in the usual way. An easier approach would be to generalize the frequency transformation given by (4.70), by substituting $\Omega = \frac{s}{j}$ and $v = \frac{W}{j}$ to obtain

$$W^2 = \frac{s^2 + \Omega_x^2}{1 - \Omega_x^2} \tag{4.73}$$

This complex mapping can be used directly on the natural modes s_k of the Chebyshev filter to obtain the corresponding natural modes W_k of the modified Chebyshev filter.

4.4 INVERSE CHEBYSHEV FILTERS

The Inverse Chebyshev filter provides a maximally flat attenuation in the passband around the origin, and an equi-minima, or Chebyshev, response in the stopband. Therefore, it is not an all-pole or polynomial filter, but has finite loss-poles. It is optimum in the sense that, of all the approximations that provide a maximally flat passband and a certain minimum stopband attenuation, the inverse Chebyshev filter will be the most economic; that is, will have the lowest order. Obviously, the inclusion of the finite loss-poles results in a higher cut-off rate for the inverse Chebyshev filter than that obtained with a comparable Butterworth filter. Compared to the Chebyshev filter, however, the inverse Chebyshev provides no advantage from the point of view of attenuation characteristics. In fact, for a given passband distortion and stopband attenuation, both approximations have to be of equal order. This fact will be proven at a later stage. The inverse Chebyshev filter, however, has a much better group delay characteristic than the Chebyshev filter. This important point will be illustrated by an example.

The reason for including the inverse Chebyshev filter approximations at this point is that, as the name implies, they can be easily derived from Chebyshev characteristic functions. Of course the iterative stopband techniques of Chapter 5 together with the maximally flat passband formula of Chapter 3 will yield the same results. Consider the 5th order squared Chebyshev characteristic function shown in Fig. 4.13a, and generally given by,

$$|K(\Omega)|^2 = \varepsilon^2 C_N^2(\Omega) \tag{4.74}$$

Replacing Ω by $\frac{1}{\Omega}$ in (4.74) obtains the squared function $|K(\frac{1}{\Omega})|^2$,

$$|K(\tfrac{1}{\Omega})|^2 = \varepsilon^2 C_N^{\ 2}(\tfrac{1}{\Omega}) \tag{4.75}$$

which is sketched in Fig. 4.13b. We immediately note that this latter function is that of the high-pass Chebyshev where the passband extends from $\Omega = 1$ to $\Omega = \infty$. Let us now form a new squared characteristic function $|\bar{K}(\Omega)|^2$ as follows,

$$|\bar{K}(\Omega)|^2 = 1 \Big/ |K(\tfrac{1}{\Omega})|^2$$

$$\text{Thus,} \quad |\bar{K}(\Omega)|^2 = 1 \Big/ \varepsilon^2 C_N^{\,2} (\tfrac{1}{\Omega}) \tag{4.76}$$

A sketch of this latter function can easily be obtained by inverting the curve of Fig. 4.13b point-by-point. The result is shown in Fig. 4.13c.

Careful inspection of the characteristic function of (4.76) and Fig. 4.13c indicates that it is the one we are seeking. Since there are N reflection zeros at the origin, obtained as the reciprocal of the N loss-poles of the high-pass characteristic of Fig. 4.13b, then the characteristic is maximally flat in the pass band around $\Omega = 0$. This fact should also be clear from (4.76), where we note that the numerator will be Ω^{2N}. The stop-band is clearly Chebyshev in nature with a minimum attenuation of

$$\begin{aligned} A_s &= 10 \log_{10} (1 + \frac{1}{\varepsilon^2}) \\ &\simeq - 20 \log_{10} \varepsilon \end{aligned} \tag{4.77}$$

The loss-poles are the reciprocals of the reflection zeros of the original Chebyshev filter. Thus, employing (4.44), the loss-poles are given by

$$\Omega_{\ell k} = \frac{1}{\cos\left[\frac{2k+1}{N}\frac{\pi}{2}\right]}, \quad k = 0,1,2,\ldots,\frac{N-1}{2} \text{ for N odd} \quad \text{or } k = 0,1,2,\ldots,\frac{N}{2} - 1 \text{ for N even} \tag{4.78}$$

We now address ourselves to the question of the required order N of the inverse Chebyshev with respect to satisfying given tolerance requirements. The reader should note that for the inverse

Chebyshev approximation derived above, frequency is normalized such that the stopband starts at $\Omega = 1$. Let the passband edge be at a frequency Ω_p, and denote the passband distortion parameter by $\bar{\varepsilon}$ (in order to distinguish it from the parameter ε of the original Chebyshev). Since the stopband starts at $\Omega = 1$, it follows that $\Omega_p = \frac{1}{\Omega_s}$ where Ω_s denotes, as usual, the ratio of the frequencies defining the stopband edge and the passband edge.

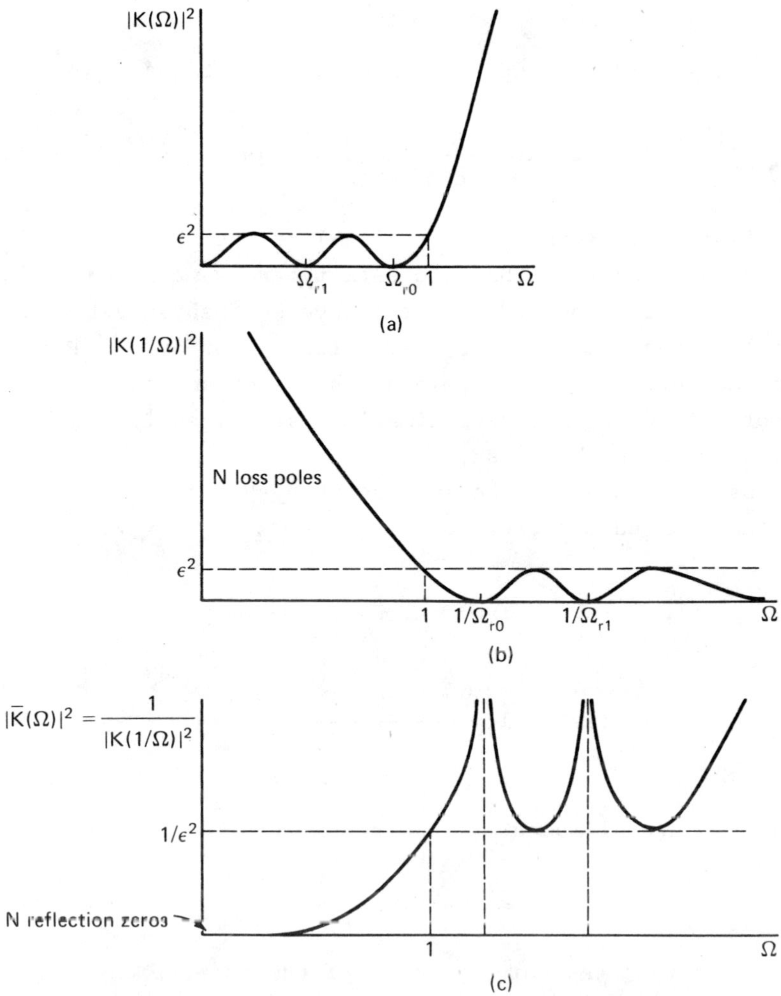

Figure 4.13 Development of the inverse Chebyshev function.

The squared characteristic function of the inverse Chebyshev in (4.76), when evaluated at $\Omega = \Omega_p = \frac{1}{\Omega_s}$, results in

$$\bar{\varepsilon}^{2} = 1 / \varepsilon^{2} \ C_N^{\ 2}(\Omega_s) \tag{4.79}$$

Thus,

$$- 20 \log \varepsilon + 20 \log \frac{1}{\bar{\varepsilon}} = 20 \log C_N(\Omega_s)$$

Substituting for the 1st term from (4.77) obtains

$$A_s + 20 \log \frac{1}{\bar{\varepsilon}} = 20 \log C_N(\Omega_s) \tag{4.80}$$

which is identical to equation (4.49) of the Chebyshev filter. We therefore conclude that, to satisfy the same given attenuation characteristics, the order of the inverse Chebyshev filter is identical to that of the Chebyshev filter. Moreover, the design chart and methods for determining the order and exercising design tradeoffs in Chebyshev approximations are directly applicable to the inverse Chebyshev case.

The natural modes of the inverse Chebyshev filter can be found from the squared transfer function,

$$|H(j\Omega)|^2 = 1 + \frac{1}{\varepsilon^2 \ C_N^{\ 2} \ (\frac{1}{\Omega})}$$

That is, $$H(s)\,H(-s) = 1 + \frac{1}{\varepsilon^2 \ C_N^{\ 2} \ (\frac{j}{s})} \tag{4.81}$$

Thus, the natural modes are the roots of

$$\varepsilon^2 \ C_N^{\ 2} \ (\frac{j}{s}) = - 1 \tag{4.82}$$

It follows that the natural modes of the inverse Chebyshev filter are the inverse of those for the corresponding Chebyshev filter, given in (4.62).

We are now in a position to summarize the procedure for generating inverse Chebyshev approximations which satisfy tolerance plots such as the one shown in Fig. 4.10. We note, at the outset, that the generated inverse Chebyshev approximation will be normalized in the manner illustrated in Fig. 4.14.

The procedure is as follows:

(1) Determine $\Omega_s = \frac{\omega_s}{\omega_p}$ and $\bar{\varepsilon} = \sqrt{10^{A_p/10} - 1}$.

(2) From A_s, $\bar{\varepsilon}$ and Ω_s, use the design chart of Fig. 4.11 and/or equation (4.30) to determine the order N. Also, if desired, the ripple parameter $\bar{\varepsilon}$ may be decreased and the stopband attenuation margin, if any, reduced to zero. (See Example 4.2)

(3) Determine the parameter ε of the corresponding Chebyshev filter from (4.79); that is,

$$\varepsilon = \frac{1}{\bar{\varepsilon}\, C_N\,(\Omega_s)}$$

(4) The loss poles of the inverse Chebyshev filter can be determined by using (4.78).

(5) The natural modes can be obtained by inverting the natural modes of the corresponding Chebyshev filter, given in (4.62).

Since Chebyshev filters of even order do not have reflection zeros at the origin, then the inverse Chebyshev filters of even order will have no loss-poles at ∞. As shown in Chapter 7, this makes an LC ladder realization impossible. Thus, we seek a modified even order inverse Chebyshev filter with loss-poles at ∞. This problem is analogous to that of the modified Chebyshev filter considered in Section 4.3. The solution consists of moving the highest loss-pole to ∞. This can be done either by applying the derivation of this section to the modified Chebyshev filter, or by finding a frequency transformation to be applied directly to the characteristic function in (4.76). This problem is left as an exercise for the reader.

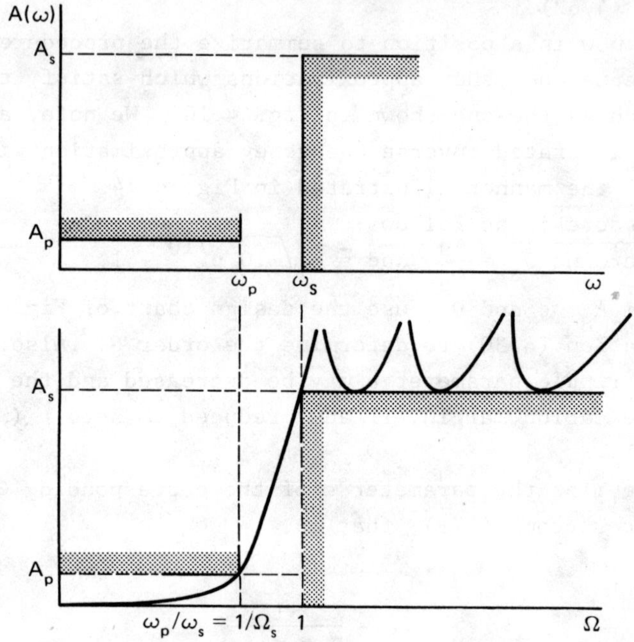

Figure 4.14 Illustrating the frequency normalization for inverse Chebyshev filters.

Example 4.4

Find the parameters of the inverse Chebyshev approximation which satisfies the requirements given in Example 4.2.

Solution

From the given tolerance plot we have

$$\Omega_s = 1.5\ ,\ A_p = 0.5\ \text{dB},\ \overline{\varepsilon} = 0.3493\ \text{and}\ A_s = 50\text{dB}$$

Using the design chart of Fig. 4.11, we obtain the order N = 8. If we keep the ripple factor $\overline{\varepsilon}$ unchanged, the stopband loss will be

$$A_s' = 51.73\ \text{dB}$$

The parameter ε of the corresponding Chebyshev filter is determined from (4.79) as

$$\varepsilon = \frac{1}{0.3493\ C_8(1.5)} = 0.002594$$

The loss poles of the inverse Chebyshev are obtained by using (4.78) for N = 8, as

$$\Omega_{\ell 0} = \frac{1}{\Omega_{r_0}} = 1.01959$$

$$\Omega_{\ell 1} = \frac{1}{\Omega_{r_1}} = 1.20269$$

$$\Omega_{\ell 2} = \frac{1}{\Omega_{r_2}} = 1.79995$$

$$\Omega_{\ell 3} = \frac{1}{\Omega_{r_3}} = 5.1258$$

The characteristic function is thus given by

$$|\bar{K}| = \frac{385.453\ \Omega^8}{(\Omega^2 - 1.039566)(\Omega^2 - 1.446463)(\Omega^2 - 3.23983)(\Omega^2 - 26.2741)}$$

The natural modes are obtained by inverting the natural modes of an 8th order Chebyshev with ε = 0.002594. The latter are obtained using (4.62). The result for the inverse Chebyshev is,

$$S_{0,7} : \Omega_o = 0.739888 \qquad Q = 3.7249$$

$$S_{1,6} : \Omega_o = 0.801642 \qquad Q = 1.2073$$

$$S_{2,5} : \Omega_o = 0.923155 \qquad Q = 0.7005$$

$$S_{3,4} : \Omega_o = 1.052444 \qquad Q = 0.5209$$

Note that the natural modes of the inverse Chebyshev have lower Q-factors than those for the Chebyshev filter satisfying the same requirement (Example 4.3). As will be seen in later chapters, the lower the natural mode Q-factor, the less sensitive its network realization. The disadvantage here is that the inverse Chebyshev filter has finite loss poles. Filters with finite loss poles require more complex network realizations than all-pole or polynomial filters.

We may now write the transfer function of the inverse Chebyshev as

$$H(s) = 385.452 \frac{s^2 + 0.1986307\ s + 0.547436}{s^2 + 1.039566}$$

$$\times \frac{s^2 + 0.664016\ s + 0.642631}{s^2 + 1.446463}$$

$$\times \frac{s^2 + 1.317873\ s + 0.852214}{s^2 + 3.23983}$$

$$\times \frac{s^2 + 2.02046\ s + 1.107637}{s^2 + 26.2741}$$

where the multiplicative constant has been determined from $|H(0)| = 1$. All the functions above are normalized relative to a frequency $f = 15$ KHz. To obtain the denormalized H(s), we replace s by $\dfrac{s}{2\pi \times 15 \times 10^4}$.

As mentioned at the beginning of this section, the inverse Chebyshev has group delay characteristics which are far superior to those of the Chebyshev filter of equal order. This fact is illustrated in Fig. 4.15, which shows plots of the group delay function τ for both the Chebyshev filter of example 4.3 and the inverse Chebyshev filter of the present example. Recall that both filters are of order $N = 8$ and satisfy identical attenuation requirements. Note also that the ripples in $\tau(\Omega)$ are much more evident for the Chebyshev case. This is due to the higher natural modes Q-factors. We may therefore conclude that, whenever the constant delay is important, an inverse Chebyshev filter will be a better choice. The disadvantage will be a more complex circuit realization.

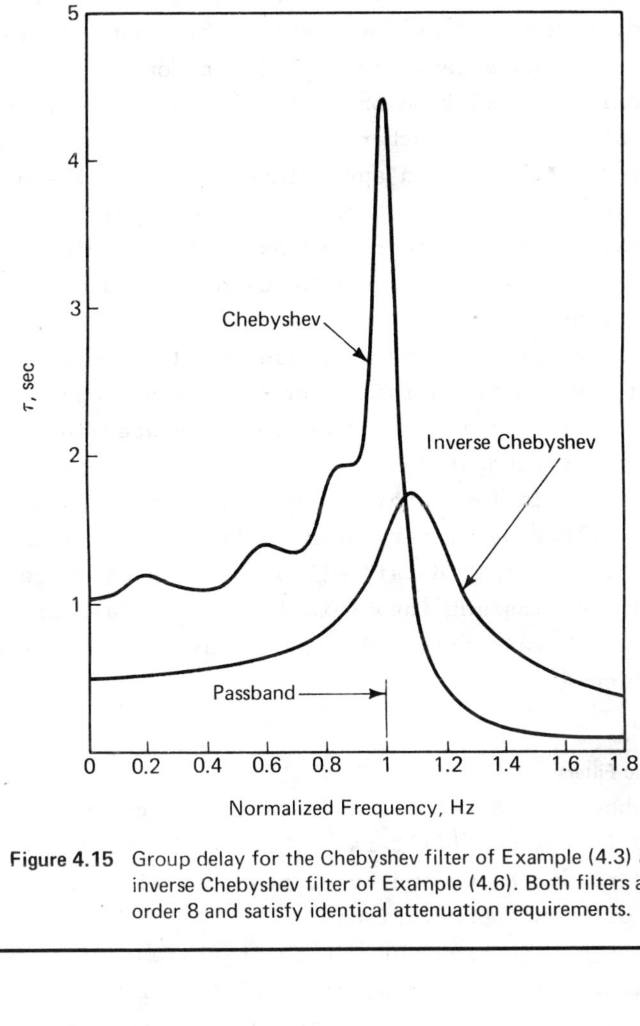

Figure 4.15 Group delay for the Chebyshev filter of Example (4.3) and the inverse Chebyshev filter of Example (4.6). Both filters are of order 8 and satisfy identical attenuation requirements.

4.5 ELLIPTIC FILTERS

Both the Butterworth and Chebyshev filter approximations provide easy-to-realize polynomial filters. They are, however, inefficient as far as the stopband is concerned, more attenuation is provided than is usually needed. Moving some of the loss-poles from infinity closer to the passband edge will increase the cut-off rate, a very desirable goal. This has already been demonstrated in the case of inverse Chebyshev filters. Their maximally flat passband, however, does not enable a high cut-off rate. Thus, if the stopband attenuation specification is constant, an optimum approximation will be one that is equiripple in both the passband and stopband. In other words, it is an approximation whose characteristic function approximates zero in the passband and infinity in the stopband, both in an equiripple fashion. Such an approximation is referred to as Elliptic, since the doubly periodic elliptic functions may be used to obtain the characteristic functions.

In Chapter 5, a general method is presented for placing the loss-poles in a way that satisfies an arbitrary stopband attenuation characteristic. This same method can be used in the special case where the stopband attenuation is constant, resulting in the elliptic filters. In fact, this is usually more economical than computing the filter parameters using elliptic functions. In the following, we consider in detail elliptic low-pass filters, with the aim of gaining insight into this important class of filter approximations. Toward this end, we make use of the formulae derived in Section 4.1.

Properties of Elliptic Filters

Fig. 4.16 shows the attenuation characteristics for order 5 and order 6 elliptic low-pass filters. These are representatives of elliptic filters for N odd and even, respectively. From Fig.4.16 we make the following observations:

(1) The elliptic filter approximation is characterized by four parameters; namely, the passband distortion or ripple A_p (or any of its equivalents such as, ε, $|\rho|$ or A_e), the selectivity factor Ω_s, the minimum stopband attenuation A_s and the order N. As will be shown, any three of these four parameters can be independently specified.

(2) The order N and the selectivity factor Ω_s uniquely determine the locations of the reflection zeros Ω_r and the loss-poles Ω_ℓ,

hence the characteristic function. Subsequently, one can trade off A_p vs. A_s; that is, increasing A_p also increases A_s, and vice versa. This point has already been made in Chapter 2 and will be further illustrated in the derivation of elliptic filters below.

(3) For an odd N, the elliptic filter has a reflection zero at the origin and a loss-pole at ∞, while for N even, these do not occur. The other reflection zeros and loss-poles occur on the jΩ axis and are all doubled. Thus, in the s-plane, the characteristic function will be of the form,

$$K(s) = k \frac{s \prod_{i=1}^{(N-1)/2} (s^2 + \Omega_{r_i}^2)}{\prod_{i=1}^{(N-1)/2} (s^2 + \Omega_{\ell_i}^2)}, \quad N \text{ odd} \tag{4.83}$$

$$\text{or, } K(s) = k \frac{\prod_{i=1}^{N/2} (s^2 + \Omega_{r_i}^2)}{\prod_{i=1}^{N/2} (s^2 + \Omega_{\ell_i}^2)}, \quad N \text{ even} \tag{4.84}$$

(4) As in the Chebyshev filters, for even order, the attenuation at $\Omega = 0$ is the maximum passband ripple A_p. Note also that, for any N the total number of peaks and valleys in the passband is equal to N.

Extracting the constant ε from K(s), we obtain the function that characterizes a family of elliptic filter approximations. This function is called the Chebyshev Rational Function and is denoted by $R_N(\Omega)$. It follows that

$$R_N(\Omega) = r \frac{\Omega \prod_{i=1}^{(N-1)/2} (\Omega^2 - \Omega_{r_i}^2)}{\prod_{i=1}^{(N-1)/2} (\Omega^2 - \Omega_{\ell_i}^2)}, \quad N \text{ odd}$$

$$\text{or, } R_N(\Omega) = r \frac{\prod_{i=1}^{N/2} (\Omega^2 - \Omega_{r_i}^2)}{\prod_{i=1}^{N/2} (\Omega^2 - \Omega_{\ell_i}^2)}, \text{ N even}$$

where r is a multiplicative constant. We will have more to say about the Chebyshev rational function at a later stage.

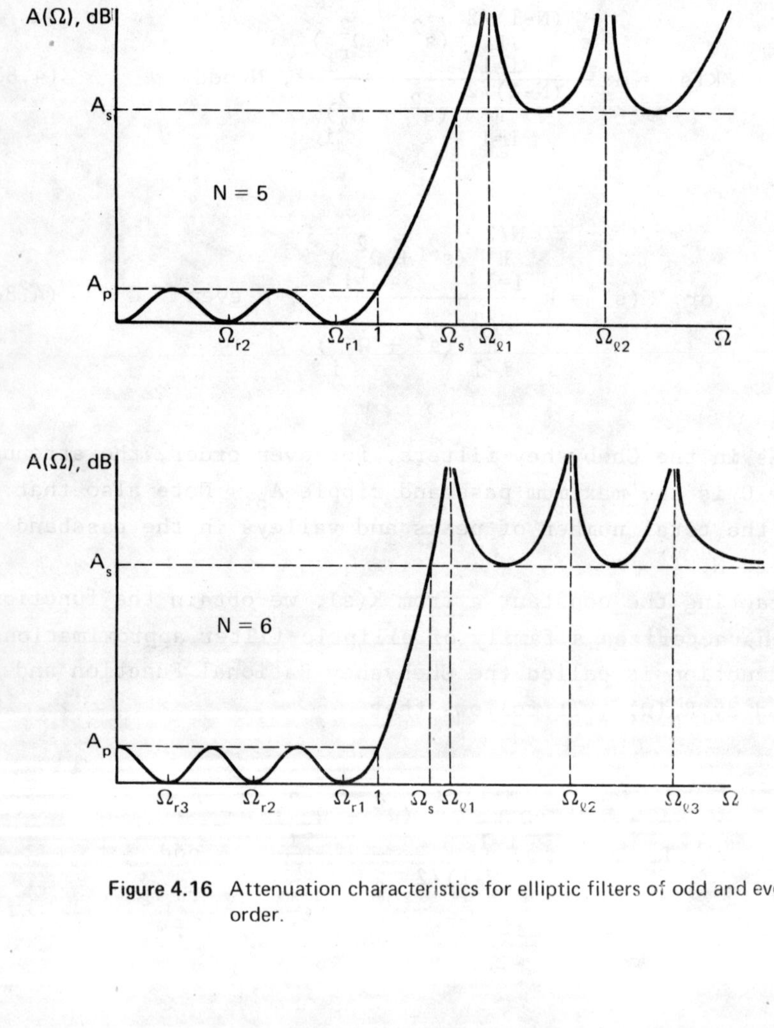

Figure 4.16 Attenuation characteristics for elliptic filters of odd and even order.

Derivation of the Characteristic Rational Function

The starting point in the derivation of an elliptic filter characteristic function is the z-plane squared characteristic function which provides an equiripple passband and an arbitrary stopband attenuation. This has been derived in section 4.1 equation (4.18), which we repeat here as

$$K'(z)\,K'(-z) = \varepsilon^2 \frac{[\text{Ev } P'(z)]^2}{P'(z)\,P'(-z)} \tag{4.85}$$

For an elliptic approximation, using the observations above, one can write an expression for the squared loss-pole polynomial

$$P'(z)\,P'(-z) = (1-z^2) \prod_{i=1}^{(N-1)/2} (z^2 - x_i^2)^2 \,, \quad N \text{ odd} \tag{4.86a}$$

$$\text{or, } P'(z)\,P'(-z) = \prod_{i=1}^{N/2} (z^2 - x_i^2)^2 \,, \quad N \text{ even} \tag{4.86b}$$

where $x_i^2 = 1 - \frac{1}{\Omega^2_{\ell i}}$. Note that $0 < |x_i| < 1$.

The problem now reduces to finding a set of loss-pole locations x_i to yield an equiripple stopband response. Toward this end, we first show that, for an elliptic approximation, the loss-poles and reflection zeros exhibit geometric symmetry around a frequency in the transition band, given by $\sqrt{\Omega_s}$. That is,

$$\Omega_{\ell 1}\,\Omega_{r1} = \Omega_{\ell 2}\,\Omega_{r2} = \ldots\ldots = \Omega_s \tag{4.87}$$

This relation together with (4.86) and (4.85) lead to the desired characteristic function $K'(z)$ or $K(s)$.

To show that (4.87) is true, consider, as an example, the 6th order elliptic filter, whose squared characteristic function $|K(j\Omega)|^2$ is shown in Fig. 4.17a. If Ω is replaced by $\frac{\Omega_s}{\Omega}$, we obtain the squared high-pass characteristic function $|K(j\frac{\Omega_s}{\Omega})|^2$ shown in Fig. 4.17b. Now, let us form a new characteristic function $|\overline{K}(j\Omega)|^2$ as

$$|\overline{K}\,(j\Omega)|^2 = 1 \Big/ |K\,(j\,\frac{\Omega_s}{\Omega}|^2$$

which is sketched in Fig. 4.17c. We now note that the new function also produces a sixth order elliptic filter. But, since the characteristic function of an elliptic filter is unique, then the zeros and poles of $\overline{K}\,(j\Omega)$ should coincide with those for $K\,(j\Omega)$, which proves (4.87).

We thus conclude that for a given order N and a given selectivity ratio Ω_s, one obtains a family of elliptic approximations characterized by a Chebyshev rational function, which we now denote by $R_{(N,\Omega_s)}(\Omega)$, and parameterized by ε; that is,

$$K(\Omega) = \varepsilon\, R_{(N,\Omega_s)}(\Omega)$$

with

$$R_{(N,\Omega_s)}(\Omega) = r\,\Omega \prod_{i=1}^{(N-1)/2} \frac{(\Omega^2 - \Omega_{r_i}^{2})}{(\Omega^2 - \Omega_{\ell_i}^{2})}, \quad N \text{ odd}$$

or

$$R_{(N,\Omega_s)}(\Omega) = r \prod_{i=1}^{N/2} \frac{(\Omega^2 - \Omega_{r_i}^{2})}{(\Omega^2 - \Omega_{\ell_i}^{2})}, \quad N \text{ even}$$

where $\Omega_{r_i}\,\Omega_{\ell_i} = \Omega_s$,

and $R_{(N,\Omega_s)}$ takes the form shown in Fig. 4.18 for N = 3 and 4.

Note that $R_{(N,\Omega_s)}(\Omega)$ oscillates in the passband around zero and

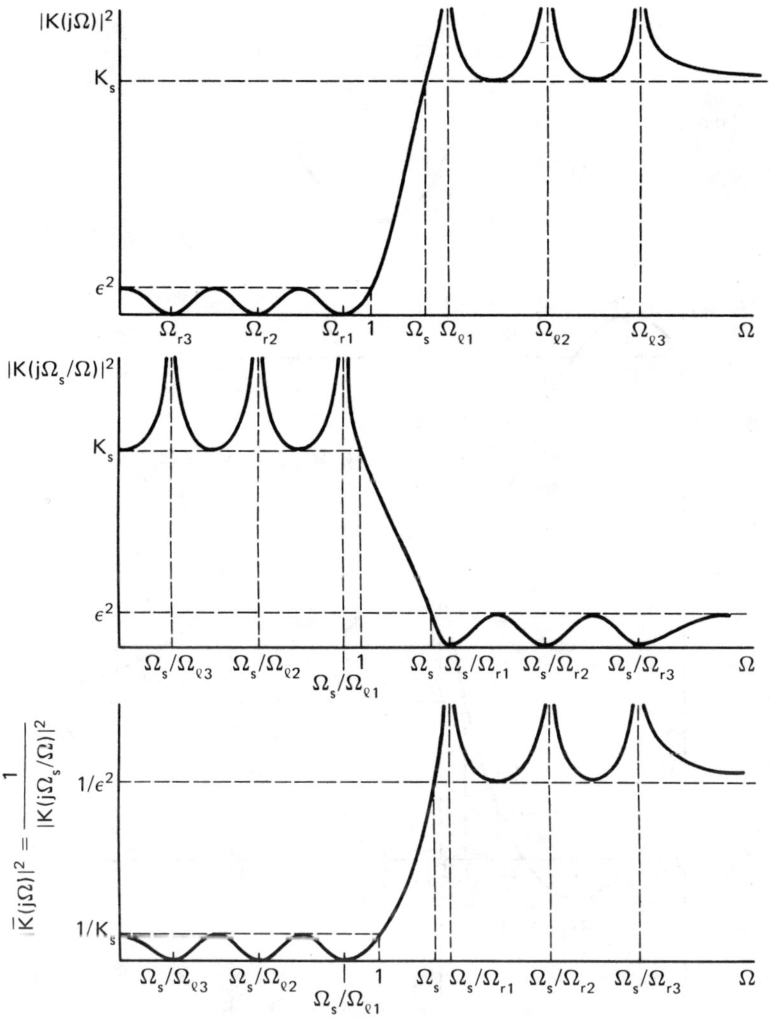

Figure 4.17 Illustrating that $\Omega_{\ell 1}\Omega_{r1} = \Omega_{\ell 2}\Omega_{r2} = \ldots = \Omega_s$.

takes on a maximum magnitude of 1. In the stopband, the magnitude of R exhibits equiminima. Moreoever, the minimum value of the magnitude of R in the stopband is equal to its magnitude at Ω_s. The frequencies of minima in the stopband also exhibit geometric symmetry with the frequencies of maxima in the passband, around $\sqrt{\Omega_s}$.

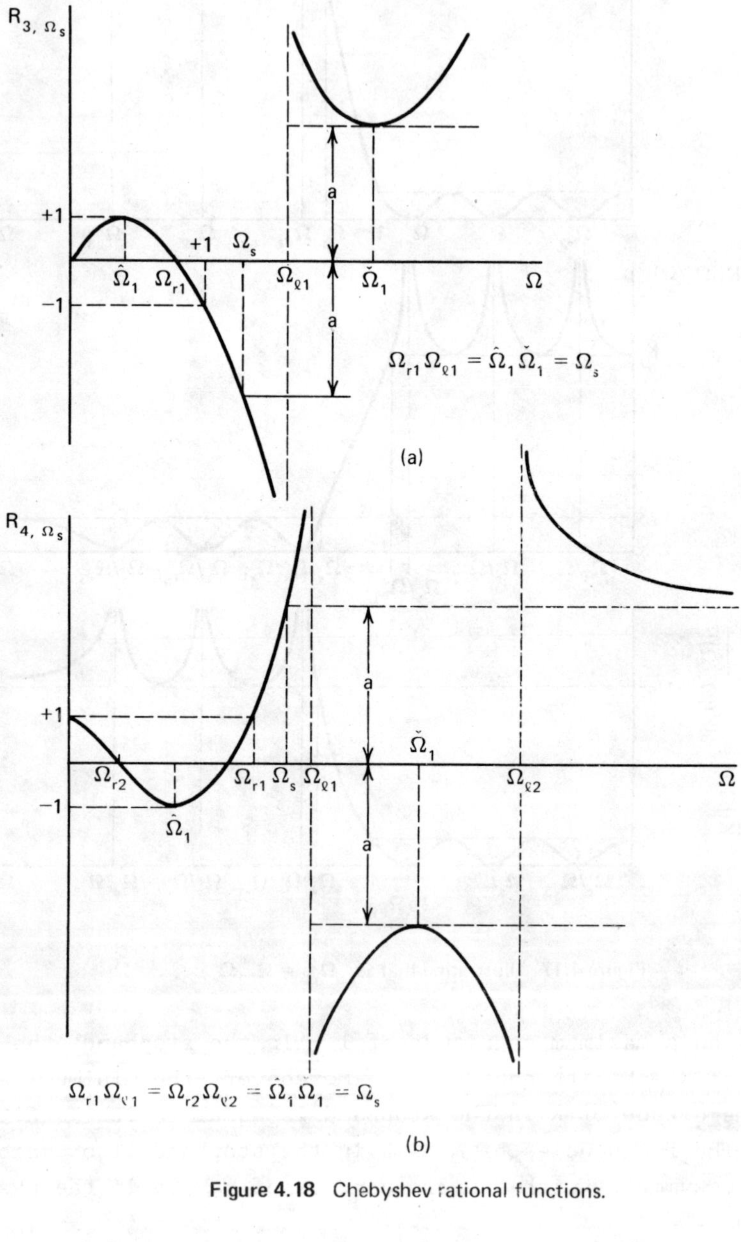

Figure 4.18 Chebyshev rational functions.

Example 4.5

Find the characteristic function for a third order elliptic filter having a selectivity factor $\Omega_s = 2$.

Solution

From (4.86) the filter will have a squared loss polynomial of

$$P'(z)\ P'(-z) = (1 - z^2)\ (z^2 - x_1^2)^2$$

where $x_1^2 = 1 - \dfrac{1}{\Omega_{\ell_1}^2}$ (4.88)

Thus, $P'(z) = (1 + z)\ (z + x_1)^2$

$$\text{Ev}\ [P'(z)] = z^2 + x_1^2 + 2x_1\ z^2 = (1 + 2x_1)\ z^2 + x_1^2$$

Substituting in (4.85) results in

$$K'(z)\ K'(-z) = \varepsilon^2\ \frac{[(1 + 2x_1)\ z^2 + x_1^2]^2}{(1 - z^2)\ (z^2 - x_1^2)^2} \tag{4.89}$$

The s-plane characteristic function K(s) can now be obtained from (4.89) by substituting $z^2 = 1 + \dfrac{1}{s^2}$, and with some simple algebra,

$$K(s) = \varepsilon\ \Omega_{\ell_1}^2\ (1 + x_1)^2\ \frac{s\ [s^2 + \dfrac{1 + 2x_1}{(1 + x_1)^2}]}{s^2 + \Omega_{\ell_1}^2} \tag{4.90}$$

Thus, $\Omega_r = \dfrac{\sqrt{1 + 2x_1}}{1 + x_1}$ (4.91)

Using the reciprocal relation $\Omega_r = \dfrac{\Omega_s}{\Omega_\ell}$ together with (4.91) and (4.88) yields the following equation in x_1,

$$x_1^4 + 2x_1^3 + x_1\left(\frac{2}{\Omega_s^2} - 2\right) + \left(\frac{1}{\Omega_s^2} - 1\right) = 0 \qquad (4.92)$$

Note that (4.92) is the general equation for determining the loss-pole location of any elliptic filter of order 3. For the special case $\Omega_s = 2$, the only positive root of (4.92) is $x_1 = 0.897745$. This results in $\Omega_{\ell_1} = 2.270068$ and $\Omega_r = 0.881031$.

Using the values of x_1 and Ω_{ℓ_1} in (4.90) we obtain

$$K(s) = \varepsilon\, 18.558946\, \frac{s\,(s^2 + 0.776216)}{(s^2 + 5.153207)}$$

Thus, the third order Chebyshev rational function having $\Omega_s = 2$ is given by

$$R_{(3,2)}(\Omega) = 18.558946\, \Omega \frac{(\Omega^2 - 0.776216)}{(\Omega^2 - 5.153207)}$$

The points of minima and maxima of $R_{(3,2)}$ can be easily obtained by differentiation, and the result is

$$\hat{\Omega} = 0.526941 \qquad \check{\Omega} = 3.795489$$

Note that $\hat{\Omega}\,\check{\Omega} = 2$

Also by direct evaluation

$$|R_{(3,2)}(\hat{\Omega})| = |R_{(3,2)}(1)| = 1$$

$$|R_{(3,2)}(\Omega_s)| = |R_{(3,2)}(\check{\Omega})| = 103.7627$$

As expected, the calculation of the singularities of the characteristic function does not depend on ε. For a given ε, one can then use the computed characteristic function in the Feldtkeller equation and proceed to determine the transfer function H(s), together with the filter natural modes.

Although in the above example we have been able to compute the loss-pole locations with relative ease, this is not the case for higher order elliptic filters. For such cases, a simple iterative procedure[2] may be applied as follows:

(1) Proceed as in Example 4.5 to obtain the expression for K(s), as (4.90).

(2) Find an initial guess for $\Omega_{\ell 1}, \Omega_{\ell 2}, \ldots,$. The initial guess could be obtained by the reciprocal relation applied to the reflection zeros of the Chebyshev filter of the same order. Find $x_1, x_2, \ldots,$ corresponding to your initial guess.

(3) Use the values for x_i in the numerator of K(s), factor it and obtain new values of Ω_{ri}, and hence, (by the reciprocal relation) find $\Omega_{\ell i}$ and x_i .

(4) Repeat step 3 a number of times until the process converges to the required accuracy.

Determining the Order N

The question now arises of the required order N of an elliptic filter which satisfies a given set of attenuation specifications. One means of estimating N is to use plots of 20 log $|R_{(N,\Omega_s)}(\Omega_s)|$ versus Ω_s for various values of N. To illustrate how such a chart can be used to determine N, we recall that

$$A_s = 10 \log_{10} [1 + |K(j\Omega_s)|^2$$

$$\simeq 20 \log_{10} |K(j\Omega_s)|$$

$$= 20 \log_{10} \varepsilon\, |R_{(N,\Omega_s)}(\Omega_s)|$$

Thus,

$$A_s + 20 \log_{10} \frac{1}{\varepsilon} = 20 \log_{10} |R_{(N,\Omega_s)}(\Omega_s)| \tag{4.93}$$

This equation is similar in form to that used in determining the order of the Chebyshev filter, equation (4.49), with the Chebyshev polynomial replaced by the Chebyshev rational function. Also note that, for small ε ($A_p \le 1$ dB), the term $20 \log_{10} \frac{1}{\varepsilon}$ can be approximated by the minimum echo loss A_{emin} resulting in

$$A_s + A_{emin} = 20 \log_{10} |R_{(N,\Omega_s)}(\Omega_s)|$$

Fig. 4.19 shows plots for $20 \log_{10} |R_{(N,\Omega_s)}(\Omega_s)|$ versus Ω_s for various values of N. This design chart can therefore be used to determine N, given A_p, A_s and Ω_s, as well as to exercise the tradeoff between A_p and A_s in a manner identical to that used in the Chebyshev case. Note that the curves marked (b) for N even are those for elliptic filters modified in the manner described below.

Example 4.6

Find the order N of the elliptic filter which satisfies the requirements given in example 4.2.

Solution

From given requirements,

$$A_s + 20 \log_{10} \frac{1}{\varepsilon} = 50 + 20 \log_{10} \frac{1}{0.3493} = 59.13 \text{ dB}$$

Locating the point $\Omega_s = 1.5$ and Attenuation = 59.13 dB on the chart of Fig. 4.19, reveals that the required order is N = 5. It is also apparent that a fifth order elliptic meets the specifications with no margins left. Thus no tradeoff is possible between stopband and passband performance.

Modified Elliptic Filters

Since elliptic filters of even order do not have loss-poles at infinity it is not possible to realize them with an LC ladder void of magnetic coupling. Through frequency transformations, however, one can move the highest loss-poles to ∞ resulting in a modified elliptic filter. In this process some of the stopband attenuation is sacrificed. This transformation process is analogous to the one performed in the modified inverse Chebyshev and will not be done here. The curves marked (b) in Fig. 4.19 correspond to these modified elliptic filters.

A further modification of the even order elliptic filters consists of moving the lowest reflection zeros to the origin in order to allow the LC ladder realization to be equally terminated.

Many excellent tabulations for elliptic filter data exist [2]-[5]. In addition one can easily design elliptic filters using computer routines[1] that implement the general approximation methods given in this book.

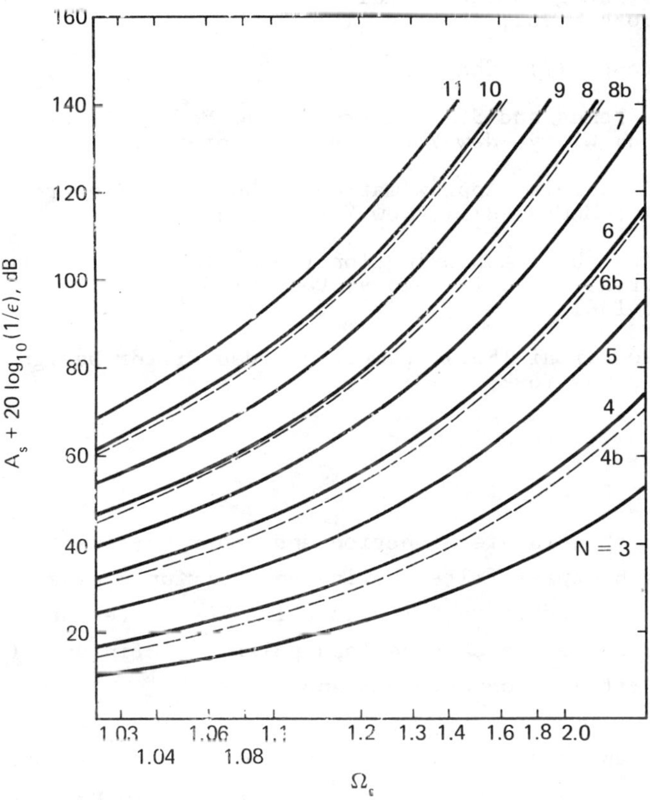

Figure 4.19 Chart for determining the required order of elliptic filters.

4.6 REFERENCES AND SOURCES FOR FURTHER READING

(1) W. M. Snelgrove and A. S. Sedra, FILTOR 2 - A Computer Aided Filter Design Package, Matrix Publishers, Champaign, Ill.

(2) D. S. Humphreys, The Analysis, Design and Synthesis of Electrical Filters, Prentice-Hall, Englewood Cliffs, N.J., 1970, Chapter 8.

(3) A. I. Zverev, Handbook of Filter Synthesis, Wiley, New York, 1967.

(4) J. K. Skwirzynski, Design Theory and Data for Electrical Filters, D. Van Nostrand, Princeton, N.J., 1965.

(5) E. Christian and E. Eisenmann, Filter Design Tables and Graphs, Wiley, New York, 1966.

(6) L. Weinberg, Network Analysis and Synthesis, McGraw-Hill, New York, 1962, Chapter 11.

(7) Reference (2), Chapter 6.

(8) G. C. Temes and S. K. Mitra (Eds), Modern Filter Theory and Design, Wiley, New York, 1973, Chapter 2.

(9) R. W. Daniels, Approximation Methods for Electronic Filter Design, McGraw-Hill, New York, 1974.

(10) H. Watanabe, "Approximation theory for filter networks", IRE Trans. Circuit Theory, Vol. CT-8, No. 3, pp. 341-356, Sept. 1961.

(11) G. Szentirmai (Ed.), Computer Aided Filter Design, IEEE Press, New York, 1973.

PROBLEMS

P4.1 Find the transfer function and other pertinent parameters of a bandpass filter having an equiripple passband extending from $\Omega=0.5$ to $\Omega=1$ with $\varepsilon=1$. The filter has one loss pole at zero and three loss poles at infinity. Calculate the attenuation at $\Omega=0.3$ and $\Omega=1.2$.

P4.2 Find the transfer function and other pertinent parameters of a low-pass filter having an equiripple passband from $\Omega=0$ to $\Omega=1$ with $\varepsilon=1$. The filter has a pair of finite loss poles at $s=\pm j2$ and one loss pole at infinity. Calculate the minimum stopband attenuation and the frequencies at which this value of attenuation is obtained. Sketch $A(\Omega)$.

P4.3 Give the Chebyshev polynomials $C_5(\Omega)$ and $C_6(\Omega)$ and sketch their values for $-1\leq\Omega\leq 1$. Also calculate the location of their zeros.

P4.4 Consider a normalized fifth order Chebyshev filter with $\varepsilon=1$:

(a) Give expressions for $|K(\Omega)|^2$ and $|H(\Omega)|^2$.

(b) Sketch $A(\Omega)$ and give the location of the reflection zeros.

(c) Calculate the attenuation at $\Omega=2$.

(d) Give the transfer function $H(s)$ together with Ω_o and Q of the natural modes.

P4.5 Find the order N, the characteristic function, the transfer function and the natural modes ω_o and Q of a Chebyshev filter that meets the following low-pass specifications:

$$f_p = 1 \text{ KHz},\ f_s = 2 \text{ KHz},\ A_p \leq 1 \text{ dB and } A_s \geq 60 \text{ dB}$$

Calculate the group-delay at $\Omega=1$ (see Example 1.1).

P4.6 Derive equation (4.50)

P4.7 Derive equation (4.51)

P4.8 Show that the Chebyshev natural modes, given by equation (4.62), reduce to the natural modes of a Butterworth filter as $\varepsilon\to o$.

P4.9 Using the fact that for an LC ladder the reflection coefficient is defined as $\rho(s) = (Z_1-R)/(Z_1+R)$, find the values of L and C so that the network of Fig. P4.9 realizes a Chebyshev filter with $\varepsilon=1$ and where $R=1\Omega$.

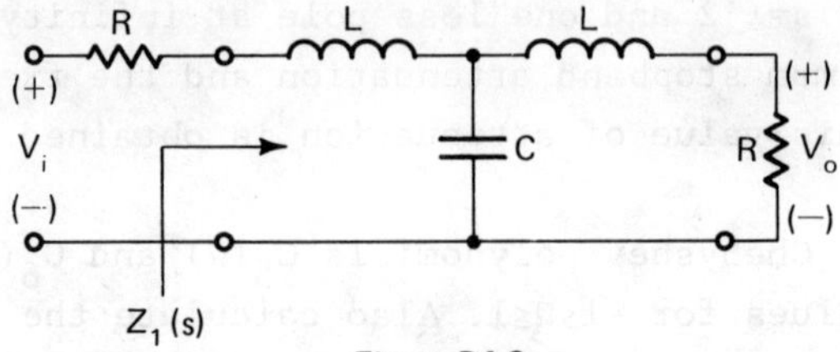

Figure P4.9

P4.10 Modify the eighth order Chebyshev filter of Examples 4.2 and 4.3 to introduce a pair of attenuation zeros at the origin. Find all the relevant parameters of the modified approximation. Also calculate the reduction in stopband attenuation at Ω_s suffered as a result of the modification.

P4.11 Show that the 3 dB frequency Ω_{3dB} of the inverse Chebyshev filter normalized as in Fig. 4.14 is given by

$$\Omega_{3dB} \simeq 2\left(\frac{\varepsilon}{2}\right)^{\frac{1}{N}}$$

where $\varepsilon \ll 1$ is the ripple parameter of the corresponding Chebyshev approximation.

P4.12 Consider a Chebyshev and an inverse-Chebyshev filter both of third order and hence meet identical specifications. Let the passband edge be at $\Omega=1$ A_p = 3dB, and Ω_s=1.5. Find the natural modes of each filter and hence plot its group delay function. Give the value of the maximum variation in $\tau(\Omega)$ over the passband for both filters.

P4.13 Find the transfer function and all pertinent parameters for a modified inverse Chebyshev low-pass filter with N=6, A_p=0.5dB, f_p=1KHz and Ω_s=2.

P4.14 Find the parameters of the inverse Chebyshev filter that satisfies the specifications given in Problem 4.5. Calculate the group delay at Ω=1.

P4.15 Find the transfer function and all pertinent parameters of a third-order elliptic low-pass filter having $\Omega_s=1.5$ and $\varepsilon=1$. Sketch $A(\Omega)$ and give the location of the reflection zeros and the loss poles. Calculate the minimum stopband attenuation (see Example 4.5).

P4.16 Consider a fifth-order elliptic filter with $\Omega_s=2$. If the finite loss poles are at 2.0892 and 3.2508 find the location of the reflection zeros. Give $|K(\Omega)|^2$ and find the minimum stopband attenuation for $A_p=0.5$dB and $A_p=1$dB.

P4.17 Using the fact that the reflection coefficient $\rho(s)$ for an LC ladder filter is defined as $\rho(s)=[Z_1(s)-R]/[Z_1(s)+R]$ find component values for the network in Fig. P4.17 (assume $R=1\Omega$) to realize

(a) An inverse Chebyshev filter

(b) An elliptic filter with $\Omega_s=2$ and $\Omega_\ell=2.2701$

For both filters $A_p=1$dB and the passband edge is at $\Omega=1$ rad/sec.

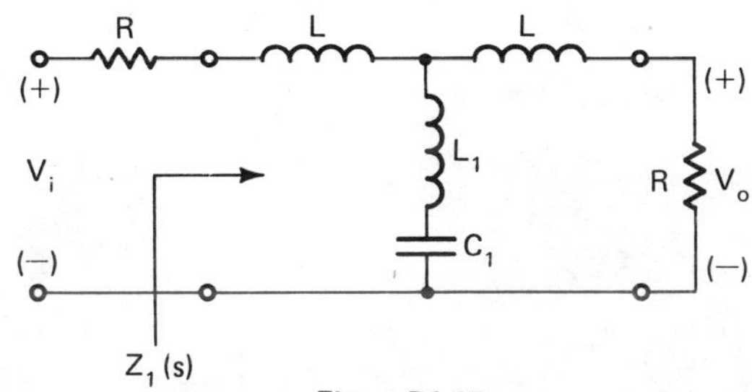

Figure P4.17

P4.18 Determine the order N of the elliptic filter that meets the specifications given in P4.5.

7.15 Find the transfer function and all pertinent parameters of a third-order elliptic low-pass filter having [illegible] and [illegible]. Sketch A(ω) and give the location of the reflection zeros and the loss poles. Calculate the minimum stopband attenuation (see Example [illegible]).

7.16 Consider a fifth-order elliptic filter with [illegible]. If the three loss poles are at [illegible] and [illegible], find the location of the reflection zeros [illegible] and (b) the minimum stopband attenuation for [illegible] and [illegible].

7.17 Using the fact that the reflection coefficient [illegible] for an LC ladder filter is defined as [illegible], find numerical values for the [illegible] of the [illegible]:

(a) [illegible] filter

(b) an elliptic filter with [illegible] and [illegible].

For both filters [illegible] and the [illegible] as [illegible] [illegible].

Figure 7.17

7.18 Determine the order N of the elliptic filter that [illegible] the specifications given in [illegible]

CHAPTER 5

ITERATIVE STOPBAND APPROXIMATION

In this Chapter, we complete our study of approximation methods for filter attenuation functions by describing an iterative technique for obtaining a set of loss poles which allow the filter attenuation specifications to be met. The method to be described depends upon and makes use of the formulae developed in the two previous Chapters for the characteristic functions of filters having maximally flat or equiripple passbands. The reader will recall that these latter formulae were given explicitly in terms of an arbitrary set of loss poles. Thus, the method we shall describe here allows an optimum placement of the loss poles over the stopbands to satisfy the stopband specifications, while simultaneously producing the desired passband response via the explicit formulae for the characteristic functions. At this point, the reader is urged to review Sections 2.4 and 2.6 of Chapter 2, dealing with approximation criteria and strategy.

5.1 PRESCRIBED LOSS POLES

Before proceeding to the stopband approximation algorithm, we should like to place all of the preceding Chapters in perspective and indicate that, even at this point, the reader has quite an extensive arsenal of tools available for filter approximation.

In the previous Chapters, we have obtained explicit formulae for the characteristic functions of filters having prescribed equiripple or maximally flat passband responses. These formulae are expressed as functions of an arbitrary set of loss poles. If the loss poles as well as the type of passband are prescribed, then the formulae allow the direct calculation of the reflection zeros, and the natural modes can be found by solving Feldtkeller's equation in the z-plane, i.e. by finding all the roots of

$$E'(z)E'(-z) = P'(z)P'(-z) + \varepsilon^2 F'(z)F'(-z) \tag{5.1}$$

We have then only to transform the natural modes and loss poles back to the s-plane. The transfer function $H(s) = E(s)/P(s)$ is then formed by selecting all of the left-half-plane roots of $E(s)E(-s)$ for $E(s)$ and appropriately selecting the desired roots of $P(s)P(-s)$ for $P(s)$. The approximation problem is thus solved.

The above procedure of using prescribed loss poles is indeed the case for many practical problems; for instance, if one requires conventional "all-pole" designs where all loss poles lie at either zero or infinity in the s-plane. Simply by programming the formulae on a computer and making use of a root finding routine, the approximation problem for prescribed loss poles can be easily automated. Such a simple program could release the designer from reliance on the tabulated and catalogued filters, which provide results only for discrete steps of A_p and A_s.

Fig. 5.1 is a flow chart which illustrates the logic of a computer program to solve the approximation problem when the loss poles and passband response are prescribed. A complete set of FORTRAN SUBROUTINES that may be used in such a program have been included in the FILTOR 2 package[1], and their names given in the flow chart of Fig. 5.1.

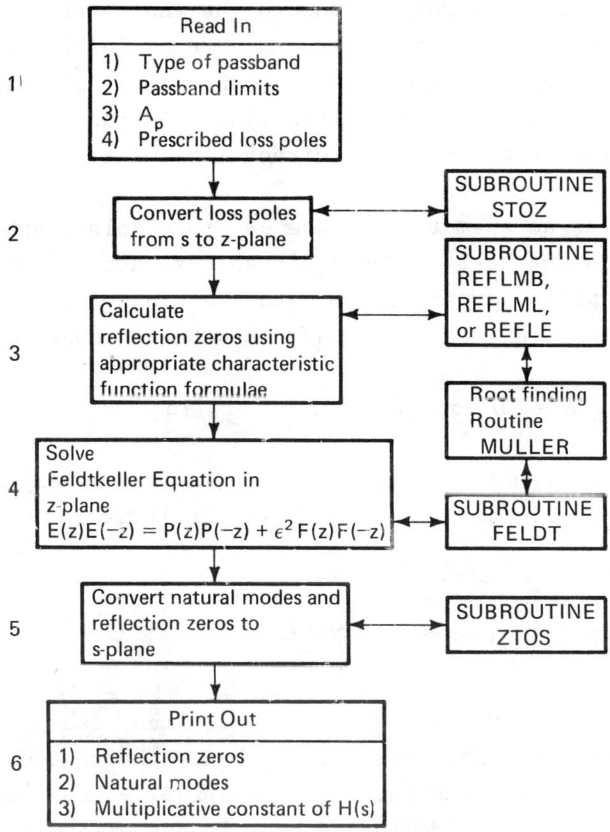

Figure 5.1 Flow chart for filter approximation with prescribed loss poles.

5.2 THE LOGARITHMIC AXIS

In this section, we introduce a logarithmic transformation of the frequency axis which has been found to enhance the numerical accuracy of the procedures. The transformation

$$\gamma = \ell n(z) = \frac{1}{2} \ell n \left(\frac{s^2 + 1}{s^2 + a^2}\right) \tag{5.2}$$

is applicable to bandpass filters, and to low pass filters by setting a = 0. It is interesting to note that before the widespread use of computers this transformation was used, in conjunction with certain approximations to the characteristic function formulae, to obtain simple graphical "templates" which enabled early designers to find by hand suitable loss pole placements, satisfying the stopband specifications. In fact, the early computer programs consisted of a direct implementation of these graphical template methods.

Consider as an example the case of equiripple passband filters described in Chapter 4. The first approximation made in obtaining the template formula is to note that in the stopband of most filters, where $|K|^2 >> 1$, we may take the attenuation to be

$$A = 10 \log (1 + |K|^2) \simeq 10 \log |K|^2. \tag{5.3}$$

Next, noting from (4.13) that $K'(z)K'(-z)$ may be expressed in terms of the all pass function $[Q(-z)/Q(z)]$ as

$$K'(z)K'(-z) = \frac{\varepsilon^2}{4} \left[\frac{Q(z)}{Q(-z)} + 2 + \frac{Q(-z)}{Q(z)}\right], \tag{5.4}$$

where $Q(z)Q(-z) = P'(z)P'(-z)$ is the squared loss pole polynomial it is easily seen that over the stopband (the region in which the loss poles lie) the variable z has real values and the formula (5.4) is usually dominated by only one of the terms in the bracketed expression. Thus, the second approximation is to take

$$K'(z)K'(-z) \simeq \frac{\varepsilon^2}{4} \left[\frac{Q(z)}{Q(-z)}\right], \tag{5.5}$$

so that the attenuation along the stopband may now be taken as

$$A \simeq 10 \log \frac{\varepsilon^2}{4} \left|\frac{Q(z)}{Q(-z)}\right| \tag{5.6}$$

Noting that the all pass function $[Q(z)/Q(-z)]$ can be expressed as a product of factors involving the loss poles x_k as

$$\frac{Q(z)}{Q(-z)} = \prod_k \left(\frac{x_k + z}{x_k - z}\right) \tag{5.7}$$

We can now write the approximate expression for the stopband attenuation in (5.6) as the sum

$$A \simeq 20 \log \left(\frac{\varepsilon}{2}\right) + \sum_k 10 \log \left|\frac{x_k + z}{x_k - z}\right| , \tag{5.8}$$

of which a typical term is 10 log $|(x_k+z)/(x_k-z)|$. Now, introducing the logarithmic transformation $\gamma = \ell n(z)$, we may find a "template" for the typical term and plot the attenuation over the γ-axis in the same manner as the well known Bode plots are obtained. Introducing the notation $\gamma_k = \ell n(x_k)$, we may express the typical term as

$$\log \left|\frac{x_k + z}{x_k - z}\right|_{z=e^{\gamma}} = \log \left|\frac{e^{\gamma_k} + e^{\gamma}}{e^{\gamma_k} - e^{\gamma}}\right| = \log \coth \frac{|\gamma-\gamma_k|}{2} , \tag{5.9}$$

resulting in the "template" formula for the approximate stopband loss of an equiripple passband filter

$$A(\gamma) \simeq 20 \log \left(\frac{\varepsilon}{2}\right) + \sum_k 10 \log \coth \left[\frac{|\gamma-\gamma_k|}{2}\right] . \tag{5.10}$$

Fig. 5.2 illustrates the "template" shape for the typical term (5.9). This template may be constructed of clear plastic and used with appropriately scaled graph paper, on which the specifications have been drawn, and a set of drawing instruments to place loss poles by hand.

The approximations made in arriving at the template formula are reasonable for many filter specifications, but in certain cases the results are almost useless. Of course, with the advent of the computer there is no need to make such approximations, and the exact formulae may be programmed into a routine. SUBROUTINE REMEM of FILTOR 2[1] uses the exact formulae for both equiripple and maximally flat filters.

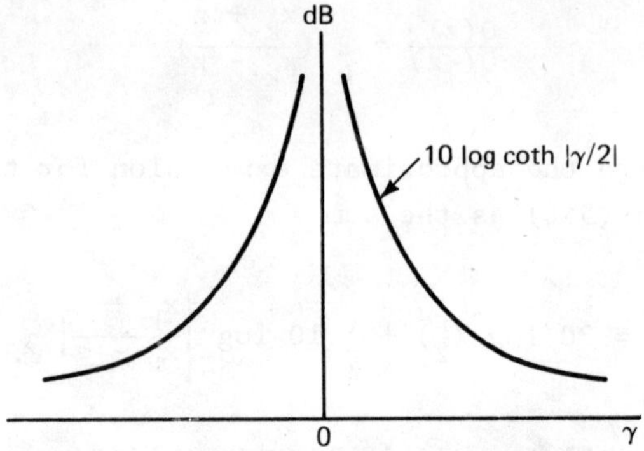

Figure 5.2 Illustration of the template curve.

We shall later derive certain expressions for $A(\gamma)$ in an example and leave the remaining derivations as exercises. Therefore, the reader should become familiar with the relationship between the ω-axis and the γ-axis as illustrated in Fig. 5.3. We note that the upper stopband region is mapped into $[-\infty < \gamma < 0]$, and the lower stopband region is mapped into $[\ell n(1/a)<\gamma<\infty]$. The real axis of the s-plane is mapped into the region $[0<\gamma<\ell n(1/a)]$ lying between the two stopbands. The point at infinity in the s-plane corresponds to $\gamma = 0$, while the origin of the s-plane corresponds to $\gamma = \ell n(1/a)$. The reader should note that the frequency variable in the γ-plane stopbands increases from left to right, in the same manner as the ω variable, even though the relative positions of the two stopbands is reversed. Of course, the same mapping is applicable to low pass filters by setting $a = 0$, in which case the lower stopband vanishes.

5.3 THE MODIFIED REMEZ METHOD

There exists a well known procedure for the gradual equalization of a set of function extremum by an iterative process. This procedure, known as "Remez second method", is described in the references at the end of this Chapter. Remez second method has been adapted to the stopband approximation problem we are discussing. The method, as described in some references, makes use of

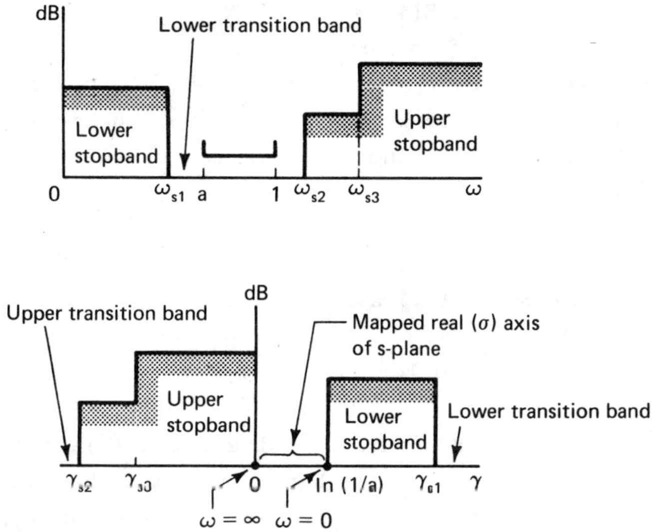

Figure 5.3 Illustrating the relationship between the γ-axis and the ω axis.

the "template" formulae and so is not exact although it provides excellent results for many types of specifications. If the procedure is to be computerized, however, it is no problem to program the exact formulae.

Outline of the Procedure

We begin the procedure by first transforming the filter stopband specifications to the γ-axis. A typical set of γ-axis tolerance contours and a typical attenuation function are illustrated in Fig. 5.4. An initial estimate of the number and placement of the loss poles is also required. The accuracy of this initial estimate is not critical, since it may be changed if required after the algorithm has converged to an unsatisfactory answer. Of course, a good initial estimate will be repaid by an optimum placement of the loss poles in the very few iterations.

The error or "margin" between the specifications contour $A_s(\gamma)$ and the filter attenuation $A(\gamma)$ is denoted by $E(\gamma)$. Thus, $E(\gamma)$ is expressed as

$$E(\gamma) = A(\gamma) - A_s(\gamma) \tag{5.11}$$

Clearly, we require expressions for the attenuation over the γ-axis, $A(\gamma)$, and these we may obtain from the z-plane formulae for the characteristic functions via the substitution

$$z = e^{\gamma}. \tag{5.12}$$

Thus, we may express $A(\gamma)$ as

$$A(\gamma) = 10 \log [1 + K'(z)K'(-z)]\big|_{z=e^{\gamma}} \tag{5.13}$$

where we use the appropriate formulae for $K'(z)K'(-z)$ corresponding to the type of filter passband desired. The initial loss poles will be transformed to the γ-axis positions γ_k, and $A(\gamma)$ may be expressed in terms of these γ_k.

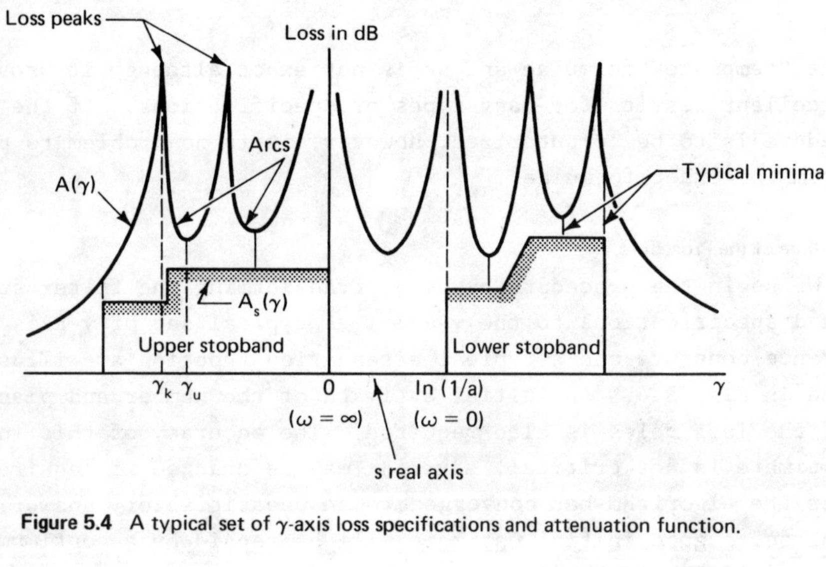

Figure 5.4 A typical set of γ-axis loss specifications and attenuation function.

Example 5.1

Find an expression for $A(\gamma)$ for the case of a maximally flat low pass filter having an arbitrary set of imaginary axis loss poles.

Solution

From Chapter 3, we find the expression (3.9) for $K'(z)K'(-z)$ as

$$K'(z)K'(-z) = \varepsilon^2 \frac{C}{P'(z)P'(-z)} ,$$

where, for our case of $j\omega$-axis loss poles only, we obtain $P'(z)P'(-z)$ from (2.39) as

$$P'(z)P'(-z) = (-1)^{n_i} (z^2-1)^{n_i} \prod_{k=1}^{n_f} (z^2-x_k^2)^2 ,$$

and C is given by (3.11) as

$$C = \prod_{k=1}^{n_f} x_k^4$$

with the order of the filter given by $N = (n_i + 2n_f)$. Thus, for our problem,

$$K'(z)K'(-z) = \frac{(-1)^{n_i} \varepsilon^2 \prod_{k=1}^{n_f} x_k^4}{(z^2-1)^{n_i} \prod_{k=1}^{n_f} (z^2-x_k^2)^2} .$$

Now transforming to the γ-axis, the following three functions may be defined.

$$R(\gamma_k) = \prod_{k=1}^{n_f} e^{4\gamma_k} ,$$

$$T(\gamma) = \frac{\varepsilon^2}{e^{2N\gamma} [e^{-2\gamma}-1]^{n_i}} ,$$

$$Q(\gamma,\gamma_k) = \prod_{k=1}^{n_f} [e^{2(\gamma_k-\gamma)}-1]^2 .$$

Thus, we have

$$|K(\gamma)|^2 = \frac{T(\gamma)\ R(\gamma_k)}{Q(\gamma,\gamma_k)} ,$$

with the attenuation given by

$$A(\gamma) = 10 \log [1 + |K(\gamma)|^2] ,$$

or

$$A(\gamma) = 10 \log [1 + \frac{T(\gamma)\ R(\gamma_k)}{Q(\gamma,\gamma_k)}] ,$$

which is the required result, expressed in a form suitable for use in a computer routine.

In the case of all jω-axis loss poles, the minima of the error function $E(\gamma)$ are easily characterized. Clearly, these minima must occur in the "arcs" or "loops" of the attenuation function, which appear as the attenuation dips downward between the loss poles. In addition, there will be a minimum in each end region between the stopband edge and the first loss pole encountered inside that stopband edge. Thus, if the number of loss poles placed in one of the stopbands, excluding those placed at the transforms of zero and infinite frequency, is denoted by m, then we have (m + 1) possible minima in that stopband. These stopband minima locations, or points of closest approach of $A(\gamma)$ to $A_s(\gamma)$, are denoted by γ_u and are illustrated in Fig. 5.4. Using this notation, we may express the stopband minima as

$$E(\gamma_u) = A(\gamma_u,\gamma_k) - A_s(\gamma_u), \tag{5.14}$$

where we have shown explicitly the dependence of $A(\gamma)$ on the loss pole set γ_k.

As described in Chapter 2, we wish to find a placement of the loss poles so as to force all of the stopband minima $E(\gamma_u)$ to become equal to an as yet unknown constant margin which we shall denote by M. Of course, if all of the minima are equal we have an optimum solution. Thus, we wish to find the incremental margin $\Delta E(\gamma_u)$ that is required at each minima point to make all of the margins equal to M. This is expressed as

$$\Delta E(\gamma_u) = M - E(\gamma_u) \tag{5.15}$$

We can obtain an approximate expression for $\Delta E(\gamma_u)$ by expanding in a Taylor's series and retaining only first order terms. Thus we have

$$\Delta E(\gamma_u) \simeq \sum_{k=1}^{n_f} \frac{\partial E(\gamma_u)}{\partial \gamma_k} \Delta\gamma_k\,, \tag{5.16}$$

where n_f is the total number of finite non-zero loss poles which we are treating as variables in the approximation. These n_f loss poles are to be placed in such a way as to force all $E(\gamma_u)$ to become equal to M. Equation (5.16) provides a relation from which we could possibly solve for increments $\Delta\gamma_k$ to be added to the γ_k.

The partial derivatives in (5.16) are readily available analytically from (5.14) as

$$\frac{\partial E(\gamma_u)}{\partial \gamma_k} = \frac{\partial A(\gamma_u, \gamma_k)}{\partial \gamma_k} \tag{5.17}$$

and we already have explicit expressions for $A(\gamma, \gamma_k)$ which we may differentiate. We shall develop one such expression as an example, and leave the remainder as exercises.

Example 5.2

Find an expression for $(\partial A/\partial \gamma_k)$ for the case of a maximally flat low pass filter having an arbitrary set of imaginary axis loss poles.

Solution

Straightforward differentiation of the expression for $A(\gamma,\gamma_k)$ found in Example 5.1 yields the result

$$\frac{\partial A(\gamma,\gamma_k)}{\partial \gamma_k} = \frac{40 \log (e) \, |K(\gamma)|^2}{[1 + |K(\gamma)|^2]} \left[\frac{1}{1 - e^{2(\gamma_k-\gamma)}}\right]$$

where $|K(\gamma)|^2$ is as given in that Example.

Now let us rewrite (5.16), using (5.15) and (5.17) as

$$\sum_{k=1}^{n_f} \frac{\partial A(\gamma_u,\gamma_k)}{\partial \gamma_k} \Delta\gamma_k - M = -E(\gamma_u) . \tag{5.18}$$

This relation provides a set of $(n_f + 1)$ equations for the low pass (single stopband) case, or $(n_f + 2)$ equations for the band-pass case, in the $(n_f + 1)$ unknowns $(\Delta\gamma_1, \Delta\gamma_2, \ldots . \Delta\gamma_{n_f}, M)$. This set of equations, which we write in symbolic matrix form as

$$\left[\left[\frac{\partial A}{\partial \gamma_k}\right]_{\gamma=\gamma_u} \;\vdots\; -1\right] \left[\begin{array}{c}\Delta\gamma_k \\ \hline M\end{array}\right] = -\left[E(\gamma_u)\right] , \tag{5.19}$$

may be solved for the loss pole increments $\Delta\gamma_k$ and constant margin M. The resulting increments may then be added to the initial loss poles $\gamma_k^{(i)}$ to produce a new set of loss poles $\gamma_k^{(i+1)}$ as

$$\gamma_k^{(i+1)} = \gamma_k^{(i)} + \Delta\gamma_k . \tag{5.20}$$

Of course, due to the approximation involved in (5.16), we do not expect these new loss poles to be correct, but they certainly might be better than the original estimate. At this point, the procedure is repeated. The present set of $\gamma_k^{(i+1)}$ are treated as an initial estimate, all of the (n_f+1) or (n_f+2) margin minima are searched out and calculated and a new set of equations (5.19) are set up and solved for the new $\Delta\gamma_k$. This process is carried out repeatedly until the desired convergence is obtained. This procedure has quadratic convergence and usually only few iterations are required for a typical problem, depending of course on the initial estimate of loss pole locations. Fig. 5.5 is a flow chart illustrating the modified Remez algorithm we have just described.

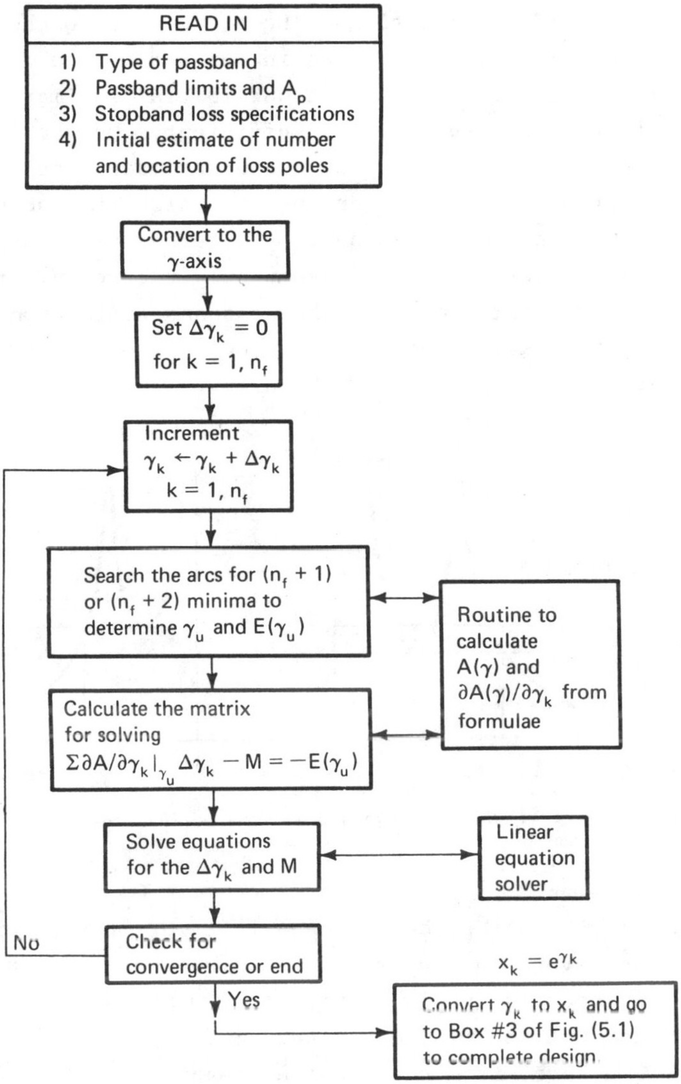

Figure 5.5 Flow chart for iterative stopband approximation.

At this point, we should note that once the Remez procedure has converged, we have merely equalized the minima and we may not in fact have satisfied our specifications, since the constant margin may be either positive or negative (it is a highly unlikely occurrence that it will be exactly zero). If the margin M is negative then we have the situation depicted in Fig. 5.6, and we have not met our specifications. Here, we have no choice but to increase the number of loss poles in our estimate and begin again. This means, of course, that our initial estimate of the order of filter required was not sufficient. On the other hand, if M is positive, then we may have overestimated the order of filter required. This may mean that the algorithm should be repeated again with fewer loss poles. Alternatively, as suggested in previous chapters, we may subsequently reduce any excess margin to zero by simply reducing the passband distortion A_p, before calculating the natural modes. Thus, we can easily trade off excess margin for an improved passband.

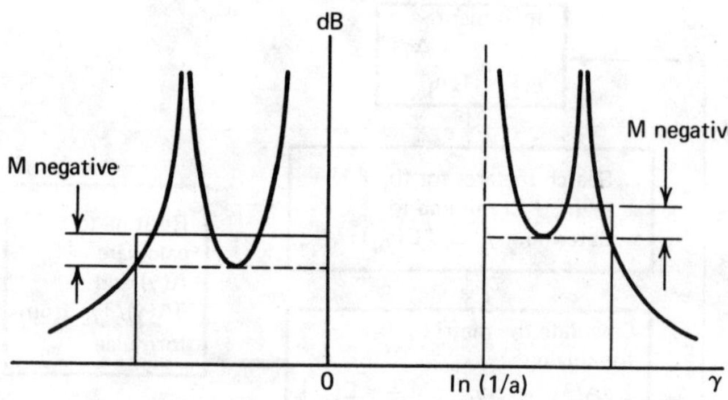

Figure 5.6 All minima equalized but M is negative.

Details of the Procedure

In the above description, we have left out many details, and we now discuss several of these. We require explicit expressions for $A(\gamma,\gamma_k)$ and its partial derivatives with respect to the γ_k, in order to find the minima and set up the equations to be solved for $\Delta\gamma_k$. We have, from the previous chapters, a collection of formulae for the z-plane characteristic functions of different classes of filters. For reasons of generality, these formulae included the most general form of loss pole polynomial, while the movable loss poles considered here are a special class; namely, those that produce infinite attenuation peaks on the physical frequency ($j\omega$) axis. The remaining loss poles, if any, will be treated as fixed parameters in the Remez iterations. (As mentioned previously, these poles may find use as variable parameters for approximating other functions, such as phase or delay).

Of course, an efficient search procedure for locating the minima in each of the arcs must be arranged so that an inordinate amount of time is not spent in locating these minima, since the search must be repeated at each iteration. We require, as well, an efficient algorithm for solving the set of linear equations established during each iteration.

The reader will find details in the literature on the convergence properties of this modified Remez procedure, at least for the low pass case. In fact, for most practical problems the iteration converges geometrically to the equiripple (equi-margin) solution and usually no more than 5 to 10 iterations are required to obtain satisfactory results. This, of course, depends upon the initial estimate of the loss pole locations.

Let us now consider the low pass case with n_f loss poles to be used as variable parameters in the approximation. A typical initial set of γ_k is illustrated in Fig. 5.7. We note that there are $(n_f + 1)$ minima along the stopband and we have exactly $(n_f + 1)$ unknown parameters to solve for. Thus, the linear equations to be solved are given by Equation A shown on page 188. These equations have to be solved for the increments $\Delta\gamma_k (k=1, n_f)$ and the unknown constant margin of MdB, at each iteration. Eventually, the loss pole increments $\Delta\gamma_k$ become very small and all of the minima $E(\gamma_u)$ will become equal to the unknown value M. Now, if M is positive, we have exceeded our specification and we may try again with a lower order function or use a lower passband ripple A_p. In fact, we may increase the minimum echo loss specification A_{emin} by very nearly the amount MdB. If, however, M is negative, then we have not satisfied the specification and we must try again with a higher order.

$$
\underbrace{\begin{bmatrix}
\dfrac{\partial A(\gamma_{u_1})}{\partial \gamma_1} & \dfrac{\partial A(\gamma_{u_1})}{\partial \gamma_2} & \cdots & \dfrac{\partial A(\gamma_{u_1})}{\partial \gamma_{n_f}} & -1 \\
\dfrac{\partial A(\gamma_{u_2})}{\partial \gamma_1} & \cdots & \cdots & \dfrac{\partial A(\gamma_{u_2})}{\partial \gamma_{n_f}} & -1 \\
\vdots & & & \vdots & \\
\dfrac{\partial A(\gamma_{u_{n_f+1}})}{\partial \gamma_1} & \cdots & \cdots & \dfrac{\partial A(\gamma_{u_{n_f+1}})}{\partial \gamma_{n_f}} & -1
\end{bmatrix}}_{n_f + 1}
\begin{bmatrix}
\Delta\gamma_1 \\ \Delta\gamma_2 \\ \vdots \\ \Delta\gamma_{n_f} \\ M
\end{bmatrix}
=
\begin{bmatrix}
-E(\gamma_{u_1}) \\ -E(\gamma_{u_2}) \\ \vdots \\ -E(\gamma_{u_{n_f+1}})
\end{bmatrix}
\quad (n_f + 1 \text{ rows})
$$

EQUATION A

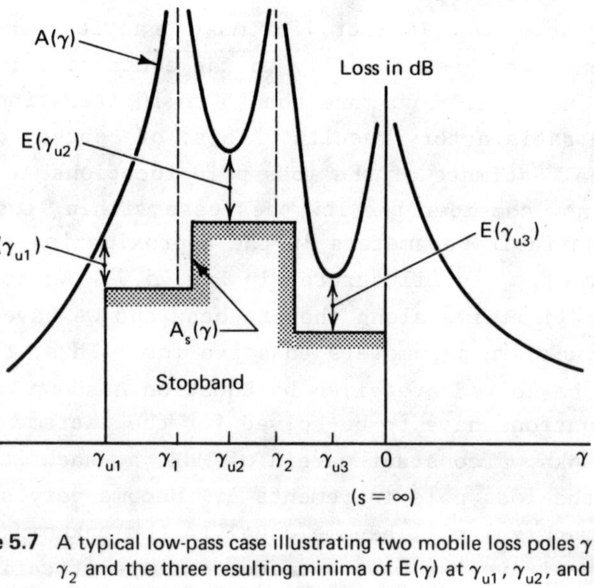

Figure 5.7 A typical low-pass case illustrating two mobile loss poles γ_1 and γ_2 and the three resulting minima of $E(\gamma)$ at γ_{u1}, γ_{u2}, and γ_{u3}.

At this point, we should note that if the low pass specification had been the conventional straight line stopband, the procedure described would result in the conventional Elliptic function or Inverted Chebyshev low pass response with an equiripple stopband, as discussed in Chapter 4. Thus, the modified Remez procedure may be used to design elliptic low pass filters, and is very nearly as efficient as direct calculations using the elliptic functions, which must, in any case, be evaluated by power series computations on a computer.

Let us now go on to the bandpass case. Here, we again have n_f variable loss poles $\gamma_k (k=1, n_f)$, of which some are placed in the lower stopband and the remainder in the upper stopband. A typical situation is illustrated in Fig. 5.8. As discussed earlier, we now find that we have $(n_f + 2)$ minima to equalize. Thus, there are $(n_f + 2)$ equations to be solved for $(n_f + 1)$ unknowns,

$$
n_f + 2 \left\{ \overbrace{\begin{bmatrix} \frac{\partial A(\gamma_{u_1})}{\partial \gamma_1} & \frac{\partial A(\gamma_{u_1})}{\partial \gamma_2} & \cdots & \frac{\partial A(\gamma_{u_1})}{\partial \gamma_{n_f}} & -1 \\ \vdots & \vdots & & \vdots & \vdots \\ \frac{\partial A(\gamma_{u_{n_f+2}})}{\partial \gamma_1} & \cdots & \cdots & \frac{\partial A(\gamma_{u_{n_f+2}})}{\partial \gamma_{n_f}} & -1 \end{bmatrix}}^{n_f + 1} \right. \begin{bmatrix} \Delta\gamma_1 \\ \Delta\gamma_2 \\ \vdots \\ \Delta\gamma_{n_f} \\ M \end{bmatrix} = \begin{bmatrix} -E(\gamma_{u_1}) \\ -E(\gamma_{u_2}) \\ \vdots \\ -E(\gamma_{u_{n_f+2}}) \end{bmatrix}
$$

EQUATION B

This system of equations is overdetermined, since there is one more equation than there are unknowns and we are faced with the dilemma that we cannot equalize all of the $(n_f + 2)$ minima. We can only obtain solutions in which $(n_f + 1)$ margins are equalized and the remaining one is arbitrary. This can be accomplished by discarding one of the equations, and solving the reduced set. One must take care to do this in such a manner that the convergence of the algorithm is not upset. In order to ensure that the specifications will be satisfied, the equation corresponding to the most positive minimum should be eliminated at each step, but this strategy may not be compatible with the convergence of the algorithm. Usually, a compromise is made where the most positive minimum of the two nearest $s = 0$ ($\gamma = \ln 1/a$) and $s = \infty$ ($\gamma = 0$) is thrown out. This latter strategy allows convergence in most cases, and places the "large" minimum far from the passband, avoiding the possibility of unnecessarily high selectivity in a transition band.

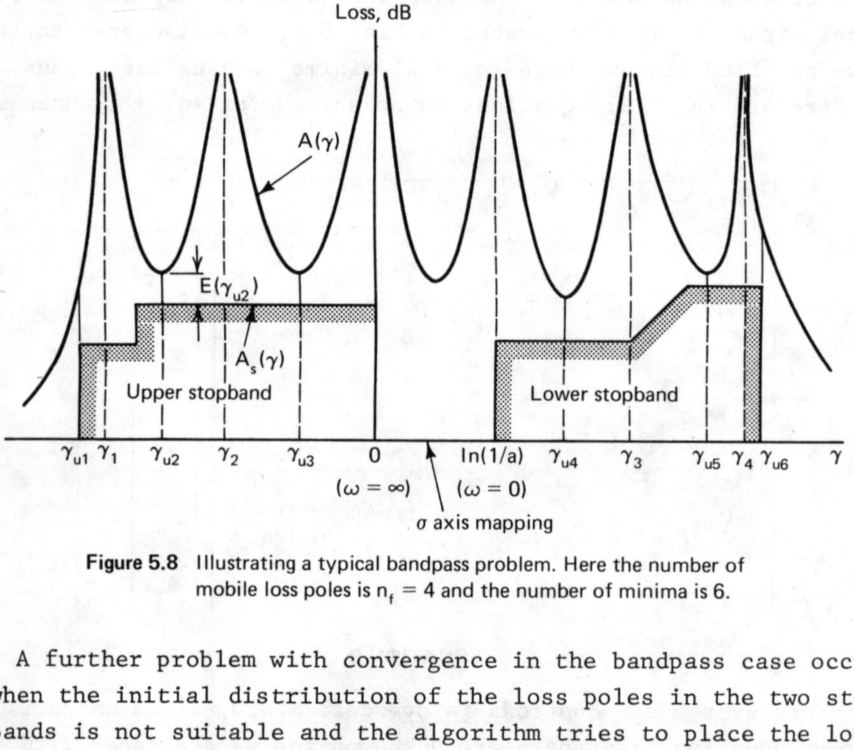

Figure 5.8 Illustrating a typical bandpass problem. Here the number of mobile loss poles is $n_f = 4$ and the number of minima is 6.

A further problem with convergence in the bandpass case occurs when the initial distribution of the loss poles in the two stopbands is not suitable and the algorithm tries to place the loss poles in the "real axis region", $[0 < \gamma < \ln(1/a)]$ or across the "real axis region" in the opposite stopband. In both of these cases, the algorithm generally will not converge. As an example, consider the case illustrated in Fig. 5.9 where initially all loss poles are placed in only one stopband, and the single minimum in the other stopband is less than any of those in the opposite stopband. Clearly, a solution of this kind cannot exist, since the algorithm will attempt to place one or more loss poles in the other stopband. Thus, we see that in general for the bandpass case, convergence is not assured, but fortunately, for many practical (narrow band) problems, convergence is obtained.

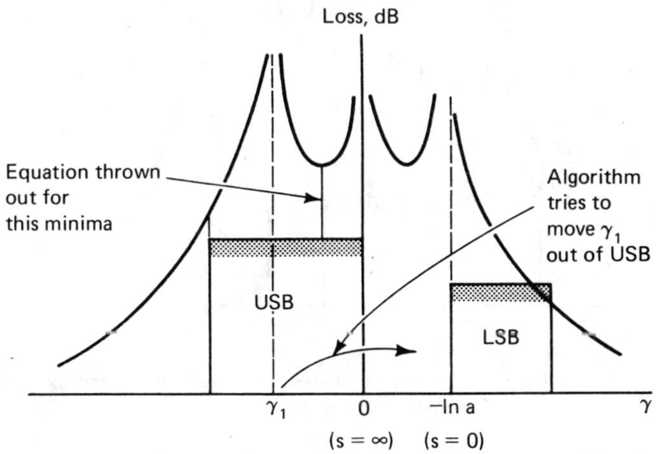

Figure 5.9 Illustrating a case of difficult convergence for the bandpass problem.

An interesting point revealed by the above discussion is that, in general, a bandpass specification with identical constant stopband attenuation, as illustrated in Fig. 5.10, cannot have an equiripple solution. Nevertheless, we know that if a conventional low pass filter with equiripple stopband (Elliptic or Inverted Chebyshev) is transformed to the bandpass (via the low pass to bandpass transformation as described in Chapter 6), just such a function is produced. In this case, however, the rather severe requirement of geometric symmetry is imposed on the function by the low pass to bandpass transformation. Thus, we conclude that the only cases in which the modified Remez iterative procedure can produce a perfectly equiripple approximation to a constant stopband attenuation of a bandpass filter are those in which the specifications have geometric symmetry. In these cases, the minimum corresponding to the equation being eliminated at each step will eventually become equal to the remaining minima.

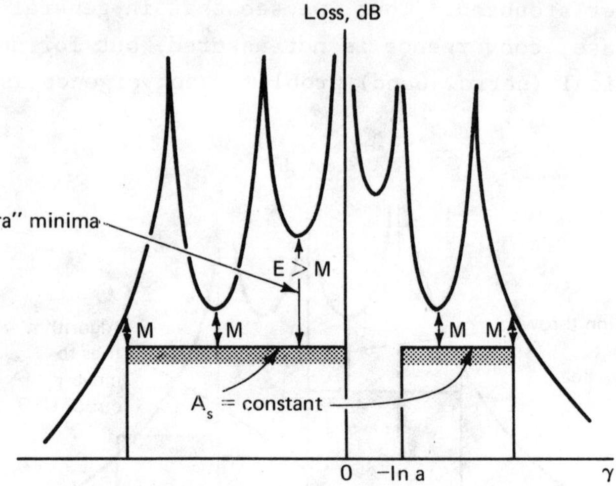

Figure 5.10 Illustrating the impossibility of an equiripple stopband approximation in the general bandpass problem.

There are, however, at least two methods of producing another unknown for the set of bandpass equations which the authors have found useful, and which usually eliminate any convergence difficulties. The first method is to allow the unknown constant margin in each stopband to be different. Thus, instead of having a single unknown parameter M we introduce the two parameters, M_u for the upper stopband and M_ℓ for the lower stopband. In this case, the equations to be solved at each stage of the iteration take on the form shown as Equation C on page 193, and the iterations proceed to a solution of the kind illustrated in Fig. 5.11. Here, one may find that either the result is satisfactory or, if not, it is immediately clear which stopband will require more, or less, attenuation and thus indicate the proper loss pole distribution between the two stopbands.

The second method which we have found useful but which is not applicable to passive LC ladder networks without mutual inductance, is to take as the extra unknown parameter another variable loss pole which is placed in the region between the two stopbands on the γ-axis, as illustrated in Fig. 5.12. This region corresponds to the real axis of the s-plane and the loss poles

$$\overbrace{}^{n_f + 2}$$

$$n_f + 2 \left\{ \begin{bmatrix} \dfrac{\partial A(\gamma_{u_1})}{\partial \gamma_1} & - - - & \dfrac{\partial A(\gamma_{u_1})}{\partial \gamma_{n_f}} & -1 & 0 \\ & & & -1 & 0 \\ \cdot & & & \vdots & \vdots \\ & & & 0 & -1 \\ \cdot & & & & \\ & & & 0 & -1 \\ \cdot & & & \vdots & \vdots \\ \dfrac{\partial A(\gamma_{u_{n_f+2}})}{\partial \gamma_1} & - - - & \dfrac{\partial A(\gamma_{u_{n_f+2}})}{\gamma_{n_f}} & 0 & -1 \end{bmatrix} \right. \begin{bmatrix} \Delta\gamma_1 \\ \cdot \\ \cdot \\ \cdot \\ \partial\gamma_{n_f} \\ M_u \\ M_\ell \end{bmatrix} = \begin{bmatrix} -E(\gamma_{u_1}) \\ \cdot \\ \cdot \\ \cdot \\ \cdot \\ \\ -E(\gamma_{u_{n_f+2}}) \end{bmatrix}$$

EQUATION C

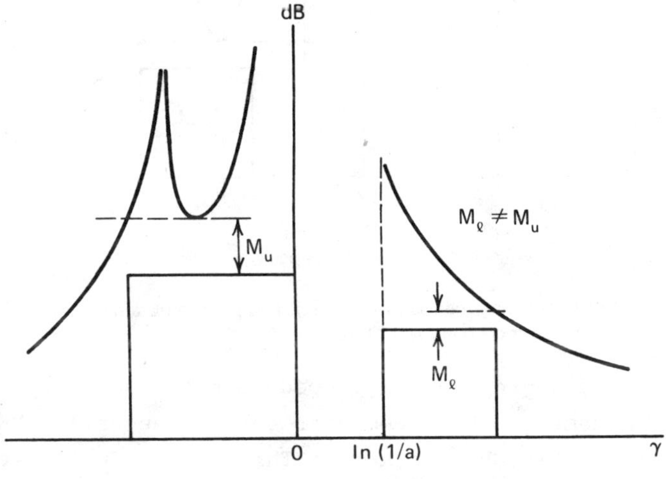

Figure 5.11 Illustrating the use of separate stopband margins to guarantee convergence in a difficult bandpass approximation.

here result in a mirror image symmetrical pair of real axis s-plane loss poles, which belong to Group 4 of the loss pole classification of Table 2.1. This type of loss pole can be produced by active circuitry or by LC networks which include mutual inductance. The reader will note that these real-axis loss poles provide an additional unknown parameter but do not increase the number of margin minima, since they do not produce a loss peak over either stopband. In fact, such a loss pole contributes loss to both stopbands and, by proper placement, can produce a perfectly equiripple result having all minima (of a bandpass filter) equalized. The authors have used this technique to design bandpass filters having almost perfect arithmetic symmetry, a necessary requirement for many types of communications systems. Such an example will be described below.

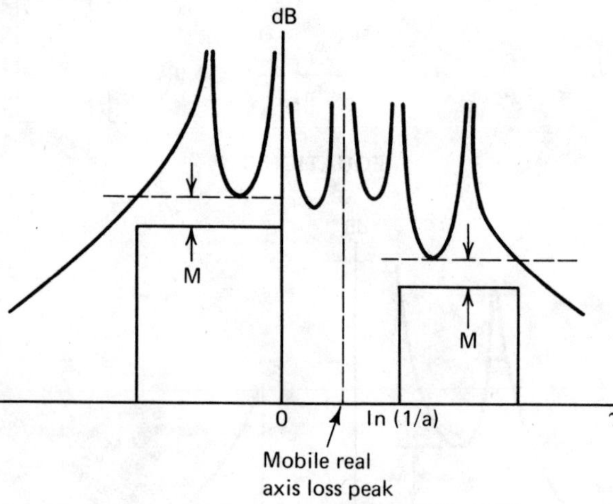

Figure 5.12 Illustrating the use of a real axis loss pole to obtain an equiripple result for a bandpass filter.

The stopband approximation procedure described in this chapter has been implemented by a computer subroutine called REMEM, which is part of the FILTOR 2 package. This subroutine, together with the ones mentioned in previous chapters, such as REFLE, REFLMB, FELDT, enable the filter designer to obtain a complete solution to the approximation problem, given arbitrary attenuation specifications. The FILTOR 2 package contains a "driver" program that "calls" the various routines at the appropriate phases of the design process. For program listings, detailed description as well as step-by-step operation procedure, the reader is referred to the FILTOR 2 manual [1]. It should be noted that many of the filters used as examples throughout this book were designed using FILTOR 2. The only exception to this are the classical designs discussed in Chapter 6.

Example 5.3

It is required to find an approximation for a bandpass filter having the following arithmetically symmetric attenuation specifications:

Passband : equiripple with $f_1 = 1$ KHz, $f_2 = 1.4$ KHz and $A_p = 0.4$ dB

Stopbands: $A_s = 40$ dB for $f \le 0.7$ KHz and $f \ge 1.7$ KHz

Solution

We note at the outset that, unless real-axis loss poles are used, we will not be able to obtain an approximation whose attenuation characteristics is arithmetically symmetric. To demonstrate this, we will present two different approximations: first without and then with real-axis loss poles.

The approximation without real axis loss poles has $n_i = 2$, $n_o = 2$ and $n_f = 2$. Other parameters of this eighth-order approximation were given in Example 2.2. The attenuation and group delay functions are plotted in Fig. 5.13. Note that, as expected, the stopbands attenuation margins are not all equal. As a result the attenuation function does not exhibit arithmetic symmetry. This filter approximation will be used as an example to illustrate the various passive and active realization methods described in subsequent chapters.

The second approximation was obtained by using only one loss pole at zero frequency, and allowing the program to place one loss pole on the real axis in the s-plane. The decision to use one (rather than two) loss pole at zero was prompted by the observation that the attenuation obtained in the first design in the lower stopband far exceeded specifications. Fig. 5.14 shows the attenuation and group delay characteristics of the new approximation. As may be seen, the response is very close to being arithmetically symmetric.

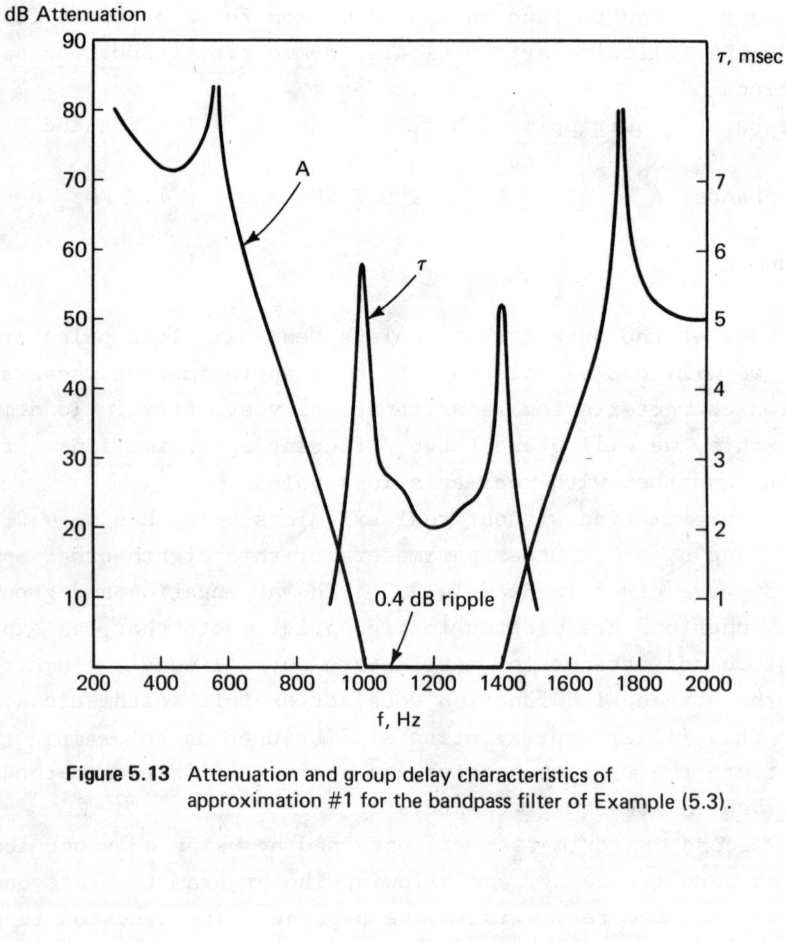

Figure 5.13 Attenuation and group delay characteristics of approximation #1 for the bandpass filter of Example (5.3).

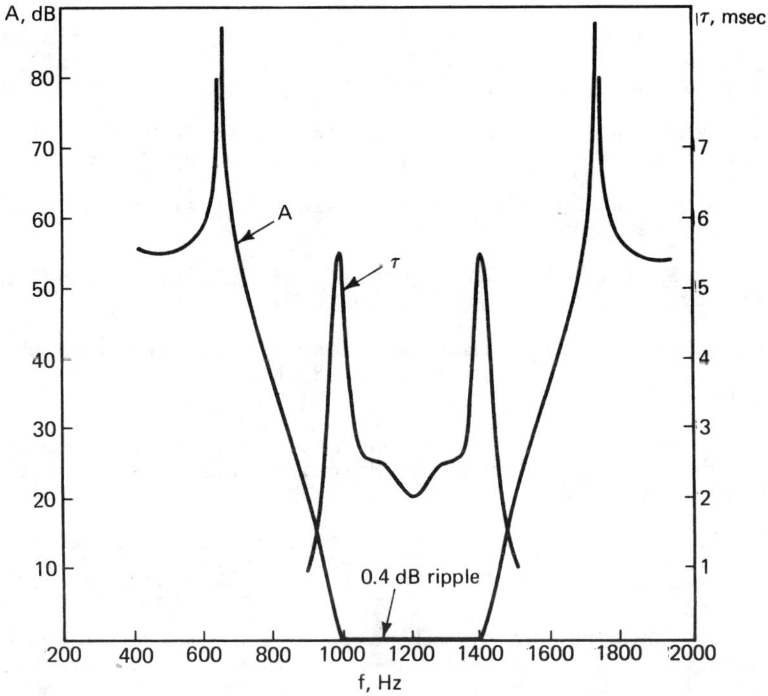

Figure 5.14 Attenuation and group delay of the second approximation (with one real loss-pole) for the bandpass filter of Example (5.3).

5.4 REFERENCES AND SOURCES FOR FURTHER READING

(1) W. M. Snelgrove and A. S. Sedra, FILTOR 2 - A Computer Aided Filter Design Package, Matrix Publishers, Champaign, Ill.

(2) Y. Ishizaki and H. Watanabe, "An iterative Chebyshev approximation method for network design", IEEE Trans. Circuit Theory, Vol. CT-15, pp. 326-336, Dec. 1968.

(3) G. Szentirmai, "Theoretical basis of a digital computer program for filter synthesis", Proc. 1st Allerton Conf. Circuit and System Theory, pp.37-49, 1963.

(4) G. Szentirmai, "Computer aids in filter design: a review", IEEE Trans. Circuit Theory, Vol. CT-18, pp.35-40, Jan. 1971.

(5) G. Szentirmai (Ed.), Computer Aided Filter Design, IEEE Press, New York, 1973.

(6) J. K. Skwirzynski, "On synthesis of filters", Trans. Circuit Theory, Vol. CT-18, pp.152-163, 1971.

(7) H. Watanabe, "Approximation theory for filter networks", IRE Trans. Circuit Theory, Vol. CT-8, No.3, pp. 341-356, Sept. 1961.

(8) F. F. Kuo and W. G. Magnuson, Jr., Eds., Computer-Oriented Circuit Design, Chapter 6, Prentice-Hall, Englewood Cliffs, N.J., 1969.

(9) B. R. Smith and G. C. Temes, "An iterative approximation procedure for automatic filter synthesis," IEEE Trans. Circuit Theory, Vol. CT-12, No. 1, pp. 107-112, March 1967.

(10) G. C. Temes and D. A. Calahan, "Computer aided network optimization - The State-of-the-art", Proc. IEEE, Vol. 53, pp. 1832-1863, Nov. 1967.

(11) R. W. Daniels, Approximation Methods for Electronic Filter Design, McGraw-Hill, New York, 1974.

(12) E. Remez, "Sur le calcul effectif des polynomes de Tchebicheff", Compt. Rend. Acad. Sci. (France), Vol. 199, pp. 337-340, July 1934.

(13) G. C. Temes and S. K. Mitra (Eds.), Modern Filter Theory and Design, Wiley, New York, 1973, Chapter 2.

(14) F. F. Kuo and J. F. Kaiser, (Eds.), System Analysis by Digital Computer, Chapter 5, Wiley, New York, 1966.

(15) H. J. Orchard and G. C. Temes, "Filter design using transformed variables", IEEE Trans. Circuit Theory, Vol. CT-15, pp. 385-408, 1968.

(16) W. M. Snelgrove, Computer Aided Filter Design, M.A.Sc. Thesis, University of Toronto, 1977.

(17) Special Issue on Computer-Aided Design, Proc. IEEE Vol. 55, No.11, 1967.

(18) Special Issue on Modern Filter Design, IEEE Trans. Circuit Theory, Vol. CT-5, No. 4, Dec. 1958.

(19) Special Issue on Modern Filter Design. IEEE Trans. Circuit Theory, Vol. CT-15, No. 4, Dec. 1968.

PROBLEMS

P5.1 A bandpass filter has the following attenuation specifications:

Passband: 1 - 1.25 KHz, with $A_p = 0.5$ dB

f = 0 to f = 400 Hz : 20 dB

f = 400 to f = 500 Hz : 50 dB

f = 500 to f = 800 Hz : 30 dB

f ≥ 1.5 KHz : 40 dB

Sketch the stopband specifications versus γ.

P5.2 Consider a third-order low-pass elliptic approximation with $|\rho|_{max} = 25\%$ and $\Omega_s = 2$. It is required to use Remez method to solve the stopband approximation problem. For simplicity, you may use the template formula given in the text. Carry out your solution in a step-by-step manner and iterate 5 times. As a result, give the location of the loss poles and the values of the stopband margins. Assume that $A_s = 25$ dB.

Note that this problem is tedious, but it illustrates the method very nicely. You may wish to write a computer program to aid in solving the problem.

P5.3 Consider a bandpass filter whose passband extends from $\omega = a$ to $\omega = 1$. The lower stopband extends from $\omega = 0$ to $\omega = \omega_{s1}$ with a specified attenuation of A_{s1} dB, and the upper stopband extends from $\omega = \omega_{s2}$ to $\omega = \infty$ with a specified attenuation of A_{s2} dB. Assume that in solving the stopband approximation problem you decided to locate one pair of loss poles in the lower stopband at $\omega = \omega_1$, one pair of loss poles in the upper stopband at $\omega = \omega_2$, and one loss pole at each of zero and infinity.

Sketch a possible $A(\gamma)$ and clearly label its minima and the attenuation margins. Write down the set of linear equations that will have to be solved in every Remez iteration. Use two different values for M; M_1 in the lower stopband and M_2 in the upper stopband.

CHAPTER 6

FREQUENCY AND NETWORK TRANSFORMATIONS

In the preceding chapters, a number of classical filter approximations have been presented. A considerable amount of design data in the form of graphs, formulae and tables have also been given. Using these data together with a calculator enables the reader to find classical approximations to many filtering problems, as has been already demonstrated in the examples. More efficient approximations can be obtained using the general approximation theory along with implementing computer routines, such as those of FILTOR 2[1].

Many of the classical approximations have been extensively tabulated in the literature. At the end of this Chapter, a list of some of these tables and design manuals is given. One of the objectives of this Chapter is to present filter design methods using classical approximations. In addition to finding transfer functions, we will also present realizations in the form of doubly terminated LC ladders. However, the theory of passive realization as well as general LC ladder design methods are treated in detail in Chapter 7. In the present Chapter, we will present closed form expressions for the elements of ladders realizing Butterworth and Chebyshev filters. Elements of LC ladders that realize other classical approximations will be obtained from filter tables. The computer routines of Chapter 7 can, of course, be used to design all such networks.

All the classical approximations presented in this book, as well as those tabulated in filter design manuals, are for low-pass filters. The design data, however, can be applied to the design of other filter types through the use of the frequency band transformations presented in this Chapter.

Since filters are usually designed using normalized frequencies and impedance levels, the final network has to be appropriately denormalized. The first section of this Chapter is devoted to the subject of frequency and impedance scaling. This material will also be used in Chapter 7, in conjunction with the study of passive filter sensitivities.

Another section of this Chapter presents some network transformations which are usually required in order to obtain practical element values for the final filter network.

The approximation methods of this and the previous Chapters can be applied to the design of discrete-time filters through the use of the bilinear transform discussed in Section 6.8.

6.1 NETWORK SCALING

In this section, we shall consider the relationship between the network functions of two networks which are related to each other by simple impedance and frequency scaling operations on their elements.

Consider the two networks η and η', which are identical in form and topology but have different element values. In particular, the element values of η', which are denoted by R_K', C_K', and L_K', are related to the element values of η denoted by R_K, C_K and L_K, by

$$R'_K = bR_K$$
$$C'_K = \frac{C_K}{ab} \tag{6.1}$$
$$L'_K = \frac{bL_K}{a}$$

Here, the factor "b" is referred to as the impedance scaling factor, since the impedance level of the elements of η' are clearly b times the impedance level of the elements of η. The factor "a" is referred to as the frequency scaling factor, since the effect of the reactive elements C'_K and L'_K of η' may be obtained from those of η with b = 1, by merely substituting (s/a) for s in the network impedances.

Now, consider a nodal analysis of network η. The matrix equations are

$$\underline{Y}(s)\ \underline{V} = \underline{I}$$

where $\underline{Y}$ is the nodal admittance matrix and $\underline{V}$, $\underline{I}$ are the corresponding node voltage and current vectors. A typical element $y_{jk}(s)$ of $\underline{Y}(s)$ has the form

$$y_{jk}(s) = \frac{1}{R_{jk}} + \frac{1}{sL_{jk}} + C_{jk}s.$$

We note that for the network η' we have

$$\underline{Y}'(s)\ \underline{V}' = \underline{I}' ,$$

and the typical element $y'_{jk}(s)$ of $\underline{Y}'(s)$ is related to $y_{jk}(s)$ of $\underline{Y}$ by (6.1)

$$y'_{jk}(s) = \frac{1}{bR_{jk}} + \frac{a}{sbL_{jk}} + \frac{C_{jk}s}{ba} = \frac{1}{b}\, y_{jk}\ (s/a)$$

so that $\underline{Y}'(s) = (1/b)\ \underline{Y}(s/a)$.

The reader will recall that the various network functions such as impedances, admittances, transfer impedances and admittances, and transfer voltage and current ratios may be computed as ratios of cofactors and determinants of $\underline{Y}$. In the case of dimensionless ratios, such as voltage and current transfer functions, the function is obtained as the ratio of cofactors of $\underline{Y}$ of identical order so that the impedance scaling factor "b" arising in both numerator and denominator will cancel. Thus, for any transfer ratios, say (V_x/V_y) or (I_x/I_y) of η, the corresponding ratios (V_x'/V_y') or (I_x'/I_y') of η' are related to the former by

$$\frac{V_x'}{V_y'}(s) = \frac{V_x}{V_y}\left(\frac{s}{a}\right)$$

or,

$$\frac{I_x'}{I_y'}(s) = \frac{I_x}{I_y}\left(\frac{s}{a}\right) .$$

In the case of dimensioned network functions (with dimensions of Ohms or mhos) such as impedances, admittances, or transfer impedances and admittances, the function is obtained as a ratio of a cofactor of $\underline{Y}$ to the determinant of $\underline{Y}$ which are of different order. In this case, the impedance scaling factor "b", as well as the frequency scaling factor, affect the relationship. Consider an impedance or admittance function (say Z_{mn} or Y_{mn}) of η, and the corresponding function of η'. We now have the relationships,

$$Z_{mn}'(s) = bZ_{mn}(s/a)$$

or

$$Y_{mn}'(s) = \frac{Y_{mn}(s/a)}{b} .$$

We shall make use of such scaling techniques in the following Sections and Chapters.

6.2 DESIGN OF LOW-PASS FILTERS

LC Ladders for All-Pole Filters

We consider first the realization of all-pole transfer functions. Fig. 6.1 shows doubly terminated LC ladders that realize such functions. Networks are given for both even and odd order functions. In each case we have two possible networks; the minimum inductance ladder, and the minimum capacitance ladder, with the two networks being duals of one another. One normally chooses the minimum inductance ladder for practical reasons. The dual ladder will, however, be of interest to us in the design of certain classes of active filters (Chapter 11).

LC ladders that are designed to deliver the maximum possible power to the load exhibit low-sensitivity to the inevitable changes in element values. This will be discussed at great length in Chapter 7. For the time being, suffice it to say that we will be specifically interested in ladders that are terminated in a manner which effects this maximum power transfer. The condition of maximum power transfer implies that, at a reflection zero, the input impedance of the ladder network should be real and equal in value to the source resistance. For filter functions that have reflection zero(s) at $\omega=0$, this condition is very easy to enforce. At $\omega=0$, the input impedance of the ladder is equal to R_2 and all that we need to do is to make $R_2 = R_1$. Thus, Butterworth filters, Chebyshev filters of odd order, modified Chebyshev filters of even order etc. will be realized with equally terminated ladders.

Filter functions which do not have reflection zeros at DC cannot be realized with equally terminated ladders. To ensure maximum power transfer at the reflection zeros, we have to choose the termination as follows. Consider as an example an even order Chebyshev filter. From the filter transfer function we have

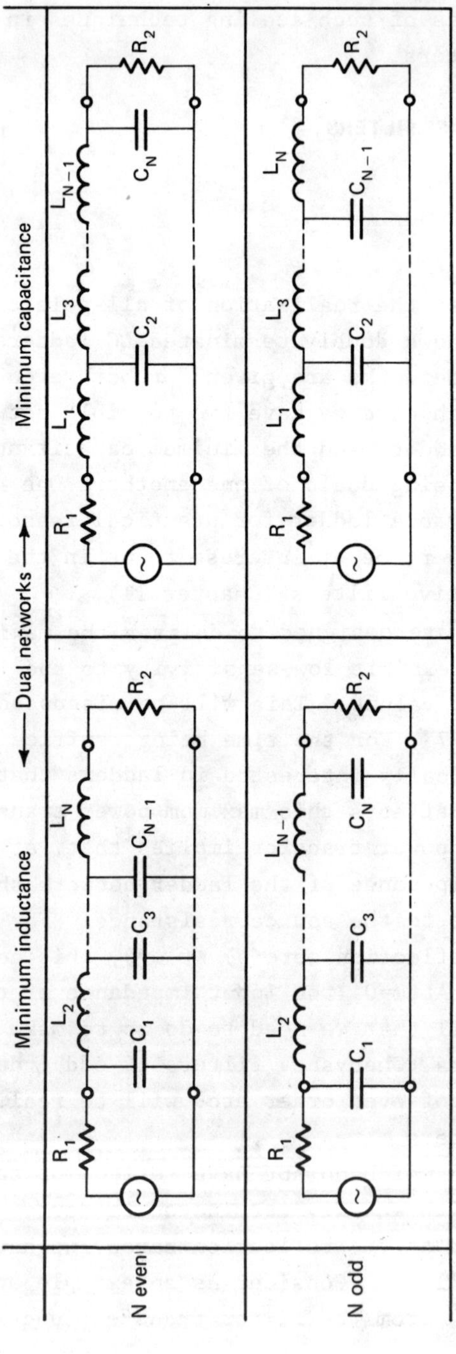

Figure 6.1 LC ladders for all-pole low-pass filters.

$$\frac{|H(0)|}{|H(j\omega_r)|} = \sqrt{1 + \varepsilon^2} \tag{6.2}$$

where ω_r denotes a reflection zero and ε is the passband ripple factor related to the ripple A_p dB by,

$$\varepsilon = \sqrt{10^{A_p/10}-1} \quad .$$

Consider next the ladder realization of this filter. For maximum power transfer at ω_r, we require the input impedance of the ladder to be matched to R_1. Thus, the power entering the ladder at this frequency is given by

$$P_{in} = \frac{|V_i(j\omega_r)|^2}{4R_1} \quad .$$

Since the LC ladder is lossless all this power will be delivered to the load. Thus, we may write,

$$\frac{|V_o(j\omega_r)|^2}{R_2} = \frac{|V_i(j\omega_r)|^2}{4R_1} \quad ,$$

and obtain,

$$\left|\frac{V_i}{V_o}(j\omega_r)\right| = 2\sqrt{\frac{R_1}{R_2}} \tag{6.3}$$

At DC, the ladder has the transfer function,

$$\left|\frac{V_i}{V_o}(0)\right| = \frac{R_1 + R_2}{R_2} \tag{6.4}$$

From (6.3) and (6.4), we obtain

$$\frac{|H(0)|}{|H(j\omega_r)|} = \frac{1}{2}\left(\sqrt{\frac{R_2}{R_1}} + \sqrt{\frac{R_1}{R_2}}\right) \tag{6.5}$$

Without loss of generality, we can assume that the ladder is impedance normalized such that $R_1 = 1\Omega$. Substituting this in (6.5) and using (6.2) gives the value of R_2 for maximum power transfer as,

$$R_2 = 1 + 2\,\varepsilon^2 \pm 2\,\varepsilon\,\sqrt{1+\varepsilon^2} \tag{6.6}$$

The two values for R_2 obtained from (6.6) are reciprocally related; that is, their product equals unity. Thus, the two values correspond to the two reciprocal ladders.

Explicit Formulae for the Element Values of Butterworth and Chebyshev Ladders

We now present, without proof, explicit formulae for the element values of ladders that realize Butterworth and Chebyshev filter functions. These formulae, together with a modern scientific calculator, enable the design of any low-pass filter of these types. Furthermore, the design of other filters, such as band-pass, high-pass and band-reject filters, can be obtained through the application of the transformations presented in the following sections. Thus, for Butterworth and Chebyshev filters, there is no need to rely on filter tables.

The formulae presented are for the minimum inductance ladders of Fig. 6.1. The minimum capacitance ladders can be obtained as duals. The values are given in terms of the terminology of Fig. 6.1, and the low-pass function is assumed to be frequency normalized such that the ripple bandwidth is unity. We also assume that $R_1 = 1\Omega$. Note that, except for the even-order Chebyshev case, the value of R_2 is also unity.

Butterworth Filters

The formula for the Nth order Butterworth filter is surprisingly simple and is given by

$$\left.\begin{array}{ll} k \text{ odd}, & C_k \\ k \text{ even}, & L_k \end{array}\right\} = 2(\varepsilon^{1/N}) \sin\left(\frac{2k-1}{2N}\pi\right) \tag{6.7}$$

Example 6.1

Find the element values for the dual networks which realize the 5th and 6th order Butterworth filters having 1Ω terminations and a unity 3 dB frequency.

Solution

Directly from the formulae (6.7), we use a scientific calculator and construct the table below.

N	C_1	L_2	C_3	L_4	C_5	L_6
5	0.6180	1.6180	2.000	1.6180	0.6180	
6	0.5176	1.4142	1.9319	1.9319	1.4142	0.5176
	L_1	C_2	L_3	C_4	L_5	C_6

Chebyshev Filters

The formulae for the Chebyshev case are also simple, but not quite so straightforward. First, we use the ripple factor ε to compute a constant h,

$$h = \sqrt[N]{\frac{1}{\varepsilon} + \sqrt{1 + \frac{1}{\varepsilon^2}}} \tag{6.8}$$

from which we obtain the quantity ξ as

$$\zeta = (h - \frac{1}{h}) \quad . \tag{6.9}$$

The formulae are then given by

$$C_1 = \frac{4 \sin (\frac{\pi}{2N})}{\xi R_1} \tag{6.10}$$

$$C_{2k-1} L_{2k} = \frac{16 \sin (\frac{4k-3}{2N})\pi \sin (\frac{4k-1}{2N})\pi}{(\xi^2 + 4 \sin^2(\frac{2k-1}{N})\pi)} \tag{6.11}$$

$$C_{2k+1} L_{2K} = \frac{16 \sin (\frac{4k-1}{2N})\pi \sin (\frac{4k+1}{2N})\pi}{(\xi^2 + 4 \sin^2 (\frac{2k}{N}) \pi)} \tag{6.12}$$

for k = 1, 2, ... N/2. The end elements are also given by

$$C_N = \frac{4 \sin (\frac{\pi}{2N})}{\xi R_2} \text{ (odd N)} , \tag{6.13}$$

or

$$L_N = \frac{4 R_2 \sin (\frac{\pi}{2N})}{\xi} \text{ (even N)} , \tag{6.14}$$

so that the formulae may be used in the reverse direction as well.

Example 6.2

Find the element values for the 3rd and 4th order Chebyshev filters with Ap = 0.1 dB, unity cutoff frequency and source resistance of 1Ω.

Solution

Here Ap = 0.1 dB so that $\varepsilon = 0.1526204$ and $\varepsilon^2 = 0.0232930$.

a) <u>3rd order Chebyshev</u> Using (6.8), we compute h = 2.36215, and (6.9) yields $\xi = 1.938812$. Now, with $R_1 = 1\Omega$,

$$C_1 = \frac{4 \sin (\frac{\pi}{6})}{\xi} = 1.031560 \text{ F}$$

$$C_1 L_2 = \frac{16 \sin(\frac{\pi}{6}) \sin(\frac{3\pi}{6})}{\xi^2 + 4 \sin^2(\frac{\pi}{3})} = 1.183609$$

$$L_2 = 1.1473966 \text{ H}$$

$$C_3 L_2 = \frac{16 \sin(\frac{3\pi}{6}) \sin(\frac{5\pi}{6})}{\xi^2 + 4 \sin^2(\frac{2\pi}{3})} = 1.183609$$

$$C_3 = 1.031560 \text{ F}$$

b) <u>4th order Chebyshev</u> Using (6.8), we compute $h = 1.905378$, and (6.9) yields $\xi = 1.380548$. Now, with $R_1 = 1\Omega$,

$$C_1 = \frac{4 \sin(\frac{\pi}{8})}{\xi} = 1.108787 \text{ F}$$

$$C_1 L_2 = \frac{16 \sin(\frac{\pi}{8}) \sin(\frac{3\pi}{8})}{\xi^2 + 4 \sin^2(\frac{\pi}{4})} = 1.448280$$

$$L_2 = 1.306184 \text{ H}$$

$$C_3 L_2 = \frac{16 \sin(\frac{3\pi}{8}) \sin(\frac{5\pi}{8})}{\xi^2 + 4 \sin^2(\frac{2\pi}{4})} = 2.31240$$

$$C_3 = 1.770351 \text{ F}$$

$$C_3 L_4 = \frac{16 \sin(\frac{5\pi}{8}) \sin(\frac{7\pi}{8})}{\xi^2 + 4 \sin^2(\frac{3\pi}{4})} = 1.448280$$

$$L_4 = 0.818075 \text{ H}$$

Now, using (6.14) to evaluate the element values from the ladder opposite end, we obtain L_4 in terms of R_2 as

$$L_4 = R_2 \frac{4 \sin(\frac{\pi}{8})}{\xi} = 1.108787\ R_2$$

Thus, R_2 is given by

$$R_2 = \frac{0.818075}{1.108787} = 0.737811\ \Omega$$

We can easily check that this value for R_2 is identical to that obtained using the condition for maximum power transfer in (6.6).

Undoubtedly, the reader can appreciate the convenience of these explicit formulae. The networks are obtained directly with no computation of poles and zeros or extensive polynomial manipulations. A great many of the simpler filter design problems can be solved by the use of these simple formulae and a scientific calculator, thus freeing the designer of reliance on extensive tables and computer programs. Unfortunately, no such formulae are yet known for the Inverse Chebyshev and Elliptic low pass prototypes, even though much is known analytically about the functions involved.

LC Ladders for Elliptic and Inverse Chebyshev Filters

Fig. 6.2 shows LC ladders for elliptic and inverse Chebyshev filters. Again, for the odd-order case, the ladder can be equally terminated. For the even-order case, we use equation (6.6) to calculate the value of R_2, assuming $R_1 = 1\Omega$, that results in maximum power transfer. Note that the ladders given for N even are for elliptic and inverse Chebyshev functions, modified such that two loss poles are placed at infinite frequency. Of course, a ladder realization is not possible unless at least one loss pole is at $s = \infty$.

For a given order N, two possible ladder network realizations, which are duals of one another, are possible. In the minimum inductance ladder, loss poles are realized by parallel resonant circuits placed in the series arms. These tank circuits are, of course, resonant at the loss pole frequencies, a fact that makes tuning or aligning passive filters relatively easy. In the case of minimum capacitance ladders, loss poles are realized by series resonant circuits placed in the shunt arms and tuned to the frequencies of the loss poles.

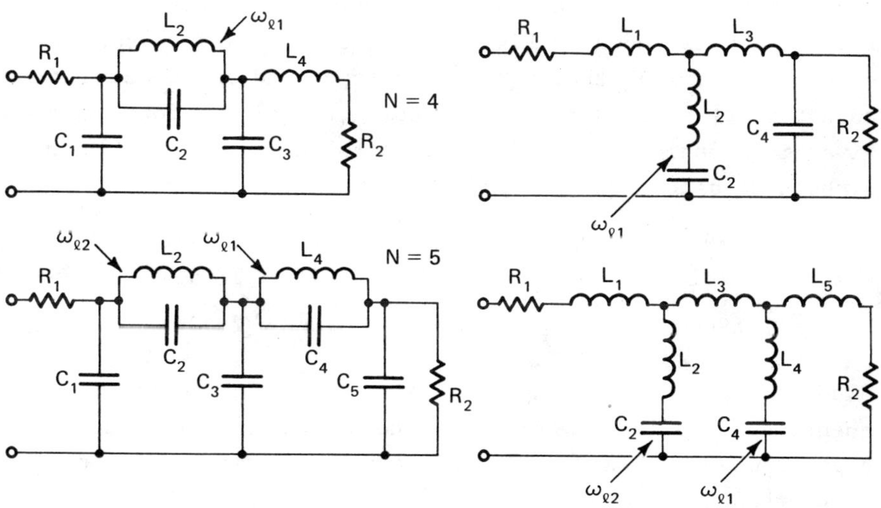

Figure 6.2 LC ladder realizations for elliptic and inverse Chebyshev filters. The even-order functions are modified and have two loss-poles at s = ∞.

As mentioned before, no closed form expressions exist for elliptic and inverse Chebyshev filter realizations. Ladder element values may be obtained using appropriate computer routines such as those of FILTOR 2. Also, elliptic function filters have been extensively tabulated[2]-[4].

Design Examples

Example 6.3

Find the passive ladder realization of the Chebyshev filter considered in Examples 4.1 and 4.2. The specifications for this filter are:

LP with $f_p = 10$ KHz , $f_s = 15$ KHz ,
$A_p \le 0.5$ dB and $A_s \ge 50$ dB

Solution

The various parameters of this eighth order filter were determined in Examples 4.1 and 4.2. Its minimum inductance ladder realization is shown in Fig. 6.3. The component values, frequency normalized such that $\Omega_p = 1$ rad./sec. and impedance normalized such that $R_1 = 1\Omega$, are obtained using the closed form expressions given in equations (6.8)-(6.14), as follows. (We use lower case symbols to denote the normalized values.)

$\varepsilon = 0.3493$ $\quad h = 1.2483$ $\quad \xi = 0.4472$

$c_1 = 1.7451$ $\quad \ell_2 = 1.2647$ $\quad c_3 = 2.6564$ $\quad \ell_4 = 1.3591$

$c_5 = 2.6964$ $\quad \ell_6 = 1.3389$ $\quad c_7 = 2.5092$ $\quad \ell_8 = 0.8796$

Also, $\ell_8 = 1.7451$ leading to $R_2 = 0.5040$. The filter may be frequency denormalized by dividing the values of all inductances and capacitances by $2\pi x10^4$. If the source impedance R_1 is to be, say, 1 KΩ, then the components should be impedance denormalized by multiplying all inductance values by 10^3 and dividing all capacitance values by 10^3. These operations result in the values shown in Fig. 6.3.

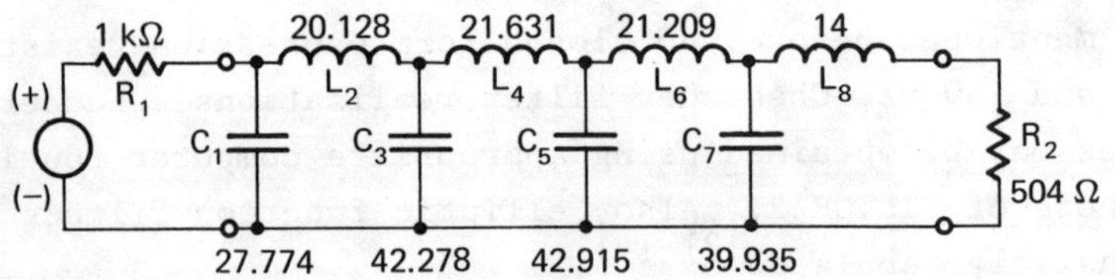

Figure 6.3 Ladder realization for the eighth order low-pass filter of Example (6.3). (Inductance in mH and capacitance in nF).

Example 6.4

Design an elliptic function filter that meets the specifications given in Example 6.3 above.

Solution

In Example 4.6 we have determined the order N to be 5. Using [2], we find the closest entry to be for $\rho = 25\%$ ($A_p = 0.28$ dB), which for $\Omega_s = 1.5243$ provides only 48.92 dB. This does not quite meet the specifications. If we have to use tables, we find ourselves forced to increase the order to 6. For N = 6, the closest entry is $\Omega_s = 1.47584$, which gives $A_s = 61.74$ dB, which is 11.74 dB higher than required. Selecting this approximation, the attenuation poles are at

$$\Omega_{\ell 1} = 1.516862 \qquad \Omega_{\ell 2} = 2.007733$$

Note that this is a modified elliptic (type b) with the highest loss poles moved to ∞.

The attenuation zeros are not given in[2], but can be obtained from[3] as

$$\Omega_{r1} = 0.293657 \qquad \Omega_{r2} = 0.755514 \qquad \Omega_{r3} = 0.97434$$

The natural modes can be obtained from either [2] or [3]. Using the latter, we obtain the complete transfer function as

$$H(s) = 51.246370 \, \frac{s^2 + 0.131886\, s + 1.044839}{s^2 + 2.300870} \cdot \frac{s^2 + 0.448076\, s + 0.709900}{s^2 + 4.0309918} \cdot (s^2 + 0.763250\, s + 0.252001)$$

In terms of Ω_o and Q we calculate:

Natural modes pair # 1: $\Omega_o = 1.022173$ Q = 7.75

Natural modes pair # 2: $\Omega_o = 0.842556$ Q = 1.88

Natural modes pair # 3: $\Omega_o = 0.501997$ Q = 0.6581

The passive ladder realization is obtained from [2]. Choosing the case of maximum power transfer, we obtain the network shown in Fig. 6.4, which has been frequency denormalized (with a denormalizing frequency of $2\pi \times 10^4$ rad/sec) and impedance denormalized (with a denormalizing resistance of 1KΩ). Note that the two tank circuits are tuned to the two loss pole frequencies.

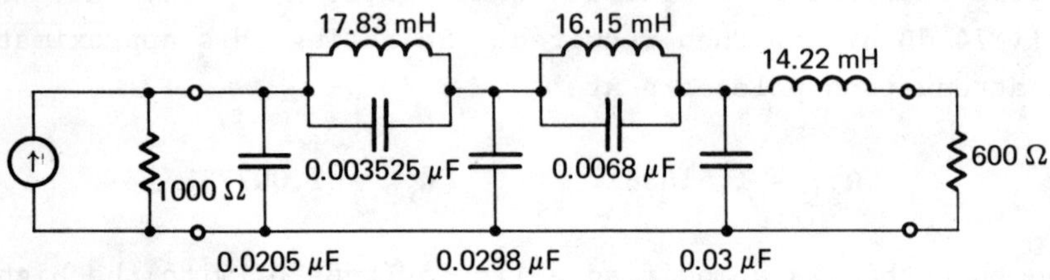

Figure 6.4 LC ladder realization of the sixth order LPF of Example (6.4).

6.3 THE LOW-PASS TO BANDPASS TRANSFORMATION

The LP to BP transformation is used to find an approximation to given bandpass attenuation specifications, by first finding an approximation to a "corresponding" low-pass filter specification. This "corresponding" low-pass filter is called a Low-Pass Prototype (LPP). Subsequently, the LP to BP transformation is applied to either the transfer function (or the network realization) of the LPP, to obtain a transfer function (or a network realization) for the bandpass filter.

The starting point in the design is a given tolerance plot for the attenuation of the bandpass filter. The first step is to find the specifications for the LPP. The method employed to accomplish this step naturally depends on the properties of the LP to BP transformation. Therefore, we will first consider these properties, and then present the method for finding the LPP specifications.

Properties of the LP to BP Transformation

Let the LPP specifications be in the form of the tolerance plot shown in Fig. 6.5. This plot can obviously be satisfied with a classical approximation. Indeed, one resorts to the use of the LP to BP transformation only if a classical approximation is sought. Otherwise, as has already been demonstrated, the approximation process can be performed directly in the bandpass domain.

Once we have obtained the parameters of the LPP tolerance plot (i.e. A_p, A_s and Ω_s), we can find an appropriate classical approximation (e.g. Butterworth, Chebyshev, elliptic, etc.) using filter tables and/or the formulae presented in preceding chapters. The result is a transfer function H(s) for the LPP ($s = j\Omega$ for physical frequencies in the LPP domain). Furthermore, we may use filter tables or explicit formulae to obtain an LC network realization for the LPP. We now seek a transformation of the frequency variable s to a new frequency variable p(with $p = j\omega$ for physical frequencies in the bandpass domain), say $s = f(p)$, such that the bandpass transfer function $H_{BP}(p) = H(s)|_{s=f(p)}$ satisfies the BP tolerance plot. It is also desirable that the transformation $s = f(p)$ be such that when applied to the network realization of the LPP, a physical network realization for the BP filter is obtained.

For the purpose of illustration, let us assume that our LPP tolerance plot has been approximated by a transfer function having the attenuation plot also shown in Fig. 6.5. Recalling from Chapter 1 that the attenuation function is an even function of Ω, we can draw the function $A(-\Omega)$ as a mirror image of $A(+\Omega)$, to obtain the complete plot of $A(\Omega)$ shown in Fig. 6.6a. A careful examination of this attenuation plot reveals that it represents a bandpass filter whose center frequency is zero, whose passband extends from -1 to +1, and whose upper stopband extends from Ω_s to ∞ and lower stopband extends from $-\Omega_s$ to $-\infty$. What we now need is a frequency transformation (Ω to ω or s to p) that will translate this attenuation plot to a predetermined location on both the positive and negative frequency axis. Such a transformation is given by

$$s = \frac{p^2 + \omega_o^2}{pB} \tag{6.15}$$

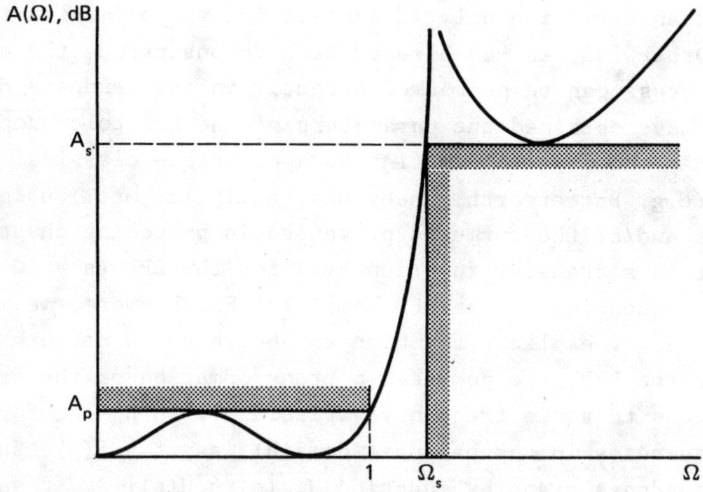

Figure 6.5 Tolerance plot for low-pass prototype (LPP), together with a typical approximation.

From (6.15), we find the following properties of the LP to BP transformation.

(1) The frequency $s = 0$ transforms to $p = \pm j\omega_o$. Thus, we refer to ω_o as the center-frequency of the BPF.

(2) Any frequency $s = j\Omega$, at which the attenuation is A dB, transforms into two frequencies; a positive one

$$\omega_2 = \frac{1}{2} B\Omega + \sqrt{\omega_o^2 + \frac{1}{4} B^2\Omega^2}$$

and a negative one

$$\omega_2' = \frac{1}{2} B\Omega - \sqrt{\omega_o^2 + \frac{1}{4} B^2\Omega^2}$$

The attenuation at both ω_2 and ω_2' is A dB.

(3) The mirror image frequency $s = -j\Omega$ also produces two frequencies having the same attenuation AdB. Again, one frequency is positive,

$$\omega_1 = -\frac{1}{2} B\Omega + \sqrt{\omega_o^2 + \frac{1}{4} B^2\Omega^2}$$

while the other is negative,

$$\omega_1' = -\frac{1}{2} B\Omega - \sqrt{\omega_o^2 + \frac{1}{4} B^2\Omega^2}$$

Note that $\omega_1' = -\omega_2$ and $\omega_2' = -\omega_1$.

(4) Since we are mostly interested in positive frequencies, we conclude that a frequency $s = \pm j\Omega$ in the LPP gives rise to two positive frequencies ω_1 and ω_2, given by

$$\omega_{1,2} = \mp \frac{1}{2} B\Omega + \sqrt{\omega_o^2 + \frac{1}{4} B^2\Omega^2} \qquad (6.16)$$

These two frequencies are related by

$$\omega_1 \omega_2 = \omega_o^2 \qquad (6.17)$$

That is, in the transformed attenuation plot (on the $+j\omega$ axis) each two frequencies with equal attenuation are geometrically symmetric with respect to the center frequency ω_o.

(5) The passband edges $s = \pm j1$ transform into four frequencies, the two positive ones can be obtained from (6.16) as

$$\omega_{p1,2} = \mp \frac{1}{2} B + \sqrt{\omega_o^2 + \frac{1}{4} B^2} \qquad (6.18)$$

Thus,

$$\omega_{p_2} - \omega_{p_1} = B ,$$

which is the bandpass filter <u>bandwidth</u>. Of course, ω_{p_1} and ω_{p_2} are also geometrically symmetric with respect to ω_o;

$$\omega_{p_1} \omega_{p_2} = \omega_o^2 .$$

(6) The frequencies $s = \pm j\Omega_s$, which define the stopband map into four frequencies; the two positive frequencies ω_{s_1} and ω_{s_2} are related by

$$\omega_{s_2} - \omega_{s_1} = B\ \Omega_s \tag{6.19}$$

From the above observations, it follows that the mapping of (6.15) produces the attenuation plot of Fig. 6.6b, with the mirror image, on the negative frequency axis, shown to reaffirm the fact that $A_{BP}(\omega)$ is an even function. The most important property to note is that the resultant BP attenuation function is geometrically symmetric. This is a fundamental limitation of the LP to BP transformation. In other words, it is not possible to design arithmetically symmetric BP filters using this LP to BP transformation.

Another property of the LP to BP transformation that is worth noting concerns the phase or delay function. It can be shown[5] that the transformation given does not preserve phase linearity. That is, a linear phase LPP will not result in a BP filter with linear phase.

Obtaining the Low-Pass Prototype Specifications

We are now in a position to outline the application of the LP to BP transformation to the design of BP filters. Let it be required to find a transfer function $H_{BP}(p)$ whose attenuation function approximates the BP tolerance plot in Fig. 6.7a in a certain classical manner, e.g. Butterworth, Chebyshev, elliptic, etc. To obtain a classical approximation, the problem specifications should be first modified as follows:

(1) The stopbands' attenuation should be made constant, since

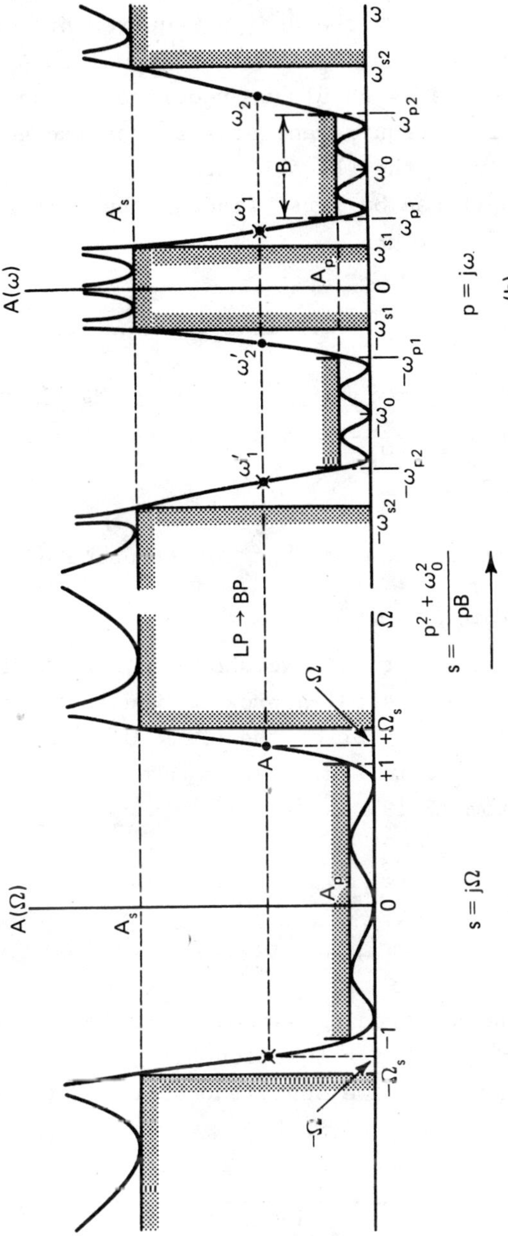

Figure 6.6 Illustrating the LP to BP transformation.

this is the only case handled with classical approximations. Obviously, to satisfy the design requirements, the minimum stopbands' attenuation A_s should be equal to A_{s2}.

(2) The stopbands' edges should be redefined to be geometrically symmetric. This requirement arises from the geometric symmetry property of the LP to BP transformation. This step can be accomplished by first finding the center frequency $\omega_o = \sqrt{\omega_{p1}\,\omega_{p2}}$. Then,

(a) evaluate $\omega_{s1} = \dfrac{\omega_o^2}{\omega_2}$. If $\omega_{s1} > \omega_1$, then the stopband edges are ω_{s1} and ω_2. However, if $\omega_{s1} < \omega_1$, then

(b) evaluate $\omega_{s2} = \dfrac{\omega_o^2}{\omega_1}$ and define the stopband edges as ω_1 and ω_{s2}.

In our example, it is assumed that the stopband edges obtained are $\omega_{s1} = \dfrac{\omega_o^2}{\omega_2}$ and $\omega_{s2} = \omega_2$.

The modified specifications are shown in Fig. 6.7b. The second step in obtaining a classical approximation is to find the parameters of the tolerance plot of the LPP. Obviously, A_p and A_s will be equal to those in Fig. 6.7b, while the selectivity factor Ω_s is obtained from (6.19) as

$$\Omega_s = \frac{\omega_{s_2} - \omega_{s_1}}{\omega_{p_2} - \omega_{p_1}} \tag{6.20}$$

We next find the required classical approximation for the LPP. The result can be a transfer function H(s), or its network realization. In either case, we apply the LP to BP transformation

$$s = \frac{p^2 + \omega_o^2}{pB}$$

where B is the passband width of the BPF,

$$B = \omega_{p2} - \omega_{p1}$$

Note that application of the transformation results directly in a frequency denormalized BP filter.

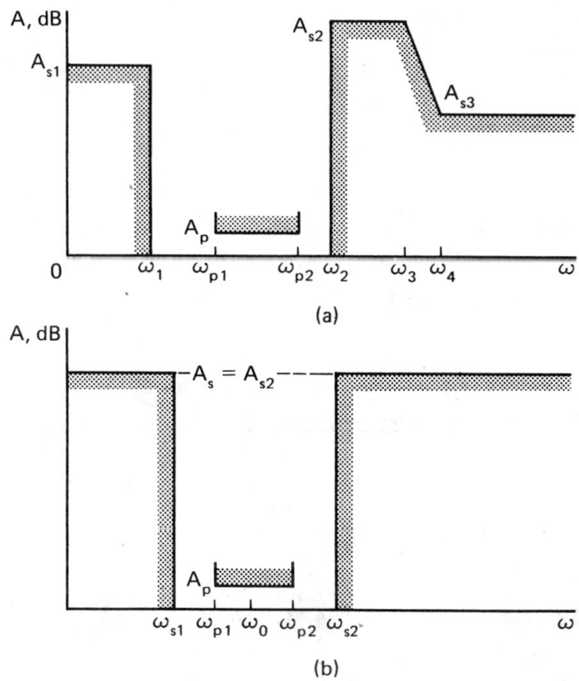

Figure 6.7 In order to be able to apply the LP to BP transformation, the tolerance requirements of the BP filter shown in (a) are modified to those in (b).

Example 6.5

It is required to design a BPF having the following specifications

Passband extends from 1900 Hz to 2100 Hz with a maximum distortion of 0.1 dB.

$A_s \geq 30$ dB at 600 Hz bandwidth.

Find an elliptic LPP.

Solution

The tolerance plot is sketched in Fig. 6.8a, from which we obtain

$$f_{p1} = 1900 \text{ Hz} \qquad f_{p2} = 2100 \text{ Hz} \qquad B = 2\pi x200 \text{ rad./sec.}$$

$$f_o = \sqrt{f_{p1}\, f_{p2}} = 1997.5 \text{ Hz}$$

$$f_{s2} - f_{s1} = 600 \text{ Hz} \qquad A_s = 30 \text{ dB}$$

$$\Omega_s = \frac{600}{200} = 3$$

$$A_p = 0.1 \text{ dB leads to } |\rho|_{max} \simeq 15\%$$

Using the parameters A_p, A_s and Ω_s of the LPP, we find from[3] the following elliptic approximation (attenuation plot is shown in Fig. 6.8b)

$$N = 3 \qquad \Omega_s = 2.9238 \qquad A_s = 34.68 \text{ dB}$$

$$\Omega_r = 0.872654 \qquad \Omega_\ell = 3.35472$$

Real natural mode at s = -1.03317
A pair of complex conjugate natural modes at,

$$s = -0.439729 \pm j1.215235$$

Thus, $\Omega_o = 1.292346$ and $q = 1.46948$*

---

*We will use the symbol q to denote the Q-factor for the LPP natural modes reserving Q for the corresponding Q factors of the BPF.

$$H(s) = 6.505517 \frac{(s + 1.033173)\,(s^2 + 0.879457\,s + 1.670157)}{(s^2 + 11.226166)}$$

An LC ladder realization of this LPP is shown in Fig. 6.8c.

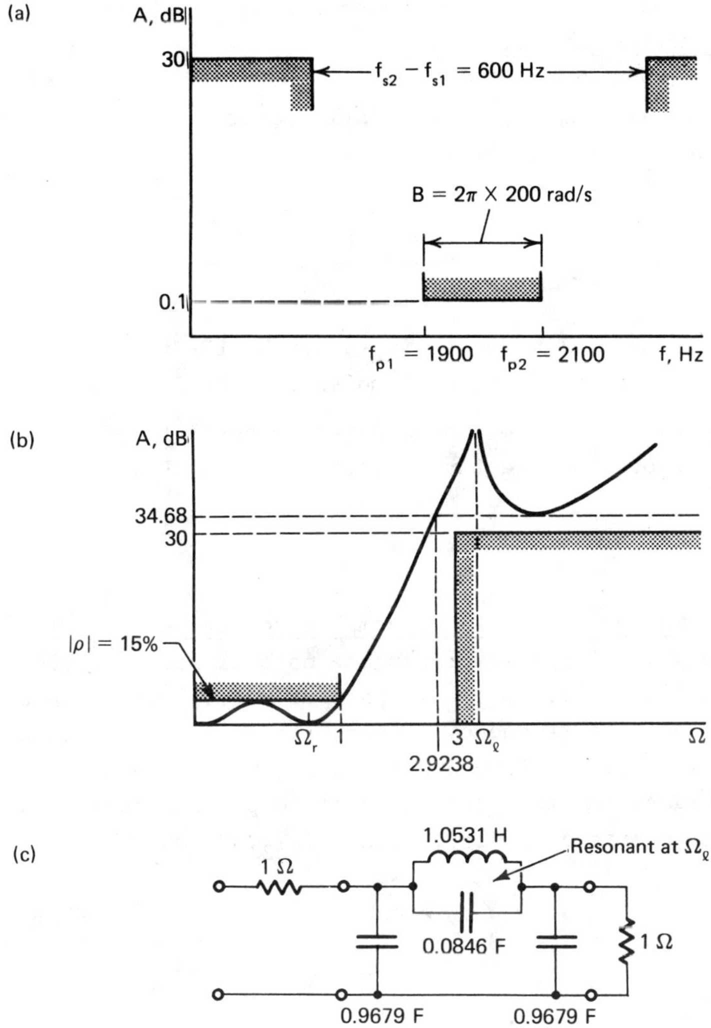

Figure 6.8 (a) Specifications for BPF of Example (6.5). (b) A third-order elliptic filter function satisfying the specifications of the LPP. (c) LC ladder realization for the LPP.

Application of the LP to BP Transformation to a Transfer Function H(s)

In this subsection, we consider the application of the LP to BP transformation to an LPP obtained in terms of its transfer function H(s). Since H(s) is usually found in factored form, we consider the application of the transformation to such factors.

A) Transformation of a real natural mode:

Consider the natural mode $s = -\sigma$, $\sigma > 0$. That is, let H(s) have a factor $(s + \sigma)$. Upon application of the LP to BP transformation, we obtain the factor

$$\frac{p^2 + \omega_o^2}{pB} + \sigma = \frac{p^2 + p\,\sigma\,B + \omega_o^2}{pB}$$

Thus, the real natural mode at $s = -\sigma$ of the LPP gives rise to a pair of complex conjugate natural modes for the BPF. This pair has a frequency equal to ω_o and a Q factor equal to $\frac{\omega_o}{\sigma B} = \frac{\hat{Q}}{\sigma}$, where we use $\hat{Q}$ to denote an effective Q-factor of the BPF;

$$\hat{Q} = \frac{\omega_o}{B} .$$

B) Transformation of a pair of complex conjugate natural modes

Consider a pair of complex conjugate natural modes in the LPP transfer function H(s). Let this pair of natural modes be defined in terms of its frequency Ω_o and Q-factor, q. It can be shown, with some algebraic manipulations, that this pair of natural modes transforms into two pairs of complex conjugate natural modes characterized by (ω_{o1}, Q) and (ω_{o2}, Q) where,

$$\omega_{o1} = \sqrt{Y}\,\omega_o \quad \text{and,} \tag{6.21}$$

$$\omega_{o2} = \omega_o / \sqrt{Y} \tag{6.22}$$

That is, ω_{o1} and ω_{o2} are geometrically symmetric. The factor Y is given by

$$Y = \frac{\Omega_o^2}{4\hat{Q}^2} + \sqrt{\left(1 + \frac{\Omega_o^2}{4\hat{Q}^2}\right)^2 - \frac{b}{4\hat{Q}^2}}$$

$$+ \frac{\Omega_o}{\sqrt{2}\,\hat{Q}} \sqrt{1 + \frac{\Omega_o^2}{4\hat{Q}^2} - \frac{b}{2\Omega_o^2} + \sqrt{\left(1 + \frac{\Omega_o^2}{4\hat{Q}^2}\right)^2 - \frac{b}{4\hat{Q}^2}}} \qquad (6.23)$$

where

$$b = \frac{\Omega_o}{q} .$$

The value of Q is given by

$$Q = \frac{\hat{Q}}{b} \left(\sqrt{Y} + \frac{1}{\sqrt{Y}}\right) \qquad (6.24)$$

C) <u>Transformation of jΩ-axis loss poles and reflection zeros</u>

The jΩ-axis loss poles and reflection zeros of the LPP can be transformed to the BP domain using (6.16), which relates Ω and ω; namely,

$$\omega = \omega_o \left[\pm \frac{\Omega}{2\hat{Q}} + \sqrt{1 + \left(\frac{\Omega}{2\hat{Q}}\right)^2}\right] \qquad (6.25)$$

Thus, a pair of loss poles in the LPP transfer function at $\pm j\Omega_\ell$ gives rise to two pairs of BP loss poles at $\pm j\omega_{\ell_1}$ and $\pm j\omega_{\ell_2}$ where,

$$\omega_{\ell_{1,2}} = \omega_o \left[\pm \frac{\Omega_\ell}{2\hat{Q}} + \sqrt{1 + \left(\frac{\Omega_\ell}{2\hat{Q}}\right)^2}\right] \qquad (6.26)$$

An identical relationship applies for the transformation of jΩ-axis reflection zeros.

Example 6.6

Obtain the transfer function H(p) and other pertinent parameters of the BPF specified in Example 6.5.

Solution

H(p) is obtained by applying the LP to BP transformation to the transfer function of the LPP obtained in Example 6.5. In this case the LP to BP transformation is given by

$$s = \frac{p^2 + \omega_o^2}{pB}$$

where $\omega_o = 2\pi \sqrt{1900 \times 2100} = 2\pi \times 1997.5$ rad/sec.,

$$B = 2\pi \times 200 \text{ rad/sec.}$$

Also, $\hat{Q} = \frac{\omega_o}{B} = 9.9875$

The stopband edge $\Omega_s = 2.9238$ transforms [using (6.25)] to the two frequencies $\omega_{s_1} = 2\pi \times 1726.4$ rad/sec. and $\omega_{s_2} = 2\pi \times 2311.2$ rad/sec. The loss poles at $\Omega_\ell = 3.35472$ transform [using (6.26)] to

$$\omega_{\ell_1} = 2\pi \times 1690.4 \text{ rad/sec. and } \omega_{\ell_2} = 2\pi \times 2360.5 \text{ rad/sec.}$$

Similarly, we obtain the two pairs of reflection zeros,

$$\omega_{r_1} = 2\pi \times 1912.1 \text{ rad/sec. and } \omega_{r_2} = 2\pi \times 2086.7 \text{ rad/sec.}$$

The real natural mode at $s = -1.033173$ produces a pair of complex conjugate natural modes with a frequency of $\omega_o = 2\pi \times 1997.5$ rad/sec. and a Q factor of $\frac{\hat{Q}}{1.033173} = 9.67$.

Finally, the pair of complex conjugate natural modes characterized by $\Omega_o = 1.292346$ and $q = 1.46948$ can be transformed using equations (6.21)-(6.24) to yield two pairs of complex conjugate modes having

$$\omega_{o1} = 2\pi \times 2112.9 \text{ rad/sec.} \qquad Q_1 = 22.53$$

$$\omega_{o2} = 2\pi \times 1879.5 \text{ rad/sec.} \qquad Q_2 = 22.53$$

Observing that the LPP has one loss pole at $s = \infty$, which transforms into a loss pole at $p = 0$ and another at $p = \infty$, we may now write the BP transfer function as

$$H(p) = 5.18 \times 10^{-3} \frac{p^2 + p\, 1.297897 \times 10^3 + 157.519 \times 10^6}{p}$$

$$\times \frac{p^2 + p\, 0.589243 \times 10^3 + 176.242 \times 10^6}{p^2 + 219.965 \times 10^6}$$

$$\times \frac{p^2 + p\, 0.524157 \times 10^3 + 139.458 \times 10^6}{p^2 + 112.8 \times 10^6}$$

Note that the multiplicative constant has been obtained as $\frac{C_{LP}}{B}$ where C_{LP} is the constant multiplying $H(s)$.

The characteristic function of the BP filter can be similarly obtained as

$$K(p) = 5.18 \times 10^{-3} \frac{p^2 + 144.344 \times 10^6}{p^2 + 219.965 \times 10^6} \times \frac{p^2 + 171.897 \times 10^6}{p^2 + 112.8 \times 10^6}$$

Fig. 6.9 shows a sketch of the BP attenuation function

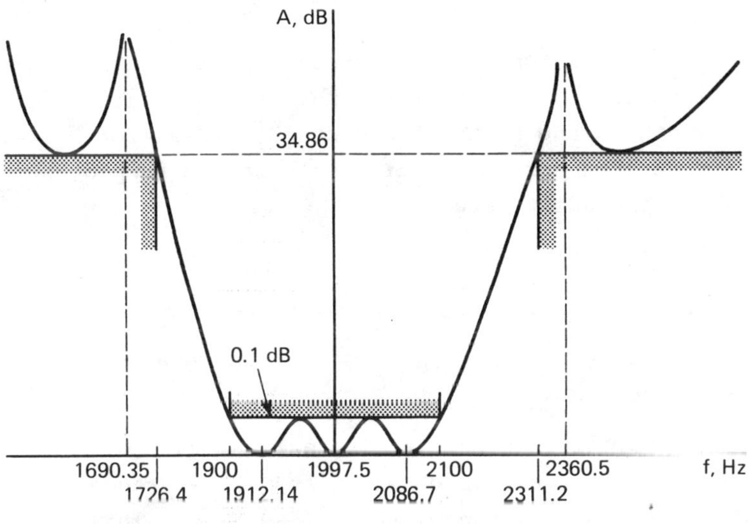

Figure 6.9 Attenuation function of the BPF in Example (6.6).

Application of the LP to BP Transformation to the LC Ladder of the LPP

The LP to BP transformation is a "reactance transformation"; that is, when applied to the LC circuit of the LPP, it results in a realizable LC circuit for the BPF. Recall that we have shown LC ladder realizations for classical LP filter approximations in Figs. 6.1 and 6.2. In the following, we shall assume that the element values of the LC ladder realizing a LPP are frequency normalized, and that R_1 and/or R_2 is equal to 1Ω. We will use lower case symbols to denote the element values of the normalized LPP ladder.

It is our purpose now to find the corresponding LC ladder realization of the BP filter obtained through the application of the LP to BP transformation to the LPP network. Toward this end, we first find the transformations of single elements and tuned circuits.

A) An inductance ℓ in the LPP transforms into the impedance

$$s\ell = \frac{p^2 + \omega_o^2}{pB}\,\ell = p\,(\frac{\ell}{B}) + \frac{1}{p\,\dfrac{B}{\omega_o^2 \ell}}$$

which is a series tuned circuit resonant at ω_o. The value of L is obtained from ℓ, by the frequency denormalization from a unity bandwidth to the actual bandwidth B. (See Fig. 6.10a)

B) In a dual manner, a capacitance c in the LP prototype transforms into the admittance,

$$sc = \frac{p^2 + \omega_o^2}{pB}\,c = p(\frac{c}{B}) + \frac{1}{p\,\dfrac{B}{\omega_o^2 c}}$$

which is a parallel tuned circuit resonant at ω_o. Again we note that the capacitance C is obtained from c by frequency denormalization by the actual bandwidth B. (See Fig. 6.10b)

Fig. 6.10c illustrates the application of the LP to BP transformation to the ladder realization of a third-order all-pole LPP. Although such a realization is easy to obtain, and to tune, a problem exists regarding the component value spread. For instance, in the resulting sixth-order bandpass filter we have,

$$\frac{L_2}{L_1} = \frac{\ell_2}{B} \omega_o^{\ 2} C_1 = \frac{\ell_2}{B} \omega_o^{\ 2} \frac{c_1}{B} = \ell_2 c_1 \hat{Q}^2$$

Since $\ell_2 c_1$ usually has a value around unity, it follows that for high $\hat{Q}$, L_2 is much larger than L_1, with the attendant practical problems in passive implementations. To overcome the component spread problems certain network transformations may be applied [2,6]. Such transformations are considered in Section 6.7.

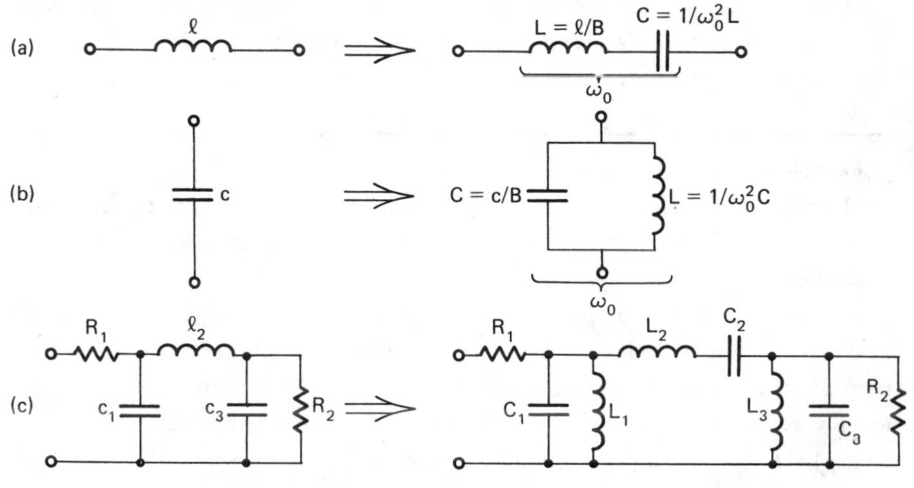

Figure 6.10 Application of the LP to BP transformation $s = \dfrac{p^2 + \omega_o^2}{pB}$ to: (a) an inductance ℓ, (b) a capacitance c, and (c) a third-order all-pole ladder network.

Consider next, the LP to BP transformation applied to the finite loss-pole ladders of Fig. 6.2. Fig. 6.11b shows the result of the direct application of the transformation for the case n=3 (N=6). Note that although the circuit in the dashed box is responsible for producing the two transmission zeros $\omega_{\ell 11}$ and $\omega_{\ell 12}$

corresponding to $\Omega_{\ell 1}$, there is no single tuned circuit resonant at either frequency. This makes tuning the filter rather difficult. A simple transformation[6] replaces this zero-producing section by two antiresonant circuits connected in series as shown in Fig. 6.11c. Each of the two parallel tuned circuits is resonant at one of the transmission zeros. The component values may be obtained from,

$$
\begin{aligned}
L_1 &= \frac{1}{\omega_o c_2 \hat{Q}} \; \frac{1}{(1 + \frac{\omega_{\ell 11}}{\omega_{\ell 12}})} = \frac{1}{\omega_o^2 C_2}, \\
L_2 &= \frac{1}{\omega_o c_2 \hat{Q}} \; \frac{1}{(1 + \frac{\omega_{\ell 12}}{\omega_{\ell 11}})} = \frac{1}{\omega_o^2 C_1}
\end{aligned}
\qquad (6.27)
$$

Finally, it should be noted that the BP ladders obtained are frequency denormalized, but should be impedance denormalized to the desired impedance level.

Example 6.7

Obtain an LC ladder realization for the BP filter specified in example 6.5. The ladder is to be terminated in 1KΩ.

Solution

In example 6.5, an LC ladder realization was obtained for the LPP and shown in Fig. 6.8c. To obtain the BP ladder, we make use of the ladder transformation formulae developed above, as well as the values of the loss pole frequencies obtained in example 6.6. After denormalization with 1KΩ impedance, we obtain the LC ladder shown in Fig. 6.12.

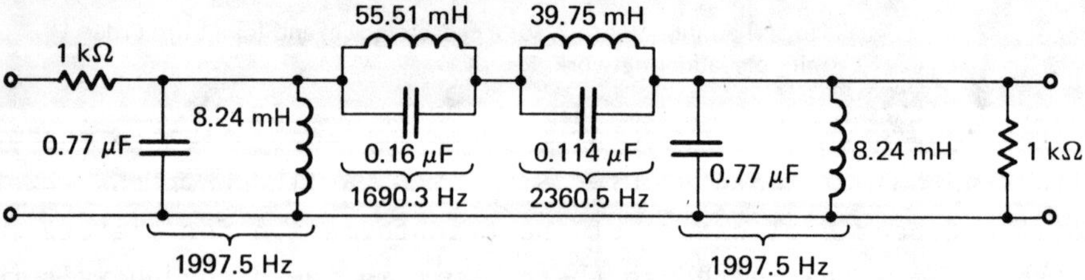

Figure 6.12 LC ladder realization of the BPF in Example (6.7).

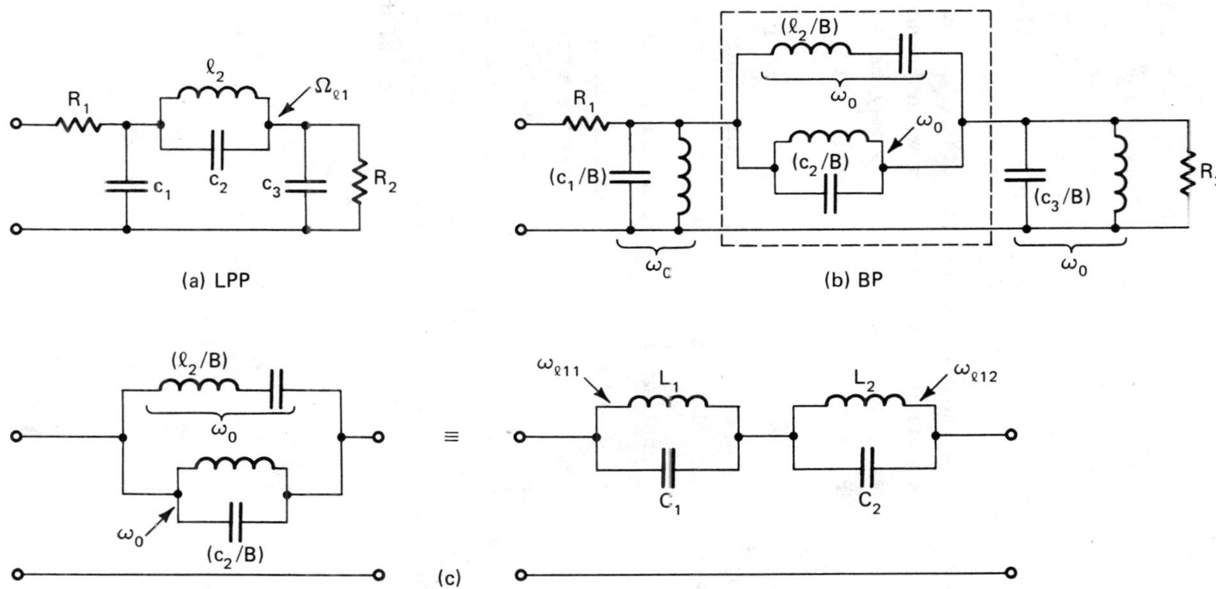

Figure 6.11 Illustrating the application of the LP to BP transformation to LC ladders with finite loss poles.

6.4 THE LOW-PASS TO HIGH-PASS TRANFORMATION

The LP to HP transformation enables the design of a HP filter by transforming the design of its LPP. This transformation is very simple. However, its importance exceeds that of the LP to BP transformation. While the latter is primarily useful for designing filters using classical approximations, the LP to HP transformation is employed in the design of any HP filter. The reader may recall that, in our general solution of the approximation problem, we concerned ourselves only with BP and LP filters. It was also mentioned that HP filters may be obtained by first designing a LPP and then transforming it using the LP to HP transformation.

To understand the LP to HP transformation, consider the attenuation function $A(\Omega)$ shown in Fig. 6.13a. This function is clearly a low-pass one. Let us find the attenuation function $A(\frac{\omega_p}{\omega})$, where ω_p is a constant frequency, by simply evaluating the argument $\Omega = \frac{\omega_p}{\omega}$ at each value of ω and then using the curve in Fig. 6.13a. The result will be the high-pass attenuation function shown in Fig. 6.13b. Note that the passband ripple is A_p dB and extends from ω_p to ∞, while the minimum stopband attenuation is A_s dB and extends from ω_s down to zero frequency. By definition, the value of ω_s is obtained from $\Omega_s = \frac{\omega_p}{\omega_s}$.

From the above, it follows that the LP to HP transformation is given by

$$\Omega = \frac{\omega_p}{\omega}$$

or,

$$s = \frac{\omega_p}{p}$$

where s is the complex frequency variable in the normalized LPP domain ($s = j\Omega$ for physical frequencies) and p is the complex frequency variable in the HP domain (p = jω for physical frequencies).

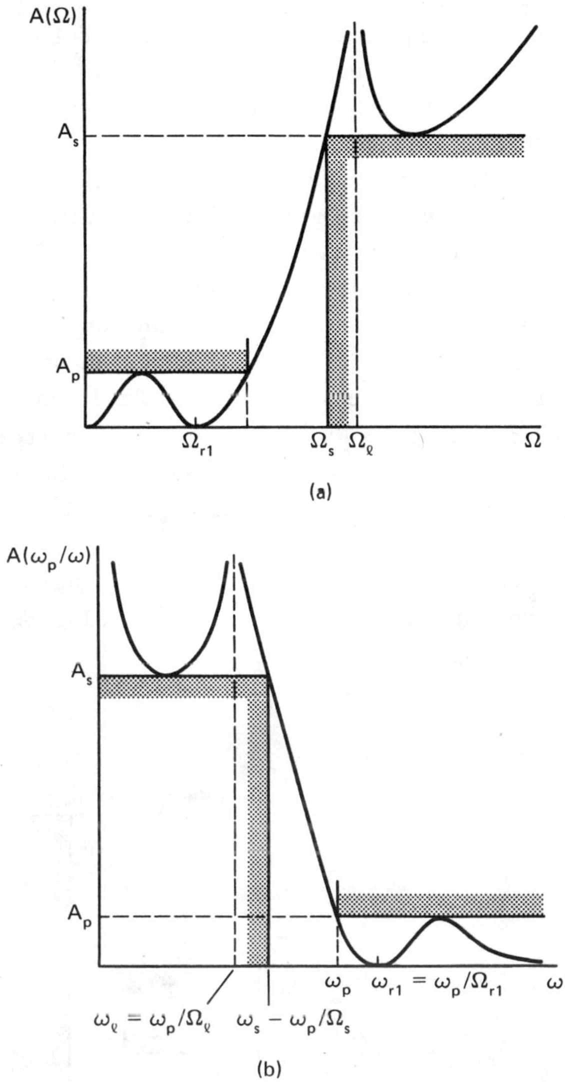

Figure 6.13 Illustrating the development of the LP to HP transformation.

Although in the development above we used a tolerance plot that can be satisfied by a classical approximation, this is only a restriction if one has to use such approximation for the LPP. The LP to HP transformation can be used to obtain approximations for HP filters with frequency-dependent stopband attenuation characteristics. The procedure for obtaining general HP designs can be summarized in the following:

(1) Given the general tolerance plot for the attenuation function of a HP filter such as the one shown in Fig. 6.14a, we determine the tolerance plot of the normalized LPP. This is shown in Fig. 6.14b. The method for obtaining the LPP tolerance plot should be obvious.

(2) The approximation methods, and associated computer programs, presented in the preceding chapters can now be used to obtain an approximation for the LPP. Such an approximation will be described by a characteristic function K(s), a transfer function H(s), etc. Furthermore, filter tables, closed form expressions, or the passive synthesis methods of Chapter 7 may be used to obtain an LC network realization of the LPP.

(3) The parameters, functions and LC network realization of the HP filter are obtained from the corresponding ones for the LPP, by simply replacing s by $\frac{\omega_p}{p}$.

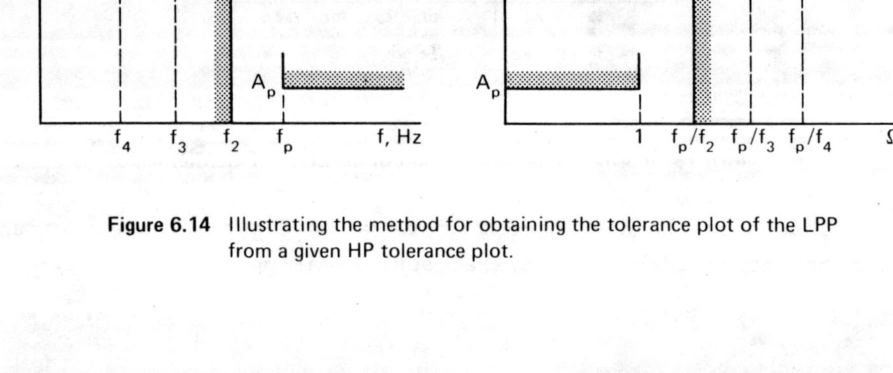

Figure 6.14 Illustrating the method for obtaining the tolerance plot of the LPP from a given HP tolerance plot.

We now consider the application of the LP to HP transformation to the different parameters, functions and network realization of the LPP.

A) Transformation of a real natural mode:

A real natural mode at $s = -\sigma$, $\sigma > 0$ will give rise to a factor

$$(s + \sigma) = \frac{\omega_p}{p} + \sigma = \sigma \left[\frac{p + (\omega_p/\sigma)}{p}\right]$$

in $H_{HP}(p)$. Thus, a real natural mode at $(-\frac{\omega_p}{\sigma})$ results.

B) Transformation of a pair of complex conjugate natural modes

A pair of complex conjugate natural modes characterized by Ω_o and q will give rise to a factor

$$(s^2 + s\frac{\Omega_o}{q} + \Omega_o^2) = (\frac{\omega_p^2}{p^2} + \frac{\omega_p}{p}\frac{\Omega_o}{q} + \Omega_o^2) = \frac{\Omega_o^2}{p^2}(p^2 + p\frac{\omega_p}{\Omega_o}\frac{1}{q} + \frac{\omega_p^2}{\Omega_o^2})$$

in the high pass transfer function. Thus, the result is a pair of complex conjugate natural modes characterized by $(\frac{\omega_p}{\Omega_o}, q)$.

C) Transformation of $j\Omega$-axis complex conjugate loss poles and reflection zeros

A pair of loss poles at $\pm j\Omega_\ell$ will transform into a pair of loss poles at $\pm j\frac{\omega_p}{\Omega_\ell}$. Similarly a pair of reflection zeros at $\pm j\Omega_r$ will produce HP reflection zeros at $\pm j\frac{\omega_p}{\Omega_r}$.

D) Transformation of LC network elements

An inductance ℓ in the LPP provides an impedance

$$s\ell = \frac{\omega_p}{p}\ell = 1 / (p/\omega_p\ell)$$

which is that of a capacitor of value $C = \frac{1}{\omega_p\ell}$. In a dual manner, a capacitance of value c transforms into an inductance $L = \frac{1}{\omega_p c}$ in the HP network. It therefore follows that a resonant circuit (series or parallel) in the LPP, tuned to a frequency Ω_o, transforms into a resonant circuit of the same configuration but tuned to a frequency $\frac{\omega_p}{\Omega_o}$.

Finally, Table 6.1 provides a summary of all the properties and pertinent formulae of the LP to HP transformation.

Table 6.1 Summary of formulae for the LP to HP transformation.*

LPP $s = j\Omega$		HP $p = j\omega$
Ω	⟶	ω_p/ω
s	⟶	ω_p/p
Real natural mode at $s = -\sigma$	⟶	Real natural mode at $s = -\omega_p/\sigma$
Pair of complex conjugate natural modes, (Ω_0, q)	⟶	Pair of complex conjugate natural modes $(\omega_p/\Omega_0, q)$
Pair of loss poles at $\pm j\Omega_\ell$	⟶	Pair of loss poles at $\pm j\omega_p/\Omega_\ell$
Pair of reflection zeros at $\pm j\Omega_r$	⟶	Pair of reflection zeros at $\pm j\omega_p/\Omega_r$
n loss poles at ∞	⟶	n loss poles at origin

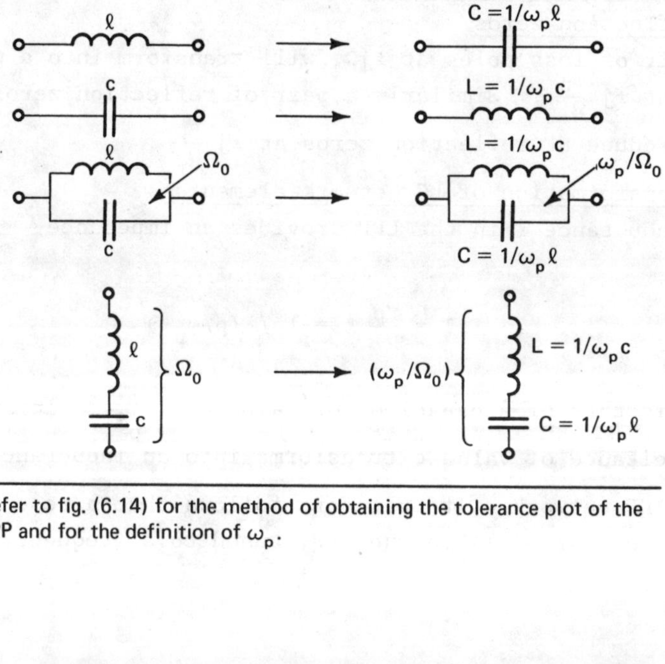

* Refer to fig. (6.14) for the method of obtaining the tolerance plot of the LPP and for the definition of ω_p.

6.5 THE LOW-PASS TO BAND-REJECT TRANSFORMATION

The LP to BR transformation is used for the design of geometrically symmetric BR filters by transforming the approximation function or network realization of the corresponding LPP. To understand the characteristics of the LP to BR transformation, consider the LP attenuation characteristic shown in Fig. 6.15a. By first applying the LP to HP transformation (with $\omega_p = 1$), we obtain the normalized HP attenuation function shown in Fig. 6.15b. If we denote the HP complex frequency variable by s', then the HP function is obtained by the transformation $s = \frac{1}{s'}$.

Since the attenuation function is an even function of frequency, we may draw the complete attenuation function $A(\Omega')$ of the HP filter as shown in Fig. 6.15c. The latter characteristic is of the BR type with the stopband extending between $-\frac{1}{\Omega_s}$ to $\frac{1}{\Omega_s}$, the upper passband from 1 to ∞ and the lower passband from -1 to $-\infty$. The "center frequency" is zero. Recalling now the logic used to develop the LP to BP transformation of section 6.3, we find that, by a parallel argument, we can transform the attenuation characteristics of Fig. 6.15c to any desired location on the positive frequency axis by applying the transformation

$$s' = \frac{p^2 + \omega_o^2}{pB} .$$

The result is shown in Fig. 6.15d. Using the properties of this LP to BP transformation, we find that

$$B = \omega_{p2} - \omega_{p1} \tag{6.28}$$

$$\omega_{s2} - \omega_{s1} = B\Omega_s' = \frac{B}{\Omega_s} \tag{6.29}$$

From the above, we conclude that the LP to BR transformation is given by

$$s = \frac{pB}{p^2 + \omega_o^2}$$

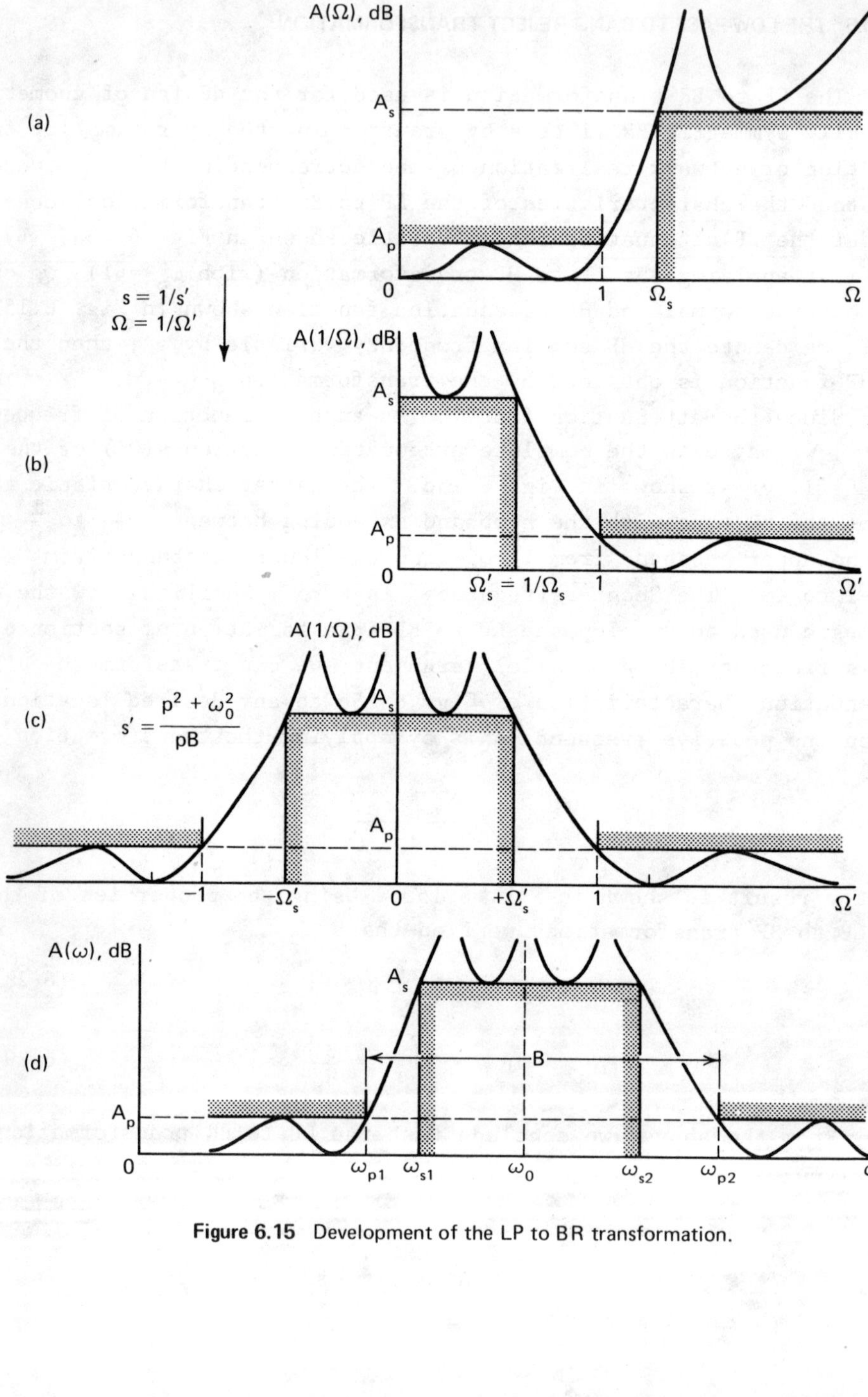

Figure 6.15 Development of the LP to BR transformation.

To obtain an approximation and/or a network realization for a given Band Reject filter specification, we perform the following steps:

(1) Modify the specifications by making the stopband attenuation constant (at its highest value) and the tolerance plot geometrically symmetric (See Section 6.3). These modifications are necessary to ensure that the LPP obtained can be approximated using classical approximations, and that the application of the LP to BR transformation of (6.30) results in a function that meets the original specifications.

(2) Let the modified BR tolerance plot be as shown in Fig. 6.16 where $\omega_{p1}\,\omega_{p2} = \omega_{s1}\,\omega_{s2}$. The corresponding LPP will have the selectivity factor Ω_s obtained from (6.28) and (6.29) as

$$\Omega_s = \frac{\omega_{p2} - \omega_{p1}}{\omega_{s2} - \omega_{s1}}$$

(3) Find an approximation and/or network realization for the LPP tolerance plot.

(4) Apply the LP to BR transformation,

$$s = \frac{pB}{p^2 + \omega_o^2}$$

where $B = \omega_{p2} - \omega_{p1}$

and $\omega_o^2 = \omega_{p1}\,\omega_{p2} = \omega_{s1}\,\omega_{s2}$

to the LPP functions or network realization to obtain the BR functions or network realization, respectively.

The details of the application of this transformation are quite similar to those developed for the BP case and will be left to the reader as an exercise.

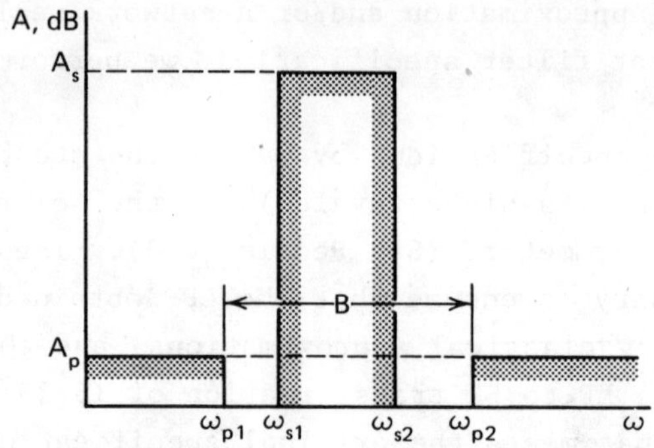

Figure 6.16 Tolerance plot for BR filter. Note that to obtain a classical approximation the plot should be geometrically symmetric, that is, $\omega_{s1}\omega_{s2} = \omega_{p1}\omega_{p2}$.

6.6 SUMMARY OF REACTANCE TRANSFORMATIONS

For quick reference, we now present a summary of the frequency band transformations discussed. Since all these transformations can be applied to an LC network and can obtain a realizable LC network, they are called reactance transformations.

Fig. 6.17 shows the tolerance plots of LP, HP, BP and BR filters. These tolerance plots are assumed to have been obtained from the original specifications, after suitable modifications to enable the use of classical approximations and to satisfy the characteristics of the reactance transformations. For this reason, the stopband(s) attenuation is made constant and the BP and BR tolerance plots are made geometrically symmetric.

Also shown in Fig. 6.17 is the tolerance plot of the normalized low-pass prototype (LPP). The specifications of the LPP are obtained from the tolerance plot in (a), (b), (c) or (d) according to Table 6.2. The next step consists of obtaining a transfer function and/or LC network realization for the LPP, using filter tables, or the methods given elsewhere in this book. Subsequently, the transfer function and/or LC network realization of the LPP is transformed, according to Table 6.2, to the original filter domain (e.g. BP, BR, ...). Note that the final transfer function H(p) will be frequency denormalized. The same holds true for the obtained LC network. However, the designer has to impedance denormalize the LC network realization.

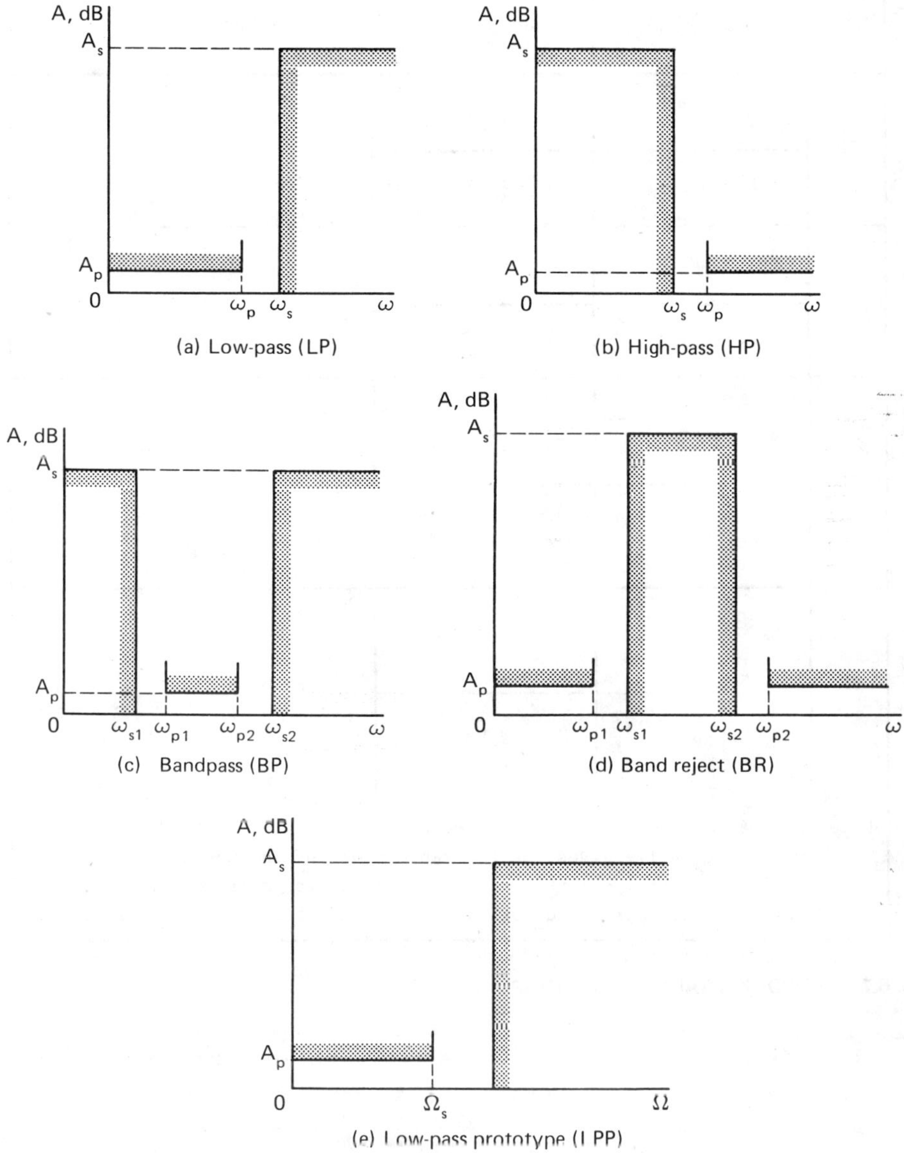

Figure 6.17 Tolerance plots for the various filter types.

Table 6.2 Frequency-band transformations

Type	Direction of Transformation	Transformation	Notes
LP	LP → LPP	$\Omega_s = \frac{\omega_s}{\omega_p}$	
	LPP → LP	$s = \frac{p}{\omega_p}$	
HP	HP → LPP	$\Omega_s = \frac{\omega_p}{\omega_s}$	
	LPP → HP	$s = \frac{\omega_p}{p}$	
BP	BP → LPP	$\Omega_s = \frac{\omega_{s2} - \omega_{s1}}{\omega_{p2} - \omega_{p1}}$	$\omega_o^2 = \omega_{p1}\omega_{p2} = \omega_{s1}\omega_{s2}$
	LPP → BP	$s = \frac{p^2 + \omega_o^2}{pB}$	$B = \omega_{p2} - \omega_{p1}$
BR	BR → LPP	$\Omega_s = \frac{\omega_{p2} - \omega_{p1}}{\omega_{s2} - \omega_{s1}}$	Same as above
	LPP → BR	$s = \frac{pB}{p^2 + \omega_o^2}$	

Note: Fig. 6.17a, d show the tolerance plots for LP, HP, BP and BR filters, while Fig. 6.17e gives the tolerance plot for the normalized ($\Omega_p = 1$) Low Pass Prototype (LPP). In the table above, s is used for the LPP domain, while p is used for the actual filter domain (denormalized).

6.7 NETWORK TRANSFORMATIONS

In this section, we consider several network transformations which are useful in the practical design of filters. The first transformation we shall consider is the introduction of an ideal transformer to allow the impedance level (and hence element values) of a portion of a network to be scaled to a different value. Such a procedure might be undertaken in order to obtain a prescribed ratio of terminating resistances, or to obtain more practical element values than those of the filter at hand. Consider the network η of Fig. 6.18a, which is partitioned into two networks η_1 and η_2. It is desired to increase the impedance level of the η_2 portion by the factor n^2. This is easily accomplished theoretically, by introducing an ideal transformer of turns ratio 1:n as shown in Fig. 6.18b. Here, the impedance of all elements of η_2' must be n^2 times the impedance level of those of η_2.

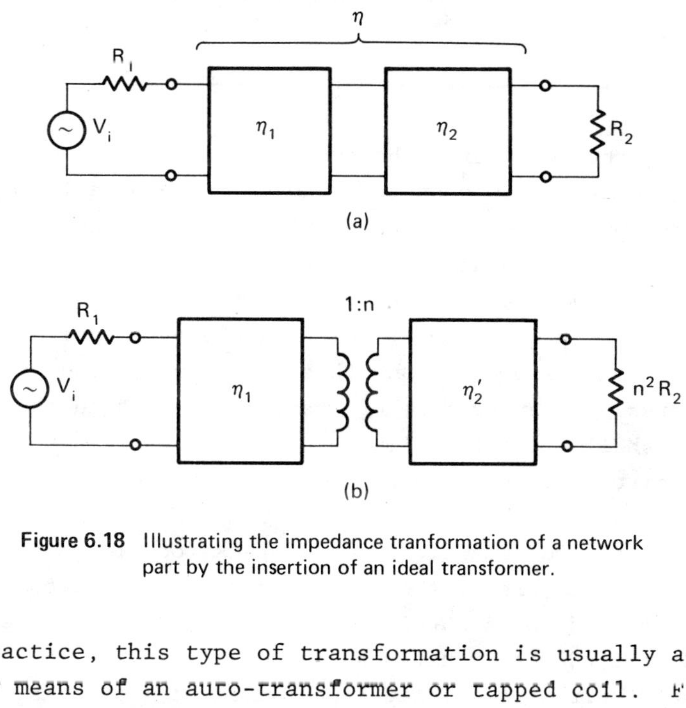

Figure 6.18 Illustrating the impedance tranformation of a network part by the insertion of an ideal transformer.

In practice, this type of transformation is usually accomplished by means of an auto-transformer or tapped coil. Fig. 6.19 illustrates an approximate equivalence between a coil-ideal transformer combination and an ideal auto-transformer or tapped coil. In practice, using ferrite cores one can obtain very close approximations to the ideal auto-transformer because of the very tight magnetic coupling between sections of the coil.

It is easily seen that one may take advantage of strategically located coils in a filter, by tapping them and making use of the impedance transformation so obtained. These procedures are best illustrated by means of an example.

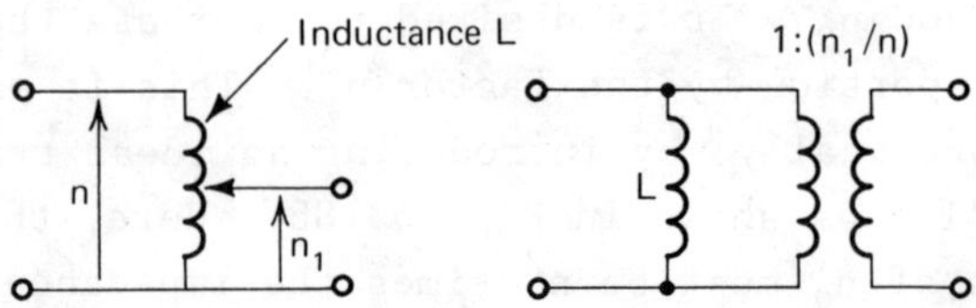

Figure 6.19 Illustrating the equivalance between an ideal tapped coil and a coil-ideal transformer combination.

Example 6.8

Consider the sixth order bandpass filter of Fig. 6.20a, which was obtained by a low pass to bandpass transformation applied to a low pass prototype. Clearly, the element values are not practical. The shunt capacitors are very large (capacitor cost and size is usually directly proportional to the capacitance) and therefore costly. The shunt coils have a very small value compared to the series coil, and this would entail the use of different cores. As an example of what can be done in these situations, we shall attempt to impedance scale the filter, such that all capacitor values are equal, and of a reasonable practical value say 10 nF.

Scaling the two (identical) shunt tanks on either end to obtain 10 nF capacitors, results in an impedance scale factor of 176.84, so that the end coils obtain a value of 281.5 mH. However, to maintain the 600Ω terminations, these coils will have to be tapped at their original impedance level of 1.5916 mH. Thus, the turns ratio for this tapping will be $\sqrt{1.5916/281.5} = 0.07519$.

In order to obtain a series capacitor of 10 nF, we must impedance scale the central branch by 0.2211, so that the 1.2732 H coil now obtains a value of 281.5 mH! Now all coils and capacitors have identical values (an economy in manufacturing). In addition, note that since the central branch must "see" an impedance of 0.2211 times the original impedance (1.5916 mH), the end coils must have an additional tap point at the impedance level of 0.2211 x 1.5916 = 0.3519 mH, which corresponds to a turns ratio of $\sqrt{0.3519/281.5}$ = 0.03536.

The schematic and element values of the scaled network are illustrated in Fig. 6.20b, for comparison with the original network. The reader should readily appreciate the practical significance of the procedures illustrated by this simple example.

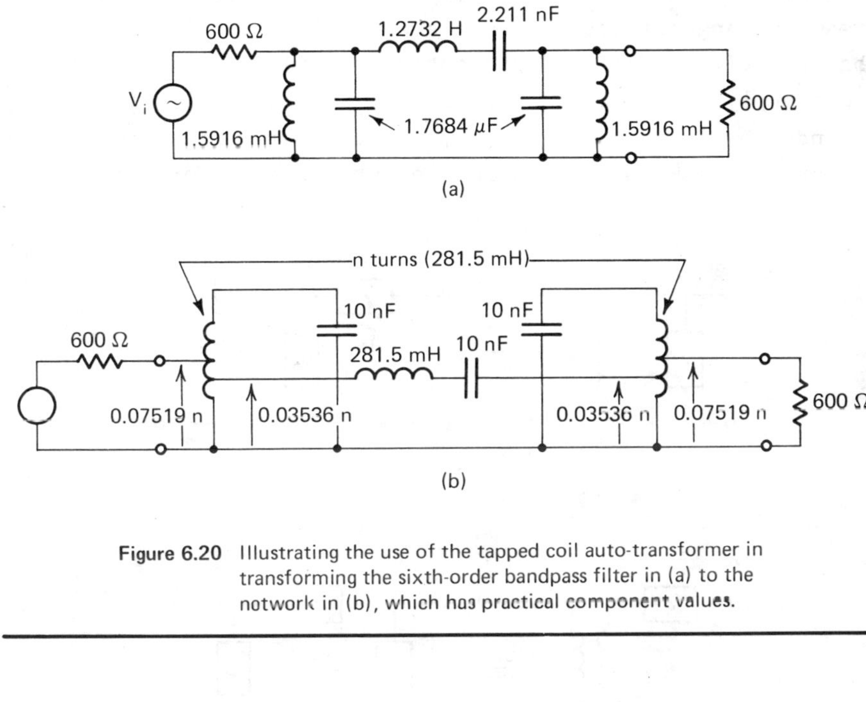

Figure 6.20 Illustrating the use of the tapped coil auto-transformer in transforming the sixth-order bandpass filter in (a) to the notwork in (b), which has practical component values.

In addition to the simple ideal transformer and its practical realization as a tapped coil, there are a number of other "tricks" or transformations which may be employed to obtain practical filter networks. The Norton transformations illustrated in Fig. 6.21 are very useful, and often allow impedance transformation without tapped coils or transformers, by operating on capacitive sub-networks. As an example, Fig. 6.22 shows the details of applying Norton's transformation to an "L" network of capacitors. Here, the "L" network shown may be partially transformed to a "Π" network or completely transformed to an "L" network with the opposite orientation, as shown in the Figure. The reader is urged to investigate how the Norton transformation has been used to obtain the equivalent networks of Fig. 6.22. One can thus use the "L" to "Π" or "L to L" transformation of capacitors to suitably raise or lower impedance levels to the left or right of the transformation section, by simply scaling impedance levels and eliminating the ideal transformer. It is clear that the transformation range is restricted (as shown in the Figure) by the fact that negative capacitors are not permitted. Similar transformations apply to inductors but are not as practical, since tapping an inductor is much more economical than winding a complete new inductor to make up the extra element for the Π or T.

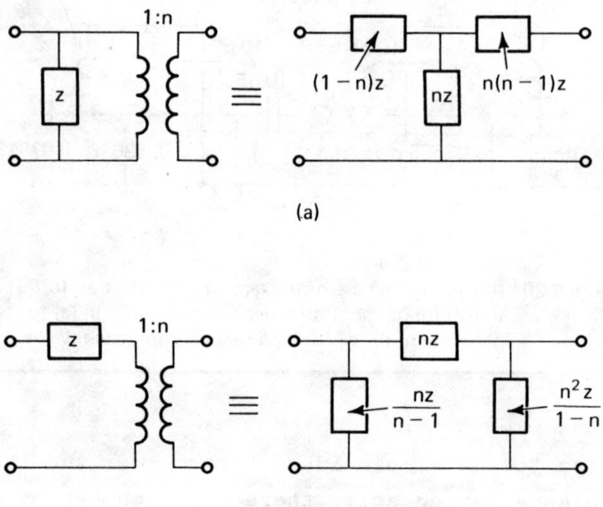

Figure 6.21 (a) Norton's shunt-T transformation, (b) Norton's series-π transformation

Figure 6.22 Applying Norton's transformation to a capacitor "L" network.

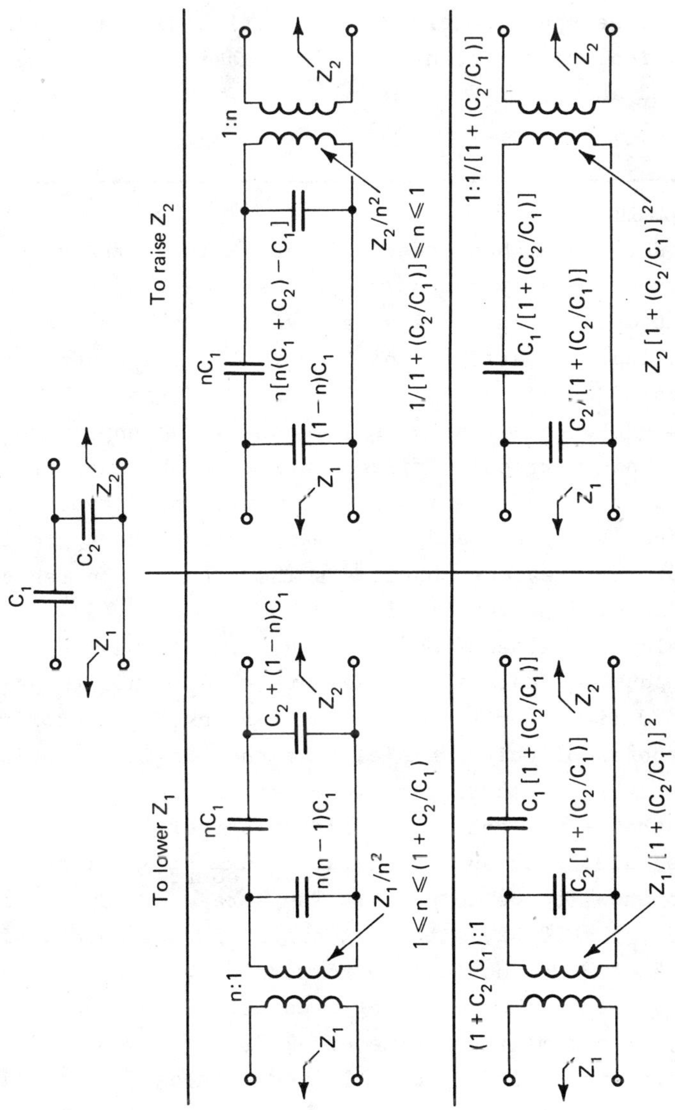

There are several other types of transformations which are helpful in designing practical passive filters and some of these are developed as problems in the exercises at the end of the Chapter. We shall conclude this Section with a further example of practical filter transformation, making use of the methods described above.

Example 6.9

Consider the eighth order bandpass filter of Fig. 6.23a. This filter, obtained by computer synthesis methods to be described in the next Chapter, appears as an example in both the approximation and realization parts of this text. It is a general parameter equiripple design having a 0.4dB passband ripple between 1 KHz and 1.4 KHz with a loss peak in each of the upper and lower stopbands. This particular filter cannot be obtained from a low-pass to bandpass transformation of a low-pass prototype and was designed directly in the bandpass domain, using the approximation methods of the previous Chapters. The attenuation response of the filter is given in Fig. 5.13.

The element values (shown in Fig. 6.23a) obtained directly from the computer synthesis routines are clearly not as practical as one would like, for the following reasons. Let us first assume, for purposes of this example, that the original specifications called for equal source and load resistances of 600Ω. The computer synthesis yielded a load resistance of 1132.8Ω, resulting in an impedance level spread of $(R_{load}/R_{source}) = 1.888$, which we must somehow reduce to unity. The spread of capacitor values is $(C_{max}/C_{min}) = 43.02$, with the largest capacitance having the value 2.115 μf. We should like to reduce this maximum value as well as the spread. The spread of inductor values is $(L_{max}/L_{min}) = 11.442$, with a minimum value of 32.826 mH. We should like to reduce this spread and obtain all inductances in a smaller range near 400 mH which we can assume, for purposes of this example, yields the highest Q-factor for the particular ferrite core we choose to use. Bringing all like elements closer to the same values will also result in a wider dynamic range, since for this condition all components will encounter their maximum ratings together (maximum capacitor voltages and maximum inductor currents) rather than having the maximum signal level limited by a single component at some lower signal level.

We begin by making use of a capacitive Norton transformation, turning the two capacitors marked with asterisks in Fig. 6.23a into a Π of capacitors, and raising the impedance level of all components to the right by a factor of 16, as shown in Fig. 6.23b. The purpose of this transformation is to reduce the value of the very large capacitor (2.115 μF) in the series tuned circuit which produces the lower stopband loss peak. Note that the new value of this capacitor is 132.19 KpF; however, the new load resistance is 18.125 KΩ, and the tank circuit inductance in the upper stopband peak producing section has risen to 2.221 H.

Our next step is to reduce the impedance to the right of the 28.47 KpF capacitor of Fig. 6.23b. Making use of a further capacitive Norton transformation, we convert the "L" network of capacitors marked with asterisks in Fig. 6.23b to an "L" network of the opposite orientation, as shown in Fig. 6.23c. This reduces the impedance to the right by a factor of 67.126, resulting in a load resistance of 270.0Ω, and reducing the 2.221 H coil to 33.08 mH. Note that we now have a 2.196 μF capacitor shunting the load.

Our next step is to reduce the value of the 2.196 μF capacitor, raise the load resistance to 600Ω, and increase the value of the 33.08 mH tank inductor to a value closer to 400 mH. We will simultaneously try to obtain standard stock values for some of the capacitors. To achieve this, we use the tapped coil auto-transformer effect in two places. First, we increase the tank inductor from 33.08 mH to 344.6 mH (near 400 mH) to obtain a 24 KpF (stock value) resonating capacitor. This coil must now be tapped at the 33.08 mH level for connection to the rest of the network as shown in Fig. 6.23d. Similarly, we raise the value of the 7.824 mH inductor shunting the 270.0Ω load in Fig. 6.23c to 520.7 mH (near 400 mH), obtaining a stock value of 33 KpF for the capacitor replacing the huge 2.196 μF value. This new 520.7 mH inductor must now be tapped at two points, the first at the 7.824 mH level to match the circuit at the left, and at the 17.387 mH level to obtain a 600Ω load on the right. The final configuration is shown in Fig. 6.23d.

Examining our final result in Fig. 6.23d, we see that we have achieved many of our objectives. We now have both 600Ω source and load resistors, the largest capacitance is now 233.3 KpF, and all inductor values are in the vicinity of 400 mH. It is, of course, possible to make a more detailed study of the possibility of obtaining more stock-valued capacitors, by considering other values than those chosen for the Norton transformations. However, due to the low element sensitivity of lossless networks (as shown in the next Chapter), one can obtain quite acceptable performance by substituting nearby stock values for those in the circuit of Fig. 6.23d. For instance, one could use 36 KpF for the 36.86 KpF value, and 12 KpF for the 12.288 KpF value, etc., depending upon the accuracy desired in the final filter response.

Due to the complexity of these transformation procedures, it is not very feasible to automate them by a computer routine. A clever design engineer with a little experience can very quickly spot the types of transformations, and places to apply them, in the wide variety of circuit configurations found in practice. It is feasible to automate a lot of the bookkeeping required to perform the transformations, and the FILTOR 2 package[1] contains a computer routine which, under the user's direction, will perform all of the numerical work involved in a transformation process, such as that illustrated by Fig. 6.23.

The examples given in this Section are representative of some of the techniques employed by filter designers in obtaining practical designs. In this area of design, there is no substitute for ingenuity and experience, and the examples have been worked out merely to indicate the nature of the design process, since a complete coverage of all of the filter designer's "bag of tricks" would require much more space than is available in this work. The reader is encouraged to obtain further examples of practical filter design, from other sources listed as references, and review as many of these as possible to help build his own back-

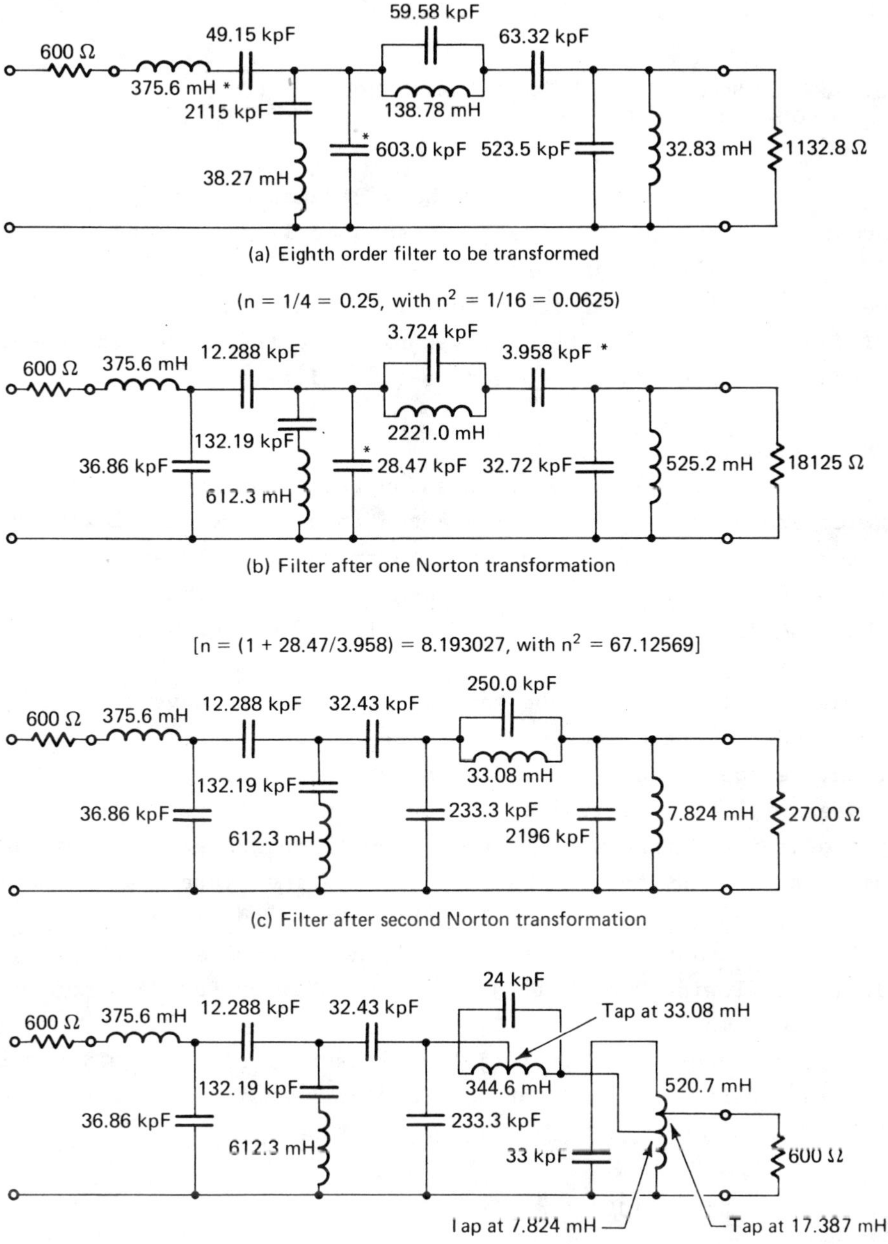

(a) Eighth order filter to be transformed

(b) Filter after one Norton transformation

(c) Filter after second Norton transformation

(d) Final configuration of filter after all transformations

Figure 6.23 Steps in the impedance transformation of a filter

ground of experience in this area.

6.8 APPLICATION OF THE APPROXIMATION METHODS TO DISCRETE-TIME FILTERS

The previous Chapters of this book discussed the approximation problem for continuous-time or nonsampled linear systems used in filtering applications. In this Section, we shall show that by means of a simple transformation of variable, all of the methods discussed to date may be easily applied to the approximation problem for sampled-data or discrete-time filters.

Discrete-Time Signals

A discrete-time system operates on the signal samples x(nT), which are obtained by sampling a continuous-time signal x(t) at intervals T. Here, n takes on integer values only and T is the sampling period, with $f_s = \frac{1}{T}$ being the sampling frequency. Note that the discrete-time signal x(nT) is defined only at the sampling times.

The signal x(nT) will be generally assumed to have a continuous amplitude. If the amplitude is discretized (quantized), we obtain the special case of a digital signal. The corresponding system will be a digital system or a digital filter. The material of this section applies equally well to analog and digital discrete-time systems. Analog discrete-time systems are becoming popular, due to the availability of Integrated Circuit analog delay lines in the form of charge transfer devices (CTD's)[7,8]. In the following, we shall use the name "discrete" to denote discrete-time.

The frequency spectrum $X_d(j\omega)$ of a discrete signal x(nT) may be obtained through the use of the Fourier transform and shown to be,

$$X_d(j\omega) = \frac{1}{T} \sum_{r=-\infty}^{\infty} X(j\omega + jr\omega_s) \qquad (6.31)$$

where the subscript "d" denotes discrete and $X(j\omega)$ is the frequency spectrum of the continuous signal x(t). It follows that $X_d(j\omega)$ is a periodic function of frequency with the period equal to the sampling frequency ω_s. In order to be able to recover the signal from its sampled form, we have to ensure that the repeated frequency spectra do not overlap. This is accomplished by bandlimiting the signal x(t) before sampling. Furthermore, if the highest frequency component of the bandlimited signal is ω_c then the sampling theorem states that the signal should be sampled at a minimum rate of $2\omega_c$ (the Nyquist rate). In practice, however, one usually samples at a much higher rate ($5\omega_c$-$10\omega_c$ is not unusual).

The Laplace transformation of the sampled signal x(nT) is given by

$$X_d(s) = \frac{1}{T} \sum_{r=-\infty}^{\infty} X(s + jr\omega_s) , \tag{6.32}$$

which is also periodic with a period of $j\omega_s$.

Discrete-Time Systems

A discrete system operates on the samples x(nT) and produces the output samples y(nT). In the time-domain, such a system may be characterized by its response to a unit-sample $\delta(nT)$, where

$$\delta(nT) = 1, \text{ for } n = 0$$
$$\delta(nT) = 0, \text{ for } n \neq 0 .$$

The unit-sample response, or more loosely, impulse response, of the system is denoted by g(nT). The output and input signals are related to g(nT) via the convolution sum.

The discrete system may also be described by a linear constant-coefficient difference equation of the form,

$$y(nT) = \sum_{k=1}^{N} a_k\, y(nT - kT) + \sum_{r=0}^{M} b_r\, x(nT - rT) \tag{6.33}$$

This equation has an easy interpretation. The output at t = nT is a function of the N previous output samples and of the present and past M input samples. Such a system will have an impulse response g(nT) of infinite duration and is referred to as an infinite-impulse response (IIR) system. If the first summation on the right-hand side of the difference equation is missing, the system will have an impulse response of finite duration, and is referred to as an FIR system. IIR systems provide more efficient filters (of the type considered in this book). Therefore, in the following, we shall only be interested in the design of IIR systems.

The difference equation description suggests that a discrete system should have the means for storing a signal sample for multiples of the sampling interval T. Thus, the basic building block of a discrete system is the unit-delay element, represented in Fig. 6.24a. Fig. 6.24b shows the representation of the unit-delay in the Laplace domain. If we substitute $e^{sT} = z$, we end up with the z-domain representation of the unit-delay, as shown in Fig. 6.24c. The transformation $z = e^{sT}$ is a fundamental one in the analysis and design of discrete systems and will be considered in more detail below.

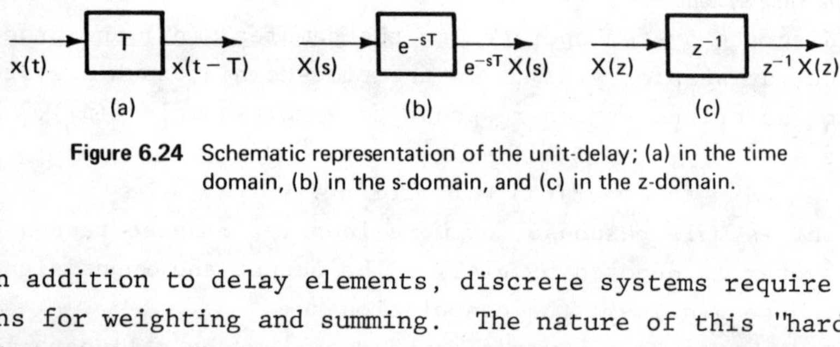

Figure 6.24 Schematic representation of the unit-delay; (a) in the time domain, (b) in the s-domain, and (c) in the z-domain.

In addition to delay elements, discrete systems require some means for weighting and summing. The nature of this "hardware" will depend on whether the system is an analog or a digital one. In an analog (continuous-amplitude) system, the weighting and summing may be performed using operational amplifier circuits. Digital systems use digital multipliers for weighting and digital adders for summing.

Realization of IIR Filters

The difference equation description of an IIR filter suggests a block diagram realization. Fig. 6.25a shows a direct implementation of equation (6.33) where, for simplicity, we have assumed that M = N. Note that the realization consists of two blocks in cascade: a "feedforward" block followed by a "feedback" block. If the filter were of the FIR type, the feedback block would not be required. Because this realization of the IIR filter contains feedback it is usually called a recursive filter.

Since the two blocks in Fig. 6.25a are in cascade, we may reverse their order and obtain the alternate realization shown in Fig. 6.25b. Both realizations, however, are non-canonic, as a total of 2N delay elements are required to realize this N^{th} order filter.

Examination of the realization of Fig. 6.25b reveals that the delay elements in both blocks have identical inputs. It is therefore possible to eliminate one set of delay elements and share the remaining N delays between the feedback and feedforward blocks. The resulting canonic realization is shown in Fig. 6.25c

Transfer Functions

The output/input transfer function $G_d(j\omega)$ of a discrete system having a unit-sample response g(nT) is given by the Fourier sum

$$G_d(j\omega) = \sum_{k=0}^{\infty} g(kT)\, e^{-jk\omega T} \tag{6.34}$$

As expected, $G_d(j\omega)$ is a periodic function of frequency with a period equal to the sampling frequency $\omega_s = \frac{2\pi}{T}$.

Since $G_d(j\omega)$ consists of an infinite sum of exponentials, some authors prefer to explicitly show this exponential dependence by writing the transfer function as $G_d(e^{j\omega T})$. From the Fourier transform of discrete-time functions, equation (6.31), we may write the transfer function of the filter with impulse response g(nT) as,

$$G_d\,(e^{j\omega T}) = \frac{1}{T} \sum_{r=-\infty}^{\infty} G[j(\omega + r\omega_s)] \tag{6.35}$$

where G(jω) is the transfer function of the filter whose impulse

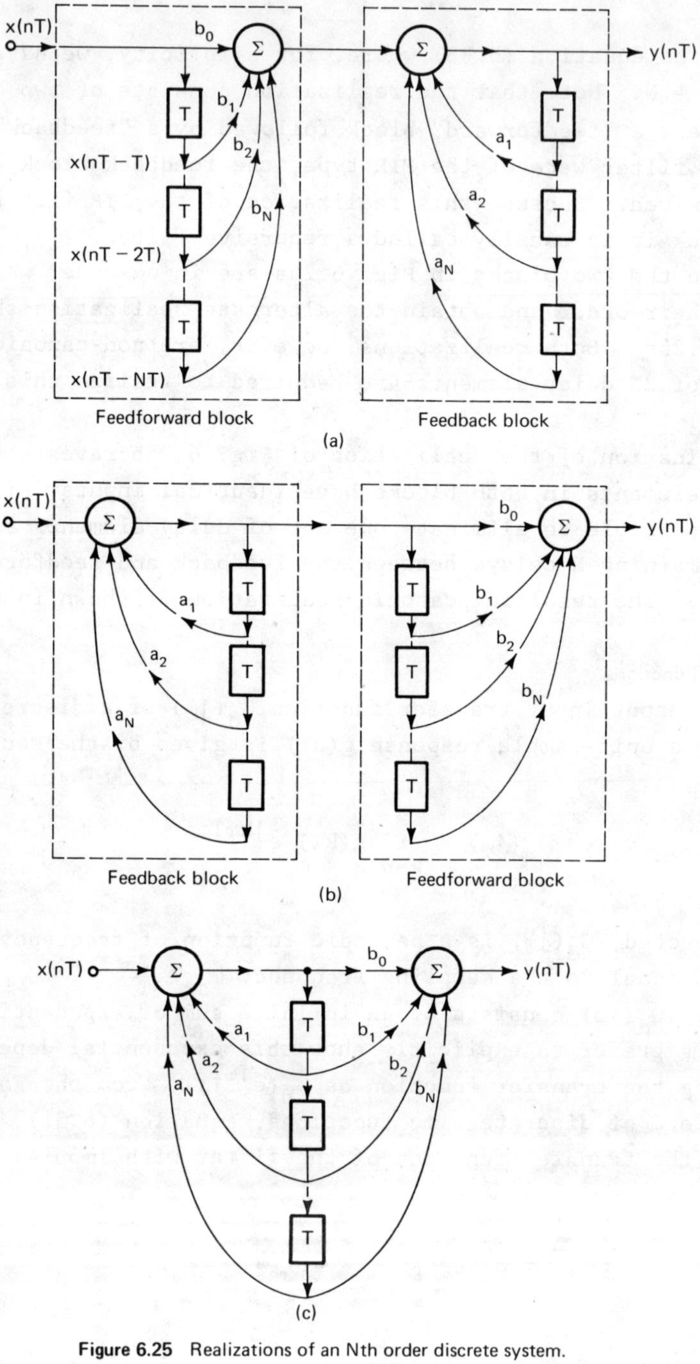

Figure 6.25 Realizations of an Nth order discrete system.

response is g(t). The summation above suggests that $G_d(e^{j\omega T})$ could be quite different from $G(j\omega)$ over the frequency band of interest, an effect usually referred to as "aliasing". This is the principal drawback of obtaining approximations for discrete filters by finding functions whose associated impulse response is the sampled version of the impulse response of a continuous filter which meets the given attenuation specifications. This method is called the impulse-invariant method and will not be discussed any further. It should be emphasized, however, that whether the discrete filter is designed by the impulse-invariant method or by any other method, its transfer function $G_d(j\omega)$ will be periodic in frequency with a period of ω_s.

The transfer function of a discrete filter in the Laplace domain is given by

$$G_d\,(e^{sT}) = \sum_{k=0}^{\infty} g(kT)\,e^{-skT} \tag{6.36}$$

Since $G_d\,(e^{sT})$ contains terms of the form e^{sT}, it has been found convenient to substitute $e^{sT} = z$ to obtain

$$G_d\,(z) = \sum_{k=0}^{\infty} g(kT)\,z^{-k}\,, \tag{6.37}$$

as the z-plane transfer function. Note that $G_d\,(z)$ is a rational function in z, and recall that the unit-delay has the transfer function z^{-1}.

In the z-plane, a transfer function is evaluated for physical frequencies by substituting $z = e^{j\omega T}$. This means that the function is evaluated over the unit-circle, as shown in Fig. 6.26. Note that the angle π in the z-plane corresponds to $\omega = \frac{\omega_s}{2}$.

The z-transform of an arbitrary discrete signal x(nT) is formally defined as

$$X(z) = \sum_{k=-\infty}^{\infty} x(kT)\,z^{-k}$$

Applying this transformation to the difference equation description of a discrete filter, that is equation (6.33), we obtain the transfer function

$$G_d(z) = \frac{Y(z)}{X(z)} = \frac{\sum_{r=0}^{M} b_r z^{-r}}{1 - \sum_{k=1}^{N} a_k z^{-k}}, \qquad (6.38)$$

which is a real rational function in z. We can use this transfer function to obtain the poles and zeros in the z-plane. Stability requirements dictate that all the poles should lie inside the unit-circle, while the zeros can be located anywhere in the z-plane.

Cascade Realization of a Given $G_d(z)$

Given a transfer function $G_d(z)$ of the form (6.38), we may realize it using the canonical form of Fig. 6.25c. It has been found, however, that the resulting circuit is quite sensitive to variations in weighting coefficients. A far more stable realization may be obtained by factoring $G_d(z)$ into the product of biquadratic and possibly one bilinear terms, realizing the transfer function of each factor separately, and then cascading the resulting networks. Assuming that the networks in the cascade do not interact, the overall transfer function will be equal to the product of the individual transfer functions, which is the required $G_d(z)$.

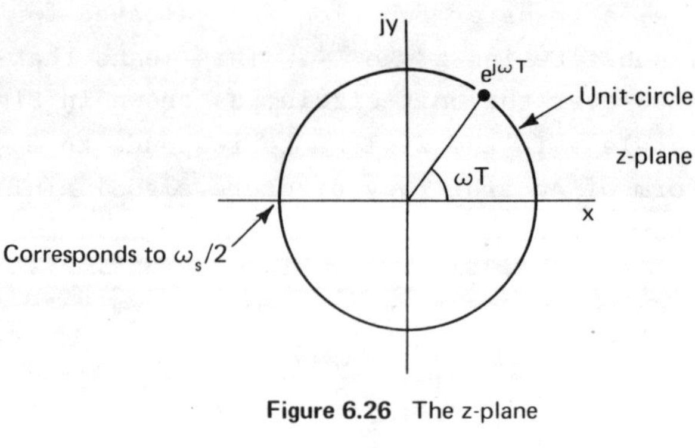

Figure 6.26 The z-plane

Fig. 6.27a shows a canonic realization of a bilinear function of the form

$$G_d(z) = \frac{b_o + b_1 z^{-1}}{1 - a_1 z^{-1}} \tag{6.39}$$

while Fig. 6.27b shows a canonic realization of a biquadratic transfer function of the form

$$G_d(z) = \frac{b_o + b_1 z^{-1} + b_2 z^{-2}}{1 - a_1 z^{-1} - a_2 z^{-2}} \tag{6.40}$$

Solving The Approximation Problem—The Bilinear Transform

We now address ourselves to the question of how to obtain a transfer function $G_d(z)$, such that its associated attenuation function

$$A_d(\omega) = -20 \log_{10}\left| G_d(e^{j\omega T})\right|$$

approximates given attenuation specifications in an optimum manner (in the sense described in previous Chapters).

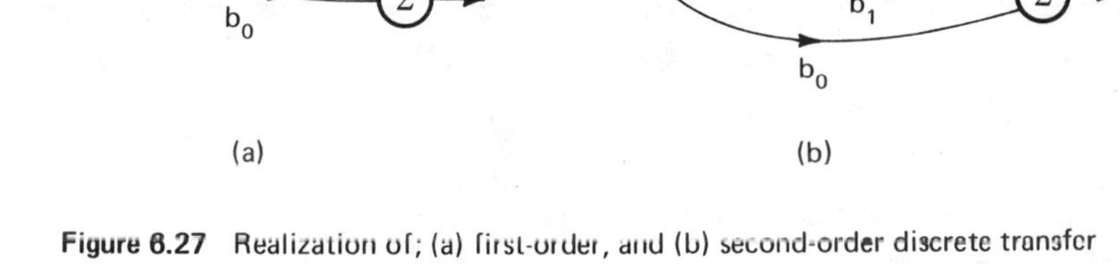

Figure 6.27 Realization of; (a) first-order, and (b) second-order discrete transfer functions.

In solving the approximation problem for discrete filters, we wish to make use of the methods and computer programs developed for continuous filters. However, there is a fundamental difficuity. The continuous filter approximation problem has been solved in a complex plane (s-plane) such that the transfer function obtained, when evaluated along a <u>vertical axis</u> (jω axis) in that plane, approximates given specifications. On the other hand, for the discrete case, we have to obtain a transfer function in a complex plane (z-plane) such that, when evaluated along <u>the unit circle</u> in that plane, it approximates the given specifications. It follows that, in order to apply the approximation methods of continuous filters to the discrete case, we need a complex mapping between the frequency variable of the continuous filter, to be denoted by S with $S = j\Omega$ for physical frequencies, and the frequency variable of the discrete filter; that is z with $z = e^{j\omega T}$ for physical frequencies. Furthermore, this transformation should map the entire jΩ axis onto the unit-circle of the z-plane. Such a mapping is given by the bilinear transformation

$$S = \frac{z-1}{z+1} \tag{6.41}$$

with its inverse being,

$$z = \frac{1+S}{1-S} \tag{6.42}$$

This transformation maps the interior of the z-plane unit-circle to the left-half of the S-plane, as illustrated in Fig. 6.28. Note that the entire jΩ-axis of the S-plane is mapped onto the unit-circle of the z-plane, resulting in a one-to-one correspondence between the points jΩ in the S-plane and the points $e^{j\omega T}$ in the z-plane. Thus for $S = j\Omega$ and $z = e^{j\omega T}$, we have

$$\omega = \frac{2}{T}\tan^{-1}(\Omega), \text{ and} \tag{6.43}$$

$$\Omega = \tan\left(\frac{\omega T}{2}\right) \tag{6.44}$$

Note that the relation between the frequency variables ω and Ω is not a linear one. This fact is usually referred to by the statement that the bilinear transformation results in "warping" the frequency axis.

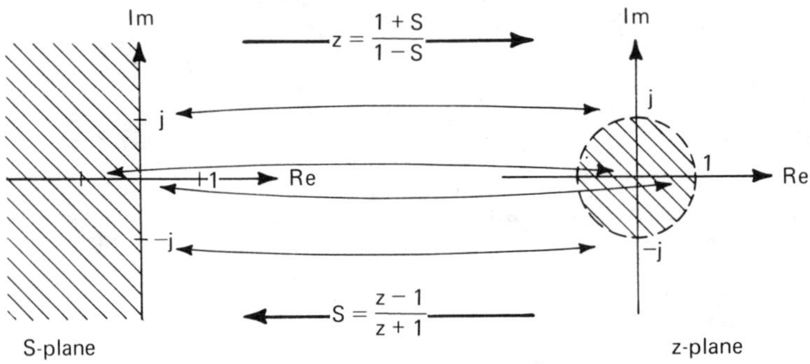

Figure 6.28 The bilinear mapping.

The procedure for solving the approximation problem for discrete filters is now straightforward. Let it be required to obtain a transfer function $G_d(z)$, such that its associated attenuation function

$$A_d(\omega) = -20 \log_{10}\left|G_d(e^{j\omega T})\right|$$

meets, for instance, the low-pass specifications given in Fig. 6.29a. The first step consists of obtaining the tolerance plot along the $j\Omega$ axis (Fig. 6.29b). This is accomplished by simply finding the frequencies Ω_p, Ω_{s1} and Ω_{s2} from the corresponding frequencies ω_p, ω_{s1} and ω_{s2} using the relation

$$\Omega - \tan\left(\frac{\omega T}{2}\right) ,$$

where T is the sampling interval chosen for the discrete filter. This step is usually referred to as "prewarping" the frequency axis. Of course, the attenuation axis remains unchanged.

Next, the methods of the previous chapters are used to obtain a suitable transfer function H(S) (input/output) such that,

$$A(\Omega) = 20 \log_{10}\left|H(j\Omega)\right|$$

meets the prewarped specifications of Fig. 6.29b.

Finally, the bilinear transformation

$$S = \frac{z-1}{z+1}$$

is used to obtain $G_d(z)$ as

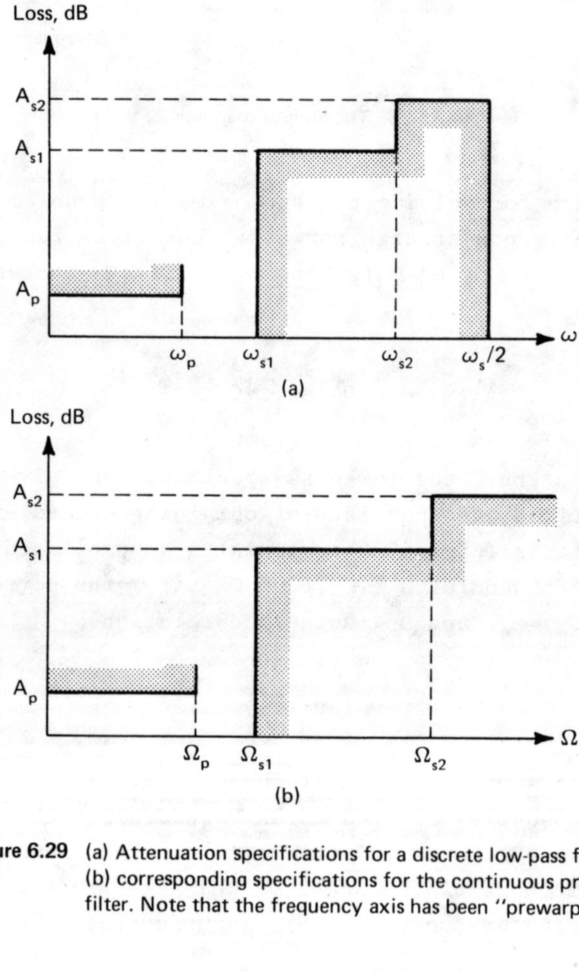

Figure 6.29 (a) Attenuation specifications for a discrete low-pass filter, (b) corresponding specifications for the continuous prototype filter. Note that the frequency axis has been "prewarped".

$$G_d(z) = [1/H(S)]_{S = \frac{z-1}{z+1}}$$

Obviously, $A_d(\omega)$ will meet the original specifications of Fig. 6.29a.

Note that the discrete filter, whose transfer function is obtained by this process, does not have the same impulse response as the prototype continuous filter. However, the frequency response of the continuous filter H(S), for measurable frequencies in the range $-\infty<\Omega<\infty$, is exactly the same as the frequency response of the resulting discrete filter in the range $-\frac{\omega_s}{2}<\omega<\frac{\omega_s}{2}$, with corresponding frequencies related by (6.43) and (6.44).

Some Useful Formulae For Applying the Bilinear Transformation

Applying the substitution (6.41) to a high-order rational polynomial in S is difficult and time consuming. A computer algorithm to implement the substitution directly is awkward to program, as well. A better procedure is to operate on the factors of G(S) [G(S) = 1/H(S)]. This procedure turns out to be most convenient when the S-plane transfer function of the continuous filter prototype is available in terms of its poles and zeros. The resulting $G_d(z)$ will then be found directly in factored form and can be realized as a cascade of first- and second-order sections. In the following, we give the formulae for the coefficients a_i and b_i of the z transformations of bilinear and biquadratic factors of G(S). The resulting z-plane transfer functions will be of the forms (6.39) and (6.40), respectively.

(A) A bilinear factor in G(S) of the form $\frac{aS+b}{S+\sigma}$ gives rise to the coefficients,

$$b_o = (a+b)/(1+\sigma)\ ,\ b_1 = (b-a)/(1+\sigma)\ ,\ \text{and } a_1 = (1-\sigma)/(1+\sigma).$$

(B) A biquadratic factor in G(S) of the form

$$\frac{aS^2 + bS + c}{S^2 + S(\frac{\Omega_o}{q}) + \Omega_o^2}\ ,$$ gives rise to the coefficients,

$$\begin{aligned} b_o &= k(a+b+c) \\ b_1 &= 2k(c-a) \\ b_2 &= k(a-b+c) \\ a_1 &= 2k(1-\Omega_o^2) \\ a_2 &= -k(1-\frac{\Omega_o}{q}+\Omega_o^2) \end{aligned} \tag{6.46}$$

where,

$$k = 1/(1+\frac{\Omega_o}{q}+\Omega_o^2)$$

Example 6.10

It is required to design a discrete low-pass Butterworth filter having the following specifications:

Passband: $f_p = 20$ Hz, $A_p = 3$ dB
Stopband: $A_s \geq 30$ dB for $f \geq 30$ Hz
Sampling frequency: 100 Hz

Solution

First, we determine the tolerance requirements of the continuous prototype filter as follows:

Passband: $\Omega_p = \tan(\frac{\omega_p T}{2}) = \tan(\frac{2\pi \times 20 \times 0.01}{2}) = 0.7265$, $A_p = 3$dB

Stopband: $\Omega_{s_1} = \tan(\frac{\omega_{s_1} T}{2}) = \tan(\frac{2\pi \times 30 \times 0.01}{2}) = 1.37638$,

where ω_{s_1} and Ω_{s_1} denote the stopband edge.

The selectivity factor Ω_s of the corresponding low-pass prototype will be given by,

$$\Omega_s = \frac{\Omega_{s_1}}{\Omega_p} = 1.8945$$

The required order of the Butterworth approximation may be determined using (3.21) to be $N \geq 5.4$. We therefore use $N = 6$. The three pairs of natural modes of the normalized (Ω_p=1) LPP have

$$\Omega_{o_1} = \Omega_{o_2} = \Omega_{o_3} = (\frac{1}{\varepsilon})^{\frac{1}{N}} = 1$$

which, when denormalized to $\Omega_p = 0.7265$, become,

$$\Omega_{o_1} = \Omega_{o_2} = \Omega_{o_3} = 0.7265 \text{ rad/sec.}$$

The natural mode Q-factors are given by (3.24),

$$Q_1 = 1.932 \qquad Q_2 = 0.707 \qquad Q_3 = 0.518$$

For a unity DC gain we may now write the transfer function G(S) as

$$G(S) = \frac{(0.7265)^2}{S^2 + S(\frac{0.7265}{1.932}) + (0.7265)^2} \times \frac{(0.7265)^2}{S^2 + S(\frac{0.7265}{0.707}) + (0.7265)^2}$$

$$\times \frac{(0.7265)^2}{S^2 + S(\frac{0.7265}{0.518}) + (0.7265)^2}$$

The transfer function $G_d(z)$ of the discrete filter can now be obtained using the transformation relationships in (6.46). The result is

$$G_d(z) = \frac{0.2772 + 0.5545z^{-1} + 0.2772z^{-2}}{1 - 0.4960z^{-1} + 0.6050z^{-2}} \times$$

$$\frac{0.2065 + 0.4131z^{-1} + 0.2065z^{-2}}{1 - 0.3696z^{-1} + 0.1958z^{-2}} \times$$

$$\frac{0.1801 + 0.3602z^{-1} + 0.1801z^{-2}}{1 - 0.3223z^{-1} + 0.0428z^{-2}}$$

This transfer function may be realized as the cascade of three sections, of the type shown in Fig. 6.27b. The reader is urged to sketch the complete block diagram realization.

In conclusion, we have presented a method by which all of the approximation methods and data developed in the previous Chapters for continuous filters may be applied to the approximation of discrete filter specifications. The technique of using the bilinear transformation is straightforward and may be easily implemented on a computer. All that is necessary to adapt a set of continuous filter approximation programs for discrete-time filters is a short program to prewarp the given specifications via equation (6.44), and another short program to transform the resulting S-plane singularities of the prototype continuous-time filter back to the z-plane via equations (6.45) and (6.46). Programs to accomplish this are available in FILTOR 2 [1].

Although we have shown how to realize a discrete filter transfer function using the cascade method, there are enormous other design details [9,10] that are beyond the scope of this book.

6.9 REFERENCES AND SOURCES FOR FURTHER READING

(1) W. M. Snelgrove and A. S. Sedra, FILTOR 2 - A Computer Aided Filter Design Package, Matrix Publishers, Champaign, Ill.

(2) A. I. Zverev, Handbook of Filter Synthesis, Wiley, New York, 1967.

(3) E. Christian and E. Eisenmann, Filter Design Tables and Graphs, Wiley, New York, 1966.

(4) J. K. Skwirzynski, Design Theory and Data for Electrical Filters, D. Van Nostrand, Princeton, N.J., 1965.

(5) R. W. Daniels, Approximation Methods for Electronic Filter Design, McGraw-Hill, New York, 1974.

(6) D. S. Humphreys, The Analysis, Design and Synthesis of Electrical Filters, Prentice-Hall, Englewood Cliffs, N.J., 1970.

(7) D. A. Smith, W. J. Butler and C. M. Puckette, "Programmable bandpass filter and tone generator using bucket-brigade delay lines", IEEE Trans. on Communications, Vol. COM-22, pp. 921-925, 1974.

(8) D. A. Smith, C. M. Puckette and W. J. Butler, "Active band-pass filtering with bucket-brigade delay lines", IEEE J. Solid-State Circuits, Vol. SC-7, pp. 421-425, 1972.

(9) A. V. Oppenheim and R. W. Schafer, Digital Signal Processing, Prentice-Hall, Englewood Cliffs, N.J., 1975.

(10) L. R. Rabiner and B. Gold, Theory and Application of Digital Signal Processing, Prentice-Hall, Englewood Cliffs, N.J., 1975.

PROBLEMS

P6.1 The LC ladder network of Fig. P6.1 realizes a LP filter which satisfies the normalized LPP tolerance plot. Find the component values such that f_p=100 KHz and R_s=R_L=1KΩ.

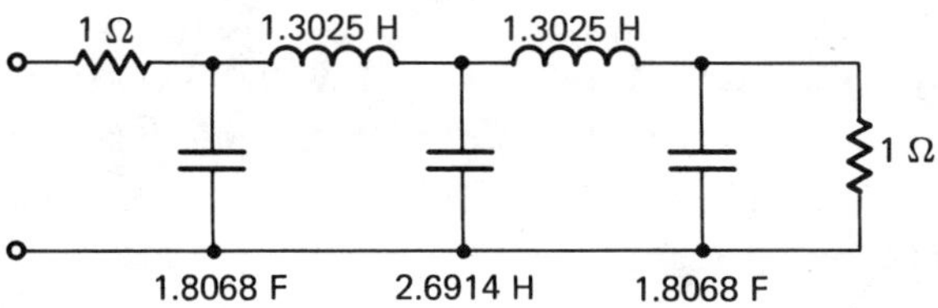

Figure P6.1

P6.2 A sixth order Chebyshev low-pass filter has A_p=1dB. The LC ladder realization has R_s=1Ω, find the value of R_L to obtain maximum power transfer.

P6.3 Find the dual of the ladder network in Fig. P6.3

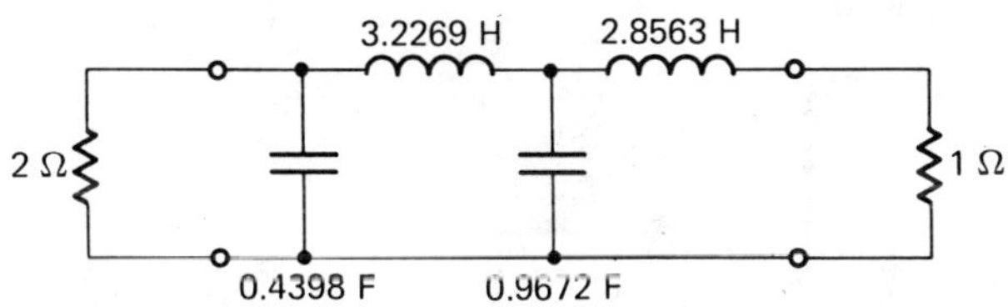

Figure P6.3

P6.4 Design a minimum inductance ladder network that realizes a 7th order Butterworth low-pass filter with f_p=1MHz, A_p=0.5dB and R_s=1KΩ.

P6.5 Design an LC ladder network that realizes a sixth order Chebyshev low-pass filter with f_p=100KHz, A_p=0.2dB and R_s=1KΩ. The ladder should be designed for maximum power transfer.

P6.6 Design an elliptic function filter that meets the LP specifications given in P4.18 and P4.5. The source resistance is to be 10KΩ. (Use filter tables)

P6.7 Find the natural modes (in terms of ω_o and Q), the transfer function and the LC ladder realization of a sixth-order bandpass filter having a maximally-flat response around 10 KHz and a 1dB bandwidth of 200 Hz. The terminations should be equal to 1KΩ. Find the passband edges and the attenuation at 12KHz.

P6.8 Find all pertinent parameters and an LC ladder realization for a bandpass filter that meets the specifications shown in Fig. P6.8. Use an elliptic function approximation and consult filter tables.

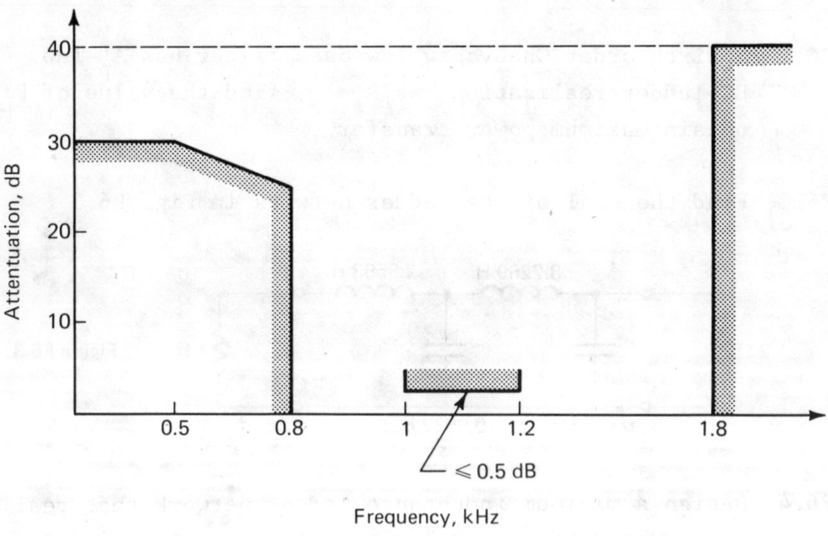

Figure P6.8

P6.9 Derive equations 6.21-6.24.

P6.10 Derive equation 6.27

P6.11 Find the reflection zeros and natural modes of a high-pass filter derived from a fifth order Chebyshev approximation. The passband edge is at 1 rad/sec and $A_p = 1$ dB.

P6.12 Design an LC ladder realization of a fifth order high-pass filter derived from a Chebyshev LPP. f_p=10KHz, A_p=1dB and the termination is 1KΩ.

P6.13 (a) Find the networks resulting from the application of the LP to BR transformation to (i) an inductance ℓ, and (ii) a capacitance c of the LPP.

(b) Find the LC ladder realization of a BR filter derived from a third-order Chebyshev LPP. The stopband extends from 3 to 4 KHz with a minimum attenuation of 30 dB. The ripple in the passbands is 1 dB. Sketch $A(\omega)$ and find the edges of the passbands.

P6.14 Use autotransformers to make all the capacitors in the LC ladder network of P6.7 equal to 10nF. Give a complete circuit schematic and specify all components.

P6.15 Verify the Norton's transformations of Fig. 6.21 by evaluating the z-parameters for the networks in (a) and the y-parameters for the network in (b).

P6.16 Verify the equivalences shown in Fig. 6.22.

P6.17 Use the Norton transformation of Fig. 6.21(b) at the line AA' in the filter shown in Fig. P6.17 to eliminate one capacitor. Give the value of the transformation constant n if:

(a) capacitor C_1 is to be eliminated, or

(b) capacitor C_3 is to be eliminated.

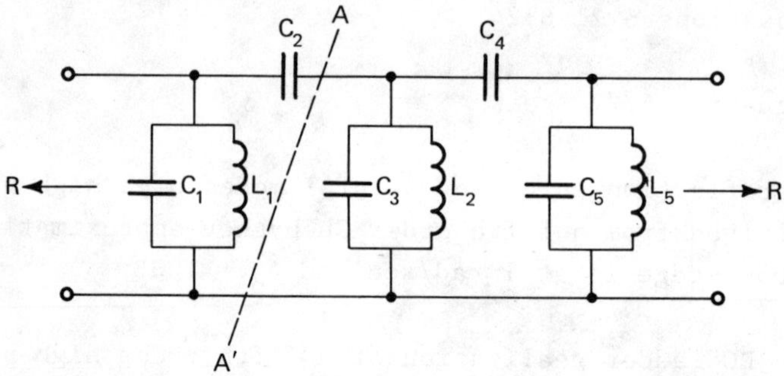

Figure P6.17

P6.18 The LC filter of Fig. P6.18 has $C_2 >> C_1$ and $L_2 << L_1$. It is required to reduce this spread in component values, without changing the impedance levels, through application of the Norton transformation of Fig. 6.21(b) as follows:

(a) Apply the transformation at AA`, with the transformer to the right of AA` and n=2, then

(b) apply the transformation at BB`, with the transformer to the right of BB` and n=0.5.

Find the final network.

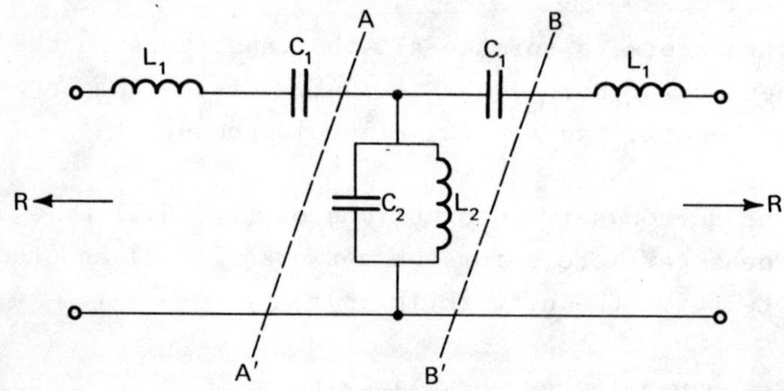

Figure P6.18

P6.19 Verify the Π to T transformations illustrated in Fig.P6.19 where:

$$\sum CC = C_1C_2+C_2C_k+C_kC_1$$

$$\sum L = L_1+L_2+L_k$$

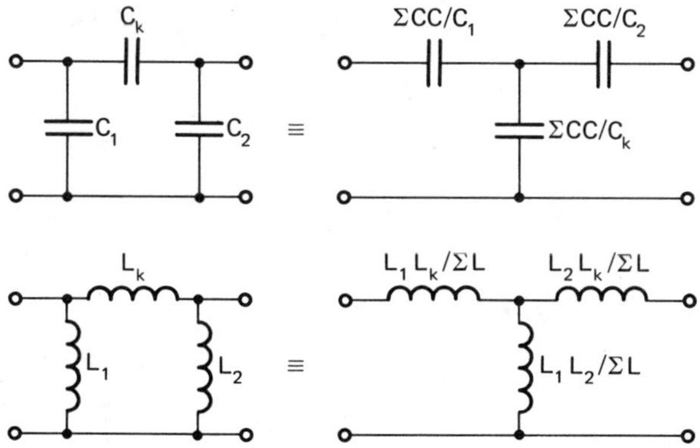

Figure P6.19

P6.20 A second order discrete filter uses a sampling frequency of 10 KHz. The z-plane poles and zeros are given, in polar form, by:

Zeros: $1 \angle 0$, $1 \angle 180^\circ$

Poles: $0.9 \angle 30^\circ$, $0.9 \angle -30^\circ$

Give $G_d(z)$ and sketch its magnitude for the frequency range 0 to 10KHz. Give a block diagram realization of the filter.

P6.21 A discrete-time low-pass filter has a fifth-order Chebyshev response with f_p=1000Hz and A_p=1dB. Find the reflection zeros and the transfer function $G_d(z)$. What is the attenuation at 1500Hz? Sketch $A_d(\omega)$. The sampling frequency is 5000Hz.

P6.22 Give a complete design of a discrete filter that meets the specifications in Fig. P6.8 using an elliptic function approximation. Sketch the resulting attenuation function. Use a sampling frequency of 5KHz.

P8.20 A one order discrete filter uses a sampling frequency of 10 kHz. The [illegible] poles and zeros are given, in polar form, by:

zeros: [illegible]

poles: [illegible]

Give G(z) and sketch its magnitude for the frequency range 0 to 10kHz. Give a block diagram realization of the filter.

P8.21 Find a [illegible] low-pass filter that has a fifth-order Chebyshev response [illegible] 1000Hz [illegible] dB. Find the reflection [illegible] points and the attenuation [illegible] is the attenuation [illegible] the sampling frequency is 5000Hz.

P8.22 Give a complete design of a discrete filter that meets the specifications in Fig. P8.6 using an elliptic function approximation. Sketch the resulting attenuation functions. Use a sampling frequency of [illegible].

CHAPTER 7

THE DESIGN OF REACTIVE LADDER FILTERS

7.1 INTRODUCTION

It is our purpose in this Chapter to present a concise treatment of the design techniques of LC ladder filter networks. Most general network synthesis courses and texts relegate this material to a small section because of the many other theoretically interesting synthesis methods which must be covered in an in-depth treatment. However, since most practical passive filter problems are solved by means of the ubiquitous doubly terminated reactive ladder, and since the present text emphasises the practical viewpoint, we choose to amplify much of the pertinent material. The reader having no previous contact with this material, or those requiring a more rigorous treatment of passive network synthesis, should consult the references at the end of this Chapter.

There are a number of reasons for the widespread use of the reactive (lossless or LC) ladder network as a filter realization method. The reactive ladder is a network structure which has a topology resembling that of a ladder, with the alternating series and shunt arms or branches made up of simple L or C elements, or simple combinations of these. The reader will recall that in a ladder network such as we are discussing here, the loss-poles (or transmission zeros) are directly related to the behaviour of the impedances of the ladder arms. In fact, the frequencies of the loss-poles usually occur at the frequencies of the impedance poles of the series arms and the frequencies of the admittance poles of the shunt arms. This one-to-one correspondence between the ladder arms and loss-poles is one reason for the popularity of ladders, since it greatly simplifies the tuning and adjustment of such networks.

A further advantage, to be explained in detail later, is that when a reactive network is designed to deliver maximum power from a resistive source to a resistive load over a given passband, then the sensitivity of the transfer function to changes in the element values of the lossless network is minimized. This is due to the fact that the network is lossless and any changes in the network elements can only cause the power delivered to the load at passband frequencies to decrease. This fact implies that, at passband frequencies, where distortion is usually critical in filter applications, the rate of change of average load power with respect to element changes is near zero.

The two advantages of reactive ladders pointed out above have led to the great popularity of such networks as a filter realization technique. The major disadvantage of the technique in general insertion loss designs is the difficulty in calculating the required element values from the transfer function parameters. In general, such calculations require the use of quite sophisticated computer programs. It is the design or synthesis of such reactive ladders, and the prediction of their imperfections, which is the subject of most of this Chapter. The material presented here will be utilized in the design of active filters which are derived from LC ladder prototypes, as will be seen in Chapters 11 and 12.

7.2 THE PASSIVE ELEMENTS R, C AND L

We begin our study by considering some of the characteristics of the passive elements of resistance R, capacitance C and inductance L. In particular, we shall stress the practical aspects of using these elements in filter networks, in terms of the parasitic effects encountered in practical devices. For the reader actually contemplating the physical construction of high quality filter networks, we most emphatically recommend consulting the component manufacturer's literature and specifications for his components.

Of the three passive elements to be described here, it is probably the resistance element which in practice can approach most closely the ideal behaviour assumed for it. The ideal resistance R is a device which presents in one lump a prescribed amount of resistance, expressed in Ohms, to electric current flow; the resistor is of course characterized in circuit theory terms by Ohm's Law as

$$R = \frac{v}{i}$$

where v is the instantaneous voltage across the device and i is the current flowing through it.

The next element which can, in practice, approach its ideal behaviour fairly closely is the capacitor. The ideal capacitor C is a device which provides in one lump a prescribed amount of charge storage proportional to the impressed voltage. Here, in circuit theory terms, we have the characterization

$$C = \frac{q}{v}$$

where q is the instantaneous amount of charge stored in the capacitor, v is the instantaneous voltage impressed upon it, and C is the capacitance in Farads. Of course, we may make use of the relation $i = (dq/dt)$ and characterize the ideal capacitor C in terms of an integral equation as

$$v = \frac{1}{C}\int i\,dt$$

which involves the more usual variables of current and voltage.

Finally, the inductance element is in practice by far the "worst" of the three elements in that it usually departs quite markedly from its assumed ideal behaviour. Here, in circuit theory terms, the ideal inductor is characterized by the differential equation

$$v = L\,\frac{di}{dt}$$

which describes the counter e.m.f. or voltage v produced by an ideal inductance of L Henries in response to the rate of change of current (di/dt) in amperes per second.

Returning now to discuss some aspects of the non-ideal behaviour of these passive devices, we first consider the resistance element. Of course, as with all three elements, the nominal value of the component will change with changing environmental conditions (e.g. temperature, humidity, pressure, etc.). Of these effects, probably the most important is the temperature dependence. This will usually be described by manufacturers in terms of a temperature coefficient which allows one to extrapolate the resistance value linearly from the nominal temperature to another. In high quality resistors, such temperature dependence is quite accurate, while in other types the temperature dependence of resistance is very nearly unpredictable.

At higher frequencies one must beware of resistors which have a physical construction that produces a considerable amount of reactance. In particular, wire-wound types and certain types of metal film resistors have a rather large inductive component.

Capacitors of various types are available from manufacturers and, in general, one can usually obtain (for a price!) capacitors that appear to be very nearly ideal for most filter applications. Of course, just as with resistors, good quality capacitors will have a well characterized temperature dependence, usually described by a linear temperature coefficient in the component specifications.

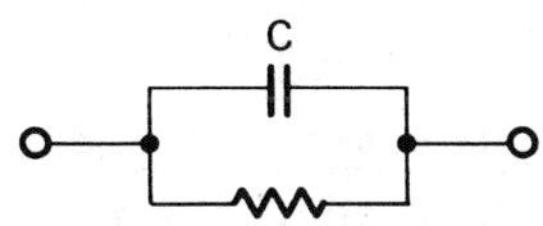

Figure 7.1 A dissipative capacitor.

Over most frequency ranges below 30 MHz one can obtain capacitors which have a negligible inductive component of reactance, but usually there is some resistive component in parallel with the capacitance. This resistance, depicted in Fig. 7.1, is usually caused by leakage across the capacitor dielectric. This resistance causes the capacitor to be slightly dissipative instead of being a purely lossless component. It has been noted that to a first approximation the parallel resistance is inversely dependent upon the size of the capacitor. Because of this, it has become common to express this "parasitic" dissipation in terms of the so called dissipation factor d_c where

$$G = d_c\, C$$

yields the conductance G in mhos of the parasitic resistance appearing in parallel with the capacitor. Clearly, the ideal capacitor will have $d_c = 0$. It is also common to express this dissipative effect in terms of a "quality factor" Q_c which is defined in terms of the ratio of the susceptance $B(\omega)$ at a particular frequency to the conductance G, thus

$$Q_c = \frac{B(\omega)}{G} = \left(\frac{\omega}{d_c}\right).$$

Finally, we should note that as of this writing, there are commercially available capacitors with dissipation factors as low as $d_c = 0.001$, which should indicate the degree of ideality that can be achieved in practice.

Inductors of good quality, for filter applications, are not generally available as "off-the-shelf" items and one must generally wind (or have manufactured) coils to the desired specifications. In this regard it is interesting to note that many neophite passive filter designers either are not aware of or have forgotten that in physical terms the inductance of a coil is directly proportional to the linear dimensions and proportional to the square of the number of turns. The general form for the inductance formula is

$$L = A_L n^2$$

where n is the number of turns and A_L is the so called "inductance factor" which is a function of the shape and dimensions of the coil as well as the magnetic permeability of the surrounding medium. Usually manufacturers of ferrite cores quote the value of A_L in Henries (or mH) per turn; this leads to erroneous results if one takes this number and calculates turns in direct proportion resulting in a filter which "won't work"! Some manufacturers also use the so-called "turns factor" α where

$$n = \alpha \sqrt{L}$$

yields the number of turns required.

In general, it is difficult if not impossible to derive analytic formulae for the inductance of coils of arbitrary shapes and winding details. The reader will recall that deriving such formulae from fundamental principles involves the determination of surface integrals which become nearly impossible to evaluate for any but the simplest of shapes (e.g. a straight wire or loop). However, much empirical work has been done to obtain approximate formulae and tables of correction factors for a wide variety of coils which have physical shapes and characteristics falling into specific and commonly used categories. For instance, the approximate inductance of a single-layer air-core coil may be calculated from the simple formula

$$L = \frac{a^2n^2}{9a + 10b}, \quad \text{(note here that } A_L = \frac{a^2}{9a + 10b}\text{)} ,$$

where L is the inductance in microHenrys (μH), a is the coil radius in inches, b is the coil length in inches, and n is the number of turns. Here, the wire diameter must be small relative to the coil diameter, the length b should be greater than 80% of the radius a, and the turns should be spaced out to fill the specified length.

For more complicated shapes and other core materials, the reader is referred to the references at the end of this Chapter. In special and unique cases, the only quick method of obtaining inductance formulae is by experimentally winding several coils of similar physical characteristics but having different numbers of turns, and measuring the inductance to obtain an experimental value for A_L which can then be used to wind coils of specified values. This latter problem explains the popularity of the commercially available ferrite cores for which the manufacturer supplies the required A_L or α values from which the coil design proceeds easily.

As mentioned previously, practical inductors are the least ideal of the three passive elements and they exhibit several parasitic effects which we shall briefly discuss. First, we note that, as with the other elements, there will be a temperature dependence of inductance since the physical dimensions as well as magnetic permeability, if a magnetic core is used, will vary with temperature. Again here in the case of ferrite cores (torroid and pot types), the manufacturer will usually specify a temperature coefficient which may be used to linearly extrapolate the inductance values for temperature variations from the nominal. The reader might note the possibility of choosing capacitors with a temperature coefficient complementary to that of inductors so that a measure of temperature independence might be obtained for LC filter networks.

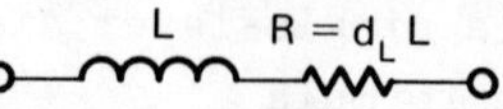

Figure 7.2 A dissipative inductor.

Usually, the most non-ideal property of an inductor is the effect of parasitic dissipation or resistance. In circuit terms this effect appears as a resistance (in general frequency dependent) in series with the coil as depicted in Fig. 7.2. Here the resistance is usually the result of two phenomenae. The first is the actual "copper loss" or resistance of the wire used to wind the coil. This loss will be frequency dependent, increasing with frequency due to the well known "skin effect". This component of resistance is not very level-dependent and will remain constant, at a given frequency, independent of the current level in the inductor. The second component is the so-called "core loss" which is due to hysteresis in the ferrite core material. This effect exhibits both frequency and level dependence, the loss increasing with both frequency and level. In addition, both the core loss and the copper loss are an increasing function of the number of turns used to produce the inductance. This, of course, depends in turn on the A_L factor chosen for the coil design.

The effect of this parasitic series resistance can be described, over a small frequency range, for a given coil configuration by means of a "dissipation factor" d_L where the series resistance R is very nearly proportional to the inductance, thus

$$R = d_L\ L\ .$$

Again here the "quality factor" Q_L of the inductor is given by the ratio of the inductive reactance to the effective resistance, so that

$$Q_L = \frac{X(\omega)}{R} = \left(\frac{\omega}{d_L}\right)\ .$$

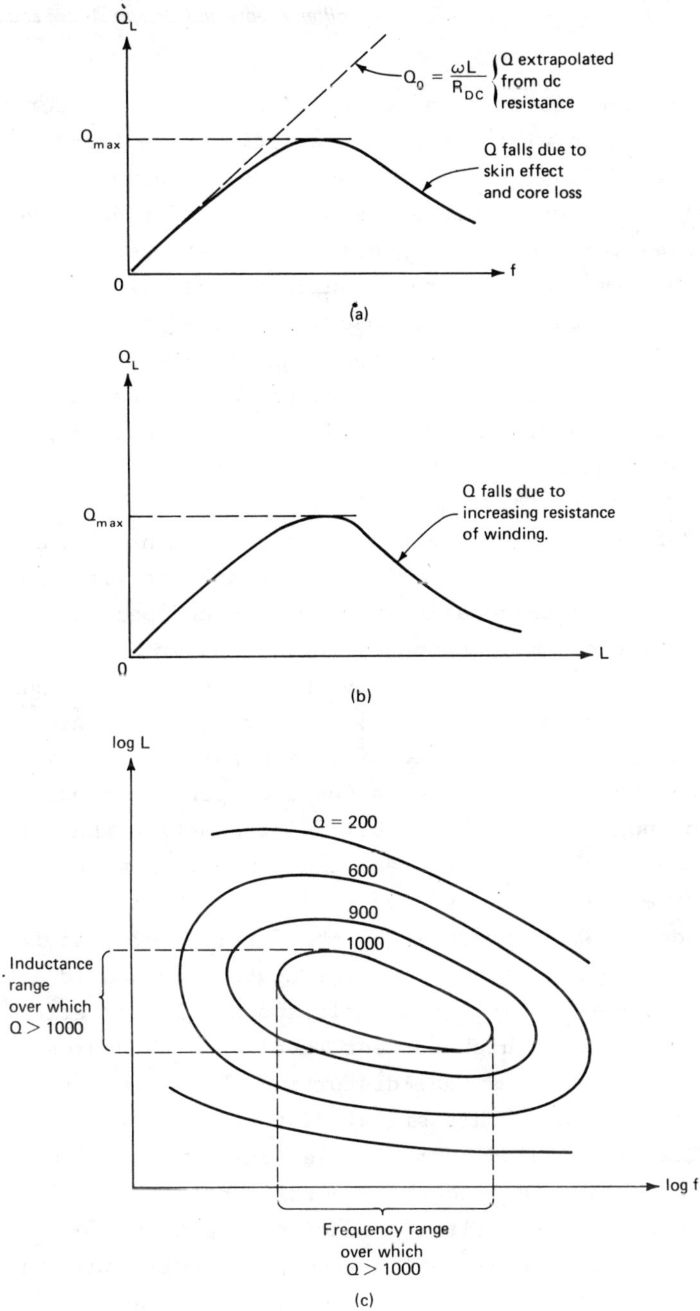

Figure 7.3 Illustrating: (a) the dependence of inductor Q upon frequency; (b) the dependence of inductor Q upon inductance; and (c) a set of constant – Q curves for a practical inductor with ferrite core

From our previous discussion, we should note that Q_L for a given coil configuration will exhibit a variation with frequency as shown in Fig. 7.3(a), and with inductance as shown in Fig. 7.3(b). Many ferrite core manufacturers supply sets of such curves with various cores for use by designers. The most useful design aid (supplied by some manufacturers) which may be fairly easily prepared by the designer are the so-called "iso-Q curves" or curves of constant Q versus both inductance and frequency. An example of such a set of curves having Q values as a parameter are illustrated in Fig. 7.3(c). As the reader can see, such a presentation of design variables for each type of ferrite core enables one to determine rapidly and efficiently the "best" core for a particular filter design. Finally, we might mention that as of this writing there are commercially available ferrite cores (on which coils having reasonable inductances and operated at reasonable current levels may be wound) for the upper and super audio frequency ranges, which can achieve Q values in the neighbourhood of 1500 to 2000. Of course, these values are only possible under a relatively small range of design parameters which are dictated by the filter configuration and specifications. The reader can appreciate from our brief introduction that inductor design is an involved subject and those desiring more information are directed to the references and to the excellent handbooks and design aids produced by the ferrite core manufacturers.

Before leaving our discussion of inductors we should like to point out another highly undesirable and non-linear effect caused by core saturation at high current levels, which causes both a rapid Q collapse and a signal distortion which tends to flatten out the signal peaks. This signal distortion is fairly well characterized by a third harmonic distortion measurement. The filter designer should note that overdriving his "passive" filter will result in the filter producing output at the third harmonic of the input signals which might very well cause him embarrassment since the "passive" filter is not intended to do this.

Another parasitic effect in inductors is that of stray capacitance caused by the capacitance between adjacent turns and layers of the coil winding. Special winding techniques may minimize this phenomena, but it is always present to some extent. A simple circuit model of the parasitic dissipation and capacitive effects taken together is shown in Fig. 7.4. In the case of narrow-band bandpass filters, the effect of the stray shunt capacitance may be compensated for by suitably offsetting the design value of inductance so that the "apparent" inductance across the narrow passband approximates the desired value. In the case of wideband filters, this effect becomes difficult to deal with unless one obtains a network design in which there are actually capacitances present across the inductors into which the parasitic capacitances may be absorbed.

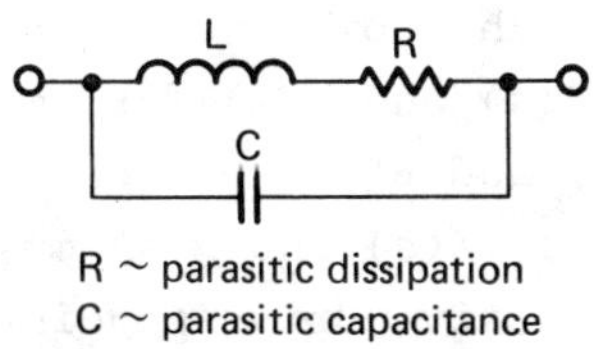

Figure 7.4 A suitable model for a practical inductor.

Before closing this section we should like to introduce some terminology regarding the effect of parasitic dissipation. In case all reactive components in a network have identical dissipation factors $d_c = d_L = d$, the network is said to be "uniformly dissipative". The reader will note that in this case replacing the Laplace transform variable s everywhere by (s + d) in all network functions of the ideal non-dissipative network will result in the network functions for the uniformly dissipative network. The case where all inductors have identical dissipation factors d_L, while all capacitors have identical dissipation factors d_c but $d_L \neq d_c$ is described as "semi-uniform dissipation". In this case, one can often make the assumption of uniform dissipation with $d = (d_L + d_c)/2$ and obtain a reasonably close approximation to the network functions for the dissipative network by substituting (s + d) for s in the ideal network functions. The case where all components have different dissipation factors (the most realistic situation) is described as "non-uniform dissipation".

7.3 REACTANCE FUNCTIONS

It will be assumed that the reader has a passing familiarity with reactance (and susceptance) functions and the associated synthesis methods. Our intent here is to briefly review this material and to organize it in a fashion helpful to the understanding of the synthesis of lossless two-port ladder networks.

Properties of LC Driving Point Functions

The driving point impedance (reactance) and the driving point admittance (susceptance) of lossless (LC) networks are rational polynomial functions of the complex frequency variable. We shall use the usual symbols Z(s) and Y(s) for impedance and admittance, respectively, with X(s) and B(s) for reactance and susceptance, respectively. Of course, X(s) and B(s) must be positive real functions, as are all passive driving point functions. In addition to this, lossless driving point functions have the following restrictions. The lossless driving point functions must be odd rational functions; that is, they must satisfy the relations

$$X(s) = -X(-s)$$

$$B(s) = -B(-s)$$

This in turn implies that the functions X(s) and B(s) are ratios of odd to even polynomials or vice versa. Of course, the numerator and denominator polynomial degrees can differ by at most one since the functions are positive real, and since they are ratios of even to odd or odd to even polynomials, the degrees of numerator and denominator must always differ by one. Also, all of the poles of these functions are simple with real and positive residues, and occur only on the imaginary axis of the s-plane. Since $B = 1/X$ has similar properties, all of the zeros of these functions are simple and occur only on the imaginary axis. The positive slope property, $\frac{\partial X(\omega)}{\partial \omega} > 0$, and $\frac{\partial B(\omega)}{\partial \omega} > 0$, and the fact that the poles and zeros must always be interlaced on the imaginary axis easily follow from the simple pole and the positive real residue conditions.

The Four Classes of Reactance (Susceptance) Functions

We now introduce a classification of reactance functions based upon the behaviour of the functions at zero and infinite frequency. First, we indicate the two general forms in which a reactance (susceptance) function may occur in terms of the parity of its numerator and denominator. It must be either the ratio of an odd to an even polynomial as indicated by

$$X(s) = \frac{sP(s^2)}{Q(\varepsilon^2)}$$

or the ratio of an even to an odd polynomial represented by

$$X(s) = \frac{P(s^2)}{sQ(s^2)} .$$

Here, P and Q must both be even polynomials as indicated by showing their arguments explicitly as s^2. In each of the above cases, we have two different possibilities for the relative degrees of the numerator and denominator. This gives rise to four forms or classes of reactance (susceptance) functions, classified according to the parity and degree of the numerator and denominator polynomials. These classes may in turn be related to the behaviour of X(s) at the extreme frequencies of zero and infinity; i.e. X(0) and X(∞).

<u>Class 0:</u>

Class 0 has zeros at both s = 0 and s = ∞, so that X(0) = X(∞) = 0. The function must be of the form

$$X_0(s) = k \frac{s \prod_{m=1}^{N-1} (s^2 + \omega_{zm}^2)}{\prod_{n=1}^{N} (s^2 + \omega_{pn}^2)}$$

which has an odd numerator of a degree less than the degree of the even denominator. Using H and L for higher and lower degrees, respectively, and E and O for even and odd respectively, we can mnemonically represent the Class 0 function as,

$$X_0 = (\frac{O,L}{E,H}).$$

Class 1:

A class 1 function has a pole at zero, and a zero at infinity; i.e. $X(\infty) = 0$, and $X(0) = \infty$. The reader may now recognize the binary numbering scheme we are using here for the four classes. The Class 1 function is of the form

$$X_1(s) = k \frac{\prod_{m=1}^{N} (s^2 + \omega_{zm}^2)}{s \prod_{n=1}^{N} (s^2 + \omega_{pn}^2)}$$

which has an even numerator of a degree less than the degree of the odd denominator. Using our mnemonic scheme, we have

$$X_1 = (\frac{E,L}{O,H})$$

Class 2:

Class 2 has a pole at infinity, and a zero at zero; i.e. $X(\infty) = \infty$, and $X(0) = 0$. Here the form of X(s) is

$$X_2(s) = k \frac{s \prod_{m=1}^{N} (s^2 + \omega_{zm}^2)}{\prod_{n=1}^{N} (s^2 + \omega_{pn}^2)}$$

which has an odd numerator of a degree greater than the degree of the even denominator. This may be expressed in our mnemonic scheme as

$$X_2 = \left(\frac{O,H}{E,L}\right)$$

Class 3:

The final class has a pole at each of zero and infinity; i.e. $X(\infty) = \infty$, and $X(0) = \infty$. The general form is

$$X_3(s) = k \frac{\prod_{m=1}^{N} (s^2 + \omega_{zm}^2)}{s \prod_{n=1}^{N-1} (s^2 + \omega_{pn}^2)}$$

which has an even numerator of a degree greater than the degree of the odd denominator. In our mnemonic scheme this may be expressed as

$$X_3 = \left(\frac{E,H}{O,L}\right)$$

These four classes (0 through 3) represent the four forms into which reactance (or susceptance) functions may be classified according to the parities and degrees of numerator and denominator or, equivalently, their asymptotic behaviour at extremes of frequency. Table 7.1 provides a summary of the four classes of reactance (susceptance) functions.

Canonic Realizations of LC Driving-Point Functions

Any reactance (susceptance) function may be expanded in partial fractions in terms of the poles of the function. The general form of the expansion for any LC driving point function $F_{LC}(s)$ is

$$F_{LC}(s) = k_\infty s + \frac{k_o}{s} + \sum_{n=1}^{M} \frac{k_n s}{(s^2 + \omega_n^2)}$$

where k_∞ is the residue of the pole at infinity, k_o is the residue of the pole at zero, k_n is twice the residue of the pole at $s = j\omega_n$, and $F_{LC}(s)$ may be either X(s) or B(s). We note from this expansion that when $F_{LC}(s)$ is a reactance, we may consider it to be the sum of elementary reactances (or impedances); i.e.

$$X(s) = z_1 + z_2 + z_3 + \ldots + z_m .$$

Table 7.1 Summary of the four classes of reactance (susceptance) functions.

Class	Form	Behaviour at Extreme Freq's. F(∞)	F(0)	Parities and Degrees
0	$\dfrac{ks \prod_{n=1}^{N-1} (s^2 + \omega_{zn}^2)}{\prod_{m=1}^{N} (s^2 + \omega_{pm}^2)}$	0	0	$(\frac{O,L}{E,H})$
1	$\dfrac{k \prod_{n=1}^{N} (s^2 + \omega_{zn}^2)}{s \prod_{m=1}^{N} (s^2 + \omega_{pm}^2)}$	0	∞	$(\frac{E,L}{O,H})$
2	$\dfrac{ks \prod_{n=1}^{N} (s^2 + \omega_{zn}^2)}{\prod_{m=1}^{N} (s^2 + \omega_{pm}^2)}$	∞	0	$(\frac{O,H}{E,L})$
3	$\dfrac{k \prod_{n=1}^{N} (s^2 + \omega_{zn}^2)}{s \prod_{m=1}^{N-1} (s^2 \quad \omega_{pm}^2)}$	∞	∞	$(\frac{E,H}{O,L})$

where the individual impedances are given by the terms of the partial fraction expansion. This in turn may be recognized as the driving point impedance of a series connection of sub-networks as illustrated in Fig. 7.5.

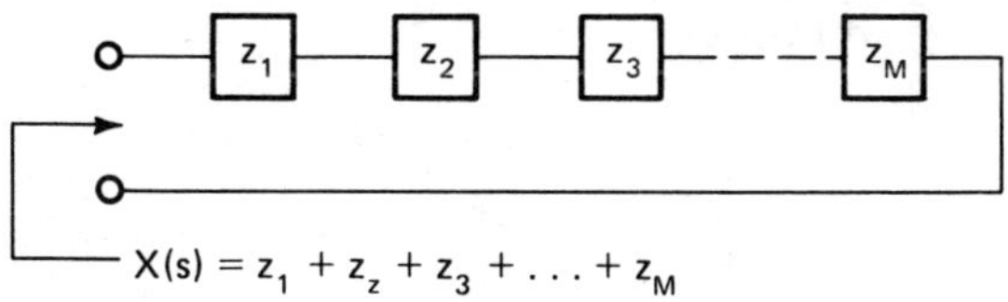

Figure 7.5 A series connection of subnetworks.

Similarly, if $F_{LC}(s)$ is a susceptance, we may consider the partial fraction expansion to be the sum of elementary susceptances (or admittances); i.e.

$$B(s) = y_1 + y_2 + y_3 + \ldots + y_m$$

which may be recognized as the driving point admittance of a parallel connection of sub-networks, as illustrated in Fig. 7.6. The reader may recognize these two forms as the well known Foster Forms of canonic LC driving point function realization.

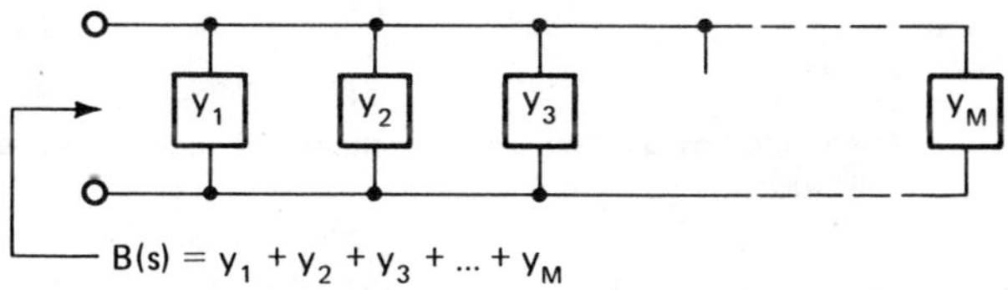

Figure 7.6 A parallel connection of subnetworks.

In particular, we may write the impedance form as

$$X(s) = L_\infty s + \frac{1}{C_o s} + \sum_{n=1}^{M} \frac{1}{[C_n s + \frac{1}{L_n s}]} .$$

which corresponds to the series connection of a coil of value L_∞, a capacitor of value C_o, and M tank circuits each of which is a parallel one resonant at one of the poles of X(s). This is the general character of Foster's First Form for a canonic LC driving point function realization, and it is illustrated in Fig. 7.7. Similarly we may write B(s) as

$$B(s) = C_\infty s + \frac{1}{L_o s} + \sum_{n=1}^{M} \frac{1}{[L_n s + \frac{1}{C_n s}]} .$$

which is the general character of Foster's Second Form for a canonic LC driving point function realization, illustrated in Fig. 7.8. The most interesting aspect of these Foster forms for realizing a driving point function is that they place the sub-networks in a one-to-one correspondence with the poles of the network driving point function.

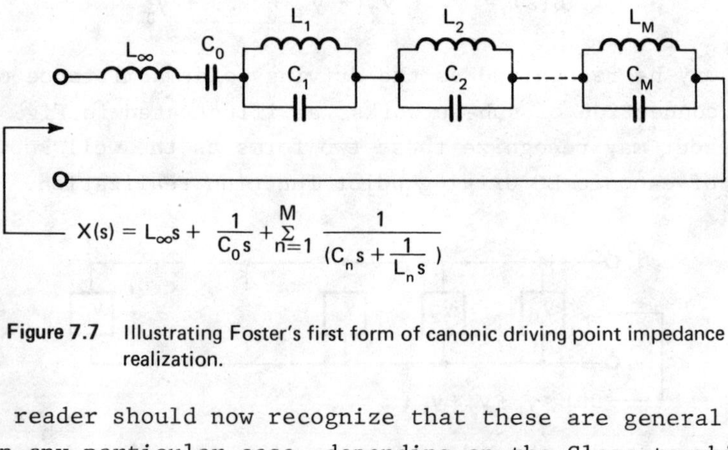

$$X(s) = L_\infty s + \frac{1}{C_0 s} + \sum_{n=1}^{M} \frac{1}{(C_n s + \frac{1}{L_n s})}$$

Figure 7.7 Illustrating Foster's first form of canonic driving point impedance realization.

The reader should now recognize that these are general forms and in any particular case, depending on the Class to which X or B belongs, some or all of the elements L_∞, C_∞, C_o, and L_o may be missing.

Since Class 0 has no poles at zero or infinity its impedance expansion will be

$$X_0(s) = \sum_n \frac{1}{[C_n s + \frac{1}{L_n s}]} .$$

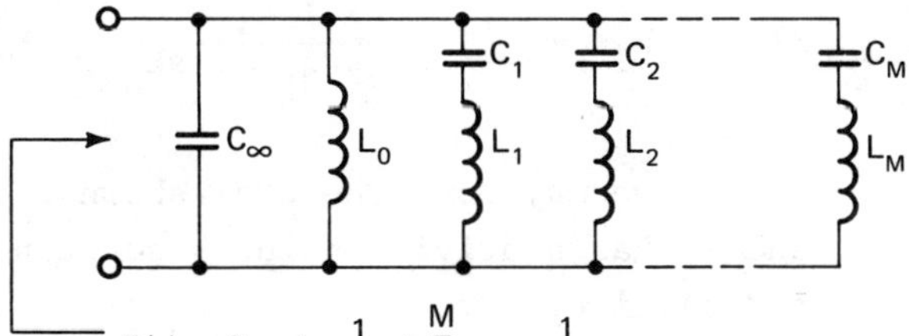

$$B(s) = C_\infty s + \frac{1}{L_0 s} + \sum_{n=1}^{M} \frac{1}{L_n s + \frac{1}{C_n s}}$$

Figure 7.8 Illustrating Foster's second form of canonic driving point admittance realization.

The reader may easily verify that the impedance expansions for the remaining three classes are of the forms

$$X_1(s) = \frac{1}{C_o s} + \sum_n \frac{1}{[C_n s + \frac{1}{L_n s}]}$$

$$X_2(s) = L_\infty s + \sum_n \frac{1}{[C_n s + \frac{1}{L_n s}]}$$

$$X_3(s) = L_\infty s + \frac{1}{C_o s} + \sum_n \frac{1}{[C_n s + \frac{1}{L_n s}]}$$

Similarly, there are four possible forms for Foster's Second Form of (admittance) expansion. In particular, if we invert a reactance of Class 0, we obtain a susceptance of Class 3, a reactance of Class 1 leads to a susceptance of Class 2, a reactance of Class 2 leads to susceptance of Class 1, and finally a reactance of Class 3 leads to a susceptance of Class 0.

The reader will recall that the driving point functions of two networks which are reciprocal are related by $z_1 \; z_2 = 1$ or $y_1 \; y_2 = 1$, and that dual networks are reciprocal. We may obtain the dual of a given network by inspection without the need for any computation. For example, given Z(s) in the form

$$Z(s) = L_1 s + \frac{1}{sC_2} + \frac{1}{[sC_3 + 1/sL_4]}$$

shown in Fig. 7.9(a), we may form the dual shown in Fig. 7.9(b) by inspection, and it has a driving point susceptance equal to the reactance of Z(s); i.e.

$$Y(s) = C_a s + \frac{1}{L_b s} + \frac{1}{[sL_c + 1/sC_d]}$$

where $C_a = L_1$, $L_b = C_2$, $L_c = C_3$, and $C_d = L_4$.

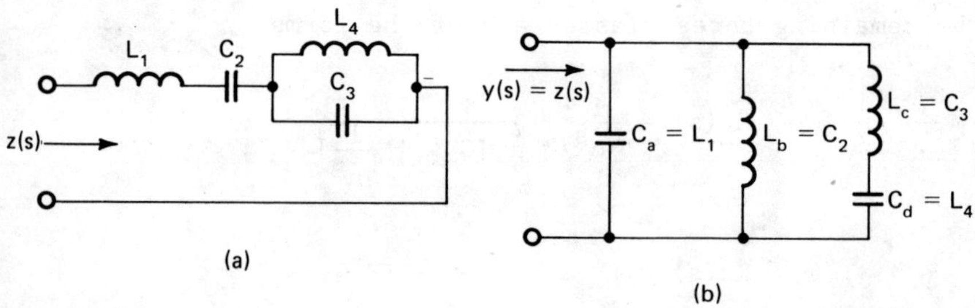

Figure 7.9 Illustrating dual networks.

The reader may now note that the Second Foster Forms of Fig. 7.8, are duals of the First Foster Forms of Fig. 7.7, and that there are in fact only four distinct classes of driving point functions when characterized in terms of the behaviour of the reactance at the extremes of frequency $s = 0$ and $s = \infty$. However, there may be many different realizations of these four classes, and the reader may recall the Cauer forms and the so-called fifth canonical form of LC driving point realization. The general character of Cauer's first and second forms, and the so-called fifth canonical form are illustrated in Fig. 7.10. These forms may be obtained by continued fraction expansion of X(s) and B(s) about $s = 0$, $s = \infty$ or alternately $s = 0$, and $s = \infty$. Alternatively, they may be obtained by a pole removal process described below.

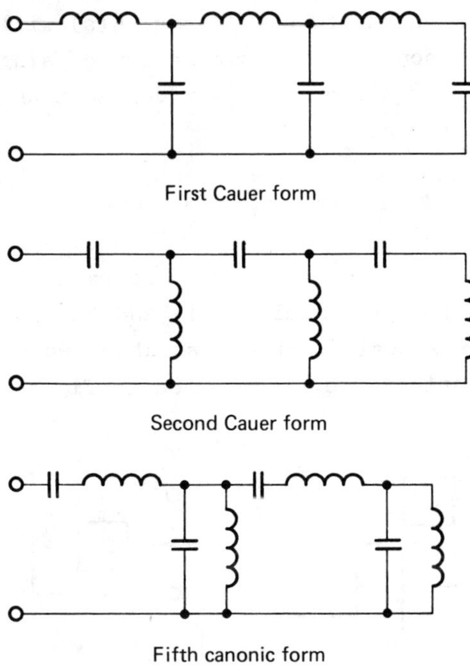

Figure 7.10 Illustrating the general character of some canonic forms of LC driving point function realizations.

7.4 POLE REMOVAL OPERATIONS

The fundamental operation of passive network synthesis is the pole removal operation. This operation consists of recognizing that a given driving point function $F_1(s)$ may be expressed as the sum of an "elementary" function $E(s)$, which represents a particular term in the partial fraction expansion of $F_1(s)$, and a "remainder" function $F_2(s)$. Thus,

$$F_1(s) = E(s) + F_2(s)$$

may be interpreted as the series connection of sub-networks of impedances $E(s)$ and $F_2(s)$ if $F_1(s)$ is an impedance, or as the parallel connection of sub-networks of admittances $E(s)$ and

$F_2(s)$ if $F_1(s)$ is an admittance. Here, $E(s)$ will represent a single element or some simple combination of single elements, and usually $F_2(s)$ is a function of lower degree than $F_1(s)$. We naturally compute $F_2(s)$ as

$$F_2(s) = F_1(s) - E(s),$$

and refer to this as a "pole removal" operation. Next, a pole is removed from $F_2(s)$ or its reciprocal, and the process is continued until the residual function is exhausted. The corresponding network operations are illustrated in Fig. 7.11.

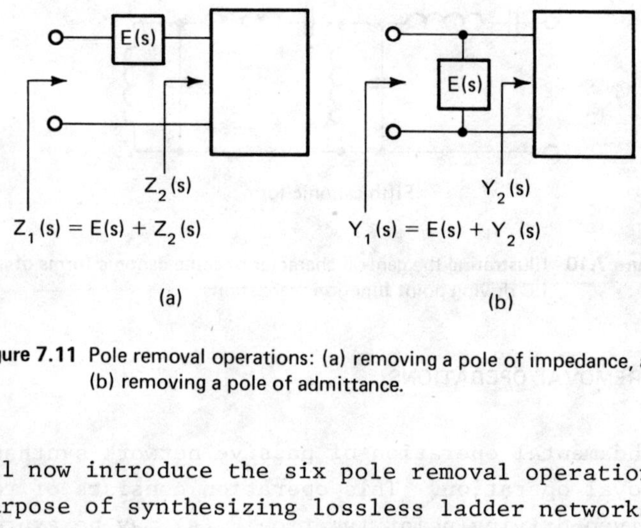

Figure 7.11 Pole removal operations: (a) removing a pole of impedance, and (b) removing a pole of admittance.

We shall now introduce the six pole removal operations central to our purpose of synthesizing lossless ladder networks.

(A) <u>The removal of a pole of impedance at s = 0.</u>

This is accomplished by removing a series capacitor from the pertinent reactance. The removal equation of interest is*

$$X_1(s) = \frac{1}{sC} + X_2(s).$$

---

*Here the subscripts 1 and 2 in X_1 and X_2 <u>do not</u> denote reactance classes.

The network operation is illustrated in Fig. 7.12. Clearly, as s approaches zero, the reactance $X_1(s)$ approaches that of the capacitance so that taking limits on both sides we may write

$$\lim_{s\to 0} X_1(s) = \lim_{s\to 0} \frac{1}{sC} .$$

Multiplying both sides through by s, and inverting, we obtain the expression for C as

$$C = \lim_{s\to 0} \left[\frac{1}{s\,X_1(s)}\right] .$$

Obviously, C is the reciprocal of the residue of the pole of $X_1(s)$ at $s = 0$. The remaining reactance $X_2(s)$ is given by

$$X_2(s) = X_1(s) - \frac{1}{sC} .$$

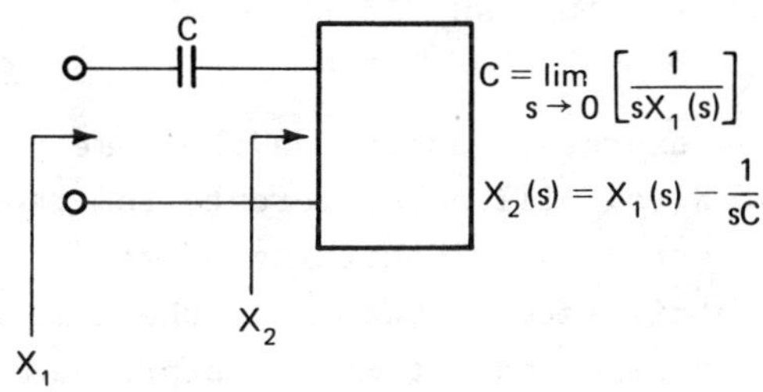

Figure 7.12 The removal of a pole of impedance at s = 0.

This removal operation may only be applied to reactances of Class 1 or Class 3, which are the only two classes with poles at zero (see Table 7.1). These classes of functions have in common the property that they are both ratios of an even to an odd poly nomial. Using the notation of Table 7.1, we have for Class 1

$$X_1(s) = \frac{k \prod_{n=1}^{N} (s^2 + \omega_{zn}^2)}{s \prod_{m=1}^{N} (s^2 + \omega_{pm}^2)}$$

Here the value of C is obtained as

$$C = \lim_{s\to 0} \frac{s\prod_{m=1}^{N}(s^2+\omega_{pm}^2)}{ks\prod_{n=1}^{N}(s^2+\omega_{zn}^2)} = \frac{\prod_{m=1}^{N}\omega_{pm}^2}{k\prod_{n=1}^{N}\omega_{zn}^2} ,$$

and X_2 is computed as

$$X_2(s) = \frac{k\prod_{n=1}^{N}(s^2+\omega_{zn}^2)}{s\prod_{m=1}^{N}(s^2+\omega_{pm}^2)} - \frac{1}{sC} ,$$

which when brought to a common denominator leaves

$$X_2(s) = \frac{Ck\prod_{n=1}^{N}(s^2+\omega_{zn}^2) - \prod_{m=1}^{N}(s^2+\omega_{pm}^2)}{sC\prod_{m=1}^{N}(s^2+\omega_{pm}^2)} .$$

Clearly, $X_2(s)$ as expressed above has the same poles as $X_1(s)$, but the zeros of $X_2(s)$ will be different from those of $X_1(s)$. This is often referred to as "shifting of zeros" and will be discussed in more detail later. Examining the numerator of $X_2(s)$ in more detail, we find that it may be expressed as

$$\left[\frac{\prod_{m=1}^{N}\omega_{pm}^2}{\prod_{n=1}^{N}\omega_{zn}^2}\prod_{n=1}^{N}(s^2+\omega_{zn}^2) - \prod_{m=1}^{N}(s^2+\omega_{pm}^2)\right] .$$

The lowest order coefficient here will be the constant term which we find, by letting s = 0, is zero; i.e.

$$\left(\frac{\prod\omega_{pm}^2}{\prod\omega_{zn}^2}\right)\left(\prod\omega_{zn}^2\right) - \prod\omega_{pm}^2 = 0$$

Thus, since the numerator of $X_2(s)$ is an even polynomial with no constant term and the highest coefficient of s is of degree 2N, we may factor s^2 from the numerator. The remaining expression is of the form

$$X_2(s) = \frac{\alpha s^2 \prod_{\ell=1}^{N-1} (s^2 + \omega_{z\ell}^2)}{s \prod_{m=1}^{N} (s^2 + \omega_{pm}^2)}$$

where the $\omega_{z\ell}$ are the new roots of the numerator polynomial, and α is the new constant into which we have incorporated the value of C. Clearly, we may divide numerator and denominator by s leaving $X_2(s)$ as a function with no pole at s = 0. Thus, $X_2(s)$ may be expressed as

$$X_2(s) = \frac{\alpha s \prod_{\ell=1}^{N-1} (s^2 + \omega_{z\ell}^2)}{\prod_{m=1}^{N} (s^2 + \omega_{pm}^2)} ,$$

which we note is a Class 0 reactance. $X_2(s)$ is a function of order 2N which, upon comparison with the order of $X_1(s)$ which is 2N + 1, we note is of one degree lower.

Thus we find that if we remove a pole of impedance by means of a series capacitor from a Class 1 reactance, we are left with a Class 0 reactance of one degree less. We shall refer to this as a class <u>converting</u> removal since it <u>converts</u> the class of the reactance from which it is removed to another class.

Let us now examine what happens when we remove a pole of impedance at s=0 from the only other allowed class which is Class 3. Here X_1 is of the form (E,H/O,L), which may be expressed as (see Table 7.1),

$$X_1(s) = \frac{k \prod_{n=1}^{N} (s^2 + \omega_{zn}^2)}{s \prod_{m=1}^{N-1} (s^2 + \omega_{pm}^2)}$$

We compute C as

$$C = \lim_{s\to 0} \frac{s \prod_{m=1}^{N-1} (s^2 + \omega_{pm}^2)}{ks \prod_{n=1}^{N} (s^2 + \omega_{zn}^2)} = \frac{\prod_{m=1}^{N-1} \omega_{pm}^2}{k \prod_{n=1}^{N} \omega_{zn}^2}$$

In this case, X_2 is given by

$$X_2(s) = \frac{k \prod_{n=1}^{N} (s^2 + \omega_{zn}^2)}{s \prod_{m=1}^{N-1} (s^2 + \omega_{pm}^2)} - \frac{1}{sC} .$$

Which when reduced to a common denominator yields

$$X_2(s) = \frac{Ck \prod_{n=1}^{N} (s^2 + \omega_{zn}^2) - \prod_{m=1}^{N-1} (s^2 + \omega_{pm}^2)}{sC \prod_{m=1}^{N-1} (s^2 + \omega_{pm}^2)} .$$

Again, the lowest order coefficient of the numerator of $X_2(s)$ is found to be zero. Subsequently factoring out s^2 in the numerator, and dividing s from both numerator and denominator leaves us with the form

$$X_2(s) = \frac{\alpha s \prod_{\ell=1}^{N-1} (s^2 + \omega_{z\ell}^2)}{\prod_{m=1}^{N-1} (s^2 + \omega_{pm}^2)}$$

We note from Table 7.1 that $X_2(s)$ is of Class 2, and comparing degrees with X_1 we find that X_2 is of one degree lower than X_1.

Summarizing the above results, we find that we may extract a series capacitor from either a Class 1 or a Class 3 reactance, and after the removal we are left with a remainder reactance of lower degree, and of Class 0 or Class 2. Thus, removing a series capacitor <u>converts</u> a Class 1 reactance to a Class 0 reactance or it converts a Class 3 reactance to a Class 2 reactance, i.e. it decreases the class number by one. Finally, we should note that if the series C branch is part of a two-port ladder network, the series C will produce a loss-pole (or transmission zero) at s=0, which is the s-plane location of the pole of impedance removed by the series C.

(B) The removal of a pole of admittance at s = 0.

This removal is accomplished by removing a shunt coil from the pertinent susceptance. The network operation is depicted in Fig. 7.13. It is left as an exercise for the reader to show that L is obtained as

$$L = \lim_{s\to 0}\left[\frac{1}{sB_1(s)}\right] = \lim_{s\to 0}\left[\frac{X_1(s)}{s}\right]$$

where $X_1(s) = 1/B_1(s)$ is the reactance from which the shunt coil is being removed, and to show that in terms of reactances, the removal of a pole of admittance by means of a shunt coil may be applied only to a reactance of Class 2 or Class 0 which is then converted to a remainder reactance of Class 3 or Class 1, i.e. it increases the class number by one. Finally we should note that if the shunt coil were part of a lossless ladder two-port network, it would produce a loss-pole or transmission zero at s = 0, the s-plane location of the pole of admittance being removed. Thus the two removal operations just described, namely, (A) the removal of a series C, and (B) the removal of a shunt L will produce loss-poles at zero in a two-port ladder network.

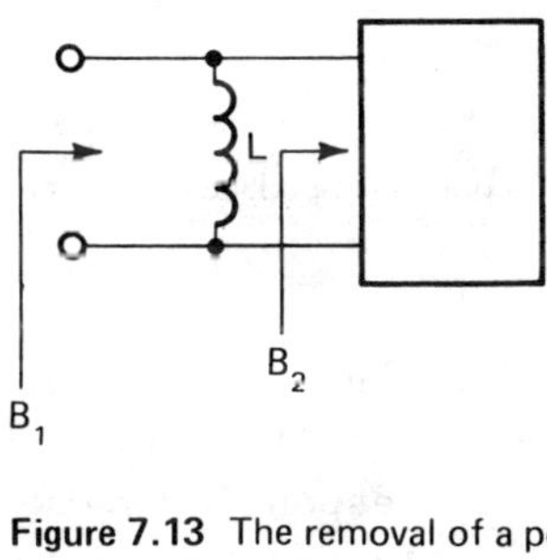

$$B_1 = \frac{1}{X_1};\ B_2 = \frac{1}{X_2}$$

$$L = \lim_{s\to 0}\left[\frac{X_1(s)}{s}\right]$$

$$B_2(s) = B_1(s) - \frac{1}{sL}$$

Figure 7.13 The removal of a pole of admittance at s = 0.

(C) The removal of a pole of impedance at $s = \infty$.

This removal is accomplished by removing a series coil from the pertinent reactance. The network operation is illustrated in Fig. 7.14. It is left for the reader to show that

$$L = \lim_{s\to\infty} [\frac{X_1(s)}{s}] ,$$

and that this removal operation may only be applied to reactances of classes which have poles at infinity, namely Class 2 and Class 3. We note that these two classes have in common the property that their numerator is of higher degree than their denominator.

It can also be shown that the removal of a series coil converts a Class 2 reactance to a remainder reactance of Class 0, and the same removal applied to Class 3 converts the reactance to a remainder of Class 1 (i.e. this removal operation decreases the Class number by two). Finally, we should note that if the series L were part of a two-port ladder network, it would produce a loss-pole at $s = \infty$, the s-plane location of the pole of impedance removed by the extraction of the inductance.

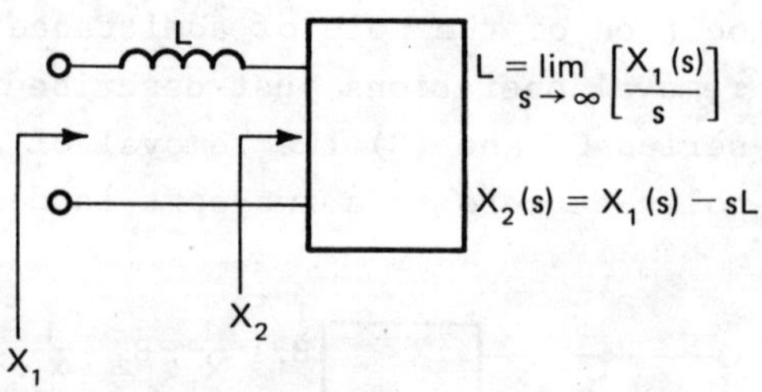

Figure 7.14 The removal of a pole of impedance at $s = \infty$.

(D) The removal of a pole of admittance at $s = \infty$.

This removal is accomplished by extracting a shunt C from the pertinent susceptance. The corresponding network operation is shown in Fig. 7.15. It can easily be shown that

$$C = \lim_{s\to\infty} \left[\frac{B_1(s)}{s}\right] = \lim_{s\to\infty} \frac{1}{s\,X_1(s)}$$

where $B_1(s) = 1/X_1(s)$.

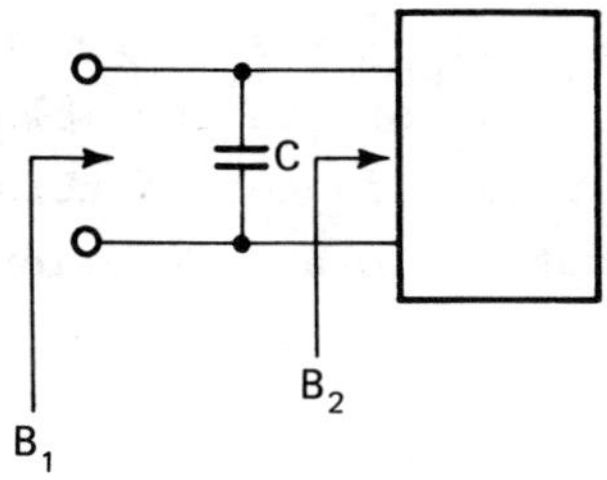

Figure 7.15 The removal of a pole of admittance at $s = \infty$.

Clearly, as with cases (A) and (B), the shunt capacitor case (D) is just the dual of case (C) described above, so that $B_1(s)$ must be of either Class 2 or Class 3 and the corresponding classes of $B_2(s)$ are Class 0 and Class 1 respectively. On a reactance basis, this means that the shunt C removal converts reactance Classes 1 and 0 to Classes 3 and 2 respectively, i.e. the class numbers are increased by two. In addition we should note that the shunt C will produce a loss-pole at infinity in a two-port ladder network.

(E) The removal of a pole of admittance at $s = j\omega_{pn}$.

This removal is accomplished by removing a series LC tank in a shunt arm, as a pole of the pertinent susceptance. The removal equation is

$$B_1(s) = \frac{k_n\, s}{(s^2 + \omega_{pn}^2)} + B_2(s)$$

where the inductance L_n is given by

$$L_n = 1/k_n \quad ,$$

and the capacitance C_n is given by

$$C_n = k_n/\omega_{pn}^2 \quad .$$

The corresponding network operation is shown in Fig. 7.16. Obviously this sub-network may be extracted from any susceptance function which has a pole at $s = j\omega_{pn}$, i.e. it must contain a factor of the form $(s^2 + \omega_{pn}^2)$. Any of the four classes of reactance or susceptance may contain a factor of this form so that we may apply this removal operation to any function which has a pole at the required s-plane location.

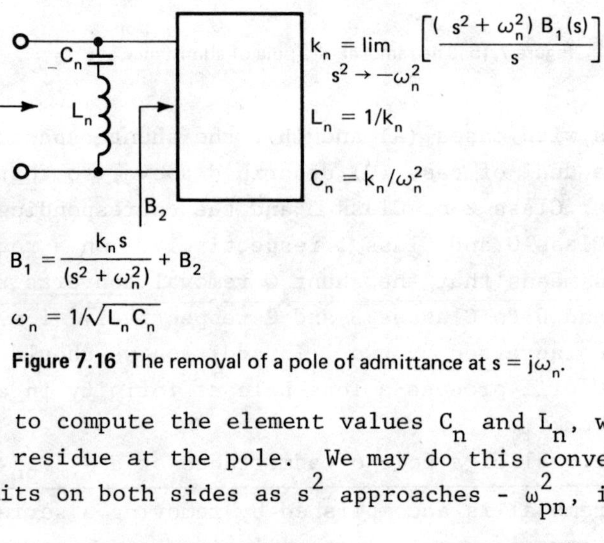

Figure 7.16 The removal of a pole of admittance at $s = j\omega_n$.

In order to compute the element values C_n and L_n, we must evaluate the residue at the pole. We may do this conveniently by taking limits on both sides as s^2 approaches $-\omega_{pn}^2$, i.e.

$$\lim_{s^2 \to -\omega_{pn}^2} B_1(s) = \lim_{s^2 \to -\omega_{pn}^2} \left[\frac{k_n s}{s^2 + \omega_{pn}^2}\right]$$

Multiplying both sides by $(s^2 + \omega_{pn}^2)/s$, we obtain k_n as

$$k_n = \lim_{s^2 \to -\omega_{pn}^2} \left[\frac{(s^2 + \omega_{pn}^2)\; B_1(s)}{s}\right] = \lim_{s^2 \to -\omega_{pn}^2} \left[\frac{(s^2 + \omega_{pn}^2)}{s\, X_1(s)}\right]$$

where $X_1(s) = 1/B_1(s)$, and the reader should recall that the factor $(s^2 + \omega_{pn}^2)$ will cancel a similar factor that must be present in the denominator of $B_1(s)$. In addition the factor s will either cancel with a similar factor, or multiply a similar factor, producing an s^2 factor. In all cases, the expression will be a function only of s^2, and no complex arithmetic will be involved. Once the value of k_n has been computed, L_n and C_n follow.

The remainder function $B_2(s)$ is then computed as

$$B_2(s) = B_1(s) - \frac{k_n s}{(s^2 + \omega_{pn}^2)} .$$

Here we will find that the pole at $s = j\omega_{pn}$ is no longer present in $B_2(s)$, and $B_2(s)$ will be of a degree two less than $B_1(s)$. It is left as an exercise for the reader to show that the class number of the remainder reactance will be the same as the class number of the initial reactance, i.e. the class number will be conserved. Finally, the reader should note that if the shunt LC sub-network is a part of a ladder two-port, a loss-pole will be produced at $s = \pm j\omega_{pn}$.

(F) The removal of a pole of impedance at $s = j\omega_{pn}$.

This removal is accomplished by removing a parallel LC tank circuit in a series arm as illustrated in Fig. 7.17. This sub-network is removed as a pole of the pertinent reactance. The removal equation is

$$X_1(s) = \frac{k_n s}{s^2 + \omega_{pn}^2} + X_2(s)$$

where the inductance is given by

$$L_n = k_n/\omega_{pn}^2$$

and the capacitance by

$$C_n = 1/k_n \quad .$$

Here k_n is evaluated by the formula

$$k_n = \lim_{s^2 \to -\omega_{pn}^2} \left[\frac{X_1(s)\,(s^2 + \omega_{pn}^2)}{s}\right]$$

where again we will find the expression on the right to be a function only of s^2 so that no complex arithmetic is involved. This operation may be applied to any class of reactance and the class will be conserved, i.e. the class of X_2 will be the same as that of X_1. The degree of the remainder reactance will again be two less than that of the initial reactance. Finally, the reader should note that in a two-port ladder network this sub-network will produce a loss-pole at $s = \pm j\omega_{pn}$.

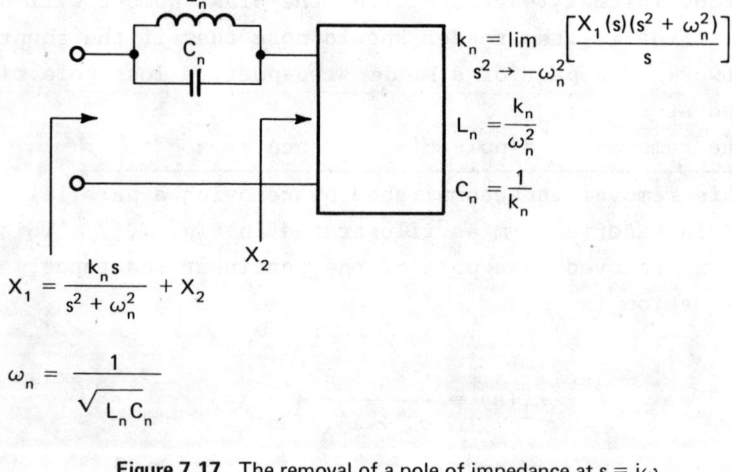

Figure 7.17 The removal of a pole of impedance at $s = j\omega_n$.

In summary, we have presented six pole removal operations. The first four are removals of poles of impedance and admittance at the extreme frequencies of zero and infinity. We found these to be class converting removals which produce a remainder reactance of a different class than the initial reactance from which they are removed. The last two were pole removals at finite non-zero imaginary axis locations, and we found that these are class conserving removals which do not change the class of the reactance from which they are removed. In the first four instances, the remainder reactance was found to be of a one degree less than the initial reactance, while one element was realized. In the last two instances the remainder reactance was found to be of two degrees less than the initial reactance, while two elements were realized.

If we start with a given reactance or susceptance function, we may always synthesize it by a sequence of these removal operations each one of which removes a pole of a remainder function reducing its degree and yielding one or two elements. When the remainder function has been exhausted or reduced to degree zero, the complete network will have been realized with as many elements as the degree of the initial function. This is a minimal element or canonic realization of the driving point function. These removal operations may be used to synthesize any one of the five canonical forms described previously, or mixtures of these forms.

Table 7.2 provides a summary of these removal operations and their properties. The reader should note that given any reactance or susceptance function it is not possible to perform these removal operations in any arbitrary sequence. The class structure outlined here and summarized by the class converting and conserving operations as shown in Table 7.2 prevent this. The sequences that are allowed may be neatly summarized by means of a state transition diagram as shown in Fig. 7.18. Here, the nodes of the graph represent the state or class of the reactance function, and the directed branches labeled A through F represent the removal operations that may be performed on a particular reactance class leading to the next class which is the class of the remaining reactance after that removal is performed. The interested reader might note that the nodes are arranged according to a "Gray Code" similar to the arrangement of a Karnaugh map and that, if the classes are numbered in binary, the removal operat-

Table 7.2 Summary of the six pole removal operations

Type	Removal Operation	Pole Type	Class	X_1	X_2
A	Series C	Z at s = 0	Converting Decrease by 1	Class 1 Class 3	Class 0 Class 2
B	Shunt L	Y at s = 0	Converting Increase by 1	Class 0 Class 2	Class 1 Class 3
C	Series L	Z at s = ∞	Converting Decrease by 2	Class 2 Class 3	Class 0 Class 1
D	Shunt C	Y at s = ∞	Converting Increase by 2	Class 0 Class 1	Class 2 Class 3
E	Shunt LC	Y at $s = j\omega_p$	Class Conserving		
F	Series LC	Z at $s = j\omega_p$	Class Conserving		

ions A to D may only be performed in an order for which the class number changes by one bit position with each removal operation.

7.5 REACTIVE TWO-PORTS

In the previous two sections we have reviewed and organized the properties of reactance functions and lossless one-port networks. In this section we consider the properties of lossless two-port networks and reactance matrices.

The reader will recall that two-port networks may be characterized by the open circuit impedance parameters or simply z-parameters which constrain the port variables V_1, V_2, I_1 and I_2 as,

$$\begin{bmatrix} V_1 \\ V_2 \end{bmatrix} = \begin{bmatrix} z_{11} & z_{12} \\ z_{21} & z_{22} \end{bmatrix} \begin{bmatrix} I_1 \\ I_2 \end{bmatrix} .$$

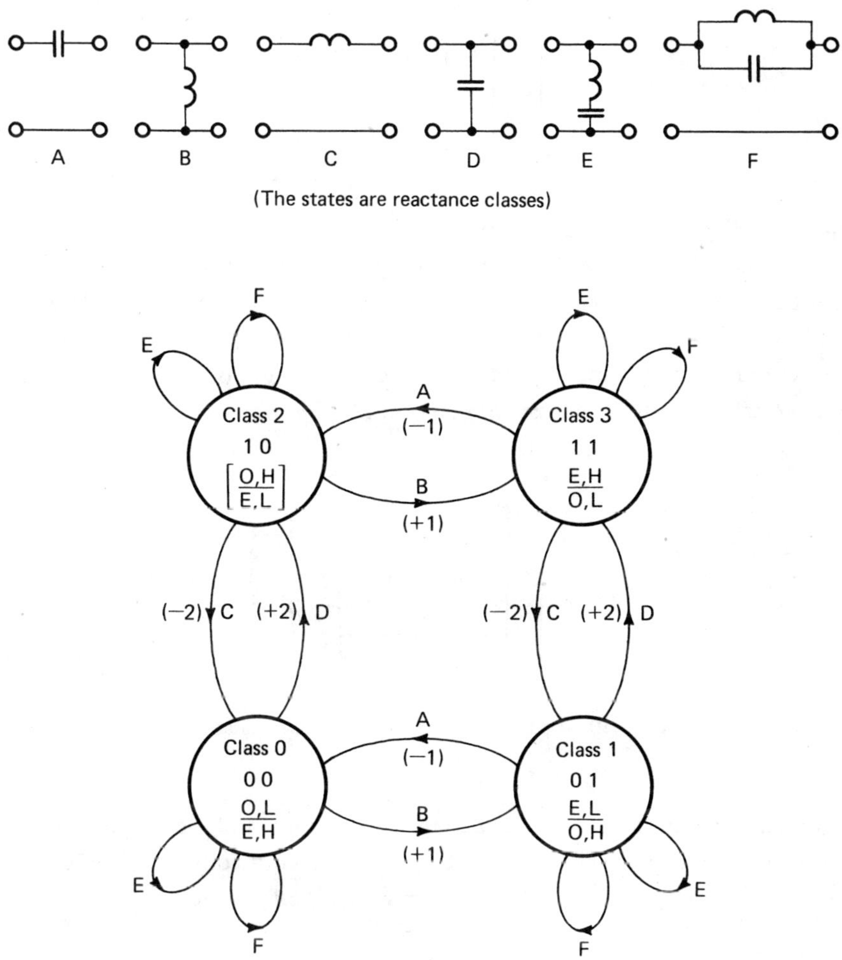

Figure 7.18 State transition flow graph illustrating allowed sequences of removal operations.

Here $z_{11}(s)$ is the driving point impedance at port 1 with port 2 open circuited, and $z_{22}(s)$ is the driving point impedance at port 2 with port 1 open circuited. Thus for a passive network $z_{11}(s)$ and $z_{22}(s)$ are positive real functions. In addition the reader will recall that for a passive network $z_{12}(s) = z_{21}(s)$. The short circuit admittance parameters or y-parameters are defined by

$$\begin{bmatrix} I_1 \\ I_2 \end{bmatrix} = \begin{bmatrix} y_{11} & y_{12} \\ y_{21} & y_{22} \end{bmatrix} \begin{bmatrix} V_1 \\ V_2 \end{bmatrix}$$

where the y-matrix is the inverse of the z-matrix. Here again for a passive network $y_{11}(s)$ and $y_{22}(s)$ must be positive real functions, and $y_{12}(s) = y_{21}(s)$. The reader will note that $y_{11}(s)$ is the driving point admittance at port 1 with port 2 short circuited, and $y_{22}(s)$ is the driving point admittance at port 2 with port 1 short circuited.

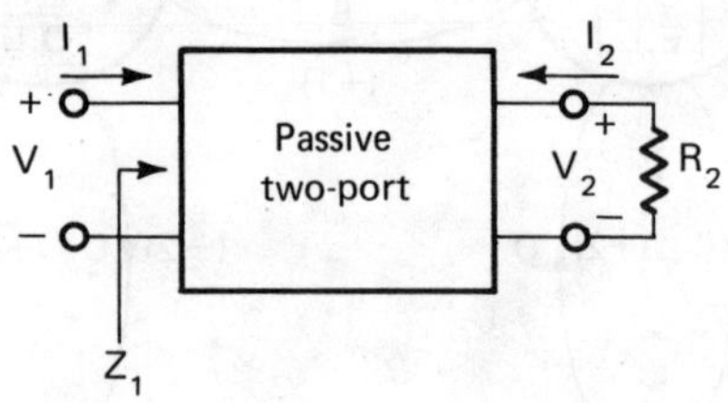

Figure 7.19 A passive two-port network terminated at port 2.

We shall now review certain definitions and develop some relations which will prove useful in later sections. Consider the network of Fig. 7.19, which is terminated at port 2 by a resistance R_2. Using the polarity convention for I_2 as shown, the output voltage across R_2 is

$$V_2 = - I_2 R_2$$

Substituting this into the z-parameter equations, we obtain an expression for the impedance at port 1 as

$$Z_1 = \frac{V_1}{I_1} = z_{11} - \frac{z_{12}^2}{R_2 + z_{22}}$$

Placing a resistor R_1 at port 1 yields a similar expression for the port 2 impedance as

$$Z_2 = \frac{V_2}{I_2} = z_{22} - \frac{z_{12}^2}{R_1 + z_{11}} ,$$

Networks for which $Z_1 = Z_2$ are termed "symmetric" networks, and networks for which $Z_1 = 1/Z_2$ are termed "antimetric". In a similar manner, we can obtain an expression for the transfer impedance Z_{12} as

$$Z_{12} = \frac{V_2}{I_1} = \frac{z_{12}\, R_2}{R_2 + z_{22}}$$

Expressions for the current transfer ratio M and the voltage transfer ratio N may be derived in a similar fashion. The results are

$$M = \frac{I_1}{-I_2} = \frac{R_2 + z_{22}}{z_{12}} ,$$

and

$$N = \frac{V_1}{V_2} = \frac{1/R_2 + y_{22}}{-y_{12}}$$

For lossless two-ports the z-parameters and y-parameters must all be odd rational functions. Thus $z_{ij}(s) = -z_{ij}(-s)$ and $y_{ij}(s) = -y_{ij}(-s)$. In particular for this case, $z_{11}(s)$ and

$z_{22}(s)$ are reactance functions and $y_{11}(s)$ and $y_{22}(s)$ are susceptance functions, having all of the properties and restrictions discussed in Section 7.3. The transfer parameters $z_{12}(s)$ and $y_{12}(s)$ are not reactance functions although they must be odd functions. In particular, the numerators of $z_{12}(s)$ and $y_{12}(s)$ must have zeros with quadrantal symmetry in the s-plane. If in addition to being lossless we restrict the topology to be a ladder structure with all positive elements, the zeros of $z_{12}(s)$ and $y_{12}(s)$ are restricted to the imaginary axis (including zero and infinity). The reader will recall that the zeros of $z_{12}(s)$ and $y_{12}(s)$ are the transmission zeros or loss poles of the network so that for a lossless ladder network we shall be restricted to transfer functions containing only imaginary axis loss poles. In addition we should recall that the loss poles occur at the zeros of shunt arm impedances and the poles of series arm impedances, and that in certain cases where the remaining network has a pole or zero coincident with that of the arm in question a loss pole may not be caused by a particular arm. We shall discuss this point in more detail in a later section.

We shall now define the four port reactances as the open circuit reactance at port 1 $X_{1o} = z_{11}$, the open circuit reactance at port 2 $X_{2o} = z_{22}$, the short circuit reactance at port 1 $X_{1s} = 1/y_{11}$, and the short circuit reactance at port 2 $X_{2s} = 1/y_{22}$.

The reader should not make the mistake of assuming that X_{1o} and X_{1s} are identical since in general $z_{11} \neq 1/y_{11}$. In fact, the order of these functions is not the same in general. The question of which of the functions X_{1o}, X_{2o}, X_{1s}, X_{2s} is of highest degree is related to the topological structure of the end arms of the ladder. This is illustrated in Fig. 7.20 where we note that the type of arm (series or shunt) at the ends of the network determine which of the four functions describes the complete network.

Using the current and voltage ratios M and N and the fact that the z and y parameters are odd rational functions for a lossless network, we may obtain formulae for these parameters as follows. Evaluating M once for s positive and then again for s negative, we obtain the two equations

$$z_{12}(s)\ M(s) - z_{22}(s) = R_2$$

$$-z_{12}(s)\ M(-s) + z_{22}(s) = R_2$$

which may be easily solved for z_{12} and z_{22} to obtain

$$z_{12}(s) = \frac{2\ R_2}{M(s) - M(-s)}$$

and

$$z_{22}(s) = R_2 \left[\frac{M(s) + M(-s)}{M(s) - M(-s)}\right]$$

We note that both z_{12} and z_{22} are odd functions as required. In a similar manner, using N we may solve for y_{12} and y_{22}. The results are

$$y_{12}(s) = \frac{-2/R_2}{N(s) - N(-s)}$$

and

$$y_{22}(s) = \frac{1}{R_2}\ \frac{N(s) + N(-s)}{N(s) - N(-s)} .$$

Again we note that the functions are odd as was assumed. Now, making use of the relations $z_{11} = -z_{12}\ y_{22}/y_{12}$ and $y_{11} = -y_{12}\ z_{22}/z_{12}$ which the reader can easily verify from the definitions, we obtain formulae for the remaining two parameters as

$$z_{11}(s) = R_2 \left[\frac{N(s) + N(-s)}{M(s) - M(-s)}\right]$$

and

$$y_{11}(s) = \frac{1}{R_2}\left[\frac{M(s) + M(-s)}{N(s) - N(-s)}\right]$$

Using these relations, the open circuit and short circuit reactances are found as

$$X_{1o} = z_{11} = R_2\left[\frac{N(s) + N(-s)}{M(s) - M(-s)}\right]$$

$$X_{2o} = z_{22} = R_2\left[\frac{M(s) + M(-s)}{M(s) - M(-s)}\right]$$

$$X_{1s} = \frac{1}{y_{11}} = R_2\left[\frac{N(s) - N(-s)}{M(s) + M(-s)}\right]$$

$$X_{2s} = \frac{1}{y_{22}} = R_2\left[\frac{N(s) - N(-s)}{N(s) + N(-s)}\right]$$

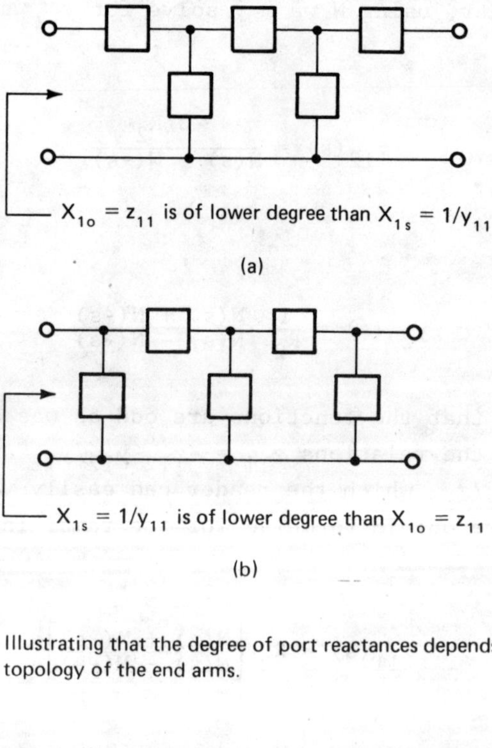

Figure 7.20 Illustrating that the degree of port reactances depends upon the topology of the end arms.

We shall use these formulae in the next section where we obtain expressions for these reactances in terms of a suitably chosen transfer function H(s) and characteristic function K(s) which we might have obtained by the methods outlined in the chapters on approximation theory.

7.6 DOUBLY TERMINATED LOSSLESS NETWORKS

For reasons which were hinted at in the introduction to this chapter, and to be fully explained in later sections, the lossless two-port network operating between a resistive source and a resistive load and designed for maximum power transfer over the passband region is very nearly an optimum realization for filter functions. In this section we shall discuss the properties of such "doubly terminated" networks, and relate certain functions of these networks to the functions H(s), K(s), $\rho(s)$ and t(s) which were introduced (rather arbitrarily) in Chapter 2 dealing with approximation theory. We will then show how the port reactances of the lossless network are related to these functions. Using these results, and extending the pole removal procedure for one-ports we shall outline the method of synthesizing a lossless two-port which realizes a given H(s) when operating between resistive source and load. Thus, at least for this case, the filter design process will have come "full circle" from specification to realization.

The Transfer Function H(s)

Consider the doubly terminated lossless two-port of Fig. 7.21. The network is operating between a resistive voltage source of V_{o1} volts behind R_1 ohms, and a resistive load R_2. We should first like to define a transfer function H(s) for this situation which has the properties of the function defined in Chapter 2. To this end we define the so-called "incident power" P_{max} to be the maximum possible power that can be delivered to the network. It is a well known theorem that this power is only supplied when the load is matched to the source. In this case, the optimum load will be R_1, and the voltage developed across it will be $V_{o1}/2$ so that the maximum power available is

$$P_{max} = \frac{|V_{o1}|^2}{4\,R_1} \text{ watts.}$$

The actual power being dissipated in the load R_2 will be

$$P_2 = \frac{|V_2|^2}{R_2} \text{ watts.}$$

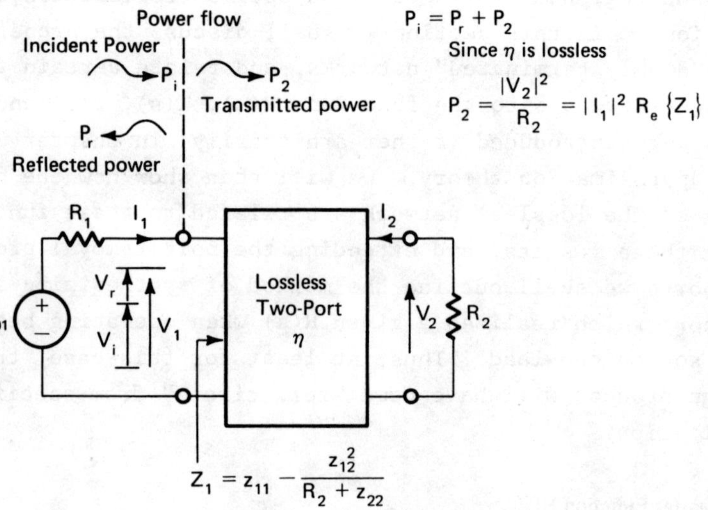

Figure 7.21 Illustrating a doubly terminated lossless two-port.

Now, since the network is both passive and lossless, P_2 must always be less than or equal to P_{max}, i.e. $P_2 \leq P_{max}$. Thus the ratio of P_{max} to P_2 must always be greater than unity. We note that this is just the property we required of the transfer function magnitude $|H|$ in Chapter 2, on approximation theory. We shall define the transfer function of the doubly terminated network by

$$|H(j\omega)|^2 = \frac{P_{max}}{P_2} = \frac{R_2}{4\ R_1} \left|\frac{V_{o1}}{V_2}\right|^2 \geq 1\ .$$

so that H(s) is

$$H(s) = \sqrt{\frac{R_2}{4\ R_1}}\ \frac{V_{o1}}{V_2}\ ,$$

the ratio of an input variable V_{o1} to an output variable V_2, and $|H| \geq 1$. Usually the "effective transmission constant" Θ is defined by

$$H(s) = e^{\Theta} = e^{\alpha + j\phi}$$

so that

$$\Theta = \alpha + j\phi = \ell n\ |H| + j\ \arg H.$$

Here $\alpha(\omega)$ is the attenuation function in Nepers, and ϕ is the phase function. We shall continue to use the attenuation function $A(\omega)$ in dB where

$$A(\omega) = 20 \log |H| \quad dB.$$

and we note that

$$A(\omega) = 8.686\alpha(\omega)$$

Since $A(\omega)$ and H are here related in the same way as $A(\omega)$ and H in Chapter 2 on approximation, we may use such suitably chosen functions $A(\omega)$ as the starting point for our filter design.

The Reflection Coefficient $\rho(s)$ and the Return Loss

To continue with our development, we define the "incident port voltage" V_i to be the voltage occurring at port 1 when the filter presents a perfect match to the source, i.e. when $Z_1 = R_1$.

In this case, $V_i = V_{o1}/2$. We then resolve the port voltage V_1 into a sum of two voltages

$$V_1 = V_i + V_r = \frac{V_{o1}}{2} + V_r$$

where V_i is the constant incident voltage, and V_r is the so-called "reflected" voltage. Note that the sign convention in this definition is arbitrary, and we might just as well have chosen $V_1 = V_i - V_r$, (which would produce sign changes in our definition for ρ later).

The port voltage V_1 is calculated in terms of the source and port impedance Z_1 as

$$V_1 = \frac{V_{o1} Z_1}{R_1 + Z_1}$$

where it should be remembered that Z_1 is a frequency dependent quantity. Using the above two equations we find that

$$V_r = \frac{V_{o1}}{2}\left[\frac{Z_1 - R_1}{Z_1 + R_1}\right] = V_i \; \rho_1(s) \; ,$$

i.e. the reflected voltage V_r is equal to the incident voltage times a factor $\rho_1(s)$, where

$$\rho_1(s) = \left[\frac{Z_1 - R_1}{Z_1 + R_1}\right]$$

is called the input reflection coefficient. Those familiar with the scattering matrix will note that this is the so-called primary scattering coefficient. We note that because of its bilinear form and the fact that Z_1 is a positive real function, the magnitude of $\rho_1(j\omega)$ is restricted to be always less than unity for any value of Z_1,

$$0 \le |\rho_1(j\omega)| \le 1$$

and that it achieves the value zero when $Z_1 = R_1$. Thus $\rho_1(s)$ satisfies the conditions we set for the function $\rho(s)$ in Chapter 2 on approximation theory.

The so-called "reflected power" is defined by

$$P_r = \frac{|V_r|^2}{R_1} = \frac{|V_{o1}|^2}{4\ R_1} |\rho_1(j\omega)|^2 \text{ watts,}$$

which we see is a fraction of the incident or maximum power; i.e.

$$P_r = P_{max} |\rho_1(j\omega)|^2 \text{ watts.}$$

We note that P_r is zero when the filter input matches the source ($Z_1 = R_1$). In this case, the maximum power P_{max} is delivered to the filter, and since the filter is lossless, the power must all be transferred to the load so that $P_2 = P_{max}$ and $|H| = 1$. In the case where $Z_1 \neq R_1$, P_r is not zero, and the reflected power P_r is "returned" to the source where it is dissipated in R_1. Also, since the filter impedance Z_1 being mismatched does not allow P_2 to obtain its maximum, the "reflected" power increases when the "transmitted" power P_2 decreases. In particular, $|\rho_1| = 1$ when $Z_1 = 0$ or ∞, which is the worst possible mismatch. In this case, $P_r = P_{max}$, and $P_2 = 0$, so that no power is transmitted through the filter. Obviously, $\rho_1(s)$ and by implication $Z_1(s)$, gives us a lot of information about the performance of the filter. We can now see the reason for the name "return loss" or "echo loss" given to $A_e(\omega)$ in Chapter 2 where

$$A_e(\omega) = 20 \log \frac{1}{|\rho_1|} = 10 \log \left[\frac{P_{max}}{P_r}\right] \text{ dB}$$

A_e tells us how much of the incident power is being lost on the "return path" from the source to the filter and back again.

Power Conservation and Feldtkeller's Equation

At this point, considering the relationships discussed above, the reader is probably wondering if a direct relationship between the "incident power" P_{max}, the "transmitted power" P_2 and the "reflected power" P_r can be established. Such a relationship can be shown quite easily, and was originally obtained by Darlington. We make use of the fact that since the network is lossless it can neither contribute or dissipate power, so that the power entering the network which is $|I_1|^2$ Re $\{Z_1\}$ must equal the power P_2 delivered to the load. Thus

$$P_2 = \frac{|V_2|^2}{R_2} = |I_1|^2 \text{ Re } \{Z_1\}$$

where Re $\{Z_1\}$ denotes the real part of Z_1. Now the input current I_1 is given by

$$I_1 = \frac{V_{o1}}{R_1 + Z_1}$$

so that we may write P_2 in terms of the source voltage as

$$P_2 = \frac{|V_{o1}|^2 \text{ Re } \{Z_1\}}{|R_1 + Z_1|^2} .$$

We have previously defined the reflected power P_r as

$$P_r = \frac{|V_{o1}|^2}{4\, R_1} |\rho_1(j\omega)|^2 = \frac{|V_{o1}|^2}{4\, R_1} \frac{|Z_1 - R_1|^2}{|Z_1 + R_1|^2} .$$

Summing P_2 and P_r we obtain

$$P_2 + P_r = \frac{|V_{o1}|^2}{4\, R_1} \left(\frac{|Z_1 - R_1|^2 + 4\, R_1 \text{ Re } \{Z_1\}}{|Z_1 + R_1|^2} \right)$$

Simply calculating the square in the numerator and adding it to $4 R_1 \operatorname{Re}\{Z_1\}$ shows that the quantity in brackets is unity. We have established the fact that

$$P_2 + P_r = P_{max} \tag{7.1}$$

i.e. the sum of the transmitted power and the reflected power equals the incident power, which is a direct consequence of the lossless property of the network. Equation (7.1) is one form of the Feldtkeller equation introduced in Chapter 2 in connection with approximation theory.

Dividing this equation through by P_{max} obtains the form

$$\frac{P_2}{P_{max}} + \frac{P_r}{P_{max}} = 1$$

where we recognize $|H|^2$ and $|\rho_1|^2$ so that we may write

$$\frac{1}{|H|^2} + |\rho_1|^2 = 1 .$$

We now define $|t|^2 = 1/ |H|^2$ to be the squared "transmission coefficient" and obtain

$$|t|^2 + |\rho_1|^2 = 1 \tag{7.2}$$

an equation relating $|t|^2$ and $|\rho_1|^2$ which is the same defining equation we proposed in Chapter 2 on approximation theory. Dividing (7.1) through by P_2 obtains

$$\frac{P_{max}}{P_2} = 1 + \frac{P_r}{P_2} \tag{7.3}$$

We recognize $|H(j\omega)|^2$ on the left as a sum of the form used to define the characteristic function $|K(j\omega)|^2$ in approximation theory. Thus we take the squared characteristic function to be

$$|K(j\omega)|^2 = \frac{P_r}{P_2} = \frac{R_2}{R_1}\left|\frac{V_r}{V_2}\right|^2$$

and rewrite (7.3) as

$$|H(j\omega)|^2 = 1 + |K(j\omega)|^2 \quad . \tag{7.4}$$

In addition, we note that $|K|^2$ is also related to $|H|^2$ through the reflection coefficient as

$$|K|^2 = |\rho_1|^2 \; |H|^2 \tag{7.5}$$

We now have satisfactory definitions for all four of the functions $|H|^2$, $|K|^2$, $|t|^2$ and $|\rho_1|^2$, which were used in approximation theory, in terms of the physics of a doubly terminated lossless network.

The Three Polynomials P(s), F(s) and E(s)

Let us digress briefly here and review the relationship of the network functions $|H|^2$, $|K|^2$ and $|\rho_1|^2$ to the polynomials P(s), F(s) and E(s) which were introduced in the chapters on approximation theory. The transfer function H(s) is the ratio of the natural mode polynomial E(s) and the loss pole polynomial P(s),

$$H(s) = \sqrt{\frac{R_2}{4\,R_1}}\;(\frac{V_{o1}}{V_2}) = \frac{E(s)}{P(s)} \quad . \tag{7.6}$$

E(s) must of course be a strictly Hurwitz polynomial (all roots in left half plane) for a stable network. Thus, given H(s) H(-s), we simply take all of the left half plane roots of the "squared" function E(s) E(-s) and assign them to E(s). The question remains as to the determination of P(s) from a given H(s) H(-s), that is from P(s) P(-s). We saw in a previous section that for a lossless network the loss pole polynomial P(s) must have all roots on the imaginary axis or else in quadrantal symmetry. In particular for a <u>lossless</u> <u>ladder</u> with no mutual inductance, P(s) may only have imaginary axis roots and is purely even or purely odd. In this case there is no choice and we simply assign one half of the roots (in conjugate pairs if not zero) of P(s) P(-s) to P(s). Thus for a lossless ladder, we may only use loss poles from the first three groups of our loss pole classification in Chapter 2. We <u>must</u> choose at least one loss pole at zero or infinity since loss poles are produced by impedance poles of series ladder arms and impedance zeros of shunt ladder arms and there must be at least one such purely inductive or capacitive arm in the ladders we are considering. The general form of loss pole polynomial being considered here then is

$$P(s) = s^{n_o} \prod_{m=1}^{n_f} (s^2 + \omega_{pm}^2)$$

where n_o is the number of loss poles at the origin and n_f is one half the number of finite non-zero loss poles. We note that the parity (even or odd) of P(s) is the same as the parity of n_o, and the degree of P(s) is $\deg(P) = n_o + 2n_f$. We recall that the degree of E(s) is $\deg(E) = n_o + n_i + 2n_f$ where by definition n_i is the number of loss poles at infinity, and $\deg(P) \leq \deg(E)$. The degree of E(s) is the order of the network N,

$$N = n_o + n_i + 2n_f = \deg(E).$$

Let us now consider the characteristic function K(s). So far we have only established a relationship for its magnitude squared. We have by definition

$$K(s)K(-s) = \frac{R_2}{R_1}\left|\frac{V_r}{V_2}\right|^2 = \frac{F(s)F(-s)}{P(s)P(-s)} \qquad \dagger$$

The same considerations discussed in the paragraph above prescribe the choice of zeros of P(s) for the denominator of K(s),

†Note that for the present Chapter we have included the passband ripple constant ε in $\frac{F}{P}$.

(i.e. they are identical to those of H(s)). The choice of assigning zeros of F(s) F(-s), the "squared" reflection zero polynomial, to K(s) is not quite so clear. In many cases all reflection zeros lie on the imaginary axis and we have no choice except that of a sign (±) to multiply F(s). In other cases, however, not all of the terms of F(s) F(-s) lie on the imaginary axis. In this case we have a choice in assigning either right-half-plane or left-half-plane (in conjugate pairs if not real) zeros of F(s) F(-s) to the numerator of K(s). These various choices of course all lead to characteristic functions having identical magnitudes but different phase angles. We should note that these choices have absolutely no effect on the transfer function H(s) since it is related to K(s) only by the Feldtkeller equation which is a relation between "squared" functions, i.e.

$$H(s)H(-s) = 1 + K(s)K(-s) \quad .$$

In what follows we assume that an appropriate choice of zeros and a sign for F(s) has been made and we simply write

$$K(s) = \sqrt{\frac{R_2}{R_1}}\left(\frac{-V_r}{V_2}\right) = \frac{F(s)}{P(s)} \; .$$

The choice of a minus sign for V_r here is arbitrary and is merely a convention.

To see the effect of different choices of zeros and signs for F(s) on the network we must use the relation involving the reflection coefficient which in turn is related to the input impedance Z_1. By definition

$$\rho_1(s) = \left[\frac{Z_1 - R_1}{Z_1 + R_1}\right] = \frac{V_r}{V_i} = -\frac{K(s)}{H(s)} \quad ,$$

so that in terms of the natural modes and reflection zeros

$$\rho_1(s) = -\frac{F(s)}{E(s)} = 2\left(\frac{V_r}{V_{o1}}\right) \quad .$$

Solving for the input impedance Z_1 we find

$$Z_1(s) = R_1 \frac{1 + \rho_1(s)}{1 - \rho_1(s)} \quad .$$

$$\text{or,} \qquad Z_1(s) = R_1 \frac{E(s) - F(s)}{E(s) + F(s)} \quad .$$

Now we can immediately see the consequences of various choices of reflection zeros and sign for F(s). Clearly these choices lead to different impedances Z_1 and hence different networks. We see that all of these impedances have identical magnitudes and differ only in their phase angles. These different impedances do not in any manner affect the transfer function H(s)! Thus in general, when the reflection zeros do not all lie on the imaginary axis, we will have at our disposal several choices of networks each of which realizes the same H(s).

7.7 DESIGN IMPEDANCES

The reader will recall that in the approximation procedures of earlier chapters, we proceeded by choosing an appropriate formula for $|K|^2$ and a set of loss poles which together are used to approximate the stopband specifications. The passband specification is automatically satisfied by the formula (maximally flat or equiripple) for $|K|^2$. The resulting "squared" loss pole and reflection zero polynomials P(s) P(-s) and F(s) F(-s) are then simply used in Feldtkeller's equation to obtain the "squared" natural mode polynomial as

$$E(s)E(-s) = P(s)P(-s) + F(s)F(-s) \quad . \tag{7.7}$$

We then choose E(s), P(s) and F(s), as described in the preceding section, to yield the functions H(s) and K(s). The remaining problem then is to synthesize the lossless ladder network from these results. We turn now to the derivation of formulae for the port reactances in terms of these functions.

Making use of $V_{o1} = V_1 + I_1 R_1$ and $V_2 = -I_2 R_2$ we may express H(s) in terms of $M(s) = (I_1/-I_2)$ and $N(s) = (V_1/V_2)$ as follows. Writing

$$H(s) = \frac{1}{2}\sqrt{\frac{R_2}{R_1}}\left(\frac{V_{o1}}{V_2}\right) = \frac{1}{2}\sqrt{\frac{R_2}{R_1}}\left[\left(\frac{V_1}{V_2}\right) + \left(\frac{R_1}{R_2}\right)\left(\frac{I_1}{-I_2}\right)\right]$$

we obtain

$$H(s) = \frac{1}{2}\left[\sqrt{\frac{R_2}{R_1}}\,N(s) + \sqrt{\frac{R_1}{R_2}}\,M(s)\right]. \tag{7.8}$$

From the definition of $\rho_1(s)$ we obtain

$$\rho_1(s) = \left[\frac{Z_1 - R_1}{Z_1 + R_1}\right] = \left[\frac{\left(\frac{V_1}{V_2}\right)\left(\frac{V_2}{I_1}\right) - R_1}{\left(\frac{V_1}{V_2}\right)\left(\frac{V_2}{I_1}\right) + R_1}\right]$$

and using $V_2 = -I_2 R_2$ again, we may write

$$\rho_1(s) = \left[\frac{\left(\frac{V_1}{V_2}\right)\left(\frac{-I_2}{I_1}\right)R_2 - R_1}{\left(\frac{V_1}{V_2}\right)\left(\frac{-I_2}{I_1}\right)R_2 + R_1}\right] = \left[\frac{\left(\frac{N}{M}\right)R_2 - R_1}{\left(\frac{N}{M}\right)R_2 + R_1}\right]$$

This may be rewritten in the more convenient form

$$\rho_1(s) = \frac{\left[\sqrt{\frac{R_2}{R_1}}\,N(s) - \sqrt{\frac{R_1}{R_2}}\,M(s)\right]}{\left[\sqrt{\frac{R_2}{R_1}}\,N(s) + \sqrt{\frac{R_1}{R_2}}\,M(s)\right]}.$$

Now, making use of $\rho_1 = -K/H$ we obtain K(s) as

$$K(s) = \frac{1}{2}\left[\sqrt{\frac{R_1}{R_2}}M(s) - \sqrt{\frac{R_2}{R_1}}N(s)\right] \qquad (7.9)$$

Thus we have expressions for H and K in terms of M and N. We now solve (7.8) and (7.9) for M and N in terms of H and K. The result is

$$\begin{aligned} N(s) &= \sqrt{\frac{R_1}{R_2}}\,[H(s) - K(s)] \\ M(s) &= \sqrt{\frac{R_2}{R_1}}\,[H(s) + K(s)] \end{aligned} \qquad (7.10)$$

Recall now the formulae developed in Section 7.5 for the open and short circuit reactances in terms of M and N. Using the subscript (ev) for the even part of a function, and (od) for the odd part of a function and recalling that, by definition, for any function G(s)

$$G_{ev}(s) = \frac{1}{2}[G(s) + G(-s)] \text{ , and}$$

$$G_{od}(s) = \frac{1}{2}[G(s) - G(-s)] \text{ ,}$$

we may rewrite the reactance formulae in the more compact form

$$X_{1o} = R_2\frac{N_{ev}(s)}{M_{od}(s)} \quad ; \quad X_{2o} = R_2\frac{M_{ev}(s)}{M_{od}(s)}$$

$$X_{1s} = R_2\frac{N_{od}(s)}{M_{ev}(s)} \quad ; \quad X_{2s} = R_2\frac{N_{od}(s)}{N_{ev}(s)}$$

Now from (7.10) we find the even and odd parts of M and N in terms of H and K as

$$
\begin{aligned}
N_{ev}(s) &= \sqrt{\frac{R_1}{R_2}}\,[H_{ev}(s) - K_{ev}(s)] \\
N_{od}(s) &= \sqrt{\frac{R_1}{R_2}}\,[H_{od}(s) - K_{od}(s)] \\
M_{ev}(s) &= \sqrt{\frac{R_2}{R_1}}\,[H_{ev}(s) + K_{ev}(s)] \\
M_{od}(s) &= \sqrt{\frac{R_2}{R_1}}\,[H_{od}(s) + K_{od}(s)]
\end{aligned}
\tag{7.11}
$$

Simply substituting these expressions in the formulae for the reactances we obtain the reactances in terms of H(s) and K(s) as

$$
\begin{aligned}
X_{1o} &= R_1\left[\frac{H_{ev}(s) - K_{ev}(s)}{H_{od}(s) + K_{od}(s)}\right] \\
X_{2o} &= R_2\left[\frac{H_{ev}(s) + K_{ev}(s)}{H_{od}(s) + K_{od}(s)}\right] \\
X_{1s} &= R_1\left[\frac{H_{od}(s) - K_{od}(s)}{H_{ev}(s) + K_{ev}(s)}\right] \\
X_{2s} &= R_2\left[\frac{H_{od}(s) - K_{od}(s)}{H_{ev}(s) - K_{ev}(s)}\right]
\end{aligned}
\tag{7.12}
$$

Since H and K are rational polynomials, and we will be using the polynomials E, F and P, we require similar expressions in terms of these design polynomials. Recalling that H = E/P, K = F/P and that P must be purely even or purely odd (i.e. the same parity as n_o) we have two cases to consider since a rational polynomial expression is even when the parity of numerator and denominator are alike and odd when they have different parities. The two sets of formulae; one for P(s) even and one for P(s) odd, are given in Table 7.3.

Table 7.3 Design impedances

P(s) is even (n_o even)	P(s) is odd (n_o odd)
$X_{1o} = R_1 \left[\dfrac{E_{ev}(s) - F_{ev}(s)}{E_{od}(s) + F_{od}(s)}\right]$	$X_{1o} = R_1 \left[\dfrac{E_{od}(s) - F_{od}(s)}{E_{ev}(s) + F_{ev}(s)}\right]$
$X_{2o} = R_2 \left[\dfrac{E_{ev}(s) + F_{ev}(s)}{E_{od}(s) + F_{od}(s)}\right]$	$X_{2o} = R_2 \left[\dfrac{E_{od}(s) + F_{od}(s)}{E_{ev}(s) + F_{ev}(s)}\right]$
$X_{1s} = R_1 \left[\dfrac{E_{od}(s) - F_{od}(s)}{E_{ev}(s) + F_{ev}(s)}\right]$	$X_{1s} = R_1 \left[\dfrac{E_{ev}(s) - F_{ev}(s)}{E_{od}(s) + F_{od}(s)}\right]$
$X_{2s} = R_2 \left[\dfrac{E_{od}(s) - F_{od}(s)}{E_{ev}(s) - F_{ev}(s)}\right]$	$X_{2s} = R_2 \left[\dfrac{E_{ev}(s) - F_{ev}(s)}{E_{od}(s) - F_{od}(s)}\right]$

We have now completed the development of doubly terminated two-port networks. Given polynomials P(s), F(s) and E(s) from the approximation methods and choosing appropriate zeros for F(s) we use the formulae of Table 7.3 to calculate the open and short circuit reactances for the lossless two-port network which will produce the required transfer function H(s) when operated between the terminations R_1 and R_2. It now remains to show that such a synthesis is possible for the polynomials under consideration, and how such a synthesis is to be obtained.

7.8 SYNTHESIS OF LOSSLESS TWO-PORTS

In sections 7.3 and 7.4 we reviewed the synthesis of lossless one-ports and introduced the reader to the concept of pole removal operations. In the last section we developed formulae for the port reactances of a lossless two-port network. Clearly, we could quite easily realize any one of these reactances as a one-port LC network which has the same reactance as the two-port we wish to synthesize. This however is not our objective, for not only do we require the two-port to have the proper port reactance, but in addition we require it to exhibit loss poles(transmission zeros) at specified frequencies. It would be quite optimistic of us to expect that using the pole removal operations to develop a ladder, we should find that the impedance and admittance poles being removed coincide exactly with the desired loss poles. Thus we must modify the pole removal procedure so as to guarantee that we can remove poles (of impedance and admittance) which do coincide with the loss poles. This leads us to the concepts of zero shifting and partial pole removals which were briefly hinted at in section 7.4.

Zero Shifting by Partial Pole Removals

The reader will recall that in section 7.4 when a pole was removed from a reactance (or susceptance) function, the poles of the remainder function were identical (apart from the one removed) to those of the original function but that the zeros of the remainder function were different (shifted) from those of the original function. This effect is termed zero shifting, and we now introduce a method of controlling this shift so as to place a zero of the remainder function at a loss pole location. The object here of course is that we may subsequently invert the remainder function and it will then have a pole (of impedance or admittance) at the desired loss pole location. In the next removal step we may then remove this impedance or admittance pole and thus produce the required loss pole.

The procedure for controlling the amount of zero shift is to "partially" remove a pole, by subtracting a fraction of elementary function which is not equal to the residue at that pole. Recall that the most general partial fraction expansion of a reactance or susceptance function is

$$F_{LC}(s) = K_\infty s + \frac{K_o}{s} + \sum_n \frac{K_n s}{s^2 + \omega_n^2}$$

where K_∞ is the residue of the pole at infinity, K_o is the residue of the pole at zero, and K_n is twice the residue of the pole at $s = j\omega_n$. In the pole removal procedures of section 7.4 we completely removed one of these poles from the function resulting in a remainder function of lower degree, having identical poles (except for the one removed) but different zeros than those of the original function. Here, we propose to remove a fraction of such a function and since the pole will not be completely removed, the remainder function will still contain the partially removed pole and, of course, its order will not be reduced. In terms of networks this implies that we have removed an extra element (or elements) which will produce a non-canonic network (not a minimal element realization). This apparently is the price we have to pay in order to shift zeros.

The process is best explained by means of examples. Let us assume that we have a Class 1 reactance ($X(\infty) = 0$, $X(0) = \infty$) and that we wish to partially remove the impedance pole at the origin by removing a series capacitor (type A removal in section 7.4). This is illustrated in Fig. 7.22 which also shows a pole-zero plot of the reactance function. In particular, the function we are considering here is given by

$$X_1(s) = \frac{k(s^2 + \omega_{z1}^2)(s^2 + \omega_{z2}^2)}{s(s^2 + \omega_{p1}^2)(s^2 + \omega_{p2}^2)}$$

for which the partial fraction expansion would be of the form

$$X_1(s) = \frac{K_o}{s} + \frac{K_1 s}{s^2 + \omega_{p1}^2} + \frac{K_2 s}{s^2 + \omega_{p2}^2}$$

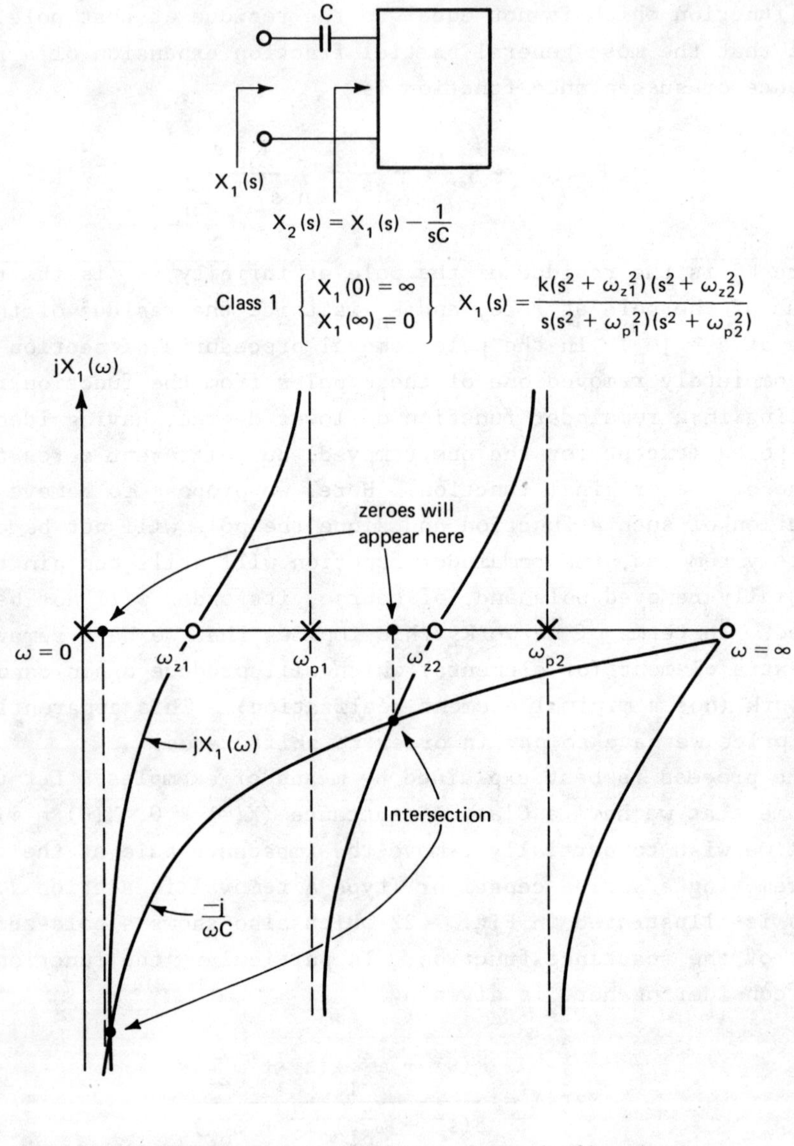

Figure 7.22 Preparatory to partially removing a pole of impedance as a series capacitor.

Fig. 7.22 illustrates the interlacing property of the poles and zeros and the positive slope property of $X(\omega)$. Clearly, $X_1(s)$ contains a series capacitive reactance. We note that if we completely removed the capacitor, it would have the value $C = 1/K_o$. In general, however, if we remove a value C, the remainder reactance becomes

$$X_2(s) = [X_1(s) - \frac{1}{sC}] \quad .$$

The zeros of this function are solutions of

$$[X_1(s) - \frac{1}{sC}] = 0$$

and occur whenever

$$X_1(s) = \frac{1}{sC} \quad .$$

The curves for both $jX_1(\omega)$ and $-j/\omega C$ are plotted in Fig. 7.22, and their points of intersection define the zeros of $X_2(s)$. We see that the zeros of $X_2(s)$ will indeed be different from those of $X_1(s)$. The poles of $X_2(s)$ will be identical to those of $X_1(s)$ (unless we completely remove the pole). We note that (at least for a positive C) the zeros of $X_2(s)$ will be shifted and lie closer to the origin (the pole being removed) than those of $X_1(s)$. We note from the alternating character of $jX_1(\omega)$ that no zero can be shifted farther than the nearest pole in the direction of the origin. We also note that if a negative capacitor were removed, the curve of the capacitive reactance would be inverted, and the zeros would then shift in the opposite direction.

As a second example we illustrate the partial removal of a pole of impedance at $s=\infty$ from a Class 2 function ($X(0) = 0$, $X(\infty) = \infty$). This is accomplished by removing a series coil (type C removal in section 7.4). The particular reactance we are considering is

$$X_1(s) = \frac{k\ s(s^2 + \omega_{z_1}^2)(s^2 + \omega_{z_2}^2)}{(s^2 + \omega_{p1}^2)(s^2 + \omega_{p2}^2)}$$

which has the partial fraction expansion

$$X_1(s) = K_\infty s + \frac{K_1 s}{s^2 + \omega_{p2}^2} + \frac{K_2 s}{s^2 + \omega_{p2}^2} .$$

Clearly, $X_1(s)$ contains a series inductive reactance produced by an inductor of value K_∞. Let us now partially remove an inductor of value L as illustrated in Fig. 7.23. The remainder reactance is

$$X_2(s) = X_1(s) - sL ,$$

and the zeros of $X_2(s)$ are given by the intersection of the curves for $X_1(s)$ and sL as shown in Fig. 7.23. We see that the zeros of $X_2(s)$ are the zeros of $X_1(s)$ shifted towards infinity (the pole being removed). Again note that if we had removed a negative inductance, the zeros would shift in the opposite direction since the line describing the reactance $j\omega L$ would be inverted in that case.

These examples illustrate the zero shifting procedure by partial pole removal. We can, of course, partially remove (or even "over remove") any type of pole and we can remove both positive and negative elements. However, if we are interested in a strictly passive network, we will never over-remove a pole or remove negative elements in an attempt to shift zeros.† It is possible to partially remove finite non-zero poles but since they require two elements and one usually tries to keep the number of elements to a minimum, these are not used in practice. Finally, we should note that partially removing a pole does not change the Class of the reactance from which it is removed since the poles and/or zeros at zero and infinity are not affected by the partial removal.

---

†In certain cases, however, we may be able to incorporate negative inductors into ideal transformers, which involve mutual inductance, but usually this is to be avoided for practical reasons.

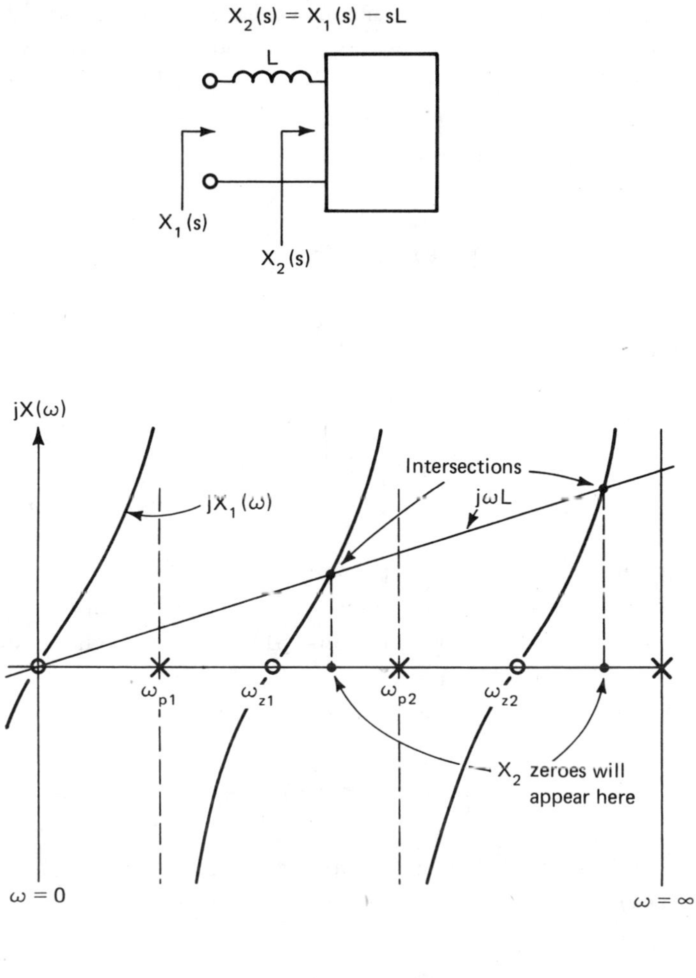

Figure 7.23 Partially removing a pole of impedance at infinity.

Implementing the Partial Pole Removal Process

From the above we conclude that there are four types of partial pole removals, corresponding to types A, B, C and D of section 7.4, in which we are interested. The remaining question is how to compute the value of element to be removed so that a zero occurs in the remainder function at the required loss pole frequency.

Consider first the partial removal of a pole of impedance at $s = 0$ by the extraction of a series capacitor. The initial reactance $X_1(s)$ must be of Class 1 or Class 3. We require the remaining reactance $X_2(s)$ to have a zero at some frequency say ω_p corresponding to a loss pole frequency. The removal equation reads

$$X_1(s) = X_2(s) + \frac{1}{sC} \quad .$$

At $s = j\omega_p$ $(s^2 = -\omega_p^2)$ we require that $X_2(s)$ be zero, so that we simply calculate C from

$$C = \left. \frac{1}{s\, X_1(s)} \right|_{s^2 = -\omega_p^2}$$

Hopefully this will yield a positive value for C, if not we must try to remove some other element or try another loss pole frequency. We shall discuss the possibilities for the occurrence of negative elements later.

At this point we calculate $X_2(s)$ as

$$X_2(s) = X_1(s) - \frac{1}{sC} \quad ,$$

where we note that $X_2(s)$ must have a zero at $s = j\omega_p$. We now invert $X_2(s)$ to obtain the susceptance $B_2(s)$

$$B_2(s) = \frac{1}{X_2(s)}$$

which has a pole (of admittance) at $s=j\omega_p$. Next we completely remove this pole of admittance from B_2 by removing a series LC circuit in a shunt arm. This results in a remainder susceptance $B_3(s)$ where

$$B_3(s) = B_2(s) - \frac{K_p\, s}{s^2 + \omega_p^2}$$

The element values for the coil and capacitor in terms of the residue K_p and frequency ω_p are calculated just as for removal type E of section 7.4. The procedure is illustrated in Fig. 7.24 where we see that the series LC circuit completely removed in shunt produces a loss pole (transmission zero) at $s = j\omega_p$.

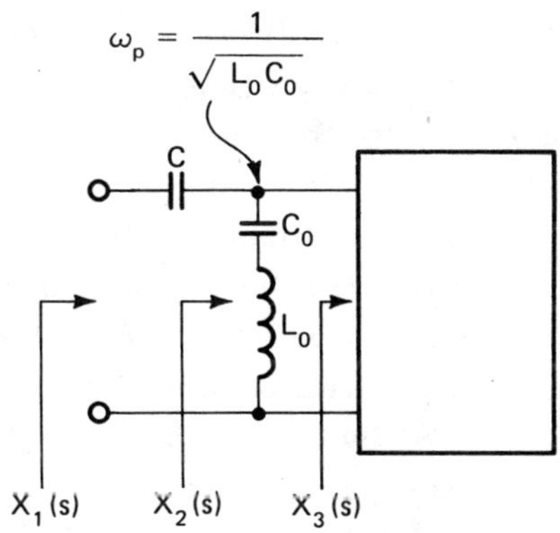

(1) $C = \left.\frac{1}{sX_1(s)}\right|_{s^2=-\omega_p^2}$ Calculate C to place a zero of X_2 at $s = j\omega_p$

(2) $X_2(s) = X_1(s) - \frac{1}{sC}$ has a zero at $j\omega_p$

(3)' $B_2 = 1/X_2$ has a pole at $j\omega_p$

(4) $B_3 = B_2 - \frac{k_p s}{s^2 + \omega_p^2} = B_2 - \frac{1}{sL_0 + \frac{1}{sC_0}}$

Figure 7.24 Partial removal of a pole of impedance at $s = 0$, followed by complete removal of a pole of admittance at $s = j\omega_p$.

We should point out here that, when we shift a zero by partial pole removal in order to place a pole of the reciprocal of the remainder function at a specific frequency, the partially removed element <u>does not</u> produce a loss pole (transmission zero). This should be clear, since the remainder reactance (or susceptance) at this point still contains a residual reactance of the kind partially removed, and thus must be completely removed at a later point in the network. The voltage (or current) divider effect produced by the partial element and the residual element (to the right of it) of the remainder reactance allows a finite transmission through the network at the point of partial removal (at the frequency of the pole that has been partially removed). Thus partial removal branches of themselves do not cause loss-poles in the two-port network, but in conjunction with the "complete" removal, which must eventually follow, a loss-pole is produced.

There are four possibilities for partial removals of a single element: in shunt or series followed by an LC circuit in series or shunt. We have already discussed one of these cases. The remaining three are the partial removal of a shunt coil followed by the full removal of a parallel LC circuit, the partial removal of a series coil followed by the full removal of an LC circuit in a shunt arm, and the partial removal of a shunt capacitor followed by the full removal of a parallel LC circuit in a series arm.

It is also possible to perform two partial removals in sequence followed by two full removals at two different frequencies. Thus, counting the four full removal operations (types A through D of section 7.4) and the partial-full removal sequences discussed above, we have a total of ten removal operations which can be performed during an LC ladder synthesis.

The detailed development of the design equations follow those outlined here and in section 7.4. The details are left to the reader and we simply present the ten removal operations and the corresponding equations in Table 7.4. We have used a numbering scheme (1 through 10) to correspond with the convention established in previous literature.

Ladder Synthesis Using the Ten Removal Operations

We have now established a set of removal operations which are at least potentially capable of realizing a driving point reactance and a set of arbitrary loss poles for a lossless ladder network. Due to the properties of the Classes of reactance functions, not all sequences of these ten removal operations are possible. At any point in a ladder development the removals that may be performed are dictated by the Class of reactance at that point which in turn depends upon the original reactance and the removals already performed up to that point. We may construct a state diagram similar to that of section 7.4, to guide us in selecting possible sequences.

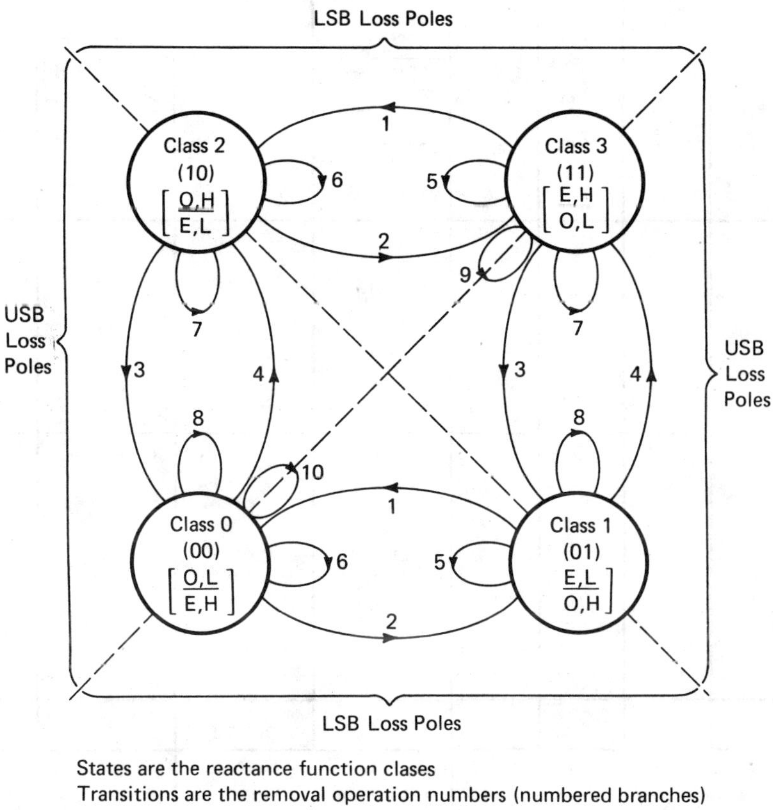

Figure 7.25 State diagram of ladder two-port removal operations.

Fig. 7.25 gives such a state diagram showing the four reactance Classes as states and the ten removal operations are marked beside the branch arrows indicating the transitions from state to

Section Removed	#	Partial Removal	Remainder Function	Full Removal	Remainder Function
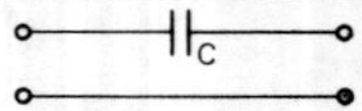	1	-	-	$C = [\frac{1}{sX}]_{s=0}$	$X_1 = X - \frac{1}{sC}$
$\omega_\ell = 0$ 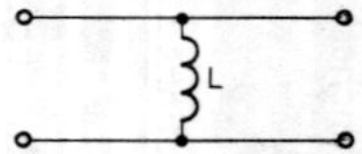	2	-	-	$L = [\frac{1}{sB}]_{s=0}$	$B_1 = B - \frac{1}{sL}$
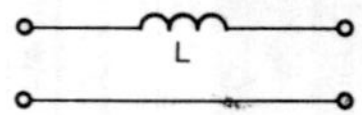	3	-	-	$L = [\frac{X}{s}]_{s=\infty}$	$X_1 = X - sL$
$\omega_\ell = \infty$ 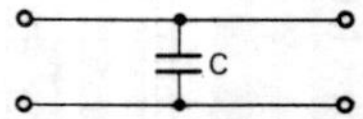	4	-	-	$C = [\frac{B}{s}]_{s=\infty}$	$B_1 = B - sC$
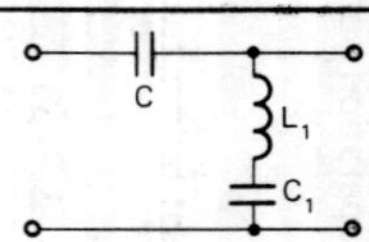	5	$C = [\frac{1}{sX}]_{s^2 = -\omega_\ell^2}$	$X_1 = X - \frac{1}{sC}$	$L_1 = \left[\frac{sX_1}{s^2 + \omega_\ell^2}\right]_{s^2 = -\omega_\ell^2}$ $C_1 = \frac{1}{\omega_\ell^2 L_1}$	$\frac{1}{X_2} = \frac{1}{X_1} - \frac{s(\frac{1}{L_1})}{s^2 + \omega_\ell^2}$
$0 < \omega_\ell < a$ 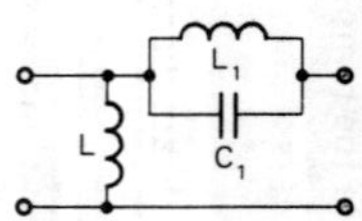	6	$L = [\frac{1}{sB}]_{s^2 = -\omega_\ell^2}$	$B_1 = B - \frac{1}{sL}$	$C_1 = \left[\frac{sB_1}{s^2 + \omega_\ell^2}\right]_{s^2 = -\omega_\ell^2}$ $L_1 = \frac{1}{\omega_\ell^2 C_1}$	$\frac{1}{B_2} = \frac{1}{B_1} - \frac{s(\frac{1}{C_1})}{s^2 + \omega_\ell^2}$
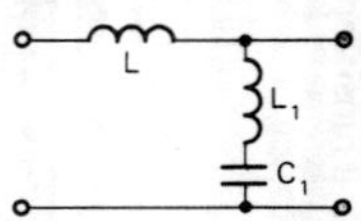	7	$L = [\frac{X}{s}]_{s^2 = -\omega_\ell^2}$	$X_1 = X - sL$	$L_1 = \left[\frac{sX_1}{s^2 + \omega_\ell^2}\right]_{s^2 = -\omega_\ell^2}$ $C_1 = \frac{1}{\omega_\ell^2 L_1}$	$\frac{1}{X_2} = \frac{1}{X_1} - \frac{s(\frac{1}{L_1})}{s^2 + \omega_\ell^2}$
$1 < \omega_\ell < \infty$ 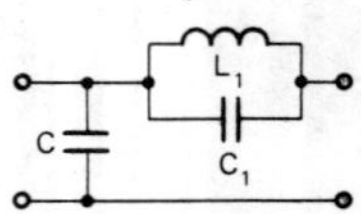	8	$C = [\frac{B}{s}]_{s^2 = -\omega_\ell^2}$	$B_1 = B - sC$	$C_1 = \left[\frac{sB_1}{s^2 + \omega_\ell^2}\right]_{s^2 = -\omega_\ell^2}$ $L_1 = \frac{1}{\omega_\ell^2 C_1}$	$\frac{1}{B_2} = \frac{1}{B_1} - \frac{s(\frac{1}{C_1})}{s^2 + \omega_\ell^2}$

Circuit	No.	Element values	Remainder	Element values	Recursion
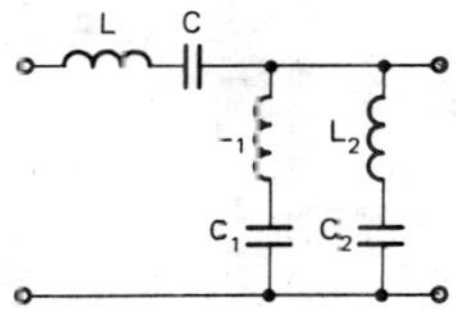	9	$A_\nu = [sX]_{s^2=-\omega_{\ell\nu}^2};\ \nu=1,2$ $L = \frac{A_1 - A_2}{\omega_{\ell 2}^2 - \omega_{\ell 1}^2}$ $C = \frac{1}{A_1 + \omega_{\ell 1}^2 L}$ $= \frac{1}{A_2 + \omega_{\ell 2}^2 L}$	$X_1 = X - sL - \frac{1}{sC}$	$L_\nu = [\frac{sX_\nu}{s^2 + \omega_{\ell\nu}^2}]_{s^2=-\omega_{\ell\nu}^2}$ $C_\nu = \frac{1}{\omega_{\ell\nu}^2 L_\nu}$ $\nu = 1,2$	$\frac{1}{X_{\nu+1}} = \frac{1}{X_\nu} - \frac{s(\frac{1}{L_\nu})}{s^2 + \omega_{\ell\nu}^2}$ $\nu = 1,2$
$0 < \omega_{\ell_1} < a$ $1 < \omega_{\ell_2} < \infty$ 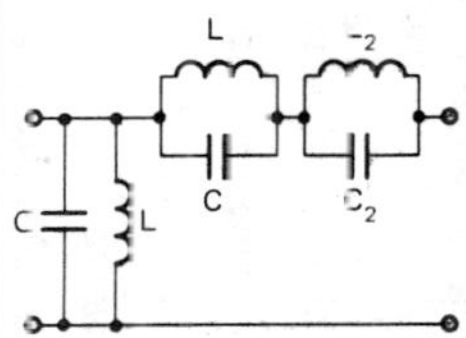	10	$A_\nu = [sB]_{s^2=-\omega_{\ell\nu}^2};\ \nu=1,2$ $C = \frac{A_1 - A_2}{\omega_{\ell 2}^2 - \omega_{\ell 1}^2}$ $L = \frac{1}{A_1 + \omega_{\ell 1}^2 C}$ $= \frac{1}{A_2 + \omega_{\ell 2}^2 C}$	$B_1 = B - sC - \frac{1}{sL}$	$C_\nu = [\frac{sB_\nu}{s^2 + \omega_{\ell\nu}^2}]_{s^2=-\omega_{\ell\nu}^2}$ $L_\nu = \frac{1}{\omega_{\ell\nu}^2 C_\nu}$ $\nu = 1,2$	$\frac{1}{B_{\nu+1}} = \frac{1}{B_\nu} - \frac{s(\frac{1}{C_\nu})}{s^2 + \omega_{\ell\nu}^2}$ $\nu = 1,2$

Table 7.4 Ladder removal operations

state. Given a reactance function and a set of loss poles, the diagram of Fig. 7.25 may be used to qualitatively sketch a number of "possible" circuit diagrams for a ladder network. Of course, the last removal for any network must of necessity be a full removal (1 through 4) since any partial removal (5 through 10) implies that there still exists at least one pole to be removed. Following the sequences allowed by Fig. 7.25 and the fact that the last removal must be a full removal we see that for any reasonable order of filter there exists a large number of "possible" circuit configurations.

There remains the question of the realizability of these "possible" configurations, since we have seen that sometimes one may encounter negative elements in the zero shifting procedure. If we do not allow negative LC elements (passive design), then not all of the "possible" removal sequences may be realizable. This problem was first solved by Fujisawa for the low pass case and subsequently by Watanabe in the bandpass case. They give a complicated set of conditions on the relative locations of the loss poles with respect to the roots of even and odd parts of the driving point impedance Z_1, which are necessary and sufficient to guarantee that at least one of the possible removal sequences will result in all positive elements (or what is equivalent, no mutual inductance). If these conditions are not satisfied then we are out of luck. Fortunately, for most types of practical specifications these conditions are satisfied. Watanabe has shown that for the equiripple case with a prescribed set of loss poles and reflection zeros there always exists a minimum passband ripple A_p for which all elements will be positive when A_p exceeds the minimum. Unfortunately these conditions do not indicate which of the many possible removal sequences will work, they only guarantee that at least one will. We are left with the problem of simply testing at each removal operation for the smallest possible non-negative element value which we can obtain and proceeding in this fashion to the end.

In practice, however, the conditions of Watanabe and Fujisawa are rarely if ever checked and certain "rules of thumb" may be used in selecting the order of removal operations and their corresponding loss pole frequencies so that one usually obtains a "good" network in one or two attempts. Clearly one should never attempt to remove numbers 1,2,5 and 6 from a lowpass filter, or numbers 3,4,7 and 8 from a highpass filter. Similarly, removals 5 and 6 should only be applied at frequencies in the lower stopband of a bandpass filter, and removals 7 and 8 should only be used at frequencies in the upper stopband of a bandpass filter. In addition, it is usually advisable to perform partial removals in the "center" of the network leaving the full removals for the "ends". The frequencies of loss poles which are the closest to the passband should be removed in the center of the network moving outwards in both directions for farther removed loss poles. Usually the above "rules" give good results but the designer is free to try any other sequence he may like for his particular reasons.

The reader should note here that in general for an Nth order network where $N = n_o + n_i + 2n_f$, the number of removal operations of Table 7.4 to be performed (for types 1 through 8) is

$$N_{rem} = (n_o + n_i + n_f) \leq N \quad .$$

While the number of elements in the resulting network will be

$$N_{el} = (n_o + n_i + 3n_f) \geq N \quad .$$

Thus if there are any partial removals ($n_f \neq 0$) the network will be non-canonic and will contain exactly n_f more elements than the order of the network, provided cases in which the loss pole frequencies lie exactly at an impedance or admittance pole do not occur. These "extra" elements of course represent the price we pay in order to obtain the finite non-zero loss poles at their proper locations on the imaginary axis.

Terminating the Ladder

We have seen that using these zero-shifting pole removal operations we may expand any given reactance (X_{1o}, X_{2o}, X_{1s} or X_{2s})

into a ladder network which realizes all of the required loss poles (and hopefully contains no negative elements). Clearly from the reactance formulae of Table 7.3, not all reactances may be of highest degree (the degree N of the network) and in order to synthesize the complete network we should select a reactance of highest degree. Synthesizing one of these high degree reactances will result in a proper reactance as seen from the end at which the synthesis was initiated but in general the resulting reactance produced at the other port will not equal the corresponding reactance computed from the formulae of Table 7.3, since we have done nothing to ensure this. In fact the ladder synthesis realizes only two of the parameters* (say z_{11} and z_{12} or y_{11} and y_{12}) and we must be satisfied with the remaining parameter (say z_{22} or y_{22}) which we obtain from the synthesis. The resulting value of z_{22} or y_{22} will be related to the desired value by an unknown constant multiplier. In other words, the value of the terminating resistance R_2 will be different from the specified value. To find the value of R_2 that we should use, a complete synthesis is first done from the left-hand-side. Let us assume that this synthesis results in a value ℓ_ℓ for a particular inductance in the ladder. Considering Table 7.3, the true value L_ℓ will be proportional to the value of R_1 used. We may express this proportionality as

$$L_\ell = \ell_\ell R_1$$

Subsequently another synthesis is performed from the right-hand-side. We need only carry this synthesis to the point of determining the value of the particular inductance under consideration. Let this value be ℓ_r. Again, the true value L_r will be proportional to the value of R_2 used; that is,

---

*There are synthesis methods (e.g. Cauer's partial fraction method) which will simultaneously realize all three of a given set of parameters (z_{11}, z_{12} and z_{22} or y_{11}, y_{12} and y_{22}) but these networks are not practical to construct.

$$L_r = \ell_r R_2 .$$

Now, since L_ℓ and L_r represent values for the same element but calculated at different impedance levels, then L_ℓ and L_r should be equated to find the ratio of impedance levels. Thus, the value of R_2 which we are forced to use is given by

$$R_2 = R_1 \left(\frac{\ell_\ell}{\ell_r}\right) . \tag{7.13}$$

In most practical situations one would use the value of R_2 obtained from (7.13). If, however, the filter is required to operate with a different value of terminating resistance say R_2', then a transformer should be added at the end of the LC ladder to transform R_2' into the value R_2 obtained from (7.13). Other techniques such as tapped coils or impedance transformations may be used to avoid the addition of a transformer.

Before leaving this point it should be mentioned that although the above analysis was performed assuming a fixed value for R_1, and correspondingly a complete synthesis from left to right, the reverse process is obviously equally acceptable. Also, we used the value of an inductance to determine the terminating resistance value. If desired, one may equally well use the value of a particular capacitor resulting in the equation

$$R_2 = R_1 \left(\frac{c_r}{c_\ell}\right) \tag{7.14}$$

As an example, a bandpass filter of order N = 10 with $n_o = 3$, $n_i = 1$, $n_f = 3$; two loss peaks in the lower stopband and one in the upper stopband, has the attenuation curve sketched in Fig. 7.26. A possible removal sequence for this filter is illustrated where we have synthesized by removals from the left. The sequence of removal operations is (2-5-8-5-1-4-2) and there are $N_{rem} = (n_o + n_i + n_f) = 7$ of these removal operations. The network has $N_{el} = (n_o + n_i + 3n_f) = 13$ elements. If assuming $R_1 = 1\ \Omega$, the last element removed from the left is ℓ_ℓ (or c_ℓ)

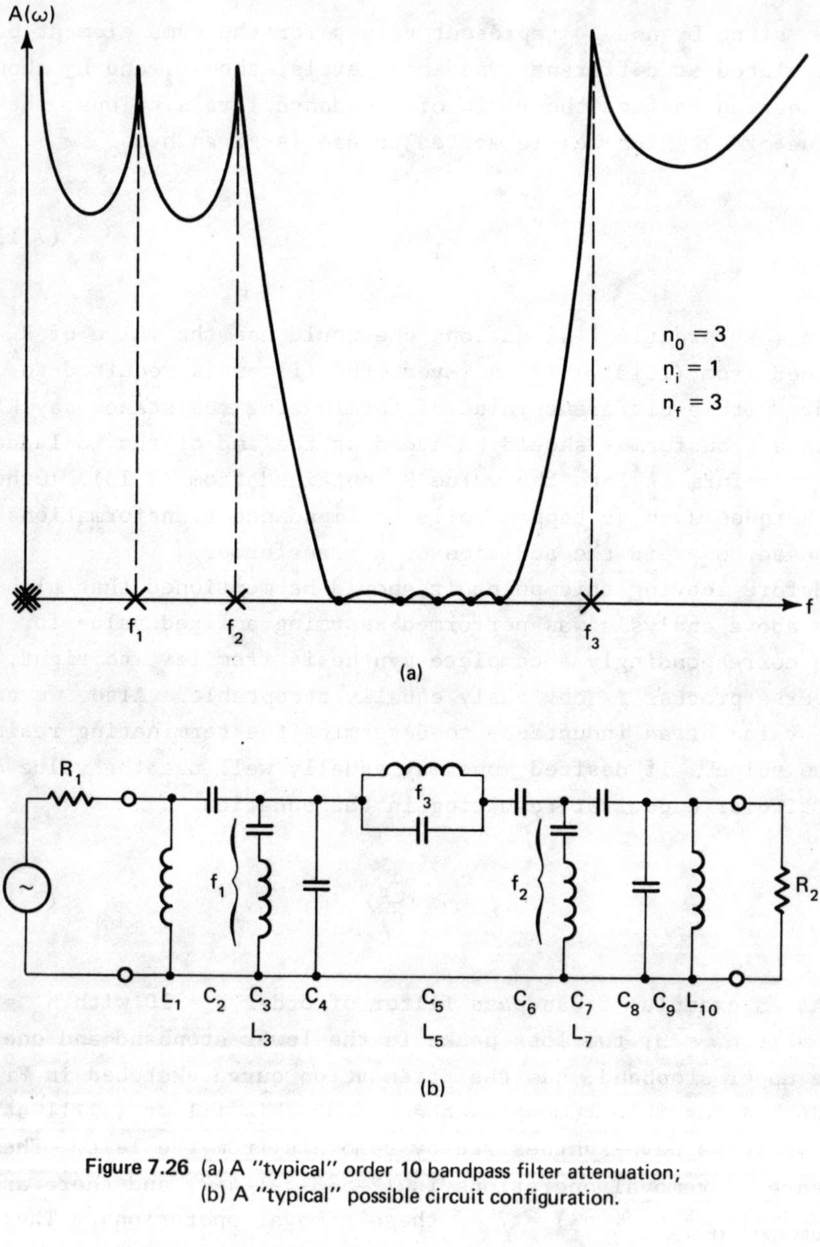

Figure 7.26 (a) A "typical" order 10 bandpass filter attenuation; (b) A "typical" possible circuit configuration.

and the same element removed from the right assuming $R_2 = 1\ \Omega$ has the value ℓ_r (or c_r), then a terminating resistance of

$$R_2 = \left(\frac{\ell_\ell}{\ell_r}\right) R_1 = \left(\frac{c_r}{c_\ell}\right) R_1$$

ohms should be used with the network if no other impedance transformations are contemplated. If the reactance functions were obtained from a set of H(s) and K(s) which were designed to give maximum power transfer ($A(\omega) \simeq 0$) in the passband then, of course, the network realized will deliver maximum power to the load R_2, over the passband region.

We now illustrate many of our results by numerical examples.

Example 7.1

All the bandpass filters considered in the approximation Chapters are of even order. Odd-order bandpass filters, though not very common, exist and are called parametric filters. A parametric filter will of necessity have at least one reflection zero on the real axis of the s-plane.

We choose to use an odd order (N=3) parametric bandpass filter to illustrate the general situation when all reflection zeros are not on the imaginary axis. Taking $n_o = 1$, $n_i = 2$, $n_f = 0$, we have a filter of order

$$N = n_o + n_i + 2n_f = 3.$$

The s-plane loss pole polynomial is odd and is simply

$$P(s) = s \quad .$$

We assume reflection zeros at $s = \pm\, j0.8$, and one at $s = -0.8$. Thus, a possible reflection zero polynomial is

$$F(s) = (s^2 + 0.64)\ (s + 0.8) = s^3 + 0.8s^2 + 0.64s + 0.512$$

The "squared" natural mode polynomial is obtained from Feldtkeller's equation; i.e.

$$E(s)\ E(-s) = P(s)\ P(-s) + F(s)\ F(-s)$$

which we evaluate as

$$E(s)\ E(-s) = -s^6 - 0.64s^4 - 0.5904s^2 + 0.262\ 144$$

Calculating all of the roots of E(s) E(-s) yields the following set of natural modes

$$s = \pm\ 0.470\ 170\ 291 \pm j0.807\ 317\ 988$$

$$\text{and } s = \pm\ 0.548\ 033\ 377$$

The complete set of "squared" singularities are plotted in the s-plane on Fig. 7.27. The loss poles are represented as crosses, the reflection zeros as small circles, and the natural modes as small squares. Fig. 7.28 illustrates the form of attenuation function exhibited by this third order bandpass filter.

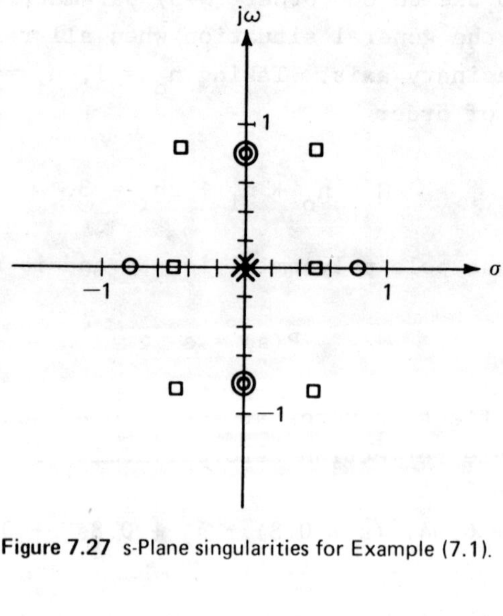

Figure 7.27 s-Plane singularities for Example (7.1).

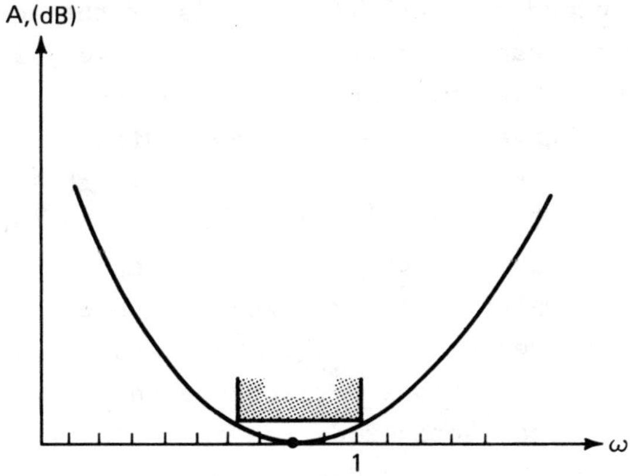

Figure 7.28 Attenuation characteristics for the bandpass filter of Example (7.1).

Selecting all of the left half plane factors, we calculate E(s) as

$$E(s) = s^3 + 1.511\ 443\ 334s^2 + 1.462\ 230\ 477s + 0.512$$

Note that in this case both the leading and trailing coefficients of E(s) and F(s) are equal. We now calculate the reactances using the formulae of Table 7.3 to obtain

$$\frac{X_{1o}}{R_1} = \frac{E_{od} - F_{od}}{E_{ev} + F_{ev}} = \frac{0.822\ 230\ 477s}{2.311\ 443\ 334\ s^2 + 1.024}$$

$$\frac{X_{2o}}{R_2} = \frac{E_{od} + F_{od}}{E_{ev} + F_{ev}} = \frac{2s^3 + 2.102\ 230\ 477s}{2.311\ 443\ 334\ s^2 + 1.024}$$

$$\frac{X_{1s}}{R_1} = \frac{E_{ev} - F_{ev}}{E_{od} + F_{od}} = \frac{0.711\ 443\ 334\ s}{2s^2 + 2.102\ 230\ 477}$$

$$\frac{X_{2s}}{R_2} = \frac{E_{ev} - F_{ev}}{E_{od} - F_{od}} = \frac{0.711\ 443\ 334\ s}{0.822\ 230\ 477}$$

The reader is urged to check the details of the above computations and note the cancellations that have taken place. Note, in particular here, that the true parities of the short circuit reactances are not given by the parities indicated in the formulae because of the cancellations that occur. In this case we have obtained only one reactance (X_{2o}) which has the degree of the network (N=3) and in fact only this reactance may be used to synthesize the complete network. We note that, X_{1o} is a Class 0 reactance and has degree two, X_{2o} is a Class 2 reactance and has degree three, X_{1s} is a Class 0 reactance and has degree two, and X_{2s} is a Class 2 reactance of degree one. Thus, port 1 is a Class 0 port and port 2 is a Class 2 port. For this network we must perform

$$N_{rem} = n_o + n_i + n_f = 3$$

removal operations, and we will obtain

$$N_{el} = n_o + n_i + 3n_f = 3$$

elements, which is a canonic network.

We can synthesize from port 2 using X_{2o} of Class 2 and we choose to use the removal sequence 3-4-2. We shall assume that $R_2 = 1\Omega$ and subsequently calculate R_1 by removing elements from port 1. The network corresponding to this removal sequence is illustrated in Fig. 7.29.

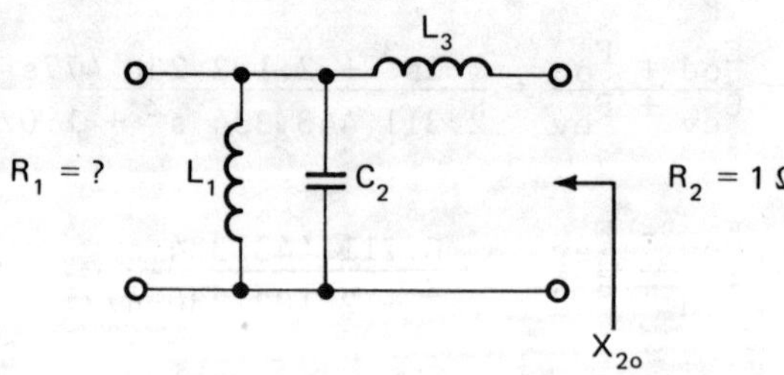

Figure 7.29 Synthesis of the third-order bandpass filter of Example (7.1) starting from X_{2o} and using the removal sequence (3–4–2).

First, removing the series inductance L_3 by removal number 3 we obtain L_3 as

$$L_3 = \left.\frac{X_{2o}(s)}{s}\right|_{s\to\infty} = \frac{2}{2.311\ 443\ 334} = 0.8653\ \text{H} .$$

We then calculate the remainder reactance as

$$X_r(s) = X_{2o}(s) - sL_3 = \frac{1.216\ 203\ 997\ s}{2.311\ 443\ 334 s^2 + 1.024}$$

which we immediately recognize as a parallel LC tank so that we need not continue through the detailed steps of removals 4 and 2. Thus, dividing through numerator and denominator we obtain

$$X_r(s) = \left[\frac{1}{1.900\ 539\ 169 s + \dfrac{1}{1.187\ 699\ 215 s}}\right]$$

and elements C_2 and L_1 are then

$$C_2 = 1.901\ \text{F}$$

$$L_1 = 1.188\ \text{H}$$

Now let us synthesize from the left. We will use X_{1o} since X_{1s} will contain L_3 and thus will not allow us to calculate the correct value for L_1 which appears in parallel with L_3 when port 2 is short circuited. We have now to perform the synthesis using (X_{1o}/R_1) from the left with the reversed removal sequence i.e. (2-4-3). However, since (X_{1o}/R_1) is only a second degree function we need not go through all of the details and we simply recognize it as a tank circuit, thus

$$X_{1o} = \frac{0.822\ 230\ 477\ R_1 s}{2.311\ 443\ 334\ s^2 + 1.024}$$

$$= \left[\frac{1}{\dfrac{2.811\ 186\ 642\ s}{R_1} + \dfrac{1}{0.802\ 959\ 450\ R_1 s}}\right]$$

yields L_1 and C_2 as

$$L_1 = 0.8030\ R_1\ \text{H}$$

$$C_2 = 2.811/R_1\ \text{F}$$

Thus we have, assuming $R_2 = 1\Omega$, the following element values synthesized from the left and from the right.

from left	from right
$L_{1\ell} = 0.8030\ R_1$	$L_{1r} = 1.188$
$C_{2\ell} = 2.811/R_1$	$C_{2r} = 1.901$

We can now compute the correct terminating resistance for port 1 by setting $L_{1\ell} = L_{1r}$ and/or $C_{2\ell} = C_{2r}$. The correct resistance is

$$R_1 = \left(\frac{1.188}{0.8030}\right) = \left(\frac{2.811}{1.901}\right) = 1.479\Omega$$

The complete network with all element values indicated is shown in Fig. 7.30. The reader is urged to calculate the input impedance Z_1 of this simple network and verify that $Z_1(s)$ indeed equals $R_1 = 1.479\Omega$ at the reflection zeros ($s = -0.8$ and $s = \pm\ j0.8$).

Let us now examine the effects of different choices of F(s), the reflection zero polynomial, on our design. In the above we have taken reflection zeros at $s = \pm\ j0.8$ and $s = -0.8$ and formed the polynomial F(s). We could equally as well have chosen F(s) to be simply the negative of this polynomial, since this also satisfies Feldtkeller's equation and has no effect on the transfer function. Please note however that this choice will change the reactances calculated. In particular, changing the sign of F(s) can be seen to exchange the roles of these reactances, such that X_{1o} becomes $1/X_{1s}$ and vice versa, while X_{2o} becomes $1/X_{2s}$. Hence by simply changing the sign of F(s), X_{2s} becomes the highest degree reactance and we obtain a different circuit configuration. In addition to this sign choice, we also have a choice in selecting the single real axis reflection zero from either the right half (s = +0.8) or left half (s = -0.8) plane. Thus, for our example, we have a choice of four different reflection zero polynomials which we might have used, these are given below.

$$\left.\begin{aligned} F_1(s) &= s^3 + 0.8s^2 + 0.64s + 0.512 \\ F_2(s) &= -s^3 - 0.8s^2 - 0.64s - 0.512 \end{aligned}\right\} \quad \text{using } s = -0.8$$

$$\left.\begin{aligned} F_3(s) &= -s^3 + 0.8s^2 - 0.64s + 0.512 \\ F_4(s) &= s^3 - 0.8s^2 + 0.64s - 0.512 \end{aligned}\right\} \quad \text{using } s = +0.8$$

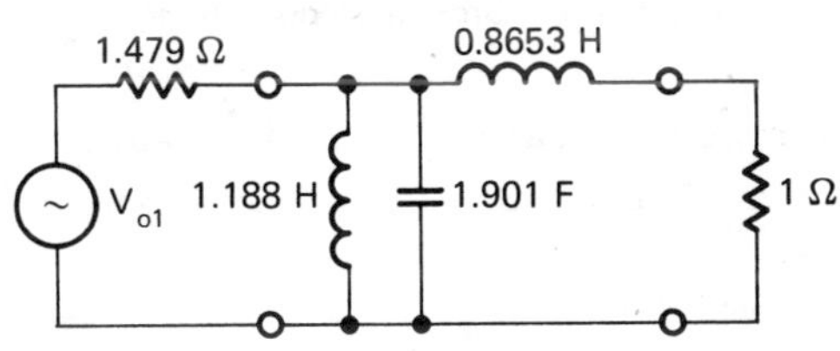

Figure 7.30 Complete realization for Example (7.1).

The reader is urged to work out the detailed synthesis of networks using these choices, as an exercise.

Example 7.2

The first step in the synthesis of an LC ladder realization is to sketch a possible circuit configuration using the removal operations and the rules for their application as discussed in the above. In this example we will illustrate this phase of the synthesis process by considering the bandpass filter whose transfer function was obtained in Example 5.3. In the following example we will discuss the results of carrying out the second phase of the synthesis process; namely, finding the values of the circuit elements. It should also be mentioned that this same filter transfer function will be realized in subsequent Chapters by a variety of active design techniques, thus allowing comparisons to be made.

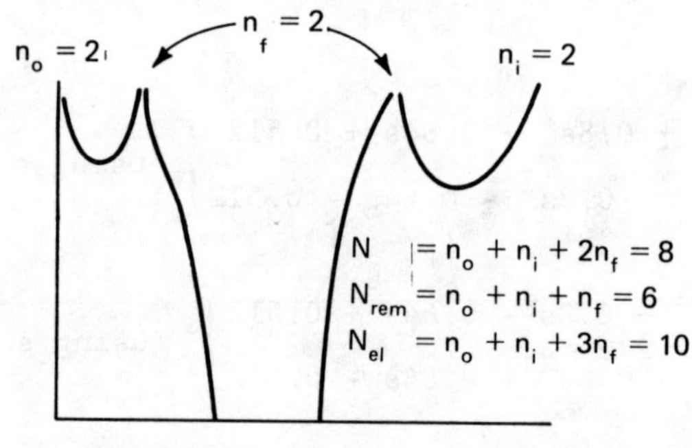

Figure 7.31 Order 8 bandpass filter.

The filter under consideration has the bandpass attenuation function sketched in Fig. 7.31. For our purpose here the pertinent parameters are:

$n_o = 2$, $n_i = 2$ and $n_f = 2$
One loss-peak in each of the two stopbands
All reflection zeros on the $j\omega$-axis

The order N may be computed as

$$N = n_o + n_i + 2\,n_f = 8 .$$

The required number of removal operations of types 1 to 8 (Table 7.4) is given by

$$N_{rem} = n_o + n_i + n_f = 6,$$

which result in the number of reactive (L and C) elements given by,

$$N_{e\ell} = n_o + n_i + 3\,n_f = 10 \quad .$$

In order to sketch possible circuit realizations, we first have to investigate the character of the various design impedances for which formulae are given in Table 7.3. Towards this end we make the following observations:

(1) Since all the reflection zeros lie on the $j\omega$-axis the only choice in calculating F(s) is its sign. Thus F(s) is an even polynomial of order eight to which we may assign either a positive or a negative sign.

(2) Since P(s) is a sixth order polynomial and since the transfer function has two loss poles at $s = \infty$, it follows from Feldtkeller's equation

$$H(s)\,H(-s) = 1 + K(s)\,K(-s) \quad ,$$

that the leading coefficients of E(s) and F(s) are equal.

Using the above two observations together with the expressions of the design impedances given in the left side column of Table 7.3 enable us to determine the class and degree of the design reactances as follows:

Since we have to select the reactance of highest order (eight) for synthesis it follows that there are four different possibilities. These are X_{1s} and X_{2o} for the case a positive sign is assigned to F(s), and X_{1o} and X_{2s} for the case a negative sign is assigned to F(s). The question now arises as to which of the four reactances should one choose for developing the LC ladder.

The answer to this question is quite simple; namely, any one of the four will do. Specifically, if we develop a ladder from one of the four possible reactances, the three ladders corresponding to the remaining three reactances can be easily obtained from the original ladder. This will be obvious by examining the expressions in the above table. For the positive F(s) case note that X_{1s} is the inverse function of X_{2o}. Thus, taking the dual of the network synthesized from X_{1s} and turning it end-for-end results in the network corresponding to X_{2o}. Similar comment applies to the networks obtained from X_{1o} and X_{2s}, in the case F(s) is negative.

$F(s) = +F_{ev}$			$F(s) = -F_{ev}$		
Formula	Degree	Class	Formula	Degree	Class
$\frac{X_{1o}}{R_1} = \frac{E_{ev} - F_{ev}}{E_{od}}$	7	1	$\frac{X_{1o}}{R_1} = \frac{E_{ev} + F_{ev}}{E_{od}}$	8	3
$\frac{X_{1s}}{R_1} = \frac{E_{od}}{E_{ev} + F_{ev}}$	8	0	$\frac{X_{1s}}{R_1} = \frac{E_{od}}{E_{ev} - F_{ev}}$	7	2
$\frac{X_{2o}}{R_2} = \frac{E_{ev} + F_{ev}}{E_{od}}$	8	3	$\frac{X_{2o}}{R_2} = \frac{E_{ev} - F_{ev}}{E_{od}}$	7	1
$\frac{X_{2s}}{R_2} = \frac{E_{od}}{E_{ev} - F_{ev}}$	7	2	$\frac{X_{2s}}{R_2} = \frac{E_{od}}{E_{ev} + F_{ev}}$	8	0

We also note from the above table that X_{1s} for positive F(s) is the dual of X_{1o} for the negative F(s). Similarly, X_{2o} for positive F(s) produces a dual network to that developed from X_{2s} for the negative F(s) case.

To summarize, we need only use one of the four possible reactances to synthesize the LC ladder. Three other realizations may be obtained from this original ladder by turning the network end-for-end, taking its dual, and taking the dual turned end-for-end.

Once we decide on a particular reactance for developing the

ladder, we can follow the pole removal rules with the assistance of the flow chart of Fig. 7.25 to sketch a number of possible circuit configurations. It should be obvious that a large number of possible circuits exist. In Fig. 7.32 we show three possible circuits (obtained from X_{1o} for the case $F(s) = -F_{ev}$) together with their duals (corresponding to X_{1s} for the case $F(s) = +F_{ev}$). Also shown are the removal sequences as well as the corresponding paths on the flow chart. The reader is urged to check the details of the removal sequences on the state diagrams.

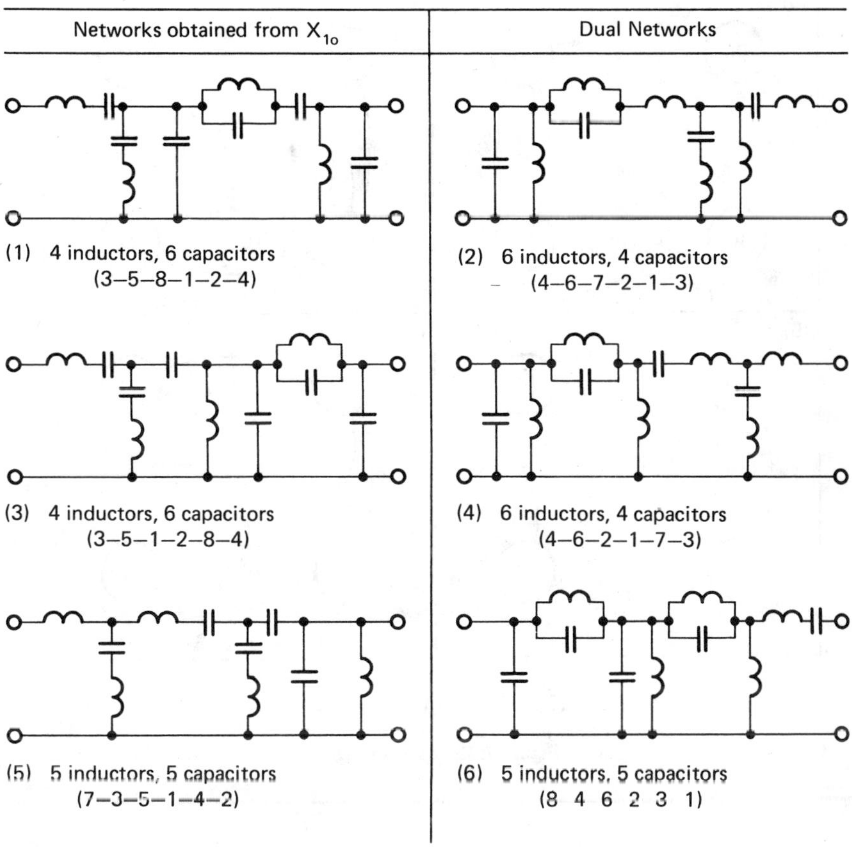

(a)

Figure 7.32 (a) Six network realizations of the order 8 bandpass filter of Example 7.2; (b) the removal sequences corresponding to the networks in (a).

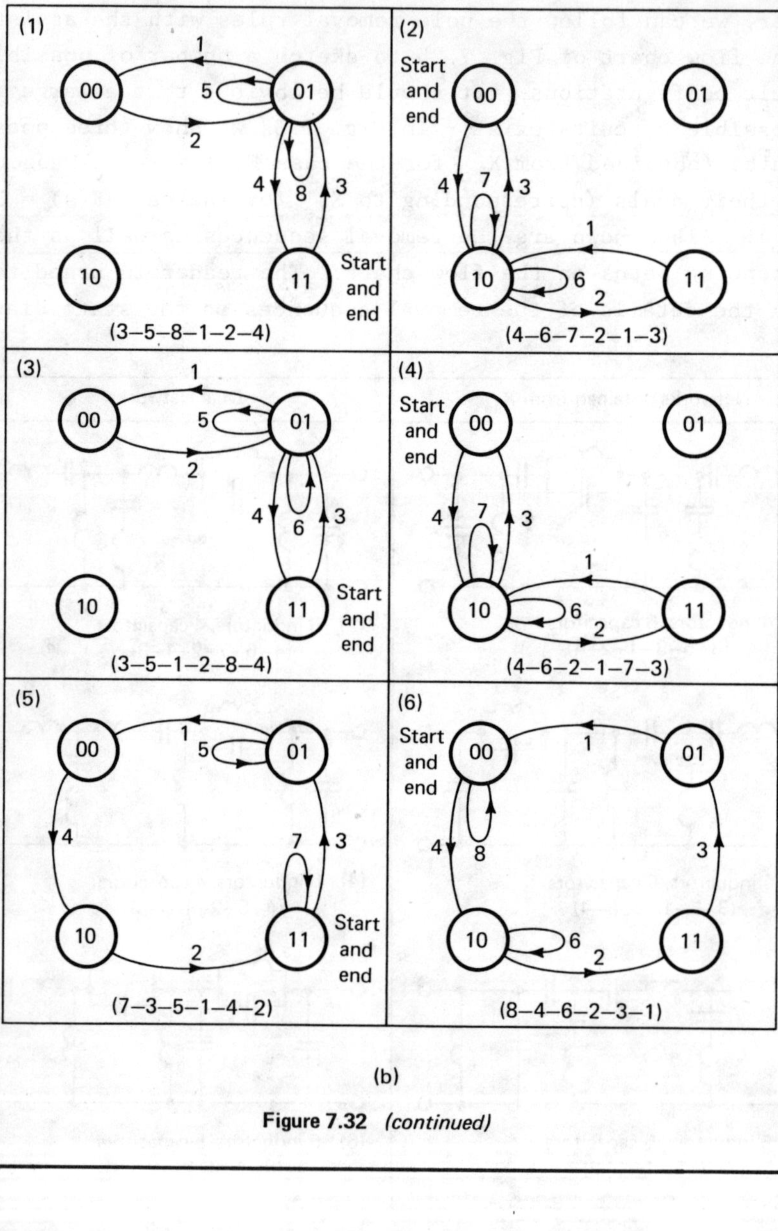

(b)

Figure 7.32 *(continued)*

Example 7.3

In this example we wish to complete the synthesis of an LC ladder realization of the bandpass filter considered above. We have seen that there exist a large number of possible circuit configurations. The designer can sketch these possible circuits using the techniques developed and illustrated in the above. Subsequently, the designer must make some kind of choice as to the "best" configuration. The problem here being the determination of what is meant by "best".

Here we should recall that in general there is no guarantee that the synthesis of any of the possible configurations will result in all positive elements. Clearly, one criterion for a passive filter design is that the best configuration should have the fewest number of inductors. A further criterion for passive design is, of course, that all elements should be nonnegative. (Such is not the case for active designs.) With the considerations of the number of inductors and positive valued elements aside, there will still remain, for a high order network, a large number of possible configurations. Each of these remaining filter realizations will have a different set of element values. The designer will usually choose the design which contains the smallest "spread" or ratio of like element values to minimize the cost, parasitic effects, and the "spread" of maximum inductor currents and maximum capacitor voltages which tend to reduce the dynamic range of the filter.

To carry out the synthesis process, a computer program that implements the procedure described earlier in this section is normally required. Such a program is available in the FILTOR 2 package. For the example in hand the approximation problem has been already solved in Example 5.3 and many of the transfer function parameters were also given in Example 2.2. We now discuss two different realizations for this bandpass filter, obtained using an LC synthesis program.

The two complete realizations are given in Fig. 7.33. The first realization was synthesized using the guidelines given earlier in this Section. It is a minimum inductance design with four inductors and six capacitors, the removal sequence is (3-5-8-1-4-2). The impedance level spread is $(R_2/R_1) = 1.888$, the inductance spread is $(L_{max}/L_{min}) = 11.442$, and the capacitance spread is $(C_{max}/C_{min}) = 43.02$ with no negative elements.

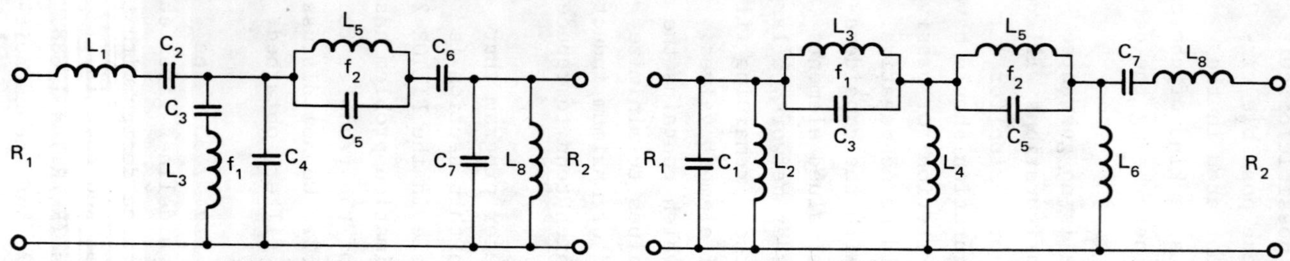

Removal sequence 3–5–8–1–4–2

$R_1 = 1.000\ \Omega$ $R_2 = 1.888\ \Omega$

$L_1 = 626.0\ \mu H$ (max)
$L_3 = 63.79\ \mu H$
$L_5 = 231.3\ \mu H$
$L_8 = 54.71\ \mu H$ (min)

$C_2 = 29.49\ \mu F$ (min)
$C_3 = 1268.7\ \mu F$ (max)
$C_4 = 361.8\ \mu F$
$C_5 = 35.75\ \mu F$
$C_6 = 37.99\ \mu F$
$C_7 = 314.1\ \mu F$

Impedance spread = 1.888
Inductor spread = 11.442
Capacitor spread = 43.02

$f_1 = 559.4$ Hz

Removal sequence 4–6–6–2–1–3

$R_1 = 1.000\ \Omega$ $R_2 = 225.6\ \Omega$

$L_2 = 294.9\ \mu H$ (min)
$L_3 = 1268.7\ \mu H$
$L_4 = 361.8\ \mu H$
$L_5 = 3661.0\ \mu H$
$L_6 = -7467.0\ \mu H$ (negative)
$L_8 = 1412600.0\ \mu H$ (max)

$C_1 = 626.0\ \mu F$
$C_3 = 637.9\ \mu F$ (max)
$C_5 = 2.259\ \mu F$
$C_7 = 0.12843\ \mu F$ (min)

Impedance spread = 225.6
Inductor spread = 4790.0
Capacitor spread = 4967.0

$f_2 = 1750.1$ Hz

Figure 7.33 Two realizations of an order 8 bandpass filter.

This design is a reasonable and practical starting point for further transformations. This design (scaled up to a 600 Ω impedance level) is the circuit used in Example 6.9 and shown in Fig. 6.23, where the design was completed to a practical conclusion. The reader should return now to that example and review the details.

The second realization of Fig. 7.33 did not follow all the guidelines of this Section. The removal sequence is (4-6-6-2-1-3) resulting in six inductors and four capacitors (the dual network would be more desirable here), and in particular, the removal operation number 6 is used to produce a loss peak in the upper stopband. This latter operation violates our guidelines and the result is that L_6 (marked by an asterisk) is negative. The impedance level spread $(R_2/R_1) = 225.6$ is excessive, with very large inductance and capacitance spreads of $(L_{max}/L_{min}) = 4790.0$ and $(C_{max}/C_{min}) = 4967.0$. Clearly this is a very unsuitable design! It is not possible in the space allowed here to examine all of the other possibilities for this filter. Suffice it to say that the designer of passive filters has not completed his design once the approximation problem is solved, and there remain many pitfalls for him to avoid before completing a successful design.

From the discussions above the reader can appreciate the fact that good passive filter design is somewhat of an art and that to date there has not yet appeared a step-by-step scientific method of arriving at "the optimum" design. In fact, there is much room for improvement of the design techniques, and filter designers will welcome any further contributions to the area which might eliminate much of the trial-and-error and dependence upon past experience required to produce successful designs.

Numerical Considerations

In our discussions to date, we have neglected the very importand numerical conditioning problem inherent in the synthesis methods described here. The synthesis problem is, if anything, more difficult than the approximation problem in terms of the numerical ill-conditioning that may be encountered. This is due to the large number of subtractions of nearly equal numbers required by the pole removal operations on a high order filter. This numerical ill-conditioning is also a function of the filter selectivity, and it is not uncommon to encounter numerical difficulties with highly selective filters of moderate order. Much research has gone into the solution of this problem, and at present there seems to be two competing techniques for overcoming the problem. As with the approximation problem, it has been found that the use of either a transformation technique making use of the z-variable throughout the synthesis removal operations, or the so-called product method where all manipulations are done on polynomials in factored form, is required for the successful computation of the element values of most practical filters. When a computer program for synthesis is used, the results should always be cross-checked by an analysis program or else the network should be resynthesized from another reactance before the element values are accepted with confidence. It is common to synthesize all filters from both ends (from left and then right) and compare the ratios of like elements from each synthesis as a check of accuracy.

A complete set of synthesis routines based on the product method has been prepared and one is available in FILTOR 2, a description of the routines and their use is provided in the FILTOR 2 manual. All of the synthesis examples of this book have been computed by these routines. At the present time it is almost unthinkable to design any practical non-tabulated filter without computer aid.

7.9 SENSITIVITY ANALYSIS

A factor of great importance to any filter designer is the effect of non-zero element tolerances and parasitic elements on the performance of a particular filter. It is of little use to expend time and effort to obtain an ideal transfer function which "optimally" meets a set of design specifications, such as that illustrated by the solid line curve of Fig. 7.34, only to find, after the filter has been constructed using practical components, that the actual filter response, shown by the dotted curve of Fig. 7.34, does not satisfy the specifications. The dotted curve illustrates a "typical" measured filter attenuation response, the attenuation deviation denoted by ΔA in Fig. 7.34, is a quantity of great interest to the designer.

If the designer had been able to predict in advance of the actual assembly of the filter the expected deviation ΔA or even a bound on the deviation, then it might have been possible for him to "overdesign" the original ideal transfer function by an appropriate amount to allow for the variation. In the example we have been discussing, it has been tacitly assumed that ΔA is due only to inexact initial element values and parasitic effects, in practice a filter must satisfy its specifications over a range of ambient conditions as well. Of course, if the network meets the specifications at some nominal set of ambient conditions, it may not do so when such conditions change, causing the filter components to change, and hence the transfer function to vary. In this Section we shall study transfer function variability and present several results which are very useful in predicting such variations as well as indicating some methods of designing around these effects.

The Sensitivity Function

Before proceeding to more specific matters we shall introduce the so-called "sensitivity function" which the reader will again encounter in later Chapters. The sensitivity function plays an important practical and conceptual role in transfer function variability studies. Sensitivity is a measure of the change of the over-all transfer function* with respect to the change of an

---

*Of course sensitivity functions may be defined for any type of network function such as impedance or admittance as well as transfer functions.

element or parameter of interest in the network. To introduce these ideas, consider the Taylor's series expansion of a transfer function $H(s) = e^{\Theta}$ with respect to some parameter denoted by x, here we have

$$H(s, x) = H(s, x_o) + \left.\frac{\partial H(s, x)}{\partial x}\right|_{x=x_o} (x - x_o) + \left.\frac{\partial^2 H(s, x)}{\partial x^2}\right|_{x=x_o} (x - x_o)^2 + \ldots \tag{7.15}$$

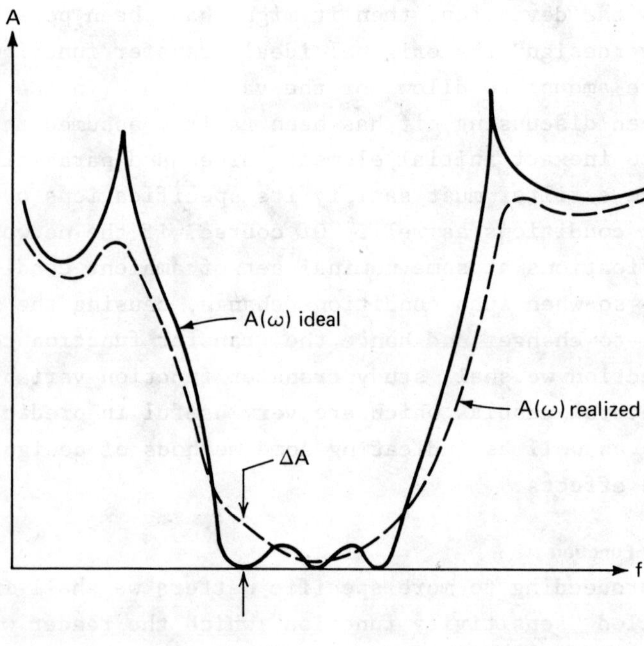

Figure 7.34 Ideal and actual filter responses.

as an infinite series, where x_o is the nominal or ideal parameter value, and $H(s, x_o)$ is the nominal or ideal transfer function. Taking $\Delta x = (x - x_o)$ as the parameter deviation, and $\Delta H = H(s, x) - H(s, x_o)$ as the transfer function deviation, we may rewrite (7.15) in the form

$$\Delta H = \left.\frac{\partial H(s,\ x)}{\partial x}\right|_{x=x_o} (\Delta x) + \left.\frac{\partial^2 H(s,\ x)}{\partial x^2}\right|_{x=x_o} (\Delta x)^2 + \ldots \quad . \tag{7.16}$$

Of course (7.16) involves an infinite number of terms containing powers of the deviation Δx and higher order derivatives, but for sufficiently "small" Δx, and "well behaved" high order derivatives we may make the first order approximation

$$\Delta H \simeq \left.\frac{\partial H(s,\ x)}{\partial x}\right|_{x=x_o} (\Delta x) \tag{7.17}$$

to obtain the transfer function deviation. The reader is cautioned here again that (7.17) depends upon Δx being "small" and the higher order derivatives being "well behaved". This may not always be the case and first order sensitivity studies, such as those to be presented here, may not be relied upon when such pathological cases arise.

The first order derivative in (7.17) is of course a measure of the variability or sensitivity of H to the parameter x, but for many purposes it has been found more convenient to deal with a normalized measure which relates per-unit changes in H to per-unit changes in x. Thus, the "first order sensitivity function" or simply sensitivity function $S_x^H(s)$ is introduced where by definition

$$S_x^H(s) = \lim_{\Delta x \to 0} \frac{[\Delta H/H]}{[\Delta x/x]} = \frac{x}{H}\frac{\partial H}{\partial x} = \frac{\partial[\ln H]}{\partial[\ln x]} \quad .$$

We note that multiplying S_x^H by the per-unit or percentage change in the parameter $(\Delta x/x)$ yields the corresponding approximate per-unit or percentage change in the transfer function $(\Delta H/H)$. The reader will note that $S_x^H(s)$ is in general a function of the complex frequency variable s, and that as such it is difficult to relate $S_x^H(s)$ to the practical measurements of attenuation and phase or group delay.

Recalling the definitions of attenuation and phase lag through the effective transmission constant Θ, where $H = e^{\Theta}$, we have

$$\Theta = \ln H = \alpha + j\phi \quad .$$

Where $\alpha(\omega) = \ln|H(j\omega)|$ is the attenuation in Nepers, while $\phi(\omega) = \arg[H(j\omega)]$ is the phase lag function in radians. Now using Θ we may express $S_x^H(j\omega)$ as

$$S_x^H(j\omega) = x\frac{\partial \ln H}{\partial x} = x\frac{\partial \Theta}{\partial x} = x[\frac{\partial \alpha}{\partial x} + j\frac{\partial \phi}{\partial x}] \quad . \tag{7.19}$$

Equating real and imaginary parts in (7.19), we obtain

$$R_e[S_x^H(j\omega)] = x\frac{\partial \alpha(\omega)}{\partial x} \tag{7.20}$$

and

$$I_m[S_x^H(j\omega)] = x\frac{\partial \phi(\omega)}{\partial x} \tag{7.21}$$

We also note that since $\alpha(\omega) = \ln|H(j\omega)|$, we may express (7.20) as

$$R_e[S_x^H(j\omega)] = x\frac{\partial \ln|H(j\omega)|}{\partial x} = S_x^{|H|}(j\omega) \quad . \tag{7.22}$$

Now (7.20) through (7.22) relate the practical quantities α and ϕ to the sensitivity function. Using these expressions we may find expressions for S_x^{α} and S_x^{ϕ} as

$$S_x^{\alpha} = (\frac{x}{\alpha})(\frac{\partial \alpha}{\partial x}) = \frac{1}{\alpha} R_e[S_x^H] = \frac{S_x^{|H|}}{\alpha} \tag{7.23}$$

and

$$S_x^{\phi} = (\frac{x}{\phi})(\frac{\partial \phi}{\partial x}) = \frac{1}{\phi} I_m[S_x^H] \quad . \tag{7.24}$$

However, since Θ is a logarithmic measure of H, the convenience associated with a normalized sensitivity relating per-unit quantities is not of direct consequence here and the absolute deviations $\Delta\alpha$ and $\Delta\phi$ in terms of per-unit parameter variations are more useful. Thus, as a first order approximation we may write

$$\Delta\alpha(\omega) = R_e[S_x^H(j\omega)](\frac{\Delta x}{x}) \; N_p \tag{7.25}$$

or,

$$\Delta\alpha(\omega) = S_x^{|H|}(j\omega)(\frac{\Delta x}{x}) \; N_p \tag{7.26}$$

and,

$$\Delta\phi(\omega) = I_m[S_x^H(j\omega)](\frac{\Delta x}{x}) \; \text{rad} \quad , \tag{7.27}$$

for the absolute deviations in α and ϕ. Finally, recalling the definition of group delay $\tau(\omega) = (d\phi/d\omega)$, we may formally write the deviation $\Delta\tau(\omega)$* as

$$\Delta\tau(\omega) = \frac{\partial}{\partial\omega}\{I_m S_x^H(j\omega)\}(\frac{\Delta x}{x}) \quad . \tag{7.28}$$

General Sensitivity Relations For Passive RLC Networks

In this section we present some general sensitivity relations for passive RLC networks which provide valuable insight into the effects of parameter variations and which will be used later to derive bounds on filter performance. Consider a general RLC network, depicted in Fig. 7.35, containing $n = r + \ell + c$ elements where r, ℓ and c denote the total number of resistors, inductors and capacitors, respectively. The kth element of each type will

---

*This expression is subject to the validity of interchanging the pertinent differentiations.

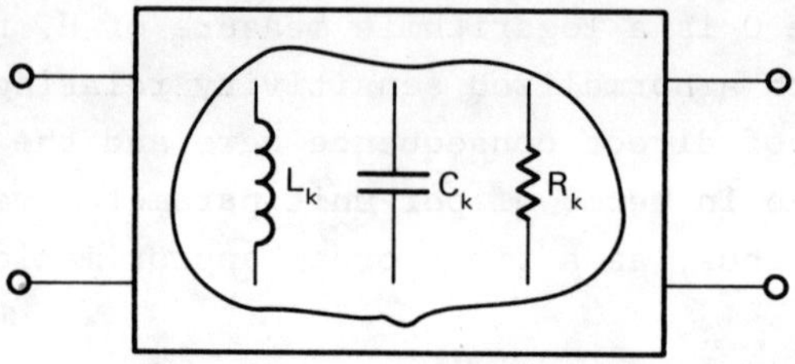

Figure 7.35 General RLC network.

be denoted by R_k, L_k and C_k, respectively, and where it is unimportant to differentiate elements we shall simply denote the elements by x_k. Consider any voltage (or current) transfer function defined for this network and let it be denoted by H(s). As an example, H(s) might be the transfer function of a doubly terminated network as defined in previous sections. We now let all components have identical per-unit variation such that the impedance of all components is multiplied by the same scale factor $\gamma = (1 + \Delta)$. Thus we obtain a new set of element values denoted by primes and related to the original set by $R_k' = \gamma R_k$, $L_k' = \gamma L_k$ and $C_k' = C_k/\gamma$. By the impedance scaling property of networks described in Chapter 6, the transfer function of the perturbed network is identical to that of the original network, thus we can write

$$H(R_k', L_k', C_k', s) = H(R_k, L_k, C_k, s) \quad .$$

Differentiating both sides with respect to the scale factor γ yields

$$\frac{\partial H(R_k', L_k', C_k', s)}{\partial \gamma} = 0.$$

Writing the left hand side explicitly in terms of the separate element types and using the chain rule results in

$$\sum_{k=1}^{r} \frac{\partial H(R_k^\backprime, L_k^\backprime, C_k^\backprime, s)}{\partial R_k^\backprime} \frac{dR_k^\backprime}{d\gamma} + \sum_{k=1}^{\ell} \frac{\partial H(R_k^\backprime, L_k^\backprime, C_k^\backprime, s)}{\partial L_k^\backprime} \frac{dL_k^\backprime}{d\gamma}$$

$$+ \sum_{k=1}^{c} \frac{\partial H(R_k^\backprime, L_k^\backprime, C_k^\backprime, s)}{\partial C_k^\backprime} \frac{dC_k^\backprime}{d\gamma} = 0 \quad .$$

Noting that $dR_k^\backprime/d\gamma = R_k$, $dL_k^\backprime/d\gamma = L_k$ and $dC_k^\backprime/d\gamma = -C_k/\gamma^2$, we substitute and then set $\gamma = 1$. Finally, dividing the result throughout by H yields the interesting sensitivity relationship*

$$\sum_{k=1}^{r} S_{R_k}^{H}(s) + \sum_{k=1}^{\ell} S_{L_k}^{H}(s) - \sum_{k=1}^{c} S_{C_k}^{H}(s) = 0 \quad . \qquad (7.29)$$

Equation (7.29) has an intuitive appeal since it confirms the designer's feeling that in a lossless network choosing inductive and capacitive elements with complementary temperature coefficients should yield a first order compensation of H(s) for temperature drift.

Another interesting relationship may be derived by considering the frequency scaling property of networks, described in Chapter 6. We now let the reactive elements of the network be scaled such that the impedance level is unchanged but the frequency response is scaled. Specifically we take the new set of elements as $L_k^\backprime = \gamma L_k$, $C_k^\backprime = \gamma C_k$ and $R_k^\backprime = R_k$. Now the transfer function of the perturbed and nominal networks are related by

$$H(L_k^\backprime, C_k^\backprime, s) = H(L_k, C_k, \gamma s) \quad .$$

Differentiating both sides with respect to γ yields,

*Note that if H(s) is not a voltage or current ratio, then the result (7.29) would not be correct. The corresponding results for impedance or admittance functions are easily derived.

$$\sum_{k=1}^{\ell} \frac{\partial H(L_k', C_k', s)}{\partial L_k'} \frac{dL_k'}{d\gamma} + \sum_{k=1}^{c} \frac{\partial H(L_k', C_k', s)}{\partial C_k'} \frac{dC_k'}{d\gamma}$$

$$= \frac{\partial H(L_k, C_k, \gamma s)}{\partial(\gamma s)} \frac{d(\gamma s)}{d\gamma}$$

Using $x_k' = \gamma x_k$, setting $\gamma = 1$ and dividing through by H results in the interesting sensitivity relation

$$\sum_{k=1}^{\ell} S_{L_k}^{H}(s) + \sum_{k=1}^{c} S_{C_k}^{H}(s) = \frac{s}{H}\frac{dH}{ds} = \frac{d(\ln H)}{d(\ln s)} \tag{7.30}$$

Here for $s = j\omega$, we have $ds = jd\omega$ with $\ln H = \alpha + j\phi$ and $\tau(\omega) = d\phi/d\omega$ so that (7.30) may be rewritten in terms of $\alpha(\omega)$ and $\tau(\omega)$ as

$$\sum_{k=1}^{\ell} S_{L_k}^{H}(j\omega) + \sum_{k=1}^{c} S_{C_k}^{H}(j\omega) = [\frac{\omega d\alpha(\omega)}{d\omega} + j\omega\tau(\omega)] \tag{7.31}$$

This relation is significant because it relates the sensitivity functions to the network transfer function parameters $\alpha(\omega)$ and $\tau(\omega)$ which are independent of the specific topology or set of element values used in the realization. In fact, bounds on $d\alpha/d\omega$ and $\tau(\omega)$ are usually set by the filter specifications prior to any filter design effort. In addition, (7.31) has an intuitive appeal since it implies that the sensitivities are somehow larger in regions of "steep" attenuation characteristics where generally both $\tau(\omega)$ and $d\alpha/d\omega$ are large. This effect is of course born out by the practical experience of any network designer.

Termination Sensitivities

In the case of a passive network operating between resistive loads, there is a special coupling relationship between the network and its terminations. This in turn leads to a special relationship between the termination sensitivities and the reflection coefficients.

Consider a passive two-port network, characterized in terms of its open-circuit impedance parameters, and terminated at port-1 in a voltage source V_{o1} having an internal resistance R_1 and at port-2 in a resistance R_2. It can be easily shown that the transfer impedance Z_{21} is given by

$$Z_{21} = \frac{V_2}{I_1} = \frac{z_{12}\, R_2}{z_{22} + R_2} \quad .$$

Also, the transfer function H(s) is given by

$$H(s) = \frac{1}{2}\sqrt{\frac{R_2}{R_1}}\ \frac{V_{o1}}{V_2}$$

These two equations, together with the relationship $V_{o1} = I_1(R_1 + Z_1)$, where Z_1 is the input impedance at port-1, enable us to write

$$H(s) = \frac{1}{2}\sqrt{\frac{R_2}{R_1}}\ \frac{R_1 + Z_1}{Z_{21}} \quad .$$

Noting that Z_{21} and Z_1 are independent of R_1, we may differentiate to obtain

$$\frac{\partial H}{\partial R_1} = \frac{1}{4}\sqrt{\frac{R_2}{R_1}}\ \frac{R_1 - Z_1}{R_1\, Z_{21}} \quad ,$$

and the sensitivity function is easily obtained as

$$S^H_{R_1} = \frac{\partial H}{\partial R_1}\ \frac{R_1}{H} = \frac{1}{2}\ \left(\frac{R_1 - Z_1}{R_1 + Z_1}\right)$$

We recognize the quantity on the right-hand-side as directly related, by definition, to the front-end reflection coefficient ρ_1 since

$$\rho_1 = \frac{Z_1 - R_1}{Z_1 + R_1}$$

The sensitivity expression may, therefore, be expressed as

$$S_{R_1}^{H} = -\frac{\rho_1}{2} \tag{7.32}$$

By an analogous procedure we can obtain

$$S_{R_2}^{H} = -\frac{\rho_2}{2} \tag{7.33}$$

where ρ_2 is the back-end reflection coefficient.

Equations (7.32) and (7.33) are the required results from which we note that the only latitude the designer has in controlling these sensitivities is through the reflection coefficients. Clearly, the designer should attempt to make $|\rho|$ as small as possible over the frequency band of interest (usually the passband) which implies as small a passband distortion (A_p in dB) as possible. In addition, we note that if the designer has the ability to design the network for maximum power transfer to the load in the passband, resulting in a small $|\rho|$, then such a choice leads to small termination sensitivities throughout the passband.

Sensitivity Analysis For Doubly Terminated Lossless Networks

In general for an arbitrary RLC network there is not much more that can be determined easily about the variability of the transfer function or about the sensitivity functions. However, in the special case of a lossless network terminated in resistive loads it is possible to obtain further interesting and practical results. In addition, if the filter network has been designed for maximum power transfer from source to load, it can be shown that the transfer function sensitivities are in some sense as "small" as possible. We shall introduce this section with an heuristic argument due to Orchard and then proceed to the derivation of estimates and bounds on the network performance.

Consider the lossless filter operating between resistances and designed for maximum power transfer throughout the passband. Here at the reflection zeros, which are distributed throughout the passband and denoted by ω_{rk}, the attenuation is, by design, zero. That is, $\alpha(\omega_{rk}) = 0$, and elsewhere in the passband $\alpha(\omega)$ is bounded by the design parameter A_p. Now consider the effect on $\alpha(\omega)$ of perturbing any of the lossless elements from its ideal or nominal value. Since the network is lossless, the only effect that a change in any inductance L_k or capacitance C_k can have on the loss at a reflection zero is to increase it, since the lossless network can have no power gain. Thus, at least at reflection zeros, we can heuristically assume that

$$\frac{\partial\alpha(\omega_{rk})}{\partial L_k} = \frac{\partial\alpha(\omega_{rk})}{\partial C_k} = 0.$$

This situation is illustrated in Fig. 7.36 which shows the manner in which $\alpha(\omega_r)$ might vary with a parameter value.

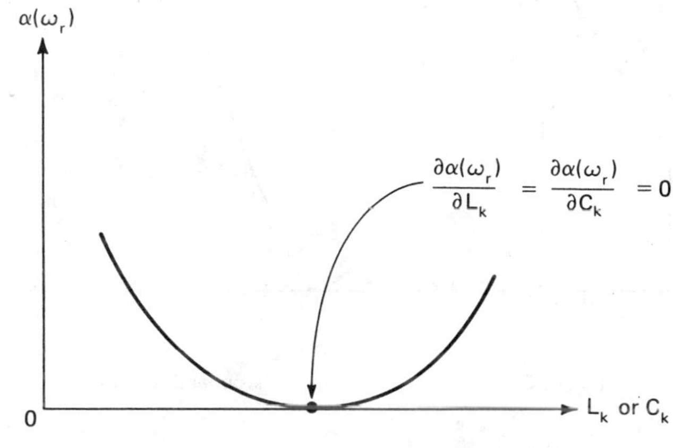

Figure 7.36 Illustrating the "small" passband sensitivities of doubly terminated lossless networks designed for maximum power transfer.

From this heuristic argument we see that $\Delta\alpha(\omega)$ must somehow be "small" throughout the passband of a properly designed doubly terminated lossless filter network. This, however, is not the case in the stopband of such a filter. In particular, at a loss pole, denoted by ω_ℓ, the deviation $\Delta\alpha(\omega_\ell)$ becomes infinite as depicted in Fig. 7.37. However, since for "filtering" applications the object is to pass undistorted signals in the passband while completely rejecting stopband signals, the effect of variations in the stopband are usually of far less importance than variations in the passband. In effect the designer does not care if the signals he is rejecting are distorted as long as the level of rejection is high enough.

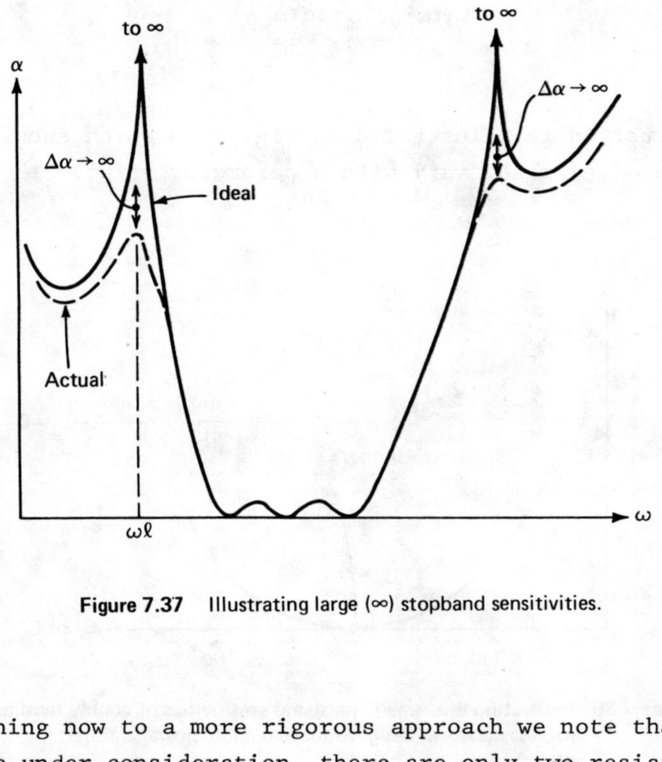

Figure 7.37 Illustrating large (∞) stopband sensitivities.

Returning now to a more rigorous approach we note that, for the case under consideration, there are only two resistors involved; namely, R_1 and R_2, so that we can write using (7.32) and (7.33)

$$\sum_{k=1}^{r} S_{R_k}^{H} = (S_{R_1}^{H} + S_{R_2}^{H}) = -\frac{1}{2}(\rho_1 + \rho_2) \quad . \tag{7.34}$$

We may now substitute this expression in (7.29) yielding

$$\sum_{k=1}^{\ell} S_{L_k}^{H} - \sum_{k=1}^{c} S_{C_k}^{H} = \left(\frac{\rho_1 + \rho_2}{2}\right) \quad , \tag{7.35}$$

which together with (7.30) which we repeat here,

$$\sum_{k=1}^{\ell} S_{L_k}^{H} + \sum_{k=1}^{c} S_{C_k}^{H} = \frac{d(\ln H)}{d(\ln s)} \tag{7.36}$$

may be solved simultaneously for the separate sums $\sum S_{L_k}^{H}$ and $\sum S_{C_k}^{H}$. The resulting expressions are

$$\sum_{k=1}^{\ell} S_{L_k}^{H} = \frac{1}{2}\left[\frac{d(\ln H)}{d(\ln s)} + \frac{(\rho_1 + \rho_2)}{2}\right] \tag{7.37}$$

and

$$\sum_{k=1}^{c} S_{C_k}^{H} = \frac{1}{2}\left[\frac{d(\ln H)}{d(\ln s)} - \frac{(\rho_1 + \rho_2)}{2}\right] \tag{7.38}$$

This is a very significant result since the inductive and capacitive sensitivity sums are now explicitly known in terms of transfer function and reflection parameters for a lossless doubly terminated network.

We may now write, as a first order approximation to the transfer function variability for the doubly terminated lossless network, the expression

$$\frac{\Delta H}{H} \sim \left[S_{R_1}^{H}\left(\frac{\Delta R_1}{R_1}\right) + S_{R_2}^{H}\left(\frac{\Delta R_2}{R_2}\right)\right] + \sum_{k=1}^{\ell} S_{L_k}^{H}\left(\frac{\Delta L_k}{L_k}\right) + \sum_{k=1}^{c} S_{C_k}^{H}\left(\frac{\Delta C_k}{C_k}\right).$$

Assuming that each element type has a constant variation or tolerance given by $\varepsilon_R = \Delta R_1/R_1 = \Delta R_2/R_2$, $\varepsilon_L = (\Delta L_k/L_k)$, $\varepsilon_C =$

$(\Delta C_k/C_k)$, for all k, which approximates the situation of similar components subjected to a changing environment, we may write

$$\frac{\Delta H}{H} \simeq -\varepsilon_R \left(\frac{\rho_1 + \rho_2}{2}\right) + \frac{\varepsilon_L}{2} \left[\left(\frac{\rho_1 + \rho_2}{2}\right) + \frac{d(\ln H)}{d(\ln s)}\right]$$

$$+ \frac{\varepsilon_C}{2} \left[\frac{d(\ln H)}{d(\ln s)} - \left(\frac{\rho_1 + \rho_2}{2}\right)\right] . \qquad (7.39)$$

Of more interest here, as pointed out earlier, are the absolute deviations in α and ϕ (or τ) due to element variations.

For the attenuation deviation we have

$$\Delta\alpha \simeq R_e \left\{ \sum_{k=1}^{n} S_{x_k}^{H} \left(\frac{\Delta x_k}{x_k}\right) \right\}$$

from which we obtain by substitution the attenuation error

$$\Delta\alpha(\omega) \simeq -\varepsilon_R R_e \left(\frac{\rho_1 + \rho_2}{2}\right) + \frac{\varepsilon_L}{2} \left[R_e \left(\frac{\rho_1 + \rho_2}{2}\right) + \frac{\omega d\alpha}{d\omega}\right]$$

$$+ \frac{\varepsilon_C}{2} \left[\frac{\omega d\alpha}{d\omega} - R_e \left(\frac{\rho_1 + \rho_2}{2}\right)\right] . \qquad (7.40)$$

Assuming $\varepsilon_R = 0$, and $\varepsilon_L = \varepsilon_C = \varepsilon_o$ results in

$$\Delta\alpha(\omega) \simeq \varepsilon_o \omega \frac{d\alpha}{d\omega} . \qquad (7.41)$$

Note that at reflection zeros $\frac{d\alpha}{d\omega} = 0$, so that $\Delta\alpha = 0$, confirming our earlier heuristic argument. The result (7.41) is a first order approximation for $\Delta\alpha$ and is considered to be a lower bound on the worst case attenuation variation. It is possible to derive an upper bound for $|\Delta\alpha|$, which we shall denote by $|\Delta\alpha|_{max}$. For the type of networks we are discussing here (lossless, minimum phase, doubly terminated) Temes and Orchard have shown that

$$|\Delta\alpha|_{max} < \frac{|\varepsilon_o||\rho|\omega\tau(\omega)}{1 - |\rho|^2} \quad . \tag{7.42}$$

This latter relation is usually easier for the designer to use since it involves only the system functions $|\rho|$ and $\tau(\omega)$, both of which are usually known, while in (7.41) the function $\frac{d\alpha}{d\omega}$ must be determined and it is not usually readily available. In a similar manner, for the phase deviation we obtain

$$\Delta\phi \simeq I_m \{ \sum_{k=1}^{n} S_{x_k}^{H} (\frac{\Delta x_k}{x_k}) \}$$

which results in

$$\Delta\phi(\omega) \simeq -\varepsilon_R I_m (\frac{\rho_1 + \rho_2}{2}) + \frac{\varepsilon_L}{2} [I_m (\frac{\rho_1 + \rho_2}{2}) + \omega\tau]$$
$$+ \frac{\varepsilon_C}{2} [\omega\tau - I_m (\frac{\rho_1 + \rho_2}{2})] \tag{7.43}$$

Errors in the transfer function due to parasitic dissipation in the reactive elements may be treated in much the same manner. We simply consider the dissipations as complex increments of the element values. Thus, making use of the dissipation factors introduced in Section 7.2 we may express the dissipation in capacitive and inductive elements in terms of first order sensitivities by noting that

$$S_{sL_k}^{H} = S_{L_k}^{H} \quad ,$$

$$S_{sC_k}^{H} = S_{C_k}^{H}$$

and taking the element variations as

$$\frac{\Delta L_k}{L_k} = \left(\frac{d_{L_k}}{s}\right)$$

$$\frac{\Delta C_k}{C_k} = \left(\frac{d_{C_k}}{s}\right)$$

Now we may write

$$\frac{\Delta H}{H} \simeq \sum_{k=1}^{\ell} S_{L_k}^{H} \left(\frac{d_{L_k}}{s}\right) + \sum_{k=1}^{c} S_{C_k}^{H} \left(\frac{d_{C_k}}{s}\right) . \tag{7.44}$$

In the case of semi-uniform dissipation where $d_{L_k} = d_L$ and $d_{C_k} = d_C$ for all elements, we obtain a simpler relation. In particular, for the attenuation deviation the result is

$$\Delta\alpha(\omega) \simeq \left(\frac{d_L + d_C}{2}\right) \tau(\omega) + \left(\frac{d_L - d_C}{2\omega}\right) I_m\left(\frac{\rho_1 + \rho_2}{2}\right) , \tag{7.45}$$

which in the case of uniform dissipation, $d_L = d_C = d$, reduces to

$$\Delta\alpha(\omega) \simeq d\tau(\omega) . \tag{7.46}$$

These formulae may be rewritten in terms of the Q factors $Q_L = \omega/d_L$ and $Q_C = \omega/d_C$ for the semi-uniform case as

$$\Delta\alpha(\omega) \simeq \frac{1}{2} \left(\frac{1}{Q_L} + \frac{1}{Q_C}\right) \omega\tau(\omega) + \frac{1}{2} \left(\frac{1}{Q_L} - \frac{1}{Q_C}\right) I_m\left(\frac{\rho_1 + \rho_2}{2}\right) , \tag{7.47}$$

and for the uniform case $Q_L = Q_C = Q$ as

$$\Delta\alpha(\omega) \simeq \frac{\omega\tau(\omega)}{Q} . \tag{7.48}$$

Transfer Function Error Prediction

In this sub-section we summarize the results on sensitivity analysis and present some useful bounds together with examples. The estimates and bounds given in the above are extremely useful to a filter designer for they allow him to estimate attenuation deviations in terms of network functions. However, in order to make use of the formulae, the network functions $\rho(\omega)$ and $\tau(\omega)$ must be obtained. Clearly, they can be obtained from a network analysis, but this implies that the network must be designed first. It would be extremely convenient if the expected attenuation deviations could be estimated from the specifications before the network is designed thus enabling the designer to allow appropriate "margins" in his initial design. Then, hopefully the initially overdesigned network will meet its specifications when the deviations occur. Alternatively, the designer could predict before beginning a design if the design is feasible using the components he must work with. In the following, several novel methods of estimating $\Delta\alpha$ from specifications will be presented.

Those terms of the attenuation deviation formulae involving the sum $[\rho_1+\rho_2]$ may be easily bounded since for doubly terminated lossless networks $|\rho_1| = |\rho_2| = |\rho|$, and we have $|\rho_1+\rho_2| \leq 2|\rho|$. Thus, in the passband of filters where the attenuation is bounded by the specifications A_p, we have

$$|\rho|_{max} = \sqrt{1 - 10^{-\frac{A_p}{10}}}$$

so that,

$$|\rho_1+\rho_2| \leq 2|\rho|_{max} \quad .$$

Using this result with (7.40), we can write for the attenuation deviation due to errors in the terminations

$$|\Delta A|_{max} < 8.69|\varepsilon_R|\sqrt{1 - 10^{-\frac{A_p}{10}}}\ \text{dB} \quad , \qquad (7.49)$$

where ε_R is the fractional deviation in the terminating resistances and A_p is obtained from the specifications. For the remaining formulae, we require some method of estimating $\tau(\omega)$ from the filter specifications.

We shall assume in what follows that the order N and distribution of the loss poles at zero (n_o) and infinity (n_i) have been estimated from the attenuation specifications by applying insights gained from the study of filter approximation techniques. Making use of this information, and the fact that almost all of the delay area falls under the passband region of a selective filter network (actually the delay peaks near the band edges where $|d\alpha/d\omega|$ is largest and dips downward in the center of the passband), we may apply equation (1.15) of Chapter 1 to estimate the maximum group delay in the passband, denoted here by τ_{max}. A fairly good estimate for τ_{max} is $\hat{\tau}$ defined by

$$\hat{\tau}B = \int_0^\infty \tau(\omega)\,d\omega = \frac{N\pi}{2} \;,$$

which results in

$$\hat{\tau} = \frac{N\pi}{2B} = \frac{N}{4F} \;,$$

where $B = (\omega_2-\omega_1)$ is the passband width of the filter obtained from the specifications, and $F = (f_2-f_1)$ is the same bandwidth measured in Hz. Since the results are only approximate any nominal measure of bandwidth such as the 3 dB bandwidth may be used here. It is also enlightening to calculate the maximum group delay in terms of the quality factor $\hat{Q}$ of the overall filter where $\hat{Q}$ is defined as $\hat{Q} = (f_2/F)$ and is a measure of the filter selectivity. (Here f_2 is the upper passband edge for a bandpass filter or the passband edge for a low pass.) Thus we have

$$\hat{\tau} = \frac{N}{4F} = \frac{N\hat{Q}}{4f_2} \;, \tag{7.50}$$

the maximum group delay proportional to the $\hat{Q}$ of the filter, which of course is an increasing function of filter selectivity.

We may now use these results with (7.42) to obtain an approximate formula for the maximum attenuation deviation $\Delta A_{max} = 8.69\ \Delta\alpha_{max}$, in the passband of a filter. Clearly in the passband region where $|\rho| \ll 1$,

$$|\Delta A|_{max} < \max \frac{8.69|\varepsilon_o\rho|}{1-|\rho|^2}\,\omega\tau < 8.69|\varepsilon_o||\rho|_{max}\,\omega_2\,\tau_{max}\ .$$

Thus, using $\hat{\tau} \simeq \tau_{max}$ we may write

$$\left.\begin{aligned} |\Delta A|_{max} &< \frac{13.65N|\varepsilon_o|f_2}{F}\sqrt{1-10^{-\frac{A_p}{10}}}\ \text{dB} \\ &\text{or in terms of filter } \hat{Q}, \\ |\Delta A|_{max} &< 13.65\ N\hat{Q}|\varepsilon_o|\sqrt{1-10^{-\frac{A_p}{10}}}\ \text{dB} \end{aligned}\right\} \quad (7.51)$$

where all of the parameters N, f_2, F, $\hat{Q}$ and A_p may be obtained directly or indirectly in a simple manner from the attenuation specifications. Clearly, these formulae are very useful as they illustrate explicitly the relationship between selectivity, element value tolerances, and attenuation deviations.

We note that the more highly selective filters will encounter greater deviations for a given element tolerance. As an example of the use of these formulae consider their application to the eighth order filter of the previous Section, Examples (7.2, 7.3). Here F = 400 Hz, $\hat{Q}$ = 3.5, f_2 = 1.4 KHz, N = 8, and A_p = 0.4 dB. Assuming an element value tolerance of 1% (ε_o = 0.01) we can predict $|\Delta A|_{max} < 1.13$ dB.

A similar development using $\hat{\tau}$ in the formula (7.47) for $\Delta\alpha$ due to parasitic effects yields a maximum attenuation deviation

ΔA_{max} as

$$\Delta A_{max} < \frac{6.82\ Nf_2}{F}\left(\frac{1}{Q_L} + \frac{1}{Q_C}\right) + 4.34\sqrt{1 - 10^{-\frac{A_p}{10}}}\left(\frac{1}{Q_L} - \frac{1}{Q_C}\right)\ \text{dB}$$

or in terms of the filter $\hat{Q}$ (7.52)

$$\Delta A_{max} < 6.82N\hat{Q}\left(\frac{1}{Q_L} + \frac{1}{Q_C}\right) + 4.34\sqrt{1 - 10^{-\frac{A_p}{10}}}\left(\frac{1}{Q_L} - \frac{1}{Q_C}\right)\ \text{dB}$$

which again are clearly very useful formulae, using only information derived from the attenuation specifications. Usually in the passband, the second term of these formulae is negligible. Thus assuming an average $Q = (Q_L + Q_C)/2$ over all reactive elements we obtain the following formula for the maximum ΔA in the passband region of most filters.

$$\Delta A_{max} < 13.65\ N\left(\frac{\hat{Q}}{Q}\right)\ \text{dB}\ , \tag{7.53}$$

from which we note that for small deviations, the Q of the elements must be much larger than the filter $\hat{Q}$

As an application of (7.53) we use the eighth-order bandpass filter mentioned in the above. Substituting $N = 8$ and $\hat{Q} = 3.5$ it follows that the average Q of the reactive components should be >400 if the attenuation deviation is to be kept below 1 dB. If one insists on say 0.1 dB deviation then the components Q factor should be of the order of 4000.

7.10 REFERENCES AND SOURCES FOR FURTHER READING

(1) E. A. Guillemin, Synthesis of Passive Networks, Wiley, New York, 1957.

(2) N. Balabanian, Network Synthesis, Prentice-Hall, Englewood Cliffs, N.J. 1958.

(3) W. Cauer, Synthesis of Linear Communication Networks, McGraw-Hill, New York, 1958.

(4) D. F. Tuttle, Network Synthesis, Vol. 1, Wiley, New York, 1958.

(5) M. E. Van Valkenburg, Introduction to Modern Network Synthesis, Wiley, New York, 1960.

(6) L. Weinberg, Network Analysis and Synthesis, McGraw-Hill, New York, 1962.

(7) H. J. Carlin and A. B. Giordano, Network Theory, Prentice-Hall, Englewood Cliffs, N.J. 1964.

(8) D. S. Humphreys, The Analysis, Design and Synthesis of Electrical Filters, Prentice-Hall, Englewood Cliffs, N.J., 1970.

(9) G. Szentirmai (Ed.), Computer Aided Filter Design, IEEE Press, New York, 1973.

(10) G. C. Temes and S. K. Mitra (Eds), Modern Filter Theory and Design, Wiley, New York, 1973.

(11) Special Issue on Modern Filter Design, IEEE Trans. Circuit Theory, Vol. CT-5, No. 4, Dec. 1958.

(12) Special Issue on Modern Filter Design, IEEE Trans. Circuit Theory, Vol. CT-15, No. 4, December 1968.

(13) O. Brune, "Synthesis of a finite two-terminal network whose driving-point impedance is a prescribed function of frequency", J. Math and Phys. Vol. 10, pp. 191-235, 1931.

(14) J. H. Mole, Filter Design Data for Communication Engineers, Wiley, New York, 1952.

(15) S. Darlington, "Synthesis of reactance four-poles", J. Math. Phys., Vol. 18, pp. 257-353, Sept. 1939.

(16) S. Darlington, "A survey of network realization techniques", IRE Trans. Circuit Theory, Vol. CT-2, pp.291-297, 1955.

(17) T. Fujisawa, "Realizability theorem for mid-series or mid-shunt low-pass ladders without mutual induction", IRE Trans. Circuit Theory, Vol. CT-2, No. 4, pp. 320-325, Dec. 1955.

(18) H. Watanabe, "Synthesis of bandpass ladder networks", IRE Trans. on Circuit Theory, Vol. CT-5, pp. 256-264, Dec. 1958.

(19) R. Saal and E. Ulbrich, "On the design of filters by synthesis", IRE Trans. Circuit Theory, Vol. CT 5, pp. 284-327, Dec. 1958. Reprinted in (9).

(20) H. Watanabe, "On the circuit with a minimum number of coils", IRE Trans. Circuit Theory, Vol. CT-7, pp. 77-78, Mar. 1960.

(21) R. Feldtkeller, "Einführung in die Vierpoltheorie der Electrischen Nachrichtentechnik", Hirzel, Stuttgart, 1962.

(22) J. D. MacDonald and G. C. Temes, "A simple method for the predistortion of filter transfer functions", IEEE Trans. Circuit Theory, Vol. CT-10, pp. 447-450, 1963. Reprinted in (9).

(23) Y. Ishizaki and H. Watanabe, "An iterative Chebyshev approximation method for network design", IEEE Trans. Circuit Theory, Vol. CT-15, pp. 326-336, Dec. 1968. Reprinted in (9).

(24) G. Szentirmai, "On the element value of filters", IEEE Trans. Circuit Theory, Vol. CT-16, p. 577, 1969.

(25) J. K. Skwirzynski, "On Synthesis of Filters", IEEE Trans. Circuit Theory, Vol. CT-18, No. 1, pp. 152-163, Jan. 1971. Reprinted in (9).

(26) D. C. Youla, "A tutorial exposition of some key network-theoretic ideas underlying classical insertion-loss filter design", Proc. IEEE, Vol. 59, pp. 760-799, 1971.

(27) G. Szentirmai, "Theoretical basis of a digital computer program for filter synthesis", Proc. 1st Allerton Conf. Circuit and System Theory, pp. 37-49, 1963. Reprinted in (9).

(28) F. F. Kuo and J. F. Kaiser, Eds., System Analysis by Digital Computer, Chapters 5 and 6, Wiley, New York, 1966.

(29) Special Issue on Computer-Aided Design. Proc. IEEE, Vol. 55, No. 11, 1967.

(30) G. C. Temes and D. A. Calahan, "Computer aided network optimization - The state-of-the-art", Proc. IEEE, Vol. 53, pp. 1832-1863, Nov. 1967. Reprinted in (9).

(31) F. F. Kuo and W. Magnuson, Eds., Computer Oriented Circuit Design, Prentice-Hall, Englewood Cliffs, N.J. 1969, Chapter 6.

(32) G. Szentirmai, "Computer aids in filter design: a review", IEEE Trans. Circuit Theory, Vol. CT-18, pp. 35-40, Jan. 1971. Reprinted in (9).

(33) D. A. Calahan, Computer Aided Network Design, McGraw-Hill, New York, 1972.

(34) W. M. Snelgrove and A. S. Sedra, FILTOR 2 - A Computer Aided Filter Design Package, Matrix Publishers, Champaign, Ill.

(35) G. C. Temes, "First-order estimation and precorrection of parasitic loss effects in ladder filters", IRE Trans. Circuit Theory, Vol. CT-9, pp. 385-400, Dec. 1962.

(36) M. L. Blostein, "Some bounds on sensitivity in RLC networks," Proc. 1st Allerton Conf. on Circuit and System Theory, Urbana, Ill., pp. 488-501, 1963.

(37) M. L. Blostein, "Sensitivity analysis of parasitic effects in resistance terminated LC filters", IEEE Trans. Circuit Theory, Vol. CT-14, pp. 21-25, Mar. 1967. Reprinted in (9).

(38) H. J. Orchard, "Inductorless filters", Electron. Lett., Vol. 2, p. 224, Sept. 1966.

(39) G. C. Temes, "The effects of semi-uniform losses in reactance two-ports", Electron. Lett., Vol. 8, pp. 161-163, Mar. 1972.

(40) G. C. Temes and H. J. Orchard, "First order sensitivity and worst-case analysis of doubly terminated reactance two-ports", IEEE Trans. Circuit Theory, Vol. CT-20, pp. 567-571, Sept. 1973.

(41) J. A. C. Bingham, "A new method of solving the accuracy problem in filter design", IEEE Trans. Circuit Theory, Vol. CT-11, pp. 327-341, Sept. 1964.

(42) C. Norek, "Product method for the calculation of the effective loss in LC filters", in Proc. Int. Symp. Network Theory (Belgrade, Yugoslavia), 1968, pp. 353-365.

(43) H. J. Orchard and G. C. Temes, "Filter design using transformed variables", IEEE Trans. Circuit Theory, Vol. CT-15, pp. 385-408, 1968. Reprinted in (9).

PROBLEMS

P7.1 Show (by means of circuit diagrams) the general form of the networks for Cauer's first and second forms, and the "fifth" canonic form for each of the four Classes of reactance function. (12 circuit diagrams are necessary.)

P7.2 Consider a reactance function of seventh degree. Sketch the function versus frequency, for each possible Class that this function may belong to. In each case give a pole-zero diagram.

P7.3 Given the reactance function

$$X(s) = \frac{s(s^2+4)(s^2+16)}{(s^2+1)(s^2+9)},$$

use the removal operations (A,B,C,D,E,F) to synthesize X(s) in each of the five canonic forms (two Foster's, two Cauer's, and the "fifth" form). Note that five networks are required.

P7.4 In problem P7.3:

(a) What type of filtering function is realized if X(s) is X_{1s} of a doubly-terminated ladder and X(s) is synthesized in Cauer's first form?

(b) What type of filtering function is realized if X(s) is X_{1o} of a doubly-terminated ladder and X(s) is synthesized in Cauer's second form?

P7.5 X_{1s} of the minimum capacitance LC ladder realization of a third-order elliptic low-pass filter, which has $\omega_p=1$ and a loss pole at $\omega_\ell=2.2701$, is given by

$$X_{1s} = 1.3757\,\frac{s(s^2+1.2938)}{(s^2+0.7398)}$$

Use the pole removal operations to develop the LC ladder realization of this filter. The network is to be equally terminated in 1Ω resistances.

P7.6 (a) Use Table 7.3 to show that if F(s) is odd and P(s) is even, then the resulting network is symmetric. [Note: A network is called symmetric if $(X_1/R_1)=(X_2/R_2)$]

(b) Synthesize an LC ladder realization for a third-order Butterworth filter with $\varepsilon=1$ and $\omega_p=1$ rad/sec. Show that the network is symmetric and give the two dual circuits with 1Ω terminations. Show that depending on the sign of F(s), one obtains either the minimum inductance or the minimum capacitance ladder.

P7.7 (a) Show (using Table 7.3) that if both F(s) and P(s) are even, the resulting network realization is antimetric. [Note: A network is called "antimetric" if $(X_{1o}/R_1)=1/(X_{2s}/R_2)$ or $(X_{1s}/R_1)=1/(X_{2o}/R_2)$.]

(b) Synthesize an LC ladder realization of a fourth-order Butterworth filter with $\varepsilon=1$ and $\omega_p=1$ rad/sec. The network is to be equally terminated in 1Ω resistances. Show that the resulting network is "physically antimetric"; that is, if we take its dual and turn it end-for-end we obtain the same network.

P7.8 Use the pole removal operations to synthesize a doubly-terminated LC ladder realization for the fourth-order bandpass filter of Example 4.1. Use $R_1=1\Omega$.

P7.9 Use Table 7.4 and its associated flow-chart to synthesize a minimum inductance LC ladder realization for the fourth-order elliptic low-pass filter with $\Omega_p=1$ rad/sec, $\Omega_s=2.14319$ and $|\rho|=15\%$. The elliptic function, modified to have loss poles at ∞, has reflection zeros at $\Omega=0.399219$ and $\Omega=0.930605$, and finite loss poles at $\Omega=2.331071$. The natural modes are at:

$$s = -0.229437 \pm j\,1.120084$$
$$\text{and } s = -0.663491 \pm j\,0.513493$$

The ladder is to be designed for maximum power transfer with $R_1=1\Omega$. [Note: H(s) and K(s) have equal multiplicative constants.]

P7.10 Repeat all the steps of Example 7.2 on a bandpass filter that has $n_o=2$, $n_i=2$, one loss peak in the lower stopband, and two loss peaks in the upper stopband. All the reflection zeros are on the $j\omega$-axis.

P7.11 For the simple RLC network in Fig. P7.11 evaluate H(s) and its sensitivities relative to the three components. Hence, verify equations (7.29) and (7.30). Use the relationship in (7.31) to evaluate $\frac{d\alpha}{d\omega}$ and $\tau(\omega)$ at resonance.

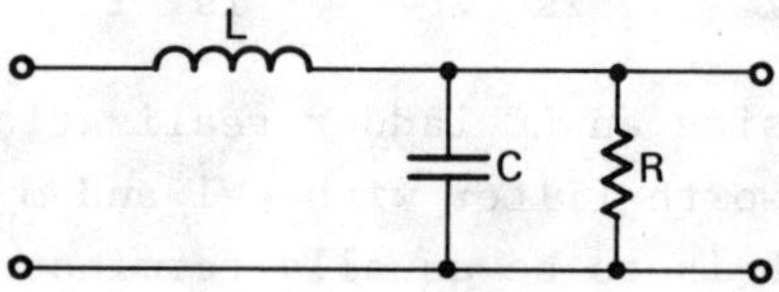

Figure P7.11

P7.12 Consider a twelfth-order bandpass filter whose passband extends from 17 to 19.7KHz with $A_p=0.2$ dB. The maximum group delay is approximately 3msec. and occurs at the upper passband edge.

(a) Find the maximum expected deviation ΔA_{max} in passband attenuation if the filter is realized by a doubly terminated LC network whose elements have 0.1% tolerance.

(b) If the LC elements have uniform dissipation with Q=2000, calculate the expected deviation in passband attenuation.

CHAPTER 8

ACTIVE FILTERS–FUNDAMENTALS

8.1 INTRODUCTION

In this and the four subsequent Chapters, we concentrate on an alternative technology for the realization of a given filter transfer function. The technology considered makes use of active devices as well as RC components, with the resulting filters termed Active-RC filters or simply, Active Filters.

The motivation behind active-RC filters lies in the desire to have inductorless filter realizations. It is well known that of the three passive RLC elements the inductor is the most nonideal one. This is especially true at low frequencies, where inductors become quite bulky and have increased losses or equivalently lower Q-factors.

Although research in active-RC filters started more than two decades ago, the demand for their existence as well as the impetus for their mature development should be credited to the rapid progress of Linear Integrated Circuit Technology in the mid and late sixties. As most of the communication and instrumentation systems became smaller in size, the filters' part appeared to be the bulkiest. More seriously, passive LC technology is, in a sense, not compatible with the IC technology that dominates most of today's systems. Attempts to produce integrated circuit inductors have generally failed. The search therefore started for an alternative technology to implement filters, especially at low frequencies. A number of alternative technologies have been suggested[1,2] but the active-RC one still appears to be the best contender.

At this point the reader might enquire about the need for active devices; that is, why not RC filters? The answer comes from a study of network theory. It is well known that RC networks have all their natural modes on the negative s-plane real axis. This implies that RC transfer functions exhibit very low selectivity. Nevertheless, it has been shown[3] that by cascading a large number of RC networks, one can obtain a frequency selective response that duplicates those realized with RLC networks. This result, however, is of theoretical interest only since the resulting RC filter will have a large amount of dissipation or loss. Of course, we can place amplifiers in cascade with the RC filter to restore the signal to a reasonable level. The question then becomes: is there a better way of utilizing such amplifiers?

From the previous Chapters, it should be clear that the closer the natural modes are to the $j\omega$-axis, the more selective the filter will be. Therefore, the active-RC filter problem is essentially that of utilization of active devices (amplifiers) to move the poles of RC networks from the negative real axis into the complex s-plane as close to the $j\omega$-axis as necessary to produce a given filter response[5]. Thus, generally speaking, active-RC filters are feedback structures, designed to realize transfer functions obtained from the solution of the approximation problem.

Active Elements

The most basic active element in the design of linear integrated networks is the bipolar junction transistor. The transistor, however, is quite a nonideal device, and it is rather difficult to formulate and solve the active filter synthesis problem using the transistor as a building block. It is considerably easier to divide the active filter design problem into two steps. The first consists of identifying and defining an idealized active element or gain block. For this element to be useful in the synthesis of active filters it has to be:

1. easy to use in the synthesis problem, and
2. easy to realize and construct using standard linear IC technology.

The second step consists of devising a method for the realization of a given filter transfer function, using the defined active element together with RC components.

Various active elements as well as synthesis procedures employing them have been proposed[6]-[14]. A partial list includes: Negative Impedance Converter (NIC), Negative Impedance Inverter (NIV), Positive Impedance Converter (PIC), Gyrator, Generalized Impedance Converter (GIC), Current Conveyor (CC) and Operational Amplifier (Op. Amp.). Although the introduction of these elements has stimulated research in the area of active network theory, very few elements have made their way to large scale production to become available as off-the-shelf items. The reason for this is mostly an economic one. For a device to become available at low cost, it has to be used in substantially large quantities. It follows that such a device has to be versatile enough to be of use in a number of applications, of which active filter design is only one. The analog circuit design area has found these attributes in the Integrated Circuit (IC) Operational Amplifier (Op. Amp.).

The IC Op. Amp. is currently the most popular linear active element. It is available from a large number of manufacturers, at reasonable cost and with good performance characteristics. Furthermore, electronic engineers have become accustomed to the use of Op. Amps. It is, therefore, only natural that the Op. Amp. would be the most popular active element in the design of active-RC filters. Section 8.2 reviews some of the characteristics and applications of Op. Amps. that are important in active filter design.

As Op. Amps. became increasingly popular, workers in the active filters area sought and found realizations of many of the other active elements using Op. Amps. For instance, there are Op. Amp. circuit realizations for NIC's, PIC's, GIC's etc. In this book, we are primarily interested in active filter design methods that employ Op. Amps. Even those design methods that are based on Generalized Immittance Converters indirectly employ op. amps. This will be demonstrated in Section 8.4, which is devoted to the study of Gyrators and Generalized Immittance Converters (GIC's). For the latter, we will present realizations using Op. Amps.

Biquadratic Filter Sections (Biquads)

In Chapter 1 section 1.5, we defined the second-order or biquadratic transfer function and presented a number of special cases of such functions. Since biquadratic filter sections, or simply Biquads, play an important role in the design of active filters, the reader is urged to review section 1.5 at this point. Chapter 9 is devoted to the design of biquads. There we present circuits containing one or more Op. Amps. together with RC elements and having biquadratic voltage transfer functions. We will also refer to such circuits as biquads. For instance, a low-pass (LP) biquad is one that implements the voltage transfer function

$$T(s) = K \frac{1}{s^2 + s(\frac{\omega_o}{Q}) + \omega_o^2}$$

Design Methods for High-order (> 2) Active Filters

There are basically two distinct approaches to the design of active filters with an order N > 2. The first approach recognizes the fact that passive LC ladder filters exhibit quite good sensitivity performance, as has been shown in Chapter 7. Furthermore, there is a great deal of accumulated knowledge and data pertaining to the design of high quality LC ladder filters. Therefore, it is logical to seek active filter design techniques that are based on the simulation of LC ladders. The starting point here is an LC ladder design obtained using the methods of Chapters 6 and 7. Subsequently, we may either simulate the inductors in the ladder or, alternatively, simulate the internal workings of the passive ladder. Chapter 11 is devoted to design methods based on inductance simulation. These methods make use of the gyrators and GIC's discussed in section 8.4. The design methods based on operational simulations of ladder networks are presented in Chapter 12. Basically, these methods employ Op. Amp. integrators (studied in Section 8.3) to simulate the current voltage relationships exhibited by the inductors and capacitors in the LC ladder. Note that for both methods; that is, inductance simulation and operational simulation, the sensitivity analysis of LC ladder networks presented in Chapter 7 is directly applicable.

The second major approach to the design of active filters starts directly from the required transfer function T(s). The latter is obtained from a solution to the approximation problem, using the methods and computer programs of the previous Chapters. Subsequently, a structure of coupled biquadratic sections is utilized to realize T(s). Thus, our basic building blocks here are the biquads of Chapter 9. The most straightforward way of coupling the biquads is by simply cascading them. Since biquads are always designed to have very low output impedance, there will be no interaction between the different sections in the cascade. Thus, the filter transfer function is realized as the product of the transfer functions of the biquads in the cascade. It is therefore evident that the cascade method is the easiest among the various active filter design methods.

Unfortunately, the simplicity of the cascade design method is not accompanied by sufficiently low sensitivity to component changes. For filters with stringent requirements, one therefore seeks other coupled biquad structures. A representative one of those, together with more details on the cascade method, are presented in Chapter 10. At this point, it is sufficient to say that, by coupling biquadratic sections through negative feedback, one is attempting to simulate the internal interactions present in an LC ladder filter. In other words, we are trying to make each natural mode dependent on more than one biquad and thus, hopefully, the natural mode will be less sensitive to variations in a single biquad. It will be noted that the circuits of Chapter 12, obtained from operational simulations of LC ladders, are also coupled biquad structures. However, the design techniques of Chapter 12 start from an LC realization, while those of Chapter 10 start directly from the transfer function T(s)*.

8.2 OPERATIONAL AMPLIFIERS (OP. AMPS.)

It is assumed that the reader has encountered the operational amplifier in his previous studies of electronic circuits. If this is not the case references [15]-[18] are recommended.† In the following, we briefly summarize those aspects of Op. Amp. terminology, specifications and circuit techniques that are important for its application in active filter design.

Figure 8.1 shows the circuit symbol for an Op. Amp. with the input terminals numbered 1 and 2 and the output terminal numbered 3. Terminal 2 is the positive or noninverting input terminal, while terminal 1 is the negative or inverting input terminal.

---

* Note that we will use T(s) to denote the output/input voltage transfer function of an active filter. Thus, T(s) is identical to the system transfer function G(s), introduced in Chapter 1. Also, T(s) can be regarded as the inverse of H(s), with any required multiplicative gain constant included.

† In addition, a great deal of useful and practical information can be found in the data and application notes available from the major semiconductor manufacturers such as Fairchild, Motorola, National, etc.

The differential gain of the Op. Amp. is denoted by A‡. Ideally, A is infinite, the common-mode rejection ratio is infinite, the input impedances (both differential and common mode) are infinite and the output impedance is zero. With the exception of the finite frequency-dependent A, the other nonideal performance characteristics of the Op. Amp. do not normally pose problems to the active filter designer.

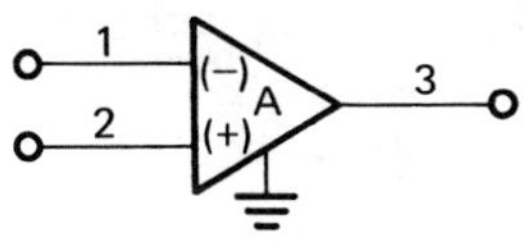

Figure 8.1 Circuit symbol for the op. amp.

In the above, we have considered only the signal parameters of the Op. Amp. Since IC Op. Amps are DC coupled, there are associated DC problems, such as the input offset voltage, the input bias current and the input offset current. The Op. Amp. user has to acquire some knowledge of how to estimate the magnitude of the expected output offset voltage and how to minimize it or, if necessary, how to "null it" out. This is generally necessary, even if the filter does not pass DC, since the offset will cause the amplifier to operate in a region where the gain might not be at its maximum. Furthermore, the dynamic range will be decreased. It goes without mention that special attention should be paid to DC problems in the design of active low-pass filters.

The Small-Signal Gain A

A major concern to an active filter designer is the finite frequency-dependent gain A. Figure 8.2 shows a typical frequency response of an Op. Amp. Note that A has a uniform 6 dB/octave or

‡ In order to conform with the literature, the symbol "A" is used to denote the op. amp. gain. This should not be confused with the attenuation function $A(\omega)$.

20 dB/decade roll-off with a 3-dB frequency f_b and a unity-gain bandwidth f_t. Such a uniform 6 dB/octave roll-off is required to ensure stable operation of the Amp. and is achieved by including frequency compensating networks either internally or externally. The first group of amplifiers are referred to as "internally compensated" and are usually easier to work with. Op. Amps. that are not internally compensated offer the flexibility of tailoring the frequency response according to the application at hand, and thus increase the useful operating frequency range. For a detailed coverage of the methods of frequency compensation in IC Op. Amps., the reader is referred to [13,17].

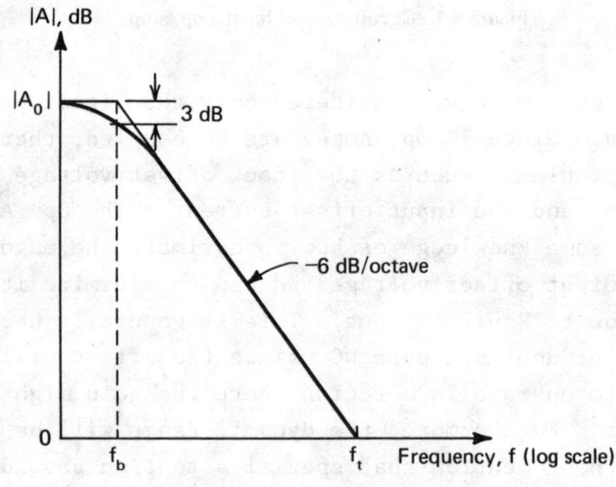

Figure 8.2 Frequency response of an op. amp. having a dominant-pole compensation.

In this book, we will assume that all Op. Amps. have the frequency response shown in Fig. 8.2. Thus, A is given by

$$A = \frac{A_o}{1 + \frac{s}{\omega_b}} \tag{8.1}$$

where A_o is the DC gain and ω_b is the 3 dB frequency. For frequencies $\omega \gg \omega_b$, (8.1) can be approximated as,

$$A \simeq \frac{A_o \omega_b}{s} \tag{8.2}$$

Note that from (8.2), and from the definition of ω_t, we have

$$\omega_t = A_o \omega_b \tag{8.3}$$

$$\text{or, } f_t = A_o f_b \tag{8.4}$$

$$\text{Thus, } A \simeq \frac{\omega_t}{s}, \quad \omega >> \omega_b \tag{8.5}$$

$$\text{or, } |A| \simeq \frac{\omega_t}{\omega}, \quad \omega >> \omega_b \tag{8.6}$$

For a typical general purpose Op. Amp. such as the 741 type, $A_o \simeq 10^5$ and f_t = 1 MHz. The tolerance on these parameters, however, is usually quite wide (50% is not unusual). Furthermore, A_o and f_t are usually temperature and power supply dependent.

Inverting and Noninverting Configurations

Figure 8.3 shows the two most common closed-loop configurations of Op. Amps. The configuration of Fig. 8.3a is referred to as the inverting configuration, while the circuit of Fig. 8.3b is the non-inverting configuration. If we assume ideal Op. Amps., the transfer function of the inverting configuration is

$$\frac{V_o}{V_i} = -\frac{Z_2}{Z_1} \tag{8.7}$$

while that of the non-inverting configuration is

$$\frac{V_o}{V_i} = 1 + \frac{Z_2}{Z_1} \tag{8.8}$$

Assuming a finite Op. Amp. gain A, equations (8.7) and (8.8) change to (8.9) and (8.10) respectively,

$$\frac{V_o}{V_i} = -\frac{Z_2/Z_1}{1 + \dfrac{1 + (Z_2/Z_1)}{A}} \tag{8.9}$$

$$\frac{V_o}{V_i} = \frac{1 + (Z_2/Z_1)}{1 + \dfrac{1 + (Z_2/Z_1)}{A}} \tag{8.10}$$

Consider now some important special cases of the configurations of Fig. 8.3. Fig. 8.4a shows an inverting amplifier with a closed loop gain

$$\frac{V_o}{V_i} = -\frac{R_2/R_1}{1 + \dfrac{1 + (R_2/R_1)}{A}} \tag{8.11}$$

Substituting for A from (8.5) obtains,

$$\frac{V_o}{V_i} = -\frac{R_2/R_1}{1 + s\dfrac{1 + (R_2/R_1)}{\omega_t}} \tag{8.12}$$

It follows from (8.12) that the inverting amplifier will have a 3 dB frequency given by

$$f_{3dB} = \frac{f_t}{1 + (R_2/R_1)} \tag{8.13}$$

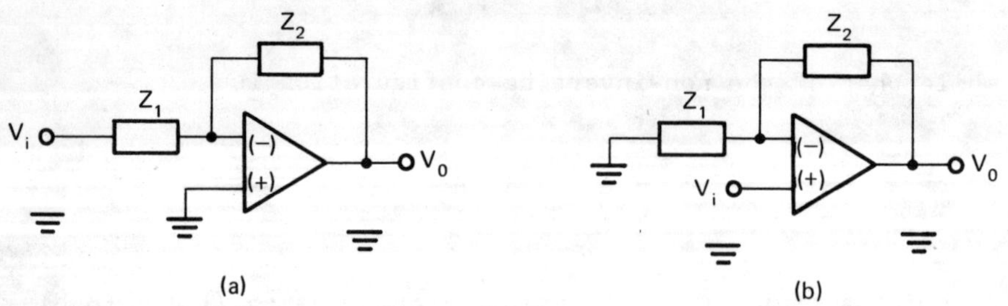

Figure 8.3 (a) The inverting, and (b) the noninverting configuration.

From Fig. 8.4a, it may be seen that the input resistance of the inverting amplifier is approximately equal to R_1. Thus, the input resistance is limited to a rather low value, especially if high closed-loop gain is required. This is easily seen if we recall that the feedback resistance should be kept to a reasonably low value, to avoid large DC offset voltages caused by the finite input bias currents. For a general purpose Op. Amp., it is a good design practice to keep the value of the feedback resistance below 1 MΩ.

To compensate for the effect of the input bias currents, a resistance R_3, equal to the parallel equivalent of R_1 and R_2, is included in the non-inverting terminal lead as shown in Fig. 8.4b. In this case, we are still left with the effect of the imbalance in the input bias currents; that is, the input offset current I_{off}. The total output DC offset voltage (worst case) is given by

$$V = V_{off} \left(1 + \frac{R_2}{R_1}\right) + I_{off} R_2$$

where V_{off} is the input offset voltage.
For a general purpose Op. Amp., $V_{off} \simeq 2$ m volts while $I_{off} \simeq$ 10 na. Although some Op. Amps. are provided with accessible terminals to which a potentiometer may be attached to trim the DC offset to zero, this entails additional cost and should be avoided in active filter applications.

For the Noninverting amplifier in Fig. 8.4c, the nominal gain is $(1 + \frac{R_2}{R_1})$, which has a minimum value of unity. The latter case is illustrated in Fig. 8.4d, which shows a unity-gain follower. An analysis similar to that carried out for the inverting configuration shows that the closed loop 3 dB frequency of the noninverting amplifier is also given by (8.13). For the unity-gain follower, it follows that,

$$f_{3dB} = f_t \tag{8.14}$$

Note that, although the result in (8.14) might have been expected, the 3dB frequency for an inverting unity-gain amplifier is not f_t but $(\frac{f_t}{2})$.

The input impedance of the noninverting amplifier is very high (typically > 100 MΩ), as it is determined by the high common-mode input resistance. All the comments above regarding the DC offset voltage of the inverting configuration apply equally well to the noninverting circuit. Although it might be obvious, we should emphasize the need for a DC path between the positive input terminal and ground. Thus, we cannot capacitively couple the input signal to the input terminal of the noninverting amplifier without first providing a resistive path to ground, such as in Fig. 8.4e. The inclusion of such a resistance, however, substantially lowers the input resistance. Bootstrapping techniques may be employed to restore the high input impedance of the noninverting amplifier[18]. This is, however, seldom used in active filter applications.

As a simple application of the inverting configuration, Fig. 8.4f shows an inverting weighted summer. Here we make use of the virtual ground present at the inverting input terminal, to provide a summer whose coefficients are easily and independently adjustable. It is easy to show that

$$V_o = -\left(\frac{R_f}{R_1} V_1 + \frac{R_f}{R_2} V_2 + \dots + \frac{R_f}{R_n} V_n\right) \tag{8.15}$$

This summer is frequently used in active filter applications. Furthermore, the virtual ground in the inverting configuration is often used to "feed in" an additional signal without disturbing the function already performed by the closed-loop amplifier. This feedforward concept will be illustrated in subsequent chapters.

A positive summer is possible using the noninverting configuration as shown in Fig. 8.4g. However, due to the absence of the "handy" virtual ground, adjustment of the summing coefficients is rather difficult.

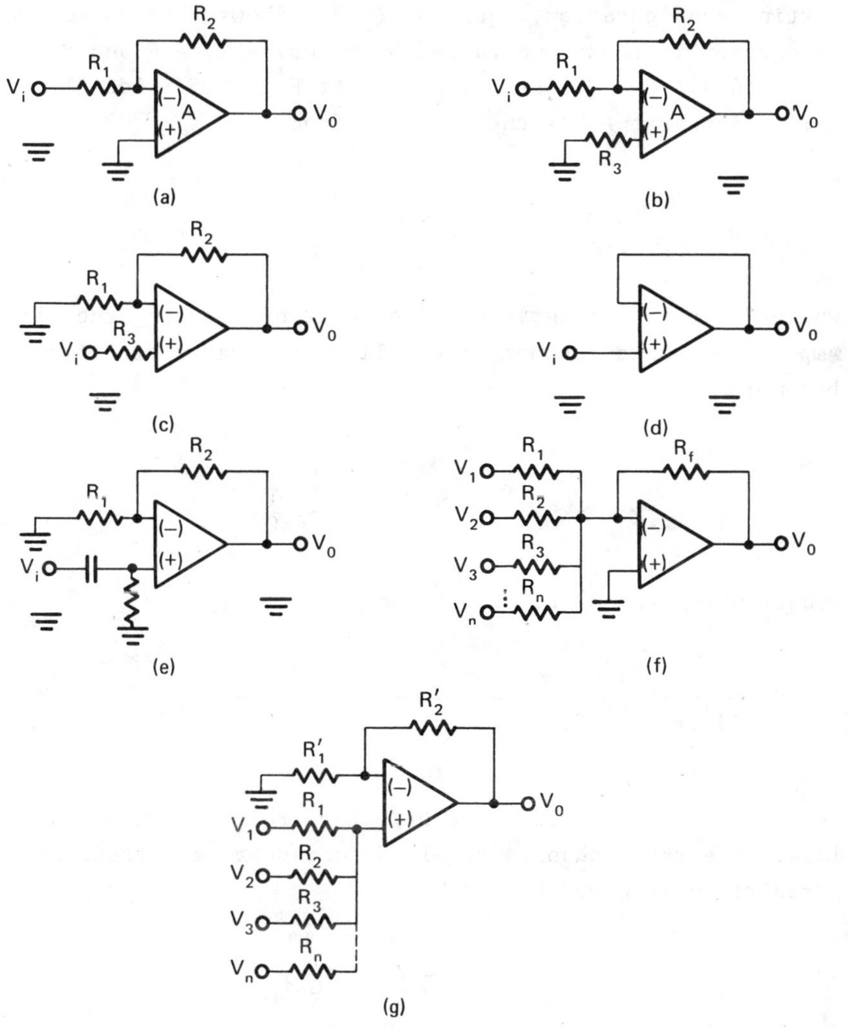

Figure 8.4 Illustrating some important op. amp. circuit configurations.

8.3 OP. AMP. INTEGRATORS

Op Amp integrators play an important role in the design of active filters. Examination of the transfer function of the inverting configuration, equation (8.7), shows that an inverting integrator is readily obtained by choosing $Z_1 = R$ and $Z_2 = \frac{1}{sC}$, resulting in the Miller integrator of Fig. 8.5. Ideally the Miller integrator has the transfer function

$$\frac{V_o}{V_i} = -\frac{1}{sCR}$$

where CR is the integrating time-constant. Taking the finite Op. Amp. gain A into account, the Miller integrator transfer function becomes

$$\frac{V_o}{V_i} = -\frac{1}{sCR}\,\frac{1}{1 + (1 + \frac{1}{sCR})/A}$$

Substituting for A from (8.5) obtains,

$$\frac{V_o}{V_i} = -\frac{1}{sCR}\,\frac{1}{(1 + \frac{1}{\omega_t CR}) + \frac{s}{\omega_t}}\,, \quad \text{for } \omega >> \omega_b\,. \tag{8.16}$$

It follows that the ideal -6 dB/octave roll-off expected from an ideal integrator changes to -12 dB/octave at the frequency of the parasitic pole given by

$$s = -(\omega_t + \frac{1}{CR})$$

which may be approximated by,

$$s \simeq -\omega_t \text{ for } \omega_t >> \frac{1}{CR}\,. \tag{8.17}$$

The response of a real integrator therefore takes the shape shown in Fig. 8.6. Note that we have also shown the low frequency pole that arises from the finite Op. Amp. gain A_o.

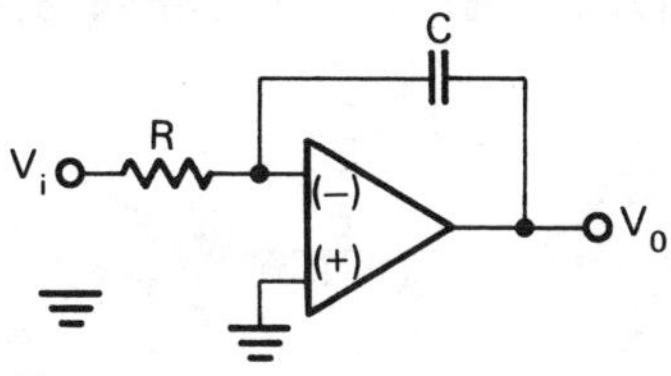

Figure 8.5 The inverting or Miller integrator.

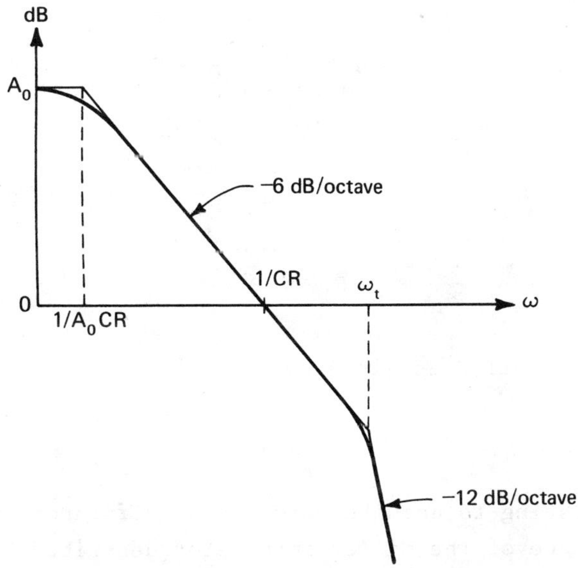

Figure 8.6 Frequency response of a Miller integrator.

The Integrator Q-factor

As we will see in Chapter 12, the integrator will be used to simulate the operation of a reactance. For instance, consider the relationship between the voltage V_L and the current I_L established by an inductor L in a passive network,

$$I_L = \frac{1}{sL}V_L \tag{8.18}$$

Scaling the current by an arbitrary resistance R_1, equation (8.18) becomes

$$(I_L R_1) = \frac{1}{s(L/R_1)} V_L \tag{8.19}$$

The operation described by (8.19) can be simulated by an integrator with a time constant of (L/R_1). If the coil L is nonideal, its associated losses can be modelled by a series resistance r. The inclusion of this resistance modifies (8.19) to,

$$(I_L R_1) = \frac{1}{s\frac{L}{R_1} + \frac{r}{R_1}} V_L$$

For $s = j\omega$,

$$(I_L R_1) = \frac{1}{\frac{j\omega L}{R_1} + \frac{r}{R_1}} V_L . \tag{8.20}$$

Note that the coil Q is given by

$$Q_L = \frac{\omega L}{r} . \tag{8.21}$$

It is interesting to enquire into the significance of the nonideal performance of the Miller integrator described by (8.16); that is

$$\frac{V_o}{V_i} = - \frac{1}{sCR + s^2 \frac{CR}{\omega_t}} \quad \text{for } CR >> \frac{1}{\omega_t} \tag{8.22}$$

Substituting $s = j\omega$ in (8.22) obtains,

$$\frac{V_o}{V_i} = - \frac{1}{j\omega CR - \frac{\omega^2 CR}{\omega_t}} \tag{8.23}$$

Comparing (8.23) to (8.20), it follows that the inductance simulated by the integrator will have a Q-factor given by

$$Q_L = -\left(\frac{\omega_t}{\omega}\right) \tag{8.24}$$

It is interesting to note that this equivalent Q-factor is equal to the negative of the gain magnitude at a frequency ω.

Similarly, if the integrator is used to simulate the relationship between the voltage V_C and the current I_C of a capacitor C in a passive network, the simulated capacitor will have an equivalent Q-factor given by a relation identical to (8.24).

The above suggests that we may use the concept of the integrator Q-factor as a measure of its nonideal performance [19,20]. Specifically, if the transfer function of an integrator circuit is expressed in the form

$$T(j\omega) = \frac{1}{R(\omega) + jX(\omega)} \tag{8.25}$$

then we define the integrator Q-factor by

$$Q_I = \frac{X(\omega)}{R(\omega)} \tag{8.26}$$

which is, obviously, a function of frequency. We have already established that, for a Miller integrator, the Q-factor is given by

$$Q_I = -\left(\frac{\omega_t}{\omega}\right) = -|A(j\omega)| \tag{8.27}$$

Noninverting Integrators

In subsequent Chapters, we will require both inverting and noninverting integrators. The most obvious way of realizing a noninverting integrator is to cascade a Miller integrator by an inverting amplifier as shown in Fig. 8.7. Analysis of this Miller-Inverter integrator results in the transfer function

$$\frac{V_o}{V_i} = \frac{1}{sCR},$$

for ideal amplifiers. Taking the finite Op. Amps'. gains into account, we use

$$A_1 \simeq \frac{\omega_{t1}}{j\omega} \text{ and } A_2 \simeq \frac{\omega_{t2}}{j\omega}$$

That is, we assume operation at frequencies $\omega >> \omega_{b1}, \omega_{b2}$. If we further assume that $\omega << \omega_{t1}, \omega_{t2}$, as is normally the case, then the transfer function $\frac{V_o}{V_i}$ is given by

$$\frac{V_o}{V_i} \simeq \frac{1}{j\omega\, CR - \frac{\omega^2 CR}{\omega_{t1}}} \; \frac{1}{1 + j\; 2(\frac{\omega}{\omega_{t2}})}$$

which for $CR >> \frac{1}{\omega_{t1}}, \frac{1}{\omega_{t2}}$ reduces to

$$\frac{V_o}{V_i} \simeq \frac{1}{-\omega^2\; CR(\frac{1}{\omega_{t1}} + \frac{2}{\omega_{t2}}) + j\omega CR} \tag{8.28}$$

The Q of the Miller-Inverter integrator is now obtained as

$$Q_I = - \frac{1}{(\frac{\omega}{\omega_{t_1}}) + (\frac{2\omega}{\omega_{t_2}})} = -1/[\frac{1}{|A_1|} + \frac{2}{|A_2|}] \tag{8.29}$$

We observe that the addition of the inverter results in a considerable degradation in the value of the integrator Q. For identical amplifiers, (8.29) yields

$$Q_I = -\frac{|A|}{3} .$$

which is one-third the value obtained for the Miller integrator.

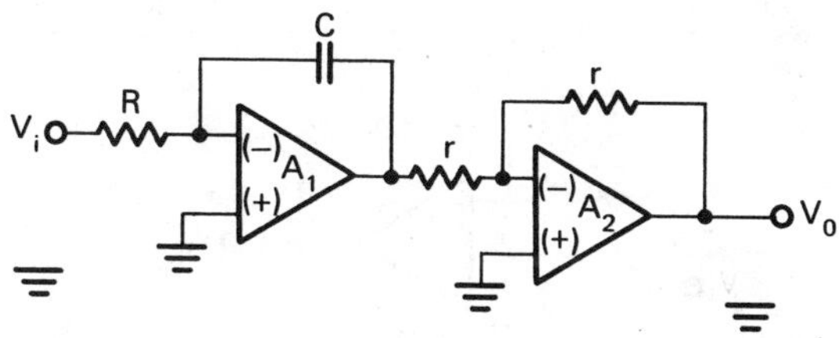

Figure 8.7 Realization of a noninverting integrator as a Miller-inverter cascade.

The Miller-inverter integrator discussed in the above requires two amplifiers. The question arises as to the possibility of realizing the non-inverting integrator using one Op. Amp. Such a circuit was reported by Deboo[21] and is shown in Fig. 8.8a, where K is the finite gain of a noninverting amplifier that can be easily implemented with an Op. Amp. in the noninverting configuration. Operation of the Deboo integrator is based on the fact that the amplifier together with R_2 provide a negative resistance in parallel with the capacitor as shown in the equivalent circuit of Fig. 8.8b. For the latter circuit to provide perfect integration, we should have

$$R_1 = \frac{R_2}{K - 1}$$

A common choice is $R_2 = R_1$ leading to K = 2 and the circuit shown in Fig. 8.8c. For the latter circuit, the ideal transfer function is given by

$$\frac{V_o}{V_i} = \frac{2}{sCR} .$$

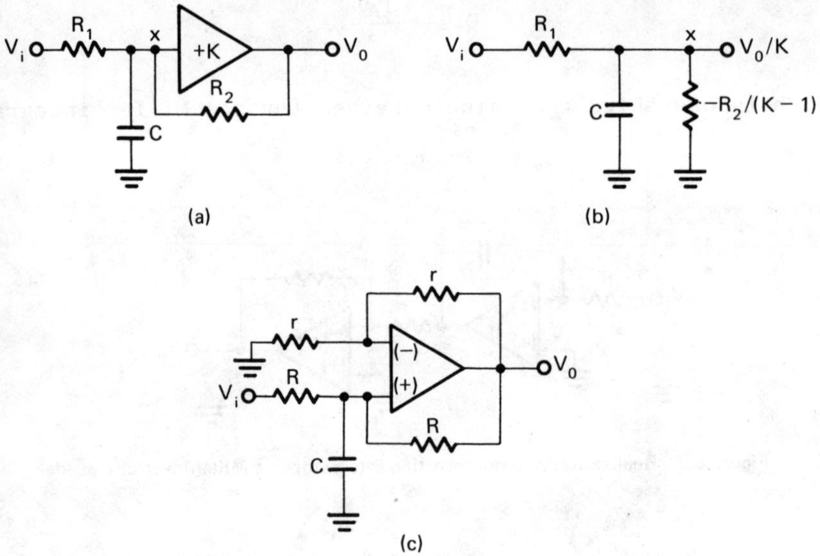

Figure 8.8 Development of the Deboo noninverting integrator.

Although the Deboo circuit provides a single Op. Amp. implementation of the noninverting integrator, as well as the convenience of a grounded capacitor, it has a serious drawback. Its operation depends critically on the matching of the two resistors labelled R, as well as on the stability of the value of K. This critical dependence stems from the required cancellation of a negative and a positive resistance, as demonstrated in the above. If the cancellation is not perfect, the integrator Q factor will be low in magnitude. As an example, if the value of K is held constant at 2, but R_1 and R_2 have a 1% mismatch, the integrator Q evaluated at $\omega=\frac{1}{CR}$ will have a value of only 100. As shown in Chapter 9, this poses a rather serious limitation on the application of the Deboo integrator to the design of biquads.

Assuming perfect passive components, but taking into account the finite bandwidth of the Op. Amp., the reader may easily verify that the Deboo integrator has a Q-factor given by

$$Q_I \simeq -\frac{1}{2}|A| ,$$

which is half that for the Miller integrator, but higher in magnitude than that of the Miller-inverter cascade.

Passive Compensation of Integrators

It should be clear that the finite high-frequency Q-factor of any of the integrators considered is due to the excess phase lag introduced by the finite Op. Amp. pole. For instance, in the Miller integrator this excess phase lag is given approximately by $(\frac{\omega}{\omega_t})$ resulting in a Q-factor of $-(\frac{\omega_t}{\omega})$. Similarly, in the Miller-inverter integrator, the excess phase lag is $(\frac{\omega}{\omega_{t1}} + 2\frac{\omega}{\omega_{t2}})$, or $(3\frac{\omega}{\omega_t})$ for identical amplifiers, resulting in a Q-factor of $-(\frac{\omega_t}{3\omega})$. This suggests that the integrator Q-factor can be restored to ∞ (ideally) by the introduction of an equal amount of phase lead. One way of accomplishing this for the Miller integrator is shown in Fig. 8.9. Here we have added a compensating capacitor C_c in parallel with R, resulting in the transfer function,

$$\frac{V_o}{V_i} = -\frac{1}{sCR}\,\frac{s\,C_c\,R + 1}{s\,\frac{1}{\omega_t}\,(1 + \frac{C_c}{C}) + (1 + \frac{1}{\omega_t CR})}$$

For $\omega_t >> \frac{1}{CR}$ this reduces to,

$$\frac{V_o}{V_i} \simeq -\frac{1}{sCR}\,\frac{s\,C_c\,R + 1}{s\,\frac{1}{\omega_t}\,(1 + \frac{C_c}{C}) + 1} \tag{8.31}$$

Thus, in addition to the parasitic real pole we have a real zero. It follows that to obtain perfect integrator the zero and pole should be made to cancel, resulting in a compensating capacitor value of

$$C_c \approx \frac{1}{\omega_t R}\,,\; C_c << C. \tag{8.32}$$

Rather than thinking of compensation in terms of a pole-zero cancellation process, we may interpret the effects in terms of phase. Specifically, the addition of a compensating capacitor creates a real zero at $\frac{1}{C_c R}$ and hence a corresponding phase lead of $\phi_c \approx \omega C_c R$, for small angle ϕ_c. To obtain compensation, we force ϕ_c to equal the excess phase lag; that is,

$$\omega C_c R = \frac{\omega}{\omega_t} ,$$

which results in the value of C_c given by (8.32).

Following a similar argument, the Miller-inverter integrator can be compensated by a capacitor C_c across the input resistance R with the value of C_c given by

$$C_c = \frac{3}{\omega_t R}$$

The above compensation technique is called "passive compensation." There is one serious drawback to passive compensation. As ω_t is usually both temperature and power supply dependent, it is difficult to maintain compensation with changes in these parameters. For this reason, we seek an alternative compensation scheme, referred to as "Active Compensation."

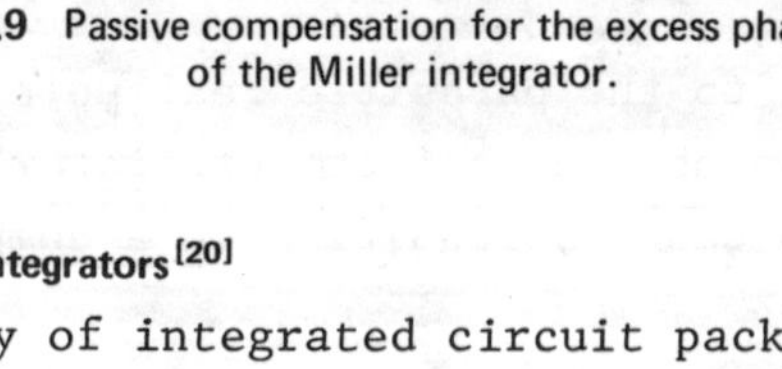

Figure 8.9 Passive compensation for the excess phase shift of the Miller integrator.

Active Compensation of Integrators[20]

The availability of integrated circuit packages containing a number of matched Op. Amps. on the same silicon chip suggests the use of matching and tracking properties to obtain circuit configurations of much wider bandwidth. This technique, called "active compensation"[20], uses Op. Amps. to compensate for the limited bandwidth effects of other matched Op. Amps. in the same circuit. Although the technique is general, we are only interested here in its application to integrator circuits.

As a first example, Fig. 8.10a shows the Miller-Inverter integrator with a slight modification. The positive input to the inverter has been lifted off-ground and connected to the negative input of the first amplifier. Simple analysis shows that the "transfer function" of the inverter is now given by

$$\frac{V_o}{V_x} = \frac{1 + \frac{2}{A_1}}{1 + \frac{2}{A_2}}$$

Thus, for matched amplifiers of arbitrary frequency response, $\frac{V_o}{V_x}$ reduces to unity. Recalling that the inverter in the Miller-Inverter cascade contributes two-thirds of the excess phase lag, the Q-factor of the Modified Miller-Inverter will be

$$Q_I \simeq - |A|$$

Thus, a simple wiring change results in a three-fold increase in the magnitude of the Q-factor !

Another modification of the two-amplifier noninverting integrator is shown in Fig. 8.10b. Here, the inverter has been placed in the feedback loop with the input terminals of the integrating amplifier interchanged to maintain stability. Analysis of this integrator circuit, which has been called Phase-lead integrator [20], shows that for the usual approximations, the Q-factor is given by

$$Q_I \simeq 1/[\frac{2\omega}{\omega_{t2}} - \frac{\omega}{\omega_{t1}}] ,$$

which for matched amplifiers reduces to

$$Q_I \simeq |A|$$

Although the magnitude of Q is not higher than that obtained for the Modified Miller-Inverter above, it is important to note that it has a positive sign. The fact that the Miller integrator and the phase-lead integrator have Q factors of equal magnitude and opposite signs will be found quite advantageous in Chapters 9 and 12.

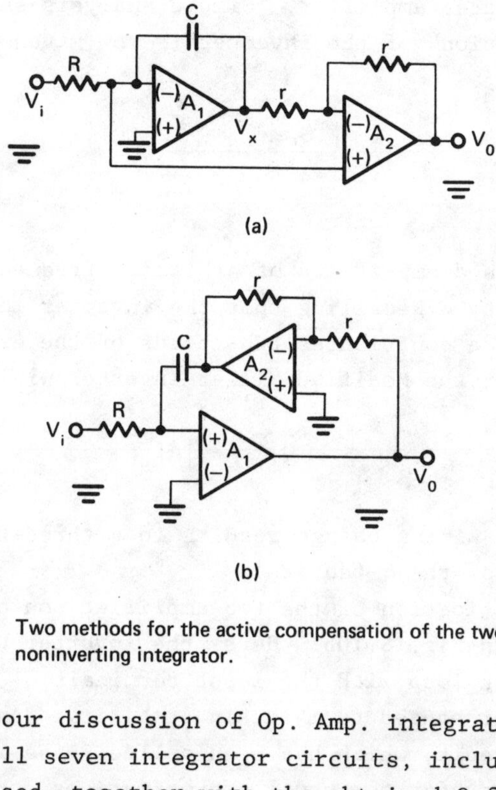

Figure 8.10 Two methods for the active compensation of the two-amplifier noninverting integrator.

We conclude our discussion of Op. Amp. integrators by presenting in Fig. 8.11 seven integrator circuits, including the ones already discussed, together with the obtained Q-factor values. The latter are evaluated assuming matched amplifiers. The two additional integrators, termed "High-Q" and "Modified High-Q", achieve very high Q values. This is obtained at the expense of one extra amplifier per integrator. Note, however, that to obtain these ultra high Q-factors, the Op. Amps. have to be very tightly matched.

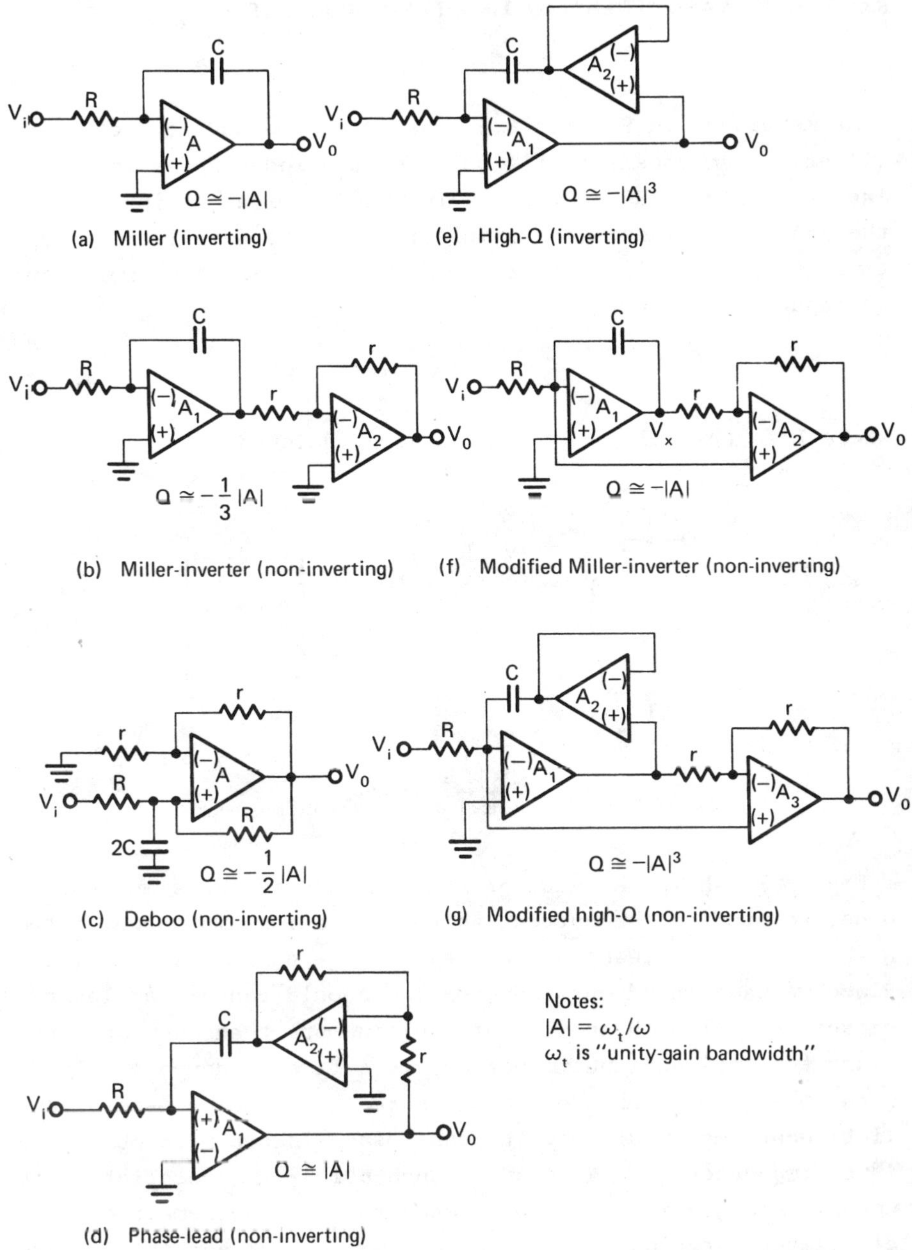

Figure 8.11 Seven integrator circuits and their Q-factors

8.4 GYRATORS AND GENERALIZED IMMITTANCE CONVERTERS

The Gyrator

As mentioned in Section 8.1, one of the methods of active-RC filter design consists of simulating the inductances in the LC ladder by active-RC networks. In this section, we investigate the problem of inductance simulation in some detail. Our objective, therefore, is to find an active-RC one-port network having an input impedance

$$Z_{11} = s\,L \tag{8.33}$$

where L is the value of the simulated inductance.

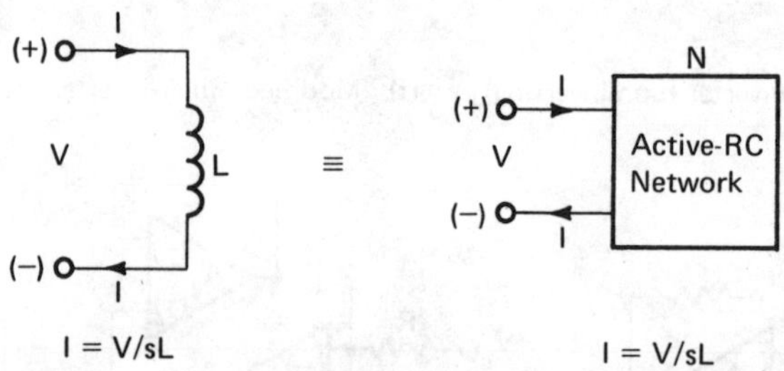

Figure 8.12 An inductance L and its active-RC simulator.

Fig. 8.12 shows an inductor L and its active-RC simulator. In order to provide the current-voltage relationship shown, network N should have at least one energy storage or reactive element. Since we have ruled out inductors, N should contain at least one capacitor. Also, N should not contain more than one capacitor, since the desired input impedance is a first order one. If N were to have more than one capacitor, then some pole-zero cancellation process would have to take place in order to obtain an input impedance of degree one. Such pole-zero cancellation is undesirable, as it invariably leads to the requirement of critical element matching and the corresponding high sensitivity problems. Consequently, network N should contain one capacitor only.

One approach to obtain a realization for the inductance simulator N proceeds as follows. We can extract the capacitor C from network N to form the two-port active-resistive network N' in Fig. 8.13a. Network N', when terminated at port-2 in a capacitor C, is required to have a purely inductive input impedance at port-1. It can be shown[5] that this is obtained if network N' is characterized by the admittance matrix

$$Y = \begin{bmatrix} 0 & g_2 \\ -g_1 & 0 \end{bmatrix} \tag{8.34}$$

where g_1 and g_2 are positive conductances.

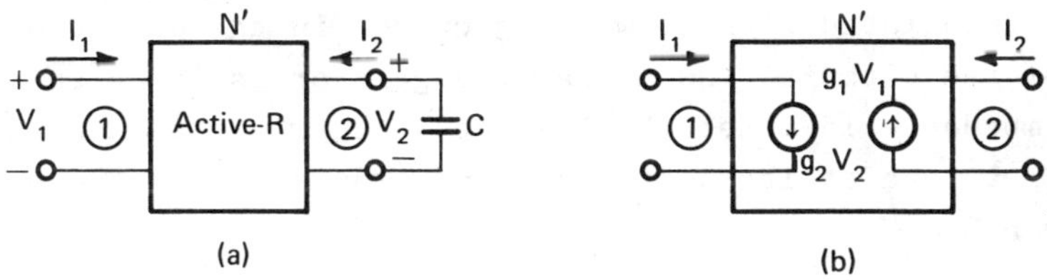

Figure 8.13 (a) Illustrating inductance simulation by an active-R network N' loaded by a capacitor C. (b) Network N' may be realized by two voltage controlled current sources.

Figure 8.13b shows a pictorial representation of network N' in the form of two ideal voltage-controlled current-sources of opposite polarity and connected back-to-back. Note that the magnitude of g_1 or g_2 by itself is immaterial; what is important is their product, which determines the value of the simulated inductance, given by

$$L = \frac{C}{g_1 g_2} \tag{8.35}$$

It is also easy to verify that if network N' is loaded at port-1 with a capacitor C, then we obtain an inductance L, of the same value as in (8.35), at port-2.

Since one of the transconductances in the admittance matrix is positive while the other is negative, the network is nonreciprocal. A special case of theoretical interest is that of $g_1 = g_2$. In this case, it can be easily verified that the two-port network is passive and lossless. This particular case was introduced by Tellegen[22] and the resulting element was termed a gyrator. Since one usually uses active elements to implement the two-port network described by (8.34), whether $g_1 = g_2$ or not, we will refer to this two-port network as a gyrator. In other words, we shall not distinguish between a passive gyrator ($g_1 = g_2$) and an active gyrator ($g_1 \neq g_2$).

A gyrator, therefore, is a two-port network described by the admittance matrix in (8.34). It has been given the symbol shown in Fig. 8.14a. The minus sign associated with g_1 can be alternatively associated with g_2 with no change in the ultimate function of the circuit. Note that the gyrator is in essence a Positive Impedance Inverter (PIV) since if terminated in an impedance Z_L at one of its ports the input impedance at the other port is given by

$$Z_{in} = 1/g_1 g_2 Z_L$$

Conductances g_1 and g_2 are called gyration conductances.

The gyrator shown in Fig. 8.14a has one of its four terminals connected to ground. Note, however, that the input inductance at port-1 is still properly floating. Such a gyrator is referred to as a floating gyrator. Since many LC ladders contain floating inductances, the need exists for floating gyrators.

Floating gyrators are difficult to realize in practice; a problem that had hindered the successful application of direct inductance replacement methods in active filter design. It is much easier to realize the three terminal or grounded gyrator shown in Fig. 8.14b. A number of transistor circuit implementations of gyrators can be found in the literature[5,9,13,14,25-28].

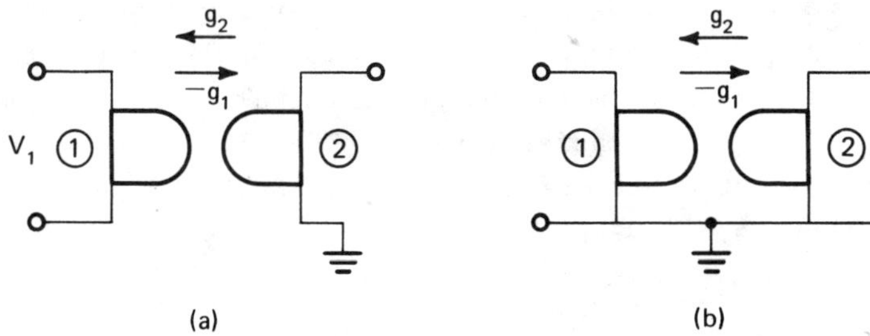

Figure 8.14 The circuit symbol for, (a) a floating gyrator, and (b) a grounded gyrator.

Inductance Simulation Using Op. Amps.

Gyrators, as we have seen in the above, require voltage-controlled current-sources for their implementation and hence are not suitable for realization using Op. Amps. The Op. Amp. is fundamentally a voltage-mode device and its utilization to implement current sources is generally unsatisfactory. Nevertheless, the economic advantages of Op. Amps. have led the proponents of the inductance simulation techniques of active filter design to seek inductance simulation circuits using Op. Amps. We should emphasize that the circuits we are about to discuss are not proper gyrators. In fact, we have to abandon the gyrator view point and seek an "inductance simulator."

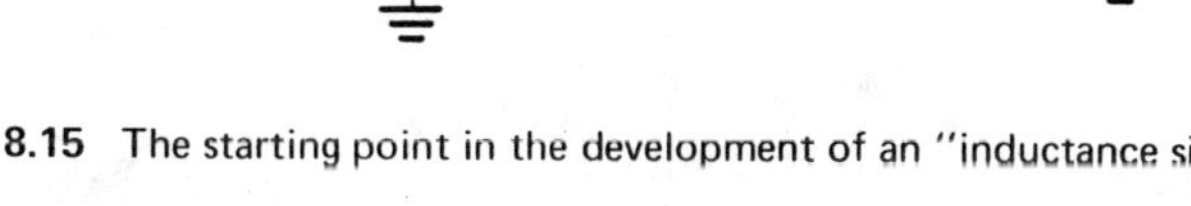

Figure 8.15 The starting point in the development of an "inductance simulator".

Consider the circuit shown in Fig. 8.15. Network N is a two-port network containing operational amplifiers, one capacitor and a number of resistors. Its input impedance is ideally infinite (that is, we are looking into the noninverting input terminal of an Op. Amp.) and its output impedance is ideally zero (that is, the output is taken from the output terminal of an Op. Amp.). We wish to find the transfer function $T(s) = \frac{V_2}{V_1}$ of N such that, at the input port of the circuit in Fig. 8.15, we have an inductance L; that is, $\frac{V_1}{I_1} = sL$. This can be easily done as follows:

$$I_1 = \frac{V_1}{sL} = \frac{V_1 - V_2}{R}$$

$$\text{Thus,} \quad \frac{V_2}{V_1} = 1 - \frac{1}{s(L/R)}$$

It follows that if we can find a network N having the voltage transfer function $T(s) = 1 - \frac{1}{s\tau}$ where τ is a time constant, then the circuit of Fig. 8.15 provides a grounded inductance $L = \tau R$ at the input.

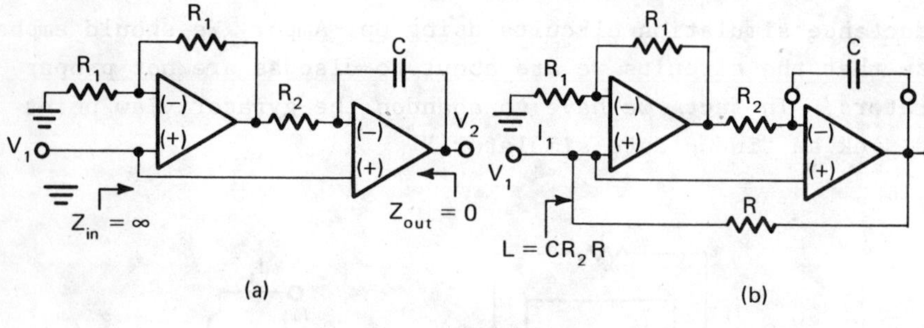

Figure 8.16 (a) An implementation of the network N of the inductance simulator of Fig. (8.15). (b) The network in (a) with a feedback resistance R has an inductive input impedance.

As an example, the circuit shown in Fig. 8.16a provides a straightforward implementation of T(s) given by,

$$T(s) \equiv \frac{V_2}{V_1} = 1 - \frac{1}{sCR_2}$$

Thus, connecting a resistance R in the feedback path, as shown in in Fig. 8.16b, results in an input grounded inductance

$$L = \tau R = C R_2 R.$$

This circuit was suggested by Riordan[29,5]. Note that although this circuit may be considered as a gyrator if port-2 is defined to be across the capacitor terminals (with the capacitor disconnected, of course), we prefer to refer to it as an inductance simulation circuit. The reason for this is that port-2 is neither grounded nor properly floating.

The circuit of Fig. 8.16b is one of a family of two-amplifier inductance simulation circuits derived by Antoniou[30,31]. Antoniou's derivation uses the theory of singular elements and will not be given here. Through the work of Antoniou[31] and subsequently of Bruton[32], a particular two-amplifier topology emerged as the "best" Op. Amp. inductance simulation circuit. This particular configuration is shown in Fig. 8.17 where we have replaced all the resistors and capacitor by general impedance branches.

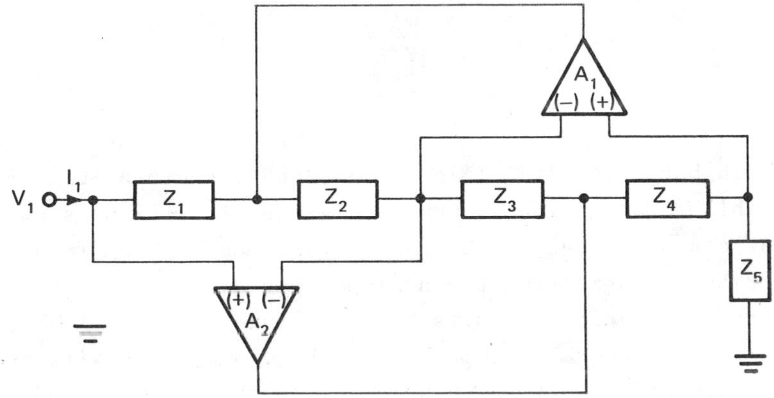

Figure 8.17 General form of the "best" op. amp. inductance simulator (the Antoniou "gyrator").

Analysis of the circuit in Fig. 8.17 yields the input impedance

$$Z_{11} \equiv \frac{V_1}{I_1} = \frac{Z_1 Z_3 Z_5}{Z_2 Z_4} \tag{8.36}$$

Thus, for Z_{11} to be of the form,

$$Z_{11} = sL,$$

we have two possible choices:

(a) $Z_1 = R_1$, $Z_2 = \frac{1}{sC}$, $Z_3 = R_3$, $Z_4 = R_4$ and $Z_5 = R_5$

leading to the circuit shown in Fig. 8.18a with

$$L = C\,\frac{R_1 R_3 R_5}{R_4}\,, \text{ or,}$$

(b) $Z_1 = R_1$, $Z_2 = R_2$, $Z_3 = R_3$, $Z_4 = \frac{1}{sC}$ and $Z_5 = R_5$

leading to the circuit shown in Fig. 8.18b with

$$L = C\,\frac{R_1 R_3 R_5}{R_2}$$

Although both circuit-A (Fig. 8.18a) and circuit-B (Fig. 8.18b) provide high quality inductance simulation, it will be shown below that circuit-B has a superior performance at high frequencies. Before considering the nonideal performance of Antoniou's inductance simulation circuits, we will take a closer look at the circuit in its general form; that is, Fig. 8.17. As will be seen this circuit provides more than just inductance simulation.

Generalized Immittance Converter (GIC)

The input impedance of the circuit in Fig. 8.17 is given in

(8.36), which we repeat as

$$Z_{11} = \frac{Z_1Z_3Z_5}{Z_2Z_4} \tag{8.37}$$

Recalling that a gyrator is essentially a Positive Impedance Inverter (PIV), we see that the circuit under consideration may be considered as a gyrator, if port-2 is defined to be that across which Z_2 is connected while all the other impedances are resistors, or if port-2 is defined to be that across which Z_4 is connected while all the other impedances are resistors. These two cases correspond directly to the two inductance simulation circuits in Fig. 8.18a and 8.18b, respectively.

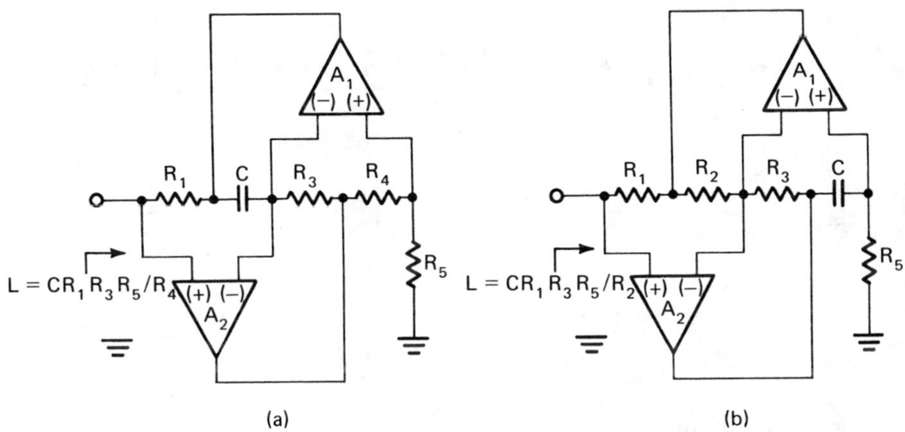

Figure 8.18 The two possible ways of using the circuit of Fig. (8.17) to simulate an inductance: (a) Type-A circuit and (b) Type-B circuit.

An examination of (8.37) reveals a more interesting fact. If we consider the circuit in Fig. 8.17 as a grounded two-port network; that is, with port-2 being that across which Z_5 is connected, then the circuit functions as an Impedance Converter with the conversion factor defined as follows

$$Z_{11} = K(s)\ Z_5$$

Thus, from (8.37) we have

$$K(s) = \frac{Z_1(s)\ Z_3(s)}{Z_2(s)\ Z_4(s)} \tag{8.38}$$

Since K(s) can be any desired function of s, the circuit of Fig. 8.19 is referred to as a Generalized Impedance Converter (GIC).

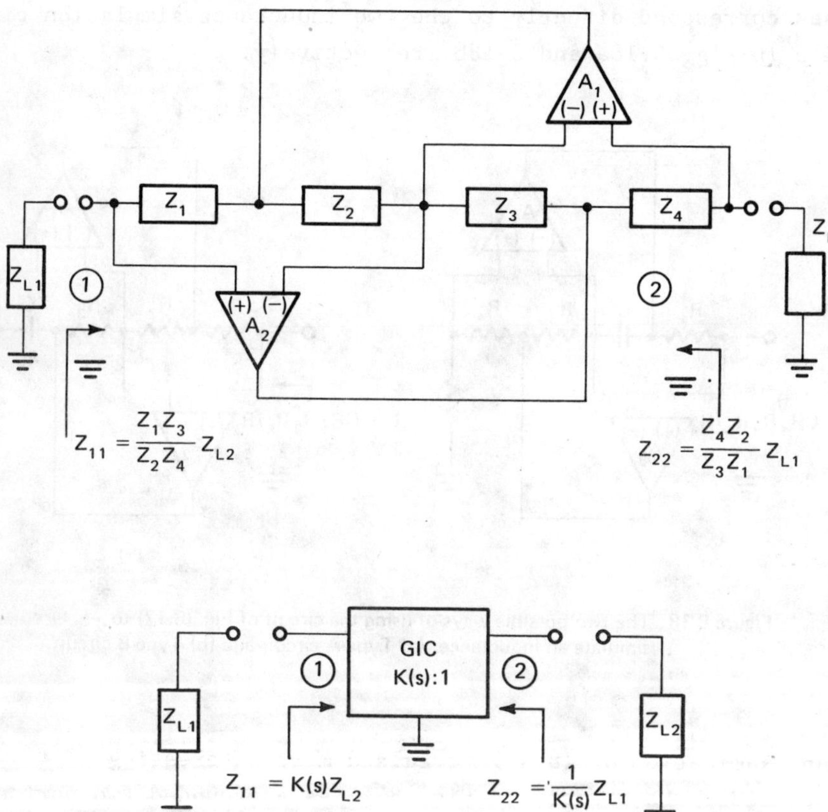

Figure 8.19 The Generalized Immittance Converter (GIC) together with its black-box representation.

To gain some insight into the operation of this GIC, consider Fig. 8.20. Assuming ideal Op. Amps., we find

$$V_2 = V_1$$

$$I_2 = -K(s)\ I_1$$

where K(s) is given by (8.38). Thus this GIC forces V_2 to equal V_1, while transforming the current by the appropriate transformation factor. For this reason, this GIC implementation has been referred to as a "current GIC". Notice that if port-1 is terminated in Z_{L1}, then the input impedance at port-2 is given by

$$Z_{22} = \frac{1}{K(s)} Z_{L1}$$

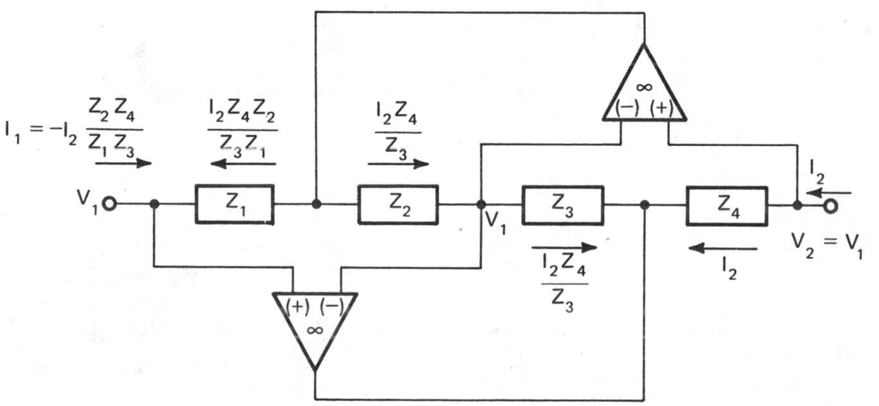

Figure 8.20 Illustrating the internal workings of the Antoniou GIC.

The symbol we shall use for the GIC is depicted in Fig. 8.19, together with the transformation relationships.

The reader might have already discovered that in both the inductance simulation circuits of Fig. 8.18, the GIC is made to have the transformation function

$$K(s) = k\, s$$

which is the one needed to convert a resistance (R_5) into an input inductance at port-1. The two GIC implementations corresponding to circuits A and B of Fig. 8.18 are shown in Fig. 8.21, together with their conversion functions. Note that if port-1 is terminated in a capacitor C, then the input impedance at port-2 is given by

$$Z_{22}(s) = \frac{1}{k\, s}\frac{1}{sC} = \frac{1}{kC\, s^2}$$

Thus,

$$Z_{22}(j\omega) = -\frac{1}{k\, C\, \omega^2}$$

which is a Frequency Dependent Negative Resistance (FDNR). The FDNR concept was introduced by Bruton[33] and used in the simulation of low-pass LC ladder filters. We will have more to say about FDNR's in Chapter 11.

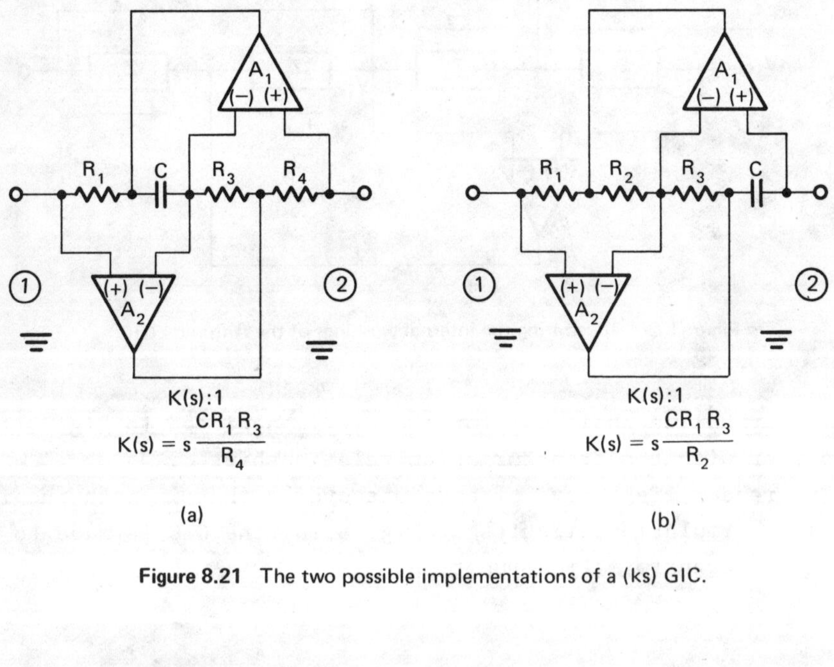

Figure 8.21 The two possible implementations of a (ks) GIC.

It should be noted that there are other ways than that discussed above for realizing FDNR's. For instance, a GIC with $K(s) = \frac{k}{s^2}$ realizes an FDNR at port-1 if loaded in a resistance at port-2. This FDNR realization, together with the one obtained using the (ks) GIC are shown in Fig. 8.22. Also shown is the circuit symbol used[33] for the FDNR element.

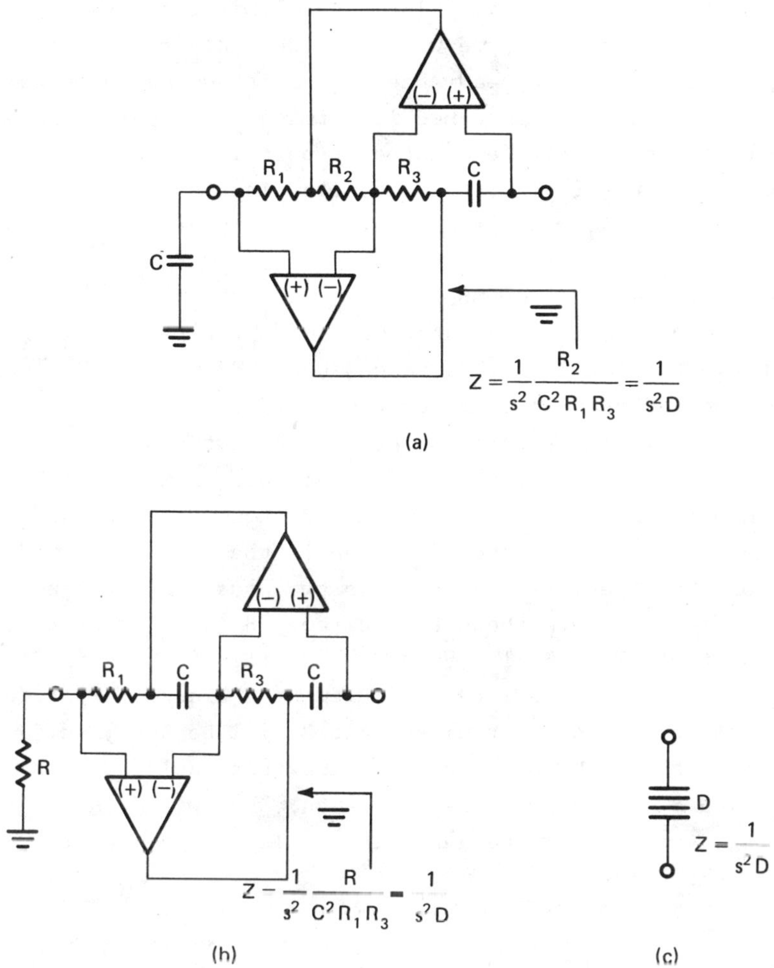

Figure 8.22 (a), (b) Two different circuits for implementing FDNR elements, (c) the circuit symbol for FDNR.

Note that whereas to resonate a GIC simulating an inductance we terminate it at the input port in a capacitor C, a GIC realizing an FDNR is resonated by terminating it at the input port in a resistance R. In both cases resonance is obtained at $\omega_o = \frac{1}{CR}$ if all the resistors are equal to R and the two capacitors are equal to C.

In the next subsection, we shall study in detail the nonideal performance of GIC circuits. For space limitations, as well as practical considerations, we will consider only the two circuits of Fig. 8.21. It has already been established that both circuits can realize an inductance if terminated at port-2 in a resistance, or an FDNR if terminated at port-1 in a capacitance. From a loop point of view, the two situations are equivalent when the other port is appropriately terminated to produce resonance.

Nonideal Performance of the Antoniou GIC

In the above, we assumed ideal Op. Amps. for realizing the GIC and hence obtained ideal inductances and FDNR's. In this subsection, we address ourselves to the nonideal performance of the GIC, due to the finite gain and bandwidth of the Op. Amps.

Evaluation of the nonideal performance of a family of GIC's has been performed by Antoniou[31] and Bruton[32]. It has been shown that the best GIC configuration is the one considered in this book. Furthermore, between its two possible arrangements, Bruton and Lim[34] have shown that circuit-B is superior to circuit-A, both of Fig. 8.21. However, as will be shown in Chapter 9, circuit-A has the ability to implement a biquad. Therefore, in the following, both circuits A and B will be analysed.

Fig. 8.17 shows the GIC configuration that includes both circuits A and B. Assuming ideal Op. Amps., but with finite gains A_1 and A_2, straightforward analysis results in the input admittance, at the port-1,

$$Y_{in_1} = \frac{Y_1Y_3Y_5}{Y_2Y_4} \frac{1 + \frac{1}{A_1}\left(1+\frac{Y_4}{Y_5}\right) + \frac{1}{A_2}\frac{Y_2}{Y_3}\left(1+\frac{Y_4}{Y_5}\right) + \frac{1}{A_1A_2}\left(1+\frac{Y_2}{Y_3}\right)\left(1+\frac{Y_4}{Y_5}\right)}{1 + \frac{1}{A_1}\frac{Y_3}{Y_2}\left(1+\frac{Y_5}{Y_4}\right) + \frac{1}{A_2}\left(1+\frac{Y_5}{Y_4}\right) + \frac{1}{A_1A_2}\left(1+\frac{Y_3}{Y_2}\right)\left(1+\frac{Y_5}{Y_4}\right)} \tag{8.39}$$

Ignoring the terms that contain $\frac{1}{A_1A_2}$, we may approximate (8.39) as,

$$Y_{in_1} \simeq \frac{Y_1Y_3Y_5}{Y_2Y_4} \frac{1 + \frac{1}{A_1}(1 + \frac{Y_4}{Y_5}) + \frac{1}{A_2}\frac{Y_2}{Y_3}(1 + \frac{Y_4}{Y_5})}{1 + \frac{1}{A_1}\frac{Y_3}{Y_2}(1 + \frac{Y_5}{Y_4}) + \frac{1}{A_2}(1 + \frac{Y_5}{Y_4})} . \quad (8.40)$$

Similarly, if Z_5 is connected across port-1, then the input admittance at port-2 will be given by,

$$Y_{in_2} \simeq \frac{Y_4Y_2Y_5}{Y_3Y_1} \frac{1 + \frac{1}{A_2}(1 + \frac{Y_1}{Y_5}) + \frac{1}{A_1}\frac{Y_3}{Y_2}(1 + \frac{Y_1}{Y_5})}{1 + \frac{1}{A_2}\frac{Y_2}{Y_3}(1 + \frac{Y_5}{Y_1}) + \frac{1}{A_1}(1 + \frac{Y_5}{Y_1})} . \quad (8.41)$$

Inductance simulation by circuit-A

To obtain an inductance simulation using the type-A circuit, we select

$$Y_1 = g_1 , Y_2 = j\omega C, Y_3 = g_3, Y_4 = g_4 \text{ and } Y_5 = g_5 .$$

Using these in (8.40), together with substituting

$$A_1 \simeq \frac{\omega_{t_1}}{j\omega} \text{ and } A_2 \simeq \frac{\omega_{t_2}}{j\omega} \text{ results in}$$

$$Y_{in_1} = \frac{1}{j\omega L_o} \frac{[1 - (\frac{\omega}{\omega_{t_2}})(\frac{\omega C}{g_3})(1 + \frac{g_4}{g_5})] + j(\frac{\omega}{\omega_{t_1}})(1 + \frac{g_4}{g_5})}{[1 + (\frac{\omega}{\omega_{t_1}})(\frac{g_3}{\omega C})(1 + \frac{g_5}{g_4})] + j(\frac{\omega}{\omega_{t_2}})(1 + \frac{g_5}{g_4})} \quad (8.42)$$

$$L_o = C\frac{g_4}{g_1g_3g_5} \quad (8.43)$$

Assuming that $\frac{\omega}{\omega_{t_1}} \ll 1$ and $\frac{\omega}{\omega_{t_2}} \ll 1$, equation (8.42) may be approximated by approximated by

$$Y_{in_1} \simeq \frac{1}{j\omega L_o} \{[1 - (\frac{\omega}{\omega_{t_1}})(\frac{g_3}{\omega C})(1 + \frac{g_5}{g_4}) - (\frac{\omega}{\omega_{t_2}})(\frac{\omega C}{g_3})(1 + \frac{g_4}{g_5})]$$

$$+ j\ [(\frac{\omega}{\omega_{t_1}})(1 + \frac{g_4}{g_5}) - (\frac{\omega}{\omega_{t_2}})(1 + \frac{g_5}{g_4})]\} \qquad (8.44)$$

where we have neglected second order terms of the form $(\frac{\omega}{\omega_t})^2$, with $\omega_t = \omega_{t_1} \simeq \omega_{t_2}$.

From (8.44), the Q-factor of the simulated inductance is given by the approximate expression,

$$Q \simeq \frac{1}{g_4 + g_5} \frac{1}{(\frac{\omega}{\omega_{t_1}})(\frac{1}{g_5}) - (\frac{\omega}{\omega_{t_2}}) \frac{1}{g_4}} \qquad (8.45)$$

Thus, very high Q-factors may be obtained if

$$g_5\ \omega_{t_1} = g_4\ \omega_{t_2}$$

Normally $\omega_{t_1} \simeq \omega_{t_2}$ and thus, we nominally choose

$$g_4 = g_5 \qquad (8.46)$$

but trim one of the two conductances to obtain the maximum Q-factor. If ω_{t_1} and ω_{t_2} have a fractional mismatch (e.g. due to lack of tracking) of ε_t; that is,

$$\left| \frac{\omega_{t_2} - \omega_{t_1}}{\omega_t} \right| = \varepsilon_t ,$$

then the inductance Q-factor is given by

$$Q = \frac{1}{2\varepsilon_t} (\frac{\omega_t}{\omega}) \qquad (8.47a)$$

i.e.

$$Q = \frac{1}{2\varepsilon_t} |A(j\omega)| \tag{8.47b}$$

Returning back to equation (8.44), we may determine an approximate expression for the actual value L_k of the inductance as,

$$L_k \simeq L_o \left[1 + \left(\frac{\omega}{\omega_{t_1}}\right)\left(\frac{g_3}{\omega C}\right)\left(1 + \frac{g_5}{g_4}\right) + \left(\frac{\omega}{\omega_{t_2}}\right)\left(\frac{\omega C}{g_3}\right)\left(1 + \frac{g_4}{g_5}\right)\right]$$

Substituting $g_4 = g_5$ and $\omega_{t_1} \simeq \omega_{t_2} = \omega_t$ yields the fractional error in the inductance value as

$$\frac{\Delta L}{L_o} = 2 \left(\frac{\omega}{\omega_t}\right)\left(\frac{g_3}{\omega C} + \frac{\omega C}{g_3}\right) \quad . \tag{8.48}$$

At a given frequency ω, the value $\frac{\Delta L}{L_o}$ is minimized if g_3 and C are related by

$$g_3 = \omega C \quad .$$

We therefore enforce this condition at a critical frequency ω_c (usually in the transition band or at the passband edge) to obtain

$$g_3 = \omega_c C \tag{8.49}$$

Thus, at ω_c the error in the value of L is

$$\left.\frac{\Delta L}{L_o}\right|_{\omega=\omega_c} = 4 \left(\frac{\omega_c}{\omega_t}\right) = \frac{4}{|A(j\omega_c)|} \tag{8.50}$$

Note that at this frequency ω_c, the deviation in the value of L is insensitive to variations in both C and g_3.

Using (8.43) and (8.49) results in the value of g_1, given by

$$g_1 = \frac{1}{\omega_c L_o} \tag{8.51}$$

Thus, (8.51) is used to determine g_1 and the remaining design equations are (8.47) and (8.49). Two degrees of freedom are still available and may be used to select all conductances equal. Finally, note that the value of g_1 may be used to adjust the value of the simulated inductance, without affecting the Q-factor. The value of the latter should be first maximized by trimming either g_4 or g_5.

Inductance simulation using circuit-B

To obtain an inductance simulation using the type-B circuit we select,

$$Y_1 = g_1 \;,\; Y_2 = g_2 \;,\; Y_3 = g_3 \;,\; Y_4 = j\omega C \text{ and } Y_5 = g_5 .$$

We then substitute these values together with $A_1 \simeq \frac{\omega_{t_1}}{j\omega}$ and $A_2 \simeq \frac{\omega_{t_2}}{j\omega}$ in (8.40). Following an analysis identical to that presented for circuit-A above, the following results are obtained

$$L_o = C\,\frac{g_2}{g_1 g_3 g_5} \tag{8.52}$$

$$Q \simeq \frac{1}{\left[\left(\frac{\omega}{\omega_{t_1}}\right) + \left(\frac{g_2}{g_3}\right)\left(\frac{\omega}{\omega_{t_2}}\right)\right]\left(1 - \frac{g_3}{g_2}\right)} \tag{8.53}$$

From (8.53), it follows that high Q-factors may be obtained by matching g_2 and g_3. Note that the Q-factor is independent of mismatches in ω_{t_1} and ω_{t_2}. Since it is much easier to ensure the matching and tracking of resistors than of op. amps., the high-frequency performance of circuit-B is superior to that of circuit-A. We therefore choose

$$g_2 = g_3 \quad (8.54)$$

and trim one of the two conductances in order to maximize the value of the Q-factor. A fractional mismatch ε_g in the values of g_2 and g_3 results in the Q-factor given by

$$Q = \frac{1}{2\,\varepsilon_g}\left(\frac{\omega_t}{\omega}\right) \quad (8.55a)$$

i.e.

$$Q = \frac{|A(j\omega)|}{2\,\varepsilon_g} \quad (8.55b)$$

The deviation in the value of the simulated inductance L due to the finite bandwidth of the op amps. is given by (for $g_2 = g_3$)

$$\frac{\Delta L}{L_o} \simeq \left(\frac{\omega}{\omega_{t_1}} + \frac{\omega}{\omega_{t_2}}\right)\left(\frac{\omega C}{g_5} + \frac{g_5}{\omega C}\right) .$$

The value of $\frac{\Delta L}{L_o}$ is minimized at a critical frequency ω_c if we select

$$g_5 = \omega_c C \quad (8.56)$$

The corresponding value of $\left(\frac{\Delta L}{L_o}\right)$, assuming $\omega_{t_1} \simeq \omega_{t_2} = \omega_t$, is given by

$$\left.\frac{\Delta L}{L_o}\right|_{\omega=\omega_c} = 4\left(\frac{\omega_c}{\omega_t}\right) = \frac{4}{|A(j\omega_c)|} \quad (8.57)$$

Equation (8.56) can be used with (8.52) and (8.54) to obtain the value of g_1 as,

$$g_1 = \frac{1}{\omega_c L_o} \quad (8.58)$$

This together with (8.54) and (8.56) constitute three design equations in the five unknown component values. The two remaining degrees of freedom may be used to make all conductances equal.

FDNR simulation using circuit-B

To realize an FDNR element, we may use the type-B circuit and terminate it at port-1 in a capacitance C. The input admittance at port-2 can be obtained by substituting in (8.41) the values,

$Y_1 = g_1$, $Y_2 = g_2$, $Y_3 = g_3$, $Y_4 = j\omega C$, $Y_5 = j\omega C$, $A_1 \simeq \frac{\omega_t}{j\omega}$ and $A_2 \simeq \frac{\omega_t}{j\omega}$. The result is,

$$Y_{in_2} = - \omega^2 D_o \frac{[1 + (\frac{\omega}{\omega_{t_2}})(\frac{g_1}{\omega C}) + (\frac{\omega}{\omega_{t_1}})(\frac{g_3}{g_2})(\frac{g_1}{\omega C})] + j\,[(\frac{\omega}{\omega_{t_2}}) + (\frac{\omega}{\omega_{t_1}})(\frac{g_3}{g_2})]}{[1 - (\frac{\omega}{\omega_{t_2}})(\frac{g_2}{g_3})(\frac{\omega C}{g_1}) - (\frac{\omega}{\omega_{t_1}})(\frac{\omega C}{g_1})] + j\,[(\frac{\omega}{\omega_{t_2}})(\frac{g_2}{g_3}) + (\frac{\omega}{\omega_{t_1}})]} \tag{8.59}$$

where

$$D_o = \frac{C^2}{g_1}\frac{g_2}{g_3} \tag{8.60}$$

If we assume that $\omega << \omega_{t_1}, \omega_{t_2}$, then second order terms of the form $(\frac{\omega}{\omega_t})^2$, where $\omega_t = \omega_{t_1} \simeq \omega_{t_2}$, may be ignored. Thus, the admittance Y_{in_2} is given by the approximate expression,

$$Y_{in_2} \simeq -\omega^2 D_o \{[1 + (\frac{\omega}{\omega_{t_1}})(\frac{\omega C}{g_1} + \frac{g_3}{g_2}\frac{g_1}{\omega C}) + (\frac{\omega}{\omega_{t_2}})(\frac{g_1}{\omega C} + \frac{g_2}{g_3}\frac{\omega C}{g_1})]$$
$$+ j\,[(\frac{\omega}{\omega_{t_1}}\frac{g_3}{g_2} + \frac{\omega}{\omega_{t_2}})(1 - \frac{g_2}{g_3})]\} \tag{8.61}$$

The Q-factor of the FDNR element, defined as the ratio of the real part of Y_{in_2} to its imaginary part, can now be obtained from (8.61). An approximate expression for Q is,

$$Q \simeq 1 \bigg/ \left(\frac{\omega}{\omega_{t_1}} \frac{g_3}{g_2} + \frac{\omega}{\omega_{t_2}}\right)\left(1 - \frac{g_2}{g_3}\right) \tag{8.62}$$

Thus, the value of Q can be made very high by matching g_2 and g_3. This high Q-factor will be obtained independent of amplifier matching. Circuit-B, therefore, provides a high quality implementation of FDNR's, provided that we select

$$g_2 = g_3 \tag{8.63}$$

If g_2 and g_3 have a fractional mismatch ε_g, then the Q-factor of the FDNR, for $\omega_{t_1} \simeq \omega_{t_2} = \omega_t$, will be given by

$$Q = \frac{|A(j\omega)|}{2\,\varepsilon_g} \tag{8.64}$$

The actual value of the realized FDNR will differ from the nominal value D_o by an amount ΔD. The latter can be easily evaluated from (8.60) and is given by (for the case $g_2 = g_3$),

$$\frac{\Delta D}{D_o} = \left(\frac{\omega}{\omega_{t_1}} + \frac{\omega}{\omega_{t_2}}\right)\left(\frac{g_1}{\omega C} + \frac{\omega C}{g_1}\right) \tag{8.65}$$

From (8.65), it follows that to minimize the value of ΔD at a critical frequency ω_c, we have to arrange that

$$g_1 = \omega_c C \tag{8.66}$$

Substituting in (8.65) obtains,

$$\left.\frac{\Delta D}{D_o}\right|_{\omega=\omega_c} = 4 \,/\, |A(j\omega_c)| \tag{8.67}$$

Using (8.60), (8.63) and (8.66) yields

$$C = \omega_c D_o \tag{8.68}$$

Thus, (8.68), (8.66) and (8.63) are three equations in the four unknown component values. The remaining degree of freedom may be used to make all conductances equal. Finally, we note that to maximize the value of Q, either one of the two conductances g_2 or g_3 should be adjusted. Subsequently, g_1 may be trimmed to yield the desired value of D_o.

8.5 SENSITIVITY FUNCTIONS

As in any design problem, the active-filter designer is concerned with how close the resulting response will be to the ideal or desired function. The reason for response deviation from the ideal is the finite tolerances of the RC elements, as well as the nonideal performance of the active elements. In the latter case, not only the gain changes or tolerances have to be considered but the effect of the limited amplifier bandwidth on the filter response must also be evaluated. Although effects of initial component tolerances may be "trimmed out" during the initial filter alignment or tuning process, a sensitive design will deviate from the required specifications as time progresses, due to component variations with temperature, aging, humidity, etc. Note also that a sensitive design might be extremely difficult to tune in the first place or, the initial adjustment will be quite uneconomical.

The answer to the tolerance question can be obtained through sensitivity studies. Considerable emphasis has been placed, in the active filter literature, on the study of sensitivity. The outcome of these efforts has been a large number of sensitivity functions and relationships, with some attendant confusion. In this section, we lay the groundwork for the sensitivity studies of the following chapters, by focussing on some of the more useful and widely accepted sensitivity functions.

Sensitivities of Active Filters Obtained from LC Ladders

In Chapters 11 and 12, two approaches are presented for the design of active filters, starting from a prototype LC ladder. The method in Chapter 11 is based on inductance simulation, while that of Chapter 12 is based on simulation of the internal workings of the ladder. In both cases, the performance of the resulting active circuits can be easily evaluated using the sensitivity analysis of LC ladder filters presented in Chapter 7. The application of the latter formulae to active filters will be demonstrated in Chapters 11 and 12. The sensitivity functions presented in the following are especially useful for evaluating the performance of the biquads of Chapter 9, as well as the performance of higher order filters designed by methods other than ladder simulation.

Magnitude Function Deviation and Classical Sensitivity

We should note at the outset that the filter designer is ultimately concerned with the expected deviation or variability in the filter function of interest, due to the known components' (passive and active) tolerances and variations. For instance, with a filter designed to meet certain magnitude characteristics, one is concerned with the deviation in $|T(j\omega)|$, that is $\Delta|T(j\omega)|$, both in the passband and the stopband. However, normally one can tolerate much larger attenuation deviations in the stopband than in the passband. It is usually desirable to express the expected deviation in decibels. The deviation $D(\omega)$dBs in the magnitude function may be evaluated as follows. Let the function $|T(j\omega)|$ change to $[|T(j\omega)| + \Delta|T(j\omega)|]$ then,

$$D(\omega) = 20 \log \frac{|T(j\omega)| + \Delta|T(j\omega)|}{|T(j\omega)|}$$

$$D(\omega) = 8.68 \ln \left[1 + \frac{\Delta|T(j\omega)|}{|T(j\omega)|}\right] \quad \text{dB} \qquad (8.69)$$

For small variability, (8.69) can be approximated as,

$$D(\omega) \simeq 8.68 \frac{\Delta|T(j\omega)|}{|T(j\omega)|} \quad \text{dB} \qquad (8.70)$$

Thus, the deviation in the magnitude response in nepers is equal to the per-unit variability in the magnitude of the transfer function. The problem now reduces to that of evaluating the per-unit change in $|T(j\omega)|$. It should be noted that this is not an easy problem, since $T(j\omega)$ is a function of many elements with different tolerances and tolerance statistics. Furthermore, the per-unit change is a function of frequency. Let us first recall the definition of the Classical Sensitivity function S_x^y, where y is an output variable of interest, usually a function of many parameters of which x is one,

$$S_x^y \equiv \frac{\partial y}{\partial x}\,\frac{x}{y} = \frac{\partial(\ell n y)}{\partial(\ell n x)} \tag{8.71}$$

Note that from the definition (8.71), S_x^y is the limit value for $\frac{(\Delta y/y)}{(\Delta x/x)}$ as $\Delta x \to o$. Thus, for small variations

$$S_x^y \simeq \frac{(\Delta y/y)}{(\Delta x/x)} \tag{8.72}$$

The usefulness of the classical sensitivity function is evident from (8.72). The per-unit or percentage change in y, due to a given per-unit or percentage change in x, can be easily obtained by multiplication with S_x^y, i.e.

$$\left(\frac{\Delta y}{y}\right) \cong S_x^y \left(\frac{\Delta x}{x}\right) \tag{8.73}$$

Since we will be making extensive use of the classical sensitivity function, a number of sensitivity identities that considerably simplify the evaluation of S_x^y are given in the following [35]. If y, u, v are single-valued differentiable functions of x and k, n are real constants, then:

$$S_x^{kx} = 1$$

$$S_x^{kx^n} = n$$

$$
\begin{aligned}
S_x^y &= S_u^y \cdot S_x^u \\
S_x^{u \cdot v} &= S_x^u + S_x^v \\
S_x^{1/y} &= - S_x^y \\
S_x^{ky} &= S_x^y \\
S_x^{y+k} &= \frac{y}{y+k} S_x^y \\
S_x^{u+v+..} &= \frac{1}{u+v+\ldots} \quad (u\, S_x^u + v\, S_x^v + \ldots) \\
S_x^{u/v} &= S_x^u - S_x^v \\
S_x^{y^n} &= n\, S_x^y
\end{aligned}
\tag{8.74}
$$

The verification of these identities is left as an exercise.

Example 8.1

Consider the non-inverting amplifier configuration of Fig. 8.4c. Evaluate the sensitivities of the closed-loop gain K with respect to R_1, R_2 and A_o (Assume the Op. Amp. gain to be a constant A_o, which is equivalent to assuming operation at frequencies $\omega << \omega_t/K$). Also, find the expected tolerance on K due to an absolute value resistor tolerance of ± 5%, a resistor ratio tolerance of ± 1% and open-loop gain tolerance of ± 20%, for K=1, 10 and 1000. Assume $A_o \simeq 10^5$.

Solution

$$
K = \frac{1 + \frac{R_2}{R_1}}{1 + \left(1 + \frac{R_2}{R_1}\right)/A_o} \tag{8.75}
$$

$$K_{nominal} = 1 + \frac{R_2}{R_1} \tag{8.76}$$

$$S^K_{R_2} = \frac{R_2/R_1}{1 + (R_2/R_1)} \tag{8.77}$$

$$S^K_{R_1} = - \frac{R_2/R_1}{1 + (R_2/R_1)} \tag{8.78}$$

$$S^K_{(R_2/R_1)} = \frac{(R_2/R_1)}{1 + (R_2/R_1)} \tag{8.79}$$

$$S^K_{A_o} = - S_{A_o}^{[1 + (1 + \frac{R_2}{R_1})/A_o]}$$

$$= + \frac{(1 + \frac{R_2}{R_1})/A_o}{1 + (1 + \frac{R_2}{R_1})/A_o} \simeq \frac{(1 + \frac{R_2}{R_1})}{A_o} = \frac{K}{A_o} \tag{8.80}$$

Note that in evaluating the passive K sensitivities, we assumed $A_o >> (1 + \frac{R_2}{R_1})$, which is a reasonable approximation. Using (8.75-8.80) , the values in the table below are obtained. Thus, if the resistors' tolerances are uncorrelated, one would expect a worst case gain variation, due to resistor variabilities, of ± 10%. However, since K is a function of $(\frac{R_2}{R_1})$ and since resistive ratios' tolerance is ± 1%, the resulting gain variability is only ± 1% for the case K = 1000.

The Gain-Sensitivity Product [36–38]

An important consideration in evaluating the sensitivity of a filter parameter with respect to the closed-loop gain is the tol-

	K = 1	K = 10	K = 1000
$S^K_{R_1}$	0	- 0.9	≃-1
$S^K_{R_2}$	0	0.9	≃ 1
$S^K_{(R_2/R_1)}$	0	0.9	≃ 1
$\frac{\Delta K}{K}\Big\vert_{\Delta R_1}$	0	± 4.5%	± 5%
$\frac{\Delta K}{K}\Big\vert_{\Delta R_2}$	0	± 4.5%	± 5%
$\frac{\Delta K}{K}\Big\vert_{\Delta (\frac{R_2}{R_1})}$	0	± 0.9%	± 1%
$S^K_{A_o}$	10^{-5}	10^{-4}	10^{-2}
$\frac{\Delta K}{K}\Big\vert_{A_o}$	± 2 ppm	± 20 ppm	± 0.2 %

erance on the closed-loop gain due to open-loop gain variability. From (8.80), we see that $S^K_{A_o}$ is directly proportional to K. Thus, the tolerance of K is directly proportional to K. We thus conclude that, in addition to evaluating the sensitivity of a certain active filter parameter with respect to a gain K, it is important to also know the nominal value of K. For instance, if we are interested in the variability of a certain performance parameter y of an active filter, that has as one of its components an amplifier with a closed-loop gain K, then

$$\left(\frac{\Delta y}{y}\right) = S^y_K \left(\frac{\Delta K}{K}\right)$$

$$= S^y_K \, S^K_{A_o} \left(\frac{\Delta A_o}{A_o}\right)$$

$$= S_K^y \frac{K}{A_o} \left(\frac{\Delta A_o}{A_o}\right)$$

$$\text{Thus, } \left(\frac{\Delta y}{y}\right) = (K\, S_K^y) \left(\frac{\Delta A_o}{A_o^2}\right) \qquad (8.81)$$

From (8.81), it follows that for a given Op. Amp. gain A_o and gain variability $\left(\frac{\Delta A_o}{A_o}\right)$, the variability of y is proportional to the gain-sensitivity product $(K\, S_K^y)$, rather than the sensitivity S_K^y. For this reason, we define the y-gain-sensitivity-product GS_K^y as

$$GS_K^y \equiv K \cdot S_K^y \qquad (8.82)$$

Note that the above development for the gain-sensitivity product has been carried out in terms of a noninverting amplifier. Similar development applies for the inverting amplifier. In fact, we can extend the definition (8.82) to the open-loop gain,

$$GS_{A_o}^y \equiv A_o \cdot S_{A_o}^y \; , \qquad (8.83)$$

and note that,

$$GS_K^y = GS_{A_o}^y$$

Thus, since the ultimate goal is the variability or tolerance rather than the sensitivity, the gain-sensitivity product is a better index for comparing different designs.

Determining the Variability of the Transfer Function Magnitude

We now return to the problem of determining the magnitude function variability $\Delta|T(j\omega)|/|T(j\omega)|$. Consider an active filter having ℓ resistors, m capacitors and n amplifiers. Let the amplifier k have an open loop gain A_{ok} and, possibly, a closed loop gain K_k. The variability of the magnitude function is given by

$$\frac{\Delta|T(j\omega)|}{|T(j\omega)|} = \sum_{i=1}^{\ell} S_{R_i}^{|T(j\omega)|} \left(\frac{\Delta R_i}{R_i}\right) + \sum_{j=1}^{m} S_{C_j}^{|T(j\omega)|} \left(\frac{\Delta C_j}{C_j}\right)$$

$$+ \sum_{k=1}^{n} GS_{K_k}^{|T(j\omega)|} \left(\frac{\Delta A_{ok}}{A_{ok}^2}\right) \qquad (8.84)$$

Note that each of the sensitivity functions in (8.84) is a function of frequency. In a high-order filter realization, the different sensitivity functions might be difficult to evaluate. Also, in evaluating the expression in (8.84), one might consider the worst case where all the sensitivity terms add up arithmetically, rather than algebraically, and where all the tolerance terms are equal to the worst case tolerance. There are other methods for establishing bounds on the variability expression [39-41].

In evaluating $S_x^{|T(j\omega)|}$, one may make use of the relationship

$$S_x^{|T(j\omega)|} = \mathrm{Re}[S_x^{T(j\omega)}] \qquad (8.85)$$

where Re denotes real part.

Note that $S_x^{|T|}$ is different from $|S_x^T|$. Furthermore, it can be easily shown that

$$S_x^{|T|} \le |S_x^T|$$

Sensitivity of the Transfer Function Relative to its Poles

In the above, we have derived a relationship between the deviation in the magnitude of filter transfer function and the transfer function variability; namely,

$$D(\omega) \simeq 8.68 \frac{\Delta|T(j\omega)|}{|T(j\omega)|} \text{ dB} \qquad (8.86)$$

Since the filter natural modes, or poles, play an important role in determining the passband response, we wish to relate the variability in the transfer function to the poles' displacements. Furthermore, since each pole pair is completely characterized by its ω_o and Q, the transfer function variability may be conveniently related to the variability of the latter parameters. Ignoring the variability of a possible real pole (for N odd), as the effect of such a change is usually negligible, we may express the transfer function magnitude variability as,

$$\frac{\Delta|T(j\omega)|}{|T(j\omega)|} = \sum_{i=1}^{m} S_{\omega_{oi}}^{|T|} \left(\frac{\Delta\omega_{oi}}{\omega_{oi}}\right) + \sum_{i=1}^{m} S_{Q_i}^{|T|} \left(\frac{\Delta Q_i}{Q_i}\right) \qquad (8.87)$$

where it is assumed that the filter has m pairs of complex poles.

To evaluate the magnitude function variability in (8.87), we need expressions for the sensitivity functions [39,42,43]. Toward this end, consider the general expression for T(s)

$$T(s) = k\frac{P(s)}{s+\sigma} \prod_{i=1}^{m} \frac{1}{s^2 + s\frac{\omega_{oi}}{Q_i} + \omega_{oi}^2} \qquad (8.88)$$

where P(s) is the loss-pole or transmission-zero polynomial, k is a multiplicative constant and $(-\sigma)$ is a real pole which is present if the filter order is odd.

Substituting $s = j\omega$, the magnitude function can be obtained from (8.88) as

$$|T(j\omega)| = k\frac{|P(j\omega)|}{\sqrt{\sigma^2 + \omega^2}} \prod_{i=1}^{m} \frac{1}{\left[(\omega_{oi}^2 - \omega^2)^2 + \frac{\omega^2\omega_{oi}^2}{Q_i^2}\right]^{1/2}}$$

The required sensitivity expressions may now be obtained,

$$S_{\omega_{oi}}^{|T|} = -\frac{[2(1-\gamma_i^2) + \frac{\gamma_i^2}{Q_i^2}]}{[(1-\gamma_i^2)^2 + \frac{\gamma_i^2}{Q_i^2}]} \tag{8.89}$$

$$S_{Q_i}^{|T|} = \frac{\frac{\gamma_i^2}{Q_i^2}}{(1-\gamma_i^2)^2 + \frac{\gamma_i^2}{Q^2}} \tag{8.90}$$

where γ_i is the normalized frequency variable,

$$\gamma_i = \frac{\omega}{\omega_{oi}}$$

Examination of (8.89) and (8.90) reveals the following properties:
(1) For high - Q poles, which are the ones of major importance, the magnitude of $S_{\omega_{oi}}^{|T|}$ reaches a peak value approximately equal to Q at the 3dB frequencies $\gamma_i = 1 \pm \frac{1}{2Q}$.
(2) The maximum value of $S_{Q_i}^{|T|}$ is unity and occurs at $\gamma_i = 1$.
(3) The ratio of $S_{\omega_{oi}}^{|T|}$ to $S_{Q_i}^{|T|}$ is given by

$$S_{\omega_{oi}}^{|T|} \Big/ S_{Q_i}^{|T|} = -[1 + 2\,Q^2\,(\frac{1}{\gamma_i^2} - 1)] \tag{8.91}$$

For frequencies close to resonance, (8.91) can be approximated (narrow-band approximation) by

$$S_{\omega_{oi}}^{|T|} \Big/ S_{Q_i}^{|T|} \simeq [4\,\Delta\,Q^2 - 1] \tag{8.92}$$

where $\Delta \equiv \frac{\Delta\omega}{\omega_{oi}}$

Thus at the 3dB frequencies, $\Delta = \pm \frac{1}{2Q}$, we see that the magnitude of the ratio of sensitivities is $\simeq 2Q$. The above properties clearly indicate the relative contribution of variations in the pole frequency and variations of the pole Q to the overall transfer function variability. For equal contributions to the deviation in the filter magnitude function, the tolerance or variability of ω_{oi} should be Q_i times smaller than the tolerance of Q_i.

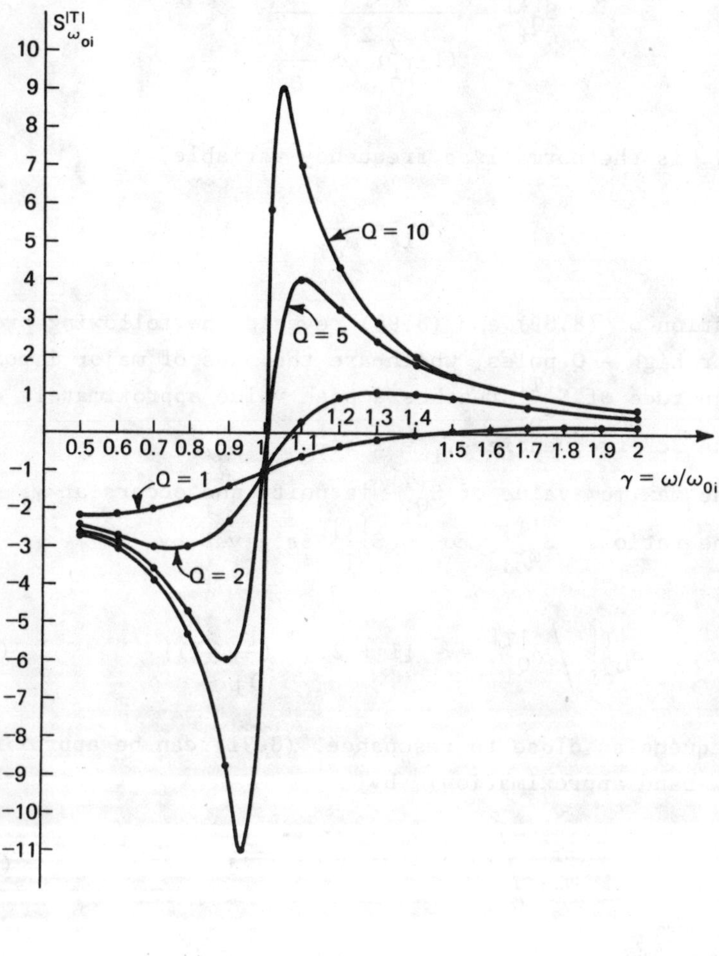

Figure 8.23 Universal curves for $S^{|T|}_{\omega_{oi}}$

This fact will guide us in our attempts to optimize designs for second-order filters or biquads. The order of magnitude of the ω_o-sensitivities in a biquad should be kept Q-times smaller than that of the Q-sensitivities. Note also that the sensitivities ratio in (8.92) becomes progressively higher as one moves away from resonance. Finally, we present in Figs. 8.23 and 8.24 plots of $S_{\omega_{oi}}^{|T|}$ and $S_{Q_i}^{|T|}$, respectively. These plots are shown versus the normalised frequency variable γ_i rendering their use universal.

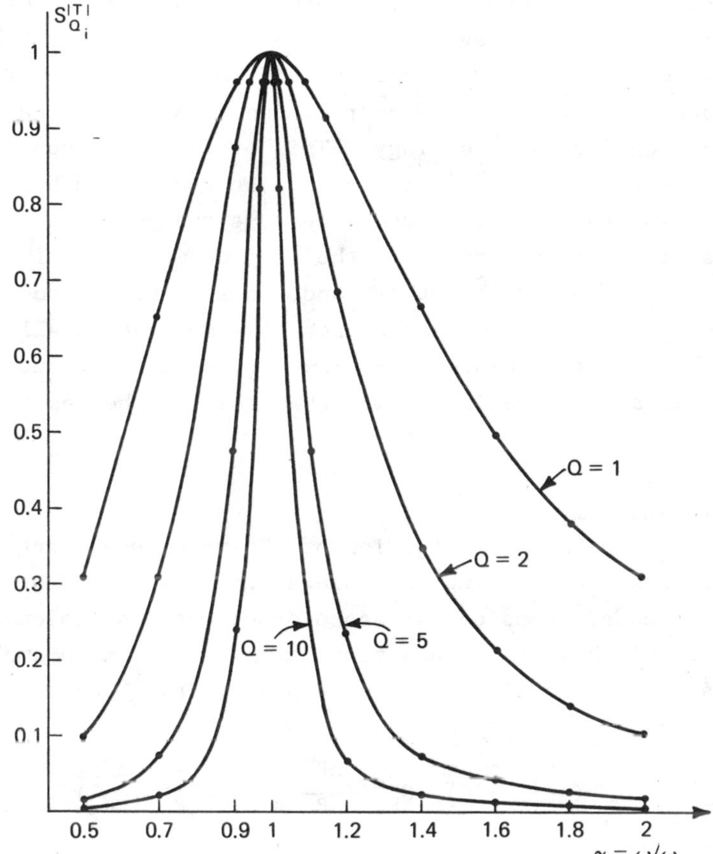

Figure 8.24 Universal $S_{Q_i}^{|T|}$ Curves.

Sensitivities of the Poles' ω_0 and Q

Having determined the sensitivities of $|T(j\omega)|$ with respect to the poles' ω_o and Q, the next step in evaluating the transfer function variability is to determine the variabilities of ω_{oi} and Q_i. In a cascade design, each pair of the filter poles is identical to the poles of a biquad in the cascade. The changes $\frac{\Delta\omega_{oi}}{\omega_{oi}}$ and $\frac{\Delta Q_i}{Q_i}$ can thus be evaluated from each biquad ω_o and Q sensitivities. In multiple-loop feedback designs, such a one-to-one correspondence between the overall transfer function poles and the poles of the second-order building blocks does not exist. One therefore needs to evaluate the sensitivities of ω_{oi} and Q_i relative to the ω_o and Q of each biquad in the filter. These latter sensitivities are, of course, functions of the particular multiple-loop feedback topology. This problem has been dealt with in the literature[43] and will not be considered here. Suffice it to say that, in cascade designs, a change in one of the biquadratic sections results in the displacement of only one of the filter poles. On the other hand, in a multiple-loop feedback design a change in one biquad causes displacement of all the filter poles. These displacements, however, tend to produce a smaller overall change in $|T(j\omega)|$ than that in the cascade design.

Sensitivities of Biquad's ω_0 and Q

The last step in the evaluation of the magnitude function deviation consists of finding the tolerances of ω_o and Q of the individual biquads. Consider a second-order section implemented with a circuit having ℓ capacitors, m resistors and n amplifiers. The ω_o and Q variabilities can be expressed as

$$\frac{\Delta\omega_o}{\omega_o} = \sum_{i=1}^{\ell} S_{C_i}^{\omega_o} \left(\frac{\Delta C_i}{C_i}\right) + \sum_{j=1}^{m} S_{R_j}^{\omega_o} \left(\frac{\Delta R_j}{R_j}\right) + \sum_{k=1}^{n} S_{K_k}^{\omega_o} \left(\frac{\Delta K_k}{K_k}\right) \tag{8.93}$$

$$\frac{\Delta Q}{Q} = \sum_{i=1}^{\ell} S_{C_i}^{Q} \left(\frac{\Delta C_i}{C_i}\right) + \sum_{j=1}^{m} S_{R_j}^{Q} \left(\frac{\Delta R_j}{R_j}\right) + \sum_{k=1}^{n} S_{K_k}^{Q} \left(\frac{\Delta K_k}{K_k}\right) \tag{8.94}$$

The passive component tolerances are dependent on the technology employed in the realization. It should be noted that in evaluating the effects of passive components' tolerances, one might wish to make use of the statistical properties of such tolerances.

The ω_o and Q sensitivities with respect to resistors and capacitors are usually referred to as passive sensitivities, while those relative to the amplifiers' gains K are called active sensitivities. The active sensitivities in (8.93) and (8.94) are evaluated with respect to the low-frequency amplifier gain. We have used K_k to denote the nominal low-frequency closed-loop gain of amplifier #k. At this point, the reader should recall our previous observation that the tolerance of the closed-loop gain is directly proportional to the value of the gain. To take this into account, we have suggested, in the above, the use of the gain sensitivity product in place of the active sensitivities. Thus, we may rewrite (8.93) and (8.94) as

$$\frac{\Delta\omega_o}{\omega_o} = \sum_{i=1}^{\ell} S_{C_i}^{\omega_o} \left(\frac{\Delta C_i}{C_i}\right) + \sum_{j=1}^{m} S_{R_j}^{\omega_o} \left(\frac{\Delta R_j}{R_j}\right) + \sum_{k=1}^{n} GS_{K_k}^{\omega_o} \left(\frac{\Delta A_k}{A_k^2}\right) \quad (8.95)$$

$$\frac{\Delta Q}{Q} = \sum_{i=1}^{\ell} S_{C_i}^{Q} \left(\frac{\Delta C_i}{C_i}\right) + \sum_{j=1}^{m} S_{R_j}^{Q} \left(\frac{\Delta R_j}{R_j}\right) + \sum_{k=1}^{n} GS_{K_k}^{Q} \left(\frac{\Delta A_k}{A_k^2}\right) \quad (8.96)$$

$$\text{where} \quad GS_{K_k}^{\omega_o} \equiv K.S_K^{\omega_o} \quad ,$$

$$GS_{K_k}^{Q} \equiv K.S_K^{Q} \quad ,$$

and $\frac{\Delta A_k}{A_k}$ is the tolerance of the low-frequency open-loop gain A_k of the $k^{\underline{th}}$ amplifier.

In comparing the sensitivity performance of the various biquad designs of Chapter 9, we will use the passive sensitivities as well as the gain-sensitivity products. We note that these sensitivity functions are particularly convenient to use, since their values are real numbers independent of ω. However, the question arises regarding the effects of the limited bandwidth of the amplifiers. Such effects will be expressed in terms of the shift induced in the pole locations. We will thus be concerned with finding the actual values of pole frequency and Q, denoted by ω_{oa} and Q_a. Specifically, expressions will be found for,

$$\delta \equiv \frac{\Delta\omega_o}{\omega_o} = \frac{\omega_{oa} - \omega_o}{\omega_o}$$

$$\text{and} \quad \eta \equiv \frac{Q_a}{Q} ,$$

for each biquad design.

Example 8.2

The circuit shown in Fig. 8.25 implements a second-order BP function, where A is ideally infinite. Find expressions for the transfer function $T(s) \equiv \frac{V_o(s)}{V_i(s)}$, assuming first an ideal op. amp. with a finite gain A. Hence, find expressions for ω_o and Q. For the case $R_1 = R_2 = R$, $C_1 = C$, and $C_2 = \frac{C}{m}$, find the time constant CR, the value of m, the ω_o and Q passive sensitivities and the gain-sensitivity products.

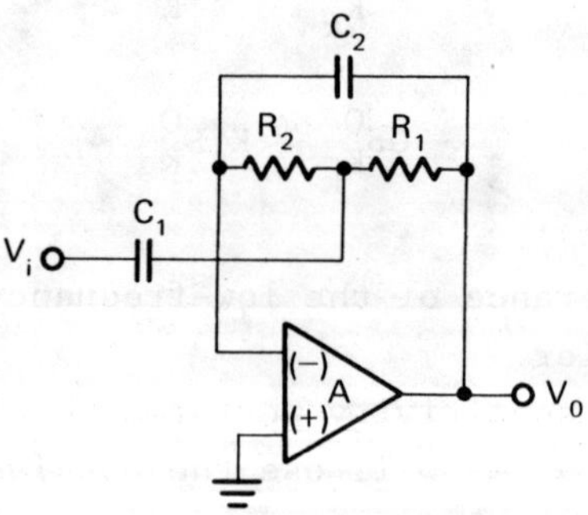

Figure 8.25 Circuit for Example (8.2).

Solution

Assuming an ideal op. amp. with infinite gain, simple circuit analysis yields

$$T(s) \equiv \frac{V_o}{V_i}(s) = \frac{-s\,\frac{1}{C_2R_2}}{s^2 + s\left(\frac{1}{C_1R_1} + \frac{1}{C_1R_2}\right) + \frac{1}{C_1R_1C_2R_2}} \tag{8.97}$$

For a finite gain A, we obtain

$$T(s) = \frac{1}{(1 + \frac{1}{A})} \frac{-s\,\frac{1}{C_2R_2}}{s^2 + s\left[\frac{1}{C_1R_1} + \frac{1}{C_1R_2} + \frac{1}{C_2R_2}\,\frac{1}{A+1}\right] + \frac{1}{C_1R_1C_2R_2}} \tag{8.98}$$

Thus, $\omega_o = \frac{1}{\sqrt{C_1R_1C_2R_2}}$

and $$Q = \frac{1}{\sqrt{C_1R_1C_2R_2}} \Big/ \left[\frac{1}{C_1R_1} + \frac{1}{C_1R_2} + \frac{1}{C_2R_2}\,\frac{1}{A+1}\right] \tag{8.99}$$

In the ideal case $A = \infty$ and the design equations reduce to

$$\omega_o = \frac{1}{\sqrt{C_1R_1C_2R_2}} \tag{8.100}$$

$$Q = \frac{1}{\sqrt{C_1R_1C_2R_2}} \Big/ \left(\frac{1}{C_1R_1} + \frac{1}{C_1R_2}\right) \tag{8.101}$$

If we choose $R_1 = R_2 = R$, $C_1 = C$, and $C_2 = C/m$, then from (8.101)

$$Q = \frac{\sqrt{m}}{2}\,.$$

$$\text{Thus, } m = 4\,Q^2\,.$$

The value of CR may now be obtained from (8.100),

$$CR = \frac{2Q}{\omega_o}$$

To evaluate the passive ω_o-sensitivities, we use (8.100) to obtain

$$S_{R_1}^{\omega_o} = S_{R_2}^{\omega_o} = S_{C_1}^{\omega_o} = S_{C_2}^{\omega_o} = -\frac{1}{2} \cdot$$

Since ω_o is independent of the amplifier gain A, then the active ω_o-sensitivity is zero;

$$S_A^{\omega_o} = 0 \text{ and } GS_A^{\omega_o} = 0 \ .$$

The passive Q-sensitivities are obtained from (8.101).

$$S_{R_1}^{Q} = -\frac{1}{2} - [\frac{R_1}{R_1 + R_2} - 1] = 0$$

$$S_{R_2}^{Q} = 0$$

$$S_{C_1}^{Q} = -\frac{1}{2} + 1 = \frac{1}{2}$$

$$S_{C_2}^{Q} = -\frac{1}{2}$$

Finally, the active Q-sensitivity is obtained from (8.99),

$$S_A^Q = \frac{A}{A+1} - \frac{A\ (\frac{1}{C_1 R_1} + \frac{1}{C_1 R_2})}{A(\frac{1}{C_1 R_1} + \frac{1}{C_1 R_2}) + \frac{1}{C_1 R_1} + \frac{1}{C_1 R_2} + \frac{1}{C_2 R_2}}$$

$$= \frac{A}{A + 1} - \frac{A(\frac{\omega_o}{Q})}{(A + 1)\ (\frac{\omega_o}{Q}) + 2\omega_o Q},$$

$$= \frac{A}{A + 1} - \frac{A}{A + 1 + 2Q^2}$$

Thus, $S_A^Q \simeq \frac{2Q^2}{A}$ for $A >> 2Q^2 + 1$

$$GS_A^Q \simeq 2Q^2$$

Note that, although the approximation $A >> (2Q^2 + 1)$ might appear unjustified, this is not the case. Since GS^Q is proportional to Q^2, this circuit is useful only for low Q applications ($Q \leq 10$), in which case $A >> 2Q^2 + 1$.

From the above example, we note that

$$\sum S_{R_i}^{\omega_o} = -1 \tag{8.102}$$

$$\sum S_{C_i}^{\omega_o} = -1 \tag{8.103}$$

$$\sum S_{R_i}^{Q} = 0 \tag{8.104}$$

$$\sum S_{C_i}^{Q} = 0 \tag{8.105}$$

which are true for every biquad realization. In fact, one usually uses (8.102)-(8.105) to check the sensitivity calculations. The proof of (8.102) and (8.103) lies in the fact that ω_o is a homogeneous function of order - 1 in the resistors R_I and in the capacitors C_i. Similarly, (8.104) and (8.105) follow since Q is a homogeneous function of order zero (i.e., dimensionless) in R_i and C_i.

8.6 TWO-PORT RC NETWORKS

In designing second-order active filters (biquads), we make use of a variety of two-port RC networks. In this section, we present some useful two-port RC networks, together with their pertinent transfer functions. We will only be concerned with RC networks that provide second-order transfer functions. Some of the networks considered, however, are third-order networks which provide second-order transfer functions through a pole-zero cancellation. As will be shown in Chapter 9, the use of these non-canonic RC networks can result in a dramatic reduction in the active sensitivity figures of the resulting biquads.

Two-port RC networks have all their poles on the negative real axis of the s-plane. On the other hand, the zeros can generally lie anywhere in the s-plane. However, if we restrict the network to have a common terminal between the input and output ports (common ground), the transmission zeros cannot lie on the positive real axis. Furthermore, to realize complex transmission zeros we shall use RC networks that have two parallel paths from input to output, such as the bridged-T and twin-T structures. With these general properties reviewed, each of the following subsections presents a particular RC two-port network. Note that the notation used below is chosen to correspond with the terminology of Chapter 9.

Ladder (LD) Networks

Fig. 8.26a shows a general two-section ladder network fed by a voltage source V_3 at terminal 3, and with the output measured at terminal 2. The open-circuit voltage transfer function is given by

$$T_{23}(s) \equiv \frac{V_2}{V_3} = \frac{y_1 y_3}{y_1 y_3 + y_1 y_4 + y_2 y_3 + y_2 y_4 + y_3 y_4} \qquad (8.106)$$

The transmission zeros are at the zeros of y_1 and y_3, the series arms, and at the poles of y_2 and y_4, the shunt arms.

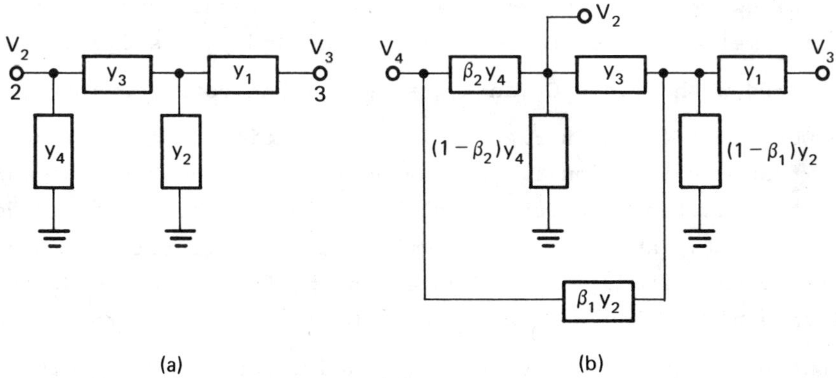

Figure 8.26 (a) General two-section ladder. (b) Realization of transmission zeros by feedforward, without altering the natural modes.

The biquad design method adopted in Chapter 9 realizes the active-filter transmission zeros by a feedforward scheme. The basis for such a scheme is that any admittance connected to ground can be partially, or fully, "lifted" and connected to the input voltage signal. This process does not alter the natural modes of the network. For the ladder network under consideration, we may lift the two admittances $\beta_1 y_2$ and $\beta_2 y_4$ and connect them to the signal V_4, as shown in Fig. 8.26b. Note that both β_1 and β_2 are ≤ 1. It should be clear that the natural modes of the ladder do not change by such a process. The open-circuit voltage transfer function between terminals 4 and 2 with 3 grounded is given by

$$T_{24}(s) \equiv \left.\frac{V_2}{V_4}\right|_{V_3 = 0}$$

$$T_{24}(s) = \frac{\beta_1 y_2 y_3 + \beta_2 y_4 (y_1+y_2+y_3)}{y_1y_3 + y_1y_4 + y_2y_3 + y_2y_4 + y_3y_4} \tag{8.107}$$

As expected, the denominators of the transfer functions in (8.106) and (8.107) are identical.

For biquad design we will be especially interested in RC ladders realizing a second-order band-pass function. From (8.106), it follows that there exist two possible canonic band-pass ladders; namely, LD 1 obtained with $y_1 = \frac{1}{R_1}$ and $y_3 = s\, C_3$, and LD 2 obtained with $y_1 = s\, C_1$ and $y_3 = \frac{1}{R_3}$. The corresponding two networks are given in Fig. 8.27. In addition, Fig. 8.27 provides the transfer function $T_{23}(s)$, as well as the numerator polynomial of the transfer function $T_{24}(s)$. The latter polynomial is denoted by $N_{24}(s)$ and is given for the indicated fractional components being lifted. For instance, in the LD 1 case the $N_{24}(s)$ given applies when a capacitance $\beta_1 C_2$ is extracted from C_2 and connected to the input voltage V_4, and a conductance $\frac{\beta_2}{R_4}$ is extracted from the conductance $\frac{1}{R_4}$ and connected to the input terminal 4. The remainder of both components are $(1-\beta_1)\, C_2$ and $\frac{1-\beta_2}{R_4}$. Note that all the β's have values smaller than or equal to unity.

In order to increase the versatility of the two dual ladder networks, we may augment them by additional components, as shown in Fig. 8.27 for the ALD 1 and ALD 2 networks. This results in more components to "lift" off ground and, correspondingly, more terms in $N_{24}(s)$, which means more flexibility in placing the zeros of $N_{24}(s)$. This point will be further illustrated when the ladder networks are applied to the design of biquads in Chapter 9.

Although only BP ladders are explicitly given in Fig. 8.27, LP and HP ladders can be easily derived from the information presented. The canonic LP ladder is obtained from LD 2 by setting $\beta_1 = 1$ and $\beta_2 = 0$ and grounding terminal 3. The resultant circuit and its transfer function are given in Fig. 8.28a. Similarly the canonic HP ladder can be derived from LD 1 by setting $\beta_1 = 1$ and $\beta_2 = 0$ and grounding terminal 3. The resultant circuit and function are given in Fig. 8.28b.

Network	*Transfer Functions*

General ladder

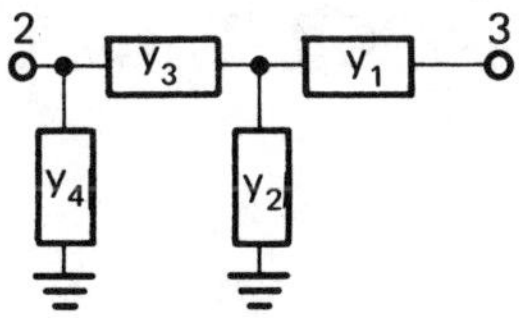

$\beta_1 : y_2 \qquad \beta_2 : y_4$

$$T_{23}(s) = \frac{y_1 y_3}{y_1 y_3 + y_1 y_4 + y_2 y_3 + y_2 y_4 + y_3 y_4}$$

$$N_{24}(s) = \beta_1 y_2 y_3 + \beta_2 y_4 (y_1 + y_2 + y_3)$$

1st BP ladder (LD1)

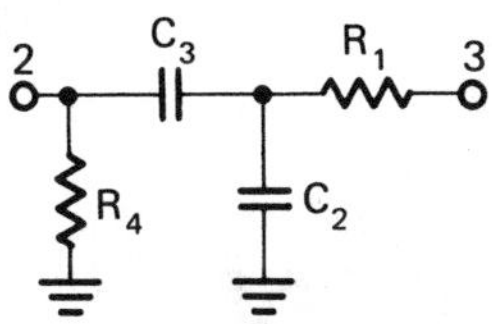

$\beta_1 : C_2 \qquad \beta_2 : 1/R_4$

$$T_{23}(s) = \frac{s\dfrac{1}{C_2 R_1}}{s^2 + s\left(\dfrac{1}{C_2 R_1} + \dfrac{1}{C_2 R_4} + \dfrac{1}{C_3 R_4}\right) + \dfrac{1}{C_2 C_3 R_1 R_4}}$$

$$N_{24}(s) = \beta_1 s^2 + \beta_2 \left[s\frac{1}{R_4}\left(\frac{1}{C_2} + \frac{1}{C_3}\right) + \frac{1}{C_2 C_3 R_1 R_4}\right]$$

Augmented 1st BP ladder (ALD1)

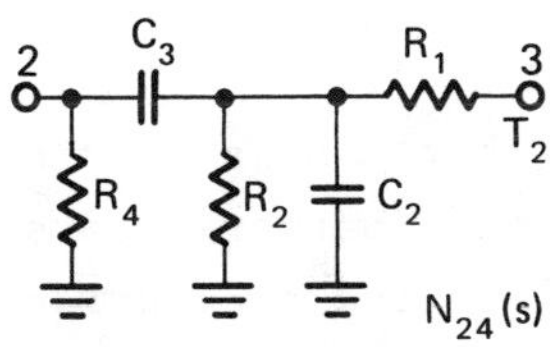

$\beta_1 : C_2 \qquad \beta_2 : 1/R_4 \qquad \beta_3 : 1/R_2$

$$T_{23}(s) = \frac{s\dfrac{1}{C_2 R_1}}{s^2 + s\left(\dfrac{1}{C_2 R_1} + \dfrac{1}{C_2 R_2} + \dfrac{1}{C_2 R_4} + \dfrac{1}{C_3 R_4}\right) + \dfrac{1}{C_2 C_3 R_4}\left(\dfrac{1}{R_1} + \dfrac{1}{R_2}\right)}$$

$$N_{24}(s) = \beta_1 s^2 + \beta_3 \frac{s}{C_2 R_2} + \beta_2 \left[s\frac{1}{R_4}\left(\frac{1}{C_2} + \frac{1}{C_3}\right) + \frac{1}{C_2 C_3 R_4}\left(\frac{1}{R_1} + \frac{1}{R_2}\right)\right]$$

2nd BP ladder (LD2)

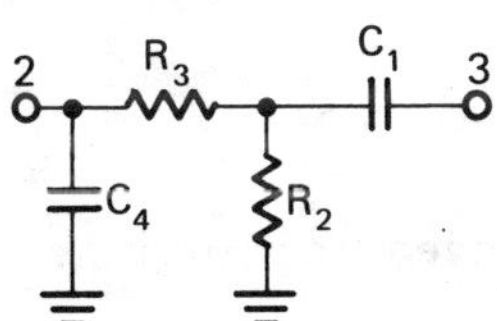

$\beta_1 : 1/R_2 \qquad \beta_2 : C_4$

$$T_{23}(s) = \frac{s\dfrac{1}{C_4 R_3}}{s^2 + s\left(\dfrac{1}{C_1 R_2} + \dfrac{1}{C_1 R_3} + \dfrac{1}{C_4 R_3}\right) + \dfrac{1}{C_1 C_4 R_2 R_3}}$$

$$N_{24}(s) = \beta_2 \left[s^2 + s\left(\frac{1}{C_1 R_2} + \frac{1}{C_1 R_3}\right)\right] + \frac{\beta_1}{C_1 C_4 R_2 R_3}$$

Augmented 2nd BP ladder (ALD2)

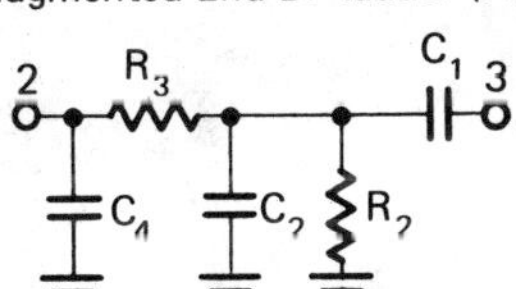

$\beta_1 : 1/R_2 \qquad \beta_2 : C_4 \qquad \beta_3 : C_2$

$$T_{23}(s) = \frac{s\dfrac{1}{[1 + (C_2/C_1)]}\left(\dfrac{1}{C_4 R_3}\right)}{s^2 + s\dfrac{1}{C_1 + C_2}\left(\dfrac{1}{R_2} + \dfrac{1}{R_3} + \dfrac{C_1}{C_4 R_3}\right) + \dfrac{1}{C_4 (C_1 + C_2) R_2 R_3}}$$

$$N_{24}(s) = \beta_2 \left[s^2 + s\frac{1}{(C_1 + C_2)}\left(\frac{1}{R_2} + \frac{1}{R_3}\right)\right] + \beta_2 s \frac{C_2}{C_1 + C_2}\frac{1}{C_4 R_3} + \frac{\beta_1}{C_4 (C_1 + C_2) R_2 R_5}$$

Figure 8.27 Second-order RC ladder networks.

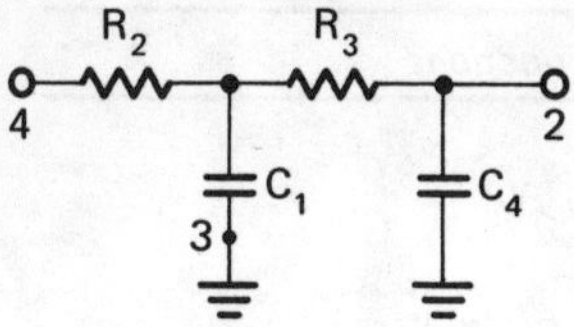

$$T_{24}(s) = \frac{1/C_1C_4R_2R_3}{s^2 + s\left(\frac{1}{C_1R_2} + \frac{1}{C_1R_3} + \frac{1}{C_4R_3}\right) + \frac{1}{C_1C_4R_2R_3}}$$

(a)

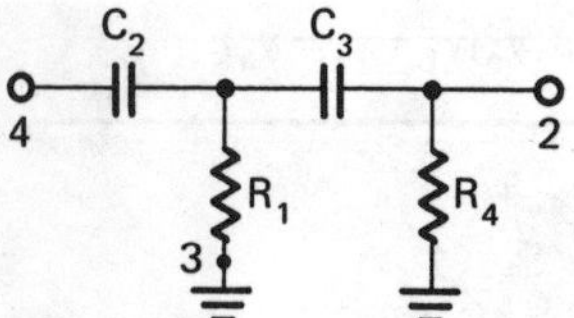

$$T_{24}(s) = \frac{s^2}{s^2 + s\left(\frac{1}{C_2R_1} + \frac{1}{C_2R_4} + \frac{1}{C_3R_4}\right) + \frac{1}{C_2C_3R_1R_4}}$$

(b)

Figure 8.28 Canonic second-order RC ladders: (a) Low-pass derived from LD2 of Fig. (8-27) (b) High-pass derived from LD1 for Fig. (8.27).

Bridged-T (BT) Networks

Some biquad designs require RC networks that have a pair of complex transmission zeros. The simplest such networks have a bridged-T structure, as shown in Fig. 8.29a. The open-circuit voltage transfer function for this general BT network is given by

$$T_{13}(s) = \frac{y_1y_2 + y_1y_3 + y_2y_3 + y_3y_4}{y_1y_2 + y_1y_3 + y_2y_3 + y_2y_4 + y_3y_4}$$

The only admittance in the BT with a grounded terminal is y_4. Hence, feedforward is possible through a fraction β of y_4, as shown in Fig. 8.29b, with the resulting transfer function given by

$$T_{14}(s) \equiv \left.\frac{V_1}{V_4}\right|_{V_3 = 0}$$

$$T_{14}(s) = \frac{\beta\, y_4y_2}{y_1y_2 + y_1y_3 + y_2y_3 + y_2y_4 + y_3y_4}$$

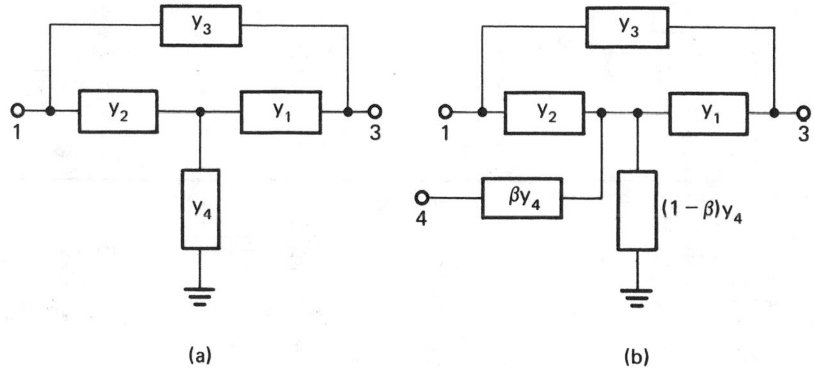

Figure 8.29 (a) Bridged—T (BT) network.
(b) BT network with feedforward.

Figure 8.30 presents the two simplest canonic BT networks namely, BT 1 and its dual BT 2, together with their augmented forms. The notation adopted is similar to that used with ladder networks. The fraction of the component used for feedforward is indicated in each case. For instance, in the augmented BT 1 case, we have given the numerator polynomial $N_{14}(s)$ of the transfer function $T_{14}(s)$ from an input terminal 4 connected to a capacitance $\beta_1 C_4$ in parallel with a conductance $\frac{\beta_2}{R_4}$. The remaining capacitance and conductance from node 5 to ground are $(1-\beta_1)C_4$ and $\frac{(1-\beta_2)}{R_4}$, respectively. Note, of course, that all the β values have to be in the range,

$$0 \le \beta \le 1 \ .$$

The augmented BT networks given provide additional components for feedforward and hence, more versatility in the use of the RC networks to generate different active-filter transmission zeros. Again we remind the reader that the potential of this technique will be fully appreciated after a study of Chapter 9.

Network	Transfer Functions

General BT

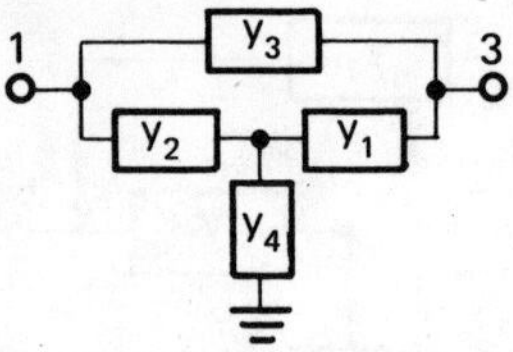

$\beta : y_4$

$$T_{13}(s) = \frac{y_1 y_2 + y_1 y_3 + y_2 y_3 + y_3 y_4}{y_1 y_2 + y_1 y_3 + y_2 y_3 + y_2 y_4 + y_3 y_4}$$

$$N_{14}(s) = \beta y_4 y_2$$

1st Bridged-T (BT1)

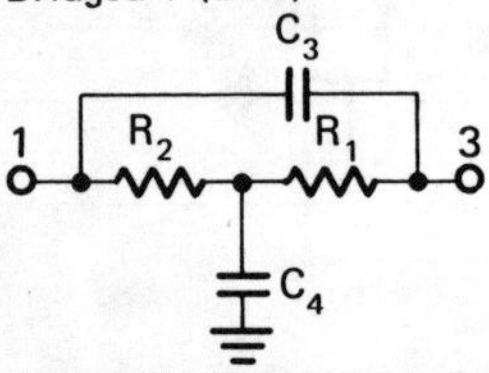

$\beta : C_4$

$$T_{13}(s) = \frac{s^2 + s\left(\frac{1}{R_1} + \frac{1}{R_2}\right)\frac{1}{C_4} + \frac{1}{C_3 C_4 R_1 R_2}}{s^2 + s\left(\frac{1}{C_4 R_1} + \frac{1}{C_4 R_2} + \frac{1}{C_3 R_2}\right) + \frac{1}{C_3 C_4 R_1 R_2}}$$

$$N_{14}(s) = \beta s \frac{1}{C_3 R_2}$$

Augmented 1st Bridged-T (ABT1)

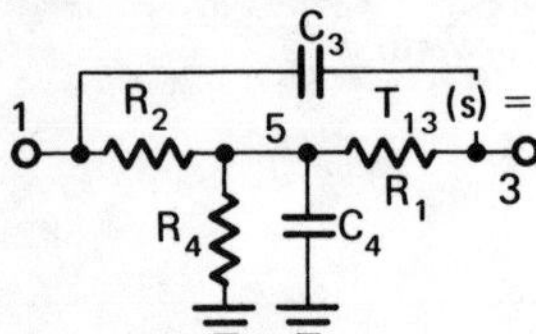

$\beta_1 : C_4 \qquad \beta_2 : 1/R_4$

$$T_{13}(s) = \frac{s^2 + s\frac{1}{C_4}\left(\frac{1}{R_1} + \frac{1}{R_2} + \frac{1}{R_4}\right) + \frac{1}{C_3 C_4 R_1 R_2}}{s^2 + s\left(\frac{1}{C_4 R_1} + \frac{1}{C_4 R_2} + \frac{1}{C_4 R_4} + \frac{1}{C_3 R_2}\right) + \frac{1}{C_3 C_4 R_2}\left(\frac{1}{R_1} + \frac{1}{R_4}\right)}$$

$$N_{14}(s) = \beta_1 s \frac{1}{C_3 R_2} + \beta_2 \frac{1}{C_3 C_4 R_2 R_4}$$

2nd Bridged-T (BT2)

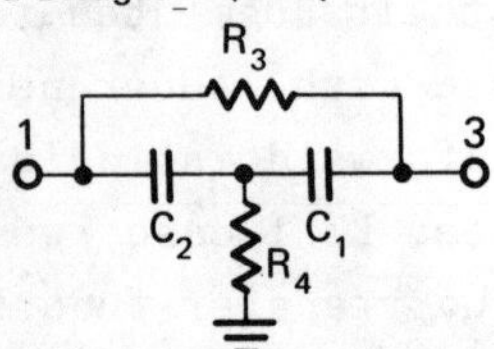

$\beta : 1/R_4$

$$T_{13}(s) = \frac{s^2 + s\left(\frac{1}{C_1} + \frac{1}{C_2}\right)\frac{1}{R_3} + \frac{1}{C_1 C_2 R_3 R_4}}{s^2 + s\left(\frac{1}{C_1 R_3} + \frac{1}{C_2 R_3} + \frac{1}{C_1 R_4}\right) + \frac{1}{C_1 C_2 R_3 R_4}}$$

$$N_{14}(s) = \beta s \frac{1}{C_1 R_4}$$

Augmented 2nd Bridged-T (ABT2)

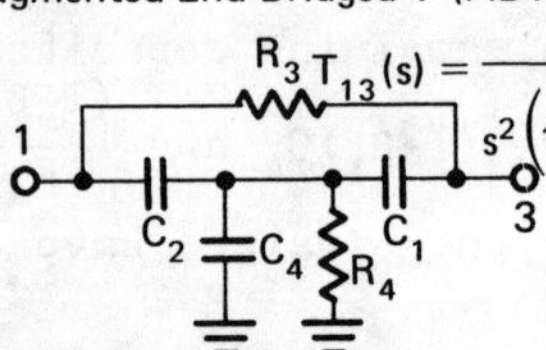

$$T_{13}(s) = \frac{s^2 + s\left(\frac{1}{C_1} + \frac{1}{C_2} + \frac{C_4}{C_1 C_2}\right)\frac{1}{R_3} + \frac{1}{C_1 C_2 R_3 R_4}}{s^2\left(1 + \frac{C_4}{C_2}\right) + s\left(\frac{1}{C_1 R_3} + \frac{1}{C_2 R_3} + \frac{C_4}{C_1 C_2 R_3} + \frac{1}{C_1 R_4}\right) + \frac{1}{C_1 C_2 R_3 R_4}}$$

$$N_{14}(s) = \beta_2 s^2 \frac{C_4}{C_1} + \beta_1 s \frac{1}{C_1 R_4} \qquad \beta_2 : C_4 \quad \beta_1 : 1/R_4$$

Loaded 2nd Bridged-T (LBT2)

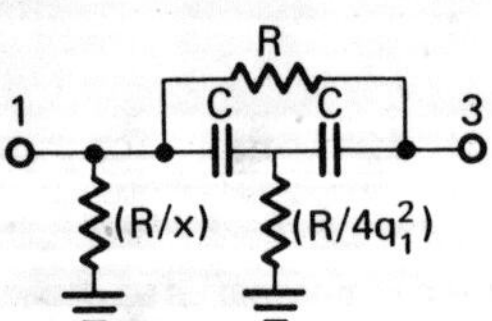

$\beta_1 : (4q_1^2/R) \qquad \beta_2 : (x/R)$

$\omega_1 = (2q_1/CR)$

$\omega_2 = \omega_1\sqrt{1+x}$

$q_2 = q_1\sqrt{1+x}\,/[2q_1^2 + 1 + x]$

$$T_{13}(s) = \frac{s^2 + s\left(\frac{\omega_1}{q_1}\right) + \omega_1^2}{s^2 + s\left(\frac{\omega_2}{q_2}\right) + \omega_2^2}$$

$$N_{14}(s) = s\beta_1(2\omega_1 q_1) + \beta_2 x \frac{\omega_1}{q_1}(s + \omega_1 q_1)$$

Figure 8.30 Bridged-T Networks

Example 8.3

The transfer function of a BT network may be expressed as

$$T_{13}(s) = k \frac{s^2 + s\frac{\omega_1}{q_1} + \omega_1^2}{s^2 + s\frac{\omega_2}{q_2} + \omega_2^2} \tag{8.108}$$

where (ω_1, q_1) characterize the pair of complex zeros $(q_1 > 0.5)$, and (ω_2, q_2) characterize the pair of real poles $(q_2 < 0.5)$ and k is a multiplicative constant. For the BT 1 case, find expressions for the parameters in (8.108) in terms of the circuit components. Also find an expression for the magnitude of transmission at the frequency $s = j\omega_1$. For the case $R_1 = R_2 = R$, $C_4 = C$ and $C_3 = C/m$, find the time constant CR, the value of m, the value of $|T_{13}(j\omega_1)|$, as well as the ω_1 and q_1 sensitivities for a given zeros' location; that is, given ω_1 and q_1.

Solution

From Fig. 8.30 and equation (8.108), we obtain

$$\omega_1 = \omega_2 = \frac{1}{\sqrt{C_3 C_4 R_1 R_2}} \tag{8.109}$$

$$q_1 = \frac{1}{\sqrt{C_3 C_4 R_1 R_2}} \Big/ \left(\frac{1}{R_1} + \frac{1}{R_2}\right)\frac{1}{C_4} \tag{8.110}$$

$$q_2 = \frac{1}{\sqrt{C_3 C_4 R_1 R_2}} \Big/ \left(\frac{1}{C_4 R_1} + \frac{1}{C_4 R_2} + \frac{1}{C_3 R_2}\right) \tag{8.111}$$

$$k = 1$$

The transmission at the zeros' frequency $j\omega_1$ is given by

$$T(j\omega_1) = \frac{q_2}{q_1} = \frac{\frac{1}{C_4}\left(\frac{1}{R_1} + \frac{1}{R_2}\right)}{\left(\frac{1}{C_4 R_1} + \frac{1}{C_4 R_2} + \frac{1}{C_3 R_2}\right)} \tag{8.112}$$

For $R_1 = R_2 = R$, $C_4 = C$ and $C_3 = \frac{C}{m}$,

$$\omega_1 = \omega_2 = \frac{\sqrt{m}}{CR}$$

$$q_1 = \frac{\sqrt{m}}{2}$$

$$\text{Thus, } m = 4\, q_1^2$$

$$\text{and} \quad CR = \frac{2q_1}{\omega_1}$$

$$q_2 = \frac{\sqrt{m}}{2 + m}$$

$$\text{Thus, } T(j\omega_1) = \frac{2}{2 + m}$$

$$T(j\omega_1) = \frac{2}{2 + 4q_1^2} = \frac{1}{1 + 2q_1^2}$$

From (8.109) ,

$$S_{R_1}^{\omega_1} = S_{R_2}^{\omega_1} = S_{C_3}^{\omega_1} = S_{C_4}^{\omega_1} = -\frac{1}{2}$$

From (8.110) ,

$$S_{R_1}^{q_1} = -\frac{1}{2} - \frac{R_1}{R_1 + R_2} + 1 = 0$$

$$S_{R_2}^{q_1} = 0$$

$$S_{C_3}^{q_1} = -\frac{1}{2}$$

$$S_{C_4}^{q_1} = +\frac{1}{2}$$

Twin-T (TT) Networks

Another class of RC two-ports that is useful in the design of biquads is found in twin-T type networks. As will be shown, there are two basic twin-T types; namely, the notch twin-T and the bandpass twin-T.

An RC twin-T network is a third-order network that produces a second-order function through a cancellation of a pole with a zero. A detailed study of twin-T networks is beyond the scope of this text and can be found elsewhere[44-47]. In the following, we shall consider only some special cases of interest.

Fig. 8.31a shows a general twin-T structure with an additional bridging admittance y_7. Analysis of this circuit yields the open circuit voltage transfer function

$$T_{13}(s) = \frac{y_7 + \frac{y_1 y_2}{Y_1} + \frac{y_4 y_5}{Y_2}}{y_2 + y_5 + y_7 - \frac{y_5^2}{Y_2} - \frac{y_2^2}{Y_1}} \tag{8.113}$$

where $Y_1 = y_1 + y_2 + y_3$ and $Y_2 = y_4 + y_5 + y_6$

It can be shown that, for the RC twin-T to provide a second-order transfer function, the time constants of the two T's should be equal. This condition can be stated as follows

$$Y_2 = k\, Y_1$$

where k is a positive real constant. Substituting in (8.113) results in

$$T_{13}(s) = \frac{k\, y_7\, Y + y_4 y_5 + k\, y_1 y_2}{k\, y_7\, Y + (y_4 + y_6)\, y_5 + k\, y_2\, (y_1 + y_3)} \tag{8.114}$$

where $Y = Y_1 = Y_2/k$

As the first special case, consider the symmetrical RC twin-T of Fig. 8.31b. In this case, $y_1 = y_2 = \frac{1}{R}$, $y_3 = s2C$, $y_4 = y_5 = sC$, $y_6 = \frac{2}{R}$ and $y_7 = 0$. Thus, $Y = Y_1 = Y_2 = \frac{2}{R} + s2C$ and $k = 1$. Substituting in (8.114) obtains

$$T_{13}(s) = \frac{s^2 + \frac{1}{T^2}}{s^2 + s\frac{4}{T} + \frac{1}{T^2}} \tag{8.115}$$

where $T = CR$.

This circuit is therefore a notch twin-T with the notch frequency being $\frac{1}{T}$.

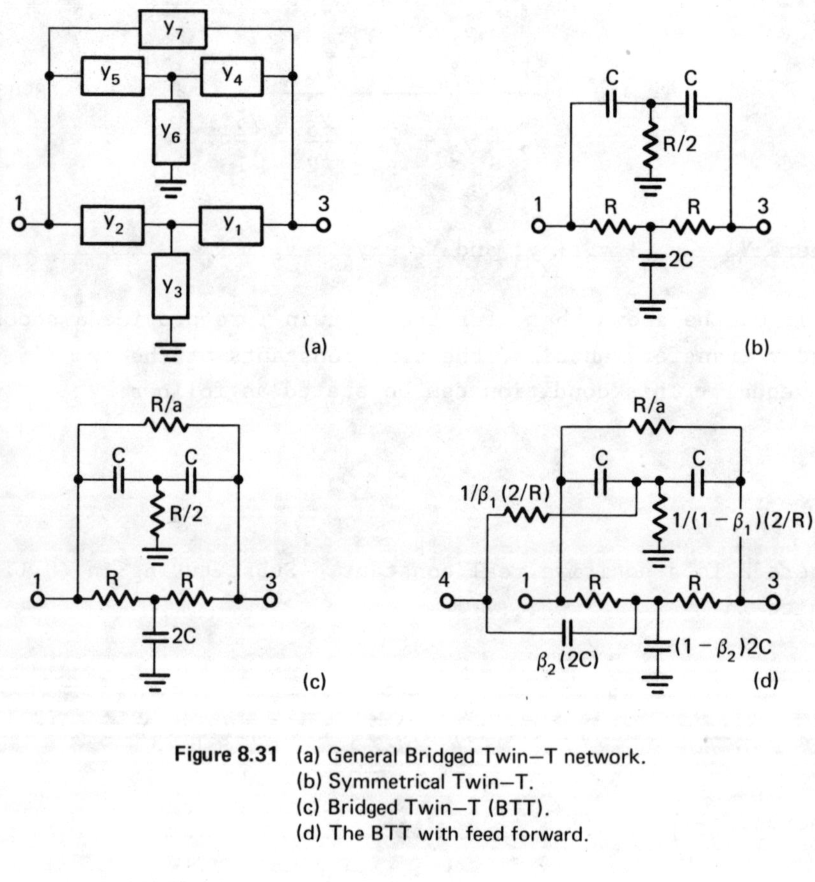

Figure 8.31 (a) General Bridged Twin–T network.
(b) Symmetrical Twin–T.
(c) Bridged Twin–T (BTT).
(d) The BTT with feed forward.

In Chapter 9 we will require twin-T networks that have a pair of complex transmission zeros in the LHP. It is thus required to find some means for shifting the transmission zeros of the symmetrical twin-T from the jω-axis to a prescribed location in the left-half of the s-plane. In other words, we require the zeros to have a finite prescribed Q-factor, rather than an infinite Q. Straightforward reasoning reveals that this task may be accomplished by the addition of a bridging resistance to the twin-T. This bridging resistance provides some controlled transmission from input to output at the frequency where the two T's provide equal and opposite polarity outputs. Note that there is an analogy between the effect of the bridging resistance in an RC twin-T and the effect of a parallel resistance in a lossless parallel-tuned LC tank circuit.

Fig. 8.31c shows the bridged twin-T (BTT) where the bridging resistance is R/a. Using (8.114) with $y_7 = \frac{a}{R}$ and with y_1 to y_6 as given above yields

$$T_{13}(s) = \frac{s^2 + s\,\frac{2a}{T} + \frac{1+2a}{T^2}}{s^2 + s\,\frac{4+2a}{T} + \frac{1+2a}{T^2}} \qquad (8.116)$$

Thus, the pair of complex zeros are characterized by a frequency ω_1,

$$\omega_1 = \frac{\sqrt{1+2a}}{T} ,$$

and a Q-factor q_1,

$$q_1 = \frac{\sqrt{1+2a}}{2a} .$$

Thus, the value of "a", and hence the value of the bridging resistance determines the Q of the zeros. For high-Q zeros, i.e. $q_1 \gg 1$,

$$q_1 \simeq \frac{1}{2a}$$

$$\omega_1 \simeq \frac{1}{T}$$

The bridged twin-T network has two components connected to ground. We may form an additional port, 4, by "lifting" fractions of these two components off ground and connecting them to terminal 4. This is shown in Fig. 8.31d with the transfer function $T_{14}(s)$ given by

$$T_{14}(s) = \frac{s\,\frac{2}{T}\,(\beta_1 + \beta_2)}{s^2 + s\,(\frac{4+2a}{T}) + \frac{1+2a}{T^2}} \;. \qquad (8.117)$$

Note that $T_{14}(s)$ is the open-circuit voltage transfer function between terminals 4 and 1 with 3 grounded. As expected, the natural modes of $T_{14}(s)$ and $T_{13}(s)$ are identical.

To obtain more terms in the numerator of $T_{14}(s)$, one has to augment the BTT network by adding more components having a ground terminal. In doing so, we should keep the time constants of the two T's equal, this being the condition for pole-zero cancellation. Fig. 8.32 shows the augmented notch twin-T (ABTT) which is generated from the BTT by the addition of a resistor (R/b) and a capacitor (fC), between each of the internal nodes and ground. Also given are the transfer functions and feedforward relationships.

Another twin-T network of interest is the bandpass twin-T. This network can be derived from the symmetrical notch twin-T of Fig. 8.32b. Specifically, if terminal 3 is grounded, and $\beta_1 = \beta_2 = 1$, then the resulting network becomes as shown in Fig. 8.33a with the transfer function being,

$$T_{14}(s) = \frac{s\,\frac{4}{T}}{s^2 + s\,(\frac{4+2a}{T}) + (\frac{1+2a}{T^2})} \qquad (8.118)$$

Note that the terminating resistance (R/a) may be eliminated, resulting in a transfer function

Network	*Transfer Functions*

(a) TT

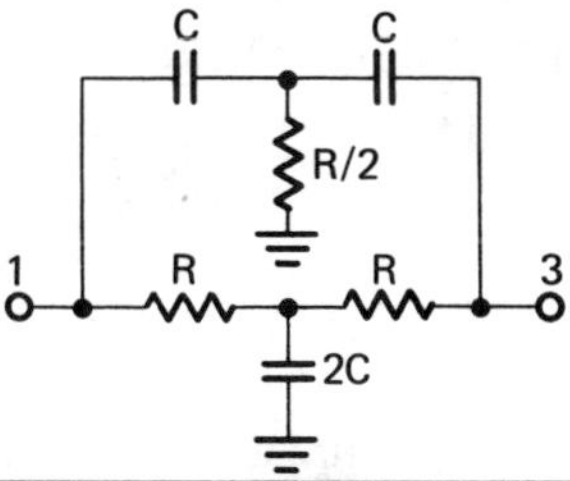

$$T_{13}(s) = \frac{s^2 + 1/T^2}{s^2 + s(4/T) + 1/T^2}$$

$$T = CR$$

(b) BTT

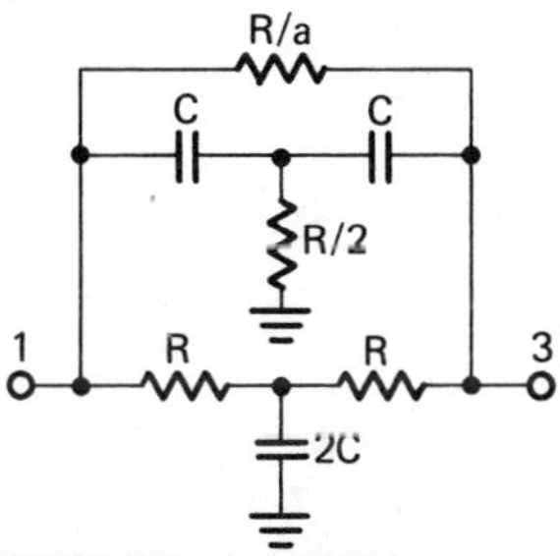

$$T_{13}(s) = \frac{s^2 + s\dfrac{2a}{T} + \dfrac{1+2a}{T^2}}{s^2 + s\dfrac{4+2a}{T} + \dfrac{1+2a}{T^2}}$$

$\beta_1 : 2/R$ $\beta_2 : 2C$

$$N_{14}(s) = s\frac{2}{T}(\beta_1 + \beta_2)$$

(c) ABTT

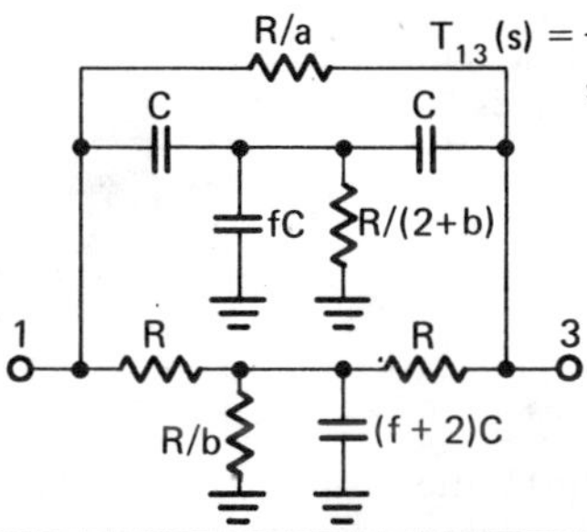

$$T_{13}(s) = \frac{s^2 + s\dfrac{a(f+2)}{T} + \dfrac{1 + a(2+b)}{T^2}}{s^2(1+f) + s\dfrac{1}{T}[4 + b + f + a(f+2)] + \dfrac{1 + b + a(2+b)}{T^2}}$$

$\beta_1 : (2+b)/R$ $\beta_3 : b/R$

$\beta_2 : (f+2)C$ $\beta_4 : fC$

$$N_{14}(s) = \beta_4 f s^2 + s\frac{1}{T}[\beta_1(2+b) + \beta_2(2+f)] + (\beta_3 b/T^2)$$

Figure 8.32 RC Twin-T notch networks.

$$T_{14}(s) = \frac{s\dfrac{4}{T}}{s^2 + s\dfrac{4}{T} + \dfrac{1}{T^2}} \tag{8.119}$$

for the network shown in Fig. 8.33b. This latter network has the unique property of unity center-frequency gain; a property impossible to obtain with bandpass RC ladders. Moreover, the value of the terminating resistance R/a, and hence the value of a, uniquely determines the center frequency gain in the loaded bandpass RC twin-T of Fig. 8.33a. We will find both these properties extremely valuable in generating high quality biquads in Chapter 9.

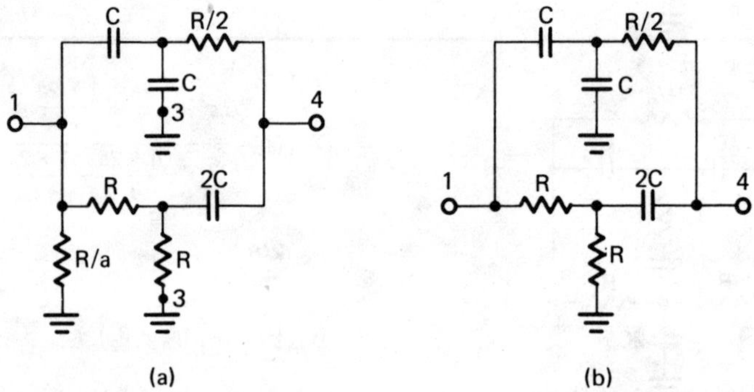

Figure 8.33 (a) Loaded bandpass Twin–T derived from the notch twin–T of Fig. (8.31C) (b) Bandpass Twin–T with unity center-frequency gain.

Returning back to the BP twin-T of Fig. 8.33b, which has a unity center- frequency gain, we note that another way to reduce the gain from unity by a controlled amount is to center-load the twin-T. In doing so, however, we have to keep the time constants of the two parallel T's equal. Thus, for any resistance we add at one of the internal nodes, we have to add a matching resistance at the other internal node. The loaded BP twin-T so obtained is shown in Fig. 8.34c. Note that, in this latter diagram, we have changed the nodes' numbering to conform with our future notation. For future reference, Fig. 8.34 includes the two other bandpass twin-T networks, along with all pertinent transfer functions.

Network	*Transfer Functions*

(a) Bandpass Twin-T

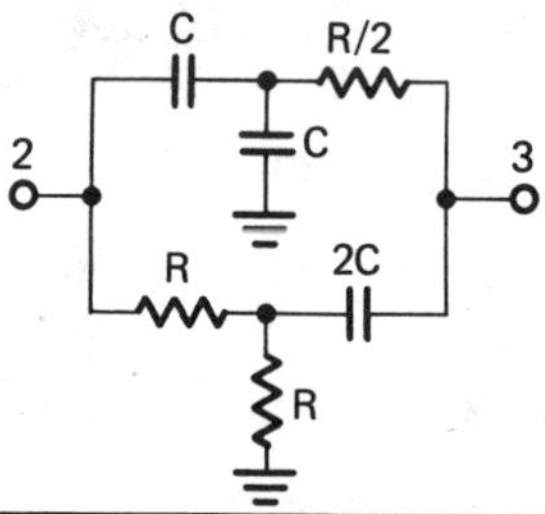

$$T_{23}(s) = \frac{s(4/T)}{s^2 + s(4/T) + (1/T^2)}$$

Center-frequency gain $G_0 = 1$

(b) End-Loaded BPTT

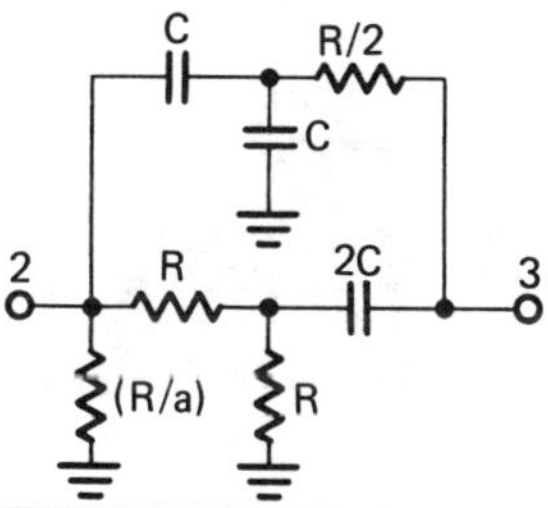

$$T_{23}(s) = \frac{s(4/T)}{s^2 + s\left(\frac{4 + 2a}{T}\right) + \frac{1 + 2a}{T^2}}$$

$$G_0 = \frac{1}{1 + (a/2)},\ \omega_2 = (\sqrt{1 + 2a})/T,\ q_2 = (\sqrt{1 + 2a})/(4 + 2a)$$

$\beta_1 : C$ $\beta_2 : 1/R$ $\beta_3 : a/R$

$$N_{24}(s) = \beta_1 s^2 + s\beta_3 \frac{2a}{T} + \frac{\beta_2 + 2a\beta_3}{T^2}$$

(c) Center-Loaded BPTT

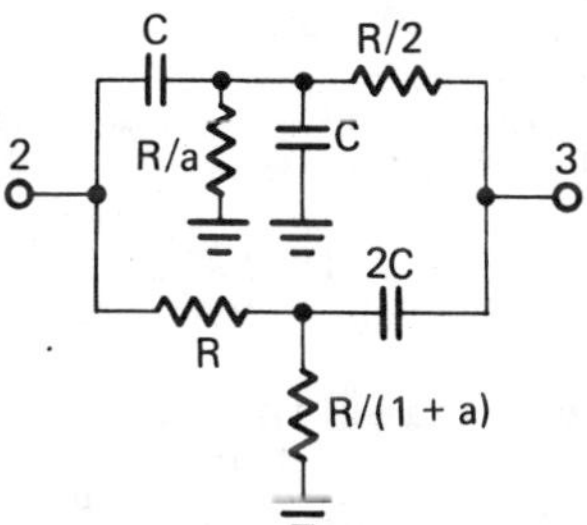

$$T_{23}(s) = \frac{s(4/T)}{s^2 + s\frac{4 + a}{T} + \frac{1 + a}{T^2}}$$

$$G_0 = \frac{1}{1 + (a/4)},\ \omega_2 = (\sqrt{1 + a})/T,\ q_2 = (\sqrt{1 + a})/(4 + a)$$

$\beta_1 : C$ $\beta_2 : a/R$ $\beta_3 : (1 + a)/R$

$$N_{24}(s) = \beta_1 s^2 + s\beta_2 \frac{a}{T} + \beta_3 \frac{(1 + a)}{T^2}$$

Figure 8.34 RC Twin-T bandpass networks.

8.7 REFERENCES AND SOURCES FOR FURTHER READING

(1) G. C. Temes and S. K. Mitra (Eds.), Modern Filter Theory and Design, Wiley, New York, 1973.

(2) G. S. Moschytz, "Inductorless filters: a survey. I. Electromechanical filters. II. Linear active and digital filters, "IEEE Spectrum, Vol. 7, No. 8, pp. 30-36 and No. 9, pp. 63 75, 1970.

(3) E. A. Guillemin, "Synthesis of RC networks", Journal of Math. Phys., Vol. 28, pp. 22-42, 1949.

(4) L. P. Huelsman, (Ed.), Active Filters: Lumped, Distributed, Integrated, Digital and Parametric, McGraw-Hill, New York, 1970.

(5) H. J. Orchard, "Gyrator Circuits", in (4), Chapter 2.

(6) K. L. Su, Active Network Synthesis, McGraw-Hill, New York, 1965.

(7) L. P. Huelsman, Theory and Design of Active RC Circuits, McGraw-Hill, New York, 1968.

(8) S. S. Haykin, Synthesis of RC Active Filter Networks, McGraw-Hill, New York, 1969.

(9) R. W. Newcomb, Active Integrated Circuit Synthesis, Englewood Cliffs, N.J., Prentice-Hall, 1968.

(10) S. K. Mitra, Analysis and Synthesis of Linear Active Networks, Wiley, New York, 1969.

(11) A. S. Sedra and K. C. Smith, "A second-generation current conveyor and its applications", IEEE Trans. Circuit Theory, Vol. CT-17, pp. 132-134, 1970.

(12) A. S. Sedra, "A new approach to active network synthesis", Ph.D. Thesis, Dept. of Elect. Eng., Univ. of Toronto, 1969.

(13) G. S. Moschytz, Linear Integrated Networks: Fundamentals, Van Nostrand Reinhold, New York, 1975.

(14) W. E. Heinlein and W. H. Holmes, Active Filters for Integrated Circuits, R. Oldenbourg Verlag, Munich, 1974.

(15) G. S. Moschytz, "The operational amplifier in linear active networks", IEEE Spectrum, Vol. 7, pp. 42-50, Jan. 1970. Reprinted in [23]

(16) J. V. Wait, L. P. Huelsman and G. A. Korn, Introduction to Operational Amplifier Theory and Applications, McGraw-Hill, New York 1975.

(17) J. K. Roberge, Operational Amplifiers: Theory and Practice, Wiley, New York, 1975.

(18) J. G. Graeme, G. E. Tobey, and L. P. Huelsman, Operational Amplifiers, McGraw-Hill, New York, 1971.

(19) L. T. Bruton and A. I. A. Salama, "Frequency limitations of coupled biquadratic active ladder structures", IEEE J. Solid-State Circuits, Vol. SC-9, pp. 70-72, Apr. 1974.

(20) P. O. Brackett and A. S. Sedra, "Active compensation for high-frequency effects in op. amp. circuits with applications to active - RC filters", IEEE Trans. Circuits and Systems, Vol. CAS-23, pp. 68-73, Feb. 1976.

(21) G. J. Deboo, "A novel integrator results by grounding its capacitor", Electron. Design, Vol. 15, pp.90, June 7, 1967.

(22) B. D. H. Tellegen, "The gyrator: A new network element", Philips Res. Rep., Vol. 3, pp. 81-101, Apr. 1948.

(23) S. K. Mitra (Ed.), Active Inductorless Filters, IEEE Press, New York, 1971.

(24) L. P. Huelsman, (Ed.), Active RC Filters: Theory and Application, in Benchmark Papers in Electrical Engineering and Computer Science, Vol. 15, Dowden, Hutchinson and Ross, Inc. Stroudsburg, Pennsylvania, 1976.

(25) B. A. Shenoi, "Practical realization of a gyrator circuit and RC gyrator filters", IEEE Trans. Circuit Theory, Vol. CT-12, pp. 374-380, Sept. 1965. Reprinted in [24].

(26) S. S. Haykim, S. Kramer, J. Schewchun, and D. H. Treleaven, "Integrated-circuit implementation of direct - coupled gyrator", IEEE J. Solid-State Circuits, Vol. SC-4, pp. 164-166, June 1969.

(27) W. H. Holmes, W. E. Heinlein, and S. Grutzmann, "Sharp-cutoff low-pass filters using floating gyrators", IEEE J. Solid State Circuits, Vol. SC-4, pp. 38-50, Feb. 1969. Reprinted in [23], [24].

(28) G. G. A. Black, R. T. Friedman and A. S. Sedra, "Gyrator implementation with integrable current-conveyors", IEEE J. Solid State Circuits, Vol. SC-6, pp. 396-399, 1971.

(29) R. H. S. Riordan, "Simulated inductors using differential amplifier", Electron. Lett., Vol. 3, pp. 50-51, Feb. 1967.

(30) A. Antoniou, "Gyrator using operational amplifiers", Electron. Lett., Vol. 3, pp. 350-352, 1967.

(31) A. Antoniou, "Realization of Gyrators using operational amplifiers and their use in RC - active network synthesis", Proc. Inst. Elec. Eng., Vol. 116, pp. 1838-1850, Nov. 1969.

(32) L. T. Bruton, "Nonideal performance of two-amplifier pos-sitive impedance converters", IEEE Trans. Circuit Theory, Vol. CT-17, pp. 541-549, Nov. 1970. Reprinted in [24].

(33) L. T. Bruton, "Network transfer functions using the concept of frequency dependent negative resistance", IEEE Trans. Circuit Theory, Vol. CT-16, pp. 406-408, Aug. 1969. Reprinted in [23], [24].

(34) L. T. Bruton and J. T. Lim, "High frequency comparison of GIC - simulated inductance circuits", Int'l J. Circuit Theory and Appl., Vol. 2, pp. 401-404, Dec. 1974.

(35) P. R. Geffe, "RC-amplifier resonators for active filters", IEEE Trans. on Circuit Theory, Vol. CT-15, pp. 415-419, Dec. 1968. Reprinted in [23].

(36) G. S. Moschytz, "Gain-sensitivity product-A figure of merit for hybrid-integrated filters using single operational amplifier", IEEE J. Solid State Circuits, Vol. SC-6, pp. 103-110, June 1971.

(37) A. S. Sedra, "Generation and classification of single amplifier filters", Int. J. Circuit Theory and Appl., Vol. 2, pp. 51-67, 1974.

(38) A. S. Sedra and J. L. Espinoza, "Sensitivity and frequency limitations of biquadratic active filters", IEEE Trans. Circuits and Systems, Vol. CAS-22, pp. 122-130, Feb. 1975.

(39) D. Hilberman, "An approach to the sensitivity and statistical variability of biquadratic filters", IEEE Trans. Circuit Theory, Vol. CT-20, pp. 382-390, July 1973.

(40) A. L. Rosenblum and M. S. Ghausi, "Multiparameter sensitivity in active RC networks", IEEE Trans. Circuit Theory, Vol. CT-18, pp. 592-594, Nov. 1971.

(41) B. A. Shenoi, "Optimum variability design and comparative evaluation of thin-film RC active filters", IEEE Trans. Circuits & Systems, Vol. CAS-21, pp. 263-267, March 1974.

(42) G. S. Moschytz, "A note on pole, frequency and Q sensitivity", IEEE J. Solid-State Circuits, Vol. SC-6, pp. 267-269, Aug. 1971.

(43) R. MacKay and A. S. Sedra, "Sensitivity evaluation and comparison for high-order active filters", Int. J. Circuit Theory and Applications, Vol. 4, pp. 345-356, 1976.

(44) G. S. Moschytz, "A general approach to twin - T design and its application to hybrid integrated linear active networks", Bell-Syst. Tech. J., Vol. 49, pp. 1105-1149, July 1970.

(45) E. Lueder, "The general second-order twin-T and its application to frequency emphasizing networks", Bell Sys. Tech. J., Vol. 51, pp. 301-316, 1972.

(46) S. A. Boctor and A. S. Sedra, "On the synthesis of RC bandpass filters", IEEE Trans. Circuit Theory, Vol. CT-19, pp. 374-376, July 1972.

(47) S. A. Boctor and A. S. Sedra, "Realization of nonminimum phase transfer functions using twin-T RC networks", IEEE Trans. Circuit Theory, Vol. CT-18, pp.471-475, July 1971.

PROBLEMS

P8.1 Consider an op amp with $A_o=10^5$, f_t=2MHz, input bias current = 1μA, input offset current = 0.1μA and input offset voltage = 3 m volts.

(a) Calculate the 3dB frequency for each of the closed-loop amplifiers whose nominal gains are +1, +10, +100 and -1.

(b) Find the output offset voltage of an inverting amplifier with a closed-loop gain of 1000 and feedback resistance of 1MΩ. What will the offset voltage be if a compensating resistor is connected in the positive input lead of the op amp. Give the value of the compensating resistor.

P8.2 Assuming ideal op amps find:

(a) Rin for the circuit in Fig. P8.2(a)

(b) I for the circuit in Fig. P8.2(b)

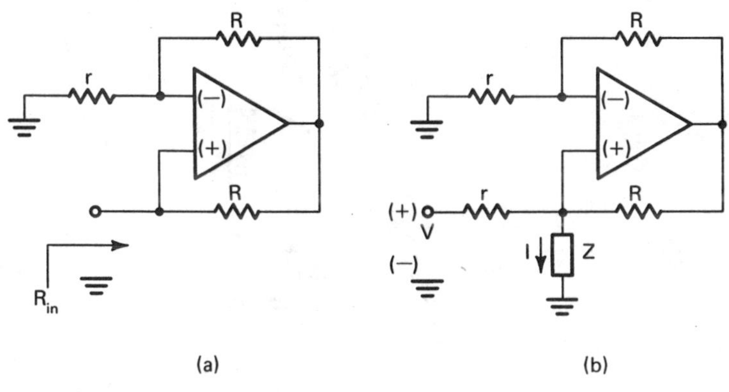

Figure P8.2

P8.3 Assuming ideal op amps, derive the transfer function $(V_o/V_i)(s)$ for each of the circuits in Fig. P8.3. Sketch a Bode plot for the magnitude of each of the transfer functions and specify the type of function.

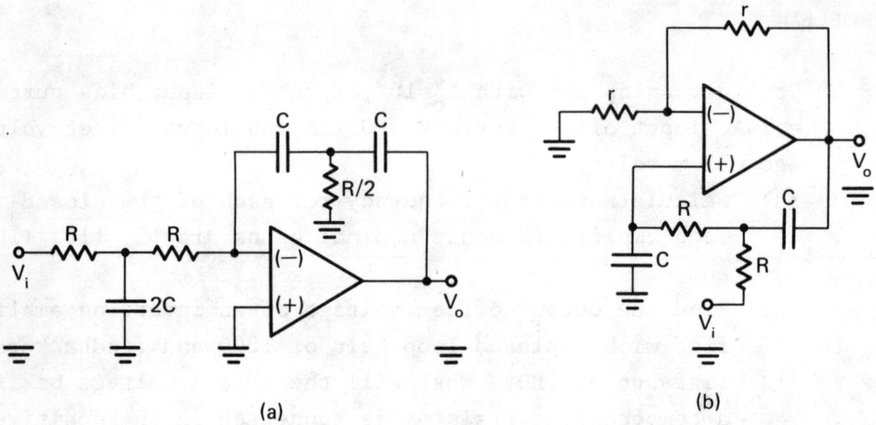

Figure P8.3

P8.4 For the circuit in Fig. P8.4 assume the op amp to be ideal and find the input impedance $Z_{in}(s)$. Give an equivalent circuit for $Z_{in}(s)$.

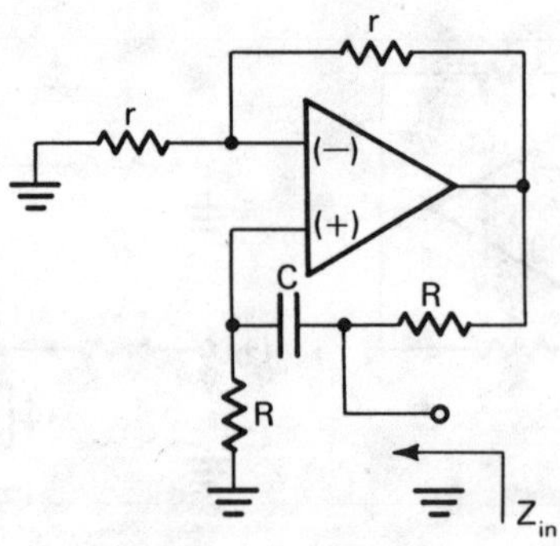

Figure P8.4

P8.5 For the circuit in Fig. P8.5, find the transfer function t(s) and its RC network realization so that $(V_o/V_i)(s)$ is a first-order all-pass function with the singularities at $s=\pm a$. Assume the op amp to be ideal.

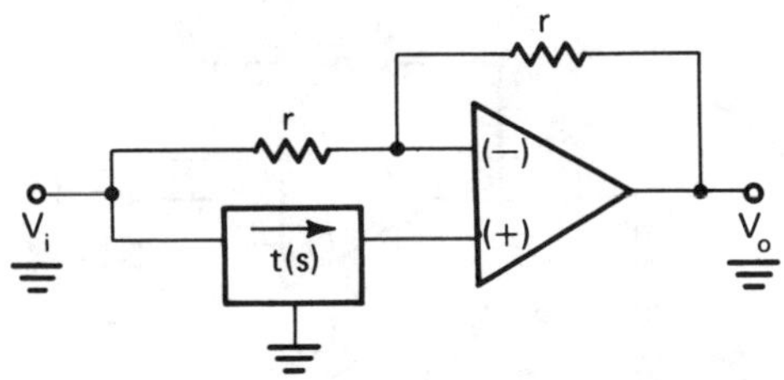

Figure P8.5

P8.6 Show that a second order all-pass function with a flat gain of -1 may be realized by the circuit in Fig. P8.5, and give the required t(s).

P8.7 Consider the inverting summer of Fig. 8.4(f) and let the parallel equivalent of the input "feed-in" resistors be denoted by R_i. Assuming a single-pole model for the op amp with a unity gain bandwidth ω_t, show that the excess phase shift (i.e. above 180°) of the transfer function from any of the inputs to the output is given by

$$\phi \simeq -\left(\frac{\omega}{\omega_t}\right)\left(1+\frac{R_f}{R_i}\right)$$

P8.8 Show that the excess phase shift (i.e. above -90°) of the Miller integrator is approximately equal to $(1/Q_I)$.

P8.9 Show that the Q-factor of the Deboo integrator is equal to $-\frac{1}{2}|A|$.

P8.10 Give the location and value of a capacitor that can be added to the Deboo integrator to provide passive compensation of the Q-factor.

P8.11 For the circuit in Fig. P8.11, find the value of R_x required to make the Q-factor equal to $+|A|$. What is the time-constant of this integrator? Assume the op amps to be identical and have a single-pole model.

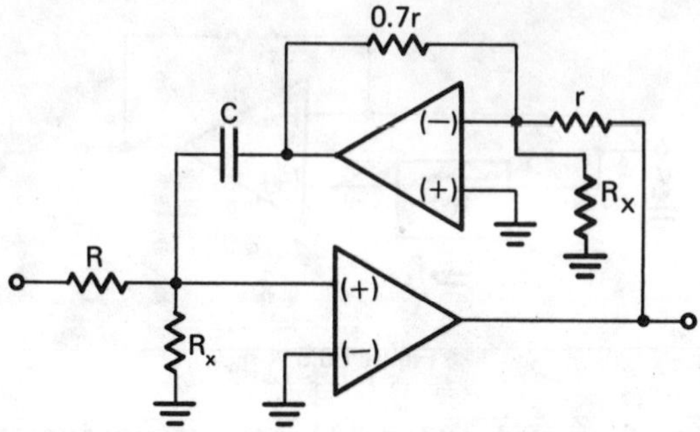

Figure P8.11

P8.12 For the circuit in Fig. P8.12 assume the op amps to be identical and have a single pole mode with a unity gain bandwidth of ω_t. Derive an expression for the excess phase shift ϕ and show that R_x may be used to control the magnitude and polarity of ϕ.

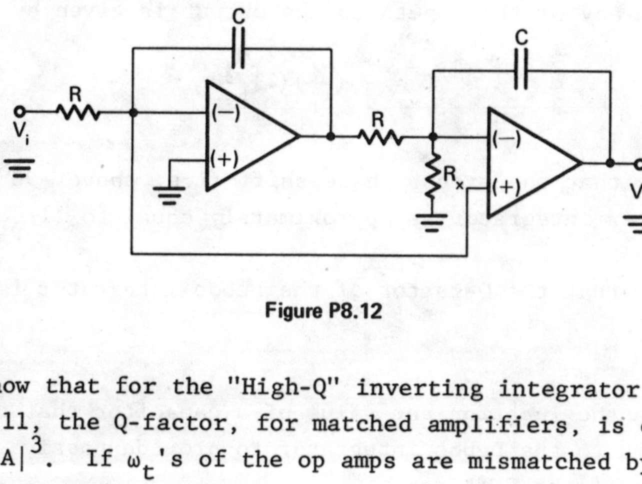

Figure P8.12

P8.13 Show that for the "High-Q" inverting integrator in Fig. 8.11, the Q-factor, for matched amplifiers, is equal to $-|A|^3$. If ω_t's of the op amps are mismatched by x% what is the value of Q?

P8.14 Derive equations (8.34) and (8.35).

P8.15 Show that the circuit in Fig. P8.15 realizes a floating inductance L, between terminals 1 and 2, and find its value.

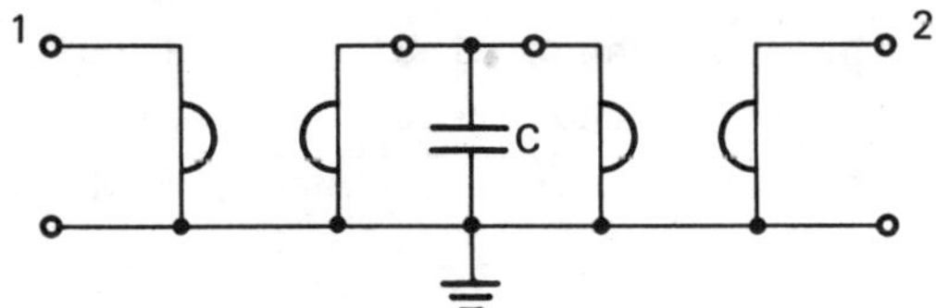

Figure P8.15

P8.16 Consider the implementation of a gyrator using two voltage controlled current sources with equal transconductances g. Let each current source have an input conductance (εg) and an output conductance (εg) where $\varepsilon \ll 1$. Derive an approximate expression for the Q-factor of the realized inductor when the gyrator is terminated in a capacitor C. Find the frequency at which the Q factor reaches a maximum and find the value of this maximum.

P8.17 Find the impedances Z_1 to Z_5 of the circuit in Fig. 8.17 so that the input is a "super inductor", that is the input impedance is $(s^2\ell)$.

P8.18 Transform the network of Fig. P8.18 by multiplying its impedances by (1/s). Show that the transfer function $(V_o/V_i)(s)$ remains constant under this transformation Give an active circuit realization of the transformed network.

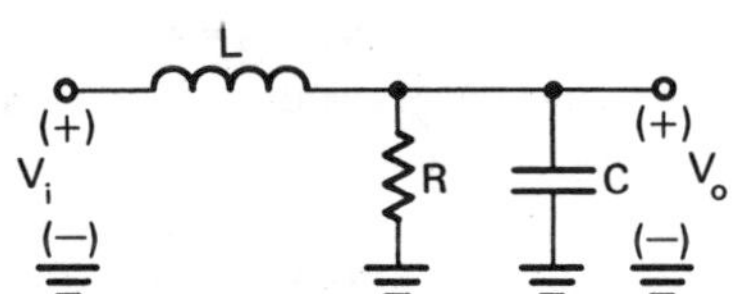

Figure P8.18

P8.19 Consider the type-A inductance simulation circuit of Fig. 8.18. If $\omega_{t1}=1.5\omega_{t2}$, find the value of (R_4/R_5) to obtain infinite Q factor (to a first order approximation).

P8.20 Consider the inductance simulation circuit of type-A in Fig. 8.18. Design the circuit to realize an inductance of 0.1H and let the critical frequency f_c be 1KHz. Let the op amps have equal f_t's of 1MHz that track to within 2% over a 50°C change in temperature. Calculate the error in the value of L at room temperature and find the value of Q after the temperature changes by 50°C.

P8.21 Verify the sensitivity identities given in Section 8.5

P8.22 (a) For the circuit in Fig. P8.22, calculate the sensitivities of ω_o and Q relative to R, C and L.

(b) If Q=10 and C changes by + 1% calculate $S_{\omega_o}^{|T|}$ and $S_Q^{|T|}$ at the 3 dB frequencies. Hence, calculate the deviation D(ω) in dB.

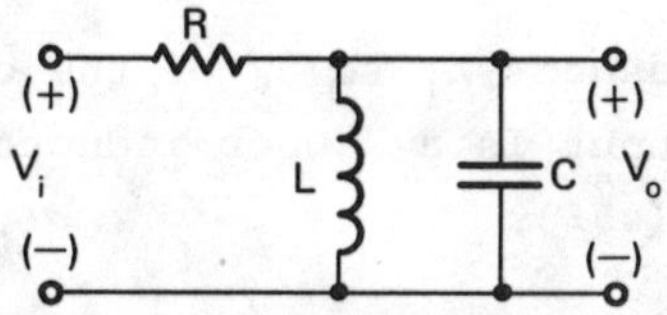

Fig. P8.22

P8.23 (a) Show that the circuit shown in Fig. P8.23 realizes a second-order bandpass function.

(b) For R_1=R, R_2=R_3=mR and C_1=C_2=C find (CR) and m in terms of ω_o and Q. Also find the gain at ω_o.

(c) Evaluate the passive ω_o and Q sensitivities and the gain sensitivity products GS^{ω_o} and GS^Q.

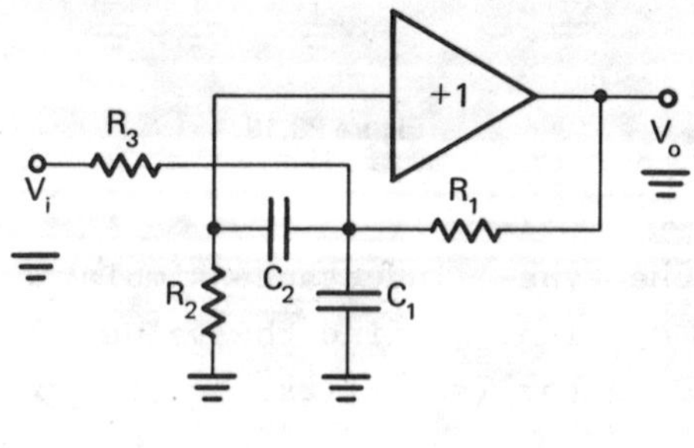

Figure P8.23

P8.24 Use the results of Fig. 8.27 to show that if the input terminal and ground exchange roles the transfer function obtained is the complement of the original function. For instance, in all networks $T_{24} = 1 - T_{23}$. (Note: this result is a general one).

P8.25 Show that exchanging input with ground in a ladder results in a bridged-T network and vice versa. Hence, identify complementary networks between Fig. 8.27 and Fig. 8.30.

P8.26 Apply the ideas of the above two problems to the twin-T networks of Fig. 8.32 and Fig. 8.34.

P8.27 Verify the relationships given in Fig. 8.32.

CHAPTER 9

SECOND ORDER SECTIONS-THE BIQUADS

9.1 INTRODUCTION

Biquadratic sections are basic building blocks in many of the structures used to realize high-order active filters. This will be clearly demonstrated in Chapters 10 and 12. In addition, there are many filtering applications that only require a second-order section. As an example, an active second-order notch filter is frequently used to eliminate a particular interfering frequency. Another example of the application of biquads is in the touch-tone telephone receiver where eight second-order band-pass filters may be connected in parallel, with each section tuned to one of the eight tones. For all these reasons, the implementation of biquads deserves a special treatment. The present Chapter is devoted to such a detailed presentation.

Scanning the active filter literature, one finds hundreds of "different" circuit implementations of biquads. This variety, however, rather than providing the filter designer with flexibility in choice, produced a state of confusion. To further compound the problem, the various inventors and analysts tended to use different criteria for comparing the large number of biquad circuits. Another symptom of the problem is found in the presentation of the subject of biquad design to students of filter theory and design. The usual practice has been to present a particular implementation, analyze it, give some sort of design equations and offer few comments about its utility. Very seldom was the student told how the circuit was arrived at in the first place or, more importantly, what relation if any does the particular circuit have to other circuits.

Recently, however, attempts have been made[3-7] to rectify this serious problem by:

(a) Devising a "somewhat general" procedure for generating or "inventing" various biquad circuits. Of course such a procedure would have been useless if the known circuits, or at least a large number of them, did not result as products of the generation process.

(b) Classifying and categorizing the various biquads into distinct groups with the members of each group sharing many common properties. This classification process is made possible by the insight gained from the procedure of (a) above.

In order to achieve the above goals, one had to, unfortunately, ignore certain particular circuits. As it turned out, however, these circuits are not of high enough practical value.

For any classification scheme to be possible, the starting point is the identification and justification of key performance indices to be used for comparison purposes. Section 9.2 introduces such performance indices, or figures-of-merit. Before studying Section 9.2, the reader is advised to review the material on sensitivity presented in Section 8.5.

As mentioned in Chapter 8, our basic active element is the popular IC op. amp. Thus, although some of the biquads presented employ finite gain amplifiers, the latter can be realized using op. amps. The number of amplifiers in a biquadratic section ranges from a minimum of one to a maximum of four. Minimizing the number of amplifiers has the obvious advantage of lower cost and power dissipation. Another not-so-obvious advantage is that properly designed single-amplifier filters can achieve a wider useful operating frequency range. Roughly speaking, adding more amplifiers to the circuit increases the excess phase shift and hence degrades the high-frequency performance. Nevertheless, as will be shown, there are multi-amplifier circuits that exhibit excellent high-frequency response. Generally speaking, the addition of more amplifiers to the circuit is motivated by two reasons:

(a) to improve some performance measure and,
(b) to increase the flexibility and manufacturability of the circuit.

We will, of course, have more to say about this in later sections.

Sections 9.3-9.7 are devoted to the subject of Single-Amplifier Biquads (SAB). There, a general theory for these minimal realizations is developed. Unfortunately, it is not easy to extend this general theory to multi-amplifier circuits. Therefore, based on our previous research[8] and experience, we present in Section 9.8 those two-amplifier circuits that have proven generally superior to single-amplifier circuits. Realizations that <u>essentially</u> use three amplifiers are presented in Section 9.9. Again, only those circuits that yield some improvement in return for the price of the extra amplifier are considered. Adding a fourth amplifier is useful only to provide flexibility in realizing finite transmission zeros. This is included in Section 9.9.

All through this chapter the emphasis is on design trade offs and comparisons. A substantial number of design tables and comparison charts are provided.

9.2 PERFORMANCE CRITERIA FOR BIQUADRATIC REALIZATIONS

A general second-order voltage transfer function is given by

$$T(s) = K \frac{s^2 + s \frac{\omega_{oz}}{Q_z} + \omega_{oz}^2}{s^2 + s \frac{\omega_{op}}{Q_p} + \omega_{op}^2} \tag{9.1}$$

where K is a gain constant, s is the complex frequency variable, ω_{oz} and Q_z characterize the transmission zeros and ω_{op} and Q_p characterize the pair of complex poles, $Q_p > 0.5$. It has been the usual practice in the active-filter literature to drop the subscript p for the pole frequency and Q, that is write T(s) as

$$T(s) = K \frac{s^2 + s \frac{\omega_{oz}}{Q_z} + \omega_{oz}^2}{s^2 + s \frac{\omega_o}{Q} + \omega_o^2} . \tag{9.2}$$

Since it is the location of the poles (natural modes) that primarily determines the filter passband performance, one is naturally concerned with the variability of ω_o and Q, due to the tolerances and variations of the various components used in the biquad circuit realization. Note, for instance, that in a cascade design the overall transfer function poles correspond directly to the poles of the biquads. Hence, the biquads' poles directly determine the filter performance in the passband where only small deviations can be tolerated.

In Section 8.5 we presented expressions for the variability of ω_o and Q in a biquad. To repeat, consider a biquad realization having ℓ capacitors, m resistors and n amplifiers. Each amplifier can be used as an infinite-gain amplifier or in a finite-gain configuration with a nominal closed-loop gain K. The variabilities in question may be expressed as

$$\frac{\Delta\omega_o}{\omega_o} = \sum_{i=1}^{\ell} S_{C_i}^{\omega_o} \left(\frac{\Delta C_i}{C_i}\right) + \sum_{j=1}^{m} S_{R_j}^{\omega_o} \left(\frac{\Delta R_j}{R_j}\right) + \sum_{k=1}^{n} GS_{K_k}^{\omega_o} \left(\frac{\Delta A_k}{A_k^2}\right) \tag{9.3}$$

$$\frac{\Delta Q}{Q} = \sum_{i=1}^{\ell} S_{C_i}^{Q} \left(\frac{\Delta C_i}{C_i}\right) + \sum_{j=1}^{m} S_{R_j}^{Q} \left(\frac{\Delta R_j}{R_j}\right) + \sum_{k=1}^{n} GS_{K_k}^{Q} \left(\frac{\Delta A_k}{A_k^2}\right) \tag{9.4}$$

where $\frac{\Delta C_i}{C_i}$ is the variability of capacitor C_i, $\frac{\Delta R_j}{R_j}$ is the variability of resistor R_j and, $\frac{\Delta A_k}{A_k}$ is the variability of the low-frequency open-loop gain A_k of op. amp. k.

The sensitivities of ω_o and Q relative to each of the RC components are evaluated using the classical-sensitivity definition. These are referred to as the passive sensitivities. For the active sensitivities, we use the gain sensitivity products, defined and explained in Chapter 8. We remind the reader that the gain-sensitivity product is a better measure of the effects of variations in amplifier gain since its value is independent of the closed-loop gain K; that is,

$$GS^{y}_{K_k} = GS^{y}_{A_k}$$

where K_k is the closed-loop gain of op. amp. k and where y denotes ω_o or Q.

In evaluating the summations in (9.3) and (9.4) a number of approaches have been suggested, such as summation of the magnitudes of the sensitivities or summation of the squared magnitudes of the sensitivities. Another approach takes into account the statistics of the component tolerances[9-11]. Although we recognize these elaborate methods, we still believe that by simply evaluating the different sensitivity figures, one is able to compare the different biquad realizations regardless of which way the numbers should be added to provide a variability figure. Therefore, in comparing biquads we will evaluate all the passive sensitivities, for both ω_o and Q, as well as the ω_o and Q gain-sensitivity products.

Since the gain-sensitivity products are evaluated at low-frequencies, the amplifiers' limited bandwidth is not taken into account. The effects of the limited bandwidth can be assessed in terms of the deviation induced in ω_o and Q. Specifically, we shall assume a single-pole model for each op. amp. (as described in Chapter 8) and find expressions for the actual values of ω_o and Q, which will be denoted by ω_{oa} and Q_a. An even better way of expressing the deviation is in terms of the parameters δ and η defined as

$$\delta \equiv \frac{\omega_{oa} - \omega_o}{\omega_o} \tag{9.5}$$

$$\eta \equiv \frac{Q_a}{Q} \tag{9.6}$$

An interesting result will emerge from our study of the expressions for δ and η for the different biquad circuits. It will be shown that in the expressions for δ and η, the only configuration-dependent quantities are the Q gain-sensitivity products. Thus, although evaluated assuming frequency-independent gain, the gain-sensitivity products predict the high-frequency performance of the biquad.

In Chapter 8, we studied in detail the relative contribution of the ω_o and Q variabilities to the variability of transfer function for which ω_o and Q characterize a pole-pair. We have seen that variations in ω_o are about 2Q times more significant than equal variations in Q. Consequently, in designing second-order sections one should attempt to reduce the ω_o passive sensitivities and the ω_o gain-sensitivity-products to their possible minima. This will be true for all the circuits presented here. In fact, many of the circuits presented have their GS^{ω_o} values identically zero, while others have GS^{ω_o} values around unity. Note that some of the circuits that have been suggested in the literature[12] have the obviously unacceptable GS^{ω_o} of the order of Q^2![8]. As far as the passive ω_o-sensitivities are concerned, these will generally be $\leq\frac{1}{2}$ in magnitude, which are the theoretical minima.

In summary, for each biquad realization we evaluate the following performance indices:

(1) the ω_o passive-sensitivities; $S_x^{\omega_o}$
(2) the ω_o gain-sensitivity-products; $GS_K^{\omega_o}$
(3) the Q passive-sensitivities; S_x^Q
(4) the Q gain-sensitivity-products; GS_K^Q

(5) the deviation in ω_o due to the amplifiers' limited bandwidth; δ

(6) the Q variation factor due to the amplifiers' limited bandwidth; η

In addition to the above-mentioned performance indices there are other practical considerations that influence the choice of a particular biquad design. These include:

(1) The required spread in component values. A good realization should not require a wide spread in the range of resistor and capacitor values.

(2) The total amount of capacitance required or, in other terms, the total amount of time-constant required. If the filter is to be produced using hybrid IC technology, it is necessary to minimize the total CR value.

(3) The versatility of the circuit configuration in realizing different types of filtering functions, that is LP, BP, HP, etc, by simply feeding the input signal through different feedforward admittances. A circuit having this property is said to have an "isotopic" structure.

(4) The ability of a circuit to provide more than one filtering function simultaneously at different output terminals. Note that properties (3) and (4) are highly desirable from the point of view of modular off-the-shelf filter design. Currently, a number of manufacturers offer universal filter modules which possess both the isotopic nature as well as the multiple-function-output property.

(5) The ease of initial adjustment or tuning. In some circuits, it is possible to adjust ω_o and Q separately in a noninteractive manner; a highly desirable feature. We will have more to say about tuning methods in this and the following Chapters.

(6) The input impedance of the biquad. For cascading purposes, it is desirable to have high input impedances. Note that this is necessary in spite of the low output impedance of biquad circuits, since general purpose op. amps. have a limited output driving capability. Moreover, the input impedance of some biquads are frequency dependent and become re-

active with small values at resonance. This might cause the preceding stage to distort or even oscillate. Raising the level of the input impedance by mere impedance scaling is not a satisfactory solution, since this usually results in some component attaining an impractical value, with the possibility of making the circuit susceptible to the influence of parasitic elements.

Needless to say, the relative importance of each of the above factors is determined by the technology chosen for construction (e.g. thin-film, thick-film or discrete), the ultimate application of the filter section (that is universal or a specific application), as well as the number of units to be made of the same filter.

9.3 A PRELIMINARY CLASSIFICATION OF SINGLE AMPLIFIER BIQUADS *

The general configuration of a Single-Amplifier Biquad (SAB) is given in Fig. 9.1a, where an op. amp. is shown embedded in a four-port RC network n. The problem then is to find an implementation for n, such that $\frac{V_o}{V_i}(s)$ realizes a given second-order function T(s). This most general problem, however, is quite difficult to solve and we shall be content with some special cases of the general configuration.

In Fig. 9.1b, we show a specialization of the general configuration where we have split the network n into two RC subnetworks: n_1 in the negative feedback path of the op. amp. and n_2 in the positive feedback path. What we have really done is eliminated the interaction between the two subnetworks. The resulting simplification is evident, as each of the subnetworks is now a three-port. Assuming that the op. amp. is ideal but has a finite gain A(s), analysis of the circuit of Fig. 9.1b results in

$$T(s) \equiv \frac{V_o(s)}{V_i(s)} = \frac{T_{24}(s) - T_{14}(s)}{T_{13}(s) - T_{23}(s) + \frac{1}{A(s)}} \tag{9.7}$$

*The material on SAB's is largely based on [6].

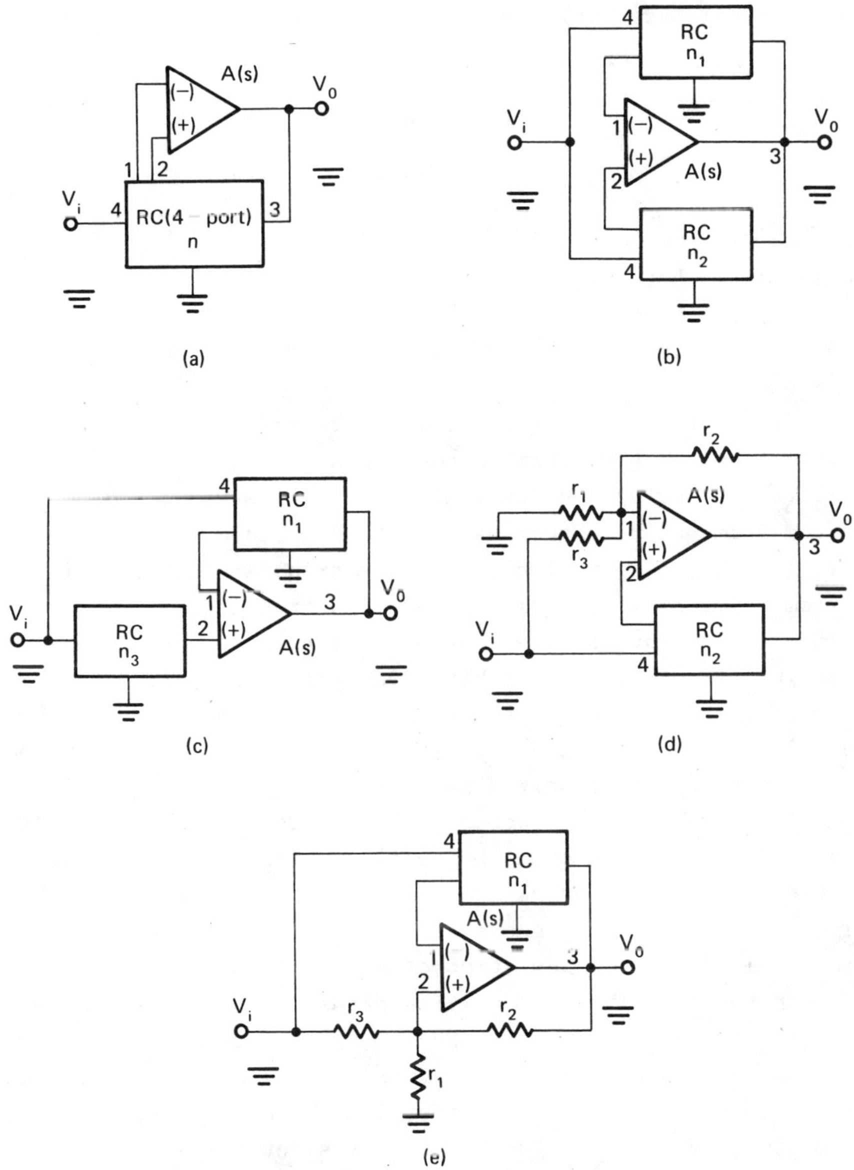

Figure 9.1 (a) General configuration of a Single-Amplifier Biquad (SAB) with special cases: (b) Mixed-Feedback configuration, (c) Negative-Feedback (NF) configuration, (d) Positive Feedback (PF) configuration, and (e) Enhanced Negative-Feedback (ENF) configuration.

In the ideal case A(s) = ∞ and (9.7) reduces to

$$T(s) = \frac{T_{24}(s) - T_{14}(s)}{T_{13}(s) - T_{23}(s)} \tag{9.8}$$

Note that in both (9.7) and (9.8):

$T_{13}(s)$ is the open-circuit voltage transfer-function of network n_1 from terminal 3 to terminal 1 with terminal 4 grounded,

$T_{14}(s)$ is the open-circuit voltage transfer-function of n_1 from terminal 4 to terminal 1 with 3 grounded,

$T_{23}(s)$ is the open-circuit voltage transfer-function of n_2 from terminal 3 to terminal 2 with 4 grounded and,

$T_{24}(s)$ is the open-circuit voltage transfer-function of n_2 from terminal 4 to terminal 2 with 3 grounded.

It is important to note that by expressing the transfer function in the form (9.8), we have, in effect, converted the problem of synthesizing three-port RC networks to the much simpler one of two-port RC networks. This should be evident from the transfer functions involved in (9.8).

Although we have considerably simplified the SAB implementation problem by going from the configuration of Fig. 9.1a to that in Fig. 9.1b, it is still quite difficult to formulate a general circuit generation procedure starting from the latter configuration. Since the configuration of Fig. 9.1b has feedback networks in both the negative and positive paths of the op. amp., we will refer to it as the Mixed Feedback (MF) configuration. Also, all circuits generated from the MF configuration will be referred to as MF circuits. We will consider some such circuits below.

To simplify our task further, we consider the three special cases shown in Figs. 9.1c, 9.1d and 9.1e. In the configuration of Fig. 9.1c we have completely eliminated the positive feedback network n_2. This configuration will therefore be referred to as the Negative Feedback (NF) configuration. It consists of an RC network n_1 connected in the negative feedback path of the op. amp. with the input signal V_i injected to node 4 in n_1 and to the noninverting input terminal of the op. amp. through another RC network n_3. Network n_3 is included only in the feedforward path and is used to increase the versatility of the NF configuration in realizing different transmission zeros.

Unlike the NF configuration, the structure of Fig. 9.1d places the RC frequency selective network in the positive feedback path and is, therefore, called the Positive Feedback (PF) configuration. Of course, some negative feedback is needed to stabilize the gain of the amplifier. This is accomplished through resistors r_2 and r_1. The input signal V_i is injected to node 4 of n_2, and to the negative input terminal of the op. amp. through resistor r_3.

Note that although both the NF and PF configurations are special cases of the more general MF circuit of Fig. 9.1b they will be treated separately. A third special case of the MF configuration which has gained considerable popularity is shown in Fig. 9.1e. Here the RC frequency selective network is placed in the negative feedback path, but some positive feedback is also applied through resistors r_2 and r_1. This circuit can, therefore, be considered as a negative feedback circuit with some additional resistive positive feedback. We will call this configuration the Enhanced Negative Feedback (ENF) circuit. Note that the input signal V_i is injected to terminal 4 of n_1, and to the noninverting input terminal of the op. amp. through resistor r_3. As will be seen below, the performance indices of the ENF class of circuits will lie in between those for the NF class and the PF class.

Now that we have concentrated on three specific topologies of SAB realizations, it remains to find methods for generating circuit configurations that belong to each of the three semi-general structures. In addition, we wish to further classify the resulting circuits according to their generic origin, which will, hopefully, also correspond to the values of their performance indices.

9.4 SINGLE AMPLIFIER BIQUADS OF THE NEGATIVE FEEDBACK (NF) TYPE

Generation Procedure

The general configuration of a negative feedback (NF) realization is shown in Fig. 9.2 The voltage transfer function may be obtained from (9.7) by setting $T_{23}(s) = 0$. Thus,

$$T(s) = \frac{T_{24}(s) - T_{14}(s)}{T_{13}(s) + \frac{1}{A(s)}} \tag{9.9}$$

where $T_{24}(s)$ is now the open-circuit voltage transfer function of n_3 between the input terminal and terminal 2. The definition of the various transfer functions in (9.9) is illustrated in Fig. 9.3. The functions of n_1 can be expressed in polynomial quotient form as,

$$T_{13}(s) = \frac{N_{13}(s)}{D(s)}$$

$$T_{14}(s) = \frac{N_{14}(s)}{D(s)}$$

where $D(s)$, the natural-mode polynomial of n_1, has its zeros on the negative real axis only. Substituting in (9.9) obtains,

$$T(s) = \frac{T_{24}(s)\ D(s) - N_{14}(s)}{N_{13}(s) + D(s)\ /\ A(s)} \tag{9.10}$$

In the ideal case $A(s) = \infty$ reducing (9.10) to

$$T(s) = \frac{T_{24}(s)\ D(s) - N_{14}(s)}{N_{13}(s)} \tag{9.11}$$

Restricting ourselves to the case,

$$T_{24}(s)\ D(s) = \text{polynomial} \tag{9.12}$$

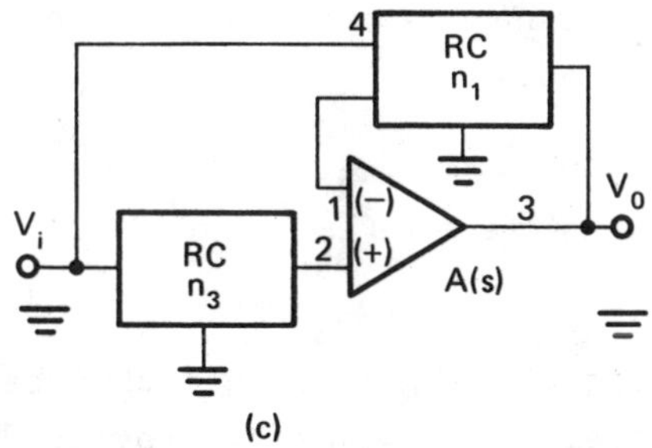

Figure 9.2 General configuration of the Negative-Feedback (NF) SAB.

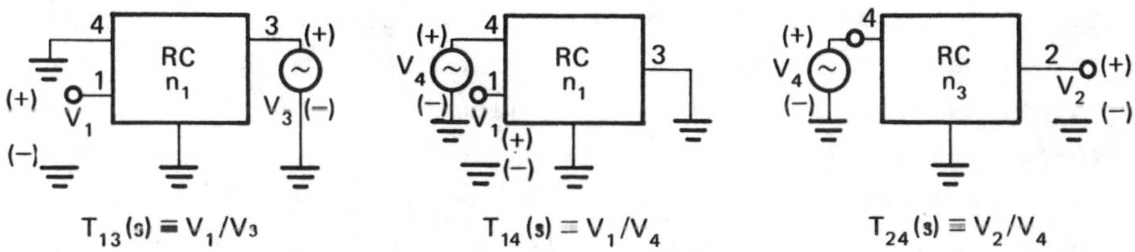

Figure 9.3 Illustrating the definition of the various transfer functions involved in equation (9.9).

it follows that:

(1) The poles of the transfer function T(s) are identical to the zeros of $N_{13}(s)$. Thus,

$$s^2 + s\left(\frac{\omega_o}{Q}\right) + \omega_o^2 = N_{13}(s) \tag{9.13}$$

Therefore, network n_1, viewed as a two-port network between terminals 3 and 1, with terminal 4 grounded, should possess a pair of complex left-half-plane (LHP) transmission zeros coincident with the desired pair of poles.

(2) The filter zeros are realized as the zeros of the numerator polynomial of (9.11); that is,

$$P(s) = T_{24}(s)\, D(s) - N_{14}(s) \tag{9.14}$$

where P(s) denotes the numerator polynomial of the biquadratic function to be realized,

$$P(s) \equiv K\left[s^2 + s\frac{\omega_{oz}}{Q_z} + \omega_{oz}^2\right] \qquad (9.15)$$

According to equation (9.14), different choices of injection nodes 4, and/or different networks n_3 provide different filter types without affecting the location of the poles.

The procedure for generating NF realizations can therefore be summarized in the following:

Step 1: A grounded RC two-port network having a pair of complex transmission zeros is chosen to implement n_1 with its input and output terminals connected to terminals 3 and 1 of the op. amp., respectively. The component values of n_1 are chosen such that its transmission zeros are at the specified pole locations. The degrees of freedom usually available in component value choices can be used to minimize the Q-sensitivities while keeping the component value spread within practical limits*.

Step 2: Any node in n_1 connected to ground can serve as a signal injection node, 4. Lifting such node off ground and connecting it to the input signal source V_i does not affect the pole locations. Different zero locations are obtained through using various combinations of such nodes 4, and/or using a network n_3 satisfying the constraint (9.12). The generation procedure can be best illustrated by an example:

Example 9.1

From Section 8.6 the simplest RC network having a pair of complex zeros is the bridged-T network, BT1 or BT2. Using BT1 to

---

* As mentioned before, the circuits are chosen such that ω_o-sensitivities are at their lowest possible values. Thus, we may exercise the available degrees of freedom to minimize the Q-sensitivities.

generate an NF realization, we first find from Fig. 8.30 that

$$N_{13}(s) = s^2 + s\left(\frac{1}{R_1} + \frac{1}{R_2}\right)\frac{1}{C_4} + \frac{1}{C_3C_4R_1R_2} \tag{9.16}$$

$$D(s) = s^2 + s\left(\frac{1}{C_4R_1} + \frac{1}{C_4R_2} + \frac{1}{C_3R_2}\right) + \frac{1}{C_3C_4R_1R_2} \tag{9.17}$$

$$N_{14}(s) = \beta s \frac{1}{C_3R_2} \tag{9.18}$$

where β is the fraction of C_4 lifted off-ground. Thus, if we use BT1 to implement n_1, we find by comparing (9.13) to (9.16) that the filter ω_o and Q are given by

$$\omega_o = 1 / \sqrt{C_3C_4R_1R_2} \tag{9.19}$$

$$Q = \frac{1}{\sqrt{C_3C_4R_1R_2}} \Big/ \left[\frac{1}{C_4}\left(\frac{1}{R_1} + \frac{1}{R_2}\right)\right] \tag{9.20}$$

The general configuration of a biquad generated from BT1 is shown in Fig. 9.4a, where the transfer function is given by

$$T(s) = \frac{T_{24}(s)\,D(s) - \beta s \frac{1}{C_3R_2}}{s^2 + s\left(\frac{\omega_o}{Q}\right) + \omega_o^2} \tag{9.21}$$

Note that network n_3, if present, should satisfy the constraint (9.12); namely,

$$T_{24}(s)\,D(s) = \text{polynomial},$$

where D(s) is given by (9.17).

The design equations are given by (9.19) and (9.20). These are two equations in four unknowns, hence there are two degrees of freedom. Usually, one choice is the impedance level of the circuit; that is, choosing a value for one of the resistors or one of the capacitors. The second possible choice can be utilized for sensitivity minimization, as shown below.

Now that we have located the filter poles at the desired positions, as determined by the given ω_o and Q, it remains to locate the zeros. To illustrate the zeros' placement operation consider the following special cases as examples:

1) If it is desired to implement a bandpass function, then from (9.21) we may set $T_{24}(s) \equiv 0$, that is eliminate network n_3 altogether and ground the positive input terminal of the op. amp. The parameter β can then be used to control the center-frequency gain of the filter. For minimum number of capacitors, we choose $\beta = 1$, leading to the circuit in Fig. 9.4b. This circuit has been referred to in the literature as the multiple-feedback realization[13]. Its' transfer function is obtained from (9.21) as,

$$T(s) = -\frac{s\frac{1}{C_3R_2}}{s^2 + s\frac{\omega_o}{Q} + \omega_o^2} \tag{9.22}$$

with ω_o and Q given by (9.19) and (9.20), respectively.

2) If it is desired to implement a high-pass function, we observe from (9.21) that one possible approach consists of setting $\beta = 0$, and selecting a network n_3 that implements a high-pass second order function. The simplest network for n_3 is the canonic high-pass ladder given in Chapter 8, Fig. 8.28b. Note that to satisfy (9.12) the denominator of $T_{24}(s)$ of the high-pass ladder should be identical to D(s). This can be achieved by selecting (See Fig. 9.4c).

$$C_1 = C_4\ ,\ C_2 = C_3\ ,\ R_3 = R_1 \text{ and } R_4 = R_2$$

The resulting high-pass circuit, shown in Fig. 9.4c, has the transfer function

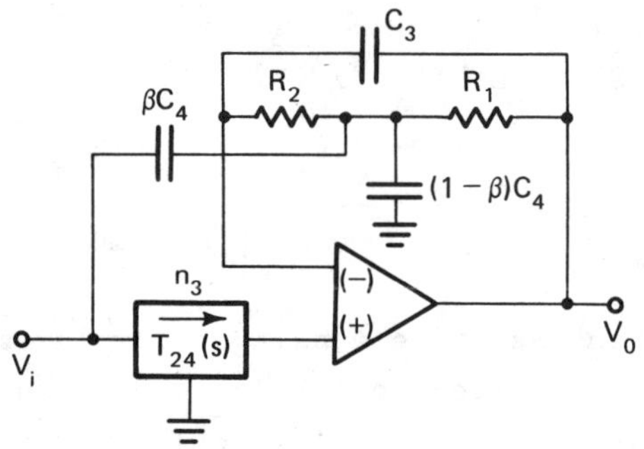

(a)

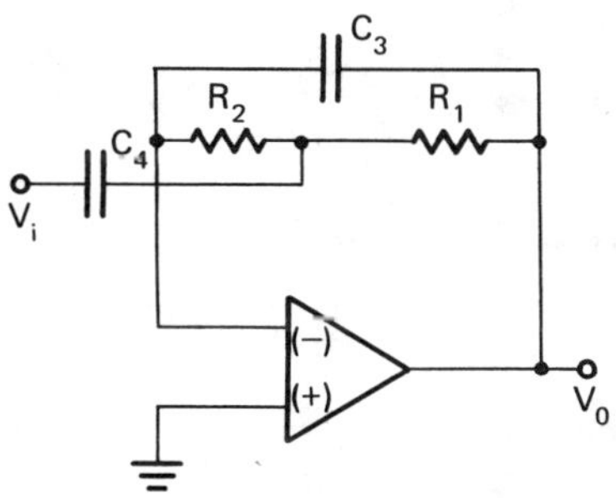

(b)

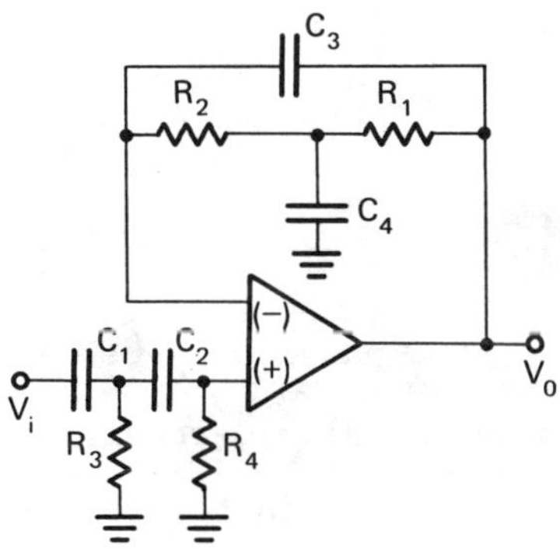

(c)

Figure 9.4 (a) The general configuration of a biquad generated from BT1; (b) A bandpass circuit obtained from the general biquad; (c) A high-pass circuit obtained from the general biquad.

$$T(s) = \frac{s^2}{s^2 + s\frac{\omega_o}{Q} + \omega_o^2} \tag{9.23}$$

This circuit is based on pole-zero cancellation and has been described in[14]. Although it utilizes more than the minimum number of components, it provides design flexibility by using different networks n_3 to obtain various zeros' locations without affecting the poles.

Sensitivity Analysis

Let $T_{13}(s)$ be given generally by

$$T_{13}(s) = \frac{s^2 + s(\omega_1/q_1) + \omega_1^2}{k\,[s^2 + s(\omega_2/q_2) + \omega_2^2]} \tag{9.24}$$

where ω_1 and q_1 characterize a pair of complex zeros, ω_2 and q_2 characterize a pair of real poles ($q_2 < 0.5$) and $k \geq 1$ is a constant. Also, let $\omega_2 = \omega_1/\gamma$, $(\gamma^2/k) \leq 1$. To evaluate the passive sensitivities, one considers an ideal op. amp.; that is, $A(s) = \infty$. In this case,

$$\omega_o = \omega_1 ,$$

$$Q = q_1 .$$

Thus, the passive ω_o-sensitivities are identical to those of ω_1 and the passive Q-sensitivities are identical to those of q_1.

If A(s) is finite, the pole-pair is obtained from the zeros of the denominator polynomial $D_v(s)$ of (9.10). Substituting for $N_{13}(s)$ and D(s) from (9.24),

$$D_v(s) = s^2 + s\frac{\omega_1}{q_1} + \omega_1^2 + \frac{k}{A(s)}\,[s^2 + s\frac{\omega_1}{\gamma q_2} + \frac{\omega_1^2}{\gamma^2}] \tag{9.25}$$

Consider first the low-frequency case, $A(s) = A_o$,

$$D_v(s) = s^2\,(1 + \frac{k}{A_o}) + s\,\frac{\omega_1}{q_1}\,(1 + \frac{q_1}{q_2}\,\frac{k}{\gamma A_o}) + \omega_1^2\,(1 + \frac{k}{\gamma^2 A_o})$$

Thus ω_o and Q of the pole-pair are given by

$$\omega_o = \omega_1 \sqrt{\frac{1 + k/\gamma^2 A_o}{1 + k/A_o}} \tag{9.26}$$

$$Q = q_1 \frac{\sqrt{(1+k/\gamma^2 A_o)\ (1+k/A_o)}}{1 + (\frac{q_1}{q_2 A_o})\ (\frac{k}{\gamma})} \tag{9.27}$$

Since k and γ are usually either unity (as in the BT1 and BT2 cases) or close in value to unity, then

$$\frac{k}{\gamma^2 A_o} << 1 \text{ and } \frac{k}{A_o} << 1 ,$$

leading to

$$\omega_o \simeq \omega_1 \tag{9.28}$$

$$Q \sim q_1 / [1 + \frac{q_1}{q_2 A_o} \frac{k}{\gamma}] \tag{9.29}$$

Thus,

$$S_{A_o}^{\omega_o} \simeq 0 ,$$

as desired. We should also note that for $\gamma = k = 1$, $S_{A_o}^{\omega_o} \equiv 0$. The active Q-sensitivity is obtained from (9.29),

$$S_{A_o}^{Q} = \frac{k\ q_1\ /\gamma\ q_2 A_o}{1 + (k\ q_1\ /\gamma\ q_2 A_o)} \tag{9.30}$$

Substituting for q_1 its nominal value of Q and noting that due to the large variability of A_o one has to keep $S_{A_o}^{Q} << 1$, then $(k\ q_1\ /\gamma\ q_2 A_o)$ should be kept << 1, leading to

$$S_{A_o}^{Q} \simeq \frac{k\ Q}{\gamma\ q_2 A_o} \tag{9.31}$$

The Q-gain-sensitivity product is given by

$$GS^Q \simeq \frac{k\ Q}{\gamma q_2} \tag{9.32}$$

From (9.32), it follows that the theoretical minimum value for GS^Q is

$$GS^Q \Big|_{min.} \simeq 2\ Q$$

Equation (9.32) applies for every NF realization. Thus, for a given Q, the value of $(\frac{k}{\gamma q_2})$ is a measure of goodness for the generating RC network n_1. To minimize the value of GS^Q, one should exercise the degrees of freedom available in choosing component values for n_1 to minimize the value of $(k\ /\gamma q_2)$. In the case $k = \gamma = 1$, the value of GS^Q is

$$GS^Q = \frac{Q}{q_2} \tag{9.33}$$

Reference to (9.24) reveals that for this case $(k = \gamma = 1)$,

$$GS^Q = 1\ /\ T_{13}(j\omega_1)\ .$$

That is, the gain-sensitivity-product is equal to the inverse of the transmission at the zero's frequency of n_1; an interesting observation.

The effect of the amplifier limited bandwidth is evaluated using the single-pole model of the op. amp.

$$A(s) = \frac{A_o}{1 + \frac{s}{\omega_b}}$$

which can be expressed as

$$\frac{1}{A(s)} = \frac{1}{A_o} + \frac{s}{\omega_t} \tag{9.34}$$

where ω_t is the unity-gain bandwidth. For frequencies $\omega >> \omega_b$, which is usually the case, equation (9.34) simplifies to

$$\frac{1}{A(s)} \simeq \frac{s}{\omega_t} \tag{9.35}$$

Substituting from (9.35) into (9.25),

$$D_v(s) = s^2 + s\,\frac{\omega_1}{q_1} + \omega_1^2 + s\,\frac{k}{\omega_t}\,[s^2 + s\,\frac{\omega_1}{\gamma q_2} + \frac{\omega_1^2}{\gamma^2}] \tag{9.36}$$

Consider for simplicity the case $k = \gamma = 1$ and substitute for ω_1 and q_1 their nominal values of ω_o and Q, respectively,

$$D_v(s) = s^3\,(\frac{1}{\omega_t}) + s^2\,(1 + \frac{\omega_o}{\omega_t}\,\frac{1}{q_2}) + s\,\frac{\omega_o}{Q}\,(1 + Q\,\frac{\omega_o}{\omega_t}) + \omega_o^2 \tag{9.37}$$

Thus, in addition to the dominant poles having changed their parameters from ω_o and Q, to say ω_{oa} and Q_a, the filter acquires a third parasitic pole, as evident from the third-order denominator polynomial in (9.37). Equation (9.37) can be written in the form

$$D_v(s) = (\frac{s}{\omega_t} + \frac{\omega_o^2}{\omega_{oa}^2})\,(s^2 + s\,\frac{\omega_{oa}}{Q_a} + \omega_{oa}^2) \tag{9.38}$$

Equating the coefficients of the like powers of s in equations (9.37) and (9.38) yields,

$$\eta \equiv \frac{Q_a}{Q} = 1 / (\frac{\omega_{oa}}{\omega_o})\,[1 + Q\,\frac{\omega_o}{\omega_t} - Q\,\frac{\omega_{oa}^2}{\omega_o\omega_t}] \tag{9.39}$$

Expressing ω_{oa} as,

$$\omega_{oa} = \omega_o (1 + \delta) \ , \tag{9.40}$$

where $\delta << 1$ so that ,

$$(1 + \delta)^2 \simeq 1 + 2\delta \ , \tag{9.41}$$

then,

$$\eta \simeq 1 \ / \ [1 + \delta - 2\delta \ Q \ (\frac{\omega_o}{\omega_t})] \ . \tag{9.42}$$

To obtain the fractional deviation δ in ω_o, we equate the coefficients of the s^2-terms in (9.37) and (9.38) to obtain

$$\frac{\omega_o^2}{\omega_{oa}^2} = 1 + \frac{\omega_o}{\omega_t} \frac{1}{q_2} - \frac{\omega_{oa}}{\omega_t} \frac{1}{Q_a} \ .$$

Making use of (9.40) and (9.41) results in

$$\delta \simeq - \frac{1}{2} (\frac{\omega_o}{\omega_t}) \ (\frac{1}{q_2} - \frac{1}{Q_a}) \ .$$

Assuming $Q_a \simeq Q$, which has a second order effect on the value of δ, we obtain,

$$\delta \simeq - \frac{1}{2} (\frac{\omega_o}{\omega_t}) \ (\frac{1}{q_2} - \frac{1}{Q})$$

$$\delta \simeq - \frac{1}{2Q} (\frac{Q}{q_2} - 1) \ \ (\frac{\omega_o}{\omega_t}) \tag{9.43}$$

Making use of the expression for GS^Q in (9.34), equation (9.43) can be expressed as,

$$\delta \simeq - \frac{1}{2Q} \ GS^Q \ \ (\frac{\omega_o}{\omega_t}) \tag{9.44}$$

Substituting for δ in (9.42) obtains,

$$\eta \simeq 1 \Bigg/ \left[1 + GS^Q \left(\frac{\omega_o}{\omega_t}\right) \left(\frac{\omega_o}{\omega_t} - \frac{1}{2Q}\right)\right] \tag{9.45}$$

Equations (9.44) and (9.45) provide a very interesting result: the only configuration dependent quantity in determining δ and η is the Q gain-sensitivity product GS^Q. Hence, minimizing GS^Q automatically minimizes the high-frequency deviations. Thus, the Q gain-sensitivity product, although evaluated at low frequencies determines the high-frequency performance of the NF circuits. As we will see later, this result applies for all the biquad circuits considered in this book. Table 9.1 provides a summary of results for the general NF SAB.

Classification and Applications

From the above, we may conclude that the performance of an NF realization depends solely on the RC network chosen to implement n_1. There are only two categories of RC networks suitable for the generating two-port n_1. From Chapter 8, these are:

(1) A second-order network of the bridged-T (BT) variety, and

(2) A third-order network that provides a second-order transfer function $T_{13}(s)$ through an internal pole-zero cancellation, such as the notch twin-T network.

Corresponding to these two types of generating RC networks, we classify the negative feedback (NF) realizations into two categories:

(1) The NF1 class generated from a BT type network in the feedback, and

(2) The NF2 class generated from a TT type network in the feedback.

As we will see shortly, members of each of the two classes share common properties. Also, the two classes are quite distinct from each other, as far as the values of the performance indices are concerned.

Table 9.1 Summary of the results for the general NF SAB of Fig. 9.2

$$T(s) \equiv \frac{V_o}{V_i} = \frac{T_{24}(s) - T_{14}(s)}{T_{13}(s)} \quad , \text{ for } A(s) = \infty$$

Let

$$T_{13}(s) = \frac{s^2 + s(\omega_1/q_1) + \omega_1^2}{k[s^2 + s(\omega_2/q_2) + \omega_2^2]}$$

$k \geq 1$,

and $\omega_2 = \omega_1/\gamma \quad , \quad \gamma^2/k \leq 1$

- Then, nominally, $\omega_o = \omega_1$ and $Q = q_1$
- Passive ω_o and Q sensitivities are identical to those of ω_1 and q_1, respectively
- $GS^{\omega_o} \simeq 0$ for values of γ and k around unity
- $GS^{\omega_o} \equiv 0$ for $\gamma = k = 1$
- $GS^Q \simeq \dfrac{kQ}{\gamma q_2}$
- $GS^Q \simeq \dfrac{Q}{q_2}$ for $\gamma = k = 1$
- $\delta \equiv \dfrac{\Delta\omega_o}{\omega_o} \simeq -\dfrac{1}{2Q} \; GS^Q \; (\dfrac{\omega_o}{\omega_t})$
- $\eta \equiv \dfrac{Q_a}{Q} \simeq 1 \; / \; [1 + \delta - 2\delta Q(\dfrac{\omega_o}{\omega_t})]$

(the last two: effects of the amplifier's limited BW, for γ=k=1)

The NF1 Class

As a representative example of the NF1 class we consider circuits generated from the BT1 network. We derived expressions for the transfer functions as well as design equations for a SAB generated from BT1 in example 9.1, with the general circuit configuration repeated in Fig. 9.5. From the results of example 9.1, we have

$$\omega_1 = 1 / \sqrt{C_3C_4R_1R_2}$$

$$q_1 = \frac{1}{\sqrt{C_3C_4R_1R_2}} / [\frac{1}{C_4}(\frac{1}{R_1} + \frac{1}{R_2})]$$

$$q_2 = \frac{1}{\sqrt{C_3C_4R_1R_2}} / (\frac{1}{C_4R_1} + \frac{1}{C_4R_2} + \frac{1}{C_3R_2})$$

$$k = \gamma = 1$$

Nominally, $\omega_o = \omega_1$ and $Q = q_1$.

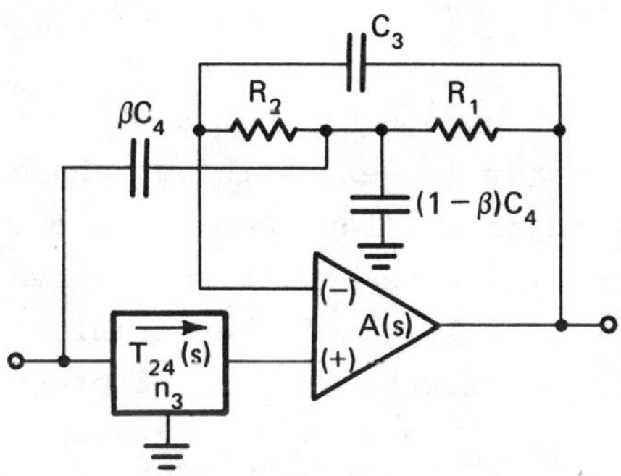

Figure 9.5 General configuration of an NF biquad generated from BT1.

The passive ω_o-sensitivities are given by

$$S_{R_1}^{\omega_o} = S_{R_2}^{\omega_o} = S_{C_3}^{\omega_o} = S_{C_4}^{\omega_o} = -\frac{1}{2}$$

Since $k = \gamma = 1$, the active ω_o-sensitivity is identically zero; i.e. $GS^{\omega_o} \equiv 0$. The passive Q-sensitivities are given by

$$S_{C_3}^{Q} = -\frac{1}{2}$$

$$S^Q_{C_4} = + \frac{1}{2}$$

$$S^Q_{R_1} = - S^Q_{R_2} = \frac{1}{2} \left(\sqrt{\frac{R_2}{R_1}} - \sqrt{\frac{R_1}{R_2}} \right) / \left(\sqrt{\frac{R_2}{R_1}} + \sqrt{\frac{R_1}{R_2}} \right)$$

Thus, all the passive Q-sensitivities are $\leq \frac{1}{2}$. The only remaining performance parameter is the Q gain-sensitivity product, given by

$$GS^Q = \frac{Q}{q_2}$$

$$= \left(\frac{1}{C_4R_1} + \frac{1}{C_4R_2} + \frac{1}{C_3R_2}\right) \quad / \quad \frac{1}{C_4}\left(\frac{1}{R_1} + \frac{1}{R_2}\right)$$

Clearly the choice of component values will affect the value of GS^Q. We are, of course, governed in our choice by the two design equations. Since there are four components, two degrees of freedom exist. One usually uses one of the degrees of freedom for selecting the impedance level of the circuit, from practical experience. Let this choice be $R_2 = R$. Expressing R_1 as a ratio $\frac{1}{m}$ of R_2; that is, $R_1 = \frac{R}{m}$, and the two capacitors as $C_4 = C$ and $C_3 = \frac{C^2}{q}$, the problem now is to determine the values of m, q and C to satisfy the two design equations

$$\omega_o = \frac{\sqrt{mq}}{CR}$$

$$Q = \frac{\sqrt{mq}}{1 + m}$$

The remaining degree of freedom may now be used to minimize GS^Q, given by

$$GS^Q = 1 + \left(1 + \frac{1}{m}\right) Q^2$$

Thus, increasing m decreases the value of GS^Q. The decrease, however, is rather small for m ranging between 1 and the theoretical upper limit of ∞. In all cases, GS^Q is proportional to Q^2, a severe limitation for this class of SABs. As a further guide to the selection of a value for m, we consider the important matter of component value spread. Clearly, as m increases, the resistors become farther apart in value. More importantly, however, is to consider the spread in capacitor values, given by

$$q = \frac{(1 + m)^2}{m} Q^2$$

which reaches a minimum for m = 1. Thus, a value of unity for m seems to be a good compromise, resulting in

$$GS^Q = 1 + 2 Q^2$$

$$q = 4 Q^2$$

$$S^Q_{R_1} = S^Q_{R_2} = 0$$

We may conclude that this circuit exhibits the desirable low passive sensitivities at the expense of intolerably high active Q-sensitivity and wide capacitor value spread. Note also that the frequency deviation δ is proportional to Q since

$$\delta \simeq - Q \left(\frac{\omega_o}{\omega_t}\right)$$

The above example illustrates in detail the process of finding "optimum" component values. The results reached using BT1 are representative of all the NF1 class of circuits. Fig. 9.6 provides a tabulation of possible generating networks for NF1 circuits. All the generating networks are of the BT type, presented in Chapter 8, Fig. 8.30. The design equations for the SAB obtained from each generating network are also given in Fig. 9.6. The component values are chosen to yield good compromise between a low value for GS^Q and a tolerable component spread. Nonethe-

	Circuit	(CR)	GS^Q	$N_{14}(s)$
BT1	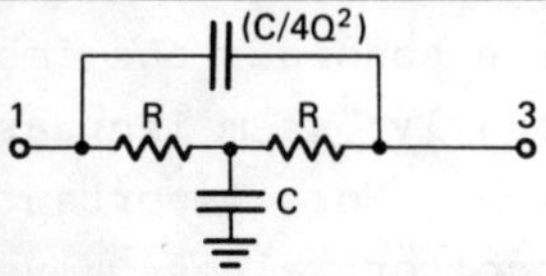	$2Q/\omega_0$	$2Q^2 + 1$	$\beta_1 : C$ $N_{14}(s) = s\beta_1(2\omega_0 Q)$
ABT1	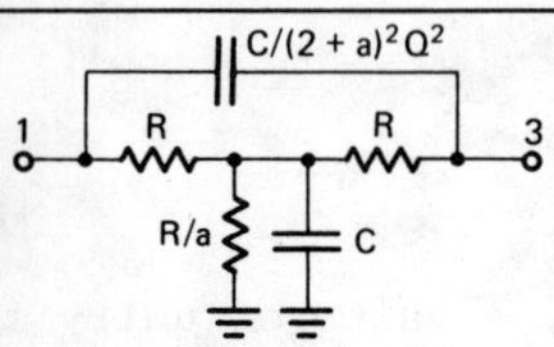	$(2 + a)Q/\omega_0$	$(2 + a)Q^2 + 1$	$\beta_1 : C \quad \beta_2 : (a/R)$ $N_{14}(s) = s\beta_1(2 + a)\omega_0 Q + \beta_2 a\omega_0^2$
BT2	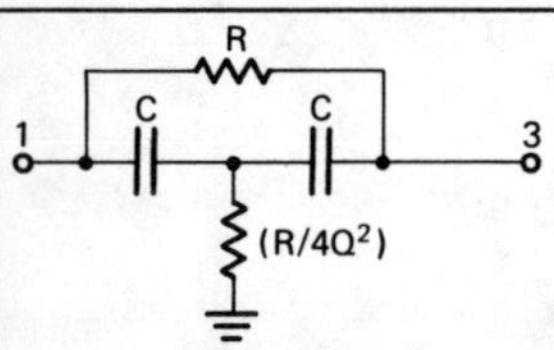	$2Q/\omega_0$	$2Q^2 + 1$	$\beta_1 : 4Q^2/R$ $N_{14}(s) = s\beta_1(2\omega_0 Q)$
LBT2	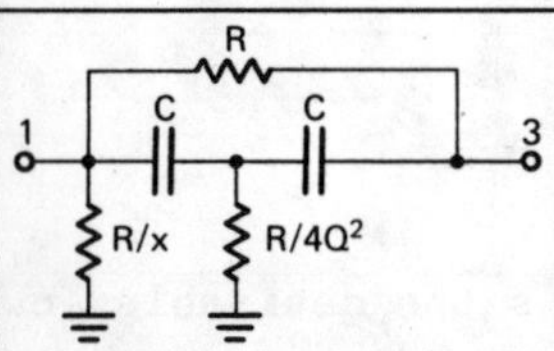	$2Q/\omega_0$	$2Q^2 + 1 + x$	$\beta_1 : \frac{4Q^2}{R} \quad \beta_2 : x/R$ $N_{14}(s) = s\beta_1(2\omega_0 Q) + \beta_2 x(\omega_0/Q)(s + \omega_0 Q)$
ABT2	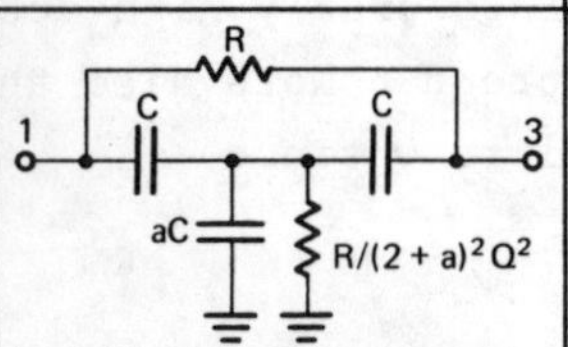	$(2 + a)Q/\omega_0$	$(2 + a)Q^2 + 1$	$\beta_1 : \frac{(2 + a)^2 Q^2}{R} \quad \beta_2 : aC$ $N_{14}(s) = a\beta_2 s^2 + s\beta_1(2 + a)\omega_0 Q$
BT3	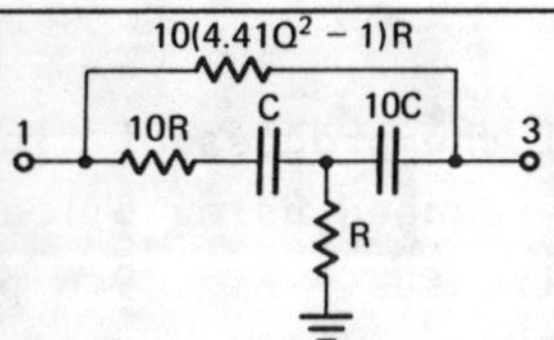	$\frac{1}{21\omega_0 Q}$	$2.1Q^2 +$ $(1.1/2.1)$	$\beta_1 : 1/R$ $N_{14}(s) = s\beta_1 \frac{\omega_0}{2.2Q}(4.41\,Q^2 - 1)$
Passive Q-Sensitivities are $\leqslant 1$				

Figure 9.6 Generating networks and design equations for NF1-type SAB's.

less, both the resulting value of GS^Q and the value of the component spread are of the order of Q^2. The NF1 circuits are, therefore, useful only for low Q(<10) applications. In the final column of Fig. 9.6, the polynomial $N_{14}(s)$ is given, assuming the signal is injected to each possible node through a designated fraction $\beta \leq 1$ of the capacitance and/or conductance connected to terminal 4, in a manner similar to that illustrated in Fig. 9.4. We finally note that network BT3 is included since it is the generating network for the popular Sallen and Key[15] negative feedback realizations.*

Fig. 9.7 shows a few examples of NF1 filters generated from the networks of Fig. 9.6. Circuit (1) is an all-pass realization originally suggested by Deliyannis[17], circuit (2) is a low-pass realization of the multiple-feedback type[13], while circuit (3) is a band-pass realization, also of the multiple-feedback type. The high-pass realization, circuit (4), is of the pole-zero cancellation type[14]. Circuit (5) provides a high-pass notch, $\omega_n < \omega_o$. Finally, circuit (6) is the familiar Sallen and Key NF realization[15]. The reader is encouraged to sketch other examples using the given network generation technique.

The loaded bridged T network (LBT2) has the attractive feature of two equal capacitors. In addition, it is quite versatile in generating various second order functions. For these reasons, we present in Fig. 9.8 the general biquad generated from LBT2. In addition, Table 9.2 gives the parameter values for the various special second-order functions realized using this biquad. Note that the only function this biquad is incapable of implementing is the low-pass one. Moreover, all other functions are realized using only two capacitors which are equal. It should be noted, however, that the notch and all-pass sections are rather sensitive to passive component values. Thus, such realizations are useful only for $Q \leq 5$.

---

* The component values for BT3 are those given by Kerwin [16].

		Circuit	Equations
(1)	BT1	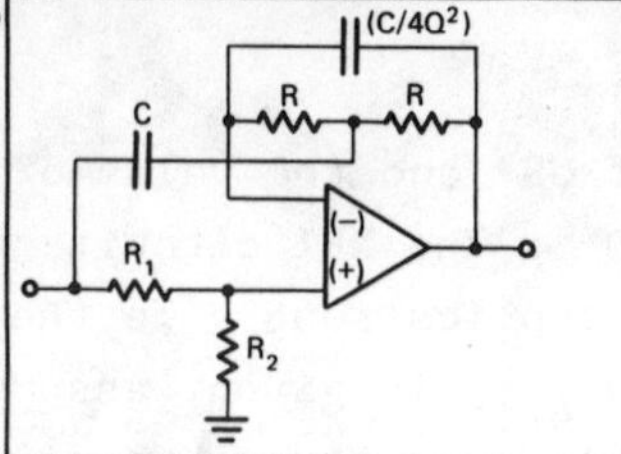	$\frac{R_2}{R_1 + R_2} = \alpha = \frac{Q^2}{Q^2 + 1}$ $T_v(s) = \alpha \frac{s^2 - s(\omega_0/Q) + \omega_0^2}{s^2 + s(\omega_0/Q) + \omega_0^2}$
(2)	ABT1	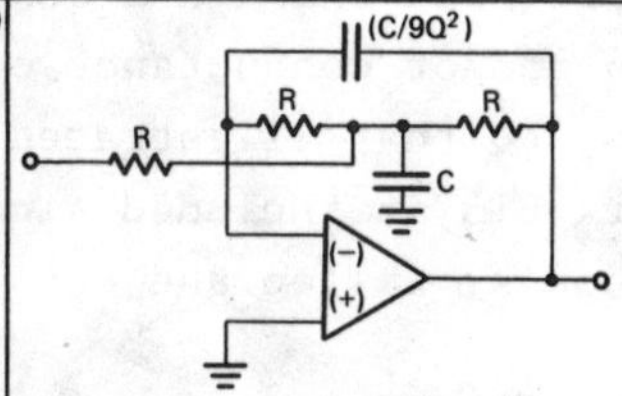	$T_v(s) = -\frac{\omega_0^2}{s^2 + s(\omega_0/Q) + \omega_0^2}$
(3)	BT2	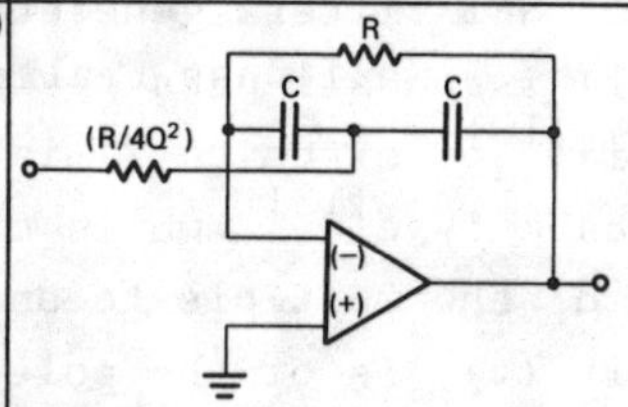	$T_v(s) = -\frac{s(2\omega_0 Q)}{s^2 + s(\omega_0/Q) + \omega_0^2}$
(4)	BT2	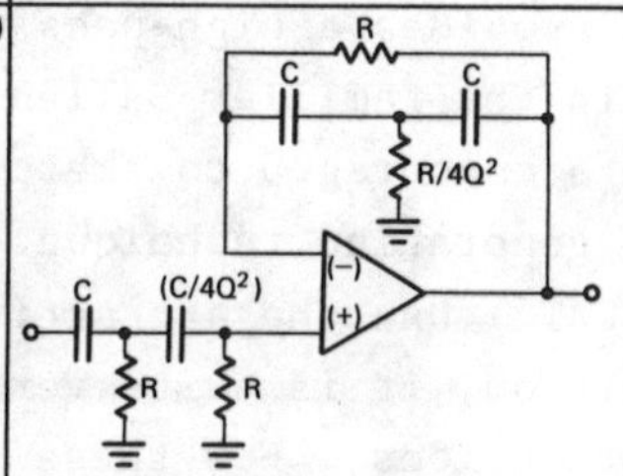	$T_v(s) = \frac{s^2}{s^2 + s(\omega_0/Q) + \omega_0^2}$
(5)	ABT2	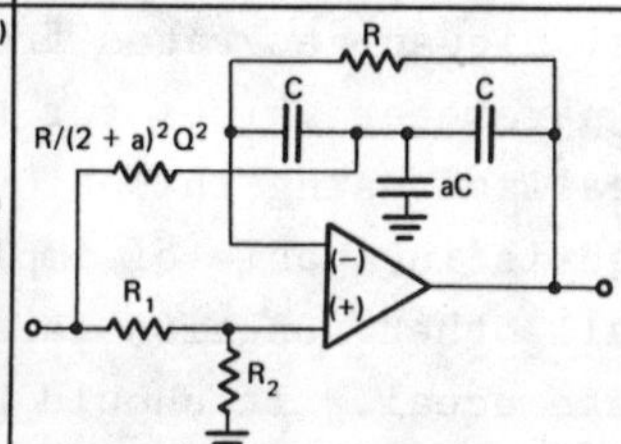	$a = \frac{\omega_0^2}{\omega_n^2} - 1, \quad \frac{R_2}{R_1 + R_2} = \alpha = \frac{(2 + a)Q^2}{(2 + a)Q^2 + 1}$ $T_v(s) = \alpha \frac{\omega_0^2}{\omega_n^2} \frac{s^2 + \omega_n^2}{s^2 + s(\omega_0/Q) + \omega_0^2}$
(6)	BT3	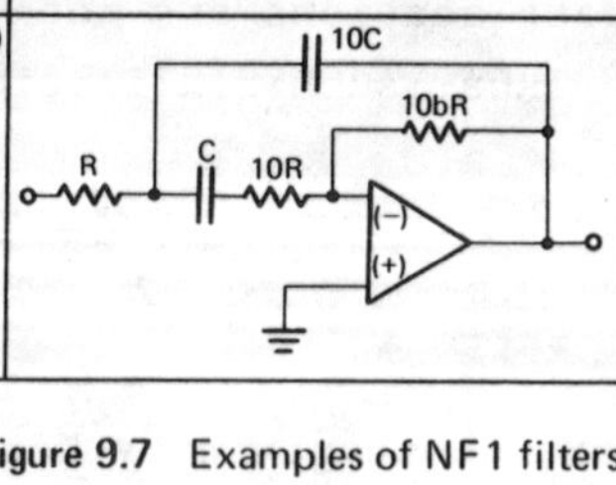	$b = (4.41\, Q^2 - 1)$ $T_v(s) = \frac{-sb\omega_0/(2.2Q)}{s^2 + s\,(\omega_0/Q) + \omega_0^2}$

Figure 9.7 Examples of NF1 filters. The first column gives the generating network.

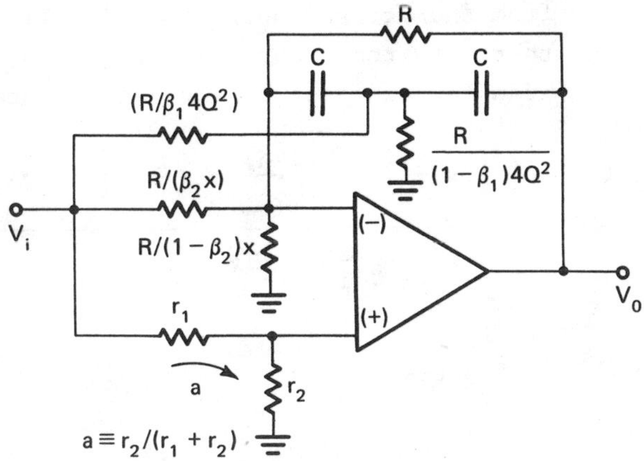

Figure 9.8 NF1 biquad generated from LBT2.

Type	β_1	β_2	a	x	Numerator
BP	Arbitrary ≤ 1	0	0	0	$s\beta_1(2\omega_o Q)$
LPN	1	0	$1\Big/[1+\frac{1}{2Q^2}\frac{\omega_{oz}^2}{\omega_o^2}]$	$[(\frac{\omega_{oz}}{\omega_o})^2 - 1]$	$a(s^2 + \omega_{oz}^2)$
HPN	1	1	$1\Big/[1+\frac{1}{2Q^2}\frac{\omega_{oz}^2}{\omega_o^2}]$	$2Q^2(\frac{\omega_o^2}{\omega_{oz}^2} - 1)$	$a(s^2 + \omega_{oz}^2)$
HP	a	1	Arbitrary, but < 1	$1\Big/(\frac{1}{a} - 1)$	$a\ s^2$
AP	1	0	$1\Big/(1 + \frac{1}{Q^2})$	0	$a(s^2 - s\frac{\omega_o}{Q}+\omega_o^2)$

Table 9.2 Special cases of the NF1 biquad of Fig. 9.8

The NF2 Class

The NF2 class of circuits is generated using the notch twin-T, presented in Section 8.6, to implement the RC feedback network n_1. As an example, Fig. 9.9 shows the general configuration of a SAB generated from the bridged notch twin-T (BTT). Using equation (9.11) together with the transfer functions of BTT given in Fig. 8.32b, the transfer function of the SAB is obtained as

$$T(s) = \frac{T_{24}(s)\ [s^2 + s\,\frac{4+2a}{T} + \frac{1+2a}{T^2}] - s\,\frac{2}{T}\,(\beta_1 + \beta_2)}{s^2 + s\,\frac{2a}{T} + \frac{1+2a}{T^2}} \tag{9.46}$$

where network n_3, if present, should satisfy the constraint,

$$T_{24}(s)\ [s^2 + s\,\frac{4+2a}{T} + \frac{1+2a}{T^2}\ ] = \text{polynomial} \tag{9.47}$$

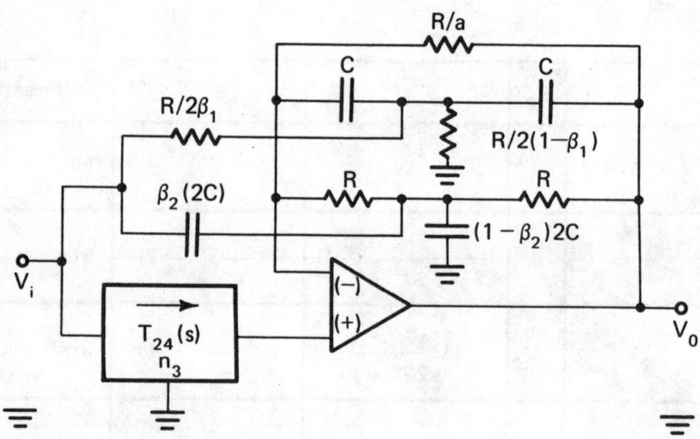

Figure 9.9 General configuration of a SAB generated from the bridged notch twin-T network.

From (9.46), it follows that the realized pole pair is characterized by

$$\omega_o = \frac{\sqrt{1+2a}}{T} \tag{9.48}$$

$$\text{and } Q = \frac{\sqrt{1+2a}}{2a} \tag{9.49}$$

Thus, given ω_o and Q, equations (9.48) and (9.49) provide the required values of a and T. It remains for the designer to select the impedance level by choosing a value for R or C using practical judgement. Finally, the transmission zeros can be located at the desired position in the s-plane by the proper choice of β_1, β_2, and/or network n_3. We consider two examples:

(1) From (9.46), it follows that if a bandpass realization is required one can dispense with network n_3 altogether and select β_1 and/or β_2 equal to unity. Three possible circuits result and are shown in Fig. 9.10[18]. Note that the three circuits differ only in the center-frequency gain and the input impedance.

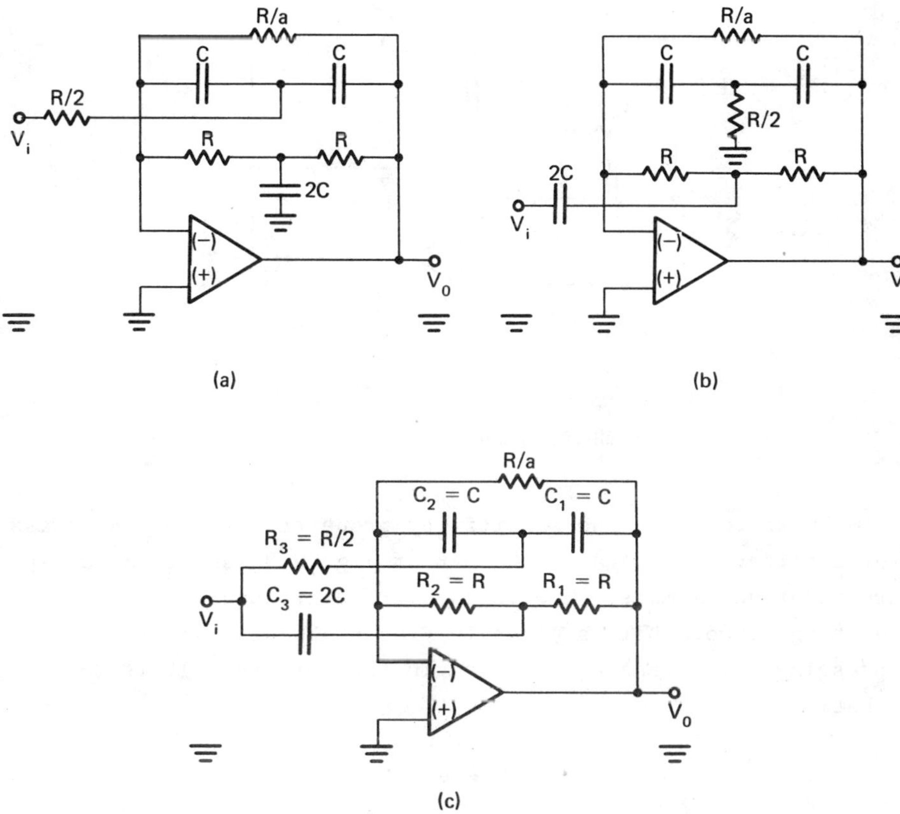

Figure 9.10 Three bandpass realizations obtained from the SAB of Figure 9.9.

(2) If it is required to implement an all-pass section, equation (9.46) suggests that $T_{24}(s)$ be a constant, say α. If we further let $\beta_1 = \beta_2 = 1$, then α may be determined from

$$(4+2a) - \frac{4}{\alpha} = -2a$$

Thus,

$$\alpha = \frac{1}{1+a}$$

where a is determined by the required Q from (9.49). The resulting all-pass circuit is shown in Fig. 9.11, with n_3 implemented as a potential divider.

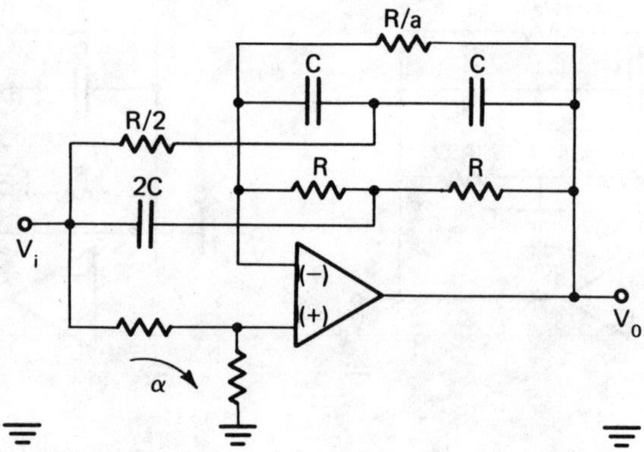

Figure 9.11 An all-pass section obtained from the SAB of Figure 9.9.

We next consider the sensitivity properties of the NF2 class of realizations. Again, the SAB of Fig. 9.9 is used as a representative example. The voltage transfer function of the generating network BTT is given in Fig. 8.32b of Chapter 8. Expressing this transfer function in the form (9.24), it follows that

$$k = \gamma = 1$$

$$\omega_1 = \frac{\sqrt{1+2a}}{T} \tag{9.50}$$

$$q_1 = \frac{\sqrt{1+2a}}{2a} \tag{9.51}$$

$$q_2 = \frac{\sqrt{1+2a}}{4+2a}$$

From these relations, as well as the sensitivity formulae of single amplifier filters, given in Table 9.1, it follows that

$$GS^{\omega_o} \equiv 0$$

$$GS^Q = \frac{2 + a}{a} \simeq \frac{2}{a} \quad \text{for } a << 1\text{; that is, } Q >> 1.$$

Thus, $GS^Q \simeq 4Q$ for $Q >> 1$.

We note the dramatic reduction in the Q-gain sensitivity-product from being proportional to Q^2 in NF1 circuits to a number proportional to Q for NF2 circuits. The price paid for this is the increased circuit complexity, as well as increased passive Q-sensitivities. The latter, however, cannot be easily evaluated from the given expressions, since we have already assumed a pole-zero cancellation in the twin-T. In fact, if one assumes the pole-zero cancellation to hold, then the passive Q-sensitivities are quite small. Although this assumption might be justified in cases where the components closely track with changes in environmental conditions, such as temperature, it is not proper to stop at this point. One reason for the circuit designer's interest in passive sensitivities is the desire to evaluate how close to the nominal values one should choose the passive components. Also, for a certain tolerance on the passive component values, what will the original error in Q be? This is particularly important for cases where initial adjustment or tuning is complex and costly.

To carry out a passive sensitivity study, we need to find expressions for the twin-T ω_1 and q_1 (which are the nominal values for ω_o and Q), assuming the pole and zero do not cancel. Such an analysis has been presented in [19] and will not be repeated here. It can be shown that for the NF2 circuit in question, the

ω_o sensitivities are at their theoretical minima while the Q-sensitivities are as follows:

$$S^Q_{R_1} = S^Q_{R_2} = \frac{Q}{4}$$

$$S^Q_{R_3} = -\frac{Q}{2}$$

$$S^Q_{C_1} = S^Q_{C_2} = -\frac{Q}{4}$$

$$S^Q_{C_3} = \frac{Q}{2}$$

where the components' numbering corresponds to that in Fig. 9.10c. Thus, the NF2 class is characterized by high passive Q-sensitivities, of the order of Q. It is fair to say that this is true for any biquad generated from a third order RC network.

Fig. 9.12 shows the BTT and ABTT generating networks together with the pertinent information required for biquad design, as well as an indication of the sensitivity performance of the NF2 class. Not shown on the diagram are the high frequency limitations of this class of SAB's. These can be easily evaluated using the values given for GS^Q together with equations (9.44) and (9.45). Note that since GS^Q is of the order of Q, the value of δ is independent of Q. The high-frequency performance of NF2 circuits is actually about the best that can be achieved in a biquad realization. This will be further collaborated below.

We conclude this section by presenting in Fig. 9.13 a SAB of the NF2 type generated from the ABTT network of Fig. 9.12. The transfer function of this versatile biquad[20] is given by,

$$T(s) = \frac{fs^2 + s(\frac{g}{CR}) + b/(C^2R^2)}{s^2 + s(\frac{\omega_o}{Q}) + \omega_o^2}$$

where $g + e = b + 2$,

	Circuit	(CR)	GS^Q	$N_{14}(s)$
BTT	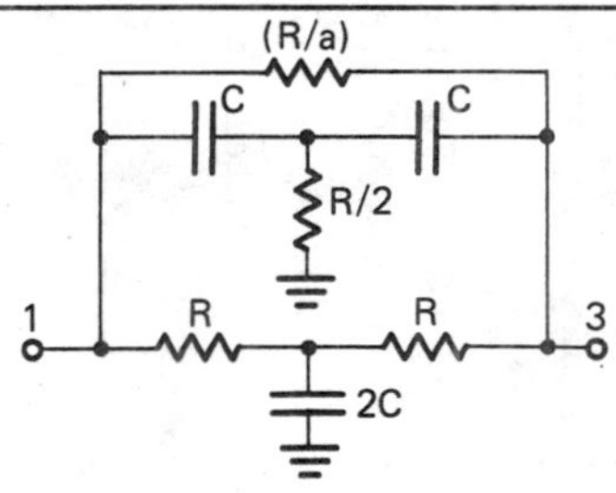 $Q = \sqrt{1+2a}/2a$	$\sqrt{1+2a}/\omega_0$	$\cong 4Q$ $(Q \gg 1)$	$\beta_1 : 2/R \quad \beta_2 : 2C$ $N_{14}(s) = s \dfrac{2\omega_0}{\sqrt{1+2a}} (\beta_1 + \beta_2)$
ABTT	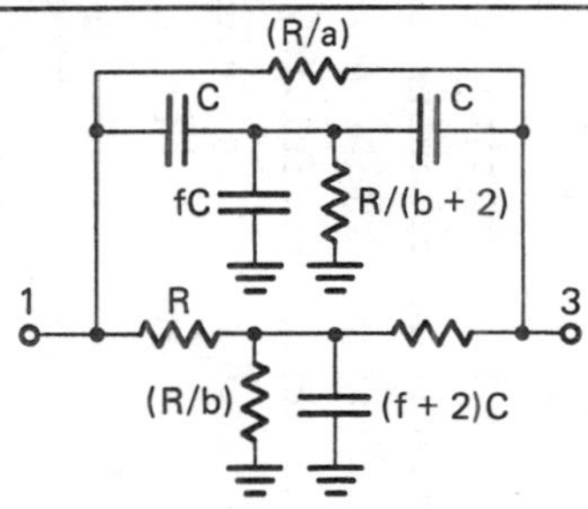 $Q = \sqrt{1 + a(b+2)}/a(f+2)$	$\dfrac{\sqrt{1+a(2+b)}}{\omega_0}$	$\cong (4+f+b)Q$ $(Q \gg 1)$	$\beta_1 : (b+2)/R \quad \beta_2 : (f+2)C$ $\beta_3 : b/R \quad \beta_4 : fC$ $N_{14}(s) = \beta_4 f s^2 + s \dfrac{\omega_0 [\beta_1 (b+2) + \beta_2 (f+2)]}{\sqrt{1+a(2+b)}} + \dfrac{\beta_3 b}{[1+a(2+b)]} \omega_0^2$
Passive Q-Sensitivities are of the order of Q				

Figure 9.12 Examples of networks generating the NF2 class of realizations.

$$\omega_o = \sqrt{1 + a(2+b)} \;/(CR)$$

$$Q = \sqrt{1 + a(2+b)} \;/[a(f+2)]$$

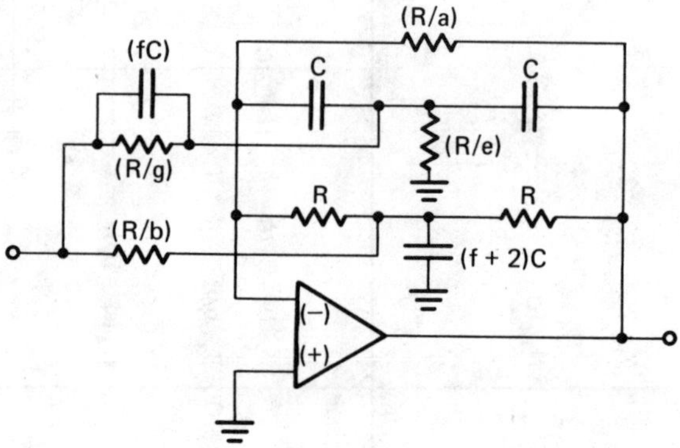

Figure 9.13 A versatile NF2-type biquad. (See text for design information).

9.5 SINGLE AMPLIFIER BIQUADS OF THE POSITIVE FEEDBACK (PF) TYPE

The second major category of single-amplifier biquads is represented by the general configuration of Fig. 9.1d, which is repeated in Fig. 9.14a. Since the RC frequency selective network is placed in the positive-feedback path of the op. amp., the class of resulting circuits is referred to as Positive Feedback (PF) realizations. In the following we present a circuit generation procedure, sensitivity analysis, and a classification scheme for PF circuits. The development is parallel to that already given for NF circuits in Section 9.4.

Generation Procedure

First, we note that in the PF general configuration of Fig. 9.14a the input signal is shown injected to network n_2, and to the inverting op. amp. terminal through resistor r_3. The latter is important, as it allows the realization of right-half-plane

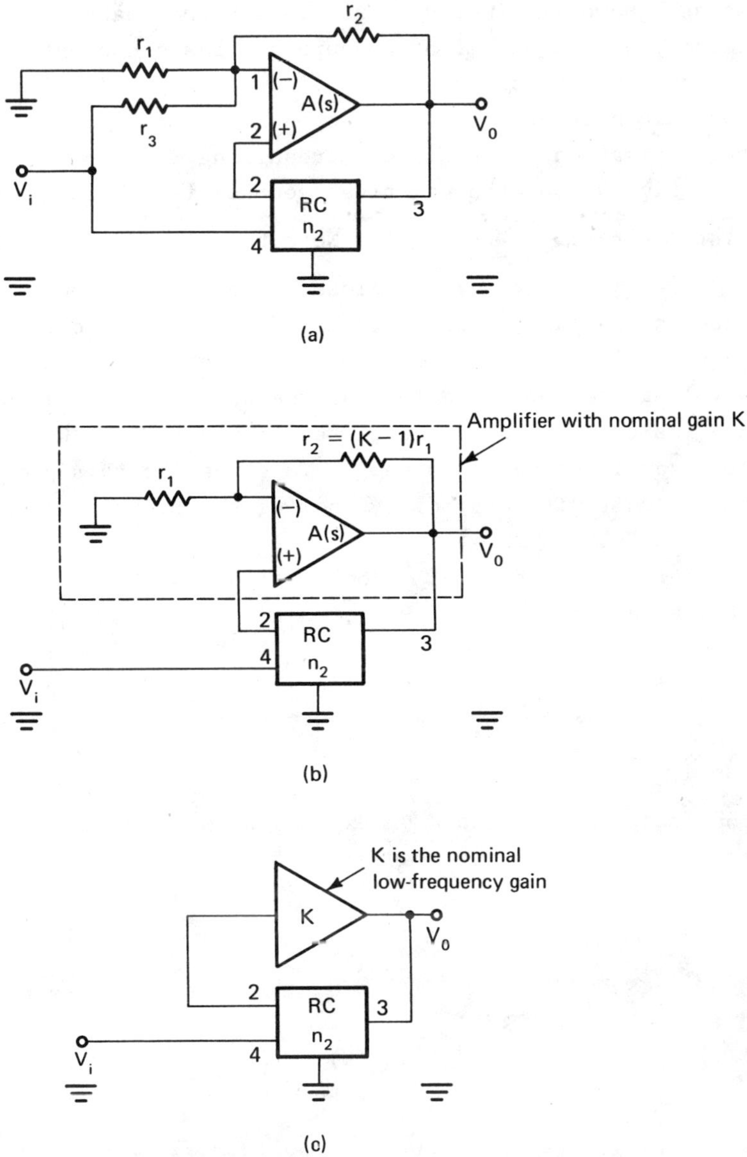

Figure 9.14 (a) General configuration of a PF SAB. (b) PF SAB without the injection resistance r_3. (c) PF SAB with the op. amp. and feedback resistors replaced by a positive-gain amplifier.

zeros. However, including r_3 renders the analysis to follow rather complex and might obscure the basic concepts. Therefore, we consider the simpler PF configuration of Fig. 9.14b, where r_3 has been omitted. The interested reader should refer to [21] for the derivation of PF circuits requiring r_3.

In Fig. 9.14b, the negative feedback is determined by the resistance ratio $\frac{r_2}{r_1} \equiv K - 1$. Note that if $A(s) = \infty$, the ideal situation, K is the nominal closed-loop gain of the amplifier enclosed in the dashed box. We therefore restrict ourselves to the case $K \geq 1$, with $K = 1$ corresponding to a unity-gain amplifier or a voltage follower. Note that the analysis to follow applies to any implementation of the closed-loop amplifier K; that is, not necessarily using an op. amp. To emphasize this point, we redraw the circuit with explicitly showing the finite gain amplifier in Fig. 9.14c.

Assuming the op. amp. to be ideal but with a finite gain A(s), it can be shown, using (9.7), that

$$T(s) \equiv \frac{V_o}{V_i} = K \frac{T_{24}(s)}{1 + K/A(s) - K\,T_{23}(s)} \tag{9.53}$$

The functions of n_2 can be expressed in polynomial quotient form as,

$$T_{23}(s) = \frac{N_{23}(s)}{D(s)}$$

$$T_{24}(s) = \frac{N_{24}(s)}{D(s)}$$

where D(s), the characteristic polynomial of n_2, has zeros on the negative real axis only. Substituting in (9.53), we obtain

$$T(s) = K \frac{N_{24}(s)}{D(s) + K\,D(s)/A(s) - K\,N_{23}(s)} \,. \tag{9.54}$$

In the ideal case $A(s) = \infty$ reducing (9.54) to,

$$T(s) = K \frac{N_{24}(s)}{D(s) - K N_{23}(s)} . \tag{9.55}$$

From (9.55), it follows that:

1. The filter zeros are realized as the zeros of $N_{24}(s)$; i.e.,

$$P(s) = K N_{24}(s) \tag{9.56}$$

2. The filter poles are realized as the zeros of the denominator polynomial of (9.55); i.e.,

$$s^2 + s\left(\frac{\omega_o}{Q}\right) + \omega_o^2 = D(s) - K N_{23}(s) \tag{9.57}$$

To render the value of ω_o independent of K and the amplifier gain A_o, two possibilities exist for the function $T_{23}(s)$:

$$\text{(a)} \quad T_{23}(s) = \frac{s\,(\omega_2/q_1)}{s^2 + s\,(\omega_2/q_2) + \omega_2^2} \equiv \frac{N_{23}(s)}{D(s)} \tag{9.58}$$

where ω_2 and q_2 characterize a pair of real poles; $q_2 < 0.5$. Thus, n_2 is a bandpass network with center-frequency gain $G_o \leq 1$ given by,

$$G_o = \frac{q_2}{q_1}$$

Using (9.57) and (9.58) leads to

$$\omega_o = \omega_2 , \tag{9.59}$$

and

$$Q = \frac{q_2}{1 - K(q_2/q_1)} = \frac{q_2}{1 - K G_o} , \quad K G_o \leq 1 . \tag{9.60}$$

$$\text{(b)} \quad T_{23}(s) = \frac{s^2 + s\,(\omega_2/q_1) + \omega_2^2}{s^2 + s\,(\omega_2/q_2) + \omega_2^2} \tag{9.61}$$

where ω_2 and q_1 characterize a pair of zeros in the left-half-plane, and ω_2 and q_2 characterize a pair of poles on the negative real axis; $q_2 < 0.5$. Use of (9.57) and (9.61) leads to

$$\omega_o = \omega_2 ,$$

and

$$Q = q_2 \frac{K - 1}{K(q_2/q_1) - 1} = q_2 \frac{K - 1}{KG_o - 1}, \quad KG_o \geq 1 \quad .$$

It is noted that this case is limited to $K > 1$. Furthermore, it can be shown that circuits generated using case (b) can become unstable during turn-on.* Therefore, in the following we consider only case (a) implementations; namely, PF circuits generated from a bandpass RC network n_2.

Equations (9.59) and (9.60) indicate that ω_o is identical to the center-frequency of n_2, while Q is realized through enhancing the passive q_2 by the amount of positive feedback $[1/(1-KG_o)]$ available at resonance. For a given value of Q, and for a particular network n_2 characterized by (9.58), the value of the resistance ratio is obtained from

$$K = \frac{1}{G_o}\left(1 - \frac{q_2}{Q}\right) . \tag{9.62}$$

In the unity-gain case, $K = 1$, network n_2 should satisfy the constraint

---

* As the power supply is switched ON, the gain K grows from zero and the poles follow a path that traverses the right-half of the s-plane.

$$G_o = 1 - \frac{q_2}{Q} \quad . \tag{9.63}$$

The procedure for generating PF realizations can be summarized in the following:

<u>Step 1</u>: A grounded RC two-port network having a bandpass characteristic is chosen to implement n_2 with its input and output terminals connected to terminal 3 and 2 of the op. amp., respectively. The component values of n_2 are chosen such that its center-frequency is coincident with the required ω_o. Also the component values and the value of K should yield the required value of Q given by (9.60). The degrees of freedom available in component value choices can be used to minimize the Q-sensitivities, while keeping the component value spread within acceptable limits.

<u>Step 2</u>: Any node in n_2 connected to ground can serve as a signal injection node, 4. Lifting such node off ground and connecting it to the input signal V_i does not affect the pole locations. Different zero locations are obtained through using different nodes, 4, with P(s) given by (9.56). The generation procedure can best be illustrated by an example.

Example 9.2

The simplest RC bandpass networks are LD1 and LD2 given in Fig. (8.27). Selecting LD1 for the implementation of n_2, we find that ω_o and Q are given by

$$\omega_o = 1 / \sqrt{C_2C_3R_1R_4}$$

and

$$Q = \frac{1}{\sqrt{C_2C_3R_1R_4}} \quad / \quad \left[\frac{1-K}{C_2\ R_1} + \frac{1}{C_2\ R_4} + \frac{1}{C_3\ R_4}\right]$$

The input signal can be injected through $\beta_1\ C_2$ and/or $\frac{R_1}{\beta_2}$, as shown in Fig. 9.15b, with the resulting $N_{24}(s)$ obtained from

Fig. 8.27. Thus, the transfer function T(s) of the biquadratic section generated from LD1 is given by

$$T(s) = K \frac{\beta_1 s^2 + \beta_2 [s \frac{1}{R_4} (\frac{1}{C_2} + \frac{1}{C_3}) + \frac{1}{C_2 C_3 R_1 R_4}]}{s^2 + s (\frac{\omega_o}{Q}) + \omega_o^2}$$

A special case of this circuit is that for which $\beta_2 = 0$ and $\beta_1 = 1$, resulting in the Sallen and Key positive feedback high-pass section[15,16].

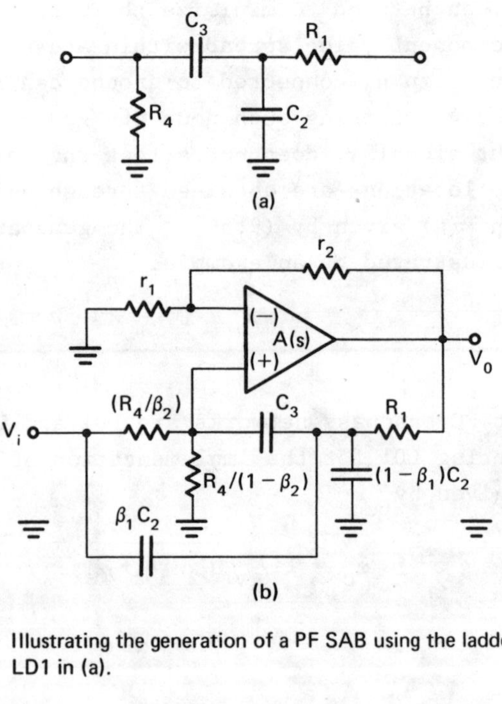

Figure 9.15 Illustrating the generation of a PF SAB using the ladder network LD1 in (a).

Sensitivity Analysis

Since the value of Q is obtained by enhancing the passive q_2 through positive feedback, the passive Q-sensitivities could be high. No specific values could be predicted at this stage, however, except for S_K^Q, given by*

$$S_K^Q = \frac{Q}{q_2} - 1 \tag{9.64}$$

Equation (9.64) is obtained using the expression for Q in (9.62). The theoretical minimum value for S_K^Q is[22]

$$S_K^Q \simeq 2Q \tag{9.65}$$

As will be shown later, the passive Q-sensitivities of PF circuits, except those for which K = 1, are rather high and of the order of Q[23].

We now turn our attention to the active sensitivities. Since network n_2 is chosen such that ω_o is independent of the amplifier gain at low frequencies, then

$$S_{A_o}^{\omega_o} = 0 \ ,$$

$$\text{and} \quad GS^{\omega_o} = 0 \ .$$

The active Q-sensitivity can be derived using (9.54), (9.58) and (9.60), with $A(s) = A_o$. The resulting Q gain-sensitivity product is given by

$$GS^Q = \frac{Q}{q_2 G_o} \left(1 - \frac{q_2}{Q}\right)^2 \tag{9.66}$$

---

* Note that K here is taken to indicate the resistive ratio not the closed-loop gain.

For $Q >> q_2$,

$$GS^Q \approx \frac{Q}{q_2 \, G_o} \, . \tag{9.67}$$

Thus to minimize the value of GS^Q, the component values of n_2 should be chosen to maximize both q_2 and G_o, with a theoretical minimum GS^Q of approximately 2Q.

At higher frequencies it can be shown, in a manner identical to that for the NF circuits above, that the filter acquires a third far removed parasitic pole, in addition to the shift in ω_o and Q characterized by

$$\delta \approx -\frac{1}{2} K \left(\frac{\omega_o}{\omega_t}\right) \left(\frac{1}{q_2} - \frac{1}{Q}\right) , \tag{9.68}$$

and

$$\eta \approx 1 \, / \, [1 + \delta - 2\delta KQ \left(\frac{\omega_o}{\omega_t}\right)] \tag{9.69}$$

Equations (9.68) and (9.69) are based on the reasonable approximations $A_o >> K$ and $\delta << 1$. Using Equation (9.68), together with the value of GS^Q derived above, it follows that

$$\delta = -\frac{1}{2Q} GS^Q \left(\frac{\omega_o}{\omega_t}\right) \tag{9.70}$$

which is an identical expression to that for the NF case, equation (9.44). Again, we note that GS^Q evaluated at low frequencies predicts the filter performance at high frequencies.

Table 9.3 provides a summary of the results for the general PF SAB of Fig. 9.14b.

Classification and Applications

As expected, classification of PF circuits will depend on the generating network chosen to implement the RC bandpass two-port n_2. From Section 8.6, there are basically two distinct types of

Table 9.3 Summary of the results for the general PF SAB of Fig. 9.14b

$$T(s) \equiv \frac{V_o}{V_i} = K\frac{T_{24}(s)}{1 - K\,T_{23}(s)} \quad , \text{ for } A(s) = \infty$$

$$\text{Let,} \quad T_{23}(s) = \frac{s(\omega_2/q_1)}{s^2 + s\left(\frac{\omega_2}{q_2}\right) + \omega_2^2} \quad , \ T_{23}(j\omega_2) \equiv G_o = \frac{q_2}{q_1}$$

$$\text{Then, nominally:} \quad \omega_o = \omega_2 \ , \text{ and } Q = \frac{q_2}{1 - KG_o} \ .$$

- Passive ω_o-sensitivities are identical to those of ω_1, that is $\leq \frac{1}{2}$
- $S_K^Q \simeq \frac{Q}{q_2}$
- Passive Q sensitivities, except for the case K=1, are large and proportional to Q
- $GS^{\omega_o} = 0$
- $GS^Q \simeq \frac{Q}{q_2\,G_o}$
- $\delta \equiv \frac{\Delta\omega_o}{\omega_o} \simeq -\frac{1}{2Q}\,GS^Q\left(\frac{\omega_o}{\omega_t}\right)$
- $\eta \equiv \frac{Q_a}{Q} \simeq 1\Big/\left[1 + \delta - 2K\delta Q\left(\frac{\omega_o}{\omega_t}\right)\right]$

(The last two items: Effects of the amplifier's limited bandwidth)

such RC networks; namely,

1) Second-order RC bandpass networks such as the ladder networks of Fig. 8.27.

2) Third-order RC networks that implement second-order bandpass functions through pole-zero cancellation. Examples of these are the twin-T bandpass networks of Fig. 8.34.

In PF realizations, there is another variable that affects the classification. This is the closed loop gain K. Specifically, it will be shown that the case K = 1 results in classes of realizations distinct from those obtained with K > 1. This is due to the constraint equation (9.63), which the RC network has to satisfy if K is chosen unity.

We thus conclude that there are four possible classes of PF realizations, characterized as follows:

Class	Network n_2	Closed-Loop Gain
PF1	Second-order	K > 1
PF2	Second-order	K = 1
PF3	Third-order	K > 1
PF4	Third-order	K = 1

The PF1 Class

Examples of generating networks for the PF1 class are given in Fig. 9.16. Shown are the ladder network (LD1), together with its dual (LD2), and the augmented form of each; namely, ALD1 and ALD2. Note that the transfer functions of these four networks have been presented in Fig. 8.27.

Fig. 9.16 also provides the design equations for biquads of the PF1 class. In each case the impedance level, R or C, is left for the designer to select. We have, however, exhausted one of the degrees of freedom available in components value choice in each case. For instance, in the LD1 case, we have chosen the time constants of the two ladder sections equal. This is done to simplify matters. The remaining degree(s) of freedom are left as parameter(s) in the design equations, and in the resulting sensitivity expressions. For instance, m is left as a design parameter in the LD1 case, with its effect on the value of GS^Q clearly visible. In this case, it is desirable to choose as small a value for m as practically feasible. Reducing the value of m below 0.1 does not, however, significantly reduce the value of GS^Q. In fact, the designer might choose m = 1, resulting in a slightly higher value for GS^Q, but with the advantage of equal values for the capacitors and equal values for the resistors.

We observe from Fig. 9.16 that PF1 circuits are characterized by a Q gain-sensitivity-product of the order of Q, rather than

	Circuit	(CR)	K	GS^Q	$N_{24}(s)$
LD1	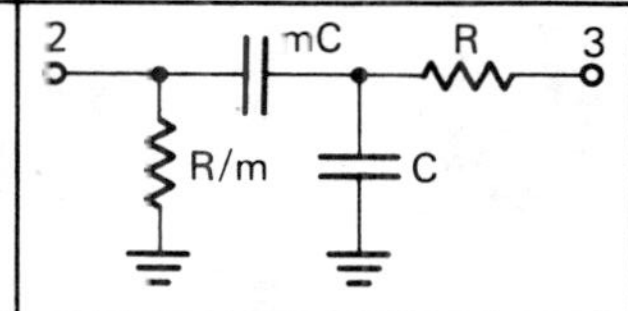	$1/\omega_0$	$2 + m - 1/Q$	$(2 + m)^2 Q$	$\beta_1 : C \qquad \beta_2 : m/R$ $N_{24}(s) = \beta_1 s^2 + s\beta_2 (1 + m)\omega_0 + \beta_2 \omega_0^2$
ADL1	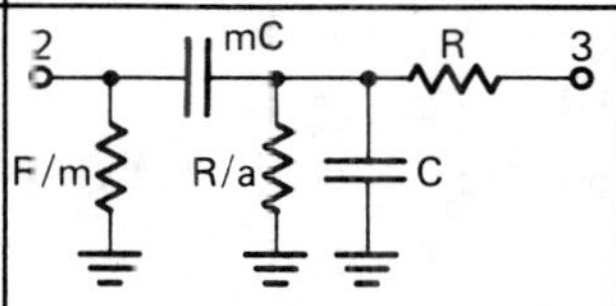	$\frac{\sqrt{1 + a}}{\omega_0}$	$2 + m + a - \sqrt{1 + a}/Q$	$\frac{(2 + m + a)^2}{\sqrt{1 + a}} Q$	$\beta_1 : C \qquad \beta_2 : m/R \qquad \beta_3 : a/R$ $N_{24}(s) = \beta_1 s^2 + s \frac{[(1 + m)\beta_2 + \beta_3 a]}{\sqrt{1 + a}} \omega_0 + \beta_2 \omega_0^2$
LD2	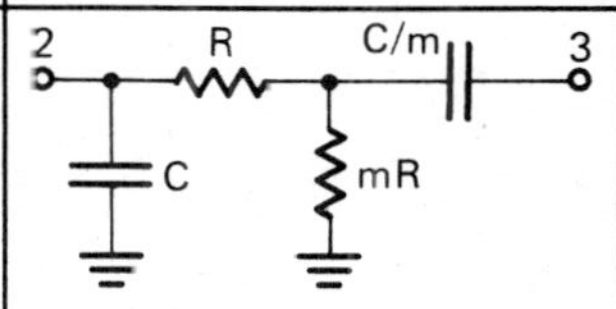	$1/\omega_0$	$2 + m - 1/Q$	$(2 + m)^2 Q$	$\beta_1 : C \qquad \beta_2 : 1/mR$ $N_{24}(s) = \beta_1 s^2 + s\beta_1 (1 + m)\omega_0 + \beta_2 \omega_0^2$
ALD2	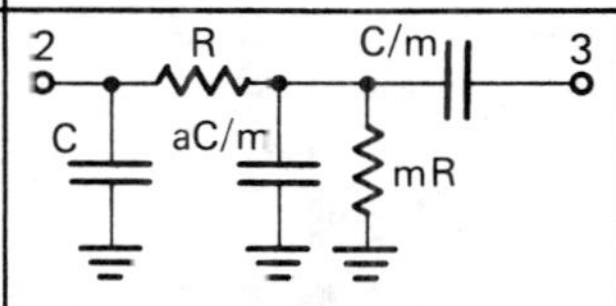	$\frac{1}{\sqrt{1 + a}\,\omega_0}$	$2 + m + a - \sqrt{1 + a}/Q$	$\frac{(2 + m + a)^2}{\sqrt{1 + a}} Q$	$\beta_1 : C \qquad \beta_2 : 1/mR \qquad \beta_3 : aC/m$ $N_{24}(s) = \beta_1 s^2 + s \frac{[(1 + m)\beta_1 + \beta_3 a]}{\sqrt{1 + a}} \omega_0 + \beta_2 \omega_0^2$

Figure 9.16 Examples of networks generating the PF1 class of realizations.

Q^2 as for NF1 circuits. This is obtained at the expense of increased passive sensitivities, also of the order of Q. Note that the ω_o deviation at higher frequencies is independent of Q.

Example 9.3

Use Fig. 9.16 to obtain a high-pass PF1 realization with Q = 10 and f_o = 1 KHz. Calculate the values of the performance indices. Assume the op. amp. to have f_t = 1 MHz.

Solution

By consulting the last column of Fig. 9.16, it follows that a high-pass realization may be generated from LD1 by setting β_2 = 0. The value of β_1 can be used to control the gain. For minimum component value spread, a value of unity is assigned to m. Thus, from Fig. 9.16 it follows that

$$CR = \frac{1}{\omega_o}$$

Selecting R = 10 KΩ then C = 0.0159 μF .

$$K = 3 - \frac{1}{Q} = 3 - 0.1 = 2.9$$

The complete circuit is shown in Fig. 9.17, which has the transfer function

$$T(s) \equiv \frac{V_o}{V_i} = \frac{2.9\ s^2}{s^2 + s(\frac{\omega_o}{Q}) + \omega_o^2}$$

where, $\omega_o = 2\pi \times 10^3$ rad/sec and Q = 10.
The performance indices may be calculated using the information presented in the above. The results are:

$$GS^{\omega_o} = 0$$

$$GS^{Q} = 9Q = 90$$

$$S_{R_1}^{\omega_o} = S_{R_2}^{\omega_o} = S_{C_1}^{\omega_o} = S_{C_2}^{\omega_o} = -\frac{1}{2}$$

$$S_{C_1}^{Q} = 9.5 \quad S_{C_2}^{Q} = -9.5 \quad S_{R_1}^{Q} = -19.5 \quad S_{R_2}^{Q} = 19.5$$

$$S_{K}^{Q} = 29, \quad \text{where } K = 1 + \frac{r_2}{r_1}$$

$$\delta \equiv \frac{\Delta\omega_o}{\omega_o} = -\frac{1}{2Q} \; GS^Q \; \left(\frac{\omega_o}{\omega_t}\right) = -0.45\%$$

$$\eta \equiv \frac{Q_a}{Q} = 1 \;/\; [1 + \delta - 2\delta KQ \left(\frac{\omega_o}{\omega_t}\right)] = 1.0042$$

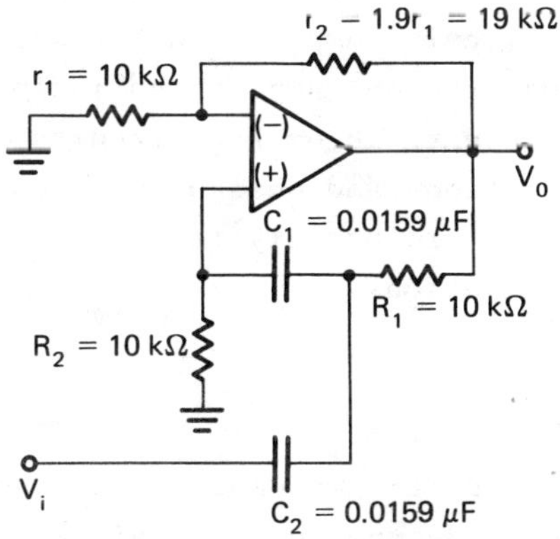

Figure 9.17 The high-pass PF1 circuit of Example (9.3).

The PF2 Class

As previously mentioned, PF2 circuits are generated using second-order RC networks and setting K = 1. In this case, the RC network has to satisfy the constraint equation (9.63). It can be shown that LD1, as an example, will have the component values given in Fig. 9.18. Note that the value of GS^Q has changed to become proportional to Q^2. The component value spread is also of the order of Q^2. The only advantage here is the low passive Q-sensitivities.

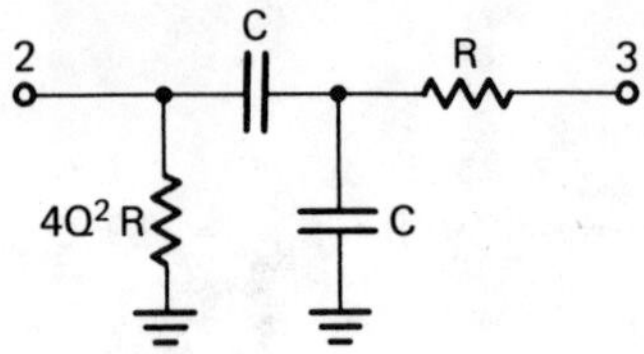

$CR = 1/2\omega_0 Q$ $\quad GS^Q \cong 2Q^2$

$\beta_1 : C$ $\quad \beta_2 : 1/4Q^2 R$

$N_{24}(s) = \beta_1 s^2 + s\beta_2(\omega_0/Q) + \beta_2 \omega_0^2$

Passive Q-sensitivities are $\leqslant 1$

Figure 9.18 Network LD1 as a generating network for the PF2 class.

Although only LD1 is shown as a generating network for PF2 circuits, the results are representative of this class of biquads. We thus conclude that PF2 circuits exhibit values for the performance indices almost identical to those obtained for NF1 circuits. While PF2 circuits employ a unity gain amplifier, the NF1 realizations use the op. amp. in an infinite gain configuration. Fig. 9.19 presents three PF2 circuits, together with their pertinent functions and design equations. We finally note that the PF2 circuits are usually referred to as the Sallen and Key unity-gain realizations.

The PF3 Class[24,26]

To generate PF3 circuits, we employ the bandpass twin-T networks given in Fig. 8.34 and set the gain $K > 1$. From our previous encounter with twin-T type networks, one should expect the passive Q-sensitivities to be high, of the order of Q. The prime motivation, therefore, for seeking PF3 realizations is to see whether one can obtain GS^Q values lower than those obtained for the PF1 class which are proportional to Q. It turns out, however, that this is not achieved. We therefore conclude that PF3 circuits do not have any sensitivity advantage over the simpler PF1 class. The only motivation for using PF3 circuits lies in their versatility in generating different filtering functions without much variation in the basic circuit topology. This has been outlined by Moschytz[24], to which the interested reader is referred.

Circuit	Transfer function $T_v = V_{out}/V_{in}$ and design equations
Low pass 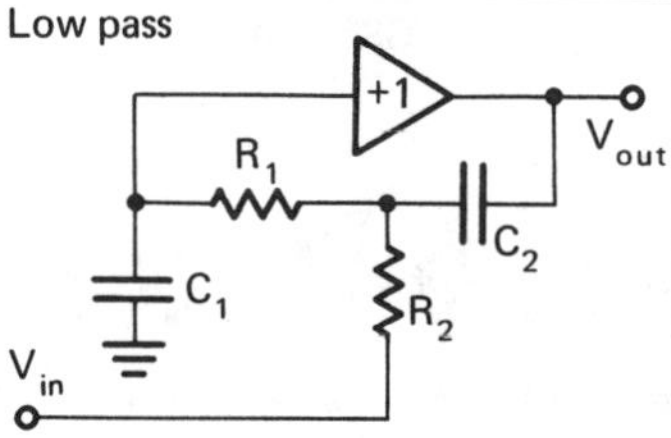	$T_v = \dfrac{\omega_0^2}{s^2 + s\omega_0/Q + \omega_0^2}$ $R_1 = R_2 = R$ $C_1 = C$ $C_2 = (4Q^2)C$ $CR = 1/(2\omega_0 Q)$
High pass 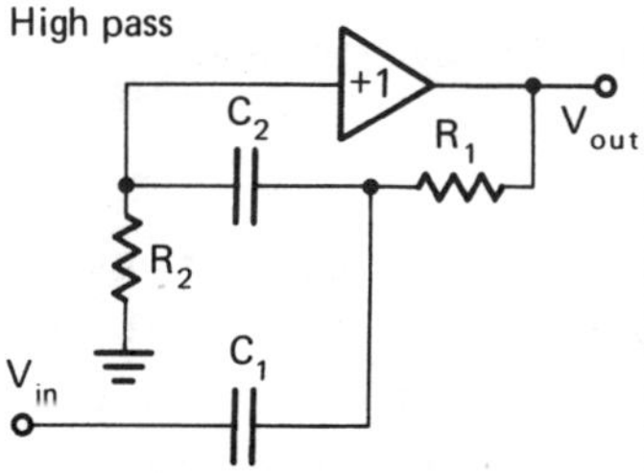	$T_v = \dfrac{s^2}{s^2 + s\,\omega_0/Q + \omega_0^2}$ $C_1 = C_2 = C$ $R_1 = R$ $R_2 = (4Q^2)R$ $CR = 1/(2\omega_0 Q)$
Bandpass 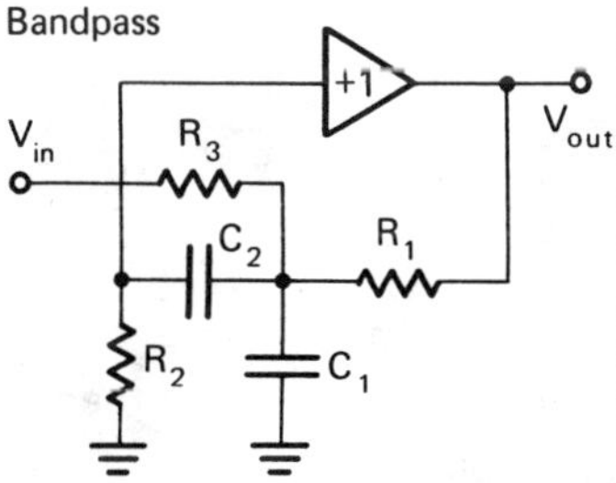	$T_v = \dfrac{s(\omega_0/3Q)}{s^2 + s(\omega_0/Q) + \omega_0^2}$ $R_1 = R$ $R_2 = R_3 = (9Q^2 - 1)R$ $C_1 = C_2 = C$ $CR = [3Q/(9Q^2 - 1)]\,(1/\omega_0)$

Figure 9.19 Sallen-and-Key unity-gain realizations (PF2).

The PF4 Class [25]

In the unity-gain case, the picture is quite different. The PF2 circuits have unacceptably high GS^Q values, of the order of Q^2. Using twin-T networks would conceivably reduce the active sensitivities at the expense of increased passive sensitivities. This is demonstrated in Fig. 9.20 where the end-loaded RC bandpass twin-T is used as a generating network. The resulting PF4 circuits have an almost identical performance to that obtained with NF2 circuits.

The reader might note that the reason for the reduced active

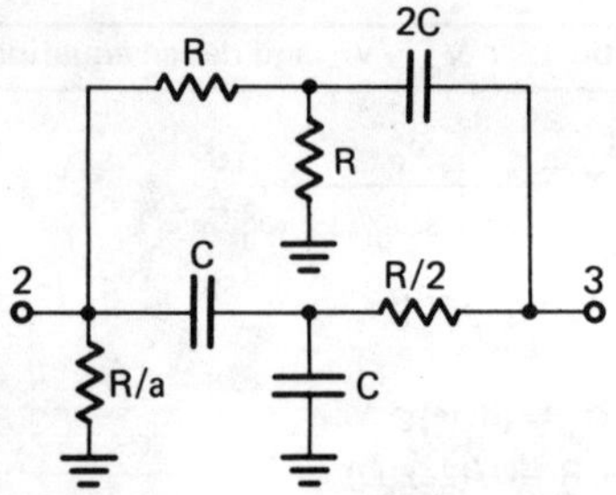

$CR = \sqrt{1 + 2a}/\omega_0$ $\quad GS^Q \cong 4Q$

$\beta_1 : C \quad \beta_2 : 1/R \quad \beta_3 : a/R$

$$N_{24}(s) = \beta_1 s^2 + s\beta_3(\omega_0/Q) + \frac{\beta_2 + 2a\beta_3}{1 + 2a}\omega_0^2$$

Passive Q-sensitivities are of the order of Q.

Figure 9.20 The end-loaded bandpass Twin—T as a generating network for the PF4 class.

sensitivities obtained for PF4 circuits is the higher center-frequency gain realized with twin-T bandpass networks. This point has been considered in Section 8.6. Although the bandpass twin-T is capable of unity center-frequency gain, the constraint equation (9.63); namely,

$$G_o = 1 - \frac{q_2}{Q} \tag{9.71}$$

forces us to load the twin-T in order to obtain a value of G_o less than unity, as determined by the required Q. From Section 8.6, it follows that there are basically two methods for loading the bandpass twin-T; namely, end-loading, such as the one shown in Fig. 8.34b, and center-loading, such as that shown in Fig. 8.34c. Correspondingly, there are two families of biquads both having identical performance and belonging to the PF4 class. These realizations are shown in Figs. 9.21 and 9.22 respectively, together with the complete design information.

Inspection of the circuits in Fig. 9.22 reveals that if the input terminal is grounded, the passive networks in all five circuits are identical to the one shown in Fig. 8.34c. Thus, it follows that all circuits can be obtained from the same building block with the associated economic advantage. Also, in all these circuits, Q is determined by only two resistors. In the notch realizations, two capacitors C_1 and C_2 determine the notch frequ-

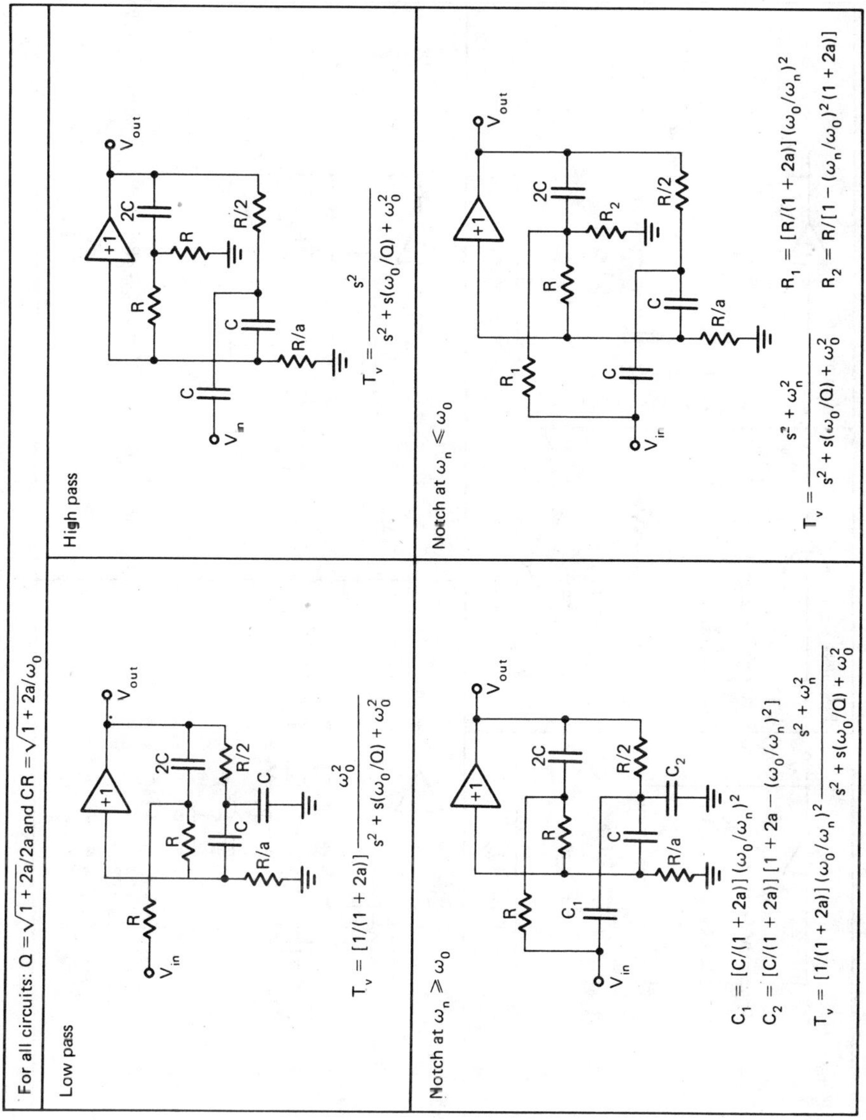

Figure 9.21 PF4 circuits generated from the end-loaded bandpass Twin–T network.

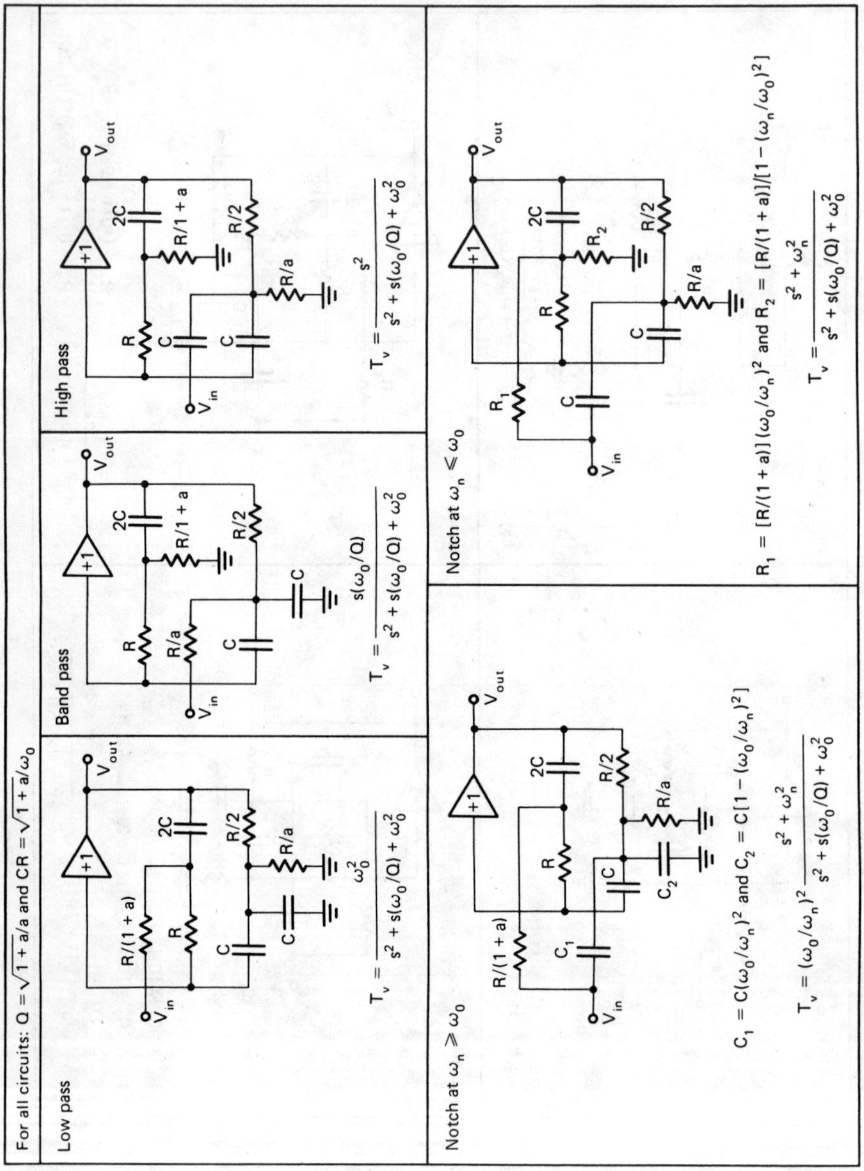

Figure 9.22 PF4 circuits generated from the center-loaded bandpass Twin–T network.

ency ω_n. Therefore, unlike many other notch realizations, the notch frequency can be varied without affecting ω_o and Q.

An interesting scheme is possible for tuning the circuits in Fig. 9.22. The passive network can be redrawn as in Fig. 9.23a. The result is the familiar notch twin-T structure having a pair of complex zeros. In Fig. 9.23b, the twin-T is redrawn with the Q determining resistors shown dotted. Without these resistors, the twin-T has a null at $\omega_o' = \frac{1}{CR}$. Methods for automatically tuning thin-film twin-T networks are described elsewhere[27]. After tuning the network in Fig. 9.23b to have a notch of sufficient depth at ω_o', the Q determining resistors can be added.

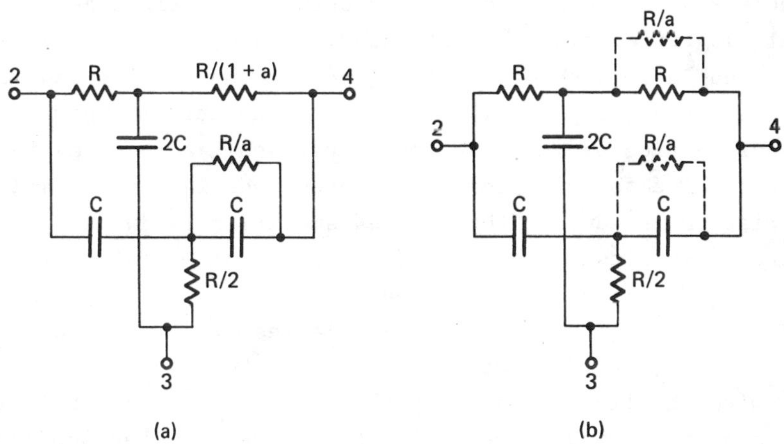

Figure 9.23 Illustrating the tuning of the PF4 circuits given in Fig. (9.22).

In the circuits of Fig. 9.21, only one resistor (R/a) determines Q. However, no bandpass realization of this type is available. All these circuits use the same passive structure and can also be obtained from one building block. Similar to the circuits of Fig. 9.22, an indirect tuning scheme is possible for the circuits of Fig. 9.21. For more details on PF4 type circuits, the reader is referred to [25].

We note that although PF4 circuits do not provide any significant sensitivity advantage over PF1 realizations, their use of unity-gain amplifiers is rather attractive. In designing PF4 circuits, one is not restricted to op. amps. and may use wider bandwidth voltage followers. The high frequency effects in PF4 circuits are also less pronounced than in PF1 circuits, since $K = 1$ in equation (9.69).

9.6 COMPARISON OF NF AND PF SAB'S

Table 9.4 provides a comparison between the six classes of single amplifier biquads considered. Since, for all the circuits presented in this chapter, the ω_o active and passive sensitivities are either very small or at their theoretical minima, only Q-sensitivities are used in the comparison. Also given in the table are the formulae for the frequency deviation, δ, and the Q variation, η, due to the limited amplifier bandwidth. It follows that circuits having GS^Q of the order of Q^2 suffer large frequency deviation (proportional to Q) and large Q variation.

Realizations of the NF1 and PF2 classes are limited to low-Q poles only, ($Q < 10$). Other classes can be applied to higher Q's, with the limitation being the high passive sensitivities. It has been claimed, however, that with certain fabrication technologies, high passive sensitivities do not present a major obstacle. Specifically, circuits employing twin-T networks have been fabricated using Tantalum-Thin film technology[7,28,29], and automatic tuning procedures have been devised for such networks. This, together with the close tracking of passive components make these high-passive-sensitivity circuits viable.

An interesting observation from Table 9.4 is that the product of the passive sensitivities and the value of GS^Q is of the order of Q^2 for all of the six classes. This will be shown to also hold true for the Enhanced Negative Feedback (ENF) class considered next.

Table 9.4 Comparison of the six classes of SAB's

Class		Feedback Network	Passive Q-sensitivities	S^Q_{Ao} or GS^Q	Element Spread	$\delta \equiv \frac{\Delta\omega_o}{\omega_o}$	$\eta \equiv \frac{Q_a}{Q}$
NF	NF1	2nd order	Low	Q^2	Q^2	$-\frac{1}{2Q}GS^Q(\frac{\omega_o}{\omega_t})$	$\frac{1}{1+\delta-2\delta Q(\frac{\omega_o}{\omega_t})}$
	NF2	3rd order	Q	Q	Q		
FF K>1	PF1	2nd order	Q	Q	Independent of Q		$\frac{1}{1+\delta-2K\delta Q(\frac{\omega_o}{\omega_t})}$
	PF3	3rd order	Q	Q	Q		
PF K=1	PF2	2nd order	Low	Q^2	Q^2		$\frac{1}{1+\delta-2\delta Q(\frac{\omega_o}{\omega_t})}$
	PF4	3rd order	Q	Q	Q		

Notes: * Only order of magnitudes are indicated
** K = 2 to 3
*** Passive ω_o sensitivities and GS^{ω_o} are low, for all circuits

9.7 SINGLE AMPLIFIER BIQUADS OF THE ENHANCED NEGATIVE FEEDBACK (ENF) TYPE*

From the previous section, we conclude that biquads of the negative feedback (NF) or positive feedback (PF) types offer either low active-sensitivities and high passive-sensitivities, or the reverse. Specifically, the NF1 class has low passive-sensitivities at the expense of high active-sensitivities, while the opposite is true for the PF1 class. It is natural, therefore, to inquire into the possibility of departing from these two extreme situations and devising circuits that feature a continuous tradeoff between active and passive sensitivities. As will be shown below, this is indeed the case for the Enhanced Negative Feedback class of SAB's depicted in Fig. 9.24.

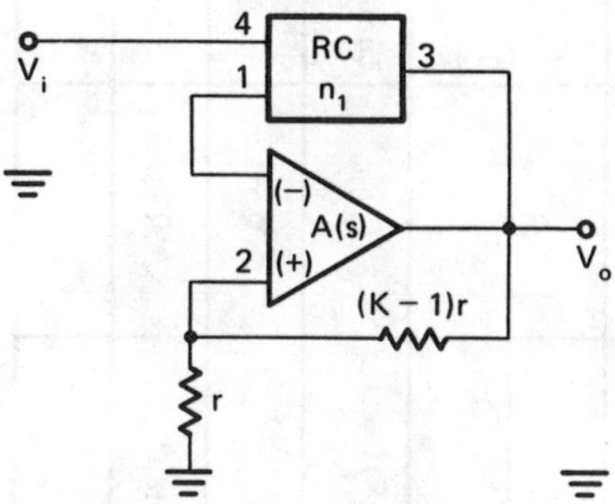

Figure 9.24 The ENF configuration.

As the name implies, the ENF configuration is similar to the NF circuit, with the exception of additional resistive feedback supplied by the two resistors r and (K-1)r in Fig. 9.24. Note that the parameter K determines the amount of positive feedback; as K is reduced more positive feedback is applied. Obviously, setting $K = \infty$ results in the NF configuration. In the following, we shall be specifically interested in ENF circuits generated

* The analysis of this Section is largely based on [30]

from a canonical (second-order) RC network implementation of n_1. In other words, we will investigate the effects of applying some resistive positive feedback to circuits of the NF1 type. It will be shown that positive-feedback reduces the high active-sensitivities encountered in the NF1 circuits, at the expense of increasing the passive sensitivities. The tradeoff between the two sensitivities can be exercised by controlling the value of the parameter K.

Generation of ENF Circuits

The ENF configuration of Fig. 9.24 is general except for the possible injection of the input signal to the noninverting op. amp. input through a resistor. This will be needed for generating notch and all-pass circuits. For simplicity of analysis and in order to focus attention on the mechanism by which ENF circuits exhibit lower active sensitivities, we will be content with the simplified configuration of Fig. 9.24. In this way, we will deal in detail with the design of the bandpass version of the ENF class of circuits. Other filter types can be readily generated by following a method similar to that employed for the NF1 class.

The transfer function of the ENF circuit of Fig. 9.24 can be obtained by using equation (9.7), and is given by

$$T(s) = \frac{V_o}{V_i}(s) = -\frac{T_{14}(s)}{T_{13}(s) - \frac{1}{K} + \frac{1}{A(s)}} . \qquad (9.72)$$

The network functions of n_1 can be expressed in a polynomial quotient form, using our usual notation, leading to

$$T(s) = -\frac{N_{14}(s)}{N_{13}(s) - \frac{D(s)}{K} + \frac{D(s)}{A(s)}} . \qquad (9.73)$$

The filter poles are determined by the denominator polynomial $D_V(s)$,

$$D_v(s) = N_{13}(s) - \frac{D(s)}{K} + \frac{D(s)}{A(s)} \quad . \tag{9.74}$$

In the ideal case $A(s) = \infty$, we have,

$$D_v(s) = N_{13}(s) - \frac{D(s)}{K} \quad . \tag{9.75}$$

We note that if positive feedback is eliminated ($K = \infty$), the realization reduces to the NF class where the filter poles are identical to the zeros of n_1. Equation (9.75) thus suggests that the addition of positive feedback (K finite) enhances the filter Q-factor by moving the roots of $D_v(s)$ from the location of the zeros of $N_{13}(s)$ closer to the $j\omega$-axis by an amount determined by K. Part of the selectivity, therefore, is obtained by the negative feedback action, while the remainder is obtained through positive feedback. Network n_1 should, therefore, be chosen to have a pair of complex transmission zeros characterized by ω_1, q_1 ($q_1 < Q$), that is,

$$T_{13}(s) = \frac{s^2 + s(\frac{\omega_1}{q_1}) + \omega_1^2}{s^2 + s(\frac{\omega_1}{q_2}) + \omega_1^2} \tag{9.76}$$

where $q_2 < 0.5$, and it has been assumed that the poles and zeros of n_1 have the same frequency ω_1. This assumption is not essential but is made to simplify the analysis to follow. The value of q_1 is chosen to be a fraction α of the desired value of Q; that is,

$$q_1 = \alpha Q \quad , \quad \alpha < 1 \quad . \tag{9.76}$$

Note that for $\alpha = 1$, $K = \infty$, and the realization reduces to the NF case.

From (9.75) and (9.76) we obtain

$$D_v(s) = (1 - \frac{1}{K}) \; [s^2 + s \; \frac{\omega_1}{q_1(1 - \frac{1}{K})} \; (1 - \frac{1}{K} \; \frac{q_1}{q_2}) + \omega_1^2] \quad .$$

Thus,

$$\omega_o = \omega_1 \quad \text{and,}$$

$$Q = q_1 \frac{1 - \frac{1}{K}}{1 - \frac{1}{K}\frac{q_1}{q_2}} \tag{9.77}$$

Thus, positive feedback moves the filter poles closer to the $j\omega$-axis on a circle of radius ω_1. The original position of the poles is characterized by (ω_1, q_1), while their final position is characterized by (ω_1, Q). One may view the action of positive feedback as that of "enhancing" the value of the Q-factor (increasing it from q_1 to Q), as determined by (9.77).

Sensitivity Analysis

From the above, we see that at low frequencies (i.e. $A(s) \simeq A_o$),

$$S_{A_o}^{\omega_o} = 0.$$

To determine GS^Q, we use (9.74) together with (9.76) and substitute $A(s) = A_o$. The result is,

$$GS^Q \simeq \left(\frac{Q}{q_2} - 1\right) / \left(1 - \frac{1}{K}\right) \quad , \quad \text{for } A_o >> K \quad . \tag{9.78}$$

Note that for $K = \infty$ (the NF case), $GS^Q = \frac{Q}{q_2} - 1$; a result obtained in (9.33).

As will be shown below, K is usually >> 1. Nevertheless, a significant reduction in GS^Q is obtained, as compared to the NF1 case. This comes about due to the higher value of q_2 possible in the ENF case. This in turn is obtained because n_1 has zeros with lower Q-factor ($q_1 = \alpha Q$). As α is decreased from unity (K decreased from ∞), the value of q_2 increases towards the maximum possible value of 0.5, thus reducing the value of GS^Q. The price paid is an increased sensitivity with respect to K and the components of n_1. The latter sensitivities will be derived for the specific example considered below, while S_K^Q is obtained from (9.77) as,

$$S_K^Q = \frac{1 - \frac{1}{\alpha}}{1 - \frac{1}{K}} , \tag{9.79}$$

which clearly increases in magnitude as α is decreased.

We next consider the effect of the amplifier limited frequency response on the obtained center-frequency ω_{oa} and Q-factor, Q_a. Using the single-pole model for the op. amp., it can be shown that,

$$\delta \simeq - \frac{1}{2} (\frac{\omega_o}{\omega_t}) \frac{1}{q_2} \frac{1}{(1 - \frac{1}{K})} \tag{9.80}$$

From (9.78) and (9.80) it follows that

$$\delta \simeq - \frac{1}{2Q} (\frac{\omega_o}{\omega_t}) \cdot GS^Q , \tag{9.81}$$

which is of the same general form obtained for the NF and PF circuits considered in previous sections. From (9.81), it follows that the reduced value of GS^Q realized in the ENF circuits under consideration results in a smaller value for the deviation in the center-frequency due to the amplifier pole.

The actual value obtained for the Q-factor is related to Q by the factor η given by

$$\eta = \frac{Q_a}{Q} \simeq 1 / (1 + \delta) [1 - 2\delta Q \frac{1}{(1 - \frac{1}{K})} (\frac{\omega_o}{\omega_t})] \tag{9.82}$$

As a check, note that for the NF case ($K = \infty$), the expression in (9.82) reduces to that given by (9.42). Bearing in mind that usually K >> 1, the smaller value obtained for δ in ENF circuits yields a value for η closer to unity.

We thus conclude that using canonical (2nd order) RC networks for n_1 in both the NF and ENF configurations results in superior performance of the ENF circuits, as far as the active sensitivities are concerned. The tradeoff is with increased passive sensitivities. However, having the flexibility of a continuous tradeoff between active and passive sensitivities is an attractive feature that enhances the manufacturability of the ENF class of circuits. This will be illustrated in the following.

The Bandpass ENF Circuit

As an example of the class of ENF realizations, we consider the design details of the bandpass version. It should be noted that this bandpass circuit is a special case of the Deliyannis-Friend biquad[31]-[33]. For the other filter functions, the reader is referred to the references.

The bandpass ENF circuit is generated using the bridged-T network, BT2, to implement n_1 as shown in Fig. 9.25. Note that the two capacitors are chosen equal in value for practical reasons and because choosing unequal capacitors has no significant effect on reducing sensitivities[33].

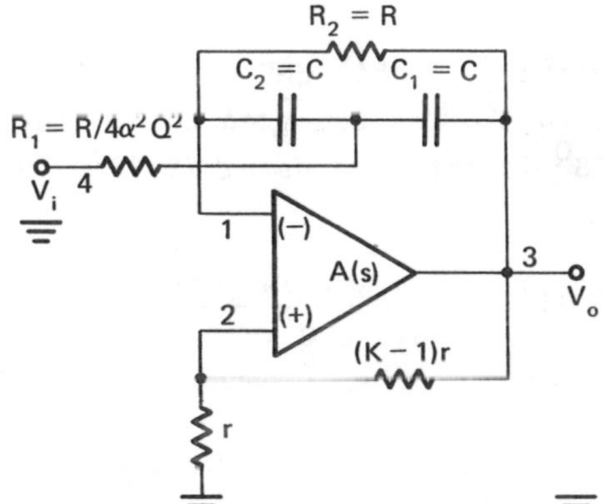

Figure 9.25 The bandpass version of the ENF class.

Since the BT network is to have transmission zeros characterized by $\omega_1 = \omega_o$ and $q_1 = \alpha Q$, then (from Fig. 9.7) its two resistors are related by $(4\alpha^2 Q^2)$ while the time constant is given by,

$$CR = \frac{2\alpha Q}{\omega_o} \tag{9.83}$$

The design thus consists of first choosing a value of α^*, from the sensitivity considerations below, and a value for the impedance level (R or C) from practical considerations. The remaining parameter is determined from (9.83) and the value of K is obtained from (9.77). For our case the value of q_2 is given by (See Chapter 8, Fig. 8.30)

$$q_2 = \frac{\alpha Q}{1 + 2\alpha^2 Q^2} \quad . \tag{9.84}$$

Substituting in (9.77) obtains,

$$K = 1 + \frac{2\alpha^2 Q^2}{1 - \alpha} \tag{9.85}$$

Note that in the circuit of Fig. 9.25, as well as in all the design equations, setting $\alpha = 1$ results in the NF1 bandpass realization.

An obvious advantage of this ENF circuit is the reduction of component spread ($4\alpha^2 Q^2$ as compared to $4Q^2$ for NF1).

The value of GS^Q is obtained using (9.78), (9.84) and (9.85) as,

$$GS^Q = \frac{(1 - \alpha + 2\alpha^2 Q^2)^2}{\alpha(2\alpha^2 Q^2)} \tag{9.86}$$

---

* Note that the design can equally well start with choosing a value for K and then determining α, etc. Both K and α determine the amount of positive feedback applied. This in turn determines the performance.

At this point, we should note that ENF circuits are used for $Q \geq 10$, since lower values of Q are easily obtained with the simpler NF1 circuits. Also, it will be shown that α is chosen such that $\alpha Q > 3$. Using these facts enables us to approximate (9.86) as,

$$GS^Q \simeq 2\ \alpha\ Q^2 \tag{9.87}$$

We now clearly see that reducing α from unity results in decreasing the value of GS^Q. The values of δ and η can be obtained from (9.81) and (9.82), using (9.85) and (9.87), as

$$\delta \simeq -\ \alpha\ Q\ (\frac{\omega_o}{\omega_t}) \quad , \tag{9.88}$$

$$\eta \simeq 1\ /\ [1 - \alpha\ Q(\frac{\omega_o}{\omega_t}) + 2\alpha Q^2(\frac{\omega_o}{\omega_t})^2 - 2\alpha^2 Q^3(\frac{\omega_o}{\omega_t})^3] \quad . \tag{9.89}$$

Again we note that decreasing α results in a more ideal performance, as far as the dependence on the amplifier is concerned. Unfortunately, decreasing α results in increasing the passive sensitivities. The latter can be easily evaluated and shown to be given by,

$$S^Q_{R_1} = -\ S^Q_{R_2} = \frac{1}{2} - \frac{1}{\alpha} \quad ,$$

$$S^Q_{C_1} = -\ S^Q_{C_2} = \frac{1}{2} + \frac{1}{2\alpha} \quad ,$$

$$S^Q_K \simeq 1 - \frac{1}{\alpha} \ , \quad \text{for } \alpha Q > 3 \ .$$

Note that for $\alpha = 1$, it is clear that the NF1 case enjoys low passive sensitivities.

To help in the decision of choosing a value for α, we tabulate approximate values for the active and passive sensitivities as functions of α as follows,

α	GS^Q	$\left\|S_R^Q\right\|$	$\left\|S_C^Q\right\|$	$\left\|S_K^Q\right\|$
$\frac{1}{Q}$	$4.5Q$	Q	$\frac{Q}{2}$	$1.5Q$
$\frac{2}{Q}$	$5Q$	$\frac{Q}{2}$	$\frac{Q}{4}$	$\frac{Q}{2}$
$\frac{3}{Q}$	$6Q$	$\frac{Q}{3}$	$\frac{Q}{6}$	$\frac{Q}{3}$
$\frac{4}{Q}$	$8Q$	$\frac{Q}{4}$	$\frac{Q}{8}$	$\frac{Q}{4}$
$\frac{5}{Q}$	$10Q$	$\frac{Q}{5}$	$\frac{Q}{10}$	$\frac{Q}{5}$
1	$2Q^2$	$\frac{1}{2}$	1	0

From this table, it appears that a value of $\alpha = \frac{4}{Q}$ or $\frac{5}{Q}$ will lead to a reasonable compromise between the values of GS^Q and the passive sensitivities. Note that once a value of αQ is decided, the value of Q can be changed using only one of the positive feedback resistors. Eliminating the positive feedback always results in a bandpass filter with a quality factor of (αQ).

Throughout the rest of this book, we will choose $\alpha Q = 5$ that is $q_1 = 5$. The resulting bandpass circuit together with the gain adjusting resistor and complete design information are given in Fig. 9.26.

9.8 TWO-AMPLIFIER BIQUADS

In the four previous sections, the limitations of single amplifier realizations were clearly outlined. It seems that a Q gain-sensitivity-product of the order of Q is acceptable and in fact, there are no currently known circuits, including multi-amplifier realizations, that provide lower values for GS^Q*. Al-

* We remind the reader that this statement applies only to circuits with low GS^{ω_o}. In fact, circuits with $GS^Q = 0$ have been reported. These, however, suffer from large values of GS^{ω_o}, rendering them impractical.

though such a value of GS^Q is obtained in a number of SAB classes the associated passive sensitivities are invariably proportional to Q. The motivation, therefore, in seeking multi-amplifier circuits is to reduce the passive Q-sensitivities, hopefully to their theoretical minima. In achieving this, however, it is not acceptable to compromise the ω_o sensitivities. The latter are quite low for all the SABs presented in the previous Sections.

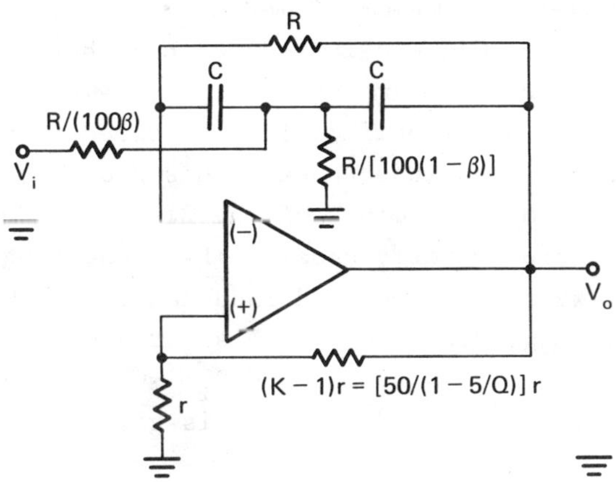

$$CR = 10/\omega_o$$

$$\frac{V_o}{V_i}(s) = -\frac{s\omega_o\beta[10.2 - (1/Q)]}{s^2 + s(\omega_o/Q) + \omega_o^2}$$

Figure 9.26 Design of the bandpass ENF circuit.

A large number of two-amplifier circuits have been reported in the literature. A comparative study[8], however, has shown that most of these circuits either obtain low Q-sensitivities at the very expensive price of high ω_o-sensitivities, or they do not have any advantage over their single-amplifier counterparts. It has been shown in [8] that some two-amplifier circuits have $C\mathcal{S}^{\omega_o}$ values of the order of Q^2 !

Only one two-amplifier resonator proved to be clearly superior to SABs and hence will be described here. This circuit is based on the Antoniou GIC presented in Chapter 8.

Before considering the circuit details, the reader should note that another motivation for seeking multi-amplifier realizations is to enhance the manufacturability of the biquad. Such a goal may be achieved by using circuits that feature easier tuning, isotopic structure, simultaneous realization of more than one filtering function and small spread in component values. We have already discussed such matters in Section 9.2.

Biquads Derived from Generalized Impedance Converters

In Chapter 8, the GIC was introduced together with two high quality op. amp. circuit implementations, namely those of Fig. 8.18. There, we were specifically interested in using the GIC to simulate a grounded inductance. This grounded inductance is obtained at the input terminal of the GIC, provided the impedances Z_1 to Z_5 are properly chosen. By connecting a capacitance C in parallel with the simulated inductance, we obtain a resonance circuit. Fig. 9.27 shows such an arrangement with the input signal fed through a resistance $R_1 = QR$ and where it is assumed that the simulated inductance L is given by *

$$L = CR^2$$

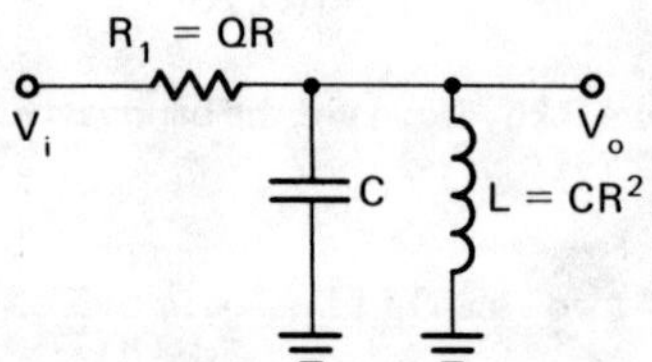

Figure 9.27 Illustrating the realization of a second-order bandpass function using a simulated inductance L.

---

* At this point, the reader is strongly urged to review Section 8.4.

It follows that, if the output is taken across the simulated inductance, then the transfer function realized is

$$\frac{V_o}{V_i} = \frac{s\,(\frac{\omega_o}{Q})}{s^2 + s\,(\frac{\omega_o}{Q}) + \omega_o^2} \tag{9.90}$$

where

$$\omega_o = \frac{1}{\sqrt{LC}} = \frac{1}{CR}$$

The transfer function in (9.90) is a bandpass one, as should be expected from the circuit in Fig. 9.27. We are thus tempted to use this circuit as a biquad. A problem exists however. The output signal is taken at a high impedance point, making loading the biquad impossible (that is, without altering the transfer function). Careful examination of the two "good" implementations presented in Chapter 8 reveals that one of them, namely circuit-A in Fig. 8.18, has the output of amplifier A_2 directly proportional to the signal at the GIC input. This simple observation allows the application of this GIC circuit as a second order biquad circuit[34]. Fig. 9.28 shows such a biquad implementing a bandpass function. It is easy to verify that the transfer function realized is

$$\frac{V_o}{V_i} = \frac{s\;2(\frac{\omega_o}{Q})}{s^2 + s\,(\frac{\omega_o}{Q}) + \omega_o^2}$$

where $\omega_o = \frac{1}{CR}$, and where the amplifiers are assumed ideal.

Straightforward analysis of this two-amplifier resonator reveals that it has the following properties.

(1) Passive ω_o-sensitivities are at their theoretical minima; namely, each is $\leq\frac{1}{2}$ in magnitude.

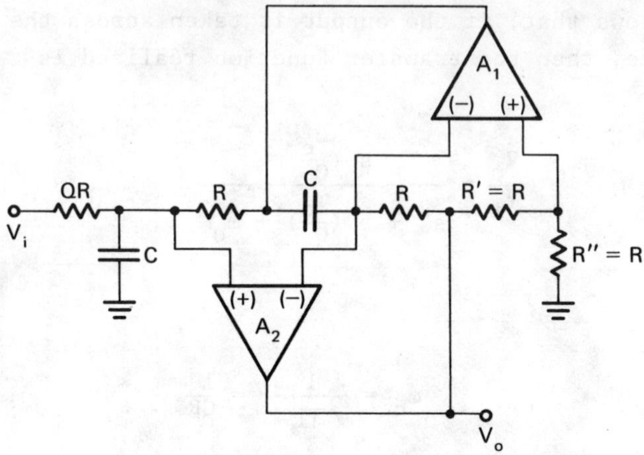

Figure 9.28 A bandpass section based on the type-A GIC circuit of Fig. (8.18).

(2) $GS_{A_1}^{\omega_o}$ and $GS_{A_2}^{\omega_o}$ are low.

(3) Passive Q-sensitivities are low.

(4) $GS_{A_1}^{Q} = GS_{A_2}^{Q} \simeq 2Q$.

(5) $\delta \simeq -\left(\frac{\omega_o}{\omega_{t_1}} + \frac{\omega_o}{\omega_{t_2}}\right)$ (9.91)

which can be written as,

$$\delta = -\frac{1}{2Q}\left[GS_{A_1}^{Q}\left(\frac{\omega_o}{\omega_{t_1}}\right) + GS_{A_2}^{Q}\left(\frac{\omega_o}{\omega_{t_2}}\right)\right] ,$$

and for identical amplifiers we have,

$$\delta \simeq -2\left(\frac{\omega_o}{\omega_t}\right) .$$

(6) $\eta \simeq 1 / (1+\delta)\left[1 - 2Q\left(\frac{\omega_o}{\omega_{t_1}}\right)\left(1 - \frac{\omega_{t_2}}{\omega_{t_1}} + 2\delta\right)\right]$ (9.92)

which for matched amplifiers reduces to

$$\eta \simeq 1 / [1 + \delta - 4\delta Q (\frac{\omega_o}{\omega_t})] \tag{9.93}$$

(7) The input impedance at resonance is infinite.

(8) All resistors are equal except for the Q determining resistor being QR. The two capacitors are equal.

We thus conclude that, from all aspects, this biquad is the "best" we have encountered this far. Not only are all the sensitivities low but in addition, for matched amplifiers, the Q variation factor η is very close to unity. In fact, the high frequency performance of this biquad is very close to that achieved with the good SAB's and certainly far superior to that exhibited by some of the state-variable biquads discussed in the next Section. The reason for the excellent high frequency performance is that the term $(1 - \frac{\omega_{t_2}}{\omega_{t_1}})$ in (9.92) vanishes when the amplifiers are matched. Matched op. amps. in dual package are readily available. The reader may recall that we have already observed this property of this GIC when we analyzed it as an inductance simulator in Chapter 8. Furthermore, if a mismatch exists in ω_{t_1} and ω_{t_2}, this can be easily corrected for by adjusting one of the resistors R' and R" in Fig. 9.28. This point was also explained in Chapter 8.

It should be obvious that initial adjustment or tuning of the circuit of Fig. 9.28 is quite easy, since Q is independently determined by the single resistor (QR).

The GIC-derived biquad has also been adapted to implement filtering functions other than the bandpass one[35-37]. The resulting circuits and their design equations are given in Fig. 9.29. Some brief comments on these circuits are:

<u>Circuit #1</u>. This bandpass circuit is based on the type-A GIC. The circuit features excellent high-frequency performance, if the two op. amps. are matched. As shown, the center-frequency gain is 2 but may be reduced by "splitting" the feed-in resistor QR.

<u>Circuit #2</u>. This high-pass circuit is based on the type-A GIC, with excellent high-frequency performance obtained for matched

op. amps. The gain can be reduced by "splitting" the input feed-in capacitor.

Circuit #3. This all-pass circuit is derived from the bandpass circuit #1 by interchanging the input and ground terminals[38]. The performance of this circuit is identical to that of circuits #1 and #2.

Circuit #4. This low-pass notch circuit is derived from the type-A GIC. Excellent high-frequency performance is obtained for matched op. amps.

Circuit #5. This high-pass notch circuit has the same topology as the type-A GIC but the component values are different. This variation, however, does not alter the excellent high-frequency performance obtained when the op. amps. are matched.

Circuit #6. This circuit is capable of realizing either a high-pass notch or a low-pass notch function. It is based on the type-B GIC topology. Unfortunately, the component values are such that the high-frequency performance is rather poor, especially if $\frac{\omega_n}{\omega_o}$ is much greater or much smaller than unity.

Circuit #7. This low-pass circuit is different from both GIC types A and B. Nevertheless, the high-frequency performance is excellent. The gain may be reduced from 2 by "splitting" the feed-in resistor R.

We should also note that it is possible to generate other circuits based on the type-B GIC at the expense of using an additional buffer amplifier[35,37]. The need for buffering arises because the type-B GIC circuits derive their useful output functions at internal high-impedance nodes.

Thus, of all the circuit classes considered the GIC derived biquad is the "optimum". Indeed, the performance indices for this class of biquads take on values that represent the state-of-the-art in biquad design.

The reader might wonder then about our reasons for seeking three-amplifier biquads. This is done mainly to achieve a structure that is more amenable to producing different types of filtering functions easily. Although feedforward techniques are applied in the present circuits to obtain different filtering functions, the design equations and hence the tuning becomes sli-

1. Bandpass

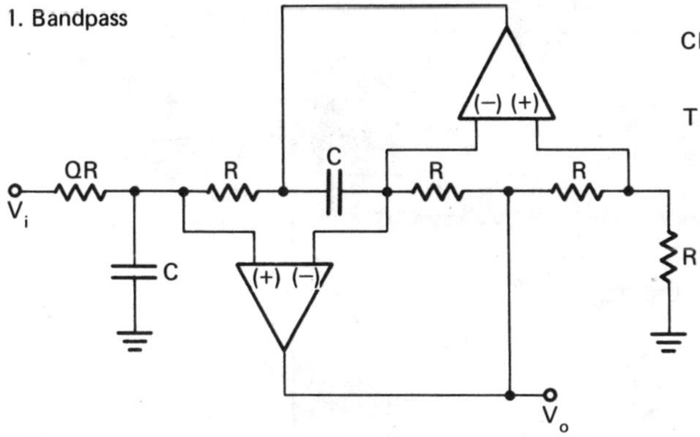

$$CR = 1/\omega_o$$

$$T(s) = \frac{s2(\omega_o/Q)}{s^2 + s(\omega_o/Q) + \omega_o^2}$$

2. High-pass

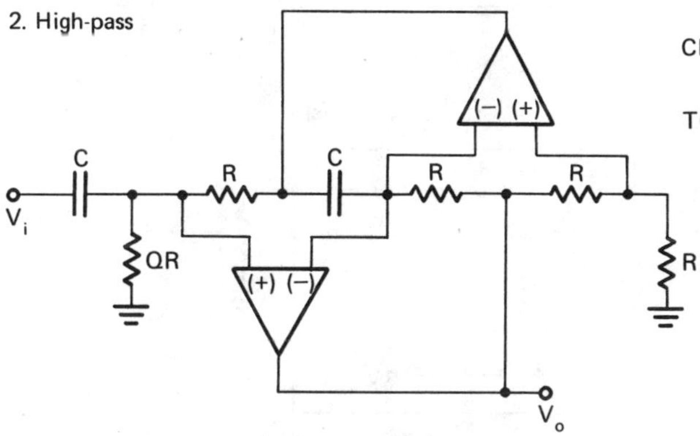

$$CR = 1/\omega_o$$

$$T(s) = \frac{2s^2}{s^2 + (\omega_o/Q) + \omega_o^2}$$

3. All-pass

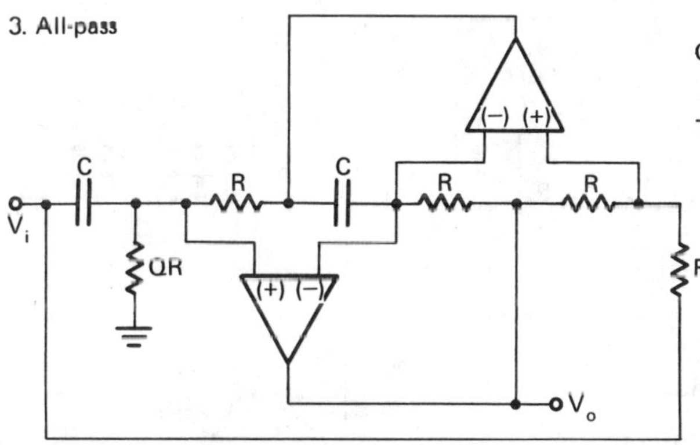

$$CR = 1/\omega_0$$

$$T(s) = \frac{s^2 - s(\omega_o/Q) + \omega_o^2}{s^2 + s(\omega_o/Q) + \omega_o^2}$$

4. Low-pass notch

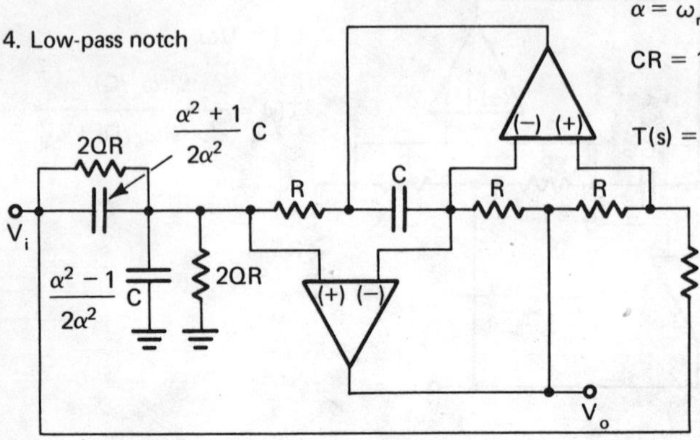

$$\alpha = \omega_n/\omega_o \geqslant 1$$

$$CR = 1/\omega_o$$

$$T(s) = (\omega_o/\omega_n)^2 \frac{s^2 + \omega_n^2}{s^2 + s(\omega_o/Q) + \omega_o^2}$$

5. High-pass Notch

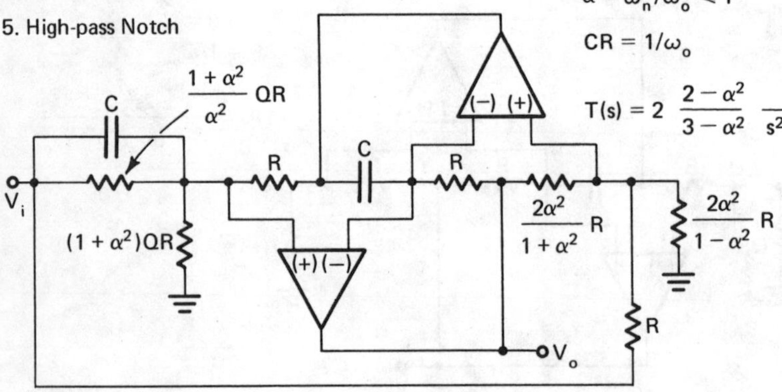

$$\alpha = \omega_n/\omega_o \leqslant 1$$

$$CR = 1/\omega_o$$

$$T(s) = 2\ \frac{2-\alpha^2}{3-\alpha^2}\ \frac{s^2 + \omega_n^2}{s^2 + s(\omega_o/Q) + \omega_o^2}$$

6. Low-pass Notch and High-pass Notch

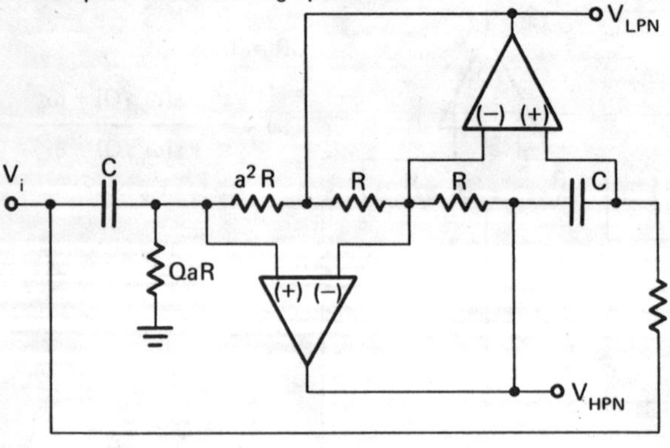

$$CR = 1/a\omega_o$$

$$a = Q|(\omega_n^2/\omega_o^2) - 1|$$

$$T(s) = \frac{s^2 + \omega_n^2}{s^2 + s(\omega_o/Q) + \omega_o^2}$$

7. Low-pass

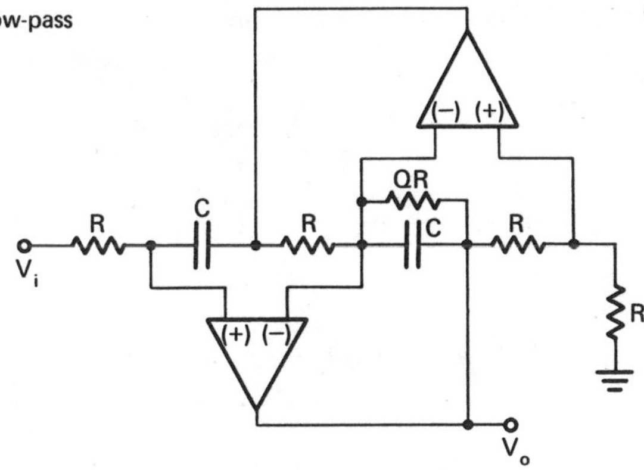

$CR = 1/\omega_o$

$$T(s) = \frac{2\omega_o^2}{s^2 + s(\omega_o/Q) + \omega_o^2}$$

Figure 9.29 Biquads based on the Antontiou GIC.

ghtly difficult. This is mainly due to the use of the op. amps. in a differential mode and the attendant absence of "handy" virtual grounds that make feedforward very simple to apply. Another, and perhaps more important reason, for using the three-amplifier biquads of the next section is their ability to provide different filtering functions at different terminals in the same circuit. This makes them viable as "universal filter modules".

9.9 THREE-AMPLIFIER BIQUADS

The biquads presented in this Section are based on the state-variable realization of transfer functions. Such realizations simulate the second-order transfer function, using integrators and summers. These methods have been used in analog computation for a long time and, in fact, our development here is based on analog simulation techniques.

Consider the transfer function

$$\frac{V_o}{V_i} = \frac{P(s)}{s^2 + s(\frac{\omega_o}{Q}) + \omega_o^2} \tag{9.94}$$

where P(s) denotes the transmission-zero polynomial. Cross multiplication of (9.94), and rearranging the terms such that only powers of $(\frac{1}{s})$ are present, obtains

$$V_o = -\frac{1}{Q}\left(\frac{V_o}{s/\omega_o}\right) - \frac{V_o}{(s^2/\omega_o^2)} + \frac{P(s)}{s^2} V_i \qquad (9.95)$$

Consider the case $P(s) = Ks^2$ and write $\frac{1}{\omega_o} = T$ then,

$$V_o = -\frac{1}{Q}\frac{V_o}{sT} - \frac{V_o}{s^2T^2} + KV_i \qquad (9.96)$$

Equation (9.96) can be implemented by the operational block diagram of Fig. 9.30. Note that since V_o is a high-pass function, then $(\frac{V_o}{sT})$, at the output of the first integrator, is a bandpass function, while $(\frac{V_o}{s^2T^2})$, at the output of the summer, is a lowpass function. The three filter functions are simultaneously available and all have the same ω_o and Q.

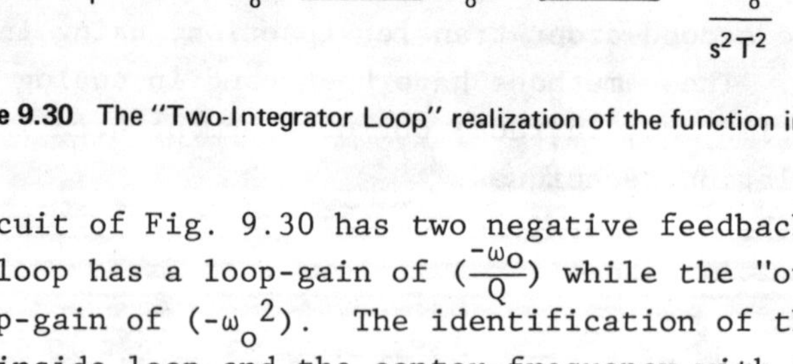

Figure 9.30 The "Two-Integrator Loop" realization of the function in (9.96).

The circuit of Fig. 9.30 has two negative feedback loops: the "inside" loop has a loop-gain of $(\frac{-\omega_o}{Q})$ while the "outside" loop has a loop-gain of $(-\omega_o^2)$. The identification of the bandwidth with the inside loop and the center-frequency with the outside loop is a powerful concept which has been employed to electronically vary the filter parameters, by controlling the appropriate loop gains[39].

All the circuits presented below are based on the "two-integrator loop" of Fig. 9.30. They differ only in the method of implementing the two integrators and the summer.

The KHN Biquad[40]

Since inverting integrators are readily implemented using the single op. amp. Miller circuit, it is natural to start by rearranging the signs of the operational blocks in Fig. 9.30, such that both integrators become inverting. The resulting circuit is shown in Fig. 9.31a, where we have also attached the appropriate sign with each of the output signals. A direct implementation of this block diagram consists of Miller integrators, and a summer having both positive and negative weights. The summer transfer function is described by

$$V_o = K\,V_i + \frac{1}{Q}\,V_{BP} - V_{LP} \tag{9.97}$$

An implementation for such a summer is shown in Fig. 9.31b and has the transfer characteristic

$$V_o = \frac{2R_1}{R_1 + R_2}\,V_i + V_{BP}\,\frac{2R_2}{R_1 + R_2} - V_{LP} \tag{9.98}$$

Comparison with (9.97) determines the ratio $\frac{R_1}{R_2}$ and the value of K to be,

$$\frac{R_1}{R_2} = 2\,Q - 1 \quad , \quad \text{and} \quad K = 2\,(1 - \frac{1}{2Q})$$

The complete circuit is given in Fig. 9.31c. As shown, the three special filter functions, HP, BP and LP, are simultaneously available. This circuit was first proposed by Kerwin, Huelsman and Newcomb[40] and is usually referred to in the literature as the KHN biquad. Note that if other values of gain are required, one may rescale the component values while keeping the loop-gains constant.

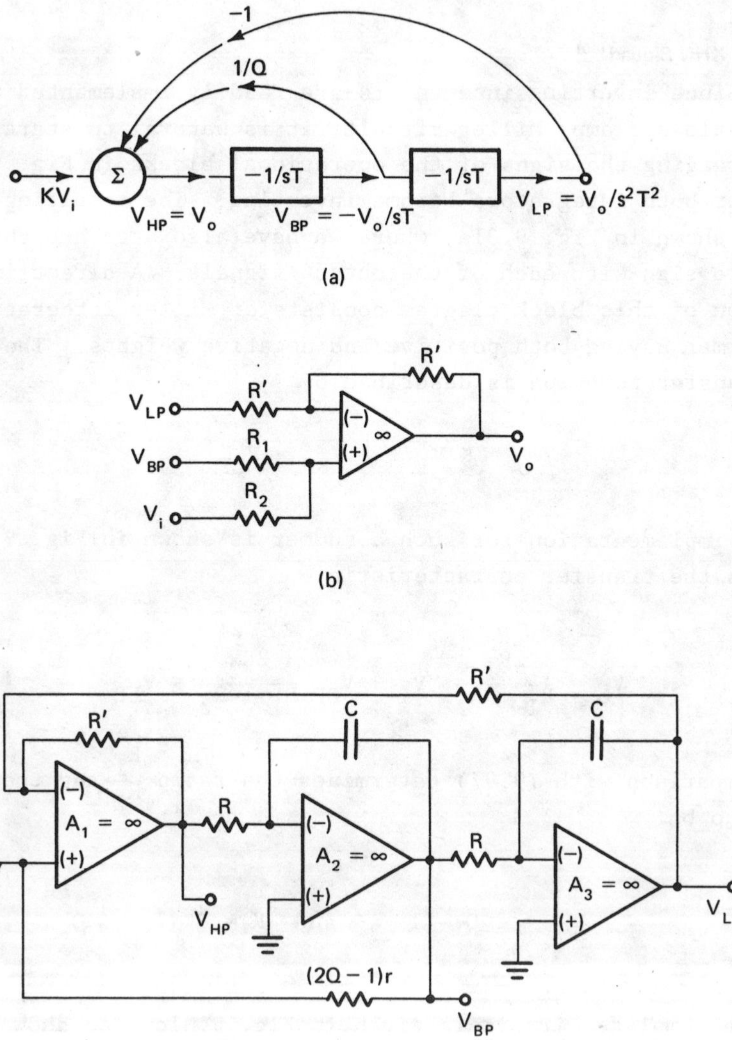

Figure 9.31 Derivation of the KHN state-variable biquad shown complete in (c).

The KHN biquad exhibits low passive and active ω_o-sensitivities, low passive Q-sensitivities and Q gain-sensitivity-products of the order of Q. It suffers, however, from severe Q-enhancement effects, due to the limited bandwidth of the amplifiers. All of these properties will be illustrated below on a slightly different form of the three-amplifier biquad. This latter circuit has the advantage of using the three amplifiers with their positive input terminals grounded.

The Tow-Thomas Biquad[41–43]

To derive the alternative three amplifier biquad, we refer the reader to the block diagram of Fig. 9.31a and note that if an inverting amplifier is added in the outer feedback path, then all the weights associated with the input summer will have positive polarity as illustrated in Fig. 9.32a. We may, therefore, perform the input summing at the virtual ground of the input integrator and obtain the circuit shown in Fig. 9.32b. This three-amplifier circuit was first proposed by Tow[41] and later developed by Thomas[42,43], and referred to as "The Biquad".

The Thomas biquad has the advantage that each of the amplifiers has a grounded input. Also, Q is determined by a resistor QR and is thus easily adjustable. Moreover, the gain K is determined by the input resistor and can be easily adjusted without affecting Q or ω_o. Compared to the KHN biquad, Thomas' biquad does not provide a high-pass output. This has been lost in the process of rearranging the block diagram. From a sensitivity and high-frequency performance point of view, both biquads are almost identical.

Sensitivity and High-Frequency Performance

The passive sensitivities of the Thomas biquad can be easily evaluated and shown to be at their theoretical minima. To eval-

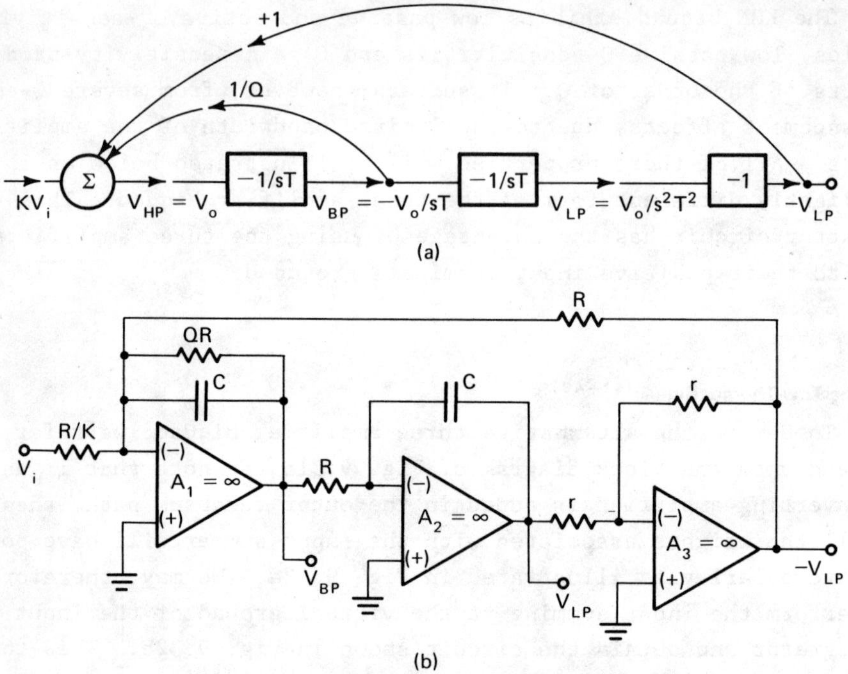

Figure 9.32 Derivation of the Tow-Thomas state-variable biquad in (b).

$$\frac{V_{BP}}{V_i}(s) = -\frac{sK\omega_o}{s^2 + s(\omega_o/Q) + \omega_o^2}, \quad \omega_o = 1/CR$$

uate the active sensitivities, we assume each op. amp. to have a finite gain and find the transfer function T(s). For K = 1, the denominator polynomial is given by

$$D_v(s) = s^2 + s\,\frac{\omega_o}{Q}\,\frac{1 + \frac{2Q+1}{A_1} + \frac{Q+1}{A_2} + \frac{3Q+1}{A_1A_2}}{1 + \frac{1}{A_1} + \frac{1}{A_2} + \frac{1}{A_1A_2}} + \omega_o^2\,\frac{\frac{1}{1+\frac{2}{A_3}} + \frac{2}{A_1A_2} + \frac{1}{QA_2} + \frac{1}{QA_1A_2}}{1 + \frac{1}{A_1} + \frac{1}{A_2} + \frac{1}{A_1A_2}} \tag{9.99}$$

where ω_o and Q denote their nominal values. Assuming real values for all the gains, we can find expressions for the actual values of ω_o and Q, that is ω_o' and Q', from (9.99). Hence, we may evaluate the various gain-sensitivity products. Assuming $Q >> 1$ and $A_i >> Q$, we find that

$$GS^Q_{A_1} \simeq 2Q + 1 \tag{9.100}$$

$$GS^Q_{A_2} \simeq Q + 1 \tag{9.101}$$

$GS^Q_{A_3}$ and all $GS^{\omega_o}_{A_i}$ are low (that is, around unity).

Thus, the three amplifier biquad has low passive sensitivities, and Q gain-sensitivity products of the order of Q. As mentioned before, this is state-of-the-art performance.

To evaluate the high-frequency limitations, we use the single-pole model of the op. amp. and substitute for each A_i in (9.99) by

$$\frac{1}{A_i} \simeq \frac{s}{\omega_{t_i}} \text{ , for } \omega >> \frac{\omega_{t_i}}{A_{o_i}} \text{ ; the range of interest.}$$

After some manipulations, in which it is assumed that $\frac{\omega_o}{\omega_{t_i}} << 1$ and that $Q >> 1$, such that terms of the form $(\frac{\omega_o}{\omega_{t_1}})(\frac{\omega_o}{\omega_{t_2}})$ and $(\frac{\omega_o}{\omega_{t_1}})(\frac{\omega_o}{\omega_{t_2}})(\frac{\omega_o}{\omega_{t_3}})$ are neglected with respect to unity, the fifth-order polynomial reduces to a third-order one given by

$$D_V(s) \simeq \frac{s^3}{\omega_o}\left(\frac{\omega_o}{\omega_{t_1}} + \frac{\omega_o}{\omega_{t_2}} + 2\frac{\omega_o}{\omega_{t_3}}\right) + s^2\left(1 + 2\frac{\omega_o}{\omega_{t_1}} + \frac{\omega_o}{\omega_{t_2}}\right) + s\frac{\omega_o}{Q}\left(1 + \frac{\omega_o}{\omega_{t_2}}\right) + \omega_o^2 \tag{9.102}$$

Thus, the filter acquires a third parasitic pole in addition to the shift in the dominant pole-pair, which is now characterized by ω_{oa} and Q_a. Equation (9.102) may be factored as

$$D_v(s) = [\frac{s}{\omega_o} (\frac{\omega_o}{\omega_{t_1}} + \frac{\omega_o}{\omega_{t_2}} + 2 \frac{\omega_o}{\omega_{t_3}}) + \frac{\omega_o^2}{\omega_{oa}^2}] [s^2 + s \frac{\omega_{oa}}{Q_a} + \omega_{oa}^2] \tag{9.103}$$

Equating the coefficients of the s^2 terms in (9.102) and (9.103), and assuming that

$$\delta \equiv \frac{\omega_{oa} - \omega_o}{\omega_o} \text{ is } << 1 \text{ and } Q_a >> 1, \text{ then}$$

$$\delta \simeq - (\frac{\omega_o}{\omega_{t_1}} + \frac{1}{2} \frac{\omega_o}{\omega_{t_2}}) \tag{9.104}$$

which by using (9.100) and (9.101), can be written in the form

$$\delta \simeq - \frac{1}{2Q} \sum_{i=1}^{3} GS_{A_i}^{Q} (\frac{\omega_o}{\omega_{t_i}}) , \tag{9.105}$$

the general form for the frequency deviation of any biquad circuit presented in this book. For identical amplifiers, (9.104) reduces to

$$\delta \simeq - \frac{3}{2} (\frac{\omega_o}{\omega_t})$$

To obtain Q_a, or alternatively the Q-variation factor $\eta \equiv \frac{Q_a}{Q}$, we equate the coefficients of the s-terms in (9.102) and (9.103) and carry out appropriate approximations to obtain

$$\eta \simeq 1 / [1 - Q (\frac{\omega_o}{\omega_{t_1}} + \frac{\omega_o}{\omega_{t_2}} + 2 \frac{\omega_o}{\omega_{t_3}})] \tag{9.106}$$

which for identical amplifiers reduces to

$$\eta \simeq 1 / [1 - 4Q (\frac{\omega_o}{\omega_t})] \qquad (9.107)$$

Equation (9.107) clearly illustrates the drastic Q-enhancement effect encountered in the standard three-amplifier biquad. In the following, we shall seek methods for compensating the biquad in order to reduce Q-enhancement. Before embarking on this task, we invoke the notion of integrator Q in order to pinpoint the cause of Q enhancement.

The concept of integrator Q-factor was introduced in Chapter 8. Specifically, if an integrator circuit has the transfer function

$$\frac{V_o}{V_i} = \frac{1}{R(\omega) + jX(\omega)} ,$$

then we define the integrator-Q as

$$Q_I = \frac{X(\omega)}{R(\omega)}$$

In the present application we will be interested in the value of Q_I evaluated at the pole-frequency of the biquad, ω_o.

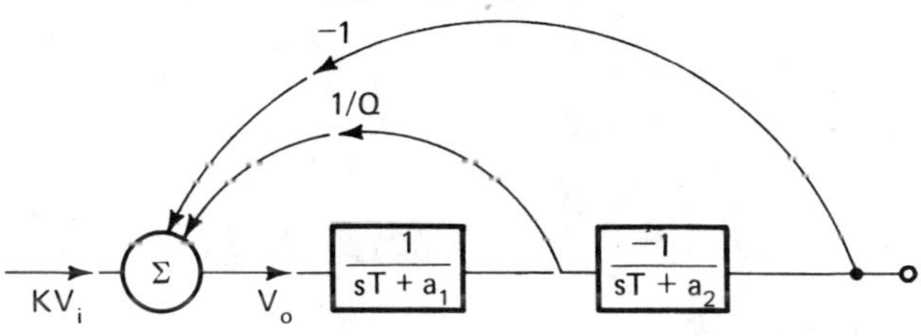

Figure 9.33 The two-integrator loop with the integrators having finite Q-factors.

Let us first evaluate the effects of finite integrator Q's on the pole-Q in a two-integrator loop. Fig. 9.33 shows a two-integrator loop where each integrator is assumed to undergo negligible change in its time constant, but a real dissipation term is introduced making the integrator Q's finite and given by

$$Q_{I1} = \frac{\omega T}{a_1}$$

$$Q_{I2} = \frac{\omega T}{a_2}$$

At the center frequency $\omega_o = \frac{1}{T}$, the Q's are

$$Q_{I1} = \frac{1}{a_1} \tag{9.108}$$

$$Q_{I2} = \frac{1}{a_2} \tag{9.109}$$

Assuming the summer to be ideal, or alternatively that its non-ideal effects are lumped inside the integrators, then the characteristic equation of the two-integrator loop is given by

$$D_V = s^2 + s\,\frac{1}{T}\,(a_1 + a_2 + \frac{1}{Q}) + \frac{1}{T^2}\,(1 + a_1 a_2 + \frac{a_2}{Q})$$

The actual realized Q is given by

$$Q_a = \frac{\sqrt{1 + a_1 a_2 + (a_2/Q)}}{a_1 + a_2 + \frac{1}{Q}}$$

$$\simeq Q \;/\; [1 + Q\,(a_1 + a_2)]$$

Thus,

$$\eta \equiv \frac{Q_a}{Q} = 1 \;/\; [1 + Q\,(a_1 + a_2)]$$

Using equations (9.108) and (9.109) results in

$$\eta = 1 / [1 + Q (\frac{1}{Q_{I_1}} + \frac{1}{Q_{I_2}})] \qquad (9.110)$$

which applies to any two-integrator loop.

In the three-amplifier biquad of Fig. 9.32b, we have two integrators: the first is an inverting one implemented by amplifier A_1, while the second is a noninverting integrator implemented by the cascade connection of the inverting integrator A_2 and the inverting amplifier A_3. For the first integrator, the Q-factor, evaluated at ω_o, is given by

$$Q_{I_1} = -\frac{\omega_{t_1}}{\omega_o} \qquad (9.111)$$

while, from the analysis in Chapter 8, the Q-factor of the Miller-inverter cascade is given by,

$$Q_{I_2} \simeq -1 / (\frac{\omega_o}{\omega_{t_2}} + \frac{2\omega_o}{\omega_{t_3}}) \qquad (9.112)$$

which for $\omega_{t_2} = \omega_{t_3} = \omega_t$ reduces to

$$Q_{I_2} = -\frac{1}{3} (\frac{\omega_t}{\omega_o}) \qquad (9.113)$$

which is one third the value of Q_{I_1} but having the same negative sign. It should be clear from (9.113) that the addition of the inverting amplifier A_3 causes a substantial reduction in the magnitude of Q_{I_2}.

Substituting (9.111) and (9.112) in (9.110) results in the Q variation factor η given in (9.106), which for identical amplifiers reduces to that in (9.107). Another point worth noting concerns the parasitic pole, or poles, associated with each integrator. In Chapter 8, we analysed the inverting Miller integrator and found that it has a parasitic real pole at a frequency ω_t, while the Miller-inverter cascade has two parasitic real poles: one caused by the Miller integrator at $\omega = \omega_t$, and the other caused by the cascade inverting amplifier at $\omega = \frac{\omega_t}{2}$. Each parasitic pole adds a corresponding amount of phase lag, with the total amount of excess phase lag being directly responsible for the Q-enhancement.

Passive Compensation for the Q-Enhancement

An approach to the compensation of the Q-enhancement consists of the addition of an amount of phase lead that cancels the excess phase lag. This process has already been applied to the compensation of a single Miller integrator in Chapter 8. There, we have added a compensating capacitor C_c in parallel with R, with the value C_c given by

$$C_c \simeq \frac{1}{\omega_t R} \tag{9.114}$$

As explained in Chapter 8, rather than thinking of compensation in terms of a pole-zero cancellation process, we may interpret the effects in terms of phase. The inverting integrator has an associated excess phase lag, determined by the parasitic pole at ω_t, and thus given by

$$\phi \simeq \frac{\omega}{\omega_t} \quad \text{for } \omega << \omega_t$$

The addition of a compensating capacitor creates a zero at $\frac{1}{C_c R}$ and hence a corresponding phase lead of $\phi_c \simeq \omega C_c R$, for small angle ϕ_c. To obtain compensation, we force $\phi_c = \phi$ which results in (9.114).

Returning to the three-amplifier biquad, the excess phase lag due to both the inverting and noninverting integrators is given by

$$\phi \simeq \frac{\omega}{\omega_{t_1}} + \frac{\omega}{\omega_{t_2}} + \frac{2\omega}{\omega_{t_3}}$$

The addition of a compensating capacitor C_c in parallel with one of the integrator resistors, R, provides a phase lead of

$$\phi_c \simeq \omega\, C_c\, R$$

For compensation, we equate ϕ_c to ϕ and obtain

$$C_c \simeq \frac{1}{R}\left(\frac{1}{\omega_{t_1}} + \frac{1}{\omega_{t_2}} + \frac{2}{\omega_{t_3}}\right)$$

For identical amplifiers, we have

$$C_c \simeq \frac{4}{\omega_t R}$$

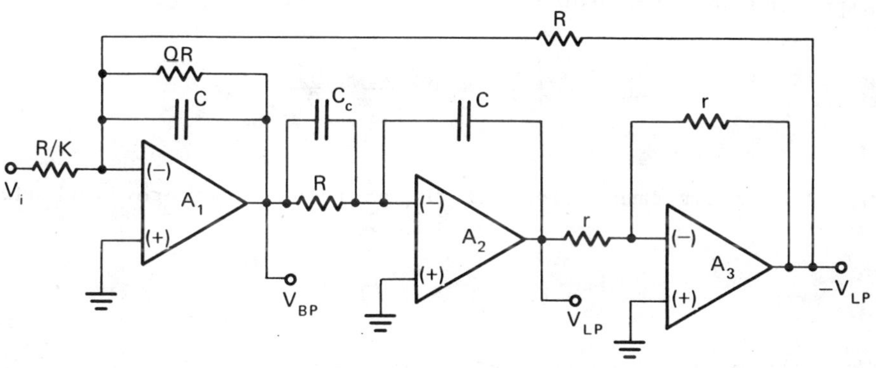

Figure 9.34 The Thomas biquad with passive compensation for the Q-enhancement. $C_c \cong 4/R\omega_t$

The biquad circuit is redrawn in Fig. 9.34, together with the compensating capacitor. This method of compensation is referred to as "Passive Compensation" and was first suggested in [42].

The technique is quite simple to apply and has been effectively employed in the design of programmable filters [39]. There is one serious drawback, however, with passive compensation. As ω_t is both temperature and power supply dependent, it is difficult to maintain compensation with changes in these parameters. For this reason, we seek an alternative compensation scheme, referred to as "Active Compensation" [44].

Active Compensation and the Akerberg-Mossberg Biquad

The availability of integrated circuit packages containing a number of matched op. amps. on the same silicon chip suggest the use of the matching and tracking properties to obtain circuit configurations with much wider bandwidth. This technique has already been discussed in Chapter 8 and called "Active Compensation". Although the active compensation method is quite general and can be applied to a variety of op. amp. circuits, we will be interested here in its application to the three-amplifier biquad.

Recalling that the Q of an inverting integrator is negative in sign, and that the Q-variation factor η is given by

$$\eta \simeq 1 / [1 + Q (\frac{1}{Q_{I_1}} + \frac{1}{Q_{I_2}})] ,$$

we seek an implementation for the positive integrator such that

$$Q_{I_2} = - Q_{I_1}$$

This obviously leads to a value of η very close to unity. Such an implementation of a noninverting integrator was given in Chapter 8 and is redrawn in Fig. 9.35. Comparing the latter circuit with the one previously used in the biquad [that is the Miller-inverter-cascade integrator], we note that both have the same number of amplifiers. Rather than placing the inverting amplifier A_3 in cascade, the compensated integrator has the inverter inside the feedback loop. For this reason, the positive input of A_2 is now used as the summing junction. The Q-factor of this modified integrator has been derived in Chapter 8 and is given by

$$Q_{I_2} \simeq 1 / [\frac{2\omega}{\omega_{t_3}} - \frac{\omega}{\omega_{t_2}}] \quad (9.115)$$

For matched amplifiers (9.115) reduces to

$$Q_{I_2} \simeq \frac{\omega_t}{\omega} \quad (9.116)$$

which is equal in magnitude and opposite in sign to the Q factor of the inverting integrator, the desired result.

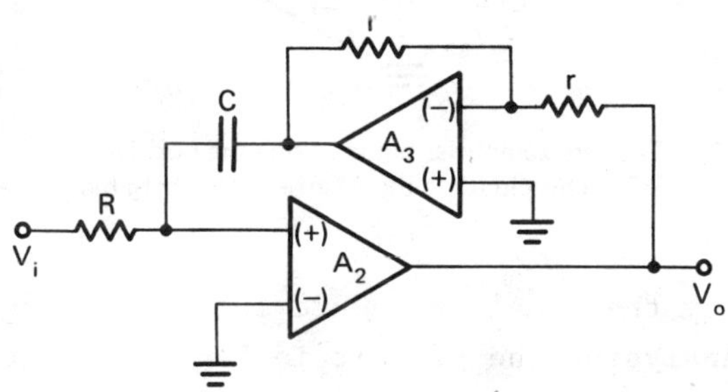

Figure 9.35 A positive integrator using the "Active Compensation" technique.

If we now use the modified positive integrator to replace the Miller-Inverter cascade integrator in Thomas biquad, the three-amplifier biquad of Fig. 9.36 is obtained. This circuit was originally introduced by Akerberg and Mossberg[45,46] and we will therefore refer to it as the AM biquad. It is interesting, however, to note that the Akerberg and Mossberg derivation of this biquad is quite different than the one presented above.

The values of Q_{I_1} and Q_{I_2} derived in the above may now be used to find an approximate expression for η in the AM biquad,

$$\eta \simeq 1 \;/\; [1 + Q\,(\frac{2\omega_o}{\omega_{t_3}} - \frac{\omega_o}{\omega_{t_2}} - \frac{\omega_o}{\omega_{t_1}})] \qquad (9.117)$$

which for $\omega_{t_1} = \omega_{t_2} = \omega_{t_3}$ results in η approximately unity! As already mentioned, dual and quad op. amps. which have matched characteristics and whose ω_t values track with changes in environmental conditions are readily available.

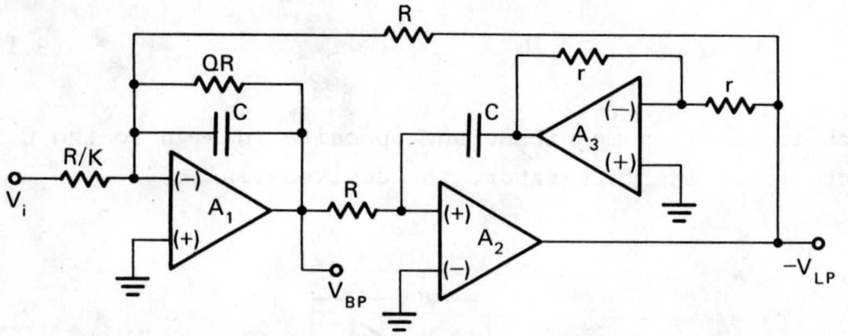

Figure 9.36 The three-amplifier biquad using the positive integrator of fig. 9.35. This circuit is the Akerberg-Mossberg biquad (AM).

We now employ the usual procedure for evaluating δ and η of the AM biquad. Analysing the circuit in Fig. 9.36 with K = 1 and using the single-pole model for each op. amp. results in the fifth-order denominator polynomial,

$$D_V(s) \simeq s^5 \frac{2}{\omega_{t_1}\omega_{t_2}\omega_{t_3}} + s^4 \frac{1}{\omega_{t_1}\omega_{t_2}} (1 + 2\frac{\omega_{t_1}}{\omega_{t_3}})$$

$$+ s^3 \frac{1}{\omega_{t_1}} (1 + \frac{\omega_{t_1}}{\omega_{t_2}}) + s^2 (1 + \frac{\omega_o}{\omega_{t_2}} + 2\frac{\omega_o}{\omega_{t_1}})$$

$$+ s\frac{\omega_o}{Q} (1 + 2Q\frac{\omega_o}{\omega_{t_3}}) + \omega_o^2 \qquad (9.118)$$

where we have assumed $Q>>1$ and $\omega_o << \omega_{t_1}, \omega_{t_2}, \omega_{t_3}$. Examination of (9.118) reveals the welcome fact that both the fifth and fourth order terms may be neglected, leading to

$$D_V(s) \simeq s^3 \frac{1}{\omega_{t_1}} (1 + \frac{\omega_{t_1}}{\omega_{t_2}}) + s^2 (1 + 2 \frac{\omega_o}{\omega_{t_1}} + \frac{\omega_o}{\omega_{t_2}})$$

$$+ s \frac{\omega_o}{Q} (1 + 2Q \frac{\omega_o}{\omega_{t_3}}) + \omega_o^2 \tag{9.119}$$

Equation (9.119) may now be used to find δ and η with the procedure identical to that already outlined for the Thomas biquad. The results are

$$\delta \simeq - (\frac{\omega_o}{\omega_{t_1}} + \frac{1}{2} \frac{\omega_o}{\omega_{t_2}}) \tag{9.120}$$

$$\eta \simeq 1 / (1 + \delta) [1 + Q (\frac{2\omega_o}{\omega_{t_3}} - \frac{\omega_o}{\omega_{t_2}} - \frac{\omega_o}{\omega_{t_1}}) - 2\delta Q \frac{\omega_o}{\omega_{t_1}} (1 + \frac{\omega_{t_1}}{\omega_{t_2}})] \tag{9.121}$$

The expression for δ also follows from the general form (9.105) where $GS^Q_{A_1} \simeq 2Q$, $GS^Q_{A_2} \simeq Q$ and $GS^Q_{A_3}$ is low. Since δ is small, expression (9.121) for η reduces to the approximate expression in (9.117), derived using the integrator Q's. It follows from (9.121) that the Q-variation can be dramatically reduced if the amplifiers are matched. In the latter case, we have

$$\delta \simeq - \frac{3}{2} \frac{\omega_o}{\omega_t} \tag{9.122}$$

$$\eta \simeq 1 / (1+\delta) [1-4\delta Q \frac{\omega_o}{\omega_t}] \tag{9.123}$$

Table 9.5 Frequency limitations of biquads

Circuit	$\delta \equiv \frac{\Delta\omega_o}{\omega_o}$ and $\eta \equiv \frac{Q_a}{Q}$	
All Circuits (in this book)	$\delta \simeq -\frac{1}{2Q}\sum_i GS^Q_{A_i}\left(\frac{\omega_o}{\omega_{t_i}}\right)$	
SAB's	$\delta \simeq -\frac{1}{2Q} GS^Q\left(\frac{\omega_o}{\omega_t}\right)$ $\eta \simeq 1/[1+\delta-2K\delta Q\left(\frac{\omega_o}{\omega_t}\right)]$	Notes: * K=1 for NF, PF2 and PF4 Circuits. ** K=1 to 3 for PF1 and PF3 circuits.
Two-Amplifier Biquad derived from GIC (in BP configuration)	$\delta \simeq -\left(\frac{\omega_o}{\omega_{t1}}+\frac{\omega_o}{\omega_{t2}}\right)$ $\eta \simeq 1/[1-2Q\left(\frac{\omega_o}{\omega_{t2}}\right)\left(1-\frac{\omega_{t2}}{\omega_{t1}}+2\delta\right)](1+\delta)$	<u>For identical Amplifiers</u>: $\delta \simeq -2\left(\frac{\omega_o}{\omega_t}\right)$ $\eta \simeq 1/[1+\delta-4\delta Q\left(\frac{\omega_o}{\omega_t}\right)]$
Thomas Biquad	$\delta \simeq -\left(\frac{\omega_o}{\omega_{t1}}+\frac{1}{2}\frac{\omega_o}{\omega_{t2}}\right)$ $\eta \simeq 1/[1-Q\left(\frac{\omega_o}{\omega_{t1}}+\frac{\omega_o}{\omega_{t2}}+2\frac{\omega_o}{\omega_{t2}}\right)]$	<u>For identical Amplifiers</u>: $\delta \simeq -\frac{3}{2}\left(\frac{\omega_o}{\omega_t}\right)$ $\eta \simeq 1/[1-4Q\left(\frac{\omega_o}{\omega_t}\right)]$
Akerberg-Mossberg Biquad	$\delta \simeq -\left(\frac{\omega_o}{\omega_{t1}}+\frac{1}{2}\frac{\omega_o}{\omega_{t2}}\right)$ $\eta \simeq 1/(1+\delta)\ [1+Q\left(\frac{2\omega_o}{\omega_{t3}}-\frac{\omega_o}{\omega_{t2}}-\frac{\omega_o}{\omega_{t1}}\right)-2\delta Q\frac{\omega_o}{\omega_{t1}}\left(1+\frac{\omega_{t1}}{\omega_{t2}}\right)]$	<u>For identical Amplifiers</u>: $\delta \simeq -\frac{3}{2}\left(\frac{\omega_o}{\omega_t}\right)$ $\eta \simeq 1/[1+\delta-4\delta Q\left(\frac{\omega_o}{\omega_t}\right)]$

Note that the value of η is close to that obtained in some SABs and in the two-amplifier circuit derived from the Antoniou GIC. For comparison purposes, Table 9.5 provides the high-frequency limitations formulae for some of the biquad implementations presented in this Chapter.

Realization of Finite Transmission Zeros

The KHN biquad realizes the LP, BP and HP functions, while the Thomas and the AM biquads realize the BP and LP functions. To obtain arbitrary transmission zeros, either the summation method or the feedforward method can be used. The summation method has been quite popular, especially with the KHN biquad. It consists of forming the weighted sum of the LP, BP and HP outputs using a fourth summing amplifier. This technique is illustrated in Fig. 9.37. Formulae for the values of the summing resistors have been derived for the Thomas biquad[47].

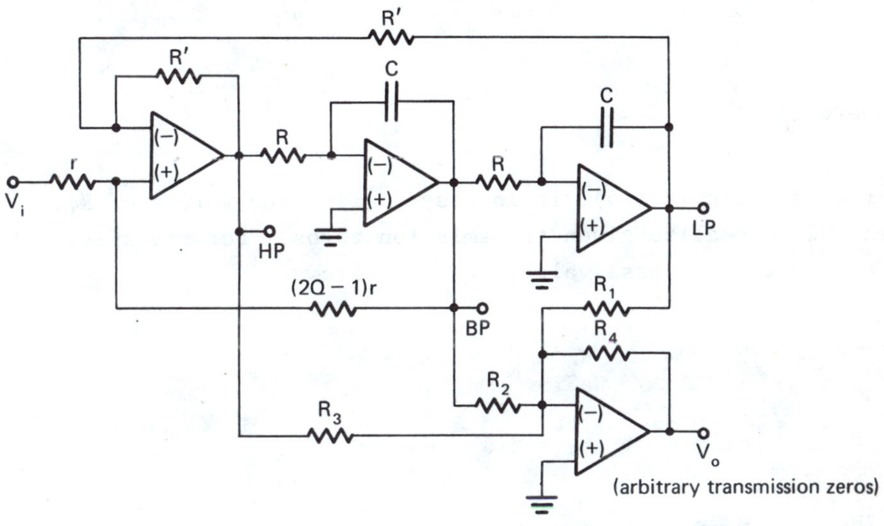

Figure 9.37 The KHN biquad with a summing amplifier used to obtain arbitrary transmission zeros.

The feedforward method utilizes the available virtual grounds as summing nodes and thus does not require any additional amplifiers. In the following, we develop the design equations for the Thomas biquad with feedforward[48]. Similar equations can be derived for the AM biquad and are left to the reader.

Fig. 9.38 shows the Thomas biquad with the order of the inverter and the second integrator interchanged, and with the input signal V_i applied to the three virtual grounds through the "feed-in" resistors R_1, R_2 and R_3. As should be expected, the values of the "feed-in" resistors will determine the location of the pair of zeros. Note that all the other circuit components have the same values as before. In particular, the time constants of the two integrators are equal, a desirable feature for minimum sensitivity.

Analysis of the circuit in Fig. 9.38 results in the transfer function

$$T(s) \equiv \frac{V_o}{V_i} = - \frac{\frac{r}{R_3} s^2 + s\left(\frac{\omega_o}{Q}\right)\left(\frac{r}{R_3} - Q\frac{R}{R_1}\right) + \omega_o^2\left(\frac{R}{R_2}\right)}{s^2 + s\left(\frac{\omega_o}{Q}\right) + \omega_o^2} \tag{9.124}$$

where $\omega_o = \frac{1}{CR}$.

From equation (9.124), it is easy to find the values of R_1, R_2 and R_3 to realize given transmission zeros. For the special filter functions, these values are as follows:

LP:	$R_1 = \infty$	$R_2 = \frac{R}{K}$	$R_3 = \infty$
BP:	$R_1 = \frac{QR}{K}$	$R_2 = \infty$	$R_3 = \infty$
HP:	$R_1 = \frac{QR}{K}$	$R_2 = \infty$	$R_3 = \frac{r}{K}$
Notch (any kind):	$R_1 = \frac{QR}{K}$	$R_2 = \frac{R}{K}\left(\frac{\omega_o}{\omega_n}\right)^2$	$R_3 = \frac{r}{K}$
All-Pass:	$R_1 = \frac{QR}{2K}$	$R_2 = \frac{R}{K}$	$R_3 = \frac{r}{K}$

Note that K is a multiplicative constant that can be chosen according to the required gain.

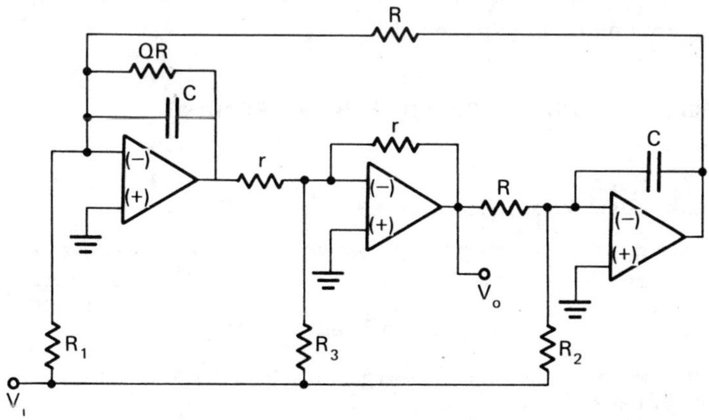

Figure 9.38 The Thomas biquad with feedforward to realize finite transmission zeros.

A disadvantage of this feedforward scheme is that notches are realized through signal subtractions, and hence tight component tolerances are required. A modified feedforward arrangement, that circumvents this problem through the use of an additional capacitor, is shown in Fig. 9.39.

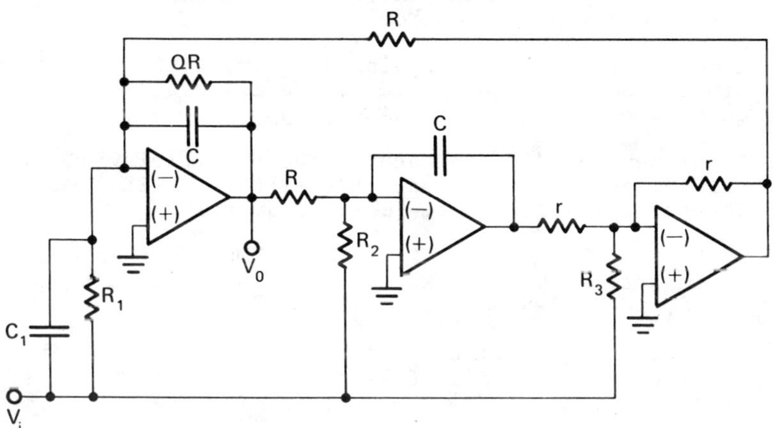

Figure 9.39 The Thomas biquad with an improved feedforward arrangement for realizing notch filters.

Finally, we should mention that the KHN biquad has been modified in order to obtain active compensation for the Q-enhancement. The resulting circuit and other design details of three-amplifier biquads are given in[37].

9.10 REFERENCES AND SOURCES FOR FURTHER READING

(1) S. K. Mirta (Ed.), "Active Inductorless Filters," IEEE Press, New York, 1971

(2) L. P. Huelsman, Ed., Active RC Filters: Theory and Application, in Benchmark Papers in Electrical Engineering and Computer Science, Vol. 15, Dowden, Hutchinson and Ross, Inc., Stroudsbury, Pennsylvania, 1976.

(3) S. K. Mitra, "Filter design using integrated operational amplifiers," 1969 WESCON Tech. papers, Session 4. Reprinted in (1).

(4) G. Daryanani, Principles of Active Network Synthesis and Design, Wiley, New York, 1976.

(5) A. Sedra, "Generation and Classification of Single Amplifier Filters," International Filter Symp., Santa Monica, Calif., Digest, pp.10-11, April, 1972.

(6) A. Sedra, "Generation and classification of single amplifier filters," Int. J. Circuit Theory and Applications, Vol. 2, pp. 51-67, 1974.

(7) G. S. Moschytz, Linear Integrated Networks: Design, Van Nostrand-Reinhold, New York, 1975.

(8) A. S. Sedra and J. L. Espinoza, "Sensitivity and frequency limitations of biquadratic active filters," IEEE Trans. Circuits and Systems, Vol. CAS-22, pp. 122-130, Feb. 1975.

(9) D. Hilberman, "An approach to the sensitivity and statistical variability of biquadratic filters," IEEE Trans. Circuit Theory, Vol. CT-20, pp. 382-390, July 1973.

(10) A. L. Rosenblum and M. S. Ghausi, "Multiparameter sensitivity in active RC networks," IEEE Trans. Circuit Theory, Vol. CT-18, pp. 592-599, Nov. 1971.

(11) B. A. Shenoi, "Optimum variability design & comparative evaluation of thin-film RC active filter", IEEE Trans. Circuits & Systems, Vol. CAS-21, pp. 263-267, March, 1974.

(12) M. A. Soderstrand and S. K. Mitra, "Active RC filters with zero gain-sensitivity product," IEEE Trans. Circuit Theory, Vol. CT-20, pp. 441-445, July, 1973.

(13) L. P. Huelsman, Theory and Design of Active RC Circuits, McGraw-Hill, New York, 1968.

(14) T. A. Hamilton and A. S. Sedra, "Some new configurations for active filters," IEEE Trans. Circuit Theory, Vol. CT-19, pp. 25-33, Jan. 1972.

(15) R. P. Sallen and E. L. Key, "A practical method of designing RC active filters." IRE Trans. Circuit Theory, Vol. CT-2, pp. 74-85, Mar. 1955. Reprinted in (1), (2).

(16) W. J. Kerwin, "Active RC network synthesis using voltage amplifiers," Chapter 2 in Active Filters: Lumped, Distributed, Integrated, Digital, and Parametric, L.P.Huelsman, Ed., McGraw-Hill, New York, 1970.

(17) T. Deliyannis, "RC active all-pass sections," Electron Lett. Vol. 5, pp. 59-60, 1969.

(18) W. J. Kerwin and C. V. Shaffer, "An integrable IF amplifier of high stability," Proc. 11th Midwest Symp. Circuit Theory, Notre Dame, Ind., pp. 78-88, 1968.

(19) M. A. Soderstrand and S. K. Mitra, "Sensitivity analysis of third-order filters," Int. J. Electron., Vol. 30, pp. 265-272, 1971.

(20) T. A. Hamilton and A. S. Sedra, "A single amplifier biquad active filter," IEEE Trans. Circuit Theory, Vol. CT-19, pp. 398-403, July 1972.

(21) A. S. Sedra and W. L. Brown, "A simple all-pass network with complex poles and zeros," IEEE Trans. Circuit Theory, Vol. CT-20, pp. 445-446, 1973.

(22) G. C. Bown, "Sensitivity in active filters," Electron. Lett. Vol. 3, pp. 298-299, 1967.

(23) W. Saraga, "Sensitivity of 2nd-order Sallen-Key active RC filters," Electron. Lett., Vol. 3, pp. 442-443, Oct. 1967.

(24) G. S. Moschytz, "A universal low-Q active filter building block suitable for hybrid-integrated circuit implementation," IEEE Trans. Circuit Theory, Vol. CT-20, pp. 37-47, Jan. 1973, Reprinted in (2).

(25) A. S. Sedra, "A class of stable active filters using unity-gain voltage followers," IEEE J. Solid-State Circuits, Vol. SC-7, pp. 311-315, Aug. 1972. Reprinted in (2).

(26) W. J. Kerwin and L. P. Huelsman, "The design of high performance active-RC bandpass filters," 1966 IEEE Int. Conv. Rec., Vol. 14, part 10, pp. 74-80. Reprinted in (1), (2).

(27) G. S. Moschytz, "Two-step precision tuning of twin-T notch filter," Proc. IEEE, Vol. 54, pp. 811-812, May 1966.

(28) G. S. Moschytz, "FEN filter design using tantalum and silicon integrated circuits," Proc. IEEE, Vol. 58, pp. 550-566, April 1970. Reprinted in (1).

(29) G. S. Moschytz and W. Thelen, "Design of hybrid-integrated filter building blocks," IEEE J. Solid-State Circuits, Vol. SC-5, pp. 99-107, 1970.

(30) A. S. Sedra and B. A. Shenoi, "Analysis of a class of mixed-feedback single amplifier filters," Unpublished report, Mar. 1974.

(31) J. J. Friend, "A single operational amplifier biquadratic filter section," presented at the 1970 IEEE Int. Sym. Circuit Theory, Atlanta, Ga., Dec. 1970.

(32) J. J. Friend, C. A. Harris and D. Hilberman, "STAR: an active biquadratic filter section," IEEE Trans. Circuits and Systems, Vol. CAS-22, pp. 115-121, Feb. 1975, Reprinted in (2).

(33) P. E. Fleischer, "Sensitivity minimization in a single amplifier biquad circuit", IEEE Trans. Circuits and Systems, Vol. CAS-23, pp. 45-55, Jan. 1976.

(34) T. A. Hamilton and A. S. Sedra, "A novel application of a gyrator-type circuit," in Fifth Asilomar Conf. Circuits and Systems, Conf. Rec., pp. 343-348, Nov. 1971.

(35) W. B. Mikhael, "Designs of high performance RC-active filters," Ph.D. thesis, Sir George Williams Univ., Montreal, Canada, May 1973.

(36) B. B. Bhattachariyya, W. B. Mikhael and A. Antoniou, "Design of RC-active networks by using generalized immittance converters," Proc. 1973 Int. Symp. on Circuit Theory, pp. 290-293, Toronto, Canada, April 1973.

(37) W. L. Brown, Contributions to state-of-the-art active filter design, MASc thesis, University of Toronto, 1977.

(38) D. Hilberman, "Input and ground as complements in active filters," IEEE Trans. Circuit Theory, Vol. CT-20, pp. 540-547, 1973.

(39) R. Sparkes and A. S. Sedra, "Programmable active filters," IEEE J. Solid-State Circuits, Vol. SC-7, pp. 93-95, Feb. 1973.

(40) W. J. Kerwin, L. P. Huelsman and R. W. Newcomb, "State variable synthesis for insensitive integrated circuit transfer

functions," IEEE J. Solid-State Circuits, Vol. SC-2, pp. 87-92, Sept. 1967. Reprinted in (1), (2).

(41) J. Tow, "Active RC filters--a state-space realization," Proc. IEEE, Vol. 56, pp. 1137-1139, 1968.

(42) L. C. Thomas, "The Biquad: Part 1--Some practical design considerations," IEEE Trans. Circuit Theory, Vol. CT-18, pp. 350-357, May 1971. Reprinted in (1), (2).

(43) L. C. Thomas, "The Biquad: Part II - A multipurpose active filtering system," IEEE Trans. Circuit Theory, Vol. CT-18, pp. 358-361, May 1971. Reprinted in (1), (2).

(44) P. O. Brackett and A. S. Sedra, "Active compensation for high-frequency effects in op. amp. circuits with applications to active-RC filters," IEEE Trans. Circuits and Systems, Vol. CAS-23, pp. 68-73, Feb. 1976.

(45) D. Akerberg and K. Mossberg, "Low sensitivity easily trimmed standard building block for active RC filters," Electron. Lett. Vol. 5, pp. 528-529, 1969. Reprinted in (2).

(46) D. Akerberg and K. Mossberg, "A versatile active RC building block with inherent compensation for the finite band width of the amplifier," IEEE Trans. Circuits and Systems, Vol. CAS-22, pp. 407-415, May 1975. Reprinted in (2).

(47) J. Tow, "Design formulae for active RC filters using operational-amplifier biquad," Electron. Lett. Vol. 5, pp. 339-341, 1969. Reprinted in (2).

(48) P. E. Fleischer and J. Tow, "Design formulae for biquad active filters using three operational amplifiers," Proc. IEEE, Vol. 61, pp. 662-663, May 1973.

(49) N. Fliege, "A new class of second-order RC-active filters with two operational amplifiers," Nachrichientech, Z., Vol. 26, pp. 279-282, 1973. Reprinted in (2).

(50) R. Tarmy and M. S. Ghausi, "Very high-Q insensitive active RC networks," IEEE Trans. Circuit Theory, Vol. CT-17, pp. 358-366, Aug. 1970. Reprinted in (1), (2).

PROBLEMS

P9.1 Use Fig. 9.6 to design an NF1 bandpass circuit with $\omega_o=10^3$ rad/sec, Q=5 and a center-frequency gain of 10. Give a circuit that uses a maximum of two capacitors. Assuming that the op amp has $\omega_t=10^6$ rad/sec, calculate the values of GS^Q, δ, η and the passive sensitivities.

P9.2 Use Fig. 9.6 to design an NF1 low-pass circuit with $\omega_{3dB}=1000$ rad/sec, $Q=1/\sqrt{2}$ and DC gain of unity.

P9.3 Use Fig. 9.6 to design an NF1 high-pass circuit with $\omega_o=1000$ rad/sec, $Q=1$ and a high-frequency gain of 2.

P9.4 Use Table 9.2 to design a low-pass notch circuit with $\omega_o=1000$ rad/sec, $\omega_n=2000$ rad/sec and $Q=2$. Give the value of the DC gain, the high-frequency gain and the maximum gain together with the frequency at which it occurs. Sketch $|T(j\omega)|$. (You may use results from Table 1.1).

P9.5 Use Table 9.2 to design a high-pass notch circuit with $\omega_n=1000$ rad/sec, $\omega_o=2000$ rad/sec and $Q=1$. Find the value of the DC gain, the high-frequency gain and the maximum gain together with the frequency at which it occurs. Sketch $|T(j\omega)|$. (You may use results from Table 1.1.)

P9.6 Use Table 9.2 to design an all-pass circuit with $\omega_o=10^4$ and $Q=2$. What is the value of the $|T(j\omega)|$?

P9.7 Consider a low-pass circuit with $Q=5$, generated from the ABT1 network of Fig. 9.6. Evaluate the values of q_2, k, γ, GS^{ω_o}, GS^Q, δ and η. Let $|A(j\omega_o)|\approx 500$. Also, evaluate all the passive ω_o and Q sensitivities.

P9.8 Consider the general NF configuration. Use the BT2 network for n_1 and a resistive divider for n_3 to design a circuit to realize a biquadratic transfer function with $\omega_{oz}=\omega_o=1$ rad/sec, $Q=1$, and (a) $Q_z=Q/2$, (b) $Q_z=2Q$. In each case give the value of the DC gain and sketch $|T(j\omega)|$.

P9.9 Design a third-order low-pass Butterworth filter using an NF1 circuit followed by an RC network. The filter is to have DC gain of unity and $\omega_{3dB}=10^4$ rad/sec.

P9.10 Design a bandpass circuit of the NF2 type to have $\omega_o=10^4$ rad/sec, Q=100 and center-frequency gain of 10. Calculate GS^Q, δ and η assuming that $\omega_t=10^7$ rad/sec.

P9.11 Use the ABTT network of Fig. 9.12 to generate a low-pass filter. Consider the case Q=100 and show that as the DC gain is increased the value of GS^Q increases. Give a design with a DC gain of unity and calculate GS^Q.

P9.12 Use the circuit of Fig. 9.13 to design a low-pass notch section with $\omega_o=1$ rad/sec, $\omega_n=2$ rad/sec, and Q=20. Give the value of GS^Q.

P9.13 Use Fig. 9.16 to obtain a bandpass PF1 circuit with Q=10 and $f_o=1$ KHz. Calculate the values of all performance indices. Assume the op amp to have $f_t=1$ MHz.

P9.14 It can be shown that interchanging the roles of the ground and input terminals in a circuit that has a transfer function T(s) results in a circuit whose transfer function is [1-T(s)].

(a) Show that this fact may be used to obtain an all-pass circuit given a bandpass circuit with a center frequency gain of 2.

(b) Use the ALD1 network of Fig. 9.16 to design a bandpass circuit with Q=1 and a center-frequency gain of 2 (For simplicity set m=1).

(c) Generate an all-pass circuit from the bandpass circuit in (b).

P9.15 (a) Show that for any SAB $S^{\omega_o}_{\omega_t}=-\delta$.

(b) Show that the per unit change in ω_o per $^\circ$C change in temperature, is given by

$$\frac{\Delta\omega_o}{\omega_o} = -(\alpha_\tau + \delta\alpha_t)$$

where α_τ denotes the temperature coefficient of the RC time constants and α_t denotes the temperature coefficient of ω_t.

(c) Consider a PF1 biquad generated from LD1. Find the value of m which will make ω_o insensitive to changes in temperature, assuming that α_τ=-100 ppm/^{o}C and α_t= - 3000 ppm/^{o}C. The op amp is assumed to have $|A(j\omega_o)| \approx 1000$.

P.9.16 Fig. P9.16 shows a method for the compensation for the excess phase lag of a voltage follower. The method consists of adding some phase lead to cancel the phase lag caused by the op amp pole. In this way, the useful frequency range of PF4 circuits can be considerably extended. Show that for $\omega \ll \omega_t$, first-order phase cancellation is obtained by selecting RC equal to $(1/\omega_t)$.

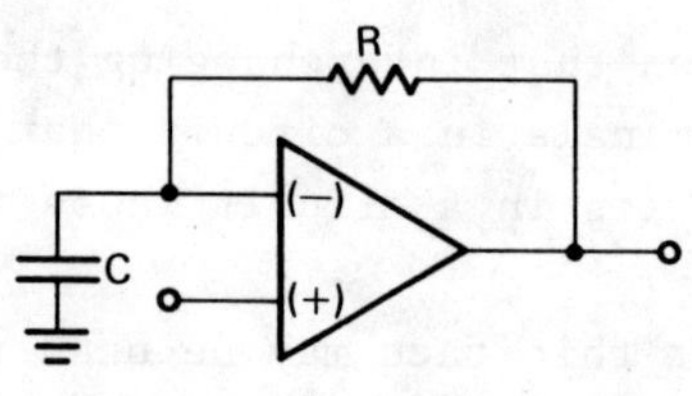

Figure P9.16

P9.17 Give the circuit and design equations for a bandpass circuit of the PF4 type with Q=∞.

P9.18 Design a high-pass notch circuit of the PF4 type with Q=50, ω_n=1 rad/sec and ω_o=1.5 rad/sec.

P9.19 Design a bandpass circuit of the ENF type to have $\omega_o=10^3$ rad/sec, Q=50 and center-frequency gain = 10. Assume $\omega_t=10^6$ rad/sec and evaluate all the performance indices.

P9.20 The circuit in Fig. P9.20 is of the PF1 type. However, this problem investigates an alternative design that features performance measures of substantially different values as compared to the design proposed in the text.

(a) Show that, given ω_o and Q, and using α as a design parameter, the design equations are:

$$CR = \sqrt{\alpha}/\omega_o$$

$$K = 1 + \frac{2}{\alpha} - \frac{1}{\sqrt{\alpha}Q}$$

(b) Derive GS^Q and plot its value versus α, for Q=10, over the range α=1-20.

(c) Derive an expression for one of the passive Q-sensitivities, say $S^Q_{R_1}$, and plot its magnitude versus α, for Q=10, over the range α=1-20.

(d) Use the results of (b) and (c) above to show that the choice of α effects a trade-off between the active and passive sensitivities. Suggest a design value for α and give reasons for your choice.

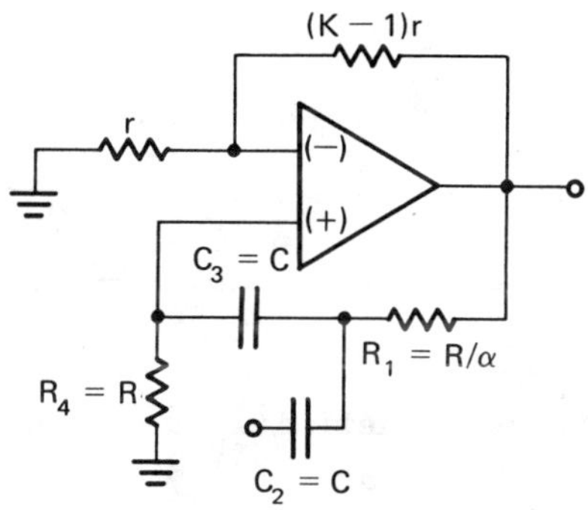

Figure P9.20

P9.21 Verify the expressions for the transfer functions given in Fig. 9.29.

P9.22 The characteristic polynomial of a general second order active filter section can be approximated by

$$D(s) \simeq s^2\left(1 + \sum_{i=1}^{n} \frac{a_i}{A_i}\right) + s\left(\frac{\omega_o}{Q}\right)\left(1 + \sum_{i=1}^{n} \frac{b_i Q}{A_i}\right) + \omega_o^2\left(1 + \sum_{i=1}^{n} \frac{c_i}{A_i}\right) \qquad \text{(P9.22.1)}$$

where ω_o and Q denote their nominal values, A_i is the gain of the ith op amp (total number of op amps is n) and a_i, b_i and c_i are coefficients that can be evaluated for a particular configuration. When amplifiers having single-pole models are used, D(s) becomes a third-order polynomial resulting in a change of the pole frequency to ω_{oa} and the quality factor to Q_a.

(a) Using the approximations $(1+x)^2 \simeq 1+2x$ and $1/(1+x) \simeq 1-x$ for $x<<1$, show that the actual value obtained for the Q-factor is given by

$$Q_a \simeq Q \Bigg/ \left[1 + Q \sum_{i=1}^{n} (c_i - a_i)\left(\frac{\omega_o}{\omega_{ti}}\right)\right] \qquad \text{(p9.22.2)}$$

and that the per unit shift in pole frequency is given by

$$\delta \equiv \frac{\omega_{oa}}{\omega_o} - 1$$

$$\delta \simeq -\frac{1}{2} \sum_{i=1}^{n} b_i \left(\frac{\omega_o}{\omega_{ti}}\right) \qquad \text{(p9.22.3)}$$

(b) Using the usual definitions of the gain-sensitivity products, show that

$$Q_a \simeq Q \Bigg/ \left[1 - 2Q \sum_{i=1}^{n} GS_{A_i}^{\omega_o} \left(\frac{\omega_o}{\omega_{ti}}\right)\right]$$

$$\delta \simeq -\frac{1}{2Q} \sum_{i=1}^{n} GS_{A_i}^{Q} \left(\frac{\omega_o}{\omega_{ti}}\right)$$

In (a) and (b) above state clearly the approximations

made (starting with P9.22.1).

P9.23 Apply the results of P9.22 in a step-by-step manner, to the two-amplifier bandpass circuit of Fig. 9.28.

P9.24 Derive the expression for δ in equation (9.91) using the expression for the input admittance of the type-A inductance simulation circuit, as given by equation (8.44).

P9.25 Design a two-amplifier low-pass notch section having $Q=20$, $\omega_n=2000$ rad/sec and $\omega_o=1000$ rad/sec.

P9.26 Design a two-amplifier low-pass filter with $Q=25$, $\omega_o=10^4$ rad/sec and DC gain = 1.

P9.27 This problem illustrates a procedure for generating circuit #4 (LPN) of Fig. 9.24. The method is based on the concept of generating complementary functions by interchanging inputs and ground (See Problem P9.14).

(a) Find the complement of the notch transfer function

$$T(s) = \left(\frac{\omega_o}{\omega_n}\right)^2 \frac{s^2+\omega_n^2}{s^2+s\left(\frac{\omega_o}{Q}\right)+\omega_o^2}, \quad \omega_n \geq \omega_o$$

i.e. find the function $[1-T(s)]$.

(b) Realize the function $\frac{1}{2}[1-T(s)]$ using the circuit of Fig. P9.27. Find the values of β_1, β_2 and CR.

(c) Use the type-A inductance simulation circuit of Fig. 8.18 to realize the inductance L in the network of Fig. P9.27. Now, taking the output at the output terminal of op. amp. A_1, the active circuit obtained will realize the function $[1-T(s)]$.

(d) Interchange the input and ground terminals in the circuit derived in (c) in order to realize $T(s)$.

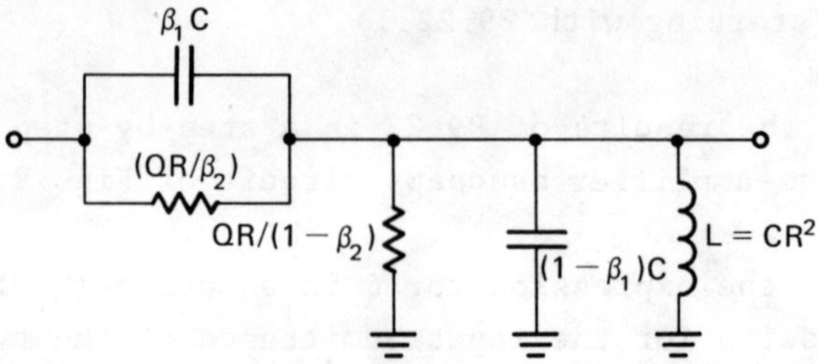

P9.28 Apply the generation procedure of P9.27 to obtain an LPN circuit using the type-B GIC. Note that here there is no low-impedance output terminal and we have to use an additional buffer amplifier. Discuss the expected performance of this LPN circuit and compare it with that given in Fig. 9.29.

P9.29 Use the Deboo positive integrator together with a Miller integrator to generate a two-amplifier state-variable realization. Give the design equations and the transfer functions realized. Discuss the expected performance of this circuit and compare it with other two- and three-amplifier circuits.

P9.30 Apply the results of problem P9.22 to the Akerberg-Mossberg biquad circuit.

P9.31 In Problem P8.12 a double-integrator circuit with a controllable excess phase shift was considered. Use this circuit together with an inverter to form a new three-amplifier biquad. Give the design equations. Derive expressions for the gain-sensitivity products, δ and η. (For deriving η, you may use the method based on excess phase shifts.) Compare this circuit with other three-amplifier biquads.

P9.32 Design a bandpass filter using the Tow-Thomas biquad with passive compensation. The filter is to have $\omega_o=10^4$ rad/sec, Q=100 and center-frequency gain = 10. Assume the op amp to have $\omega_t=10^7$ rad/sec. Give the values of the gain-sensitivity products, δ and η with and without compensation.

P9.33 Repeat Problem P9.32 using the Akerberg-Mossberg biquad. (Obviously there is no need for passive compensation here).

CHAPTER 10

CASCADE AND COUPLED BIQUAD STRUCTURES

10.1 INTRODUCTION

As mentioned in Chapter 8, there are two major approaches to the design of active filters of order N > 2. The first approach consists of first synthesizing an LC ladder realization. Subsequently, the active network is obtained as a simulation of the LC ladder. Methods based on this approach are presented in Chapters 11 and 12. The LC ladder simulation approach has the following advantages:

1. The resulting active networks retain the low-sensitivity performance of their LC ladder prototypes.
2. The wealth of knowledge in the analysis, design and manufacture of passive LC filters can be effectively utilized in the design of active filters.

On the negative side, active filters designed using LC simulation techniques tend to use a relatively large number of amplifiers. Although one might argue that this is not a serious economic disadvantage in view of the low cost of IC Op. Amps., the increased power dissipation might be objectionable. Another disadvantage is that one is restricted to functions realizable by LC ladders. The present authors believe that design methods based on LC simulation should be used whenever the filter specifications are stringent.

The other major approach is a more direct one and consists of realizing the given filter transfer function T(s) as the voltage transfer function of a feedback structure containing amplifiers and RC networks. It turns out, however, that if one attempts a direct realization of even a moderately high order transfer function in one feedback loop, the resulting circuit is quite sensitive to component variations. For this reason, a more practical and simpler technique, namely the Cascade realization method, has been developed. Here the Nth order transfer function is realized as the product of second-order (biquadratic), and possibly one first order, transfer functions. Each biquadratic function is realized in a separate block (biquad) using one of the many circuits presented in Chapter 9. The biquads, which have very low output impedances, are then connected in cascade to realize the overall transfer function. The cascade structure is the simplest and most popular for realizing active filters. Due to sensitivity considerations, however, its utility is limited to filter functions of moderate stringency. Section 10.2 is devoted to a study of the details of the cascade method of design.

One of the advantages of the cascade structure is that each separate biquad is solely responsible for realizing a pole-pair. This makes tuning and other adjustments rather easy to perform. Furthermore, the method lends itself to the "building block" approach to network design[4]. That is, one can have few standard biquad building blocks and use them in the design of a variety of filter networks. The economic advantages of this approach are obvious.

It is this "separability" or "decoupling", however, that is also responsible for the relatively high sensitivity of filters designed using cascaded biquads. A change occurring inside one of the biquads causes a shift in only one of the pole-pairs. Due to the high sensitivity of a transfer function to its poles (See Section 8.5), a change in one pair of poles causes a rather drastic alteration in the magnitude response of the filter, especially if the pole in question has a high Q-factor. On the other hand, it has been observed that the reason for the low sensitivity of LC filter is that the latter is a highly "coupled" system. Each pole-pair depends on every element in the LC ladder. As a result, a change in one element will cause all the poles to shift, with the combined shifts producing less change in the overall transfer function. Stated in another way, the transfer function sensitivity is "spread out" over more elements, thus reducing the sensitivity to each individual element.

The above observations together with the desire to retain the standard building block feature of the cascade structure have led to the development of new design methods utilizing Coupled Biquad Structures. The idea is to apply negative feedback around groups of biquads in a cascade to effect interactions similar to those encountered in an LC ladder, and thereby reduce the transfer function sensitivity to individual biquads. Two structures have resulted from this philosophy. The first is the so-called Follow-the-Leader Feedback (FLF) structure. Section 10.3 presents design methods utilizing this structure. The second is the so-called Leap-Frog (LF) structure. Design techniques utilizing the LF structure are discussed in Chapter 12 in connection with the operational simulation of LC ladders.

10.2 THE CASCADE DESIGN METHOD

Cascade design is by far the simplest and most popular method for active filter realization. The filter transfer function T(s) is realized as the voltage transfer ratio of a cascade of filter sections, each having a biquadratic voltage transfer function. An additional first-order section is required in the case the filter order N is odd. This first-order section can be easily implemented using an RC circuit. To implement the biquadratic sections (biquads), the designer may choose from the variety of circuits presented in Chapter 9.

The reader may recall that each of the biquad circuits presented derives its output from the output terminal of an op. amp. Thus, the output impedance of a biquad is very low, ideally zero. Therefore, in cascading biquads, one may assume that they do not interact; that is, the overall transfer function of a cascade of biquads is equal to the product of their individual transfer functions. The cascade method of design involves two major steps:

Step (1) Decomposing the Nth order transfer function T(s) into the product of second-order (and one first-order, for odd N) subfunctions; that is,

$$T(s) = \prod_{j=1}^{N/2} T_j(s) \quad , \text{N even where } T_j(s) \text{ has the general}$$

form,

$$T_j(s) = K_j \frac{s^2 + s\left(\frac{\omega_z}{Q_z}\right) + \omega_z^2}{s^2 + s\left(\frac{\omega_o}{Q}\right) + \omega_o^2} .$$

At this point, the reader is urged to review Section 1.5 dealing with biquadratic transfer functions.

Step (2) Realizing each biquadratic function with a circuit from the collection presented and studied in Chapter 9. Subsequently, the biquads are connected in cascade to form the complete filter.

Considerable material on biquad circuits has already been presented in Chapter 9. The merits and shortcomings of each circuit have been carefully analyzed. Therefore, the reader should find in Chapter 9 sufficient information to enable him to successfully carry out Step (2) above. In the following we consider in detail Step (1); namely, solving the decomposition problem.

Decomposing the Nth Order Transfer Function T(s)

The problem of decomposing an Nth order transfer function into the product of second-order, and possibly one first-order, functions may be split into three parts. These are:

(1) Pole-Zero Pairing:

To form second-order functions, each of the $(\frac{N}{2})$ pole-pairs has to be combined with one of the $(\frac{N}{2})$ pairs of zeros (including those zeros at $s = 0$ and $s = \infty$).* There are $(\frac{N}{2})!$ possible different combinations to choose from.

(2) Cascading Sequence:

Having determined the transfer function of each filter section, to within a gain constant, the question remains as to the sequence in which the sections should be cascaded. Note that since the sections have low output impedances, the overall transfer function of the cascade does not depend on the cascading sequence. There are $(\frac{N}{2})!$ possible different sequences to choose from.

(3) Gain Distribution:

Given a specified value of total filter gain, the question arises as to how this gain should be distributed among the different sections in the cascade. In other words, what gain level should be assigned to each biquad section? There is an infinite number of possible gain assignments.

Having defined the decomposition problem, the question now arises as to what characterizes an optimum solution. It turns out that, among the many filter performance measures which one may attempt to optimize in solving the decomposition problem, the Dynamic Range is by far the most relevant[4]. To put it in a different way, the solution of the decomposition problem affects the filter dynamic range more than any other performance measure. For this reason we shall address ourselves to the question of finding a decomposition that maximizes the filter dynamic range.

It is easy to reason that to obtain the widest possible dynamic range, two objectives have to be met. These are:

---

*For simplicity, we will usually assume that N is even. The methods presented, however, apply equally well for the case N is odd.

(1) The maximum allowable input signal amplitude should be limited only by the maximum allowable output signal swing of the op. amp. used. That is, the filter should be capable of accepting a maximum input signal of,

$$V_{in_{max}} = \frac{V_{amp}}{m} ,$$

where V_{amp} denotes the specified maximum output voltage of the op. amp. and m is the required filter gain in the passband.

(2) The signal-to-noise ratio should be maintained as high as possible throughout the cascade.

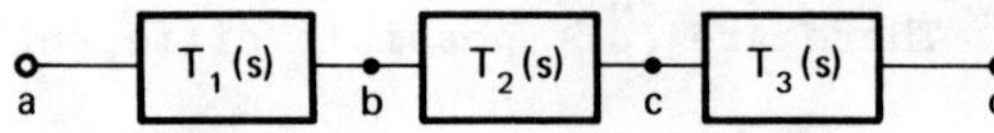

Figure 10.1 A filter designed as the cascade of three biquads. Note that the "intermediate" transfer functions are: $T_{ab} = T_1$ and $T_{ac} = T_1T_2$, while the filter transfer function $T(s) = T_1T_2T_3$.

To investigate how these objectives may be achieved, consider, as an example, a filter designed as the cascade of three biquads, as shown in Fig. 10.1. We can measure the transfer functions from the input terminal "a" to the intermediate output terminals "b" and "c", as well as the filter output terminal "d", by applying a constant-amplitude swept-frequency sine-wave signal to the filter input. The first objective stated above will be satisfied if we are able to increase the amplitude of the input signal to $V_{in_{max}}$ without any of the biquad outputs "clipping". Obviously, at this value of the input signal, the filter output will have a peak value of V_{amp}. It also follows that the maxima attained by the signals V_b and V_c should be less than or equal to V_{amp}, depending on the distribution of gains among the three filter sections. Now, to satisfy the second objective above, we should maintain the signal level as high as possible throughout the cascade. This in turn implies distributing the gain in a way that forces the peaks of V_b and V_c to be equal to V_{amp}.

We thus conclude that the optimum gain distribution is the one that makes the peak of the magnitude of each of the intermediate transfer functions equal to that of the overall transfer function.

The second objective stated above stipulates that the signal-to-noise ratio should be maintained as high as possible throughout the cascade. This implies that within the filter passband, the magnitude of the intermediate transfer functions should be made as high as possible. Otherwise, inband signals whose frequencies lie in the range of the minima of these transfer functions will suffer signal-to-noise ratio degradation. But since we have already decided to set the peak values of these magnitude functions at the maximum level specified by the filter transfer function, it follows that all we need to do now is maximize their minima. This leads to the conclusion that the pole-zero pairing and the cascading sequence should be chosen such that the magnitude of each of the intermediate transfer functions is as flat as possible, over the filter passband.

In order to obtain an optimum solution based on this flatness criterion, one needs to deal with the pairing and sequence problems simultaneously. This, however, is a rather difficult task and we shall be satisfied with a solution to the decomposition problem based on the following three steps:

(1) Solve the pairing problem such that the magnitude of the transfer function of each biquad is as flat as possible, over the filter passband.
(2) Find the sequence of cascading the biquads (obtained as a result of step-1) such that the magnitude of the intermediate transfer functions is as flat as possible over the filter passband.
(3) For a specified total filter gain, assign gain constants to the various biquads such that the peak of the magnitude of each of the intermediate transfer functions is equal to that of the filter transfer function.

Before presenting the details of the solution, it should be mentioned that there are other subjective criteria which the designer might wish to consider in performing the decomposition process. A partial list includes:

(1) A certain pole-zero pairing might be more desirable than another due to the availability of standard filter building blocks of a certain type and/or if such a pairing results in a section that is easier to tune or adjust. As an example, consider the problem of realizing the fourth-order bandpass transfer function,

$$T(s) = K \frac{s^2}{(s^2 + s \frac{\omega_{o_1}}{Q_1} + \omega_{o_1}^2)(s^2 + s \frac{\omega_{o_2}}{Q_2} + \omega_{o_2}^2)}$$

There are two possible pairings. The first involves decomposing T(s) into the product of two bandpass functions,

$$T_1(s) = K_1 \frac{s}{s^2 + s(\frac{\omega_{o_1}}{Q_1}) + \omega_{o_1}^2} \quad \text{and,}$$

$$T_2(s) = K_2 \frac{s}{s^2 + s(\frac{\omega_{o_2}}{Q_2}) + \omega_{o_2}^2}$$

The second possible decomposition consists of a low-pass function and a high-pass function; that is,

$$T_1'(s) = K_1' \frac{1}{s^2 + s(\frac{\omega_{o_1}}{Q_1}) + \omega_{o_1}^2} \quad \text{and,}$$

$$T_2'(s) = K_2' \frac{s^2}{s^2 + s(\frac{\omega_{o_2}}{Q_2}) + \omega_{o_2}^2}$$

The designer would choose the first decomposition if high-quality bandpass filter realizations are readily available. Another reason for this choice would be the fact that bandpass sections are usually easier to tune than high-pass or low-pass sections.

(2) A particular cascading sequence might be preferred. For instance, one would normally wish to place either a band-pass or a low-pass section (if available) at the start of the cascade. This will attenuate high-frequency (out-of-band) signals which otherwise might overload or cause slew-rate limiting in subsequent stages in the cascade. Similarly, one usually likes to place a bandpass or a high-pass section at the end of the cascade in order to prevent internally generated DC offsets or power-supply ripple from appearing at the output of the filter.

In any case, it is always possible to incorporate such practical criteria in the systematic decomposition procedures described below.

Before presenting optimum decomposition algorithms, a word of advice is in order. Although these methods are indispensable in high-order filters, there are many cases where the simple rules-of-thumb, also mentioned below, would lead to a good or even "optimum" decomposition.

Pole-Zero Pairing

We now present a systematic procedure for obtaining the pairing of poles and zeros which result in each biquad having the flattest possible magnitude response. The method is due to Lueder[5] and was discussed in detail by Moschytz[4].

By flattest possible magnitude response, we mean that the ratio of the maximum to the minimum transmission should be as close to unity as possible. For the maximum, we consider the entire frequency axis, $0 \le \omega \le \infty$, since a strong out-of-band signal should not be allowed to overload the filter. On the other hand, for the minimum, we consider only the filter passband, since out-of-band signals should by definition be attenuated and their signal-to-noise ratio is of no concern.

As an example, consider the transfer function $T_j(s)$ obtained by combining a pair of complex poles with a pair of jω-axis zeros (that is, a notch section),

$$T_j(s) = K_j \frac{s^2 + \omega_z^2}{s^2 + s(\frac{\omega_o}{Q}) + \omega_o^2}$$

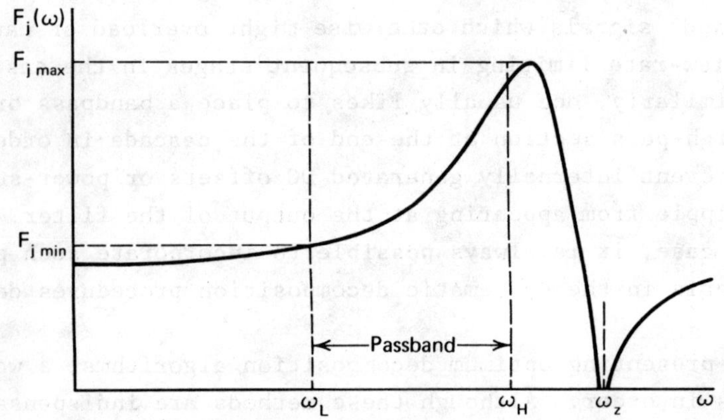

Figure 10.2 The normalized magnitude – squared function $F_j(\omega)$ of a low-pass notch section. This section is part of a cascade realization of a bandpass filter whose passband extends from ω_L to ω_H. Note the definition of $F_{j\,max}$ and $F_{j\,min}$.

Fig. 10.2 shows a sketch of the normalized magnitude squared function $F_j(\omega)$ where,

$$F_j(\omega) = \left|\frac{T_j(j\omega)}{K_j}\right|^2$$

for the case $\omega_z > \omega_o$, that is a low-pass notch (See Section 1.5). We have also indicated the filter passband extending from ω_L to ω_H. As seen, in this case the maximum of transmission happens to occur outside the passband. Nevertheless, it is this maximum that is of concern to us in optimizing for maximum dynamic range

We define $F_{j\max}$ as,

$$F_{j_{max}} = \text{Max } [F_j(\omega)] \quad , \quad 0 \le \omega \le \infty$$

The minimum of transmission occurs also outside the passband and hence is of no concern to us. In optimizing for maximum signal-to-noise ratio, we search for the minimum value of transmission within the passband. In our example, this happens to be at the lower edge of the passband. In general, we define $F_{j_{min}}$ as,

$$F_{j_{min}} = \text{Min } [F_j(\omega)] \quad , \quad \omega_L \le \omega \le \omega_H \quad .$$

Thus, our goal is to keep the ratio $[F_{j_{max}}/F_{j_{min}}]$ as close to unity as possible. For convenience we use a logarithmic measure of flatness defined as,

$$d_j = \log \left(\frac{F_{j_{max}}}{F_{j_{min}}}\right) \quad ,$$

and attempt to keep d_j as close to zero as possible.

The pole-zero pairing algorithm can be stated as follows:

(1) For a possible pole-zero assignment, find the values of d_j, $j = 1,2,\ldots,\frac{N}{2}$. Find the maximum value of d for this assignment,

$$d_{max} = \text{Max } [d_j] \quad , \quad j = 1,2,\ldots,\frac{N}{2} \quad .$$

(2) Repeat the above for every possible pole-zero assignment.

The optimum assignment is the one having the <u>smallest</u> d_{max}.

A simple method for implementing this algorithm will now be presented. For clarity, the implementation is outlined for three different cases as follows:

Case (A): Transfer Function with Complex-Poles and Complex-Zeros Only

Here, the filter transfer function has $(\frac{N}{2})$ pairs of complex poles and $(\frac{N}{2})$ pairs of complex zeros. The method consists of the following steps:

(1) Construct the $d_{\nu,\mu}$ table whose structure is shown in Fig. 10.3a. Note that the entry $d_{\nu,\mu}$ is the value of d for the biquad obtained by pairing p_ν and z_μ. It should be noted that when we speak of pairing p_ν with z_μ, we mean pairing (p_ν, p_ν^*) with (z_μ, z_μ^*), as all poles and zeros occur in complex conjugate pairs.

(2) Plot the values of $d_{\nu,\mu}$ versus $z_1, z_2, \ldots, z_{N/2}$, as shown in Fig. 10.3b. For convenience, the points $z_1, z_2, \ldots, z_{N/2}$ are located on the horizontal axis with equal spacings. Note that the numbers on the columns denote pole numbers.

(3) Draw the first assignment line which is the horizontal line of the lowest possible height, under which each pole number appears in a different column at least once.

(4) The pole closest to the first assignment line should be paired with the zero corresponding to this column. This pole-zero pair should then be excluded.

(5) If each of the remaining pole numbers appear only once, then we have obtained the optimum assignment. If some pole numbers appear more than once, then we still have some choice. Draw the second assignment line and repeat until a complete assignment is obtained.

	z_1	z_2	...	...	z_μ	...	$z_{N/2}$
p_1	$d_{1,1}$	$d_{1,2}$	...	...	$d_{1,\mu}$	...	...
p_2	$d_{2,1}$	$d_{2,2}$	...	...			
⋮	⋮	⋮			$d_{3,\mu}$		
⋮	⋮	⋮					
p_ν	$d_{\nu,1}$	$d_{\nu,2}$	...	...	$d_{\nu,\mu}$		$d_{\nu,N/2}$
⋮							
$p_{N/2}$	$d_{N/2,1}$	$d_{N/2,2}$	$d_{N/2,3}$	...	$d_{N/2,\mu}$	...	$d_{N/2,N/2}$

(a) The general $d_{\nu,\mu}$ table.

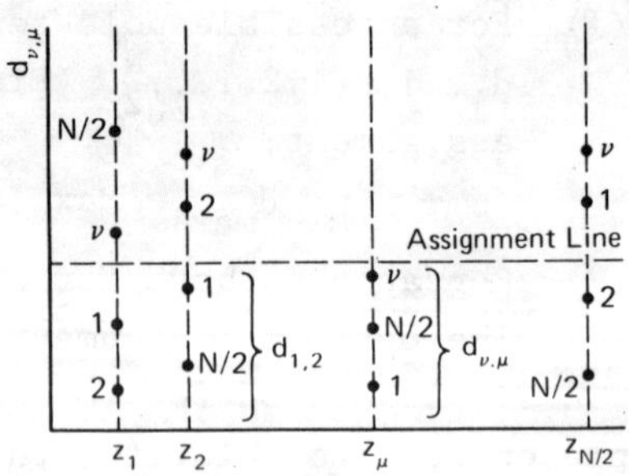

(b) The general $d_{\nu,\mu}$ graph.

Figure 10.3

The above procedure is illustrated by an example in Fig. 10.4. It should be noted that if the designer wishes to combine a particular pole-zero pair, then this pole and zero should be left-out of the above assignment procedure. On the other hand, if it is desired that a particular pole p_i should not be paired with a particular zero z_j, then a large number should be assigned to the entry $d_{i,j}$ in the $d_{\nu,\mu}$ table to guarantee the exclusion.

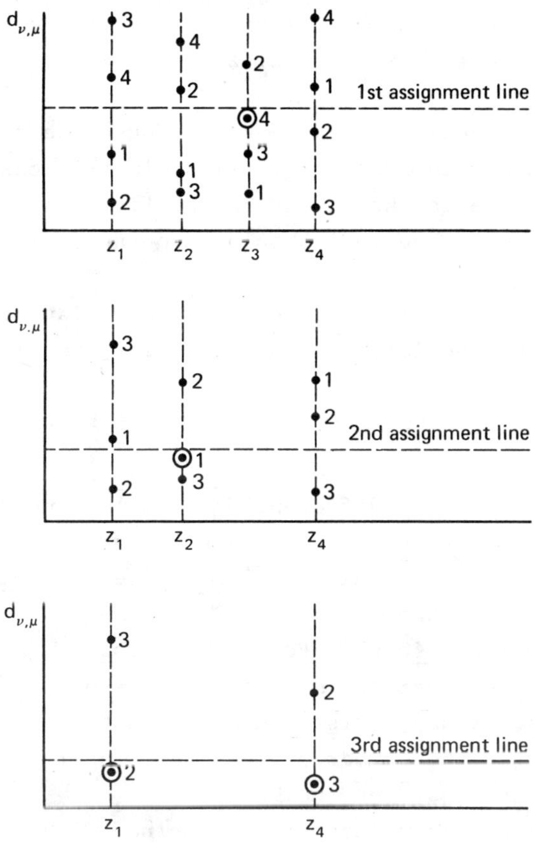

Figure 10.4 Illustrating the pole-zero pairing process for an N = 8 filter. The optimum pairing obtained is (p_4, z_3), (p_1, z_2), (p_2, z_1) and (p_3, z_4).

Case (B): Real and Complex Zeros and Complex Poles

Consider now the more general case where we have r_z real axis zeros (including those at $s = 0$ and ∞), $(N - r_z)$ complex zeros and N complex poles. It follows that both N and r_z have to be even numbers.

Before we can start the assignment procedure, we have to decide on a particular pairing of the r_z real zeros. As an example, consider the case of a sixth order filter (N = 6) with 3 pairs of complex poles, one pair of complex conjugate zeros (z_1, z_1^*) and four real-axis zeros (z_2, z_3, z_4, z_5). If all the four real zeros are distinct, then there are three different ways for pairing them. These are: $(z_2\ z_3,\ z_4\ z_5)$, $(z_2\ z_4,\ z_3\ z_5)$ and $(z_2\ z_5,\ z_3\ z_4)$. For each of these combinations, we have a $d_{\nu,\mu}$ table and the associated graph, etc. Thus, the assignment procedure has to be repeated three times, in this case. The combination that leads to the lowest-height first assignment line is obviously the only one that should be pursued and will lead to the optimum assignment.

In general, if the r_z real zeros are all distinct, there will be α_o different possible pairings of the r_z zeros where,

$$\alpha_o = (r_z-1)(r_z-3)\ldots\ldots(5)(3)(1)$$

Thus, there will be α_o different $d_{\nu,\mu}$ tables and corresponding graphs. If some of the r_z zeros are repeated (i.e. multiple real-axis zeros), then the number of different real zero pairings will be less than α_o.

Case (C): Real and Complex Poles and Zeros

Finally, we consider the most general case. Normally, only one real pole, if any, exists in addition to $\frac{(N-1)}{2}$ pairs of complex poles. Let the number of real zeros be r_z. It follows that both N and r_z are odd numbers.

One of the real zeros has to be combined with the real pole to form a first-order section. There are r_z different ways of doing this assignment. In each case, the remaining (r_z-1) real zeros have to be paired. There are α_1 different possible pairings (assuming all (r_z-1) zeros to be distinct) where,

$$\alpha_1 = (r_z-2)(r_z-4) \;\ldots\ldots\; (5)(3)(1) \quad .$$

Thus we have $(r_z\ \alpha_1)$ different possible zero configurations to start the assignment procedure. That is, the assignment procedure has to be carried out $(r_z\ \alpha_1)$ times. The case with the lowest-height 1st assignment line is obviously the only one that should be pursued to completion.

Although the systematic procedure presented is usually indispensable for high-order filters, one might be able to achieve the same results intuitively in the simpler cases. As a rule of thumb high-Q poles should be combined with the closest transmission zeros.

Example 10.1

Consider the bandpass filter* whose $|T(j\omega)|$ is shown in Fig. 10.5. The filter has an order N = 8, with equiripple passband (0.4 dB) extending from 1 KHz to 1.4 KHz. The poles and zeros are as follows (normalized with respect to 2π x 1.4 Krad/sec.)

$$z_1 \,,\, z_1^* \;=\; \pm j\ 0.3996$$

$$z_2 \,,\, z_2^* \;=\; \pm j\ 1.2501$$

$$z_3 \,,\, z_4 \;=\; 0$$

$$z_5 \,,\, z_6 \;=\; \infty$$

$$p_1 \,,\, p_1^* \;:\; \omega_{o_1} = 0.7106 \qquad Q_1 = 15.256$$

$$p_2 \,,\, p_2^* \;:\; \omega_{o_2} = 0.7911 \qquad Q_2 = 6.192$$

$$p_3 \,,\, p_3^* \;:\; \omega_{o_3} = 0.9188 \qquad Q_3 = 6.743$$

$$p_4 \,,\, p_4^* \;:\; \omega_{o_4} = 1.0048 \qquad Q_4 = 19.113$$

---

*Note that the approximation problem for this filter was considered in examples 2.2 and 5.3 where various filter parameters are given.

Note that on our normalized frequency scale the passband extends between $\omega_L = 0.7143$ and $\omega_H = 1$. It is required to find the optimum pole-zero pairing that results in maximum dynamic range for the filter.

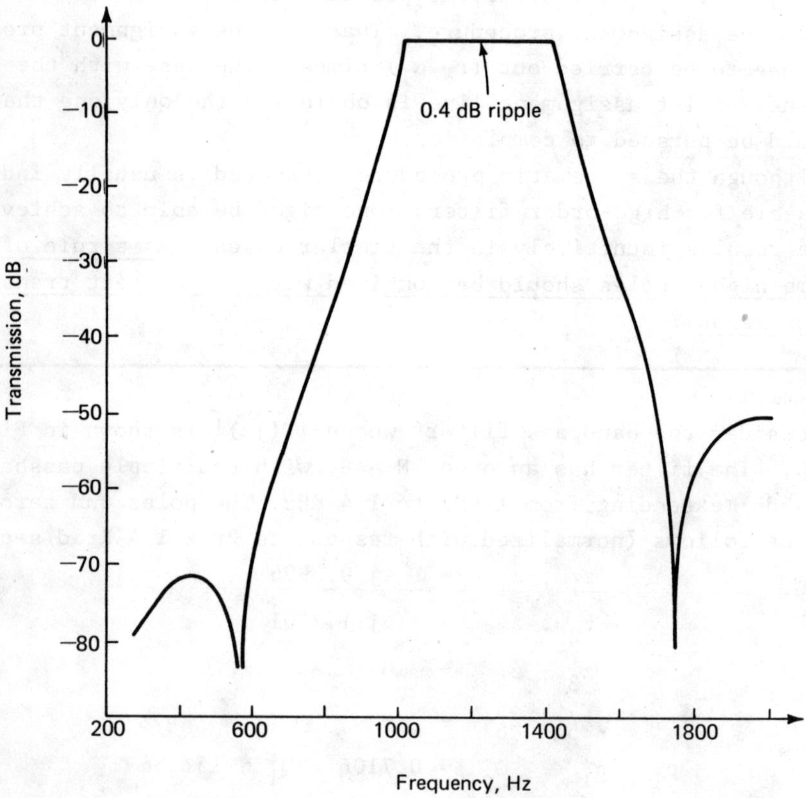

Figure 10.5 Transmission function for the bandpass filter of Example (10.1).

Solution

We have four real zeros but since two are repeated, then there are only two possible different pairings; namely, $(z_3\ z_4,\ z_5\ z_6)$ and $(z_3\ z_5,\ z_4\ z_6)$. We will denote these pairings as (z_{34}, z_{56}) and (z_{35}, z_{46}), respectively. Correspondingly, there will be two different $d_{\nu,\mu}$ tables which are combined in Table 10.1 below.

Table 10.1

	z_1	z_2	z_{34}	z_{56}	z_{35}	z_{46}
p_1	1.58	2.90	1.76	2.35	2.05	2.05
p_2	0.68	1.64	0.78	1.19	1.84	1.84
p_3	1.47	0.58	1.03	0.89	1.11	1.11
p_4	2.72	1.40	1.95	1.95	2.25	2.25

In arriving at the values given in Table 10.1 we made use of the formulae for biquad maxima given in Table 1.1.

Fig. 10.6a shows the $d_{\nu,\mu}$ graph for the first zero configuration, namely z_1, z_2, z_{34}, z_{56} while the graph for the second zero configuration is shown in Fig. 10.6b. Since the first assignment line is lower for the first case, then the optimum solution will be obtained from this case. The result is the following assignment (See Fig. 10.6a), (z_1, p_1), (z_2, p_4), (z_{34}, p_2), (z_{56}, p_3). The transfer functions of the four sections are

$$T_1(s) = K_1 \frac{s^2 + (0.3996)^2}{s^2 + s(\frac{0.7106}{15.256}) + (0.7106)^2} \qquad \text{(HPN)}$$

$$T_2(s) = K_2 \frac{s^2 + (1.2501)^2}{s^2 + s(\frac{1.0048}{19.113}) + (1.0048)^2} \qquad \text{(LPN)}$$

$$T_3(s) = K_3 \frac{s^2}{s^2 + s(\frac{0.7911}{6.192}) + (0.7911)^2} \qquad \text{(HP)}$$

$$T_4(s) = K_4 \frac{1}{s^2 + s(\frac{0.9188}{6.743}) + (0.9188)^2} \qquad \text{(LP)}$$

where we have indicated beside each transfer function the biquad type it implements.

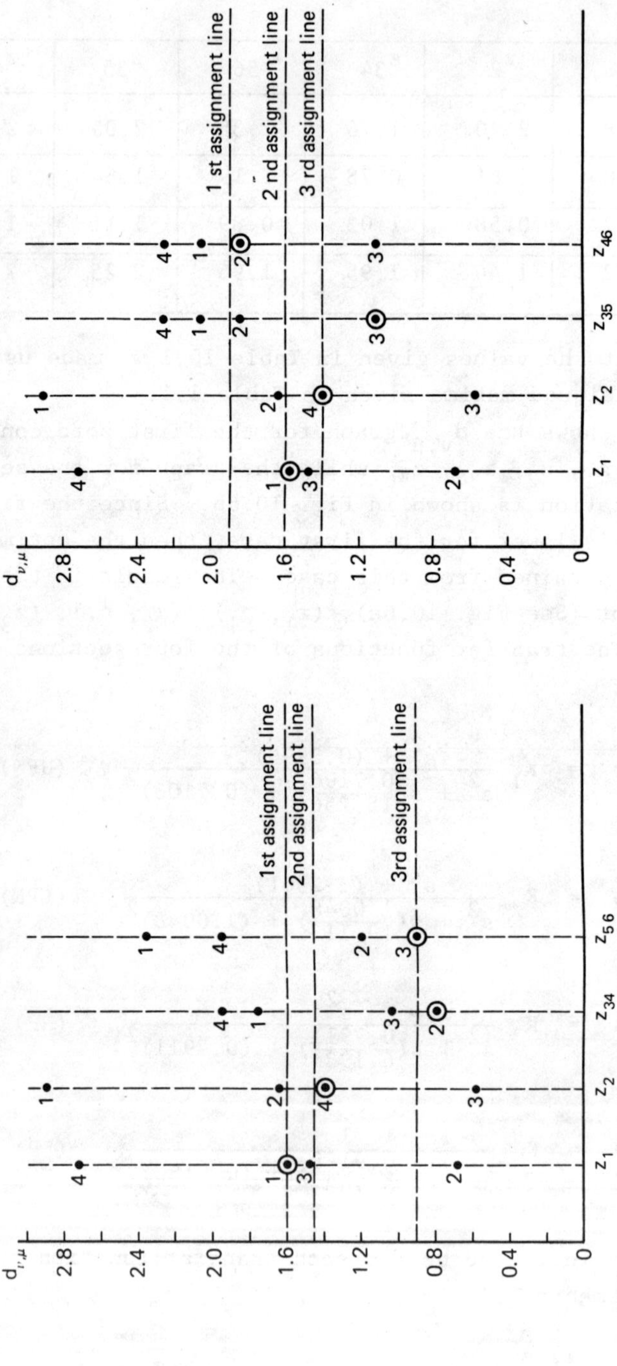

Figure 10.6 Illustrating the pole-zero assignment, for Example (10.1).

(a) The first zero configuration.

(b) The second zero configuration.

It is interesting to note that the optimization procedure resulted in an assignment that could have been, at least partly, predicted using the rule-of-thumb mentioned in the above. Specifically, note that both $T_1(s)$ and $T_2(s)$ result from combining high-Q poles with the closest transmission zeros. One can also argue that each of $T_3(s)$ and $T_4(s)$ results from combining poles and zeros that are closest to each other. It is not intuitively obvious, however, why $T_3(s)$ and $T_4(s)$ could not be replaced by two bandpass sections. In fact, if one pursues the solution to the second case, as shown in Fig. 10.6b, the resulting assignment, which is not the optimum one, has $T_1(s)$ and $T_2(s)$ identical to those above, but with $T_3(s)$ and $T_4(s)$ replaced by two bandpass functions. Finally, it should be mentioned that a filter designer might prefer this latter solution to the optimum one, since bandpass circuits are generally easier to tune.

Optimizing the Sequence of Cascading Sections

The second part in the solution of the decomposition problem concerns establishing the optimum sequence for cascading the filter sections. For an Nth order filter with an even N, there are $(\frac{N}{2})!$ different possible sequences. Our goal here is to find the particular cascading sequence that results in the transfer functions, from the filter input to the biquad outputs, having the flattest possible magnitude response. As an example, Fig. 10.7 shows one of the 24 possible sequences for cascading four biquadratic sections. Using the logarithmic measure of transmission flatness d, we obtain for this sequence

$$d_a = \log \frac{\text{Max } |T_1(j\omega)|}{\text{Min } |T_1(j\omega)|}$$

$$d_b = \log \frac{\text{Max } |T_1(j\omega)\ T_3(j\omega)|}{\text{Min } |T_1(j\omega)\ T_3(j\omega)|}$$

$$d_c = \log \frac{\text{Max } |T_1(j\omega)\ T_3(j\omega)\ T_4(j\omega)|}{\text{Min } |T_1(j\omega)\ T_3(j\omega)\ T_4(j\omega)|}$$

where for evaluating the maxima we use $0 \le \omega \le \infty$, while for the minima we use $\omega_L \le \omega \le \omega_H$. Once we evaluate the d values, we find their maximum d_{max},

$$d_{max} = \text{Max}\ [d_a, d_b, d_c]$$

The optimum cascading sequence will be the one whose d_{max} is the lowest.*

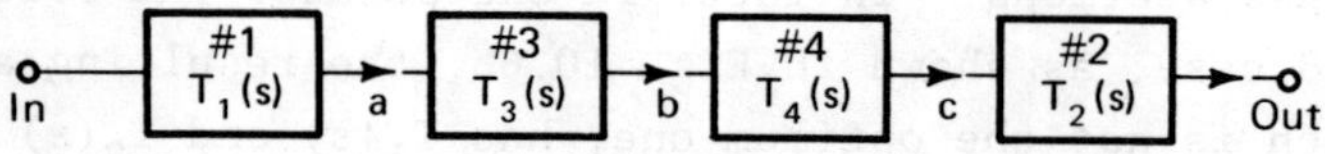

Figure 10.7 One out of the 24 possible sequences for cascading four biquads.

The question now arises as to the method for implementing the sequence optimization process outlined above. The obvious, but most tedious, method is to exhaustively search for the sequence with the smallest d_{max}. A much more efficient algorithm has been described in the literature[6,7] and will not be repeated here. The third and most commonly used method is based on trial and error aided by intuition. Although we should not expect to obtain the optimum sequence this way, the results are usually quite satisfactory.

The "trial and error" method consists of sketching the magnitude response of each individual section. We then attempt to cascade the sections in a way that "spreads out" the maxima and minima over the passband. In addition, we may wish to place a bandpass or a low-pass section (if available) at the beginning of the cascade. This will ensure that strong out-of-band signals, especially high frequency ones, will be sufficiently attenuated before reaching the second stage where they might produce overloading or slew-rate limiting. Similarly, it is usually re-

*Note that we have not evaluated d at the output of the final stage since this is the filter output and d will just be the passband ripple.

commended to place a bandpass or a high-pass section (if available) at the end of the cascade. This will help to prevent internally generated low-frequency noise from appearing at the filter output.

As a further guide in finding an "optimum" cascading sequence, one should place sections with the flattest response at the input followed by sections with lesser flatness and so on. This in turn implies that the biquads' Q-factor increases progressively from input to output in the cascade.

The Optimum Gain Distribution

The third and final part of the decomposition problem is concerned with establishing the optimum gain distribution among the various sections of the cascade. We have already shown that the optimum gain assignment is the one that results in the magnitudes of all the intermediate transfer functions having equal maxima.

The gain constants of different biquads can be evaluated as follows. Consider an Nth order transfer function T(s) and assume, for simplicity, that N is even. We assume that the pole-zero pairing and the cascading sequence have already been determined. Let the biquads in the cascade be numbered consecutively from the input side, and let $T_j(s)$ be expressed as,

$$T_j(s) = K_j t_j(s) \quad , \quad j = 1, \frac{N}{2}$$

where the K_j's denote the gain constants to be determined.

The filter transfer function T(s) is given by

$$T(s) = \prod_{j=1}^{N/2} K_j \; t_j(s) = K \prod_{j=1}^{N/2} t_j(s)$$

where K is the specified overall gain constant,

$$K = \prod_{j=1}^{N/2} K_j$$

For a uniform input signal spectrum of say 1V magnitude, the maximum of the spectrum at the filter output is easily evaluated from the given transfer function as,

$$V_{omax} = \text{Max } |T(j\omega)| \quad , \; 0 \le \omega \le \infty$$

At the output of the first section in the cascade the maximum is given by

$$V_{1max} = K_1 \text{ Max } |t_1(j\omega)| \quad , \; 0 \le \omega \le \infty$$

To make V_{1max} equal to V_{omax}, the value of K_1 should be

$$K_1 = \frac{\text{Max } |T(j\omega)|}{\text{Max } |t_1(j\omega)|}$$

Next we consider the output of the second section where the maximum signal will be

$$V_{2max} = K_1 K_2 \text{ Max } |t_1(j\omega) \; t_2(j\omega)| \quad , \; 0 \le \omega \le \infty$$

Now to make $V_{2max} = V_{1max} = V_{omax}$, we select the following value for K_2

$$K_2 = \frac{\text{Max}|t_1(j\omega)|}{\text{Max}|t_1(j\omega) \; t_2(j\omega)|}$$

Extrapolating from the above analysis, it follows that the gain constant for the jth section will be given by

$$K_j = \frac{\text{Max } |t_1(j\omega)t_2(j\omega) \; \ldots \; t_{j-1}(j\omega)|}{\text{Max } |t_1(j\omega)t_2(j\omega) \; \ldots \; t_j(j\omega)|} \; , \; j = 2, \; \ldots, \; \frac{N}{2} \quad (10.2)$$

For the last section in the cascade, it may be easier to calculate its gain constant from,

$$K_{(N/2)} = \frac{K}{K_1 K_2 \ldots K_{(N/2 - 1)}} \tag{10.2a}$$

Equations (10.1) and (10.2) provide the values of the gain constants for optimum dynamic range.

Example 10.2

We wish to continue the design of the bandpass filter of Example 10.1 by finding the optimum cascading sequence and gain assignments of the four biquads. The biquad transfer functions, obtained in Example 10.1, are,

$$T_1(s) = K_1 \frac{s^2 + (0.3996)^2}{s^2 + s(\frac{0.7106}{15.256}) + (0.7106)^2} \qquad \text{(HPN)}$$

$$T_2(s) = K_2 \frac{s^2 + (1.2501)^2}{s^2 + s(\frac{1.0048}{19.113}) + (1.0048)^2} \qquad \text{(LPN)}$$

$$T_3(s) = K_3 \frac{s^2}{s^2 + s(\frac{0.7911}{6.192}) + (0.7911)^2} \qquad \text{(HP)}$$

$$T_4(s) = K_4 \frac{1}{s^2 + s(\frac{0.9188}{6.743}) + (0.9188)^2} \qquad \text{(LP)}$$

It is required that the bandpass filter have a maximum passband gain of unity.

Solution

The value of the overall gain constant K, $K = K_1 K_2 K_3 K_4$, may be determined by evaluating the various biquad transfer functions at one of the reflection zeros (See Example 2.2 for the location of the reflection zeros), and forcing the total gain to unity. The result is,

$$K = 0.004392$$

Fig. 10.8 shows a sketch of the magnitude of the transfer functions of the four biquadratic sections. Following the discussion above, we select the low-pass section $T_4(s)$ as the first block in the cascade and the high-pass section $T_3(s)$ as the last (fourth) block. Since the peak of $|t_1(j\omega)|$ is the farthest from that of $|t_4(j\omega)|$, then it makes sense to follow $T_4(s)$ by $T_1(s)$. This leaves $T_2(s)$ as the third section in the cascade. The cascading sequence is therefore,

$$T_4(s)\ T_1(s)\ T_2(s)\ T_3(s)$$

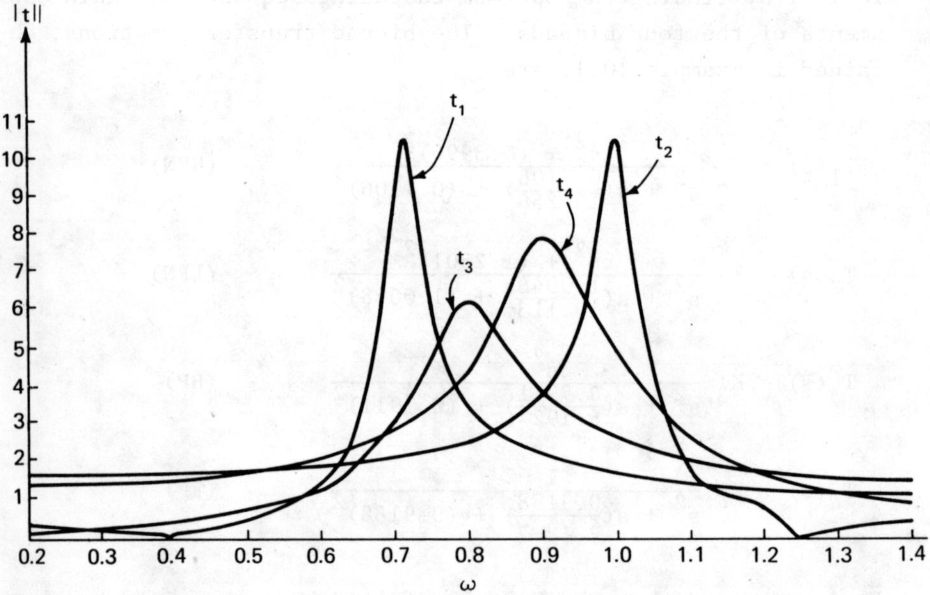

Figure 10.8 Magnitude of the transfer functions of the four biquadratic sections in the bandpass filter of Examples (10.1)–(10.3).

The gain constants K_1, K_2, K_3 and K_4 are now determined using equations (10.1) and (10.2) as follows (with the appropriate change in subscript numbers)

$$K_4 = \frac{\text{Max } |T(j\omega)|}{\text{Max } |t_4(j\omega)|} = \frac{1}{8} = 0.125$$

$$K_1 = \frac{\text{Max } |t_4(j\omega)|}{\text{Max } |t_4(j\omega)t_1(j\omega)|} = \frac{8}{30} = 0.267$$

$$K_2 = \frac{\text{Max } |t_4(j\omega)t_1(j\omega)|}{\text{Max } |t_4(j\omega)t_1(j\omega)t_2(j\omega)|} = \frac{30}{88} = 0.341$$

$$K_3 = \frac{K}{K_4K_1K_2} = 0.386$$

Note that with this gain assignment, the maximum allowable input signal is

$$V_{imax} = V_{amp}$$

Also, the signal level is kept as high as possible throughout the cascade, thus maximizing the signal-to-noise ratio.

Cascade Design Examples

Example 10.3

It is required to obtain a complete circuit implementation of the eighth-order bandpass filter considered in Examples 10.1 and 10.2.

Solution

In the two previous examples, we completely determined the biquad functions and the cascading sequence. We now address ourselves to the question of selecting circuit implementations for the four biquads. We do this with the help of the design information presented in Chapter 9. First, let us seek a realization based on the assumption that the filter will be constructed using discrete components and that little or no tuning should be required. If this is the case it is advisable to select circuits that have low passive sensitivities.

The HP and LP sections both have low Q-factors and single amplifier biquads would be appropriate. Thus, we select circuits of the PF2 type, which are given together with design equations in Fig. 9.19. To set the gain factor to the desired level, the feed-in component has to be split into two parts, as shown in the following:

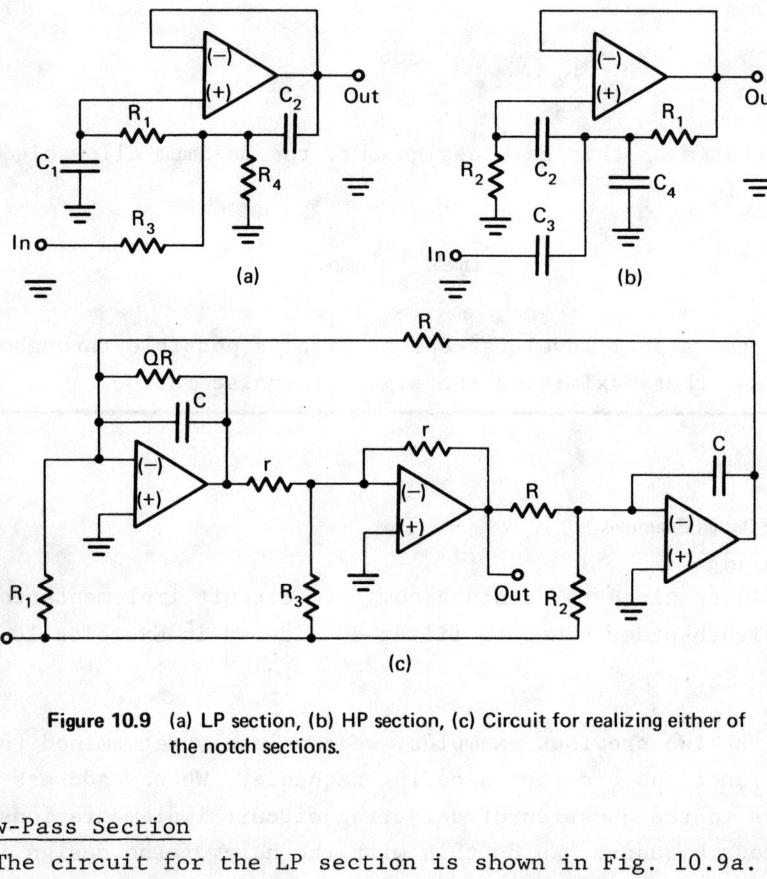

Figure 10.9 (a) LP section, (b) HP section, (c) Circuit for realizing either of the notch sections.

Low-Pass Section

The circuit for the LP section is shown in Fig. 10.9a. Note the splitting of the input feed-in resistance into the two resistances R_3 and R_4. The component values are obtained as follows

$$CR = \frac{1}{2\,\omega_o\,Q} = \frac{1}{2 \times 2\pi \times 0.9188 \times 1.4 \times 10^3 \times 6.743}$$

$$= 9.1746\ \mu\text{sec}$$

Choosing C = 360 pF, then R = 25.5 kΩ

Thus, $C_1 = C = 360$ pF

$$C_2 = 4Q^2C = 65.5 \text{ nF}$$

$$R_1 = R = 25.5 \text{ K}\Omega$$

$$R_3 = \frac{R}{[K_4/(0.9188)^2]} = 187.4 \text{ K}\Omega$$

$$R_4 = \frac{R}{[1 - \frac{K_4}{(0.9188)^2}]} = 29.93 \text{ K}\Omega$$

High-Pass Section

The circuit is shown in Fig. 10.9b

$$CR = \frac{1}{2\ \omega_o Q} = \frac{1}{2 \times 2\pi \times 0.7911 \times 1.4 \times 10^3 \times 6.192}$$
$$= 1.16 \times 10^{-5} \text{ sec.}$$

Choosing C = 2700 pF then R = 4.32 KΩ

$$C_2 = C = 2700 \text{ pF}$$

$$C_3 = K_3C = 1042\text{pF}$$

$$C_4 = (1-K_3)C = 1658 \text{ pF}$$

$$R_1 = R = 4.32 \text{ K}\Omega$$

$$R_2 = (4\ Q^2)R = 662.5 \text{ K}\Omega$$

Notch Sections

The two notch sections have relatively high Q-factors and it is therefore advisable to use either the two-amplifier circuits given in Fig. 9.29 or the three amplifier circuit given in Fig. 9.38. Although the two-amplifier circuits give excellent performance, it is rather difficult to design them for specific values of gain constants. For this reason, we will give a design using the three-amplifier biquad shown in Fig. 10.9c. This circuit applies for both types of notches. For the high-pass notch [i.e. $T_2(s)$] we obtain the following component values

$$CR = \frac{1}{\omega_o} = \frac{1}{2\pi \times 0.7106 \times 1.4 \times 10^3} = 0.16 \times 10^{-3} \text{ sec.}$$

Choosing C = 16 nF, then R = 10 KΩ ,

$$QR = 152.5 \text{ K}\Omega$$

$$R_1 = \frac{QR}{K_1} = 571.2 \text{ K}\Omega$$

$$R_2 = \frac{R}{K_1} \left(\frac{\omega_o}{\omega_n}\right)^2 = 118.4 \text{ K}\Omega$$

Also, choosing r = 10 KΩ then

$$R_3 = \frac{r}{K_1} = 37.5 \text{ K}\Omega$$

The component values for the low-pass notch section are obtained as follows

$$CR = \frac{1}{\omega_o} = \frac{1}{2\pi \times 1.0048 \times 1.4 \times 10^3} = 0.113 \times 10^{-3} \text{ sec}$$

Choosing C = 51 nF, then R = 2.21 KΩ,

$$QR = 42.24 \text{ K}\Omega,$$

$$R_1 = \frac{QR}{K_2} = 123.9 \text{ K}\Omega,$$

$$R_2 = \frac{R}{K_2} \left(\frac{\omega_o}{\omega_n}\right)^2 = 4.19 \text{ K}\Omega.$$

Also, choosing r = 10 KΩ then

$$R_3 = \frac{r}{K_2} = 39.33 \text{ K}\Omega.$$

The complete circuit diagram of the bandpass filter can be easily sketched as the cascade of the sections designed above in the order: LP-HPN-LPN-HP. A total of eight amplifiers is required.

We should note that in the interest of saving on the number of amplifiers, the designer might choose to use the two-amplifier notch circuit with its fixed gain factor. The gains of the other sections should be modified accordingly. The result is that the gain distribution is no longer optimum. The degradation in performance could, however, be very little. If the gain distribution becomes too inconvenient, then in some cases one might accept a value for the total gain somewhat different from that specified. This usually has little effect on the system design and this step could be very well justified. A design using the two-amplifier notch circuit is shown in Fig. 10.10. The component values are obtained using the design equations given in Fig. 9.29. For the sake of variety, we have replaced the LP and HP sections by BP sections implemented using two-amplifier biquads. These BP sections will exhibit better sensitivity performance than the SAB sections. The reader is urged to verify the component values shown in Fig. 10.10.

Finally, if this filter is to be fabricated using thin film technology with the attendant possibility of automated tuning schemes, then one may use biquads that employ twin-T RC networks. In Chapter 9 we studied a number of such circuits; specifically SAB's of the NF2 and PF4 types. As pointed out there, biquads generated from twin-T RC networks have excellent active sensitivities but require tight tolerances on the twin-T components. Therefore, if one has at his disposal an economic process for initial adjustment and if the circuit is constructed using hybrid IC technology, then these circuits become very strong contenders. Fig. 10.11 presents a design of our BPF using biquads of the PF4 type for the notch sections. For design and construction simplicity, we have elected to use the gain factors naturally available with these circuits. Thus, the filter does not have the optimum gain distribution. The component values have been calculated using the design data given in Fig. 9.21 and the reader is urged to verify these values as an exercise.

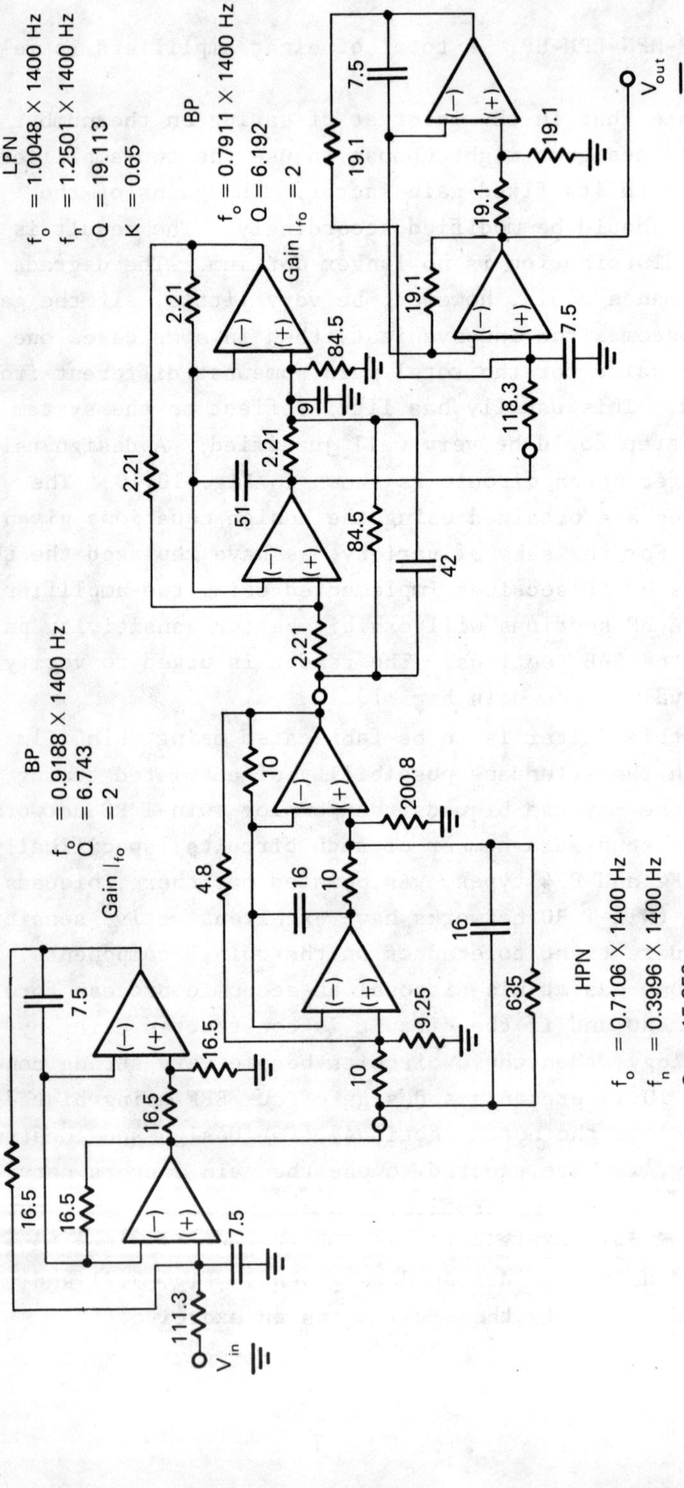

Notes: All resistors in kΩ and all capacitors in nF

Figure 10.10 An alternate realization for the BPF of Example 10.3 using two-amplifier biquads. The filter gain is approximately 20 and the gain distribution is not optimum.

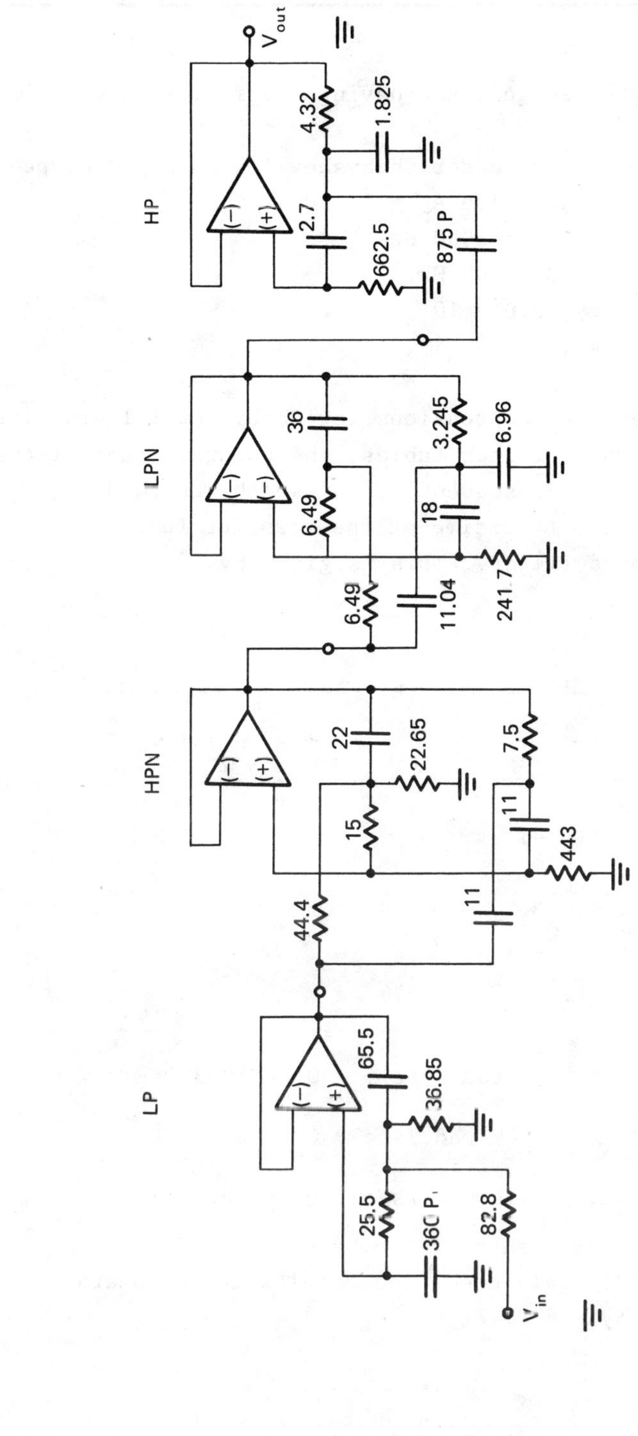

Figure 10.11 An alternate realization of the BPF of Example (10.3) using the minimum number of amplifiers. This circuit requires trimming of the notch sections. The gain distribution is not optimum. Note: Unless otherwise specified resitors are in KΩ and capacitors in nF.

Example 10.4

It is required to design a BPF having the following characteristics:

(1) Derived from a third-order Chebyshev low-pass prototype

(2) Equiripple-passband with $A_p = 0.1$ dB.

(3) Center-frequency $f_o = 1000$ Hz.

(4) Ripple bandwidth B = 100 Hz.

(5) Center-frequency gain = 10.

Give a cascade design.

Solution

Using the closed-form expressions for Chebyshev filters given in Chapter 4, or using filter tables, the natural modes of the LPP can be obtained. Subsequently, we use the LP to BP transformation of Section 6.3 to arrive at the transfer function of the sixth-order bandpass filter. This is given by

$$T(s) = K_1 \frac{s\left(\frac{\omega_{o_1}}{Q_1}\right)}{s^2 + s\left(\frac{\omega_{o_1}}{Q_1}\right) + \omega_{o_1}^2} \times K_2 \frac{s\left(\frac{\omega_{o_2}}{Q_2}\right)}{s^2 + s\left(\frac{\omega_{o_2}}{Q_2}\right) + \omega_{o_2}^2}$$

$$\times K_3 \frac{s\left(\frac{\omega_{o_3}}{Q_3}\right)}{s^2 + s\left(\frac{\omega_{o_3}}{Q_3}\right) + \omega_{o_3}^2}$$

where

$$\omega_{o_1} = 2\pi \times 1000 \text{ rad./sec.}, \quad Q_1 = 10.32$$

$$\omega_{o_2} = 2\pi \times 990.49 \text{ rad./sec.}, \quad Q_2 = 20.63$$

$$\omega_{o_3} = 2\pi \times 1009.6 \text{ rad./sec.}, \quad Q_3 = 20.63$$

and where the gain constants are to be determined such that $|T(j2\pi \times 1000)| = 10$. This condition results in the overall gain constant k given by $k = K_1 K_2 K_3 = 11.54$.

Note that we have done the pole-zero pairing such that all three sections have bandpass functions. This is practically convenient.

We choose $T_1(s)$ as the first section in the cascade since it is a bandpass function centered at the center-frequency of the filter. Moreover, $T_1(s)$ has the lowest Q and hence has the flattest response. We may follow $T_1(s)$ by either $T_2(s)$ or $T_3(s)$ with no difference at all. This is due to the geometric symmetry of the BPF. Therefore, we will select the sequence $T_1(s)\ T_2(s)\ T_3(s)$.

As far as the values of the gain constants are concerned, we use equations 10.1 and 10.2 as follows:

$$\text{Max } |T(j\omega)| = 10$$

$$K_1 = \frac{\text{Max}|T(j\omega)|}{\text{Max } |t_1(j\omega)|} = \frac{10}{1} = 10$$

$$K_2 = \frac{\text{Max } |t_1(j\omega)|}{\text{Max } |t_1(j\omega)t_2(j\omega)|} = \frac{1}{0.985} = 1.015$$

$$K_3 = \frac{k}{K_1K_2} = \frac{11.54}{10.15} = 1.137$$

For Q-factors in the range 10-20, the Friend biquad presented in Fig. 9.26 is a good choice. The desired gain factors can be easily obtained by grounding part of the input feed-in resistor. Fig. 10.12a shows the bandpass version of the Friend biquad where

$$CR = \frac{10}{\omega_o} \quad \text{and,}$$

$$\text{center-frequency gain} = \beta\ (10.2Q - 1)$$

Using these design equations, the complete realization of the order six bandpass filter is obtained and shown in Fig. 10.12b. As an exercise, the reader is encouraged to verify the component values given.

As mentioned in Section 9.7, the Friend biquad obtains low active sensitivities at the expense of increased passive sensitivities. If the latter are considered too high, we may use the two-amplifier GIC biquad described in Section 9.8. This circuit, in addition to having low sensitivities, exhibits excellent high-frequency performance. There is a slight disadvantage, however, as the maximum gain of the optimized bandpass two-amplifier configuration is only two. In the present application, the optimum gains vary from 1 to 10. Using equal gains will not obtain the maximum possible dynamiç range. As an exercise, the reader is encouraged to compute the component values and sketch the complete circuit realization of this BPF, using the two-amplifier biquad with the gain of each section equal to two. The total filter gain will be less than the specified value of 10. This, however, is a minor problem as the gain can be always restored somewhere else in the system or even by the addition of one more op. amp. to the filter circuit.

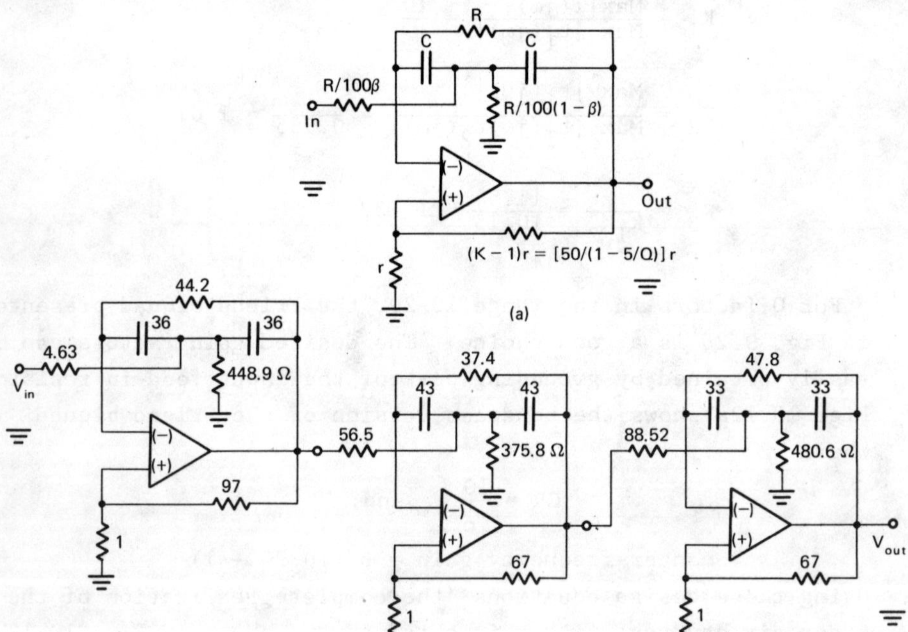

Figure 10.12 (a) The Friend bandpass circuit
(b) Complete realization for the bandpass filter of Example (10.4).
Note: Unless otherwise indicated, all resistors are in KΩ and all capacitors are in nF.

Example 10.5

It is required to design a low-pass filter having the following specifications:

(a) Equiripple passband, with $A_p \leq 0.5$ dB, extending to $f_p =$ 1 KHz.

(b) Minimum stopband attenuation of 50 dB with $f_s = 1.5$ KHz.

(c) Passband gain = 1.

Give a cascade design.

Solution

To find a transfer function T(s) satisfying the given specifications, we may either use the computer methods of Chapter 5 or alternatively use filter tables to obtain a classical approximation. In fact, we have already obtained an elliptic approximation satisfying these specifications, except for a factor of 10 in frequency, in Example 6.4. This elliptic approximation has the following transfer function

$$T(s) = K \frac{(s^2 + 2.300870)(s^2 + 4.0309918)}{(s^2 + 0.131886s + 1.044839)(s^2 + 0.448076s + 0.709900)} \times \frac{1}{(s^2 + 0.763250s + 0.252001)}$$

which is normalized with respect to the passband edge, $f_p = 1$ KHz. For a unity passband gain, the constant K should be equal to 0.019514.

Toward obtaining a cascade design, we first have to solve the decomposition problem. The pole-zero assignment will be simply done using the rule-of-thumb stating that high-Q poles should be combined with the closest transmission zeros. This results in the following biquad transfer functions

$$T_1(s) = K_1 \frac{s^2 + 2.300870}{s^2 + 0.131886\ s + 1.044839}$$

$$T_2(s) = K_2 \frac{s^2 + 4.0309918}{s^2 + 0.448076\ s + 0.709900}$$

$$T_3(s) = K_3 \frac{1}{s^2 + 0.763250\ s + 0.252001}$$

Thus we have three sections characterized by:

Section #1: LPN with $\Omega_o = 1.022173$, $\Omega_z = 1.516862$ and $Q = 7.75$
Section #2: LPN with $\Omega_o = 0.842556$, $\Omega_z = 2.007733$ and $Q = 1.88$
Section #3: LP with $\Omega_o = 0.501997$, and $Q = 0.658$.

The problem of choosing a particular sequence for cascading the three sections will be solved using the rule-of-thumb stating that the Q-factors should progressively increase throughout the cascade. This results in the sequence: $T_3(s)\ T_2(s)\ T_1(s)$. Note also that the first section is an LP one, a desirable property.

To obtain the values of the gain constants K_1, K_2 and K_3, we use equations (10.1) and (10.2) as follows:

$$K_3 = \frac{\text{Max}\ |T(j\omega)|}{\text{Max}\ |t_3(j\omega)|} = \frac{1}{3.968} = 0.252$$

$$K_2 = \frac{\text{Max}\ |t_3(j\omega)|}{\text{Max}\ |t_3(j\omega)t_2(j\omega)|} = \frac{3.968}{22.63} = 0.175$$

$$K_1 = \frac{K}{K_2 K_3} = 0.618$$

Having completely determined the transfer functions of the individual biquads, we consider their circuit implementation. Since the Q-factors involved are rather low, we may use circuits of the NF1 type discussed in Section 9.4. Specifically, for the notch sections we will use the NF1 circuit given in Fig. 9.8 with the design equations in Table 9.2, while for the LP section we use the circuit generated from the ABT1 of Fig. 9.6. The two circuits are redrawn in Fig. 10.13. The design details follow:

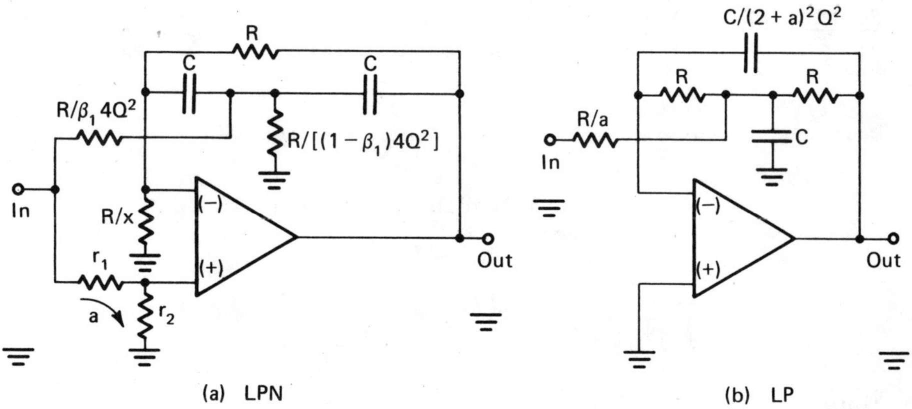

Figure 10.13 Circuits of the NF1 type to be used in the design of the LPF of Example (10.5). These circuits are discussed in detail in Section (9.4).

<u>Section #1</u>

$$CR = \frac{2Q}{\omega_o} = 2.413390 \times 10^{-3} \text{ sec.}$$

Choosing C = 4.3 nF, then R = 562 KΩ

$a = K_1 = 0.618$, leading to $r_1 = 0.618\, r_2$

$$\frac{\beta_1}{1 + \dfrac{1}{2Q^2}\dfrac{\omega_z^2}{\omega_o^2}} = 0.618 \quad .$$

Thus,

$$\frac{R}{\beta_1\, 4Q^2} = 0.006614R = 3.72 \text{ K}\Omega \ ,$$

$$\frac{R}{(1-\beta_1)4Q^2} = 0.011229R = 6.31 \text{ K}\Omega \ ,$$

$$\frac{R}{x} = \frac{R}{\left(\dfrac{\omega_z^2}{\omega_o^2} - 1\right)} = 0.831855R = 467.5 \text{ K}\Omega.$$

Section #2

$$CR = \frac{2Q}{\omega_o} = 0.710247 \times 10^{-3} \text{ sec.},$$

Choosing C = 43 nF, then R = 16.5 KΩ

$$a = K_2 = 0.175 \text{ leading to } r_1 = 4.714\, r_2$$

$$\frac{\beta_1}{1 + \frac{1}{2Q^2}\frac{\omega_z^2}{\omega_o^2}} = 0.175 ,$$

Thus,

$$\frac{R}{\beta_1 4Q^2} = 0.224141R = 3.7 \text{ K}\Omega ,$$

$$\frac{R}{(1-\beta_1)4Q^2} = 0.103347R = 1.7 \text{ K}\Omega ,$$

$$\frac{R}{x} = 0.213755R = 3.53 \text{ K}\Omega$$

Section #3

$$a\Omega_o^2 = K_3 = 0.252$$

Thus, $a = 1$,

$$\frac{R}{a} = R$$

$$CR = \frac{(2+a)Q}{\omega_o} = 0.625844 \times 10^{-3} \text{ sec.}$$

Choosing R = 19 KΩ, then C = 31.93 nF,

$$\frac{C}{(2+a)^2 Q^2} = 0.256629\, C = 8.19 \text{ nF} .$$

The complete circuit of the 6th order elliptic LPF may be easily sketched using the component values obtained in the above.

Example 10.6

In designing a bandpass filter for a communication system application, it was found necessary, in order to meet the specified maximum group delay deviation, to include an equalizer. The equalizer required is a sixth-order one with the following transfer function

$$T(s) = \frac{s^2 - s\frac{\omega_{o_1}}{Q_1} + \omega_{o_1}^2}{s^2 + s\frac{\omega_{o_1}}{Q_2} + \omega_{o_1}^2} \cdot \frac{s^2 - s\frac{\omega_{o_2}}{Q_2} + \omega_{o_2}^2}{s^2 + s\frac{\omega_{o_2}}{Q_2} + \omega_{o_2}^2} \cdot \frac{s^2 - s\frac{\omega_{o_3}}{Q_3} + \omega_{o_3}^2}{s^2 + s\frac{\omega_{o_3}}{Q_3} + \omega_{o_3}^2}$$

where $f_{o1} = 1.55$ KHz , $Q_1 = 2.55$

$f_{o2} = 2.222$ KHz , $Q_2 = 3.53$

$f_{o3} = 2.894$ KHz , $Q_3 = 4.823$

Give a complete circuit implementation for this equalizer.

Solution

In Chapter 9 we presented a number of circuits for realizing second-order all-pass functions. Since the Q-factors under consideration are relatively low, we may use an all-pass circuit of the NF1 type. Specifically, we use the all-pass circuit presented in Fig. 9.8 and redrawn in Fig. 10.14a. From Table 9.2 the design equations are given by,

$$CR = \frac{2Q}{\omega_o}$$

$$a = \frac{r_2}{r_1 + r_2} = 1 \,/\, (1 + \frac{1}{Q^2})$$

$$\text{Gain} = a$$

Using the given values for f_{o1} and Q_1 in the above design equations, the circuit of Fig. 10.14b is obtained. Note that this equalizer implementation results in signal attenuation, as the total gain is 0.77. The reader is urged to verify these results.

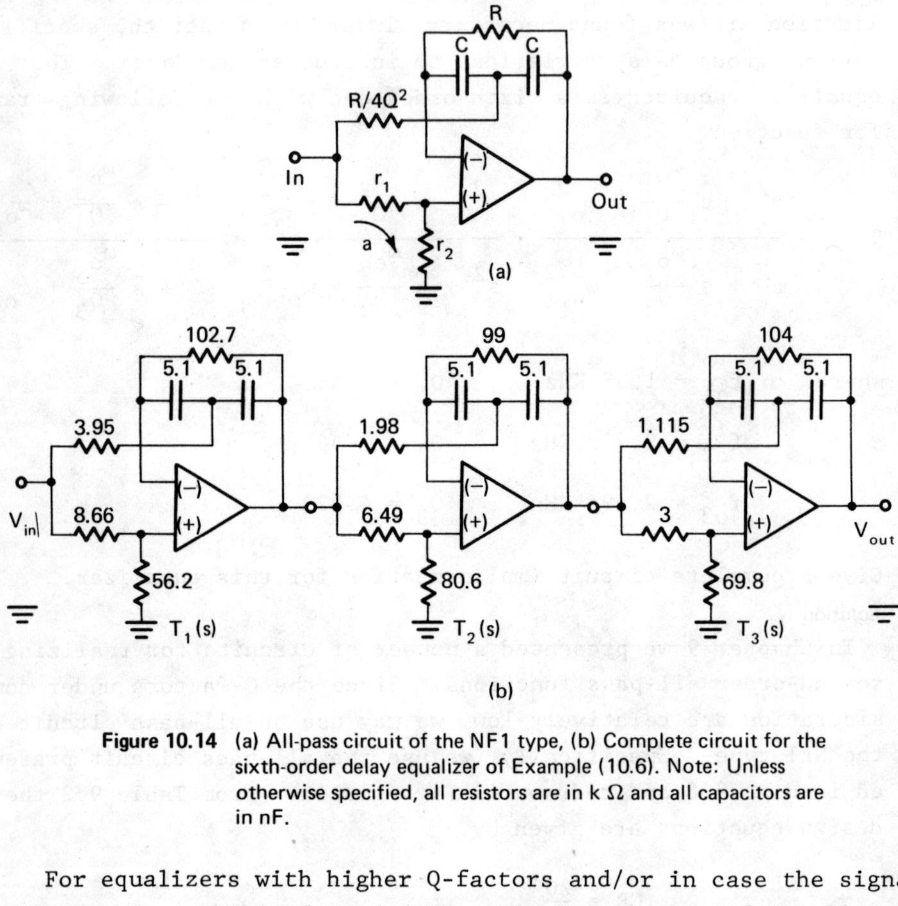

Figure 10.14 (a) All-pass circuit of the NF1 type, (b) Complete circuit for the sixth-order delay equalizer of Example (10.6). Note: Unless otherwise specified, all resistors are in k Ω and all capacitors are in nF.

For equalizers with higher Q-factors and/or in case the signal attenuation exhibited in designs using single-amplifier all-pass sections is objectionable, one may use either the two-amplifier circuit given in Fig. 9.29 or the three-amplifier circuit of Fig. 9.38. As an exercise the reader is encouraged to work out an alternative design of the sixth-order equalizer, using the two-amplifier circuit of Fig. 9.29 (circuit #3). Since the gain of each section is unity, the overall equalizer gain will also be unity.

10.3 THE "FOLLOW-THE-LEADER FEEDBACK" (FLF) STRUCTURE

A number of methods[8-16] for the design of high-order active filters attempt to provide compensatory internal interactions between the different filter sections through coupling the biquad building blocks by negative feedback paths. The resulting internal interactions resemble those that occur in LC ladder networks and thus, "Coupled-Biquad" active filters exhibit lower sensitivities to component variations than cascade filters.

In this book, we will study two coupled-biquad structures, namely the Follow-the-Leader feedback (FLF) structure and the Leap-Frog (LF) structure. The LF structure results from the operational simulation of LC ladders and hence will be discussed in Chapter 12. The present section is devoted to the study of the FLF structure and the design methods utilizing it.

A block diagram of the FLF structure is shown in Fig. 10.15. It consists of n biquad sections with negative feedback applied from their outputs back to the input of the system. Note that the feedback path around the first block is not explicitly shown as it may be considered part of the first block. The name Follow-the-Leader Feedback stems from the resemblance of the structure to the famous children's game. Also, if each block is replaced by an integrator, then the resulting structure is the state-variable system representation known as the "companion-form". For this reason, the name "shifted companion form" (SCF) has been used to refer to the FLF structure and its associated design method[16]. The reason for the word "shifted" will become apparent shortly.

There are a number of design methods based on the FLF structure. In the following, we present some of these design techniques.

Design of All-Pole Bandpass Filters

The FLF structure is particularly suited for the design of all-pole bandpass filters. The design method to be described was introduced by Hurtig[11-13], who has given this technique the name Primary Resonator Block (PRB).

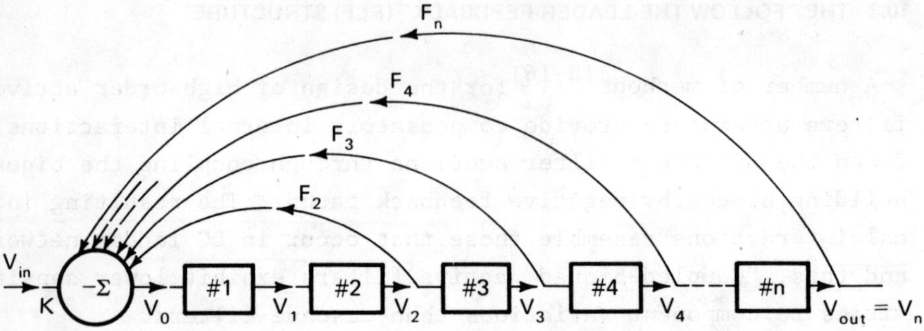

Figure 10.15 Block diagram of the FLF structure. Note that F_1 is not explicitly shown as it can be absorbed in the first block.

The starting point in the design is the transfer function T(s) of the low-pass prototype (LPP). Let T(s) be expressed in the form

$$T(s) = \frac{1}{s^n + a_{n-1} s^{n-1} + \ldots + a_o} \tag{10.3}$$

where s is the complex-frequency variable in the normalized LPP domain (See Chapter 6). To realize T(s) by the FLF structure of Fig. 10.15, we first derive an expression for its transfer function. This will be done assuming that all blocks have low output impedances (ideally zero). Thus we can write,

$$V_o = \frac{V_{out}}{T_1(s)\, T_2(s) \ldots T_n(s)} = \frac{V_{out}}{\prod_{i=1}^{n} T_i(s)}, \tag{10.4}$$

$$-V_o = K\, V_{in} + F_2 V_2 + F_3 V_3 + \ldots + F_n V_n$$

$$= K\, V_{in} + F_2 T_1 T_2 V_o + F_3 T_1 T_2 T_3 V_o + \ldots + F_n T_1 T_2 \ldots T_n V_o$$

Thus,

$$V_o = \frac{-\,K\, V_{in}}{1 + \sum_{j=2}^{n} F_j \prod_{i=1}^{j} T_i(s)} \tag{10.5}$$

Combining (10.4) and (10.5) obtains the transfer function of the FLF structure as

$$\frac{V_{out}(s)}{V_{in}(s)} = - \frac{K \prod_{i=1}^{n} T_i(s)}{1 + \sum_{j=2}^{n} F_j \prod_{i=1}^{j} T_i(s)}$$

The PRB design technique is based on all the blocks being identical. For the realization of the LPP, each block is characterized by a first-order transfer function; that is,

$$T_i(s) = \frac{k}{s+k} \quad , \qquad i = 1,2, \ldots, n \tag{10.7}$$

where k is a positive constant.* Substituting in (10.6) obtains

$$\frac{V_{out}(s)}{V_{in}(s)} = - \frac{K\ k^n}{(s+k)^n + \sum_{j=2}^{n} F_j\ k^j\ (s+k)^{n-j}} \tag{10.8}$$

For the structure of Fig. 10.15 to realize the LPP transfer function of equation (10.3), the denominators of the transfer functions in (10.3) and (10.8) should be identical; that is,

$$(s+k)^n + \sum_{j=2}^{n} F_j\ k^j\ (s+k)^{n-j} = s^n + a_{n-1}\ s^{n-1} + \ldots + a_o \tag{10.9}$$

---

*The companion-form state-variable representation is based on each block having a transfer function of $\frac{1}{s}$. Since the PRB method is based on blocks having transfer function $\frac{k}{s+k}$, we may consider k as a "shift" parameter and call the structure the shifted-companion form (SCF).

Equating the coefficients of corresponding powers of s on both sides of (10.9) results in n equations in the n unknowns; k, F_2, F_3, ..., F_n. The solution can be easily obtained as,

$$k = \frac{a_{n-1}}{n}, \tag{10.10}$$

$$F_2 = \frac{a_{n-2}}{k^2} - \frac{n(n-1)}{2!} \quad \text{and,} \tag{10.11}$$

$$F_i = \frac{a_{n-i}}{k^i} - \frac{n!}{(n-i)!\,i!} - \frac{1}{(n-i)!}\sum_{\ell=2}^{i-1} F_\ell \frac{(n-\ell)!}{(i-\ell)!}, \tag{10.12}$$

$$i=3, \ldots, n$$

Thus, given the transfer function of the LPP, all the feedback coefficients and the constant k are determined using (10.10) - (10.12).

Let the required bandpass filter have a center frequency ω_o and a bandwidth B. It follows that the LP to BP transformation takes the form (See Section 6.3)

$$s = \frac{p^2 + \omega_o^2}{pB}$$

where p is the complex frequency variable in the BP domain. Applying this transformation to the FLF block diagram realization of the LPP results in an identical block diagram, with each block characterized by a bandpass transfer function $T_{BP}(p)$ obtained as follows

$$T_{BP}(p) = \left.\frac{k}{s+k}\right|_{s = \frac{p^2 + \omega_o^2}{pB}}$$

$$\text{Thus, } T_{BP}(p) = \frac{pkB}{p^2 + pkB + \omega_o^2} \tag{10.13}$$

The feedback coefficients obviously remain unchanged.

Thus, all the building blocks of the BPF will have identical second-order bandpass functions with a center frequency ω_o and a Q-factor given by,

$$Q = \frac{\omega_o}{kB}$$

$$\text{i.e.} \qquad Q = \frac{\hat{Q}}{k} \tag{10.14}$$

where $\hat{Q} = \frac{\omega_o}{B}$ is the equivalent Q-factor of the Nth order BPF. The center-frequency gain of the complete BPF is equal to the DC gain of its low-pass prototype; that is, $\frac{Kk^n}{a_o}$ where K is the summing coefficient assigned to the input signal $V_{in}(p)$ in Fig. 10.15. Obviously, the value of K may be used to adjust the gain to any desired level.

Maximizing the Filter Dynamic Range

We now consider the problem of scaling the gains of the bandpass sections in a way that maximizes the dynamic range of the overall filter. In doing this, however, we have to change the feedback coefficients such that the gain of every loop remains constant, otherwise the filter transfer function will change.

The gain assignment process can be more easily done on the realization of the low-pass prototype. Let the ith block be assigned a gain constant K_i; that is,

$$T_i(s) = K_i \frac{k}{s+k} \quad , \qquad i = 1,2,\ldots,n$$

We also assume that the input signal has a uniform frequency spectrum; that is,

$$V_{in}(s) = 1 \quad .$$

Now, if the desired center-frequency gain of the bandpass filter is denoted by m, we can write

$$V_n(s) = \frac{m\, a_o}{s^n + a_{n-1}\, s^{n-1} + \ldots + a_o} = T(s)$$

$$V_{n-1}(s) = \frac{1}{K_n}\left(1 + \frac{s}{k}\right) T(s)$$

$$V_{n-2}(s) = \frac{1}{K_n K_{n-1}}\left(1 + \frac{s}{k}\right)^2 T(s)$$

$$V_i(s) = \frac{1}{K_n K_{n-1} \cdots K_{i+1}}\left(1 + \frac{s}{k}\right)^{n-i} T(s) \qquad (10.15)$$

$$V_o(s) = \frac{1}{K_n K_{n-1} \cdots K_1}\left(1 + \frac{s}{k}\right)^n T(s)$$

The optimum gain assignment, from the point of view of maximum dynamic range, is the one that results in the spectra of all the signals $V_i(j\Omega)$, $i=0, 1, \ldots, n$ having equal maxima over the entire frequency band, $0 \le \Omega \le \infty$. That is, we define $V_{i_{max}}$ = Max $|V_i(j\Omega)|$, $0 \le \Omega \le \infty$ and attempt to find the set of gain factors K_i, $i = 1,2,\ldots, n$ that results in

$$V_{n_{max}} = V_{n-1_{max}} = \ldots = V_{i_{max}} = \ldots = V_{o_{max}} \qquad (10.16)$$

Equations (10.15) and (10.16) result in the set of optimum gains given by

$$K_i = \frac{\text{Max}\left\{|T(j\Omega)| \left(1 + \frac{\Omega^2}{k^2}\right)^{\frac{n-i+1}{2}}\right\}}{\text{Max}\left\{|T(j\Omega)| \left(1 + \frac{\Omega^2}{k^2}\right)^{\frac{n-i}{2}}\right\}}, \quad i = 1,2,\ldots,n \qquad (10.17)$$

The values of K_i in (10.17) can be easily evaluated with the aid of a calculator.

For illustration, Fig. 10.16 shows the signal spectra at the outputs of the various biquads in a filter with idealized constant passband transmission excited by a signal with a uniform spectrum.

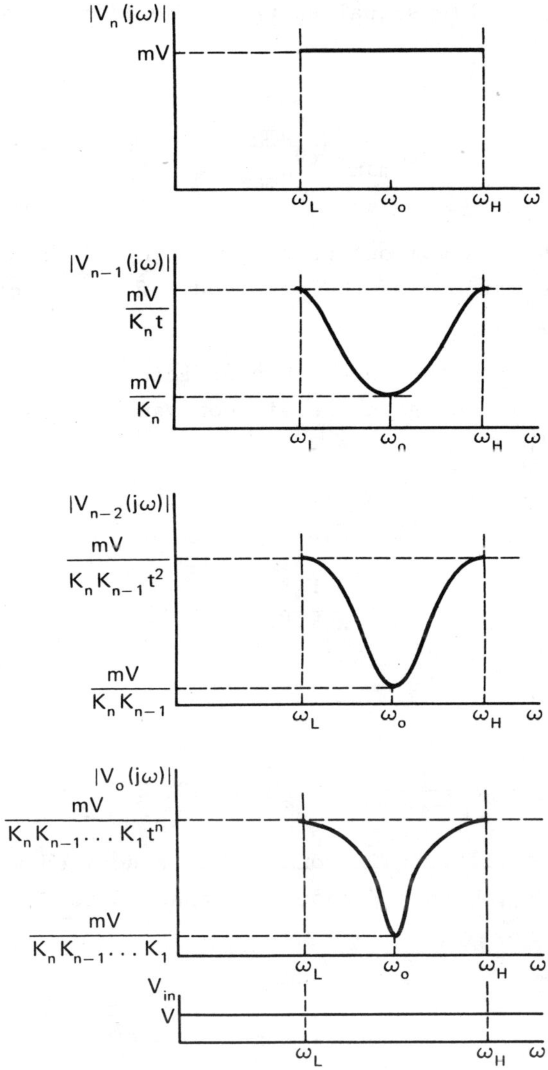

Figure 10.16 Illustrating the signal spectra at the outputs the various biquads in a PRB filter. The filter is assumed to have a constant passband transmission with a total gain of m. The excitation is assumed to have a uniform spectrum with a value V. The value of t is given by $t = 1/\sqrt{1 + (1/k^2)}$.

If the optimum gain assignment given by (10.17) is used, then the maximum permissible signal swing at the filter input will be given by

$$V_{in_{max}} = \frac{V_{amp}}{V_{n_{max}}}$$

where V_{amp} is the maximum output signal swing of the op. amp. and $V_{n_{max}}$ is the maximum of the magnitude of the specified output/input transfer function.

As mentioned above, the feedback coefficients have to be modified such that all loop gains remain constant. This leads to the modified values,

$$F'_2 = \frac{F_2}{K_1 K_2}$$

$$F'_3 = \frac{F_3}{K_1 K_2 K_3}$$

$$F'_i = \frac{F_i}{K_1 K_2 .. K_i} \quad , \quad i = 2,3, \ldots, n$$

Also, to keep the filter gain constant at a desired value m, the input multiplying constant K should be changed to K' where

$$K' = \frac{K}{\prod_{i=1}^{n} K_i}$$

Since,

$$m = \frac{K\, k^n}{a_o}$$

then,

$$K' = \frac{m\,a_o}{k^n \prod_{i=1}^{n} K_i} \tag{10.18}$$

This completes the design equations for all-pole bandpass filters using the FLF structure. For easy reference, the results are summarized in Table 10.2 where the primes are dropped from F_i and K.

Table 10.2 Design procedure for all-pole bandpass filters using the FLF structure

1. Let the LPP be characterized by

$$T(s) = \frac{m\,a_o}{s^n + a_{n-1}s^{n-1} + \ldots + a_o}$$

where m is the desired center-frequency gain of the BPF.

2. Calculate the constant k from

$$k = \frac{a_{n-1}}{n}$$

3. Evaluate the gain constants K_i to be associated with each of the bandpass sections in Fig. 10.15 from

$$K_i = \frac{\text{Max}\left\{|T(j\Omega)|\ (1 + \frac{\Omega^2}{k^2})^{\frac{n-i+1}{2}}\right\}}{\text{Max}\left\{|T(j\Omega)|\ (1 + \frac{\Omega^2}{k^2})^{\frac{n-i}{2}}\right\}} \qquad i = 1,2,\ldots,n \quad \text{and } 0 \le \Omega \le \infty.$$

4. Evaluate the feedback coefficients F_i from

$$F_2 = \left[\frac{a_{n-2}}{k^2} - \frac{n(n-1)}{2!}\right] \Big/ K_1 K_2$$

$$F_i = \frac{\left[\frac{a_{n-i}}{k^i} - \frac{n!}{(n-i)!\,i!} - \frac{1}{(n-i)!}\sum_{\ell=2}^{i-1} F_\ell\,(K_1 \ldots K_\ell)\,\frac{(n-\ell)!}{(i-\ell)!}\right]}{\prod_{j=1}^{i} K_j}$$

5. The input summing coefficient K is given by

$$K = \frac{m\,a_o}{k^n \prod_{i=1}^{n} K_i}$$

6. If the bandpass filter is required to have a center-frequency ω_o and a bandwidth B, that is an equivalent Q-factor $\hat{Q} = \frac{\omega_o}{B}$, then each of the second-order building blocks in Fig. 10.15 should have a bandpass function with a center-frequency ω_o, a Q-factor,

$$Q = \frac{\hat{Q}}{k} \quad \text{and,}$$

a center frequency gain K_i given in the above.

Note: The resulting BPF design is optimum from a dynamic range point of view.

Example 10.7

It is required to design a BPF having the following characteristics:

(1) Derived from a 3rd order Chebyshev LPP.
(2) Equiripple passband with $A_p = 0.1$ dB.
(3) Center frequency $f_o = 1000$ Hz.
(4) Ripple bandwidth B = 100 Hz.
(5) Center-frequency gain = 10.

Give a PRB design* optimized for maximum dynamic range.

*Note that we have obtained a cascade design for this filter in example 10.4.

Solution

Using the closed form expressions given in Chapter 4, or from filter tables, the following transfer function for the low-pass prototype can be obtained:

$$T(s) = \frac{16.38051}{s^3 + 1.938812\ s^2 + 2.629495\ s + 1.638051}$$

Note that we have chosen the value of the gain constant to yield a DC gain of 10, which is the desired center-frequency gain of the bandpass filter.

Following the step-by-step design procedure of Table 10.2, we obtain

$$k = \frac{1.938812}{3} = 0.646271$$

$$K_3 = \frac{\text{Max } \{|T(j\Omega)|\ (1 + \frac{\Omega^2}{k^2})^{1/2}\}}{\text{Max } \{|T(j\Omega)|\}} \simeq \frac{18.85}{10} = 1.885$$

$$K_2 = \frac{\text{Max } \{|T(j\Omega)|\ (1 + \frac{\Omega^2}{k^2})\}}{\text{Max } \{|T(j\Omega)|\ (1 + \frac{\Omega^2}{k^2})^{1/2}\}} \simeq \frac{40.63}{18.85} = 2.155$$

$$K_1 = \frac{\text{Max } \{|T(j\Omega)|\ (1 + \frac{\Omega^2}{k^2})^{3/2}\}}{\text{Max } \{|T(j\Omega)|\ (1 + \frac{\Omega^2}{k^2})\}} \simeq \frac{95}{40.63} = 2.338$$

$$F_2 = 0.654115$$

$$F_3 = 0.186667$$

$$K = 6.389697$$

$$\hat{Q} = \frac{1000}{100} = 10$$

$$Q = 15.47$$

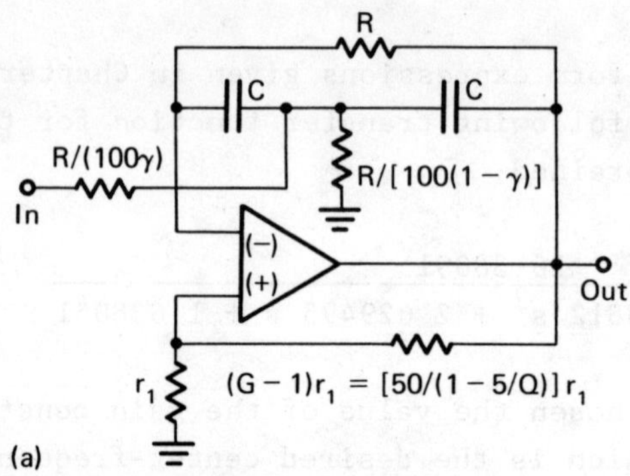

Figure 10.17 (a) The bandpass version of the Friend biquad. (b) Complete circuit for the bandpass filter of Example (10.7).

For the value of Q-factor involved, the Friend biquad presented in Section 9.7, Fig. 9.26, would be a good choice for implementing the bandpass sections. The bandpass version of the Friend biquad is redrawn in Fig. 10.17a where

$$CR = \frac{10}{\omega_o} \quad \text{and,}$$

$$\text{the center-frequency gain} = \gamma \frac{50}{1 - \frac{51}{G}}$$

For our example, all the sections are identical except for the value of γ and the two corresponding resistors. Thus we have,

$$CR = \frac{10}{2\pi \times 1000} = 1.5915 \text{ msec.}$$

$$G = 1 + \frac{50}{1 - \frac{5}{Q}} = 74.8777$$

γ for the first section = 0.0149
γ for the second section = 0.0137
γ for the third section = 0.012

The complete circuit implementation is shown in Fig. 10.17b.

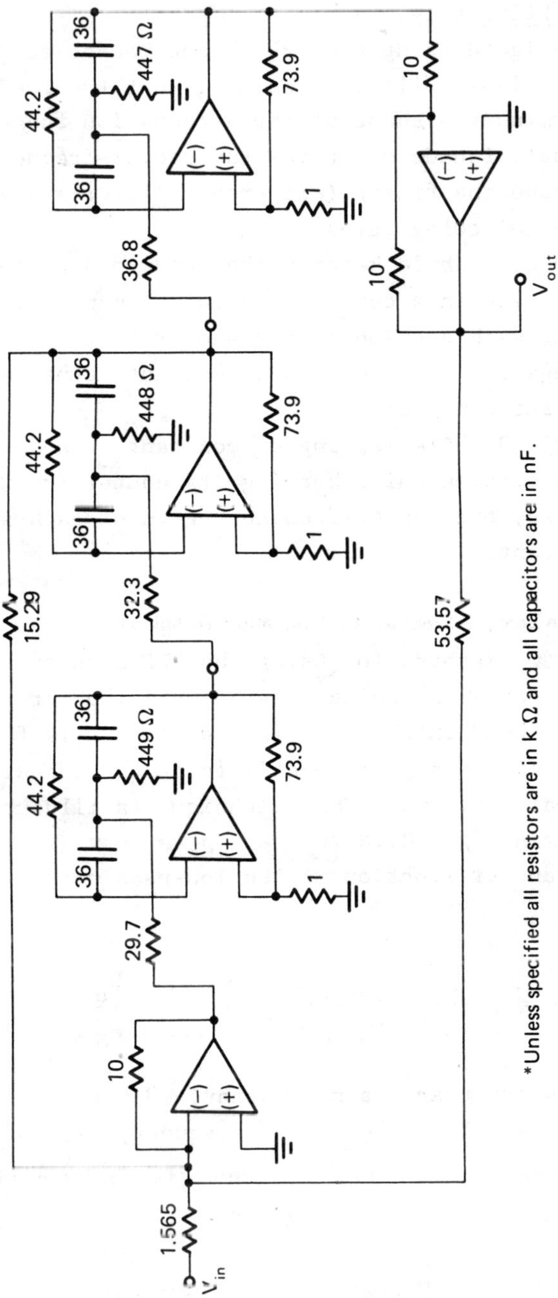

*Unless specified all resistors are in k Ω and all capacitors are in nF.

(b)

Tunable PRB Filters

Filters designed using the PRB method can be easily used as tunable or variable filters. We note that the feedback coefficients determine the shape of the attenuation function, while the individual biquads determine the center-frequency and bandwidth. To tune the filter (electronically or manually), one has to apply the following rules:

(1) To change ω_o while keeping the bandwidth B constant, tune each section in a constant bandwidth manner i.e. vary ω_o and Q of each section in the same ratio.

(2) To change ω_o while keeping $\hat{Q}$ constant, tune each section in a constant Q manner.

(3) To change B while keeping ω_o constant, change the bandwidth of each section while keeping its center frequency constant.

In all cases, the center-frequency gain of each section should be kept constant.

Finite Transmission Zeros Realized by the Summation Technique

There are two methods for using the FLF structure to realize filter functions with finite loss-poles. The first is the "Summation Method" in which the filter output is formed as the weighted sum of the signals available at the output terminals of the individual sections. This technique is illustrated in block diagram form in Fig. 10.18.

Let the transfer function of the low-pass prototype be given by

$$T(s) = K \frac{b_n s^n + b_{n-1} s^{n-1} + \ldots + b_o}{s^n + a_{n-1} s^{n-1} + \ldots + a_o} \tag{10.19}$$

where K is an arbitrary gain-constant. Hence, the DC gain, which will be equal to the center-frequency gain of the resulting BPF, is given by $m = K \frac{b_o}{a_o}$. Thus, for a specified center frequency gain m, the value of the parameter K is given by

$$K = m \frac{a_o}{b_o} \tag{10.20}$$

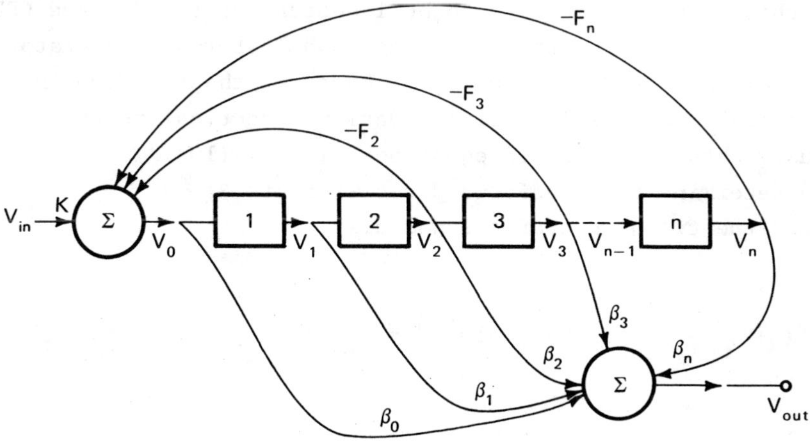

Figure 10.18 The FLF structure with finite transmission zeros realized by the summation technique.

The transfer function of the FLF structure in Fig. 10.18 can be easily derived and is given by,

$$\frac{V_{out}(s)}{V_{in}(s)} = K \frac{\beta_o + \sum_{j=1}^{n} \beta_j \prod_{i=1}^{j} T_i(s)}{1 + \sum_{j=2}^{n} F_j \prod_{i=1}^{j} T_i(s)} \tag{10.21}$$

For the PRB case all the sections are identical and have first-order low-pass transfer functions of the form

$$T_i(s) = \frac{k}{s+k} \quad , \quad i = 1,2,...,n$$

Substituting in (10.21) obtains

$$\frac{V_{out}(s)}{V_{in}(s)} = K \frac{\beta_o \; (s+k)^n + \sum_{j=1}^{n} \beta_j \; k^j \; (s+k)^{n-j}}{(s+k)^n + \sum_{j=2}^{n} F_j \; k^j \; (s+k)^{n-j}} \tag{10.22}$$

For this transfer function to be identical to the desired LPP T(s) of (10.19), the corresponding denominator and numerator polynomials should be identical. Equating the two denominator polynomials results in design equations identical to those previously obtained, that is equations (10.10)-(10.12).

To determine the feedforward coefficients $\beta_o, \beta_1, \ldots, \beta_n$, we equate the two numerator polynomials,

$$\beta_o(s+k)^n + \sum_{j=1}^{n} \beta_j k^j (s+k)^{n-j} = b_n s^n + b_{n-1} s^{n-1} + \ldots + b_o \tag{10.23}$$

Matching the coefficients of the like powers of s on both sides results in (n+1) equations in the (n+1) unknowns $\beta_o, \beta_1, \ldots, \beta_n$. Towards the solution of these equations, we note that $\beta_o = b_n$ and consider the two cases:

(a) $b_n \neq 0$

Dividing both sides of (10.23) by b_n obtains

$$(s+k)^n + \sum_{j=1}^{n} \overline{\beta_j} k^j (s+k)^{n-j} = s^n + \overline{b_{n-1}} s^{n-1} + \ldots + \overline{b_o} . \tag{10.24}$$

where $\overline{\beta_j} = \frac{\beta_j}{\beta_o}$ and $\overline{b_j} = \frac{b_j}{b_n}$.

Now equating the coefficients of the like powers of s on both sides of (10.24) results in the solution,

$$\overline{b}_{n-j} = \frac{n!}{(n-j)!\ j!} k^j + k^j \sum_{i=1}^{j} \overline{\beta_i} \frac{(n-i)!}{(n-j)!\ (j-i)!} , \quad j = 1,2,\ldots,n. \tag{10.25}$$

(b) $b_n = 0$

Equation (10.23) reduces to

$$\sum_{j=1}^{n} \beta_j \, k^j \, (s+k)^{n-j} = b_{n-1} \, s^{n-1} + \ldots + b_o$$

Equating the coefficients of like powers of s on both sides results in the solution

$$b_{n-j} = k^j \sum_{i=1}^{j} \beta_i \frac{(n-i)!}{(n-j)!\,(j-i)!} \quad , \tag{10.26}$$

$$j = 1,2,\ldots,n$$

Once the feedback and feedforward coefficients are determined, the LP to BP transformation can be applied to the circuit. The resulting BPF realization will have n identical second-order bandpass sections characterized by a center-frequency ω_o and a Q-factor,

$$Q = \frac{\hat{Q}}{k}$$

The center-frequency gain of each section is unity. This, however, may be changed in order to optimize the filter network for maximum dynamic range. We now address ourselves to the problem of assigning gain factors to each section, such that the permissible signal swing at the input is maximized. As in the all-pole case, it is more convenient to carry-out this optimization process on the low-pass prototype. Referring to the block diagram of Fig. 10.18 we assume that the input signal $V_{in}(s)$ has a constant frequency-spectrum with an arbitrary amplitude, say unity. Each section is assigned a gain factor K_i; that is,

$$T_i(s) = K_i \frac{k}{s+k} \quad , \quad i = 1,2, \ldots, n$$

Furthermore, the feedback coefficients F_I, the output summing coefficients β_i and the input gain constant K have to be modified in order to keep the filter transfer function, given by (10.19), unchanged. We shall denote the new values of these parameters, which will result in the filter having the widest possible dynamic range, by F_i'', β_i'' and K''.

The maximum dynamic range will be obtained if the signal spectra at the outputs of all sections have equal maxima, that is

$$V_{out_{max}} = V_{n_{max}} = V_{n-1_{max}} = \ldots = V_{i_{max}} = \ldots = V_{o_{max}} \tag{10.27}$$

where $V_{i_{max}}$ is defined as

$$V_{i_{max}} = \text{Max}\ \{|V_i(j\Omega)|\} \quad , \quad 0 \le \Omega \le \infty$$

Consider first the output signal, $V_{out}(s)$. The value of $V_{out_{max}}$ is specified by the transfer function T(s) and is thus readily available. On the other hand, the signal $V_n(s)$ is given by

$$V_n(s) = K \frac{k^n}{s^n + a_{n-1} s^{n-1} + \ldots + a_o} \tag{10.28}$$

and to make its maximum $V_{n_{max}}$ equal to $V_{out_{max}}$, we have to change the value of K to K' where

$$K' = K \left(\frac{V_{out_{max}}}{V_{n_{max}}}\right)$$

Denoting

$$\frac{V_{out_{max}}}{V_{n_{max}}} = q,$$

then,

$$K' = Kq = \left(\frac{ma_o}{b_o}\right) q \tag{10.29}$$

In order to keep $V_{out}(s)$ unchanged, we have to simultaneously modify the feedforward coefficients in the following manner

$$\beta_i' = (\beta_i/q)$$

Having forced $V_{n_{max}}$ to equal $V_{out_{max}}$, we now proceed to make $V_{(n-1)_{max}}, V_{(n-2)_{max}}, \ldots, V_{o_{max}}$ all equal to $V_{n_{max}}$. This problem is identical to the one we have already solved in the all-pole case. Thus, the optimum gain factor K_i for the ith section is given by

$$K_i = \frac{\text{Max}\ \{|T_n(j\Omega)|\ (1 + \frac{\Omega^2}{k^2})^{\frac{n-i+1}{2}}\}}{\text{Max}\ \{|T_n(j\Omega)|\ (1 + \frac{\Omega^2}{k^2})^{\frac{n-i}{2}}\}}, \quad i = 1,2,\ldots,n \tag{10.30}$$

where $T_n(s)$ is the transfer function from the input to the output of the nth section. This transfer function is identical to that on the right-hand-side of (10.28) except for K replaced by K'; that is,

$$T_n(s) = K' \frac{k^n}{s^n + a_{n-1} s^{n-1} + \ldots + a_o} \tag{10.31}$$

Using (10.31) together with (10.30), one can evaluate the optimum set of gain factors K_i.

The value of K' has to be modified again in order to keep the gain from input to V_n constant; that is,

$$K'' = \frac{K'}{\prod_{i=1}^{n} K_i}$$

Using (10.29) yields

$$K'' = \left(\frac{qma_o}{b_o}\right) / \prod_{i=1}^{n} K_i$$

The values of the output summing coefficients have to also be changed such that the gain from input to output remains constant; that is

$$K'' K_1 K_2 \ldots K_i \beta_i'' = K \beta_i$$

which results in

$$\beta_i'' = \beta_i \left(\frac{K_{i+1} K_{i+2} \ldots K_n}{q}\right) \quad , \quad i = 0,1,2,\ldots,n$$

Finally, the values of the feedback coefficients F_i should be changed such that the gain of each loop remains constant, resulting in

$$F_i'' = \frac{F_i}{\prod_{j=1}^{i} K_j} \quad , \quad i = 2,3,\ldots,n$$

This completes the design equations. We note that the maximum permissible signal swing at the input will be given by

$$V_{in_{max}} = \left(\frac{V_{amp}}{V_{out_{max}}}\right)$$

where V_{amp} denotes the specified maximum op. amp. signal swing. For convenience and easy reference, the design equations are summarized in Table 10.3. In the table we have dropped the prime symbols from our notation.

Table 10.3 Design procedure for bandpass filters with finite loss-poles realized by the summation technique

1. Let the LPP be characterized by

$$T(s) = m\left(\frac{a_o}{b_o}\right) \frac{b_n s^n + b_{n-1} s^{n-1} + \ldots + b_o}{s^n + a_{n-1} s^{n-1} + \ldots + a_o}$$

where m is the desired center-frequency gain of the BPF.

2. Calculate the constant k from

$$k = \frac{a_{n-1}}{n}$$

3. Calculate T_{max} where

$$T_{max} = \mathrm{Max}\ \{|T(j\Omega)|\} \quad , \quad 0 \le \Omega \le \infty$$

4. Calculate $V_{n_{max}}$ where

$$V_{n_{max}} = \mathrm{Max}\ \{|V_n(j\Omega)|\} \quad , \quad 0 \le \Omega \le \infty$$

$$V_n(s) = m\left(\frac{a_o}{b_o}\right) \frac{k^n}{s^n + a_{n-1} s^{n-1} + \ldots + a_o}$$

Hence, calculate the factor q from,

$$q = \frac{T_{max}}{V_{n_{max}}}$$

5. Define $T_n(s)$ as

$$T_n(s) = q\left(\frac{ma_o}{b_o}\right)\frac{k^n}{s^n + a_{n-1}s^{n-1} + \cdots + a_o},$$

and calculate the set of gain factors K_i,

$$K_i = \frac{\text{Max}\left\{|T_n(j\Omega)|\left(1+\frac{\Omega^2}{k^2}\right)^{\frac{n-i+1}{2}}\right\}}{\text{Max}\left\{|T_n(j\Omega)|\left(1+\frac{\Omega^2}{k^2}\right)^{\frac{n-i}{2}}\right\}}, \quad i = 1,2,\ldots,n$$

6. Calculate the feedback coefficients F_i from

$$F_2 = \left[\frac{a_{n-2}}{k^2} - \frac{n(n-1)}{2!}\right] / K_1K_2 \,,$$

$$F_i = \frac{\left[\frac{a_{n-i}}{k^i} - \frac{n!}{(n-i)!\,i!} - \frac{1}{(n-i)!}\sum_{\ell=2}^{i-1} F_\ell (K_1 \ldots K_\ell)\frac{(n-\ell)!}{(i-\ell)!}\right]}{\prod_{j=1}^{i} K_j}$$

$$i = 3,4\ldots, n$$

7. The input summing coefficient K is given by

$$K = \left(\frac{qma_o}{b_o}\right) / \prod_{i=1}^{n} K_i$$

8. Calculate the output summing coefficients as follows:

(a) If $b_n \neq 0$, then $\beta_o = (b_n/q)$,

$$b_{n-j} = \frac{k^j}{(n-j)!}\left[\frac{n!}{j!}b_n + \frac{q}{K_{j+1}\cdots K_n}\sum_{i=1}^{j}\beta_i\frac{(n-i)!}{(j-i)!}\right], \; j = 1,2,\ldots,n$$

(b) If $b_n = 0$, then $\beta_o = 0$ and,

$$b_{n-j} = \frac{k^j q}{(n-j)!(K_{j+1}\ldots K_n)}\sum_{i=1}^{j}\beta_i\frac{(n-i)!}{(j-i)!}, \; j = 1,2,\ldots,n$$

9. If the BPF is to have a center-frequency ω_o and a bandwidth B, that is an effective Q-factor $\hat{Q} = \frac{\omega_o}{B}$, then all the blocks in the FLF structure of Fig. 10.18 will have bandpass transfer functions of the form

$$T_i(p) = K_i \frac{p(\frac{\omega_o}{Q})}{p^2 + p(\frac{\omega_o}{Q}) + \omega_o^2}$$

$$\text{where } Q = \frac{\hat{Q}}{k}$$

Note: The resulting BPF design is optimum from a dynamic range point of view.

Example 10.8

It is required to find a PRB realization of the bandpass filter considered in Example 6.5. The transmission zeros are to be realized using the summation technique. The center-frequency gain of the BPF is to be equal to unity.

Solution

From the results of Example 6.5, the low-pass prototype has the transfer function

$$T(s) = 0.1537 \frac{s^2 + 11.226\,166}{s^3 + 1.91263\,s^2 + 2.578788\,s + 1.72556}$$

Note that the multiplicative constant has been chosen to result in a unity DC gain of the LPP.

We follow the step-by-step design procedure of Table 10.3 and obtain,

$$k = \frac{a_2}{3} = 0.637543$$

$$T_{max} = |T(0)| = 1$$

$$V_3(s) = \frac{0.039832}{s^2 + a_2s^2 + a_1s + a_o}$$

$$V_{3_{max}} \simeq 0.025$$

$$q = 40$$

$$T_3(s) = \frac{1.593265}{s^2 + a_2s^2 + a_1s + a_o}$$

$$K_3 = \frac{\text{Max } \{|T_3(j\Omega)| \ (1 + \frac{\Omega^2}{k^2})^{1/2}\}}{\text{Max } \{|T_3(j\Omega)|\}} \simeq \frac{2}{1} \simeq 2$$

$$K_2 = \frac{\text{Max } \{|T_3(j\Omega)| \ (1 + \frac{\Omega^2}{k^2})\}}{\text{Max } \{|T_3(j\Omega)| \ (1 + \frac{\Omega^2}{k^2})^{1/2}\}} \simeq \frac{4.33}{2} \simeq 2.165$$

$$K_1 = \frac{\text{Max } \{|T_3(j\Omega)| \ (1 + \frac{\Omega^2}{k^2})^{3/2}\}}{\text{Max } \{|T_3(j\Omega)| \ (1 + \frac{\Omega^2}{k^2})\}} \simeq \frac{10.17}{4.33} \simeq 2.349$$

$$F_2 = 0.657640$$

$$F_3 = 0.227545$$

$$K = 0.604488$$

$$\beta_o = 0$$

$$\beta_1 = 0.169792$$

$$\beta_2 = -0.216571$$

$$\beta_3 = 1.129814$$

$$f_o = 1997.5 \text{ Hz}, \qquad B = 200 \text{ Hz} \qquad \hat{Q} = \frac{f_o}{B} = 9.9875$$

$$Q = 15.67$$

Fig. 10.19 shows a block diagram realization of the BPF from which a circuit implementation using, say, the Friend biquad (See Section 9.7) for each of the bandpass sections can be easily obtained.

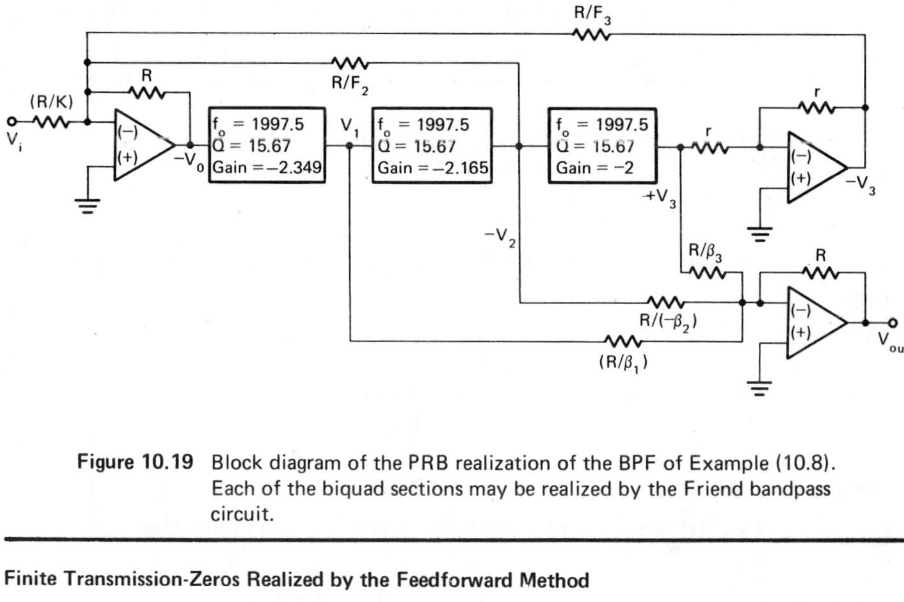

Figure 10.19 Block diagram of the PRB realization of the BPF of Example (10.8). Each of the biquad sections may be realized by the Friend bandpass circuit.

Finite Transmission-Zeros Realized by the Feedforward Method

The other method for using the FLF structure to realize filter functions with finite loss-poles makes use of feedforward, as illustrated in Fig. 10.20. Let the transfer function of the low-pass prototype be given by

$$T(s) = \frac{b_n s^n + b_{n-1} s^{n-1} + \ldots + b_o}{s^n + a_{n-1} s^{n-1} + \ldots + a_o}$$

This may be written in the form

$$T(s) = b_n + \frac{c_{n-1} s^{n-1} + c_{n-2} s^{n-2} + \ldots + c_o}{s^n + a_{n-1} s^{n-1} + \ldots + a_o} \tag{10.32}$$

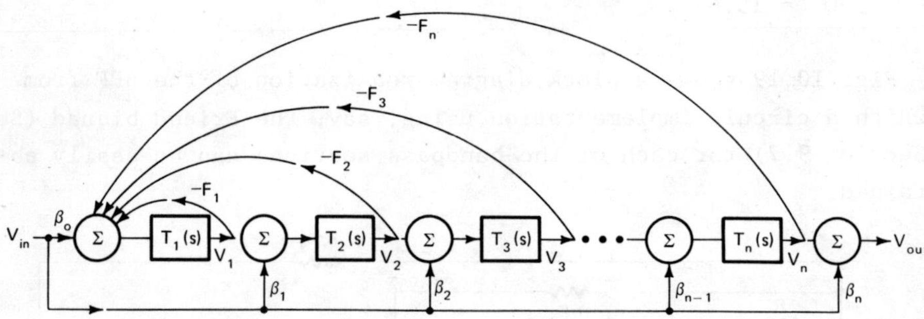

Figure 10.20 The FLF structure with feedforward for realizing finite transmission zeros.

The feedforward FLF structure of Fig. 10.20 has the transfer function

$$\frac{V_{out}(s)}{V_{in}(s)} = \beta_n + \frac{V_n(s)}{V_{in}(s)} \tag{10.33}$$

Comparing (10.32) and (10.33), it follows that for the block diagram to realize the given T(s) we have to arrange that

$$\beta_n = b_n \quad \text{and,}$$

$$\frac{V_n(s)}{V_{in}(s)} = \frac{c_{n-1} s^{n-1} + c_{n-2} s^{n-2} + \ldots + c_o}{s^n + a_{n-1} s^{n-1} + \ldots + a_o} \tag{10.34}$$

The transfer function $\frac{V_n}{V_{in}}(s)$ can be derived and expressed as

$$\frac{V_n}{V_{in}}(s) = \frac{\beta_0 T_1 T_2 \cdots T_n + \beta_1(1+F_1T_1)T_2T_3 \cdots T_n}{1 + F_1T_1 + F_2T_1T_2 + \cdots}$$

$$\frac{+\beta_2(1+F_1T_1+F_2T_1T_2)T_3T_4 \cdots T_n + \cdots}{+ F_nT_1T_2 \cdots T_n} \tag{10.35}$$

$$\frac{+ \beta_{n-1}(1+ F_1T_1+F_2T_1T_2+ \cdots + F_{n-1}T_1T_2 \cdots T_{n-1})T_n}{} .$$

As usual in the PRB case, we assume all the blocks to have identical transfer functions of the form

$$T_i(s) = \frac{k}{s+k} \quad , \; i = 1,2,\ldots,n$$

and that $F_1 = 0$. Substituting in (10.35) obtains

$$\frac{V_n}{V_{in}}(s) = \frac{\sum_{i=0}^{n-1} \beta_i \, k^{n-i} \, [(s+k)^i + \sum_{j=2}^{i} F_j \, k^j \, (s+k)^{i-j}]}{(s+k)^n + \sum_{i=2}^{n} F_i \, k^j \, (s+k)^{n-i}} \tag{10.36}$$

To obtain the feedback coefficients F_i, we match the coefficients of like powers of s in the denominator polynomials of (10.34) and (10.36). As expected, the results are identical to those obtained in the previous case; that is equations (10.10)-(10.12).

The feedforward coefficients β_i can be obtained by matching the coefficients of like powers of s in the numerator polynomials of (10.34) and (10.36). This process involves straightforward but lengthy algebra. The result may be expressed in the form

$$b_{n-i} = k^i \sum_{j=1}^{i} \beta_{n-j} \sum_{m=0}^{i-j} F_m \frac{(n-j-m)!}{(i-j-m)!(n-i)!}$$

$$F_o = F_1 = 0 \quad ,$$

$$i = 1,2,\ldots,n \tag{10.37}$$

Once the feedback and feedforward coefficients are determined, the LP to BP transformation

$$s = \frac{p^2 + \omega_o^2}{pB} ,$$

can be applied to the block diagram of Fig. 10.20 to obtain the realization of the BPF. As in the previous case, the BPF will contain identical bandpass sections characterized by a center-frequency ω_o, a Q-factor of

$$Q = \frac{\hat{Q}}{k} ,$$

and a center-frequency gain of unity. If desired, the different sections may be assigned different gain factors chosen to maximize the filter dynamic range.

More General FLF Designs

In the above, we concerned ourselves only with a special case of the FLF structure; namely, that for which $F_1 = 0$ and all the sections have equal ω_o and Q. This is the case referred to as the Primary Resonator Block (PRB) approach. As seen, the design equations and the network itself are simple and convenient to work with, especially in the all-pole case. Realizations of filters with finite loss-poles are not so convenient. In both the summation and feedforward configurations, it is rather difficult to tune the transmission zeros. This should be contrasted with cascade designs where each pair of finite transmission zeros is realized in a separate section and hence, can be easily adjusted or tuned. From this point of view, the realizations described in the following two Chapters are even more convenient.

Some sensitivity results have been reported in the literature for PRB designs[11,16,17]. Generally speaking, the passband sensitivities are much smaller than those obtained with cascade designs and are of the same order of magnitude as those encountered in Leap/Frog (Chapter 12) circuits.

It should be noted, however, that the PRB design method is limited to geometrically symmetric bandpass filters since the circuits are obtained through applying the LP to BP transformation. This is a rather severe limitation as it restricts the designer to classical approximation. The cascade method as well as the methods based on LC ladder simulations are much more general.

The question arises as to whether any advantages may be gained by allowing the different sections in the FLF structure to have different Q-factors. This question has been investigated mainly by Laker and Ghausi[14,15] and by Tow[16]. Laker and Ghausi presented a computer-aided method for obtaining a set of Q values which result in minimizing a certain sensitivity measure for bandpass filters. These designs are the most general and are referred to in the literature as the FLF design method. Although the Laker-Ghausi technique is quite interesting, the method is rather complicated when compared to the simple PRB procedure. The interested reader is referred to the references cited above.

Tow[16] addressed himself to the question of including the first feedback link F_1 in the structure and then letting all sections be identical. In this way, one gains an additional degree of freedom which can be used to minimize sensitivity. The free parameter used by Tow is the value of the constant k. Note that the case $k = 0$ corresponds to the companion-form state-variable realization of the LPP. Also, $k = \frac{a_{n-1}}{n}$ results in the PRB case. Tow's results indicate that there exist a broad range of values for k where comparable improvement over the cascade design are obtained. This range includes the PRB case. The state-variable case showed virtually no sensitivity advantage over the cascade circuit.

It is therefore concluded that for the FLF structure the PRB design technique is a good one. Again we emphasize that the method is especially suited for all-pole bandpass filters.

10.4 REFERENCES AND SOURCES FOR FURTHER READING

(1) S. K. Mitra (Ed.), "Active Inductorless Filters," IEEE Press, New York, 1971.

(2) L. P. Huelsman, Ed., Active RC Filters: Theory and Application, in Benchmark Papers in Electrical Engineering and Computer Science, Vol. 15, Dowden, Hutchinson and Ross, Inc. Stroudsburg, Pennsylvania, 1976.

(3) G. Szentirmai, (Ed.), Computer Aided Filter Design, IEEE Press, New York, 1973.

(4) G. S. Moschytz, Linear Integrated Networks: Design, Van Nostrand Reinhold, New York, 1975.

(5) E. Lueder, "A decomposition of a transfer function minimizing distortion and in band losses," Bell Syst. Tech. J., Vol. 49, pp. 455-569, 1970. Reprinted in (2).

(6) S. Halfin,"An optimization method for cascaded filters," Bell Syst. Tech. J., Vol. 44, pp. 185-190, Feb. 1970. Reprinted in (3).

(7) S. Halfin, "Simultaneous determination of ordering and amplifications of cascaded subsystems," J. Optimiz. Theory and Appl., Vol. 6, pp. 356-360, Nov. 1970. Reprinted in (3).

(8) (a) F. E. J. Girling and E. F. Good, "Active filters 12. The leap-frog or active ladder synthesis," Wireless World, Vol. 76, pp. 341-345, July 1970. Reprinted in (2).
(b) F. E. J. Girling and E. F. Good, "Active filters 13. Applications of the active ladder synthesis," Wireless World, Vol. 76, pp. 445-450, Sept. 1970.

(9) J. Tow and Y. L. Kuo, "Coupled-biquad active filters," 1972 IEEE Int. Symposium Circuit Theory, pp. 164-168, North Hollywood, Calif., 1972.

(10) G. Szentirmai, " Synthesis of multiple feedback active filters," Bell System Tech. J., Vol. 52, pp. 527-555, Apr. 1973. Reprinted in (3).

(11) G. Hurtig, III, "The primary resonator block technique of filter synthesis," Proc. Int. Filter Symposium, p.84, 1972.

(12) G. Hurtig, III, "Filter network having negative feedback loops," U.S. patent B,720,881, March 1973.

(13) G. Hurtig, III, "Voltage tunable multipole bandpass active filters," 1974 IEEE Int. Symp. Circuits and Systems, pp. 569-572, San Francisco, Calif., 1974.

(14) K. R. Laker and M. S. Ghausi, "Synthesis of a low sensitivity multiloop feedback active RC filter," IEEE Trans. Circuits and Systems, Vol. CAS-21, pp. 252-259, Mar. 1974.

(15) K. R. Laker and M. S. Chausi, "A comparison of active multiple-loop feedback techniques for realizing high-order

bandpass filters," IEEE Trans. Circuits and Systems, Vol. CAS-21, pp. 774-783, Nov. 1974.

(16) J. Tow, " Design and evaluation of shifted companion form active filters," Bell System Tech. J., Vol. 54, pp. 545-568, Mar. 1975.

(17) R. MacKay and A. S. Sedra, "Sensitivity evaluation and comparison for high-order active filters," Int. J. Circuit Theory and Applications, Vol. 4, pp. 345-356, 1976.

PROBLEMS

P10.1 A seventh-order elliptic low-pass filter that meets the normalized ($\Omega_p=1$) specifications has $|\rho|_{max}=20\%$ and $\Omega_\varepsilon=1.4142$. The poles and zeros are:

Poles: -0.4016104, $-0.0501984 \pm j\ 1.0209751$,
$-0.1712486 \pm j\ 0.8817855$, $0.3204767 \pm j\ 0.5440075$

Zeros: $\pm j\ 2.8557$, $\pm j\ 1.4397$, $\pm j\ 1.7019$

(a) Find the optimum pole-zero pairing.

(b) Find a cascading sequence to maximize dynamic range.

(c) If the DC gain is to be unity, find the gain-constant of each factor in the transfer function so that the filter's dynamic range is maximized.

(d) If the op amps have a maximum allowable signal swing of ± 5 volts, find the maximum allowable signal swing at the input of the cascade realization of this filter.

(e) Give a complete cascade design of the filter using SAB's. Design for $\omega_p=10^3$ rad/sec. Give your reasons for selecting a particular biquad circuit.

Note: To gain insight into the solution of the decomposition problem, sketch the signal spectra at the outputs of all biquads.

P10.2 Verify the design given in Fig. 10.11.

P10.3 Give a complete design of the bandpass filter of Example 10.4 using the two-amplifier GIC-based biquad.

P10.4 Give a complete design of the BPF of Example 10.6 using the two-amplifier GIC-based biquad.

P10.5 Give an optimized cascade design for a sixth-order maximally flat all-pole bandpass filter with a center frequency f_o=10KHz, a 1dB bandwidth of 200Hz and a center frequency gain of 10. Use the Friend bandpass circuit as the basic building block.
If ω_o of the highest-Q section changes by +2% calculate the resulting deviation in the passband transmission in dB.
(Note: This filter has identical specifications to that of Problem P6.7)

P10.6 Give an optimized PRB design of the bandpass filter specified in Problem P10.5. Use the Friend bandpass circuit as the basic building block. Which of the sections is the most critical? Assume that ω_o of the most critical section changes by +2% and plot the modified magnitude response of the filter. Find the maximum deviation in the passband response.
(Note: This part could be rather tedious. Try to find another way of obtaining the desired result, namely the maximum deviation in the passband response)

P10.7 In Examples 6.5 and 6.6 we derived the transfer function of a sixth-order bandpass filter. Find an optimized cascade design using the Akerberg-Mossberg biquad with feedforward.
(Note: It is preferable to use the modified feedforward scheme shown in Fig. 9.39 for the Tow-Thomas biquad. The scheme has been adapted for the Akerberg-Mossberg biquad in Problem P9.35)

P10.8 Give an alternative circuit realization of the BPF of P10.7 using SAB's. Select a suitable SAB circuit and give reasons for your selection. Comment on the expected performance of this design as compared to that obtained in P10.7.

P10.9 Use the Friend biquad to realize each of the bandpass functions in the block diagram of Fig. 10.19 and give a complete circuit schematic.

P10.10 Give an optimized PRB realization of the bandpass filter of Problem P10.7. The transmission zeros are to be realized using the summation technique. Use the Akerberg-Mossberg biquad.

P10.11 Derive equations (10.35), (10.36) and (10.37).

CHAPTER 11

COMPONENT SIMULATION OF LC LADDERS

11.1 INTRODUCTION

The methods for active filter design presented in Chapter 10 start from a given transfer function and realize this transfer function by a structure consisting of biquadratic sections coupled together by feedback and feedforward paths. The other major approach for designing active filters is based on the simulation of an LC ladder realization of the filter. Thus, the starting point for simulation methods is an LC ladder prototype which may be obtained by the use of the computer methods described in Chapter 7 or by the use of closed-form expressions or filter tables (Chapter 6). Subsequently, the designer may choose between two different simulation methods. The first is the Component Simulation method discussed in the present Chapter. Chapter 12 presents the other simulation method, called Operational Simulation.

Before getting into the details of the Component Simulation method, we will attempt to place the simulation approach to filter design in its proper perspective. We recall that the cascade method of design is the most general; that is, it is capable of realizing any filter transfer function. Moreover, cascade designs are simple to obtain and adjust as well as being flexible and economic. The resulting circuits, however, are rather sensitive to component (active and passive) tolerances and variations. On the other hand, we have seen in Chapter 7 that doubly terminated passive LC ladders designed on insertion-loss basis exhibit excellent sensitivity performance over the filter passband. One of the reasons for the reduced transfer function sensitivity with respect to each ladder component is that the ladder transfer function is dependent on all the circuit components. Hence, the transfer function sensitivity is "spread out", with each individual sensitivity being small in value. This "spreading out" of the sensitivity functions in an LC ladder may be visualized as if the ladder possesses internal negative feedback that reduces the transfer function sensitivity to each element. This point of view had led to the coupled-biquad structures, discussed in Chapter 10, which indeed have lower sensitivities than cascade designs.

The low-sensitivity performance of LC ladders, together with the wealth of knowledge that has accumulated over the years in the area of passive LC filter design, provided a strong motivation to seek methods for designing active filters based on simulating passive LC ladder prototypes[3,4]. The simplest of these methods, at least conceptually, is the component simulation method to which this Chapter is devoted. Conceptually, the component simulation approach is based on replacing the ladder inductors by simulated inductances. In Section 8.4, we studied some inductance simulation circuits. The most attractive of these circuits is the Antoniou circuit, discussed in detail in Section 8.4. This is an inductance simulator, or more generally, a Generalized Immittance Converter (GIC) that contains two op. amps. The reader is advised to review the material of Section 8.4 before proceeding with the details of this Chapter. Section 11.2 is devoted to a study of the optimum design of the Antoniou GIC when embedded in an arbitrary network. The results for the inductance simulation and FDNR cases will follow as special cases.

In the case of high-pass filters and for bandpass filters which have no finite loss-poles in the upper stopband and at most one loss-pole at $s = \infty$, the passive LC realization does not contain floating inductors. Each grounded inductor can be easily simulated by a resistively terminated (ks) GIC and the active realization is thus readily obtained. Section 11.3 presents the design detail of this special class of filters.

Low-pass filters and more general bandpass filters contain floating inductances. Although a number of circuits for simulating floating inductors have been proposed in the literature, none has been found to be practically viable. Other techniques had to be invented for simulating these classes of filters. Notable among these techniques are the Gorski-Popiel method discussed in Section 11.4 and the Bruton transformation and its applications presented in Section 11.5.

As will be seen, the component simulation approach finds its area of application in filters with stringent specifications. It is particularly suited for high-pass filters and a certain restricted class of bandpass filters, namely those that do not contain floating inductances. For low-pass filters the Gorski-Popiel and Bruton methods lead to practical designs. This is also true to a certain extent for bandpass filters. Generally speaking, however, the component simulation approach finds a strong competition, in the design of bandpass filters, in the alternative design techniques based on the Operational Simulation of LC ladders, presented in the next Chapter.

11.2 OPTIMUM DESIGN OF THE ANTONIOU GIC*

The Antoniou GIC has been studied in some detail in Section 8.4 where the design equations for optimum performance of GIC's that

*This section is based on Ref. [5]

simulate inductors and frequency dependent negative resistances (FDNR's) have been given. It will be shown in Section 11.3 that in the design of general bandpass filters, one needs to embed sections that have been impedance scaled by (1/ks), where k is a real constant, inside a network of (ks) GIC's. Therefore, the present section is devoted to the derivation of optimum design equations for the (ks) GIC embedded in an arbitrary network. The design equations for simulated inductance and FDNR elements will follow as special cases from the general equations presented here. It should be noted that the optimum circuit of the (ks) GIC will be of the type-B, presented in Section 8.4.

Terminating the GIC

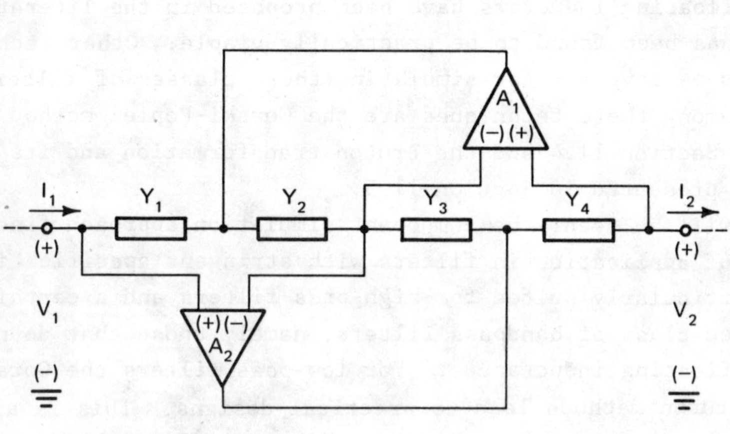

Figure 11.1 The Antoniou GIC embedded in an arbitrary network.

Figure 11.1 shows the Antoniou GIC embedded in an arbitrary network. Assuming ideal op. amps. with finite gains, A_1 and A_2, straightforward analysis results in

$$\frac{I_1}{V_1} = \frac{I_2}{V_2}\,\frac{Y_1Y_3}{Y_2Y_4}\,\frac{1 + \frac{1}{A_1}\left(1 + Y_4\frac{V_2}{I_2}\right) + \frac{1}{A_2}\frac{Y_2}{Y_3}\left(1 + Y_4\frac{V_2}{I_2}\right)}{1 + \frac{1}{A_1}\frac{Y_3}{Y_2}\left(1 + \frac{1}{Y_4}\frac{I_2}{V_2}\right) + \frac{1}{A_2}\left(1 + \frac{1}{Y_4}\frac{I_2}{V_2}\right)}$$

$$+ \frac{1}{A_1 A_2} (1 + \frac{Y_2}{Y_3})(1 + Y_4 \frac{V_2}{I_2})$$
$$\overline{+ \frac{1}{A_1 A_2} (1 + \frac{Y_3}{Y_2})(1 + \frac{1}{Y_4} \frac{I_2}{V_2})} \qquad (11.1)$$

Assuming $A_1 \simeq \frac{\omega_{t_1}}{j\omega}$, $A_2 \simeq \frac{\omega_{t_2}}{j\omega}$ and $\omega << \omega_t$, which allows the second-order terms of the form $(\frac{\omega}{\omega_t})^2$ to be ignored, gives

$$\frac{I_1}{V_1} \simeq \frac{I_2}{V_2} \frac{Y_1 Y_3}{Y_2 Y_4} [1 + j(\frac{\omega}{\omega_{t_1}})(1 - \frac{Y_3}{Y_2} + Y_4 \frac{V_2}{I_2} - \frac{Y_3}{Y_2 Y_4} \frac{I_2}{V_2}) + j(\frac{\omega}{\omega_{t_2}})(\frac{Y_2}{Y_3} - 1 + \frac{Y_2 Y_4}{Y_3} \frac{V_2}{I_2} - \frac{1}{Y_4} \frac{I_2}{V_2})] \; . \qquad (11.2)$$

For ideal performance it is desired to have,

$$\frac{I_1}{V_1} = \frac{I_2}{V_2} \frac{Y_1 Y_3}{Y_2 Y_4} \quad , \qquad (11.3)$$

independent of ω_{t_1} and ω_{t_2}. Thus from 11.2 it can be seen that we want

$$1 - \frac{Y_3}{Y_2} + Y_4 \frac{V_2}{I_2} - \frac{Y_3}{Y_2} \frac{1}{Y_4} \frac{I_2}{V_2} = 0 \qquad (11.4)$$

and

$$\frac{Y_2}{Y_3} - 1 + \frac{Y_2}{Y_3} Y_4 \frac{V_2}{I_2} - \frac{1}{Y_4} \frac{I_2}{V_2} = 0 \quad . \qquad (11.5)$$

Since in general $\frac{V_2}{I_2}$ is a function of frequency and takes on a complex value, then $(1 - \frac{Y_3}{Y_2})$ must be linearly independent of

$(Y_4 \frac{V_2}{I_2} - \frac{Y_3}{Y_2}\frac{1}{Y_4}\frac{I_2}{V_2})$. Thus from (11.4) or (11.5), it can be seen that it is necessary to have $(1 - \frac{Y_3}{Y_2}) = 0$ which implies

$$Y_2 = Y_3 \ . \tag{11.6}$$

Substituting (11.6) into either (11.4) or (11.5), we see that it is also necessary to have

$$\frac{V_2}{I_2} = \frac{1}{Y_4} \ . \tag{11.7}$$

Equation (11.6) is easily satisfied by letting $Y_2 = Y_3 = g$ where g is real. However, satisfying (11.7) is seldom possible as $\frac{V_2}{I_2}$ has in general components orthogonal to $\frac{1}{Y_4}$. Substituting (11.6) into (11.2) results in

$$\frac{I_1}{V_1} = \frac{I_2}{V_2}\frac{Y_1}{Y_4}\{1 + j(\frac{\omega}{\omega_{t_1}} + \frac{\omega}{\omega_{t_2}})(Y_4 \frac{V_2}{I_2} - \frac{1}{Y_4}\frac{I_2}{V_2})\} \tag{11.8}$$

Obviously, for best performance one would like to have the quantity $M = Y_4 \frac{V_2}{I_2} - \frac{1}{Y_4}\frac{I_2}{V_2}$ as small as possible at the most critical frequency ω_c in the filter transfer function. This frequency is the one where the greatest group delay occurs and is usually close to the passband edge.

Letting

$$\left.\frac{V_2}{I_2}\right|_{\omega=\omega_c} = A\ e^{j\theta} \ , \tag{11.9}$$

it follows that we require

$$M = Y_4\ A\ e^{j\theta} - \frac{1}{Y_4 A e^{j\theta}} \tag{11.10}$$

to be as small as possible.

Case 1:

$$\text{Let } Y_4 = g_4 \text{ ,} \tag{11.11}$$

where g_4 is real. Substituting (11.11) into (11.10) gives,

$$M = A\, g_4\, e^{j\theta} - \frac{e^{-j\theta}}{A\, g_4} \text{ ,}$$

which implies that,

$$M = (A\, g_4 - \frac{1}{A\, g_4}) \cos\theta + j(A\, g_4 + \frac{1}{A\, g_4}) \sin\theta . \tag{11.12}$$

Letting

$$X = A\, g_4 + \frac{1}{A\, g_4} \text{ ,}$$

we see that if we minimize X, then we are minimizing the last factor of the right-hand-side of (11.12). Letting $\frac{dX}{dg_4} = 0$ and solving for g_4 gives

$$g_4 = \frac{1}{A} . \tag{11.13}$$

Substituting (11.13) into (11.12) it is seen that the first factor of the right-hand-side of (11.12) goes to zero. Thus, taking

$$g_4^{\bullet} = \frac{1}{A} = \left|\frac{I_2}{V_2}\right|_{\omega=\omega_c} \tag{11.14}$$

minimizes both components of M at $\omega = \omega_c$. The corresponding GIC transfer function is obtained by substituting (11.12) and (11.13) into (11.8) as,

$$\frac{I_1}{V_1} = \frac{I_2}{V_2}\frac{Y_1}{Y_4}\,[1 - 2(\frac{\omega}{\omega_{t_1}} + \frac{\omega}{\omega_{t_2}}) \sin\theta] \quad \text{, at } \omega = \omega_c \text{ .} \tag{11.15}$$

Case 2:

Let $Y_4 = j\omega C_4$

where C_4 is real.
Therefore,

$$M = j\,\omega_c\, C_4\, Ae^{j\theta} + \frac{j\, e^{-j\theta}}{\omega_c\, C_4 A} ,$$

which implies that

$$M = j(\frac{1}{\omega_c C_4 A} + \omega_c C_4 A) \cos\theta + (\frac{1}{\omega_c C_4 A} - \omega_c C_4 A) \sin\theta \qquad (11.16)$$

The first term is minimized if one selects

$$\omega_c\, C_4 = \frac{1}{A} .$$

This selection also reduces the second term of (11.16) to zero. Thus, taking

$$\omega_c\, C_4 = \left|\frac{I_2}{V_2}\right|_{\omega=\omega_c} \qquad (11.17)$$

minimizes M. The corresponding value for the GIC transfer function is

$$\frac{I_1}{V_1} = \frac{I_2}{V_2}\frac{Y_1}{Y_4}\,[1 - 2(\frac{\omega}{\omega_{t_1}} + \frac{\omega}{\omega_{t_2}}) \cos\theta],\ \text{at}\ \omega = \omega_c . \qquad (11.18)$$

Generalizing equations (11.14) and (11.17), we may write that for most ideal performance one selects

$$\left|Y_4\right|_{\omega=\omega_c} = \left|\frac{I_2}{V_2}\right|_{\omega=\omega_c} \qquad (11.19)$$

Taking the absolute value of both sides of either (11.15) or (11.18) at $\omega = \omega_c$ and assuming $\omega \ll \omega_{t_1}, \omega_{t_2}$, we see that satisfying equation (11.19) is approximately equivalent to satisfying

$$\left| Y_1 \right|_{\omega=\omega_c} = \left| \frac{I_1}{V_1} \right|_{\omega=\omega_c} . \tag{11.20}$$

Equations (11.19) and (11.20) have an easy interpretation: at the critical frequency ω_c the magnitude of the end-admittance Y_4 on one side of the GIC must be equal to the magnitude of the admittance $\frac{I_2}{V_2}$, seen by that side of the GIC. The magnitude of the other end admittance Y_1 will then be approximately matched to the magnitude of the admittance $\frac{I_1}{V_1}$, seen by the other side of the GIC. In fact the error in this approximation is due to the non-ideality of the optimized GIC and is unavoidable. Furthermore, since we took into account both case 1 and case 2, one may design the GIC for optimum performance by exactly matching either side of the GIC. It is interesting to note that this condition implies some sort of "properly terminating" the GIC. This together with $Y_2 = Y_3$ ensures optimum GIC performance.

Special Cases

As an application of the optimum design equations derived above we consider the cases of the GIC used to simulate an inductance and FDNR element. The results obtained correspond to those given in Section 8.4.

(A) Simulated Inductance

The circuit for simulating an inductance is shown in Fig. 11.2. Setting $Y_2 = Y_3$ means that

$$R_2 = R_3 . \tag{11.21}$$

Also, setting $\left| Y_4 \right|_{\omega=\omega_c} = \left| \frac{I_2}{V_2} \right|_{\omega=\omega_c}$ implies that

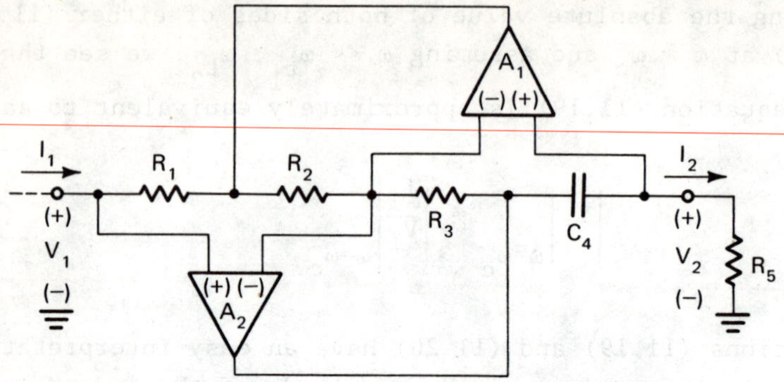

Figure 11.2 The Antoniou GIC used to realize an inductance.

$$\omega_c C_4 R_5 = 1 \ . \tag{11.22}$$

Substituting (11.9), (11.16), (11.22) and $\frac{V_2}{I_2} = R_5$ into (11.8) gives,

$$\left.\frac{I_1}{V_1}\right|_{\omega=\omega_c} = \frac{Y_1 Y_5}{Y_4}[1 - 2(\frac{\omega_c}{\omega_{t_1}} + \frac{\omega_c}{\omega_{t_2}})] \ . \tag{11.23}$$

For our case $Y_1 = \frac{1}{R_1}$ and $Y_4 = j\omega C_4$, which when substituted into (11.23) and using (11.22) result in

$$\left.\frac{V_1}{I_1}\right|_{\omega=\omega_c} \simeq j\, R_1 [1 + 2(\frac{\omega_c}{\omega_{t_1}} + \frac{\omega_c}{\omega_{t_2}})] \tag{11.24}$$

which implies that at $\omega = \omega_c$ the input inductance has infinite Q-factor but undergoes a deviation in value given by

$$\left.\frac{\Delta L}{L}\right|_{\omega=\omega_c} \simeq 2\ (\frac{\omega_c}{\omega_{t_1}} + \frac{\omega_c}{\omega_{t_2}}) \tag{11.24a}$$

Equation (11.24) also suggests that R_1 may be used to adjust or trim the value of the input inductance.

(B) Simulated FDNR

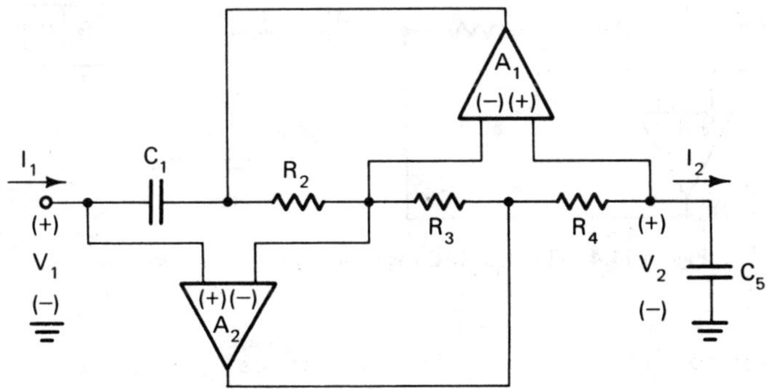

Figure 11.3 The Antoniou GIC used to realize an FDNR

The FDNR simulation circuit is given in Fig. 11.3. Following an analysis similar to that for the simulated inductance case above, it can be shown that for optimum performance one selects

$$R_2 = R_3 \qquad \text{and,}$$

$$\omega_c C_5 R_4 = 1$$

Correspondingly, the input impedance is given by

$$\left.\frac{V_1}{I_1}\right|_{\omega=\omega_c} \simeq -\frac{1}{\omega_c C_1}\left[1 - 2\left(\frac{\omega_c}{\omega_{t_1}} + \frac{\omega_c}{\omega_{t_2}}\right)\right] \qquad (11.25)$$

Thus at $\omega = \omega_c$, the FDNR element has infinite Q-factor but finite magnitude deviation of

$$\left.\frac{\Delta D}{D}\right|_{\omega=\omega_c} \simeq 2\left(\frac{\omega_c}{\omega_{t_1}} + \frac{\omega_c}{\omega_{t_2}}\right) \qquad (11.25a)$$

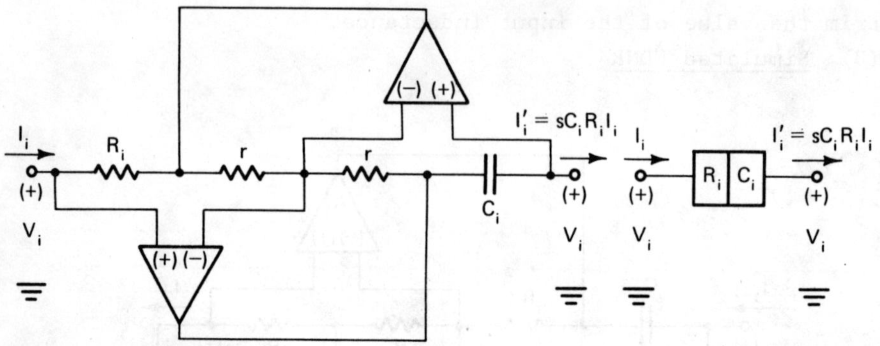

Figure 11.4 The (k_is) GIC together with its circuit symbol.

In order to simplify circuit schematics, we shall use the symbolic representation of the GIC shown in Fig. 11.4. Note that this symbol applies for a (ks) GIC of type-B with the two internal resistors equal in value.

Parasitic Effects

Before leaving this section we wish to discuss an important problem that may be encountered in using the GIC in practical filter design. This relates to the effect of the unavoidable stray capacitances that exist at the input terminals of the op. amps. in the GIC circuit. For example, with reference to Fig. 11.2, a stray capacitance C_p in parallel with R_5 will cause the simulated inductor to have a Q-factor given by

$$Q_L \simeq \frac{1}{\omega_o C_p R_5} = \frac{C_4}{C_p}, \qquad (11.26)$$

where ω_o is the frequency at which the simulated inductor is to be resonant. If one assumes a 3pF stray capacitance and substitutes $C_4 = 1.2$ nF, the quality factor will be only 400! An analogous argument applies to FDNR's and floating GIC's. To compensate for these stray capacitances a resistor R_c may be connected between the negative terminals of the op. amps. and ground. If the resistance is chosen such that (see Fig. 11.4)

$$R_c = r\frac{C_i}{C_p} \quad , \tag{11.27}$$

then the parasitic capacitance C_p at the positive terminal of the op. amp. connected to C_i will be exactly compensated for.

11.3 DESIGN OF HIGH-PASS FILTERS AND A RESTRICTED CLASS OF BANDPASS FILTERS

High-pass filters and bandpass filters that do not have loss-poles in the upper stop-band, except for possibly one at $s = \infty$, can always be realized by LC ladders void of floating inductors. Each grounded inductor may be replaced by a (ks) GIC loaded with a resistance.

The optimum design equations presented in the previous section may be used for choosing the component values of each GIC. Sensitivity analysis of the resulting network can be carried out using the formulae for passive LC filters, derived in Chapter 7.

A detailed design example of a bandpass filter, belonging to this restricted class, has been given in [6]. The design of a high-pass filter is illustrated by the following example.

Example 11.1[7]

It is required to design a high-pass filter having the following specifications:

$$A_p \le 1 \text{ dB}, \quad A_s \ge 60 \text{ dB}, \quad f_p = 1091 \text{ Hz and } f_s = 916 \text{ Hz}.$$

Using filter tables and the LP to HP transformation it was found that an order seven elliptic filter function meets these specifications. The transfer function of the high-pass filter, normalized relative to the passband edge $\omega_p = 2\pi \times 1091$ rad/sec, is

$$T(s) = \frac{s}{s + 3.948} \times \frac{s^2 + 0.2465}{s^2 + 1.2208s + 3.2701} \times$$

$$\frac{s^2 + 0.6261}{s^2 + 0.255s + 1.5072} \times \frac{s^2 + 0.8178}{s^2 + 0.05221s + 1.1933} .$$

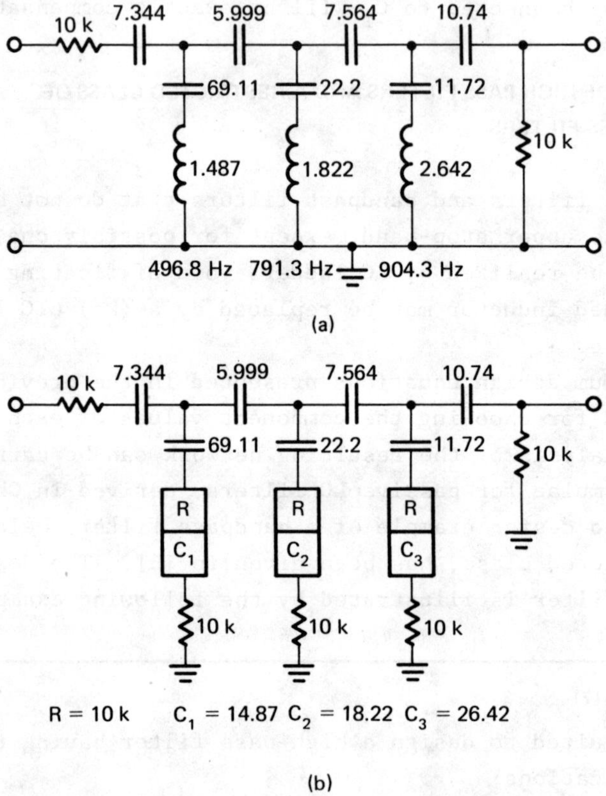

Figure 11.5 (a) LC prototype for the seventh-order elliptic HPF of Example (11.1). (b) The active realization of the LC prototype in (a). All GIC's are of the type-B circuit.
Note: All capacitances are in nF and all inductances are in H.

The denormalized LC ladder realization, obtained from filter tables together with the application of the LP to HP transformation (See Section 6.4), is shown in Fig. 11.5a. To obtain an active realization, each inductor is replaced by a type-B GIC terminated in a resistance. If we were to follow the recommended design procedure outlined in the previous section, we would have to first find the critical frequency ω_c which is the frequency at which $\frac{\omega d\alpha(\omega)}{d\omega}$ reaches a maximum. In our case this frequency would be around 1 KHz. Subsequently, each GIC would be designed such that all its resistors are equal to $R_i = \omega_c L_{o_i}$ where the subscript i corresponds to the particular inductance being simulated. For simplicity and ease of manufacture, this filter was designed[7] using 3 identical GIC's, except for the value of their capacitor C. The value of R chosen falls in the middle of the "optimum" values, namely R = 10 KΩ. Correspondingly, we obtain the three capacitance values as follows:

$$C_1 = \frac{L_1}{R^2} = \frac{1.487}{10^8} = 14.87 \quad nF$$

$$C_2 = \frac{L_2}{R^2} = \frac{1.822}{10^8} = 18.22 \quad nF$$

$$C_3 = \frac{L_3}{R^2} = \frac{2.642}{10^8} = 26.42 \quad nF$$

The active filter circuit is shown in Fig. 11.5b. The complete filter was constructed and a number of experiments were performed on it. For experimental results we refer the reader to [7].

11.4 DESIGN OF LOW-PASS AND BANDPASS FILTERS USING THE GORSKI-POPIEL TECHNIQUE

The LC ladder realizations of low-pass filters and of bandpass filters that contain finite upper-stopband loss-poles and/or more than one loss pole at s = ∞ have by necessity floating inductances. Although a number of floating transistor gyrator circuits have been reported in the literature, they lack the generality and convenience of op. amp. circuits. As mentioned before, we will concern ourselves in this book only with active filters realizable using op. amps. Although no practical op. amp. implementation of a floating inductance exists at the time of this writing, a number of rather ingenious techniques have been developed for the realization of low-pass and general bandpass filters using GIC's. All these techniques involve a complex frequency transformation of a part, or parts, of the LC ladder prototype. As will be shown, all these methods are in a sense based on the original idea introduced by Gorski-Popiel[8].

Gorski-Popiel's Ladder Embedding Technique

Gorski-Popiel's ladder-embedding method has laid the foundation for the realization of low-pass and general bandpass filters using GIC's. To understand this technique, refer to Fig. 11.6. The resistive n-port network N' is connected to the multi-GIC network N_1 to form the composite n-port network N. Note that N_1 is formed of n identical (ks) GIC's of the current-transformation type. That is, the ith GIC is described by

$$V_i = V_i' \quad \text{and,}$$

$$I_i = \frac{1}{ks} I_i' \quad .$$

Now, if the resistive network N' is characterized by an nxn impedance matrix $\underline{Z}'$, that is

$$\underline{V}' = \underline{Z}' \ \underline{I}' \ ,$$

then we may write,

$$\underline{V} = \underline{V}' = \underline{Z}' \ \underline{I}' = \underline{Z}'(ks) \ \underline{I} = (ks \ \underline{Z}') \ \underline{I} \ .$$

Thus, the composite n-port network N will be characterized by an impedance matrix $\underline{Z}$ given by

$$\underline{Z} = ks \ \underline{Z}' \ .$$

Since $\underline{Z}'$ describes a resistive network, then $\underline{Z}$ describes a topologically identical inductive network with each inductance in N

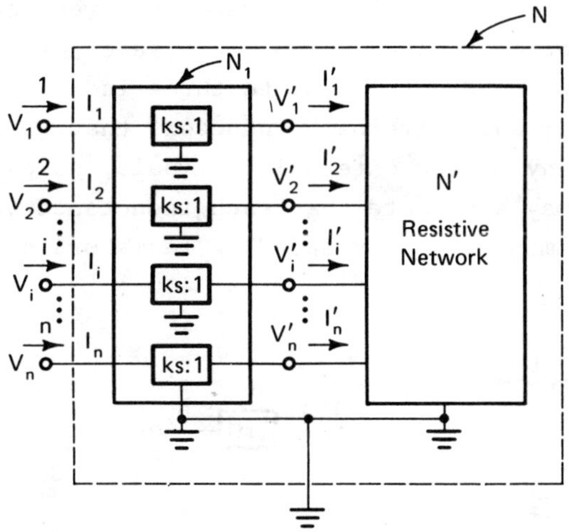

Figure 11.6 Simulation of an inductive network N through the connection of a topologically identical resistive network N′ and n GIC's.

equal to the resistance of the corresponding resistor in N' multiplied by the constant k. We thus have succeeded in simulating an arbitrary inductive subnetwork by using a resistive network connected to the rest of the system through n GIC's, where n is the number of terminals of the inductive subnetwork. Note that the number of GIC's required, and correspondingly the number of capacitors, is <u>not</u> equal to the number of inductors in the subnetwork being simulated. Rather, it is equal to the number of terminals that connect the inductive subnetwork to the rest of the LC ladder network, discounting ground.

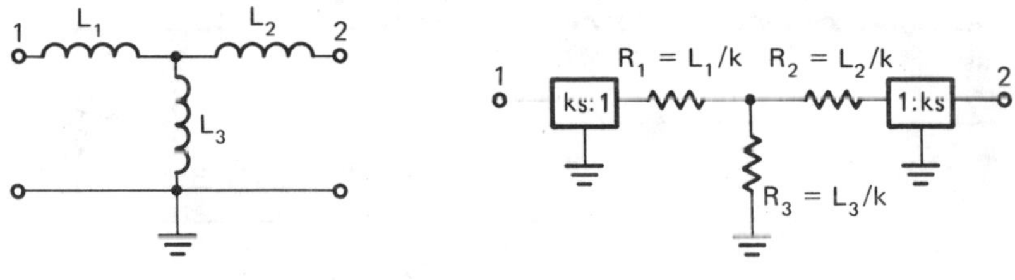

Figure 11.7 Application of the Gorski-Popiel method to the simulation of a T of inductors

As an example, Fig. 11.7 illustrates the realization of a T of inductors. Notice that only two GIC's, and correspondingly two capacitors, are needed to simulate three inductors. It can be stated that for any "internal" inductor, that is one that is not connected to any of the external terminals, no extra capacitor is needed. One may also note that each inductance value can be adjusted by trimming its corresponding resistor in the resistive network.

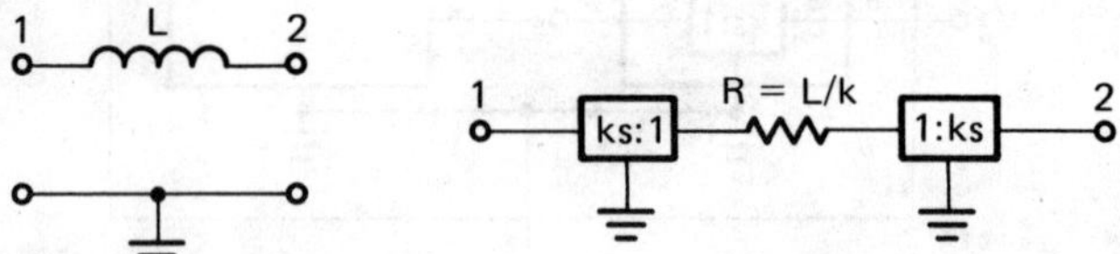

Figure 11.8 Simulation of a floating inductor L using the Gorski-Popiel technique.

A floating inductor has to be considered as a two-port network and realized using two GIC's, with two capacitors, as shown in Fig. 11.8.

Gorski-Popiel's method is further illustrated by the example in Fig. 11.9. This hypothetical LC structure is chosen to incorporate most of the ways in which inductors are used in filters. Note that since L_1 is separate from the inductance subnetwork L_2, L_3, L_4, L_5 and L_6, there is no reason to make k_1 equal to k_2. That is, the conversion factors of only the GIC's employed in a connected network need to have equal values.

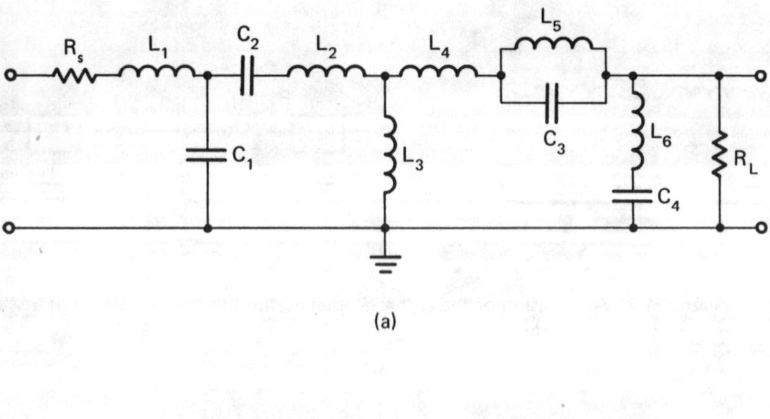

(a)

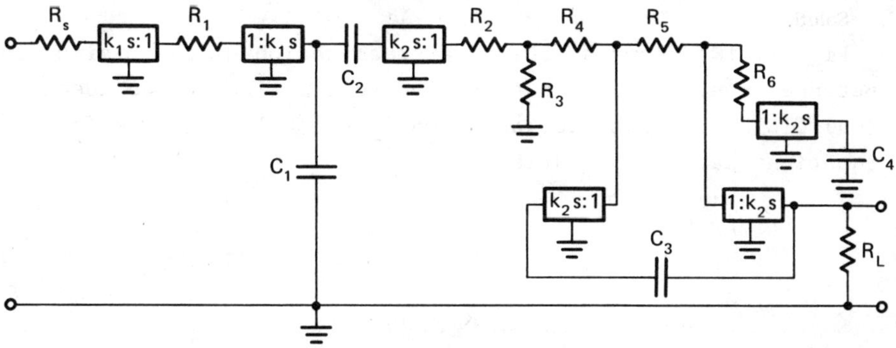

(b)

Figure 11.9 Illustrating the application of the Gorski-Popiel method to the simulation of an arbitrary LC ladder filter.

Example 11.2

Sketch the active realization of a fifth-order elliptic low-pass filter using the ladder embedding technique.

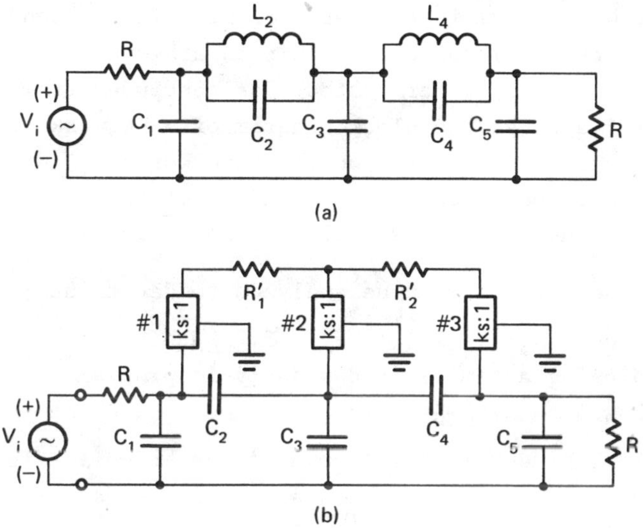

Figure 11.10 Application of the Gorski-Popiel method to the design of a fifth-order elliptic low-pass filter.

Solution

Fig. 11.10a shows an LC ladder realization of the minimum inductance type. To obtain an active realization, three identical (ks) GIC's are required, as shown in Fig. 11.10b. The following equations have to be satisfied:

$$kR_1' = L_2$$

$$kR_2' = L_4$$

The available degrees of freedom in choosing the values of the GIC's components should be used to optimize the design, as described in Section 11.2. This process will be outlined in detail below.

Design Procedure*

We now address ourselves to the optimum design of filters using the Gorski-Popiel ladder embedding technique. The resulting network will be optimum in the sense of minimized dependence on the op. amps. frequency response. This is achieved by ensuring that each GIC is "properly terminated", as concluded from Section 11.2.

Fig. 11.11a gives the block diagram of a network in which N_o is an inductive subnetwork, while subnetworks N_1, N_2 and N_3 do not contain inductances. The structure of the network has been chosen to illustrate the details of the design method.

First, the circuit must be analyzed to determine $\left|\frac{V_i}{I_i}(j\omega_c)\right|$, where for our example i = 1, ..., 5. Such an analysis can be performed using a simple ladder analysis program[9]. Next, N_o is impedance scaled by $\frac{1}{C_1R_1s}$, where C_1 and R_1 are yet to be determined, and embedded inside five GIC's as shown in Fig. 11.11b. In this circuit the ith GIC has a conversion factor $k_i = C_iR_i$.

---

*This section is based on [5]

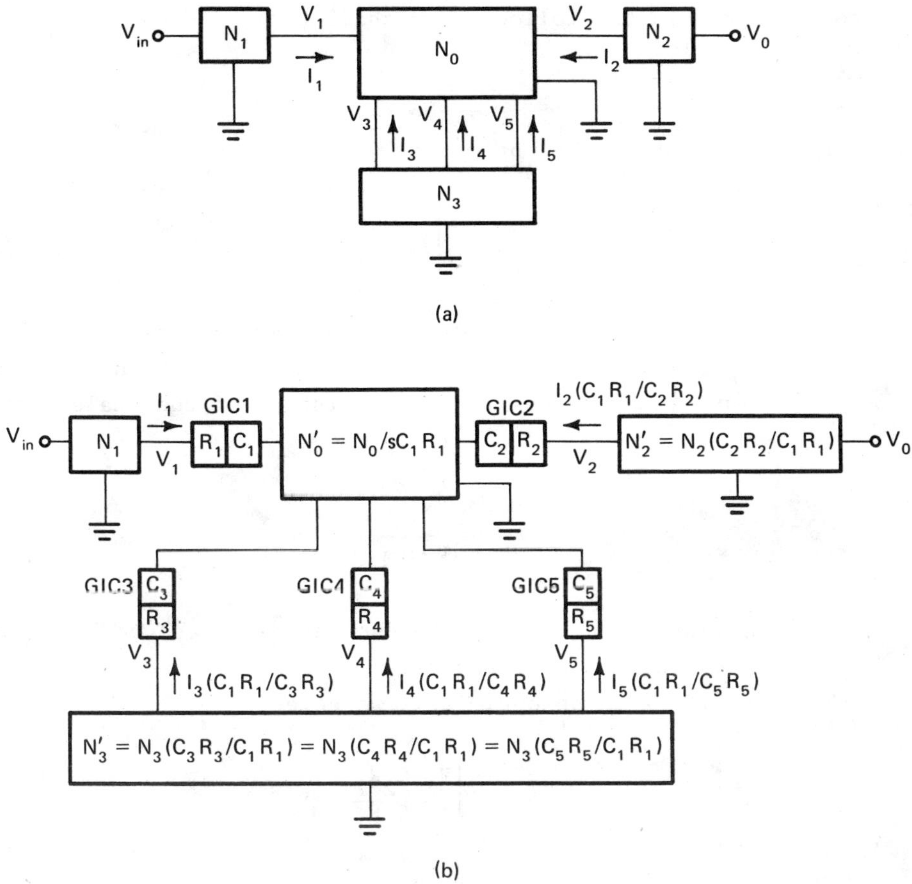

Figure 11.11 (a) Block diagram of a network containing an inductive subnetwork N_0. (b) The network of (a) after embedding the transformed subnetwork N'_0. Note that N'_2 and N'_3 are impedance scaled in order to keep the transfer function $\frac{V_0}{V_{in}}(s)$ unchanged.

Since network N_o is impedance scaled by the factor sC_1R_1 it is necessary, in order to keep the overall transfer function unchanged, to impedance scale networks N_2 and N_3 as indicated. The reader is urged to verify that these scalings indeed maintain the transfer function constant.

We now proceed to determine the element values in a step-by

step manner. In the following we shall use $\left|\frac{V_i}{I_i}\right|$ to mean $\left|\frac{V_i}{I_i}(j\omega_c)\right|$

(1) For GIC #1 to be optimized, we select

$$R_1 = \left|\frac{V_1}{I_1}\right| , \qquad (11.28)$$

while the value of C_1 can be arbitrarily chosen. It is desirable, however, to choose a value for C_1 that results in practical values for the resistors of the scaled network N_o',

$$N_o' = \frac{N_o}{sC_1R_1} ,$$

where we use N_o and N_o' to denote the impedance level of the corresponding networks.

(2) For GIC #2 to be optimized, we select

$$R_2 = \left|\frac{V_2}{I_2}\frac{C_2R_2}{C_1R_1}\right|$$

which together with (11.28) result in

$$\frac{C_2}{C_1} = \left|\frac{V_1/I_1}{V_2/I_2}\right| , \qquad (11.29)$$

while R_2 may be arbitrarily selected to yield practical values for the components of N_2'. As an aid in selecting the value of R_2 we note that,

$$R_2 = \frac{N_2'}{N_2} \left|\frac{V_2}{I_2}\right|$$

where N_2 and N_2' denote the impedance levels of the corresponding networks.

(3) Similarly, for GIC's #3, 4 and 5, we have the design equations

$$\frac{C_3}{C_1} = \left|\frac{V_1/I_1}{V_3/I_3}\right| \tag{11.30}$$

$$\frac{C_4}{C_1} = \left|\frac{V_1/I_1}{V_4/I_4}\right| \tag{11.31}$$

$$\frac{C_5}{C_1} = \left|\frac{V_1/I_1}{V_5/I_5}\right| \tag{11.32}$$

In addition, we have the constraint that all three GIC's should have equal conversion factors, as evident from Fig. 11.11b. That is,

$$C_3R_3 = C_4R_4 = C_5R_5 \quad . \tag{11.33}$$

Thus, we may select an arbitrary value for R_3 with an eye on obtaining practical component values for N_3'. Then, R_4 and R_5 are determined from

$$R_4 = R_3 \left|\frac{V_4/I_4}{V_3/I_3}\right| \tag{11.34}$$

$$R_5 = R_3 \left|\frac{V_5/I_5}{V_3/I_3}\right| \tag{11.35}$$

This completes the equations for the optimized design. The method will now be illustrated by a detailed design example.

Example 11.3

To illustrate the procedure presented above, we consider the design of a twelfth-order bandpass filter. The ideal filter characteristics are given in Fig. 11.12. The passband extends from 17 KHz to 19.7 KHz with 0.2 dB ripple. It should be noted that the highest natural mode Q-factor is 150 which makes a cascade realization very difficult.

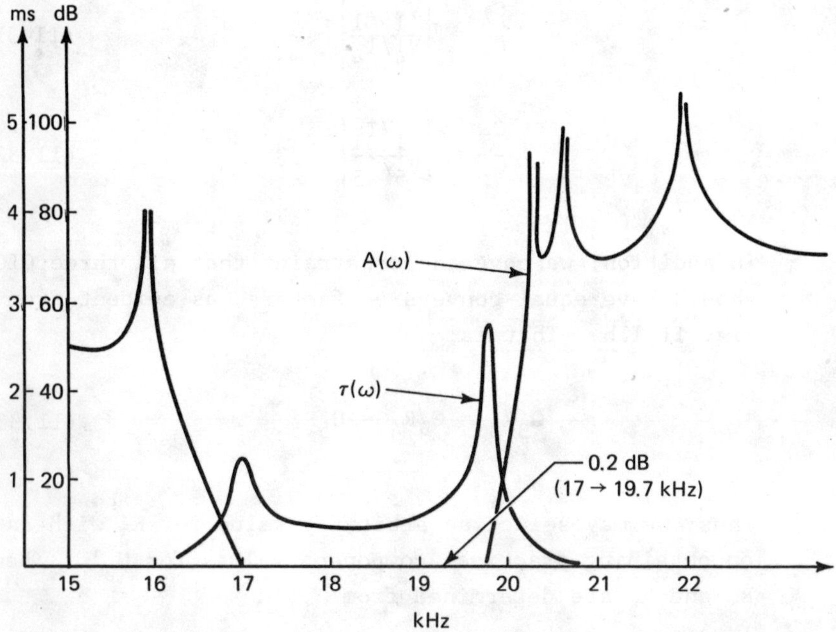

Figure 11.12 Ideal transmission characteristics of the twelfth-order bandpass filter of Example (11.3).

In order to gain an appreciation of the difficulty involved in realizing this filter, we shall evaluate the sensitivities of its passive ladder realization. It has been established in Chapter 7 that for doubly terminated lossless LC ladder filters the change in attenuation (ΔA), for given uniform element tolerance (ε) is bounded by

$$\Delta A_{max} < 8.69 \frac{|\varepsilon\rho|\omega\tau(\omega)}{1 - |\rho|^2} \text{ dB} \tag{11.36}$$

where ρ is the reflection coefficient and $\tau(\omega)$ is the group-delay. For the filter under consideration, equation (11.36) gives a deviation of 0.64 dB at 19.7 KHz if one assumes an element tolerance of 0.1% and substitutes τ = 2.8 msec. Using (7.48) it is easily shown that the attenuation deviation over the passband, due to the tank circuits having finite Q-factors (Q), is given by

$$\Delta A \simeq 4.34 \frac{\omega\tau(\omega)}{Q} \text{ dB} \tag{11.37}$$

For Q-factors of 2000, this gives $\Delta A \simeq$ 0.73 dB at 19.7 KHz. Thus, it may be concluded that the filter chosen has very stringent requirements which makes it quite sensitive to both errors in the element values and the resonant tanks having finite Q-factors.

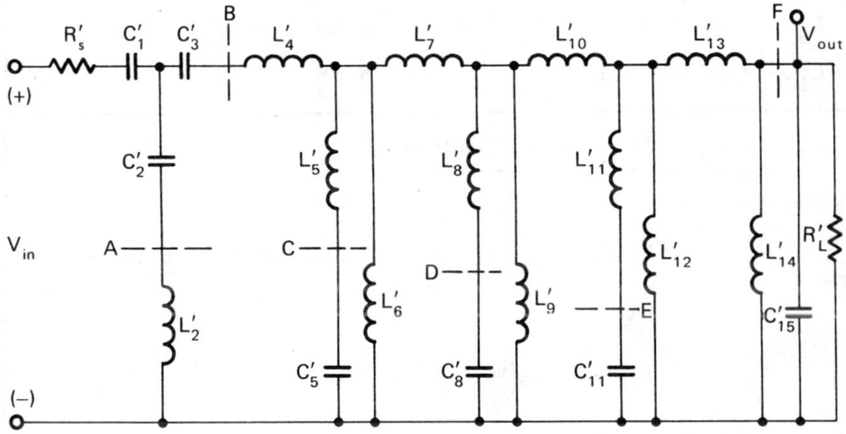

Figure 11.13 LC ladder realization of the bandpass filter. The element values are given in Table (11.1).

A passive LC realization obtained using computer synthesis programs[9] is shown in Fig. 11.13. It is easy to see that an active simulation is possible by inserting six (ks) GIC's at the points marked A, B, C, D, E and F. The passive filter was next analyzed using a simple ladder analysis program to determine

$$\left|\frac{V_i}{I_i}\right|_{\omega=\omega_c} \quad \text{where, } i = A, B, \ldots, F \text{ and } \omega_c = 2\pi \times 19.7 \times 10^3 \text{ rad/sec.}$$

The results and the element values of the LC filter are given in Table 11.1.

$R_s' = 600\ \Omega$	$R_L' = 107.414\ \text{K}\Omega$	
$C_1' = 7.5219$ nF	$C_2' = 1.6597$ nF	$C_3' = 319.5$ pF
$C_5' = 750.61$ pF	$C_8' = 750.61$ pF	$C_{11}' = 750.61$ pF
$C_{15}' = 750.61$ pF		
$L_2' = 60.69$ mH	$L_4' = 228.51$ mH	$L_5' = 82.205$ mH
$L_6' = 15.138$ mH	$L_7' = 24.048$ mH	$L_8' = 79.595$ mH
$L_9' = 39.425$ mH	$L_{10}' = 54.166$ mH	$L_{11}' = 70.043$ mH
$L_{12}' = 50.822$ mH	$L_{13}' = 260.75$ mH	$L_{14}' = 154.67$ mH
At $\omega = \omega_c = 2\pi \times 19.7$ K rad/sec:		
$\left\|\frac{V_A}{I_A}\right\| = 7.512\ \text{K}\Omega$	$\left\|\frac{V_B}{I_B}\right\| = 26.395\ \text{K}\Omega$	$\left\|\frac{V_C}{I_C}\right\| = 10.763\ \text{K}\Omega$
$\left\|\frac{V_D}{I_D}\right\| = 10.763\ \text{K}\Omega$	$\left\|\frac{V_E}{I_E}\right\| = 10.763\ \text{K}\Omega$	$\left\|\frac{V_F}{I_F}\right\| = 10.709\ \text{K}\Omega$

Table 11.1

The design procedure presented above was followed in a step-by-step manner, as illustrated in Table 11.2, to obtain the component values of the active filter shown in Fig. 11.14. Resistors R_{33} - R_{38} were added at a later time, as will be explained subsequently.

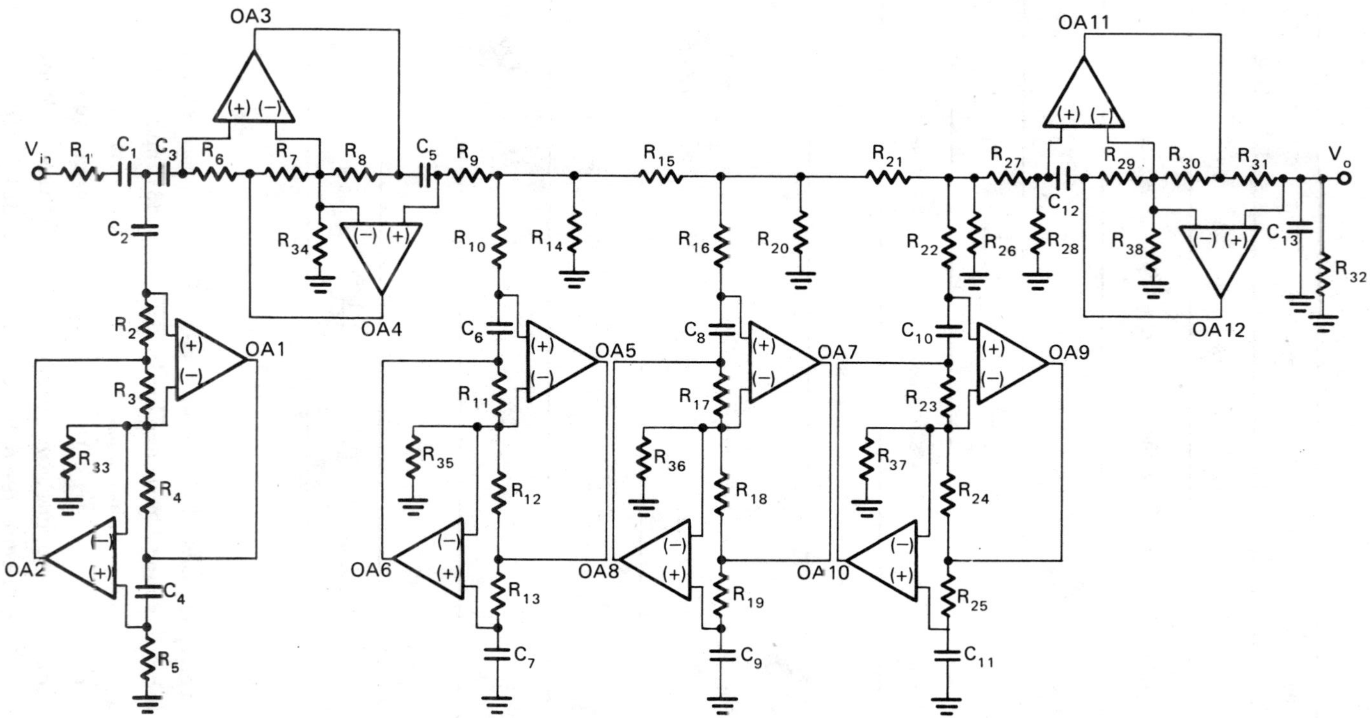

Figure 11.14 Active realization of the LC filter of Fig. (11.13) using the Gorski-Popiel technique. The component values and the design details are given in Table (11.2).

$R_1 = R_s' = 600\ \Omega,\quad C_1 = C_1' = 7.522\ \text{nF},\quad C_2 = C_2' = 1.660\ \text{nF},\quad C_3 = C_3' = 319.5\text{pF}$
LET $R_3 = R_4 = R_7 = R_8 = R_{11} = R_{12} = R_{17} = R_{18} = R_{23} = R_{24} = R_{29} = R_{30} = 3.320\text{K}\Omega$
$R_2 = \frac{V_A}{I_A} = 7.512\ \text{K}\Omega$ LET $C_4 = 2.700\ \text{nF}$ $R_5 = \frac{L_2'}{R_2C_4} = 2.992\ \text{K}\Omega$
$R_6 = \frac{V_B}{I_B} = 26.40\ \text{K}\Omega$ LET $C_5 = 504.3\ \text{pF}$
$R_9 = \frac{L_4'}{R_6C_5} = 17.17\ \text{K}\Omega$ $R_{10} = \frac{L_5'}{R_6C_5} = 6.176\ \text{K}\Omega$ $R_{14} = \frac{L_6'}{R_6C_5} = 1.137\ \text{K}\Omega$
$R_{15} = \frac{L_7'}{R_6C_5} = 1.807\ \text{K}\Omega$ $R_{16} = \frac{L_8'}{R_6C_5} = 5.979\ \text{K}\Omega$ $R_{20} = \frac{L_9'}{R_6C_5} = 2.962\ \text{K}\Omega$
$R_{21} = \frac{L_{10}'}{R_6C_5} = 4.069\ \text{K}\Omega$ $R_{22} = \frac{L_{11}'}{R_6C_5} = 5.262\ \text{K}\Omega$ $R_{26} = \frac{L_{12}'}{R_6C_5} = 3.818\ \text{K}\Omega$
$R_{27} = \frac{L_{17}'}{R_6C_5} = 19.59\ \text{K}\Omega$ $R_{28} = \frac{L_{14}'}{R_6C_5} = 11.62\ \text{K}\Omega$
$C_6 = C_5 \frac{V_B/I_B}{V_C/I_C} = 1.237\ \text{nF},$ LET $R_{13} = 2.992\ \text{K}\Omega,$ $C_7 = C_5' \frac{R_6C_5}{R_{13}C_6} = 2.700\ \text{nF}$
$C_8 = C_5 \frac{V_B/I_B}{V_D/I_D} = 1.237\ \text{nF},$ LET $R_{19} = 2.992\ \text{K}\Omega,$ $C_9 = C_8' \frac{R_6C_5}{R_{19}C_8} = 2.700\ \text{nF}$
$C_{10} = C_5 \frac{V_B/I_B}{V_E/I_E} = 1.237\ \text{nF},$ LET $R_{25} = 2.992\ \text{K}\Omega,$ $C_{11} = C_{11}' \frac{R_6C_5}{R_{25}C_{10}} = 2.700\ \text{nF}$
$C_{12} = C_5 \frac{V_B/I_B}{V_F/I_F} = 1.243\ \text{nF},$ LET $R_{31} = 883.4\ \Omega$
$C_{13} = C_{15}' \frac{R_6C_5}{R_{31}C_{12}} = 9.100\ \text{nF}$ $R_{32} = R_L' \frac{R_{31}C_{12}}{R_6C_5} = 8.860\ \text{K}\Omega$

Table 11.2

The original passive filter was designed for maximum power transfer (the passband voltage gain is 16.5 dB) and thus minimum sensitivity in the passband is expected. Any errors in the GIC conversion constants can be modelled by impedance scaling particular elements in the passive filter. Therefore, the active realization is also expected to have minimum sensitivity in the passband, with respect to the GIC conversion-constant errors.

The filter shown in Fig. 11.14 was built using polystyrene capacitors and metal film resistors. The op. amps. used were Raytheon RC 4136 quads which have a nominal unity-gain bandwidth of 3 MHz. All the passive components were chosen to within 0.5% tolerance. Each GIC was then isolated into a second-order circuit which was tuned to its proper resonance frequency. This tuning procedure is summarized in Table 11.3.

TUNING GIC INSERTED AT POINT:	GROUND THE FARTHER SIDE (FROM THE GIC) OF:	DISCONNECT:	FEED INPUT SIGNAL THROUGH 2.2 MΩ TO THE POSITIVE TERMINAL OF OP. AMP.	TO TUNE RESONANT FREQUENCY ADJUST:	SO THAT RESONANCE OCCURS AT: (KHz)
A	C2	-	OA1	R2	15.86
B	C3, R9	-	OA3	R6	18.63
C	R10	-	OA6	R13	20.26
D	R16	-	OA8	R19	20.59
E	R22	-	OA10	R25	21.95
F	R27	R32	OA12	R31	18.64

Table 11.3

The filter attenuation was then plotted on an X-Y recorder connected to an HP 3580A spectrum analyzer with the input filter set to a 100 Hz bandwidth. All measurements were taken at 0 dB (1V RMS) output (unless otherwise stated). The maximum output without noticeable distortion was measured at +5 dB which gives the filter about 100 dB dynamic range. Figure 11.15 shows a plot of the measured stopband response superimposed on the ideal response. The measured passband response together with the ideal one are shown in Fig. 11.16a.

It was felt that the deviation from the ideal attenuation, shown in Fig. 11.16a, was excessive. Upon further investigation, it was found that the filter is very sensitive to stray capacitances, as was explained in Section 11.2. The effects of these stray capacitances may be compensated for by a resistance R_c connected to each GIC as shown in Fig.11.14, resistors R_{33}-R_{38}.

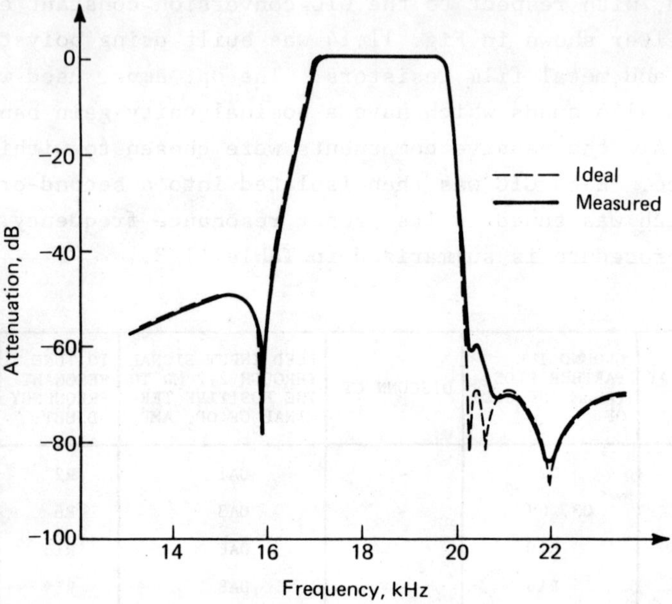

Figure 11.15 The ideal and measured responses of the filter in Fig. 11.14.

The value of these resistances can be calculated using (11.27). However, since the size of the stray capacitances are not precisely known, the tuning procedure described above was repeated adjusting R_{33} - R_{38} to give a resonance gain of unity for each of the tuned sections. The resulting passband response is plotted in Fig. 11.16b. The stopband response did not show any noticeable change.

The filter was next investigated to verify the reduced active sensitivities of the optimized design. Specifically, a number of tests were conducted on both the optimized circuit and on a circuit in which GIC #2 (composed of 0A3 and 0A4) was not designed for optimum performance. The latter design was obtained by increasing C_5 from 0.504 nF to 5.290 and then retuning GIC #2. It was found that, after tuning, the value of R_6 was different from its nominal value by 14.3%, while the corresponding difference for the optimum circuit was 4.3%; a clear indication of the reduced dependence of the optimized circuit on op. amp. imperfections.

As a further test, OA3 in both the optimum and non-optimum circuits, was replaced by a 741 package (1 MHz bandwidth) and the measured responses are shown in Fig. 11.17. For each circuit, the new response is superimposed on the previously measured graph for easy comparison. Here again, the reduced dependence on op. amp. response of the optimized design is evident.

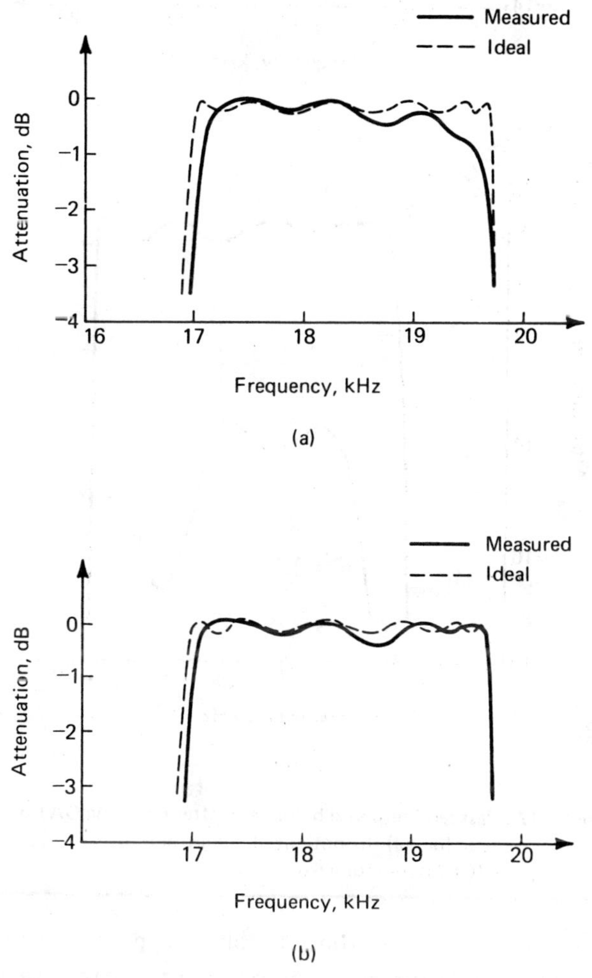

Figure 11.16 The ideal and measured passband attenuation: (a) without the Q-enhancing resistors $R_{33} - R_{38}$ and, (b) with the resistors.

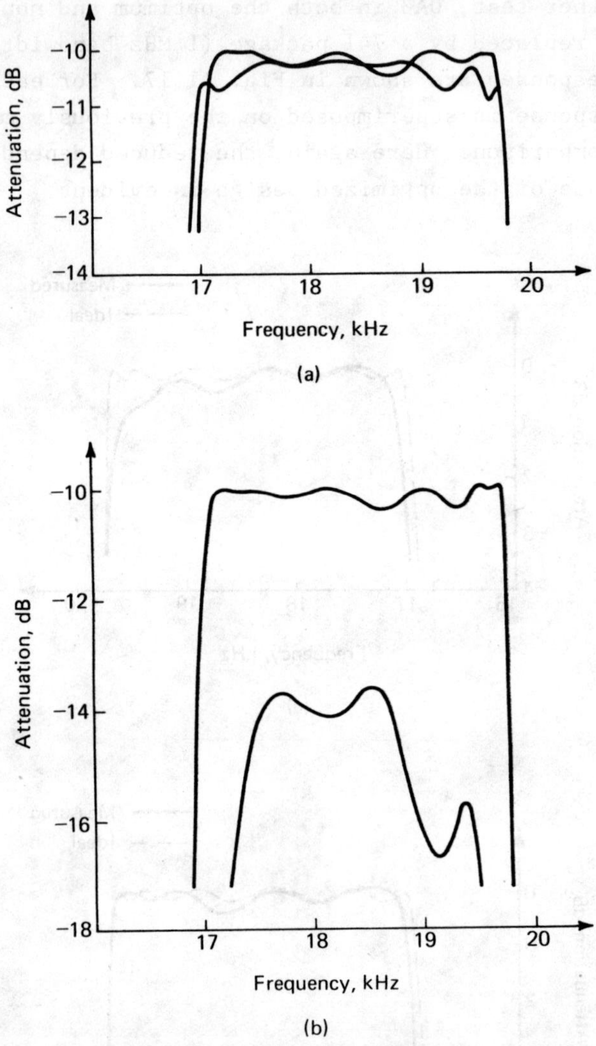

Figure 11.17 Passband response before and after replacing OA1 by a 741 type for: (a) the optimized design, and (b) the circuit with GIC#2 non-optimized.

Before leaving this section, it should be pointed out that at the time of this writing no algorithm is available for maximizing the dynamic range of filters designed using the component simulation approach.

11.5 DESIGN OF LOW-PASS FILTERS USING THE BRUTON TRANSFORMATION

Another approach to the simulation of LC filters that contain floating inductances has been proposed by Bruton[10,11]. As will be seen, Bruton's approach may be considered an extension of Gorski-Popiel's ladder embedding technique. The latter essentially involves transforming part of the LC network by the factor (1/s), where s is the complex frequency variable. Bruton's method, on the other hand, transforms the entire LC ladder by the factor $(\frac{1}{s})$. The method is particularly suited for the design of low-pass filters. The resulting designs normally require less amplifiers than those obtained by direct ladder-embedding. In the bandpass case[12], both methods usually result in similar circuits. Nevertheless, the Bruton transformation provides an alternative way of thinking that has certainly led to further developments of the component simulation approach to filter design.

The Bruton Transformation

To demonstrate how Bruton's method works, consider the fifth-order elliptic low-pass filter whose LC ladder realization is shown in Fig. 11.18a. Notice that we have used as a starting point a minimum capacitor realization. The reason for this choice will become obvious shortly. Beside each component of the LC ladder is indicated its impedance. If we divide every impedance in the LC ladder by the factor s, the voltage transfer function $\frac{V_o}{V_i}$ (s) remains unchanged. This can be easily seen if one recalls that voltage transfer functions can be expressed as ratios of network impedances, thus impedance scaling does not change the voltage transfer function.

The impedance scaling by $(\frac{1}{s})$ causes the following changes:

(a) a resistance R transforms into a capacitance of value $\frac{1}{R}$

(b) an inductance L transforms into a resistance of value L

(c) a capacitance C transforms into an impedance $\frac{1}{s^2C}$ which is the impedance of a FDNR element D equal to C in value

The transformed network is shown in Fig. 11.18b. Note that

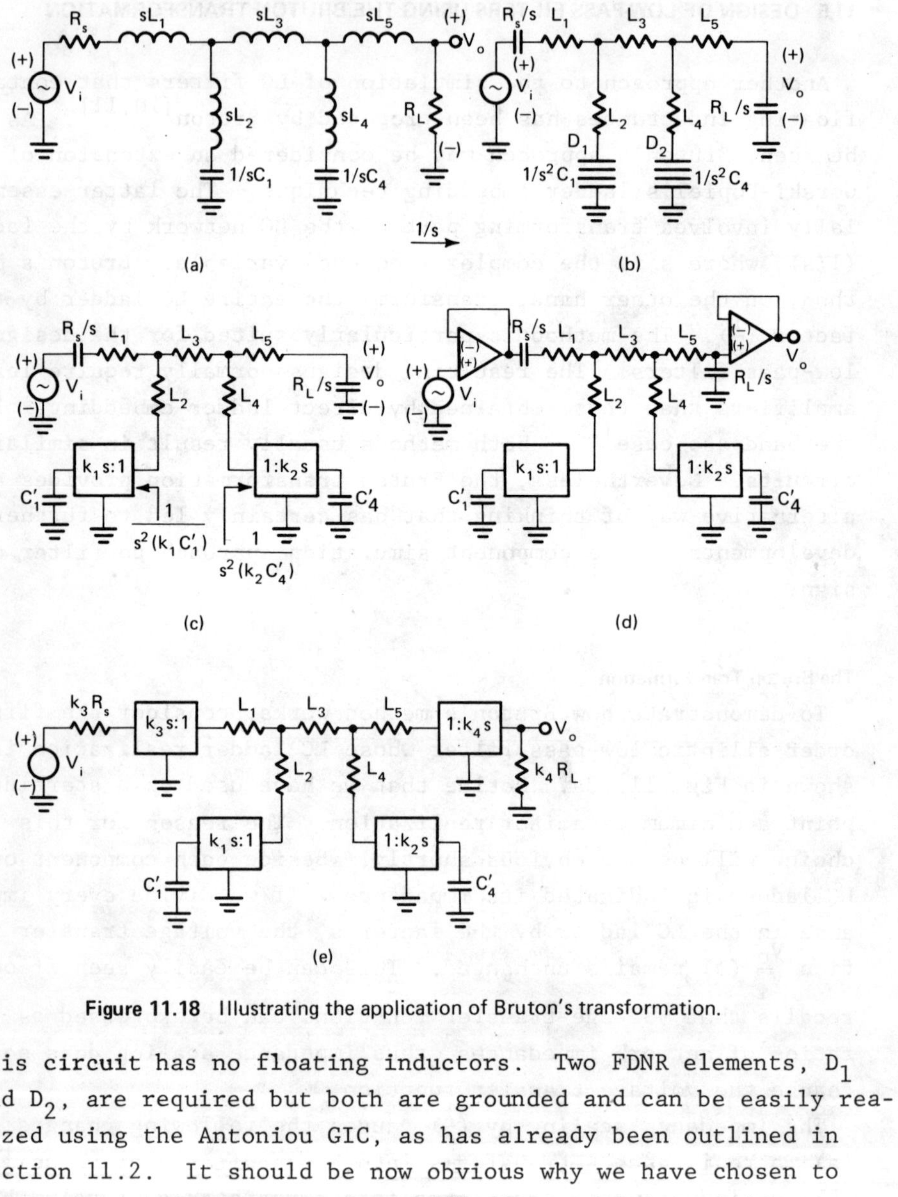

Figure 11.18 Illustrating the application of Bruton's transformation.

this circuit has no floating inductors. Two FDNR elements, D_1 and D_2, are required but both are grounded and can be easily realized using the Antoniou GIC, as has already been outlined in Section 11.2. It should be now obvious why we have chosen to start with a minimum capacitor LC ladder; a minimum inductance ladder contains floating capacitors which transform into floating FDNR's. The realization of a floating FDNR, though possible, is neither practical nor economical. For this reason, Bruton's transformation cannot be applied directly to bandpass filters. Certain modifications, however, can make this possible, with the resulting active circuits similar to those obtained by the Gorski-Popiel technique.

The reader may note that while the loss poles are realized in the LC prototype by series LC resonators, they are realized in the transformed network by series FDNR-R resonators. Fig. 11.18c shows the transformed network with each FDNR realized by a capacitively terminated $(\frac{1}{ks})$ GIC. Each GIC with the capacitance load and the series resistance on the other port forms a series resonance circuit.

Returning to the RCD ladder of Fig. 11.18b, we observe that it is no longer resistively terminated as was its LC prototype. This could be a drawback if the input source has a significant source resistance and/or if the load is resistive. These problems are easily overcome by buffering the input and/or the output as required. This is illustrated in Fig. 11.18d where we note that two additional amplifiers are required. An alternative solution, but a more expensive one, is to add one, or two, GIC's as shown in Fig. 11.18e. A close examination of this latter circuit reveals the fact that it is identical to the one we would have obtained had we directly applied Gorski-Popiel's technique to the original LC prototype. The inductance subnetwork L_1, L_2, L_3, L_4 and L_5 of the LC ladder is realized by a topologically equivalent resistance subnetwork embedded in a GIC network. The similarity between the two approaches should be apparent. Nevertheless, Bruton's method leads to a saving in the number of amplifiers if we do not require resistive terminations. The reader might also recall that in Example 11.2, the same filter is realized using 3 GIC's starting from a minimum inductance LC prototype and directly applying ladder embedding.

Circuit Implementation

Let us proceed further with our example and show a complete circuit diagram of the RCD filter of Fig. 11.18c. Using the type-B circuit to implement each of the two GIC's results in the circuit of Fig. 11.19a. Careful examination of this circuit reveals a serious problem. The noninverting input terminals of amplifiers A_1 and A_3 do not have a DC path to ground. Such a path is needed in order to supply A_1 and A_3 with the required input bias current. It should be noted that this problem will exist independent of the location of the two capacitors in each FDNR realization.

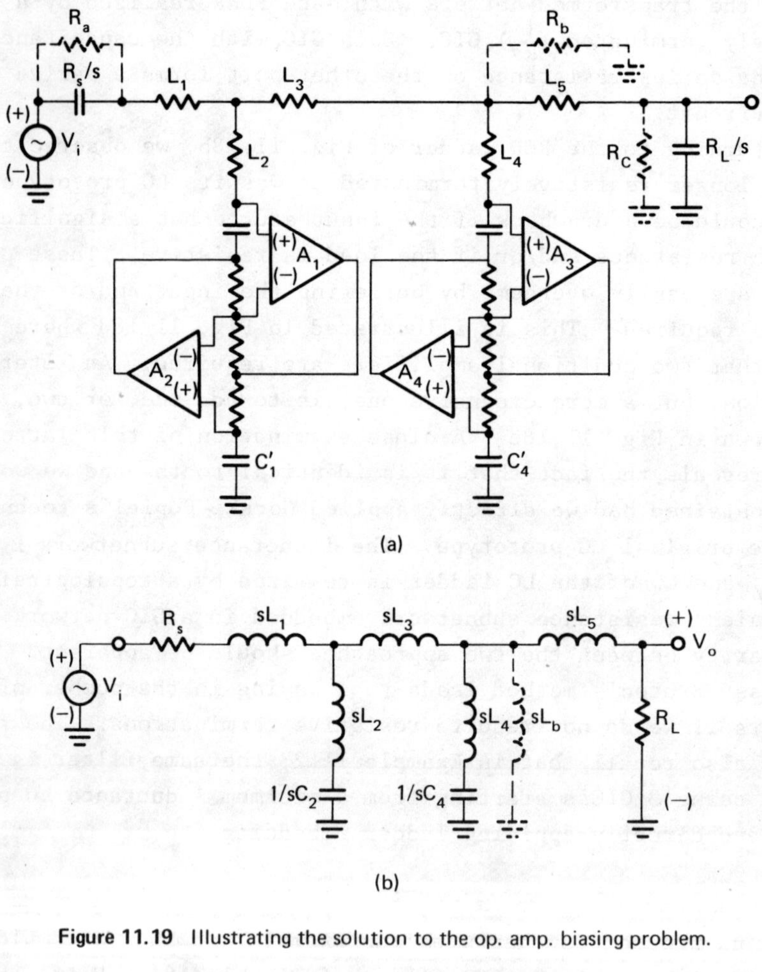

Figure 11.19 Illustrating the solution to the op. amp. biasing problem.

A straightforward solution to the bias problem consists of introducing a resistor between any node in the resistive subnetwork L_1, L_2, L_3, L_4 and L_5 to ground, such as R_b. Of course, one would choose as large a value for R_b as practically reasonable. Nevertheless, this solution is unsatisfactory. The resistance R_b corresponds to adding a parasitic inductance L_b at the corresponding location in the LC ladder, as shown in Fig. 11.19b. Such an inductance will produce a loss-pole at zero frequency. The result is a low-pass filter whose transmission falls off at low frequencies!

An alternative solution[13] to the bias problem consists of adding two resistors R_a and R_c across the terminating capacitors, as shown in Fig. 11.19a. Both amplifiers A_1 and A_3 will now have continuous paths from their non-inverting input terminals to ground. The choice of values for R_a and R_c is done as follows.

The filter transmission at DC is known from the transfer function. In our case this value is 0.5. We may therefore write

$$\lim_{\omega \to o} T(j\omega) = \frac{R_c}{R_a + L_1 + L_3 + L_5 + R_c} = 0.5 \quad , \; R_b = \infty \; .$$

Thus,

$$R_c = R_a + L_1 + L_3 + L_5 \quad .$$

The second constraint is to make R_a and R_c much larger than the other filter resistances, that is

$$R_a, R_c >> L_1 + L_3 + L_5 \quad .$$

Finally, we note that in the original LC ladder, R_a and R_c correspond to two parasitic inductances across R_s and R_L. Thus, their values clearly determine the transmission at low frequencies and eventually at DC.

Design Procedure

The design procedure for low-pass filters realized using the Bruton transformation consists of the following steps:

(1) Obtain an LC ladder prototype using closed form expressions,

filter tables or LC synthesis programs. The LC prototype is assumed to be frequency normalized ($\Omega_p = 1$) and impedance normalized.

(2) Divide the impedance of each component by s to obtain the RCD ladder. An inductance L_i H transforms into a resistance L_i Ω, a resistance R_i Ω transforms into a capacitance $\frac{1}{R_i}$F and a capacitance C_i F transforms into an FDNR element with $D_i = C_i$.

(3) Realize each FDNR by a capacitively terminated ($\frac{1}{s}$) GIC using type-B circuit as shown in Fig. 11.3. Following the design procedure developed in Section 11.2 and selecting the two capacitors equal obtains

$$R_2 = R_3$$

$$D = C^2 R_4$$

$$\omega_c C R_4 = 1$$

where ω_c is the critical frequency which is usually near the bandedge. Taking $\omega_c = \Omega_p = 1$ and selecting all the resistors equal results in

$$C = D$$

$$R_2 = R_3 = R_4 = \frac{1}{D} \quad .$$

(4) Determine the values of the bias resistances R_a and R_c (See Fig. 11.19a) from

$$\frac{R_c}{R_a + \Sigma L_j + R_c} = T(0)$$

where ΣL_j denotes the sum of all the series inductances and, R_a, R_c >> all L_j.

(5) Denormalize the component values by ω_p, which is the specified passband edge and R_n, a suitable denormalizing resistance

chosen to yield practical component values.

(6) If desired, the input and/or the output may be buffered by a voltage follower.

Example 11.4

Design a fifth-order elliptic low-pass filter with $|\rho|_{max} = 25\%$, minimum stopband attenuation = 50 dB and passband edge $f_p = 1$ KHz.

Solution

From filter tables[14] the minimum capacitor LC ladder of Fig. 11.20a is obtained. The step-by-step design procedure outlined above has been followed as illustrated in Fig. 11.20. Note that each FDNR realization is shown enclosed in a box with the op. amps. omitted for simplicity. The final circuit in Fig. 11.20d was obtained using a denormalizing resistance of 10 KΩ. If desired, buffer amplifiers can be added at the input and/or the output.

(a)

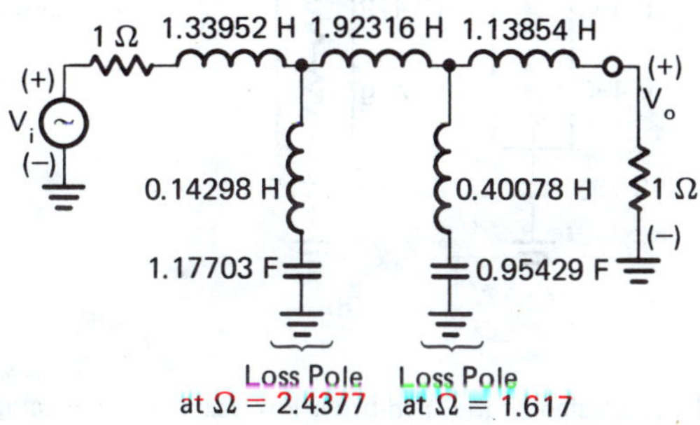

(b)

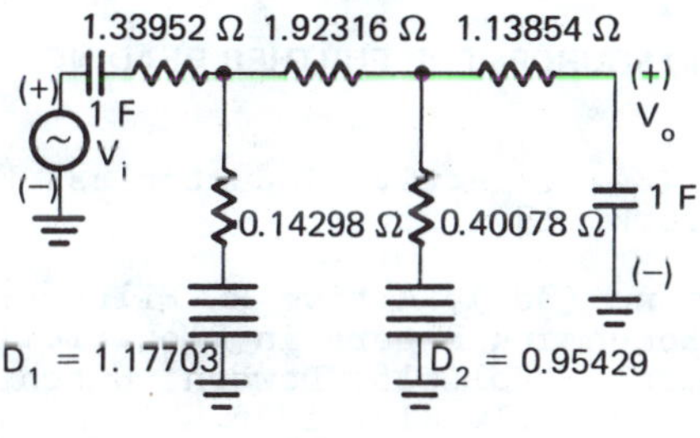

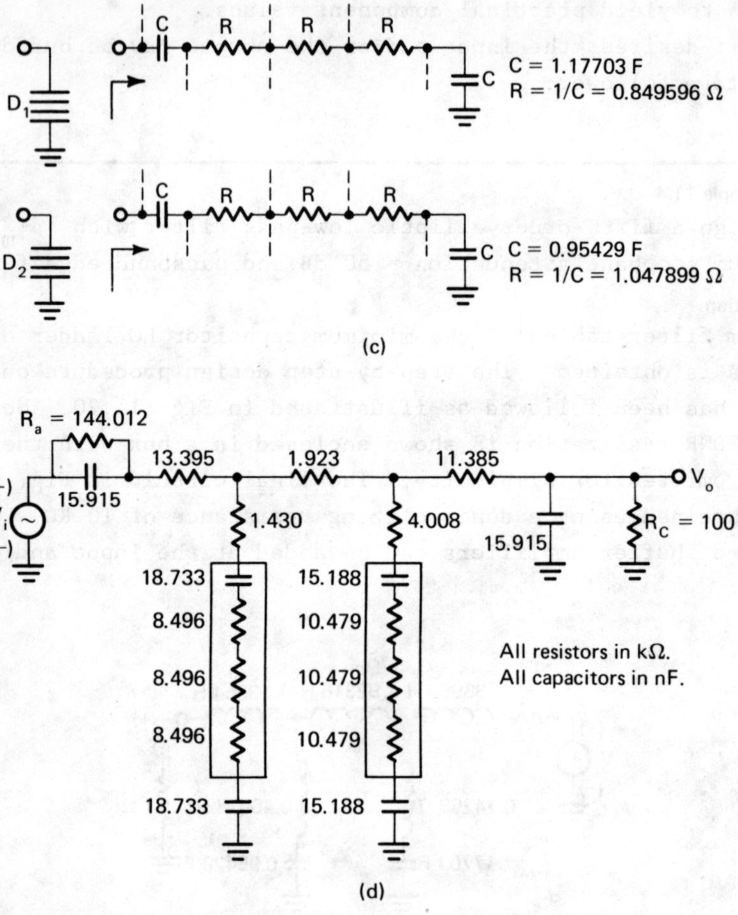

Figure 11.20 Design of the fifth-order low-pass filter of Example (11.5).

11.6 REFERENCES AND SOURCES FOR FURTHER READING

(1) S. K. Mirta (Ed.), "Active Inductorless Filters," IEEE Press, New York, 1971.

(2) L. P. Huelsman, (Ed.), Active RC Filters: Theory and Application, in Benchmark Papers in Electrical Engineering and Computer Science, Vol. 15, Dowden, Hutchinson and Ross, Inc. Stroudsburg, Pennsylvania, 1976.

(3) H. J. Orchard, "Inductorless filters," Electron. Lett., Vol. 2, p.224, Sept. 1966. Reprinted in (2).

(4) H. J. Orchard and D. F. Sheahan, "Inductorless bandpass filter," IEEE J. Solid-State Circuits, Vol. SC-5, pp. 108-118, June 1970.

(5) K. Martin and A. S. Sedra, "Optimum design of active filters using the generalized immittance converter," IEEE Trans. Circuits and Systems, Vol. CAS-24, pp. 495-503, Sept. 1977.

(6) J. Valihora, J. T. Lim and L. T. Bruton, "The feasibility of active filtering in FDM systems," Proc. 1974 IEEE Int. Symp. on Circuits and Systems, pp. 121-125, San Francisco, Calif., 1974.

(7) A. Antoniou, "Realization of Gyrators using operational amplifiers and their use in RC-active network synthesis, "Proc. Inst. Elect. Eng., Vol. 116, pp. 1838-1850, Nov. 1969. Reprinted in (2).

(8) J. Gorski-Popiel, "RC-active synthesis using positive-immittance converters," Electron. Lett., Vol. 3, pp. 381-382, August 1967.

(9) W. M. Snelgrove and A. S. Sedra, FILTOR 2-A Computer Aided Filter Design Package, Matrix Publishers, Champaign, Ill.

(10) L. T. Bruton, "Frequency selectivity using positive impedance converter-type networks," Proc. IEEE, Vol. 56, pp. 1378-1379, August 1968. Reprinted in (1).

(11) L. T. Bruton, "Network transfer functions using the concept of frequency dependent negative resistance," IEEE Trans. Circuit Theory, Vol. CT-16, pp. 406-408, Aug. 1969, Reprinted in (1), (2).

(12) L. T. Bruton and D. Treleaven, "Active filter design using generalized impedance converters," EDN, pp. 68-75, Feb. 1975.

(13) L. T. Bruton, R. T. Pederson and D. H. Treleaven, "Low-frequency compensation of FDNR low-pass filters," Proc. IEEE, Vol.60, pp.444-445, April 1972.

(14) A. I. Zverev, Handbook of Filter Synthesis, Wiley, New York, 1967.

PROBLEMS

P11.1 Consider the Antoniou CIC with

$Y_1 = sC$, $Y_2 = 1/R$, $Y_3 = (1/QR) + sC$, $Y_4 = \frac{1}{R}$, and port-2 terminated in a resistance R.

(a) Find the input admittance Y_{in} at port-1 under ideal conditions.

(b) If port-1 is terminated in a resistance R find the ω_o and Q of the resulting resonator. [Hint: The characteristic polynomial of the resonator is proportional to the total admittance at port-1.]

(c) Show that if the single pole model of each op amp is taken into account then the input admittance has a multiplicative factor which is approximately equal to $[1+4(\frac{\omega_o}{\omega_t})]$ where the two amplifiers are assumed identical, and Q>>1.

(d) Use the result of (c) to determine the actual value of ω_o and Q, denoted ω_{oa} and Q_a.

(e) Show that if the input signal is fed through the resistance R terminating port-1 and the output is taken at the output terminal of op amp A_2, then the function realized is a low-pass one. (Note: this circuit was given in Fig. 9.27.)

P11.2 Design an inductance simulation circuit to provide L=1H with infinite Q-factor at $\omega_c=10^4$ rad/sec. Assume the two op amps to be identical with $\omega_t=10^7$ rad/sec. Find the error in the realized value of L. How would you adjust the circuit to compensate for this error?

P11.3 Consider a high-pass filter designed using the inductance simulation technique. We wish to predict the changes in the attenuation function $\alpha(\omega)$ that result due to changes in temperature. The starting point is the expression for $\Delta\alpha(\omega)$, given in equation (7.40), for LC ladders.

(a) Show that since each L is given by an equation of the form $L=CR^2$ then $\varepsilon_L=2\ \varepsilon_R+\varepsilon_C$ where it is assumed that resistor values track with changes in temperature.

(b) Show that if the temperature coefficients of capacitors and resistors are equal in value and opposite in sign with an unavoidable fractional mismatch ε_τ then

$$\varepsilon_L = -\varepsilon_C \pm 2\varepsilon_\tau$$

(c) Show that

$$\Delta\alpha(\omega) = \pm\ \varepsilon_\tau\ \omega \frac{d\alpha(\omega)}{d\omega} \quad \text{Nepers}$$

(d) Show that the result in (c) may be interpreted as equivalent to a lateral shift in the frequency response given by $\Delta\omega=\pm\varepsilon_\tau\ \omega_c$ where ω_c is the critical frequency at which the group-delay peaks.

Note: Knowing the value of $\Delta\omega$ above helps the designer decide on a suitable margin in the width of the transition band.

P11.4 Derive equations (11.26) and (11.27).

P11.5 Using the Antoniou GIC for inductance simulation, design a seventh-order high-pass filter that is derived from a Butterworth LPP. The filter is to have $\omega_p=10^4$ rad/sec and $A_p=0.5$dB. You may use the closed-form expressions of Chapter 6 to compute the element values of the LPP. The GIC's should be optimized with the critical frequency being the passband edge ω_p.
Assuming that each op amp has a 3pF stray capacitance at the positive input terminal, add to your circuit suitable compensating resistors.

P11.6 (a) Sketch an active circuit realization, based on Gorski-Popiel's ladder embedding method, of a seventh-order elliptic low-pass filter using the minimum number of GIC's.

(b) For an Nth order elliptic low-pass filter designed using Gorski-Popiel's method, find the required number of amplifiers, capacitors and resistors.

P11.7 It is required to design an active circuit realization, based on Gorski-Popiel's ladder embedding technique, for the bandpass filter considered in Examples 10.4 and 10.7. The specifications of this filter are:

(1) Derived from a third-order Chebyshev low-pass prototype.
(2) Equiripple passband with A_p=0.1dB
(3) Center frequency f_o=1000Hz
(4) Ripple bandwidth B=100Hz

The method described in the text should be used for optimizing the design and for obtaining practical component values. The required |V/I| values may be obtained using manual analysis or a ladder analysis program (if available).

What is the center-frequency gain of the active filter?

P11.8 In Examples 6.5-6.7 we considered the design of a sixth-order bandpass filter derived from an elliptic LPP. Fig. 6.12 gives an LC ladder realization of this filter. Find an active realization, based on the Gorski-Popiel method, and using the design optimization method studied in the text. Note that to obtain a realization with a minimum number of GIC's one should start with an LC ladder dual to that given in Fig. 6.12. The required |V/I| values may be obtained either by manual or by computer analysis. Present your design in a systematic way and give reasons for your design decisions. Discuss practical matters such as tuning.

(Note: Cascade designs for this filter are considered in Problems P10.7 and P10.8 while a PRB design is the subject of Problem P10.10.)

P11.9 It is required to obtain an active realization based on the Gorski-Popiel method, for the eighth-order bandpass filter whose LC ladder realization is shown in Fig. P11.9. This filter is considered in a number of examples throughout this book, e.g. 5.3, 7.2, 7.3 and 10.1-10.3. The component values are given by (inductances in Henries and capacitances in Farads):

$r_s = 1\Omega$	$c_2 = 2.949 \times 10^{-5}$
$\ell_1 = 6.260 \times 10^{-4}$	$c_3 = 1.2687 \times 10^{-3}$
$\ell_3 = 6.379 \times 10^{-5}$	$c_4 = 3.618 \times 10^{-4}$
$\ell_5 = 2.313 \times 10^{-4}$	$c_5 = 3.575 \times 10^{-5}$
$\ell_7 = 5.471 \times 10^{-5}$	$c_6 = 3.799 \times 10^{-5}$
	$c_8 = 3.141 \times 10^{-4}$
	$r_L = 1.888\ \Omega$

The group-delay peaks at the frequency f_c=1KHz. Analysis of the LC ladder at this critical frequency results in

$|V_1/I_1| = 0.5288$, $|V_2/I_2| = 0.3331$, $|V_3/I_3| = 0.1254$, $|V_4/I_4| = 2.022$ and $|V_5/I_5| = 0.5832$.

Give a complete circuit schematic of the active filter with a source resistance of 10KΩ.

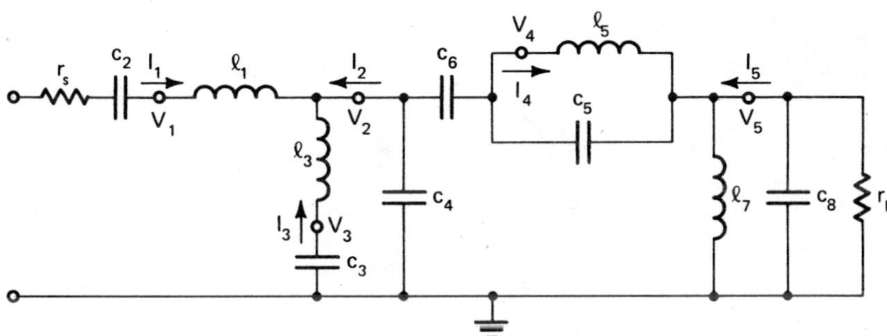

Figure P11.9

P11.10 Use Bruton's method to obtain a design of the seventh-order elliptic low-pass filter considered in Problem P10.1. The normalized (Ω_p=1) LC ladder realization is shown in Fig. P11.10. (Inductances are in Henries and Capacitances are in Farads) Design for $\omega_p=10^3$ rad/sec and consider this to be the critical frequency at which all GIC's should be optimized.

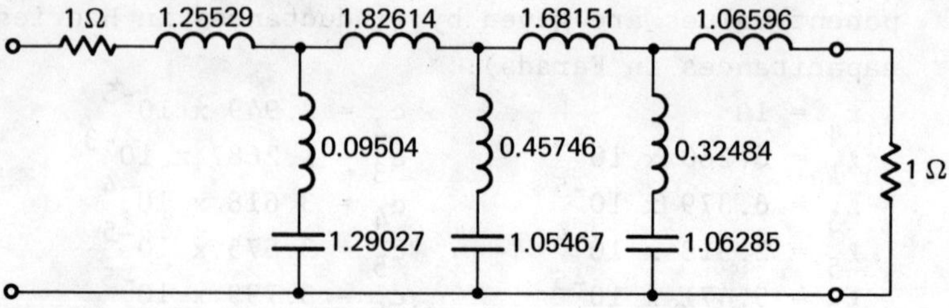

Figure P11.10

CHAPTER 12

OPERATIONAL SIMULATION OF LC LADDERS

12.1 INTRODUCTION

In the previous Chapter we studied an approach to the design of active filters starting from their passive realizations in the form of low-sensitivity LC ladders. Subsequently, the inductors in the ladder are replaced by equivalent active-RC networks; hence the name "Component Simulation" approach. The resulting active filters maintain the low-sensitivity properties of their LC prototypes.

As we have seen, the Component Simulation approach is especially suited for high-pass filters where all the ladder inductors are grounded. In the low-pass case, application of the Gorski Popiel and the Bruton transformations lead to practical designs. For the bandpass case, Gorski-Popiel's ladder embedding technique together with the optimization method presented can provide acceptable results. In general, however, it is fair to say that the Component Simulation approach is not the most convenient method for designing bandpass filters. A particular disadvantage of designs based on component simulation is the difficulty in scaling the active network to obtain maximum dynamic range. Moreover, there does not exist a systematic method for obtaining "the most convenient" LC ladder prototype for component simulation.

This Chapter presents an alternative method for the design of high-order active filters. Here, the starting point is also an LC ladder realization. However, rather than simulating the ladder inductances we seek an active circuit that simulates the "operation" of the LC ladder. To make this distinction clear, consider the simulation of an inductance L of the LC prototype. In the "Component Simulation" approach, the inductance L is replaced by a one-port network having an input impedance equal to sL. In the "Operational Simulation" method, one recognizes that the "operation" of the inductance may be described by

$$I = \frac{V}{sL} \quad ,$$

and arranges to have in the active circuit an integrator block whose time-constant is proportional to the value of L. The input voltage of this integrator will be the analogue of the inductance voltage V while the output voltage will be the analogue of the inductance current I (See Fig. 12.1).

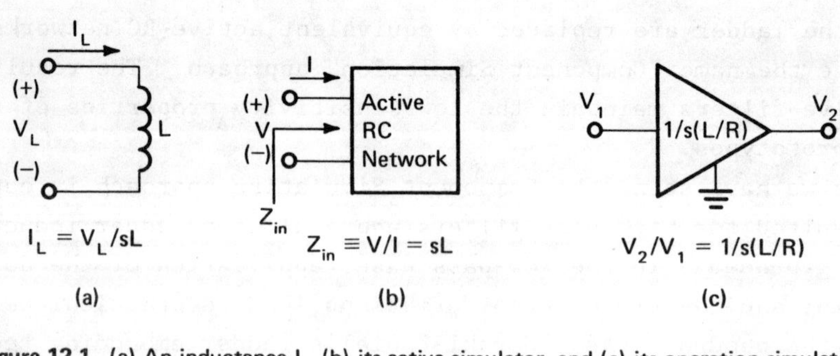

Figure 12.1 (a) An inductance L, (b) its active simulator, and (c) its operation simulator.

Thus, the basic philosophy of the "Operation Simulation" approach is to find an active circuit which "copies" the internal workings of an LC ladder prototype filter[1]. More specifically, we seek an active circuit in which the signals, say voltages, at various points are the analogues of the voltages and currents of the LC prototype. The active structure, therefore, implements all the voltage-current relationships present in the LC ladder. Hence, the active circuit will consist of Integrators that simulate the operation of inductors and capacitors, and Summers that simulate the Kirchhoff loop and node equations of the LC ladder. Since by this process a one-to-one correspondence is maintained between the integrators in the active circuit and the reactive components in the LC ladder, the resulting active filters retain the low-sensitivity properties of their LC prototype.

Section 12.2 deals with the design of all-pole filters. The resulting networks are usually referred to as Leap-Frog filters. The operational simulation of all-pole filters is quite straightforward and Section 12.2 should serve as an introduction to the general design method presented in the rest of the Chapter. The general method, which includes Leap-Frog filters as a special case, makes use of the concept of Signal-Flow Graphs (SFG's). We devote Section 12.3 to developing a method for obtaining SFG simulations of general LC ladders. The translation of these SFG simulations to actual working circuits is given in Section 12.4, where we present a number of examples to illustrate design details.

12.2 OPERATIONAL SIMULATION OF ALL-POLE FILTERS—THE LEAP-FROG STRUCTURE

The design of all-pole active filters using the operational simulation of LC ladder prototypes was discussed in detail by Girling and Good[2,3]. It also appears that Girling and Good were the first to use the name Leap-Frog for the resulting networks. This name originated from a particular way of drawing the schematic of the resulting circuits, as will be illustrated at a later stage.

Simulation of a Low-Pass Ladder Section

Fig. 12.2a shows a typical section of the LC ladder realization of an all-pole low-pass transfer function. The operation of the three reactive elements may be expressed in integral form as

$$V_1 = \frac{I_1}{sC_1}$$

$$I_2 = \frac{V_2}{sL_2}$$

$$V_3 = \frac{I_3}{sC_3}$$

In addition, the circuit is constrained by the Kirchhoff loop and node equations, namely

$$V_1 = V_2 + V_3$$

$$I_i = I_1 + I_2$$

$$I_o = I_2 - I_3$$

Since the active network we are seeking employs voltage signals, it might be helpful at this stage to rewrite the above equations using an arbitrary scaling resistance R to obtain

$$V_1 = \frac{(I_1R)}{s(C_1R)} \tag{12.1}$$

$$(I_2R) = \frac{V_2}{s(L_2/R)} \tag{12.2}$$

$$V_3 = \frac{(I_3R)}{s(C_3R)} \tag{12.3}$$

$$V_1 = V_2 + V_3 \tag{12.4}$$

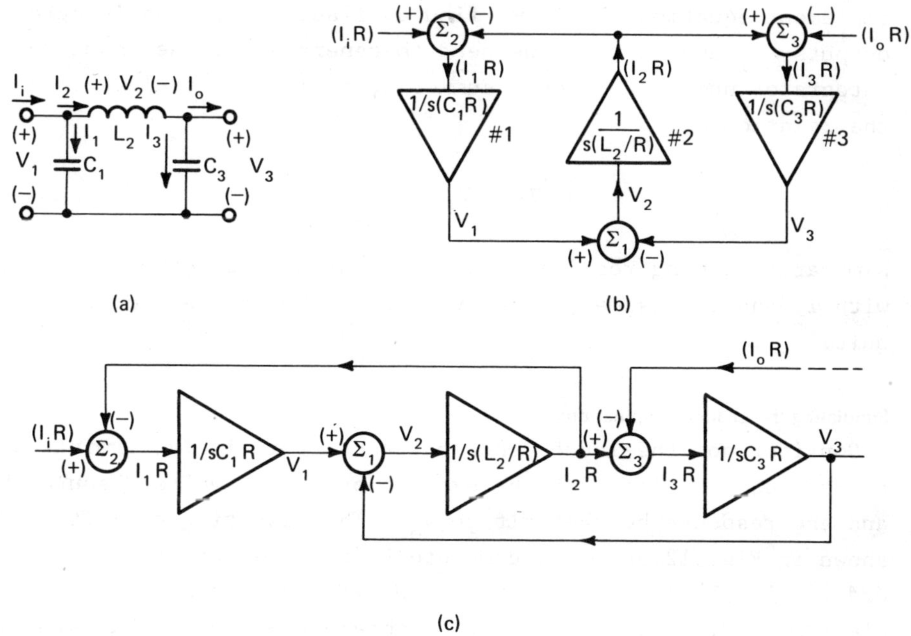

Figure 12.2 Illustrating the operational simulation of an LC ladder section.

$$(I_i R) = (I_1 R) + (I_2 R) \quad (12.5)$$

$$(I_o R) = (I_2 R) - (I_3 R) \quad (12.6)$$

To implement these equations we need three integrators and three summers, as shown in Fig. 12.2b. Note that in drawing this block diagram we kept signals that simulate ladder voltages at the bottom, while signals that simulate ladder currents are placed at the top of the diagram. This practice not only results in simple diagrams but it also speeds up the design process, as will be seen. Fig. 12.2c shows the same block diagram drawn in a different way. It is seen that the structure resembles the famous Children's game, Leap-Frog - hence the name.

Each of the integrators in the block diagram of Fig. 12.2b implements one of the three relationships (12.1) - (12.3). To implement equation (12.4) we find available to us, as integrator outputs, V_1 and V_3 while we need to generate V_2, the input to integrator number 2. Thus, summer Σ_1 is in fact implementing the relationship

$$V_2 = V_1 - V_3 \quad .$$

Similar reasoning results in the form of the two other summers, with I_i and I_o assumed to be available from the rest of the circuit.

Simulating the Ladder Terminations

Consider now the ladder section to be resistively terminated at both ends. Also, let the excitation be the voltage source V_i and the response be the voltage V_o. The resulting circuit, shown in Fig. 12.3a, is a complete LC realization of a third-order all-pole low-pass filter. We wish to augment the block diagram of Fig. 12.2b in order to obtain the operational simulation of the terminated ladder. The ladder terminations may be described by

$$(I_i R_s) = V_i - V_1$$

$$(I_o R_L) = V_3$$

Introducing the scaling resistance R modifies these two equations to

$$(I_i R) = (\frac{R}{R_s})\ V_i - (\frac{R}{R_s})\ V_1 \qquad (12.7)$$

$$(I_o R) = (\frac{R}{R_L})\ V_3 \qquad (12.8)$$

Equations (12.7) and (12.8) can be implemented as shown in Fig. 12.3b, which gives the block diagram of the operational simulation of the terminated filter of Fig. 12.3a.

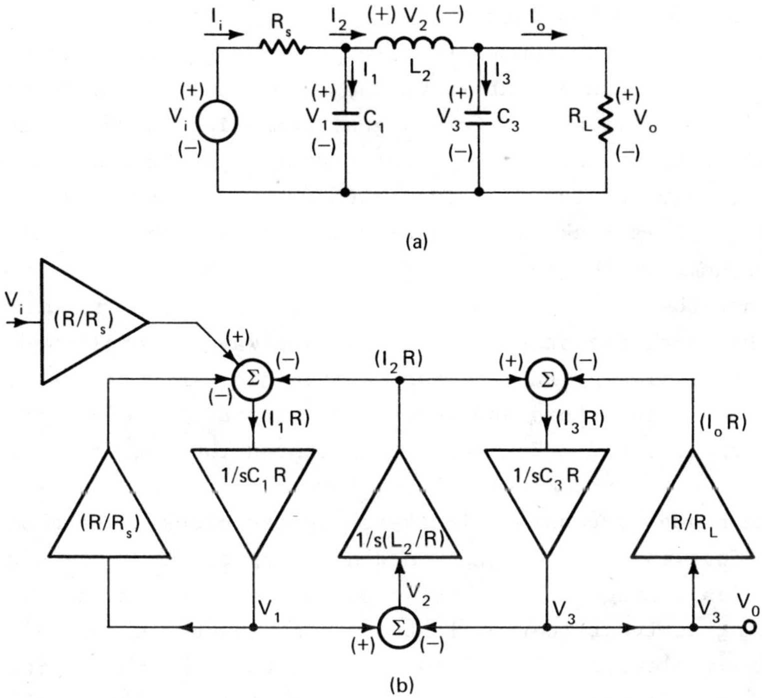

Figure 12.3 Illustrating the simulation of the terminated LC ladder.

Circuit Implementation

Before presenting a circuit implementation of the all-pole Leap-Frog filter we carefully examine the filter block diagram given in Fig. 12.3b, and make some observations. The structure consists of a number of coupled two-integrator loops (in this case two such loops exist) in addition to the two loops simulating the ladder terminations. Each loop has a negative gain - a comforting observation. The loop gains are all independent of the value of R. This is reassuring since we have chosen R arbitrarily. One can, therefore, choose R = 1Ω. To put it in a different way: what matters is the value of each loop-gain. We will, therefore, adopt the practice of starting with a prototype ladder that is impedance normalized but, of course, frequency denormalized. We will use lower case symbols to denote the component values of such a ladder.

Note that all the summations are performed prior to integration. This is convenient as we may dispense with the summer blocks altogether and perform signal summation at the virtual ground inputs of the op. amp. integrators. To be able to do this, however, requires that all inputs to a summer must have the same polarity. This task can be easily accomplished by reversing the sign of some of the integrators. In doing this, one has to ensure that the gain of each loop remains negative. It turns out that the objective stated above can be always accomplished if we adopt the following rule: "Use noninverting integrators to simulate ladder inductors and inverting integrators to simulate ladder capacitors". Following this convention together with ensuring that the termination loops have negative loop-gains is all one needs to consider in sketching the block diagram simulation. That is, the designer need not keep track of the signal polarities throughout the active structure. In some cases, the resulting active filter will have a sign inversion but this is clearly irrelevant. It should be noted that the above sign convention can be reversed. However, this will usually result in a higher number of op. amps. To appreciate this fact recall that a high quality noninverting integrator requires two amplifiers for its realization while an inverting integrator is implemented with one amplifier (See Section 8.3). This together with the fact that we usually start with minimum inductance LC ladders justifies our choice of sign convention.

Using the above observations and conclusions we obtain the modified block diagram shown in Fig. 12.4b. Note that we have included a factor k in the input signal path. Thus the DC gain of the active filter will be k times that of its LC prototype. Fig. 12.4c shows an op. amp. circuit implementation of the block diagram. Here we have used the Miller-inverter cascade (See Section 8.3) for implementing noninverting integrators. The following set of design equations is obtained by equating the gains of corresponding loops in the circuit and the block diagram.

$$C_1R_1 = c_1r_s$$

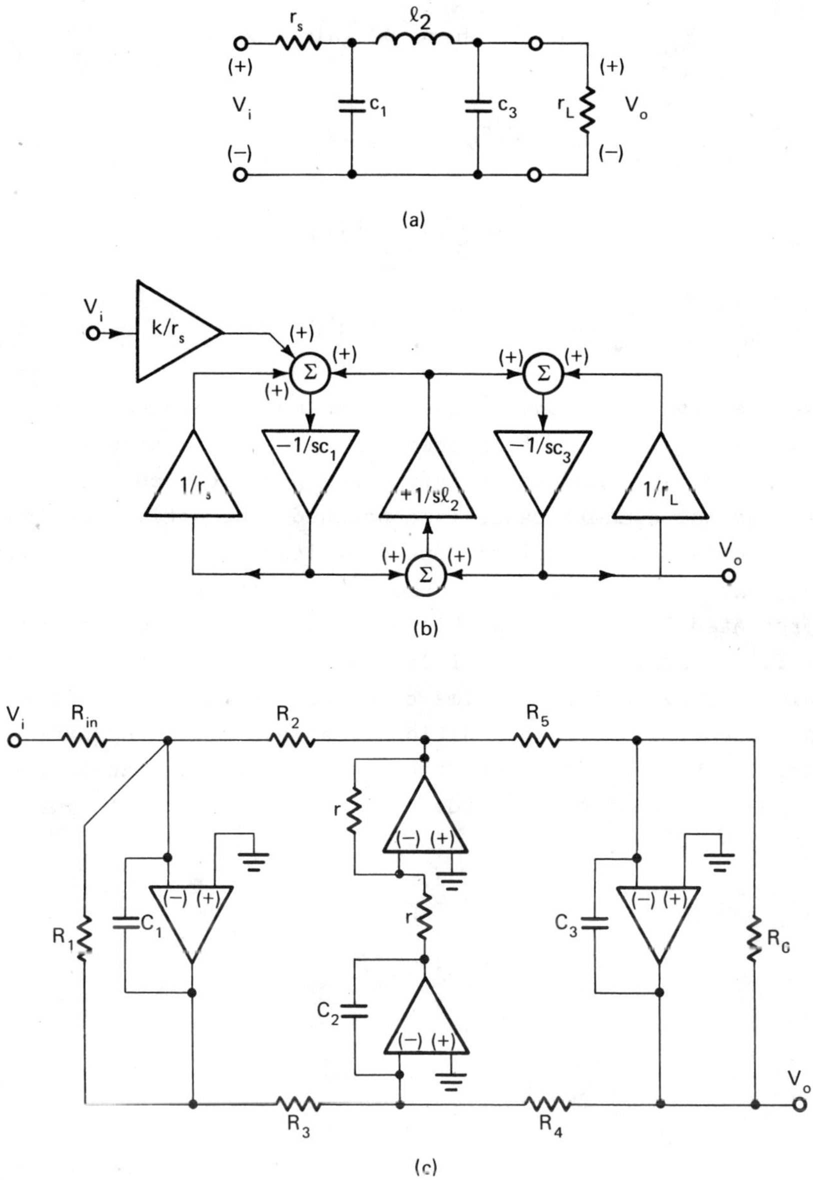

Figure 12.4 (a) Third-order all-pole LC ladder, impedance normalized; (b) Block diagram of the operational simulation of the ladder in (a); and (c) Complete circuit realization.

$$R_{in} = R_1/k \quad , \text{ where DC gain} = k\,\frac{r_L}{r_L+r_s}$$

$$C_1R_2C_2R_3 = c_1\ell_2$$

$$C_2R_5C_3R_4 = \ell_2c_3$$

$$C_3R_6 = c_3r_L$$

These are five equations in the ten unknown component values. The available degrees of freedom can be used to maximize the filter dynamic range and to satisfy other practical considerations. As far as the dynamic range is concerned, the optimum design is one in which the amplifiers' output voltages, for a given value of V_i, have equal maxima. In the next subsection, a method is presented for achieving this optimum design. For the present, we will be satisfied with a less-than-optimum solution. This is obtained by selecting the time constants of each two-integrator loop to be equal. If in addition we select all integrator capacitors to be equal to a practically convenient value C, the above design equations provide

$$R_1 = \frac{c_1r_s}{C}$$

$$R_{in} = \frac{R_1}{k}$$

$$R_2 = R_3 = \frac{\sqrt{c_1\ell_2}}{C}$$

$$R_4 = R_5 = \frac{\sqrt{\ell_2c_3}}{C}$$

$$R_6 = \frac{c_3r_L}{C}$$

Example 12.1

It is required to design a Leap-Frog active filter realizing a fourth-order Chebyshev low-pass function. The passband ripple in 0.1 dB and the ripple bandwidth is 1000 rad/sec. The maximum passband gain is to be unity.

Solution

The LC ladder prototype may be obtained using filter tables or using the closed-form expressions given in Section 6.2. We have already found an LC ladder realization for this function in Example 6.2. Denormalizing this ladder using the 1000 rad./sec. cut-off frequency leads to the circuit in Fig. 12.5a, where the component values are (capacitances in Farads and inductances in Henries):

$$r_s = 1\Omega \qquad r_L = 0.737811\ \Omega$$

$$c_1 = 1.108787 \times 10^{-3} \qquad \ell_2 = 1.306184 \times 10^{-3}$$

$$c_3 = 1.770351 \times 10^{-3} \qquad \ell_4 = 0.818075 \times 10^{-3}$$

Fig. 12.5b shows the block diagram of the operational simulation of this ladder. The value of the gain constant k is determined as follows:

$$\text{DC gain of LC ladder} = \frac{0.737811}{1 + 0.737811} = 0.42456$$

$$\text{Max. passband gain of ladder} = 0.42456 \times \sqrt{1 + \varepsilon^2}$$

$$\text{where } \varepsilon = \sqrt{10^{Ap/10} - 1} \quad , \quad Ap = 0.1 \text{ dB} \quad .$$

Thus, Max. passband gain = 0.42948

$$k = \frac{1}{0.42948} = 2.328$$

Fig. 12.5c shows the circuit diagram of the filter using Miller-Inverter-cascade for the noninverting integrators. The component values are obtained by equating the gains of corres-

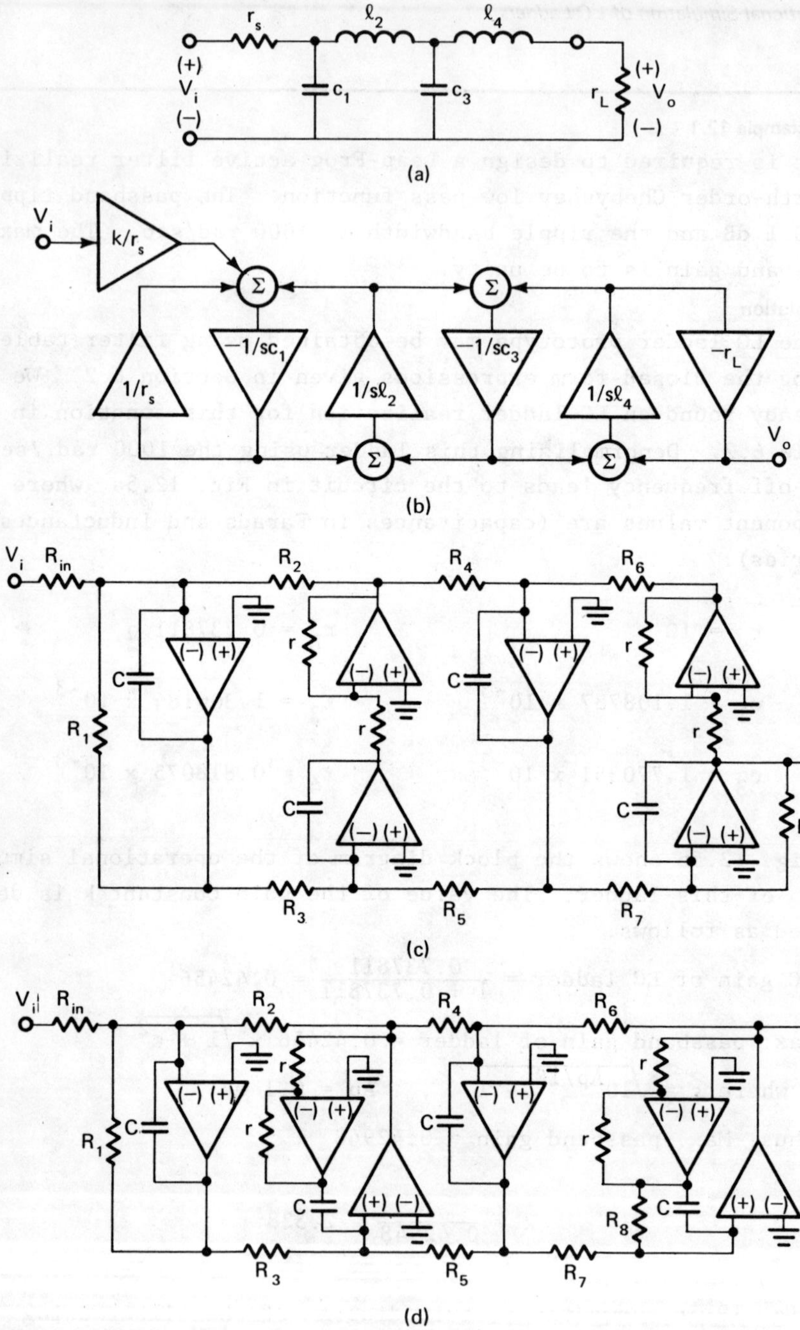

Figure 12.5 Development of the Leap/Frog realization of a fourth-order Chebyshev filter. The circuit in (d) uses the phase-lead integrator and hence has superior high-frequency performance.

ponding loops and by making the time-constants of each two-integrator loop equal. Choosing all integrator capacitors equal to C and using 0.1 μF for C results in

$$R_1 = \frac{c_1 r_s}{C} = 11.088\ \text{K}\Omega$$

$$R_{in} = \frac{R_1}{k} = 4.762\ \text{K}\Omega$$

$$R_2 = R_3 = \frac{\sqrt{c_1 \ell_2}}{C} = 12.034\ \text{K}\Omega$$

$$R_4 = R_5 = \frac{\sqrt{c_2 \ell_3}}{C} = 15.207\ \text{K}\Omega$$

$$R_6 = R_7 = \frac{\sqrt{c_3 \ell_4}}{C} = 12.034\ \text{K}\Omega$$

$$R_8 = \frac{\ell_4}{C r_L} = 11.088\ \text{K}\Omega$$

This completes the design of the filter. Note that we have three two-integrator loops of the Thomas-biquad type. In Section 9.9, we have seen that such biquads have severely limited high-frequency operation. A superior biquad is obtained by using the phase-lead integrator rather than the Miller-inverter-cascade. The resulting Leap-Frog filter is shown in Fig. 12.5d. This realization will be more tolerant to the limited op. amp. frequency response than the one given in Fig. 12.5c. The component values are identical in both cases. We will have more to say about sensitivity and high frequency limitations of filters designed by operational simulation, in a later section.

Maximizing the Filter Dynamic Range

We now consider the problem of scaling the component values such that the filter may have the widest possible signal handling capability. In Section 12.4, an algorithm is presented for accomplishing this goal for any general filter designed by the operational simulation technique. For the present, we will develop a method that applies to Leap-Frog circuits, both of the low-pass and bandpass types. We will use the third-order low-pass filter of Fig. 12.4, as a vehicle for developing the scaling method.

Fig. 12.6a shows the block diagram of the leap-frog realization of the third-order low-pass filter. This is identical to the diagram given in Fig. 12.4b except that we have absorbed the terminations' feedback paths in their corresponding integrator blocks. We will also assume that the component values have been frequency normalized. The allowable signal swing at the filter input will be maximized if the spectra of the signals at the outputs of all integrators have equal maxima. That is, if we assume that the input signal $V_i(s)$ has a constant frequency spectrum, say $V_i(s) = 1$, then the condition for maximum signal swing is

$$V_{o_{max}} = V_{1_{max}} = V_{3_{max}} ,$$

where each of these maxima is defined as

$$V_{i_{max}} = \text{Max } \{|V_i(j\Omega)|\} \quad , \ 0 \le \Omega \le \infty$$

Since $V_i(s) = 1$ then,

$$V_o(s) = T(s)$$

where T(s) is the transfer function of the normalized low-pass prototype. Thus the value of $V_{o_{max}}$ is readily available from

$$V_{o_{max}} = T_{max}$$

The value of $V_{1_{max}}$ can now be calculated from Fig. 12.6a as

$$V_1(s) = -V_o(s) \ (sc_3 + \frac{1}{r_L})$$

$$= -T(s) \ (sc_3 + \frac{1}{r_L})$$

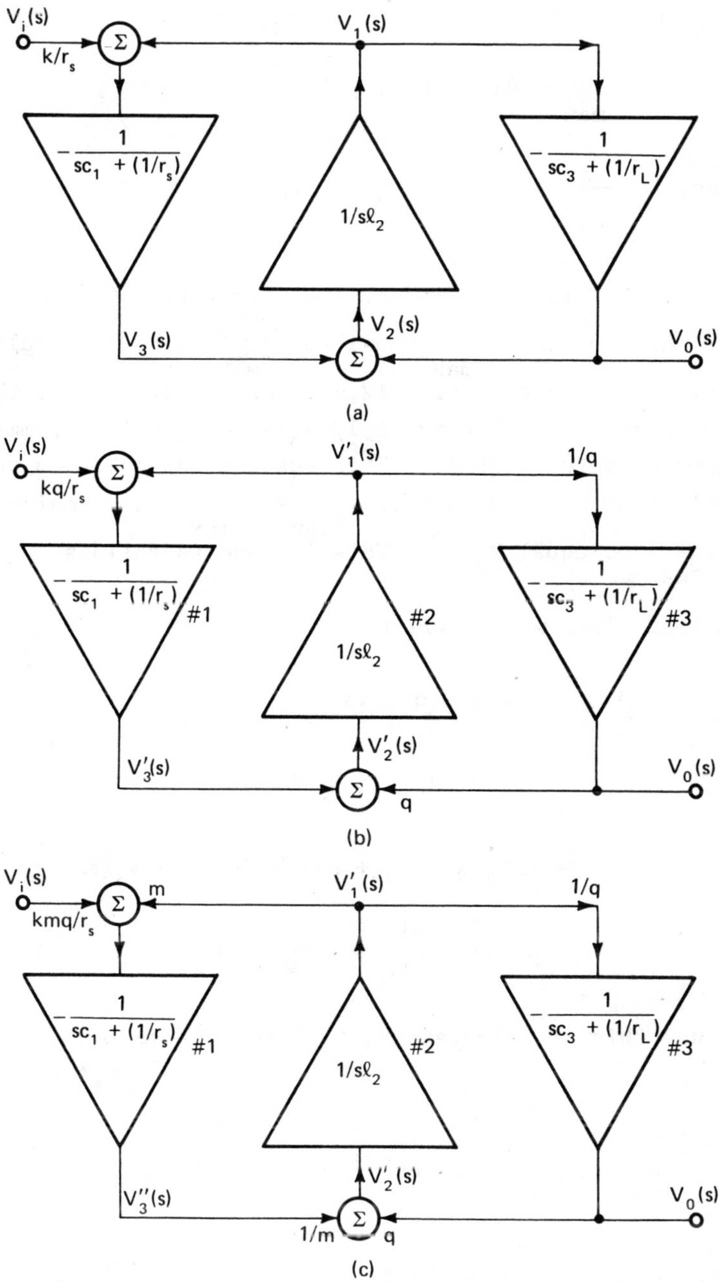

Figure 12.6 Illustrating the process of scaling for maximum dynamic range.

Thus,

$$V_{1_{max}} = \text{Max } \{|T(j\Omega)| \sqrt{(\Omega c_3)^2 + (\frac{1}{r_L})^2}\}$$

Let, $$\frac{V_{o_{max}}}{V_{1_{max}}} = q$$

Now, since $V_{o_{max}}$ is specified by the filter transfer function, it follows that to make $V_{1_{max}}$ equal to $V_{o_{max}}$ we have to multiply $V_1'(s)$ by $\frac{1}{q}$, as shown in Fig. 12.6b. Note that to keep the loop gain constant we have to multiply $V_o(s)$ by q before summing at the input of integrator #2. This scaling has achieved part of our goal; that is, now we have $V'_{1_{max}} = V_{o_{max}}$. It remains to force $V_{3_{max}}$ to equal $V_{o_{max}}$. This is done as follows:

From Fig. 12.6b we may write

$$\begin{aligned} V_3'(s) &= V_2'(s) - q\, V_o(s) \\ &= s\ell_2\, V_1'(s) - q\, V_o(s) \\ &= -s\ell_2\, q\, (sc_3 + \frac{1}{r_L})\, V_o(s) - q\, V_o(s) \\ &= -q\, (1 + \frac{s\ell_2}{r_L} + s^2\ell_2 c_3)\, V_o(s) \end{aligned}$$

We can use this to calculate $V'_{3_{max}}$. Now, if we let

$$\frac{V_{o_{max}}}{V'_{3_{max}}} = m,$$

then including the coefficients m and $\frac{1}{m}$ as in Fig. 12.6c insures that $V''_{3_{max}} = V_{o_{max}}$. Notice that in this final circuit we have changed the input summing coefficient to keep the total filter gain unchanged. The circuit of Fig. 12.6c will have the widest possible signal swing; that is, if the specified filter gain is G and the maximum allowable op. amp. output voltage is V_{amp}, then the maximum signal swing at the input is

$$V_{i_{max}} = \frac{V_{amp}}{G}$$

Extension of the above development to any filter order is straightforward. For filters of moderately high order, the calculations required can be performed on a simple calculator, as illustrated by the following example.

Example 12.2

Design a Leap-Frog circuit to realize a third-order low-pass Chebyshev filter with $A_p = 0.1$ dB, $\omega_p = 10{,}000$ rad./sec. and DC gain of 10. Optimize the circuit for maximum dynamic range.

Solution

The closed form expressions of Section 6.2 are used to obtain the following normalized component values

$$r_L = r_s = 1\Omega$$

$$c_1 = c_3 = 1.031560$$

$$\ell_2 = 1.147397$$

The block diagram realization is identical to that given in Fig. 12.6c with the scaling coefficients q and m calculated as follows

$$T_{max} = \text{Max}\ \{|T(j\Omega)|\} = |T(0)| = 10$$

Note that T(s) is given by,

$$T(s) = \frac{16.38051}{s^3 + 1.938812\ s^2 + 2.629495\ s + 1.638051}$$

$$V_{1_{max}} = \text{Max}\ \{|T(j\Omega)|\sqrt{1 + \Omega^2 c_3^2}\ \} \simeq 14.45$$

$$\text{Thus, } q = \frac{T_{max}}{V_{1_{max}}} \simeq 0.692$$

$$V'_{3_{max}} = \text{Max}\ \{q|T(j\Omega)|\sqrt{(1-\Omega^2 \ell_2 c_3)^2 + \Omega^2 \ell_2^2}\ \} = 10$$

$$\text{Thus, } m = \frac{T_{max}}{V'_{3_{max}}} = 1$$

The value of k is given by,

$$k \frac{r_L}{r_L + r_s} = 10\ .$$

Thus,

$$k = 20$$

The input gain constant is kmq = 13.84.

Using the above values together with a denormalizing frequency of 10,000 rad./sec., the circuit of Fig.12.7 is obtained. It should be noted that the time-constants of one of the loops are far from being equal. This circuit, however, will have a maximum allowable input signal swing equal to the amplifiers' specified signal capability V_{amp}, divided by the filter gain; obviously an optimum design.

Leap-Frog Realizations of All-Pole Bandpass Filters

The Leap-Frog design technique can be extended easily to the realization of all-pole bandpass filters. The starting point is an LC ladder realization of the low-pass prototype, frequency and impedance normalized. Fig. 12.8 illustrates the design process for a 6th order bandpass filter. The LC ladder realization of

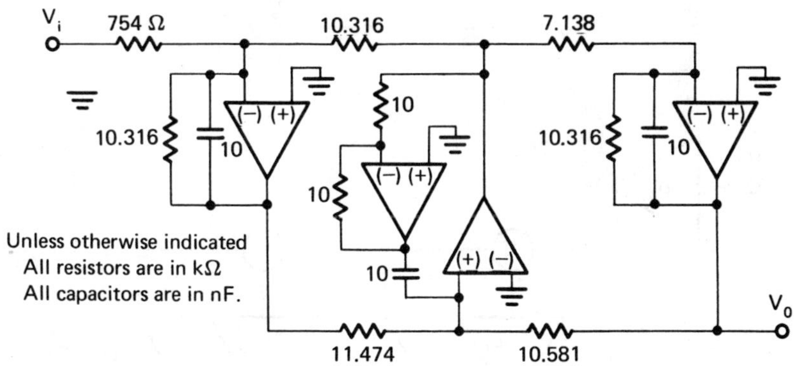

Figure 12.7 Circuit realization of the third-order low-pass filter of Example (12.2). This circuit is optimized for maximum dynamic range.

the low-pass prototype is shown in Fig. 12.8a, while Fig. 12.8b gives its operational simulation in block diagram form. The block diagram of the bandpass realization, shown in Fig. 12.8c, is obtained by applying the low-pass to bandpass transformation (See Section 6.3) to the block diagram of Fig. 12.8b. For the bandpass filter to have a center frequency ω_o and a bandwidth B, the LP to BP transformation is given by

$$s = \frac{p^2 + \omega_o^2}{pB} .$$

The negative-feedback path around each of the two end blocks in Fig. 12.8c may be included within the corresponding block, resulting in the block diagram of Fig. 12.8d. Note that the structure consists of three coupled bandpass sections all having the same center frequency, which is also equal to the center frequency of the overall filter. The three bandpass sections have alternating polarities. The two outer sections have finite Q-factor while the inner sections (one section in our example) have infinite Q-factor. The factor k determines the center frequency gain of the bandpass filter according to the relationship,

$$\text{gain of BPF at } \omega_o = k \frac{r_L}{r_L + r_s} .$$

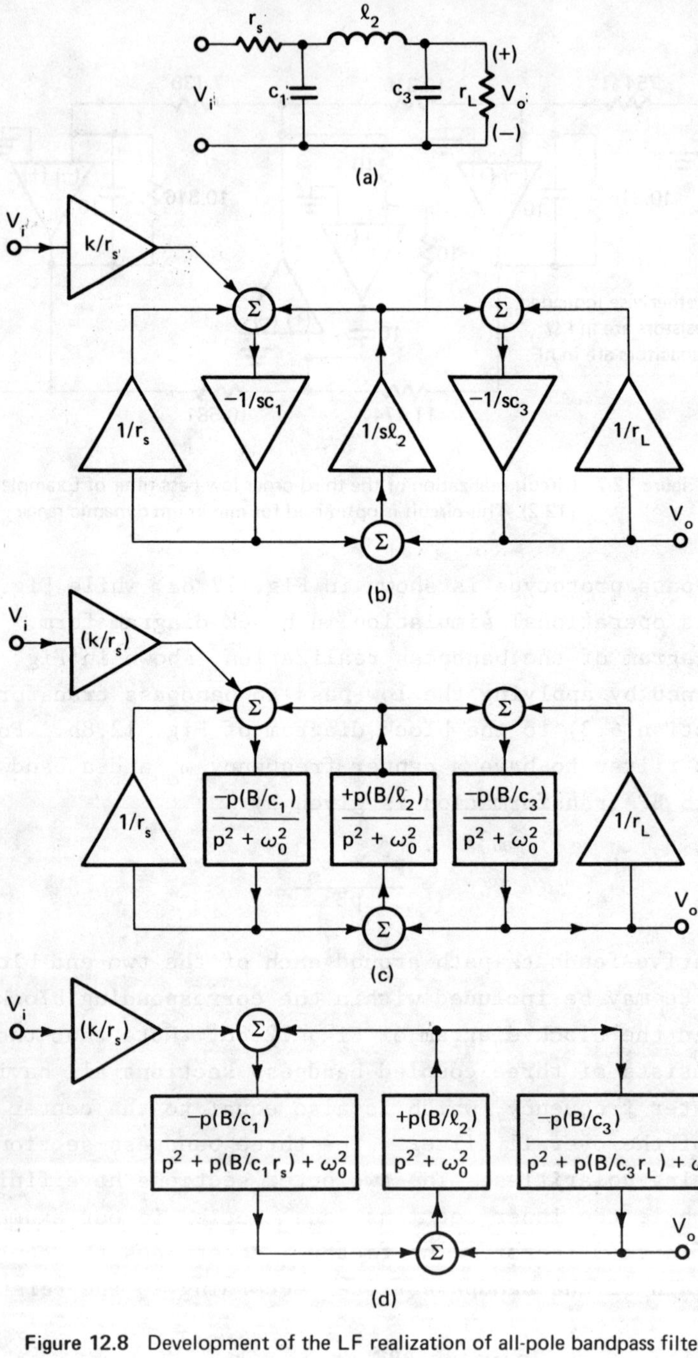

Figure 12.8 Development of the LF realization of all-pole bandpass filters.

Circuit implementation of the block diagram of Fig. 12.8d is straightforward. The bandpass sections may be implemented using any of the biquad circuits presented in Chapter 9. Note, however, that we need both inverting and non-inverting biquads. Also, the biquad circuit should be capable of implementing a bandpass function with infinite Q-factor. Finally, we note that the scaling procedure for optimum dynamic range can be directly applied to the bandpass case. In other words, we carry out the procedure on the LPP to determine the scaling factors and use them directly in the final BPF realization.

Example 12.3

It is required to design a BPF having the following characteristics:

(1) Derived from a 3rd order Chebyshev LPP.

(2) Equiripple passband with A_p - 0.1 dB.

(3) Center frequency f_o = 1000 Hz.

(4) Ripple bandwidth B = 100 Hz.

(5) Center-frequency gain = 10.

Give a Leap-Frog realization* optimized for maximum dynamic range.

Solution

We first note that the LPP for this filter is identical to that considered in Example 12.2. Thus, the values of the LPP components and the scaling factors are those obtained in Example 12.2. Using these values and following the design procedure outlined above, we obtain the block diagram realization of Fig. 12.9a.

If we choose to implement the bandpass sections with the Akerberg-Mossberg biquad, the circuit of Fig. 12.9b results. Note the modification of the biquad circuit to obtain a noninverting section.

An alternate realization using SAB's is given in Fig. 12.9c. Here, the Friend biquad is used for the inverting sections, while

---

*Note that we have obtained a cascade design for this filter in Example 10.4 and a PRB design in Example 10.7.

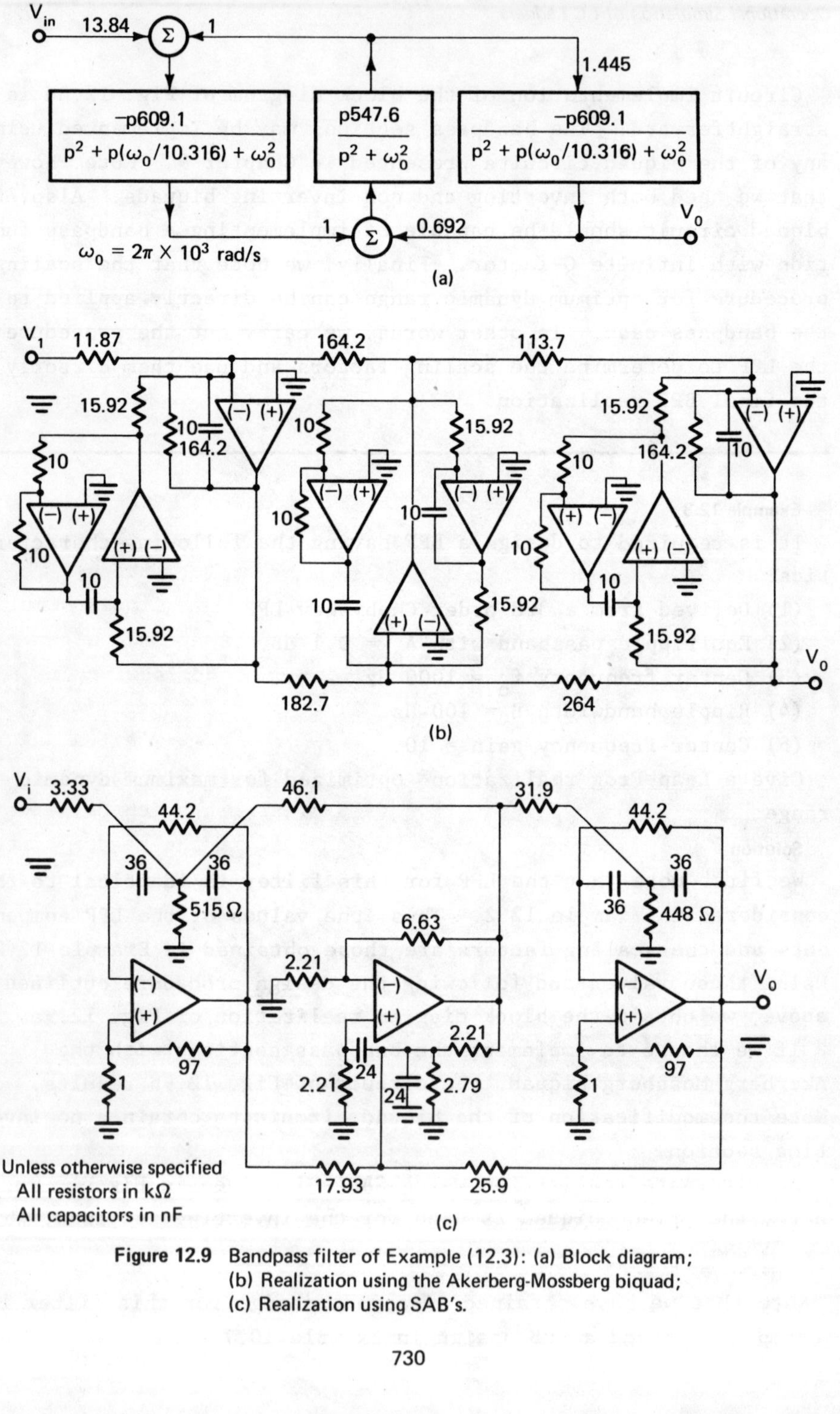

Figure 12.9 Bandpass filter of Example (12.3): (a) Block diagram; (b) Realization using the Akerberg-Mossberg biquad; (c) Realization using SAB's.

the PF1-type biquad is used for the noninverting section. Furthermore, signal summation is implemented by splitting the feed-in components; a technique discussed in detail in Chapters 8 and 9. Although this technique saves on the number of amplifiers, the resulting circuit may be difficult to adjust. Alternatively, one may use additional amplifiers to implement the summers.

12.3 SIGNAL-FLOW-GRAPH (SFG) SIMULATION OF LC LADDERS*

The Leap-Frog design technique presented in the previous section is limited to all-pole filters. Although attempts have been made[3] to extend the method to filters with finite transmission zeros, the resulting designs were quite cumbersome. Therefore, it is our objective here to present a systematic procedure for obtaining active filter designs based on the operational simulation of LC ladder prototypes. The method presented is a general one and applies for any LC ladder filter, whether derived using reactance transformations (Chapter 6) or synthesised directly using the methods of Chapter 7.

The operational simulation method to be described makes use of the concept of signal flow graph (SFG) representation of network equations. This should come as no surprise to the reader since the block diagrams employed in the previous section are in effect a form of signal flow representation. Although the material presented should be sufficiently clear, the reader who is unfamiliar with SFG's is advised to consult one of the many references dealing with the subject[7]-[11].

An LC ladder may be formed by the cascade connection of a number of the six two-port building blocks shown in Fig. 12.10. Therefore, we seek an SFG simulation for each of these building blocks and develop a set of rules for the cascade connection of the resulting SFG's. These two tasks will be accomplished concurrently.

---

*The material of this section is based on Ref. [4-6]

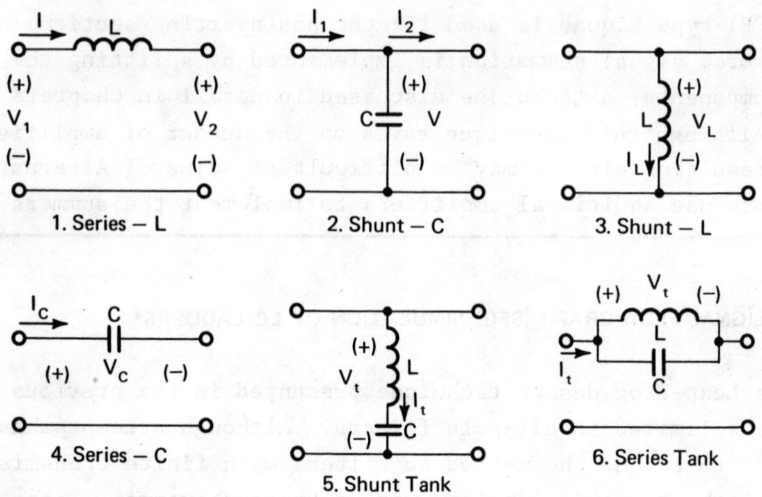

Figure 12.10 The six building blocks of LC ladders.

The Series-L

The operation of the Series-L (Fig. 12.11a) as a two-port network may be described by

$$I_1 = I_2 = \frac{V_1 - V_2}{sL}$$

In this equation we consider the series-L to be excited at its two ports by voltage signals (V_1, V_2). In response, the series-L produces the currents I_1 and I_2 which are fed to the preceding and succeeding sections, respectively, in the LC ladder. This operation is simulated by the SFG of Fig. 12.11b. As required, this SFG contains only one integrating branch which corresponds to the inductance L. In fact this SFG is unique in the sense that it is the only possible SFG simulation for the series-L that contains a single integrating branch.

Since the summing node in the series-L SFG exists at the source of the integrating branch, we will be able to perform this summation, in the circuit implementation, at the virtual ground input of the op. amp. integrator. This summing node will be called a "terminal node". A "terminal node" is defined as a node on which all oncoming signal flow lines should terminate.

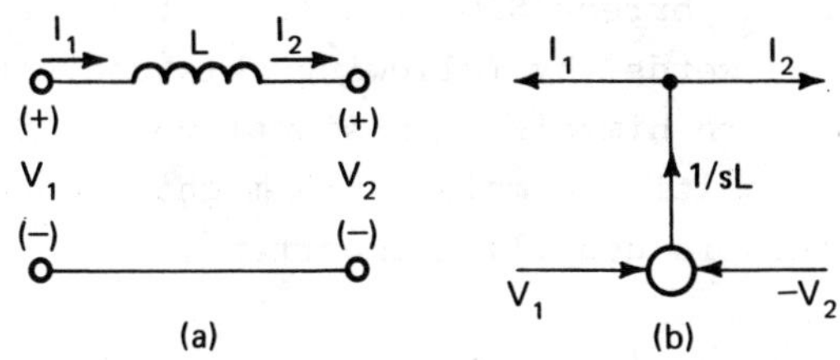

Figure 12.11 The Series – L and its simulating SFG.

The series-L SFG is a two-port SFG. Port-1 is a voltage-excited (V_1) current-response (I_1) port. Port-2 has a similar "orientation". It follows that when connecting this SFG block in cascade, the preceding SFG block should have a port-2 that is a current-excited voltage-response port. Similarly, port-1 of the succeeding SFG block, in the cascade, should have a current-excited voltage-response orientation.

The Shunt-C

The operation of the shunt-C section (Fig. 12.12a) as a two-port network may be described by

$$V_1 = V_2 = \frac{I_1 - I_2}{sC}$$

Here, we consider the excitation to be (I_1, I_2) while the response is (V_1, V_2). The simulating SFG, which is also <u>unique</u>, is shown in Fig. 12.12b. Again, we have an integrating branch preceded by a summing node. This summing node also will be designated a "terminal node".

Note that the integrating branch in the Shunt-C SFG has been assigned a negative sign. In contrast, the integrating branch in the Series-L SFG has a positive sign. We shall always adopt this sign convention. Recall that we have used this sign assignment in the previous section also. Adhering to this sign convention, together with arranging that the two end loops, which simulate the ladder terminations, have negative loop gains, is sufficient to ensure a correct SFG simulation for the complete LC ladder. In other words, by following this convention, the designer need not concern himself with signal polarities any further. Of course, the resulting complete SFG might produce a total sign inversion but this is usually unimportant.

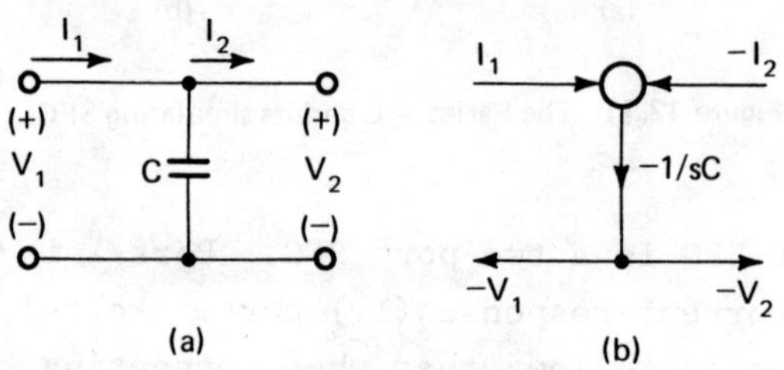

Figure 12.12 The Shunt – C and its simulating SFG.

The Shunt-C SFG is a two-port SFG with both ports having a current-excited voltage-response orientation. The reader can easily deduce the orientation that port-2 of a preceding SFG and port-1 of a succeeding SFG should have in order for their cascade connection to be possible.

Finally, it is interesting to note that while the Series-L SFG "propagates" current signals, the Shunt-C SFG "propagates" voltage signals, to both sides of the cascade.

Example 12.4

Give the SFG simulation of the fourth-order all-pole low-pass LC ladder filter shown in Fig. 12.13a.

Solution

At the outset we note that except for the resistive terminations, our LC ladder is composed of Series-L and Shunt-C sections.

The steps involved in obtaining an SFG simulation of the LC ladder are as follows:

1. Draw the SFG simulations of all sections side-by-side, as shown in Fig. 12.13b. Note that, although not necessary, we have kept track of all signal polarities.

2. Examine the interface between each two adjacent SFG blocks to determine whether a connection is possible. A connection is possible only if each two ports facing each other have dual "orientations". Since this is obviously true in our example, we simply join horizontal outgoing lines to horizontal ingoing lines. Notice that outgoing lines <u>terminate</u> on the closest "<u>terminal</u>" node.

3. Find the SFG simulation of the ladder terminations. At the source side we need to generate I_i from V_1 and the external excitation V_s. Thus, the SFG simulation of the source termination implements the relationship

$$I_i = \frac{1}{R_s}(V_s - V_1)$$

Similarly, the SFG simulation of the load termination implements

$$V_o = I_o R_L$$

The termination SFG's are then connected to the SFG of the rest of the ladder and redundant nodes are coalesced.

The complete SFG is shown in Fig. 12.13c. Although this SFG was developed in a step-by-step manner it is expected that, with some practice, the reader should be able to sketch the final SFG directly. The reader is urged to compare the resultant SFG with the Leap-Frog block diagram developed in Example 12.1, for this same LC ladder. The two structures are indeed identical, confirming our original contention that Leap-Frog designs follow as a special case of the general SFG method.

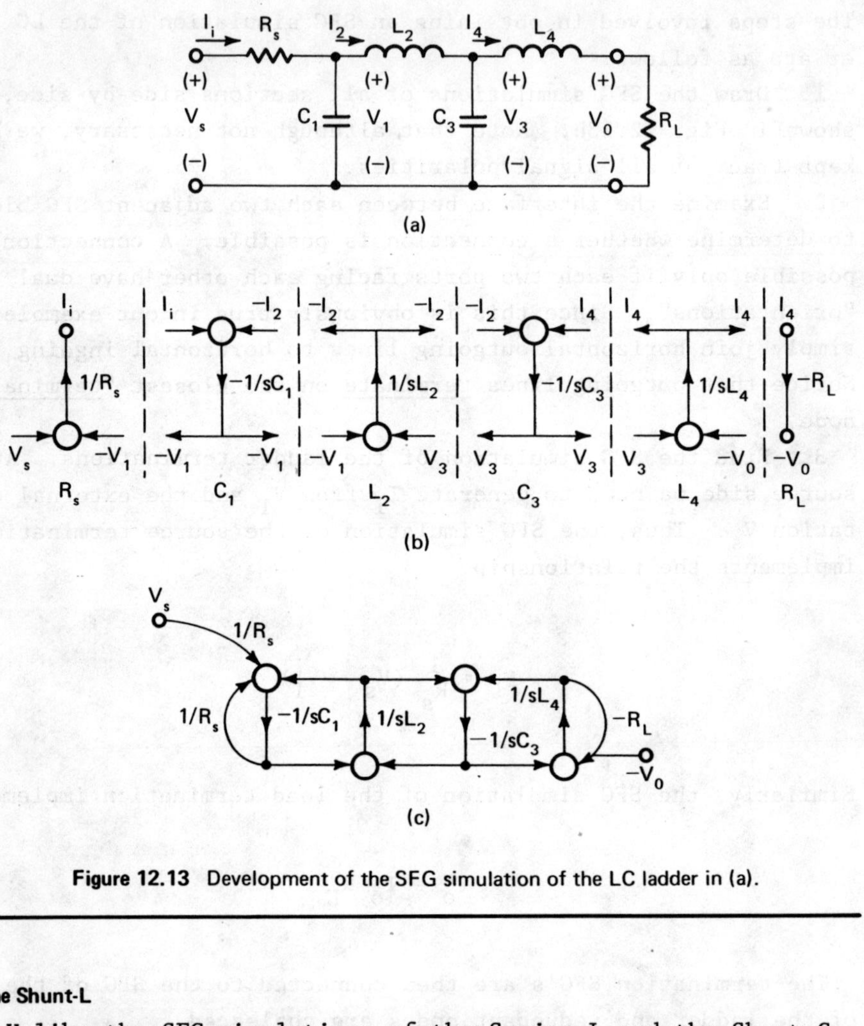

Figure 12.13 Development of the SFG simulation of the LC ladder in (a).

The Shunt-L

Unlike the SFG simulations of the Series-L and the Shunt-C, which are unique, the Shunt-L has two possible SFG simulations. As a two-port network, the operation of the Shunt-L (Fig. 12.14a) may be described by

$$I_1 = \frac{V_1}{sL} + I_2 \quad ,$$

$$V_2 = V_1$$

where the excitation is (V_1, I_2) and the response is (I_1, V_2). The SFG corresponding to this description is shown in Fig. 12.14b. Note that port-1 of this SFG is a voltage-excited current-response port, while port-2 has the dual orientation.

The alternative description of the Shunt-L operation is given by

$$V_1 = V_2 \quad ,$$

$$I_2 = I_1 - \frac{V_2}{sL}$$

Here the excitation is (I_1, V_2) while the response is (V_1, I_2) and the corresponding SFG is shown in Fig. 12.14c. Note that the ports of this SFG have the dual orientation of the corresponding ports of the SFG in Fig. 12.14b.

Which of the two Shunt-L SFG's one should choose will obviously depend on the port orientation of the adjacent SFG's in the cascade. To reflect this freedom of choice, we may draw the Shunt-L SFG omitting the signal flow directions, as in Fig. 12.14d. This SFG can obviously represent both possible SFG's of the Shunt-L.

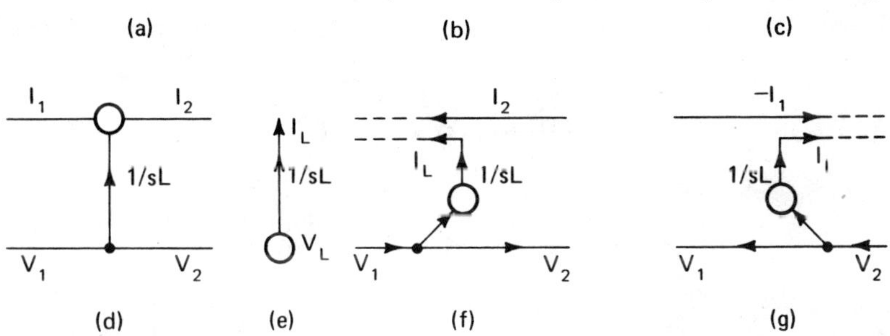

Figure 12.14 Development of the SFG simulating the Shunt − L.

The fact that in the Shunt-L SFG a summation is required "after" integration poses difficulty in circuit implementation; no longer are we able to employ the integrator op. amp. input virtual ground for signal summation. This problem is easily solved, however, by postponing the signal summation. That is, we may carry forward both signal lines, which need to be summed, to the first encountered "terminal node". This leads to the simplified Shunt-L SFG shown in Fig. 12.14e. Here we have created the node V_L, representing the inductor voltage, and left the output signal line I_L, uncommitted. Fig. 12.14f illustrates the way in which the simplified SFG may be connected in the case of (V_1, I_2) excitation while the other case is illustrated in Fig. 12.14g. In both cases, the voltage signal line feeds the V_L node and continues to travel past this node. We, therefore, call this node a "through node" to distinguish it from a "terminal node" on which all signal lines terminate. A "through node" is connected to all signal lines "passing by" it. The vertical output line, representing I_L, is bent in the direction of, and travels in parallel to, the current excitation line, I_1 or I_2.

The Series-C

The Series-C section (Fig. 12.15a) is the dual of the Shunt-L section considered above. Hence, the derivation of the Series-C SFG, shown in Fig. 12.15b, is identical to that presented for the Shunt-L and will not be repeated. Note that we have an input "through node", I_C, and an output vertical line, V_C.

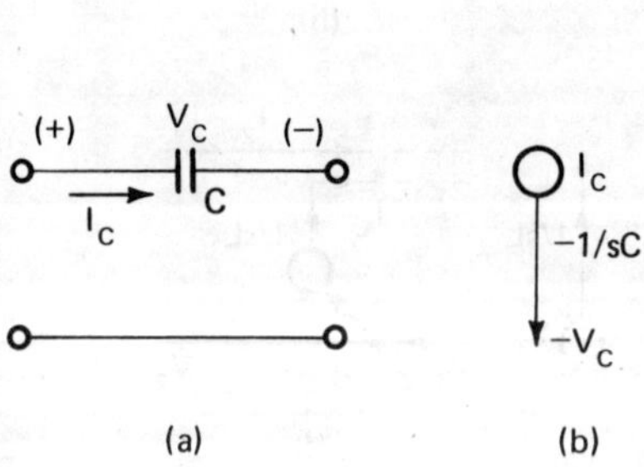

Figure 12.15 The Series − C and its SFG.

Example 12.5

Find the SFG simulation of the doubly terminated LC ladder given in Fig. 12.16a.

Solution

Fig. 12.16 illustrates the development of the complete SFG in a step-by-step manner. Note that the node representing the output variable V_o is by definition a "terminal node". Here again we have kept track of signal polarities. This, however, is an unnecessary process provided one adheres to the sign convention mentioned in the above.

The Shunt-Tank

The SFG of the Shunt-Tank (Fig. 12.17a) section is shown in Fig. 12.17b. It contains an input "through node" representing the tank voltage V_t and a "vertical" output line representing the tank current I_t. The rules for connecting this SFG, in a cascade of SFG's, are identical to those given for the Shunt-L and Series-C SFG's. Note, however, that usually the V_t node can be omitted and all summations performed at the V_L node, resulting in the simplified SFG of Fig. 12.17c. Finally, we observe that the Shunt-Tank SFG contains a loop resonant at $\frac{1}{\sqrt{LC}}$ which is the frequency of the loss-pole produced by this ladder section. The resulting active filter may therefore be aligned through tuning the two-integrator loops producing the loss-poles. This is similar to the tuning schemes employed in passive filter manufacture.

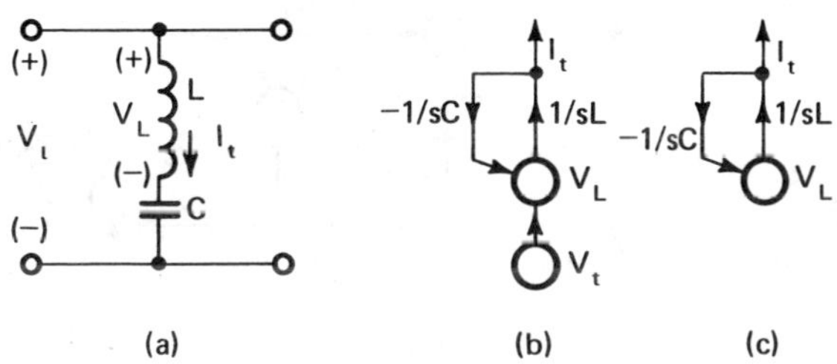

Figure 12.17 Development of the SFG simulation of the shunt-tank.

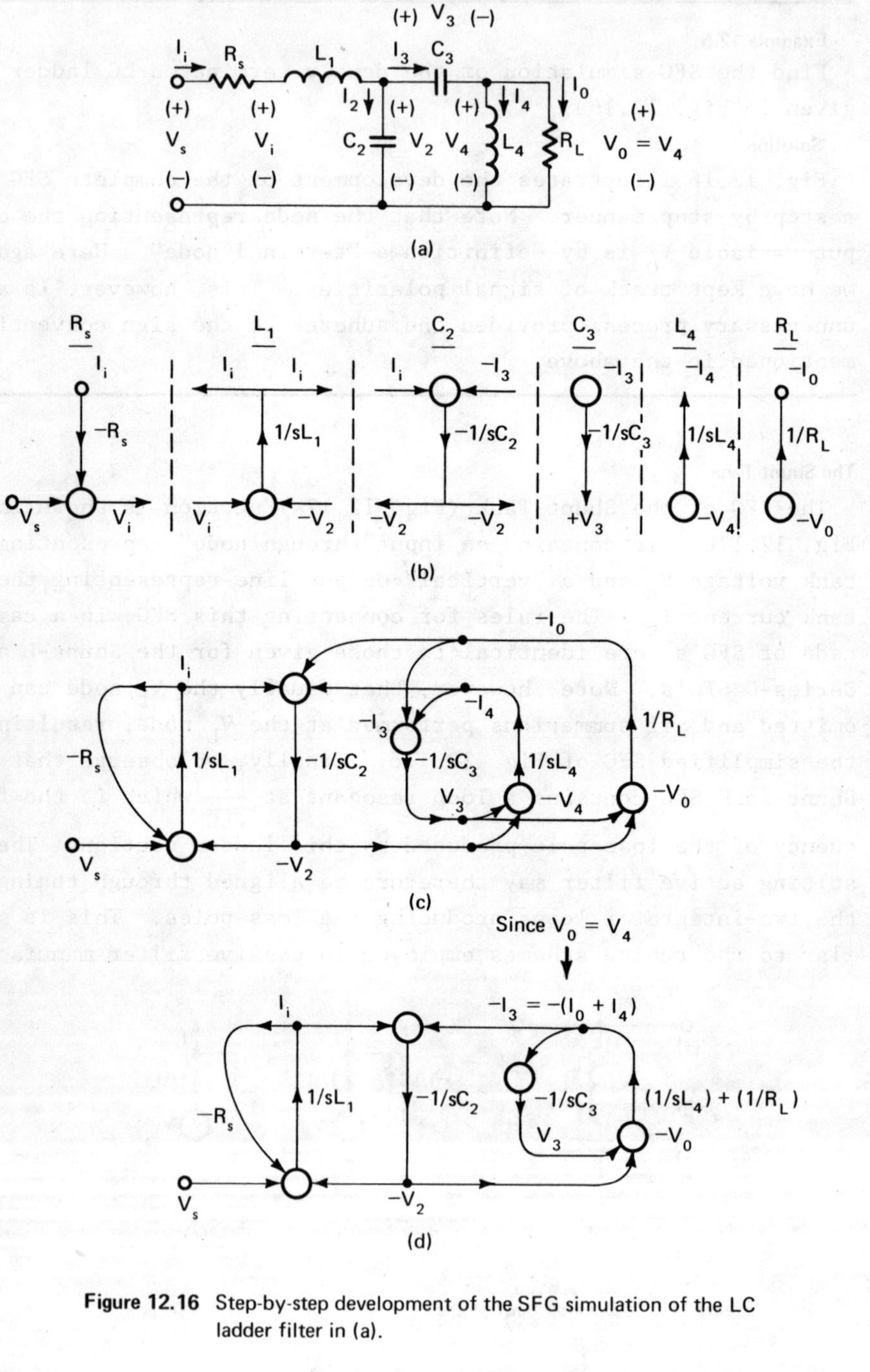

Figure 12.16 Step-by-step development of the SFG simulation of the LC ladder filter in (a).

The Series-Tank

The last of the LC ladder building blocks is the Series-Tank shown together with its SFG simulation in Fig. 12.18. The comm- ents made in the previous subsection apply equally well to the Series-Tank case.

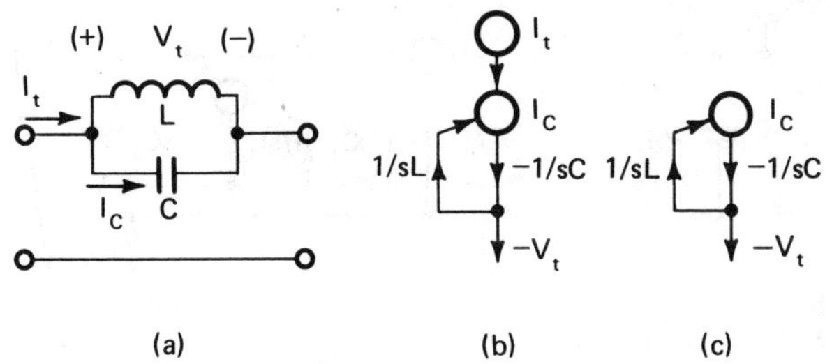

Figure 12.18 Development of the SFG simulation of the series-tank.

"Straight-Through" Sections and Port-Reciprocators

In the examples considered thus far we have encountered no difficulties in connecting the SFG blocks, following a simple set of rules. Difficulties may arise, however, if two adjacent SFG blocks have ports with the same orientation facing one another. To illustrate this situation, consider the simulation of the lad- der section shown in Fig. 12.19a. This section is commonly en- countered in LC realizations of elliptic filter functions. Fig. 12.19b shows the SFG simulation complete except for a missing in- terconnection at the interface between the tank SFG and the SFG simulating C_3. The two ports facing one another have the same current-excited voltage-response orientation and hence cannot be interconnected.

To alleviate this interconnection problem, we need to interpose a two-port SFG which can be excited by voltages and which will provide current responses, at both of its ports. The only SFG we have encountered with this ports' orientation is the Series-L SFG (Fig. 12.11b). Physically speaking, interposing this SFG corresponds to adding a series inductance L_o to our ladder cir- cuit, as shown in Fig. 12.19c. In order not to change the filter transfer function, this added inductance should be negligibly small and ideally zero. Thus the corresponding SFG will have an integrating branch with infinite transmittance as shown in Fig. 12.19d. Finally, Fig. 12.19e shows the SFG of the elliptic filter section after inserting the L_o SFG.

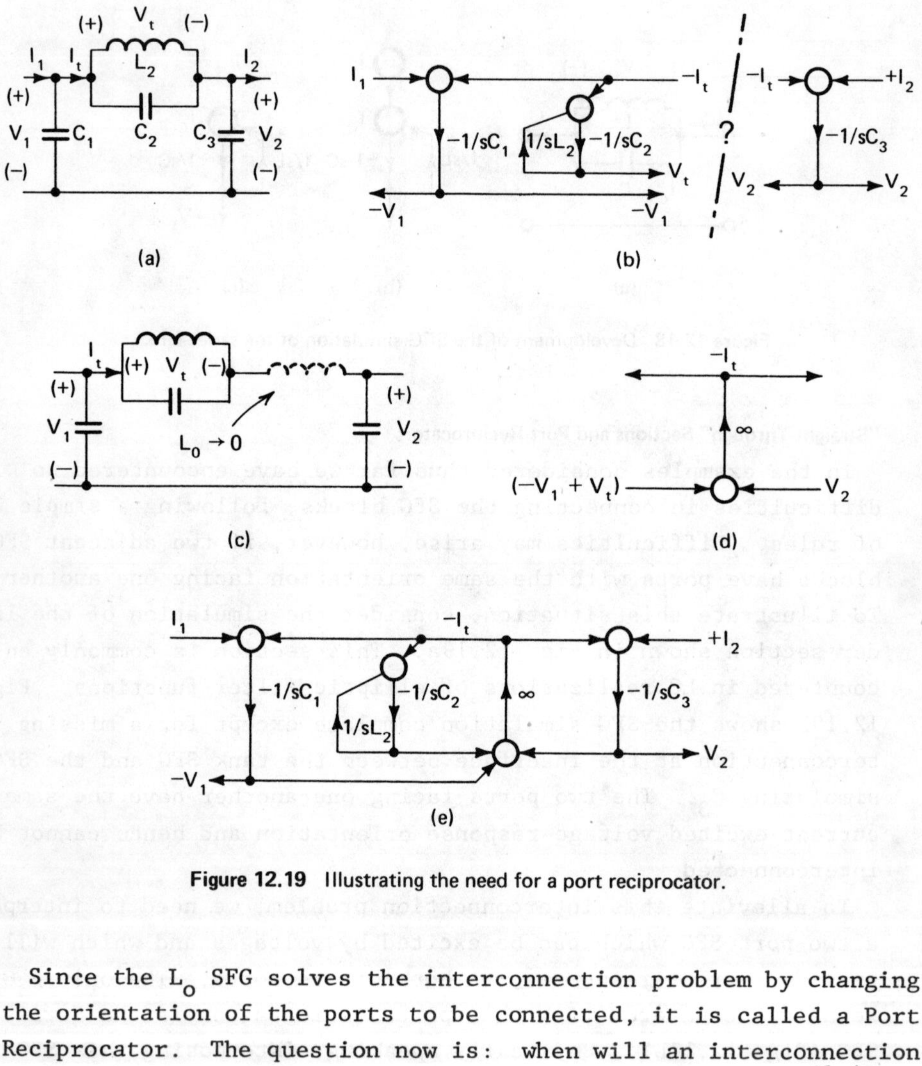

Figure 12.19 Illustrating the need for a port reciprocator.

Since the L_o SFG solves the interconnection problem by changing the orientation of the ports to be connected, it is called a Port Reciprocator. The question now is: when will an interconnection problem arise? The answer, which will not be proven here[1,4], is that interconnection problems occur whenever the LC ladder contains a Π of capacitors or a T of inductors. In the example above we had a Π of capacitors. To solve the interconnection problem we introduced a parasitic inductance $L_o \to 0$, that "breaks" this capacitive-Π. In the dual case of a T of inductors, we have to introduce a parasitic capacitance $C_o \to 0$, in parallel with the shunt arm of the T, thus "breaking" the inductive T. For this latter case we need a C_o-reciprocator which is a Shunt-C SFG with infinite transmittance.

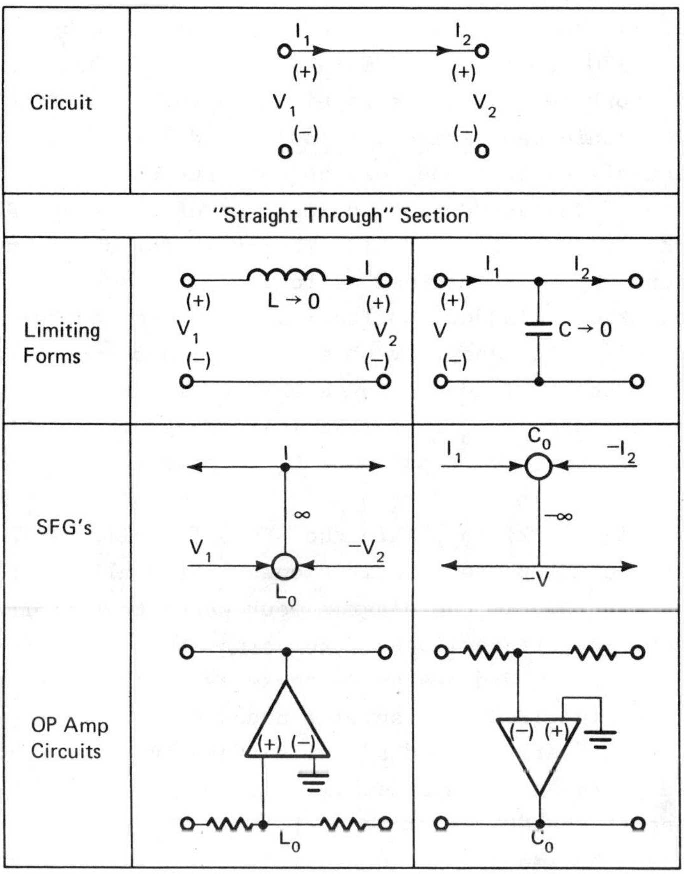

Figure 12.20 The two types of port reciprocators or SFG simulations of "straight-through" sections.

Fig. 12.20 shows both types of SFG port-reciprocators. Note that a port-reciprocator is essentially simulating a "straight-through" section or simply a pair of wires. Included in Fig. 12.20 are op. amp. circuit implementations of the port-reciprocators. Each reciprocator is in essence an integrator with zero time constant. We will have more to say about the stability of the reciprocator circuit in Section 12.4, which deals with circuit implementations.

Summary of the Simulation Rules

Fig. 12.21 shows the six LC ladder building blocks together with their SFG simulations. In addition, Fig. 12.21 includes the L_o and C_o port-reciprocators required in the simulation of ladders that contain capacitive Π's and/or inductive T's. The SFG simulation of any LC ladder can be constructed by the cascade connection of the appropriate collection of the eight SFG building blocks given in Fig. 12.21. The rules for obtaining such simulations may be summarized as follows:

1. Draw the LC ladder and label all its components.
2. Identify all capacitive Π's (if any) and inductive T's (if any) in the LC ladder. For each capacitive Π add a parasitic inductance L_o in the series arm and for each inductive T add a parasitic capacitance C_o in parallel with the shunt arm.
3. Use Fig. 12.21 to obtain the SFG's for all the ladder sections including the parasitic ones. Draw all SFG's side by side. Include in the diagram four nodes that represent the ladder two-port variables (excluding the terminations) V_i, I_i, V_o and I_o, and designate these as "terminal nodes". Note that in Fig. 12.21 the summing nodes of SFG's number 1, 2, 7 and 8 are "terminal nodes" while those in SFG's number 3, 4, 5 and 6 are "through nodes".
4. Connect the SFG sections as follows:
   (a) Each horizontal outgoing line proceeds and terminates on the first encountered "terminal node".
   (b) Each vertical outgoing line is bent in the direction of, and proceeds in parallel to, horizontal lines passing by

Section	No.
Series – L	1
Shunt – C	2
Shunt – L	3
Series – C	4
Shunt Tank	5
Series Tank	6
Straight Through L_0	7
Straigh Through C_0	8

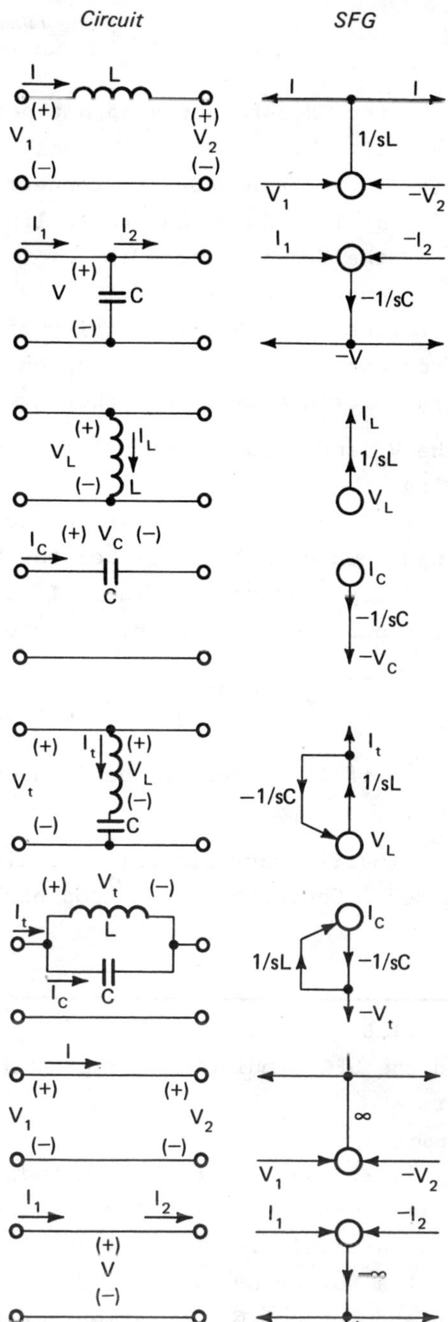

Figure 12.21 Ladder Building Blocks and Their SFG's

it. Subsequently both types of lines become indistinguishable.

(c) A "through node" is connected to each line passing by it. Note that these lines continue to travel past the "through node".

5. Connect nodes V_i and I_i by a line representing the source termination. The directivity of this line will be obvious from the diagram. Depending on its directivity, the line transmittance will be either R_s or $\frac{1}{R_s}$. Similarly, connect the V_o and I_o nodes by a line simulating the load termination.

6. The input signal V_s is fed to the summing node at the input, which will be either the V_i or the I_i node as will be clear from the diagram. If it is fed to the V_i node, it should be weighted by a factor k determined from

$$k = \frac{\text{desired gain of active filter}}{\text{gain of LC ladder}} ,$$

while if it is fed to the I_i node it should be weighted by $\frac{k}{R_s}$.

7. Simplify your diagram by coalescing unnecessary summing nodes. Check that each loop has a negative gain.

Example 12.6

Find the SFG simulation for a fifth-order elliptic low-pass filter.

Solution

Fig. 12.22a shows the minimum inductance LC ladder realization of a fifth-order elliptic low-pass filter together with two additional parasitic inductances required to "break" the two capacitive Π's formed by C_1, C_2 and C_3 and by C_3, C_4 and C_5. Fig. 12.22b shows the SFG simulations of the various ladder sections as obtained from Fig. 12.21. The different sections are subsequently interconnected and SFG's representing the terminations are added as shown in Fig. 12.22c. Finally, Fig. 12.22d presents a "cleaned up" version of the SFG where we have coalesced the nodes I_i, V_i and I_o. Note that the SFG is symmetric around the integrating branch $-\frac{1}{sC_3}$. Furthermore, there are two loops resonant at the two transmission zeros.

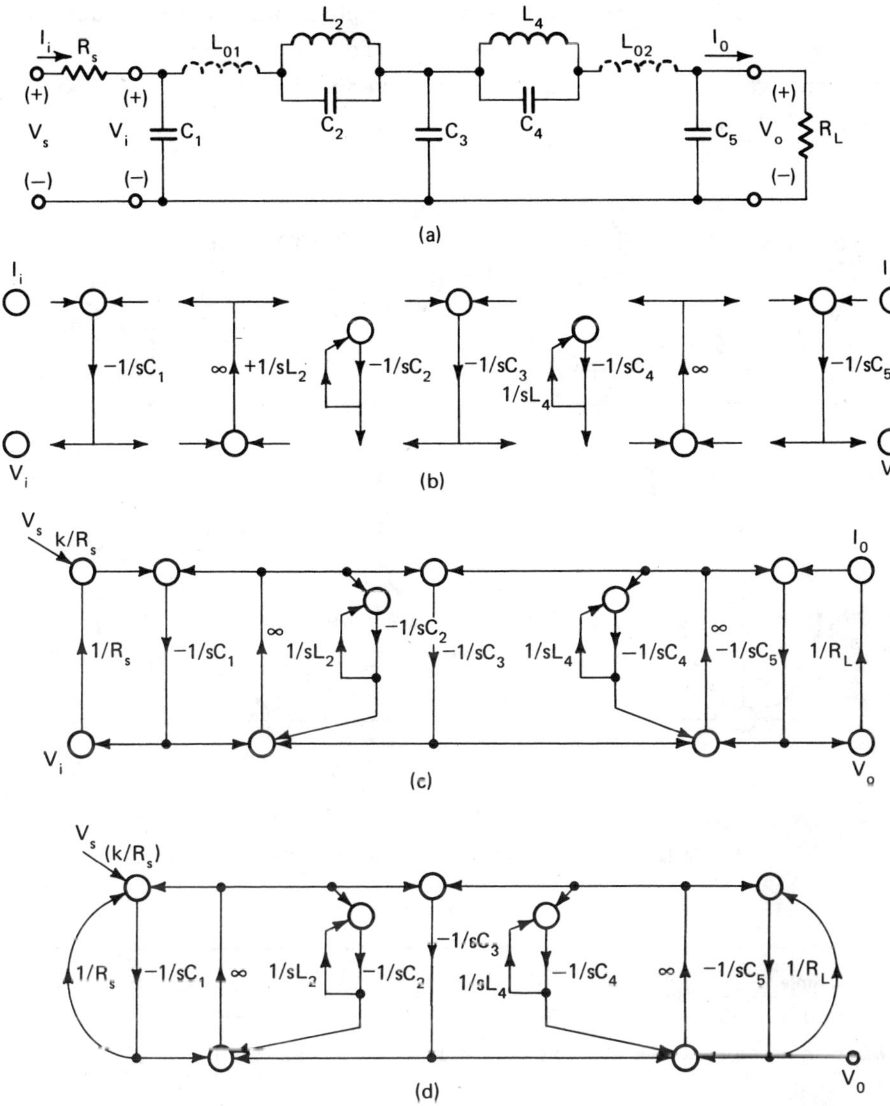

Figure 12.22 Development of the SFG simulation for a fifth-order elliptic low-pass filter.

Example 12.7

Find the SFG simulation of the eighth-order bandpass filter whose LC ladder realization is shown in Fig. 12.23a.

Solution

We identify one capacitive Π in which a parasitic inductance L_o is inserted as shown. The steps for obtaining the SFG simulation are straightforward and the resulting graph is shown in Fig. 12.23b.

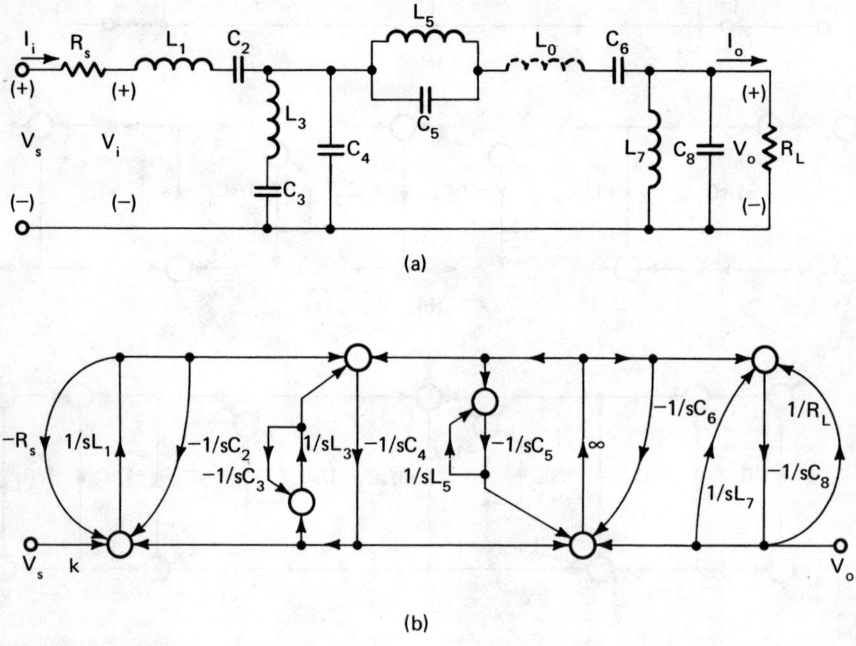

Figure 12.23 SFG simulation of an order eight bandpass filter.

Example 12.8

Find the SFG simulation of the fifth-order all-pole high-pass filter whose LC ladder realization is given in Fig. 12.24a.

Solution

There are two possible SFG simulations for this all-pole high-pass filter. The two simulations are given in Figs. 12.24b, c.

The reason for the existence of two solutions is that the SFG does not contain any horizontal outgoing lines that would determine a unique directivity for the interconnected SFG. A similar situation will thus occur for every ladder that does not contain a Series-L or a Shunt-C; that is, for every filter that does not have loss poles at infinity.

Both SFG's have a relatively large number of inputs at each summing node. In circuit terms, this implies a relatively large number of resistors. We may therefore conclude that the operational simulation approach is not well suited for the design of high-pass filters.

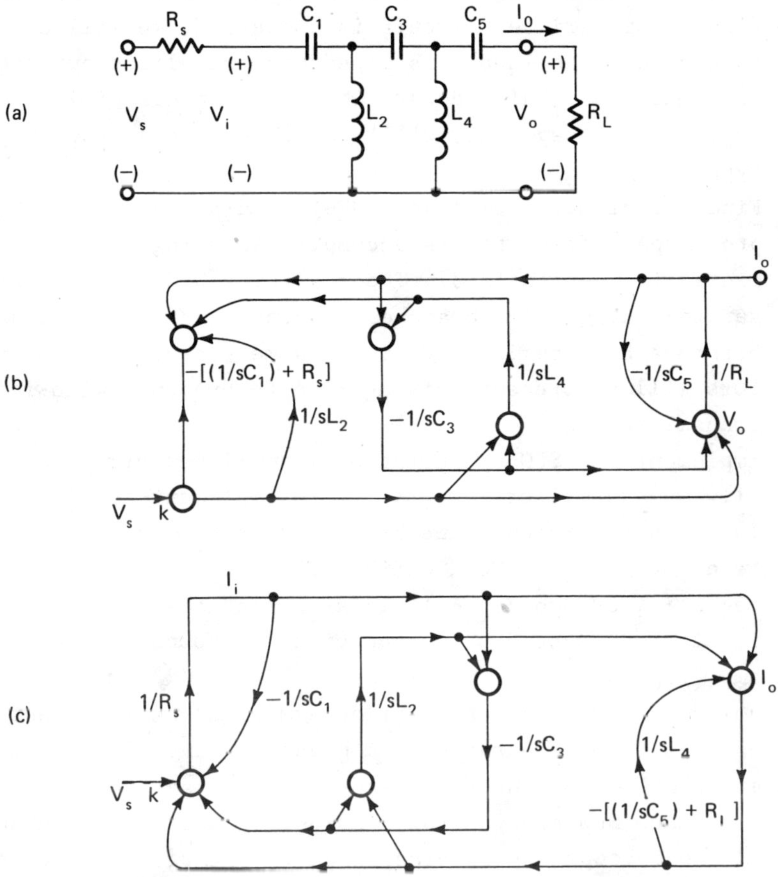

Figure 12.24 A fifth-order high-pass filter and its two possible SFG simulations.

12.4 CIRCUIT IMPLEMENTATION AND PRACTICAL DESIGN CONSIDERATIONS

Design Procedure

The design of active filters using the operational simulation approach requires the following three steps:

(1) Synthesize a doubly-terminated LC ladder that realizes the desired filter transfer function. This step may be accomplished either by using classical approximations and filter tables or, more efficiently, through application of the LC synthesis methods of Chapter 7. Although the impedance level of the resulting LC ladder is immaterial as far as the final active filter circuit is concerned, we will always assume that the ladder is impedance normalized but frequency denormalized. This ladder, whose components will be denoted by lower-case symbols, will be referred to as the LC prototype.

(2) Find the signal-flow-graph (SFG) simulation of the LC ladder prototype. This step is accomplished using the method presented in Section 12.3. The resulting SFG consists of inverting integrating branches, noninverting integrating branches and summing nodes. In some cases, port-reciprocators will be present, giving rise to branches with infinite transmittances.

(3) Implement the SFG by operational amplifier circuits. Normally we use Miller integrators to implement the inverting integrating branches and Miller-Inverter-Cascade or Phase-Lead integrators (See Section 8.3) for the noninverting integrating branches. All signal summations may be conveniently implemented using the virtual grounds present at op. amp. inputs.

We now consider the problem of determining the component values of the circuit implementation. A glance at any of the ladder simulating SFG's encountered in the previous section reveals the fact that each SFG consists of a number of negative feedback loops. These loops can be classified into three categories:

1. Two-integrator loops each having a loop gain of the

form $\frac{-1}{s^2 \ell_i c_j}$.

2. Termination loops having loop gains of the form $\frac{-r}{s\ell}$ or $-\frac{1}{scr}$.
3. Loops containing either an ℓ_o or a c_o reciprocator, where ℓ_o, $c_o \to 0$. The loop-gain of such loops take the form

$$\frac{-1}{s^2 \ell_o c_j} \text{ or } \frac{-1}{s^2 c_o \ell_j} .$$

Once we sketch the active circuit implementation of the SFG it is easy to identify the various loops and their corresponding ones in the SFG. The design equations are simply obtained by equating the gains of each two corresponding loops. For loops containing reciprocators we apply a simple artifice. Let the op. amp. implementing the reciprocator have an integrator capacitor C_r where $C_r \to 0$. This allows us to write relationships equating gains of corresponding reciprocator loops. The resulting subset of equations can be then reduced to eliminate ℓ_o, c_o and C_r.

The number of design equations, obtained by the process of equating loop gains, is usually much smaller than the number of component values to be determined. The available degrees of freedom can then be used to effect desirable practical properties of the active filter. Highest on the list is scaling for maximum dynamic range. To optimize the filter dynamic range we force all op. amps. to have equal maximum output voltages. A filter that is not scaled this way will have one of its op. amps. in saturation while all others have relatively small signals at their outputs. This particular op. amp. will then determine the maximum possible input signal level. Scaling for wide dynamic range eliminates such a problem and increases the signal handling capability of the filter. In a later subsection, we will present a method for scaling that will achieve an optimum dynamic range. For the present, however, we will be satisfied with a simpler but suboptimum approach. This consists of making the two time constants in each two-integrator loop equal. For loops containing reciprocators, we make the time constant as large as practically possible. This minimizes the effect of the parasitic element introduced by the reciprocator.

The remaining degree of freedom in component value choice can be used to make all capacitors equal, at a suitable value; a practically desirable property. The above design procedure is best illustrated by an example.

Example 12.9*

We wish to find a circuit implementation of the bandpass filter considered in Example 12.7. The ideal attenuation and delay functions of this eighth-order bandpass filter are shown in Fig. 12.25. The LC ladder prototype together with the SFG simulation are reproduced for convenience in Fig. 12.26. Pertinent specifications of the filter are:

(1) the passband, between 1 KHz and 1.4 KHz, is equiripple with a maximum ripple of 0.4 dB ($|\rho|_{max} \simeq 0.3$);

(2) the loss poles are: two at $s = 0$, two at $s = \infty$, one at $s = \pm j\ 0.3996$ (559.4 Hz), and one at $s = \pm\ j\ 1.2501$ (1750.1 Hz);

(3) the natural modes have the following frequencies (normalized with respect to 1.4KHz) and Q-factors:

ω_o	Q
1.0048	19.113
0.9188	6.743
0.7911	6.192
0.7106	15.256

Note that the transfer function does not have geometric symmetry and hence the LC design cannot be obtained using tables. The LC synthesis program provides the following ladder element values (inductances in Henries and capacitances in Farads):

$r_s = 1\ \Omega \qquad c_2 = 2.949 \times 10^{-5} \qquad r_L = 1.888\ \Omega$

$\ell_1 = 6.260 \times 10^{-4} \qquad c_3 = 1.2687 \times 10^{-3}$

$\ell_3 = 6.379 \times 10^{-5} \qquad c_4 = 3.618 \times 10^{-4}$

---

*We have considered this same filter in a number of examples in the previous chapters.

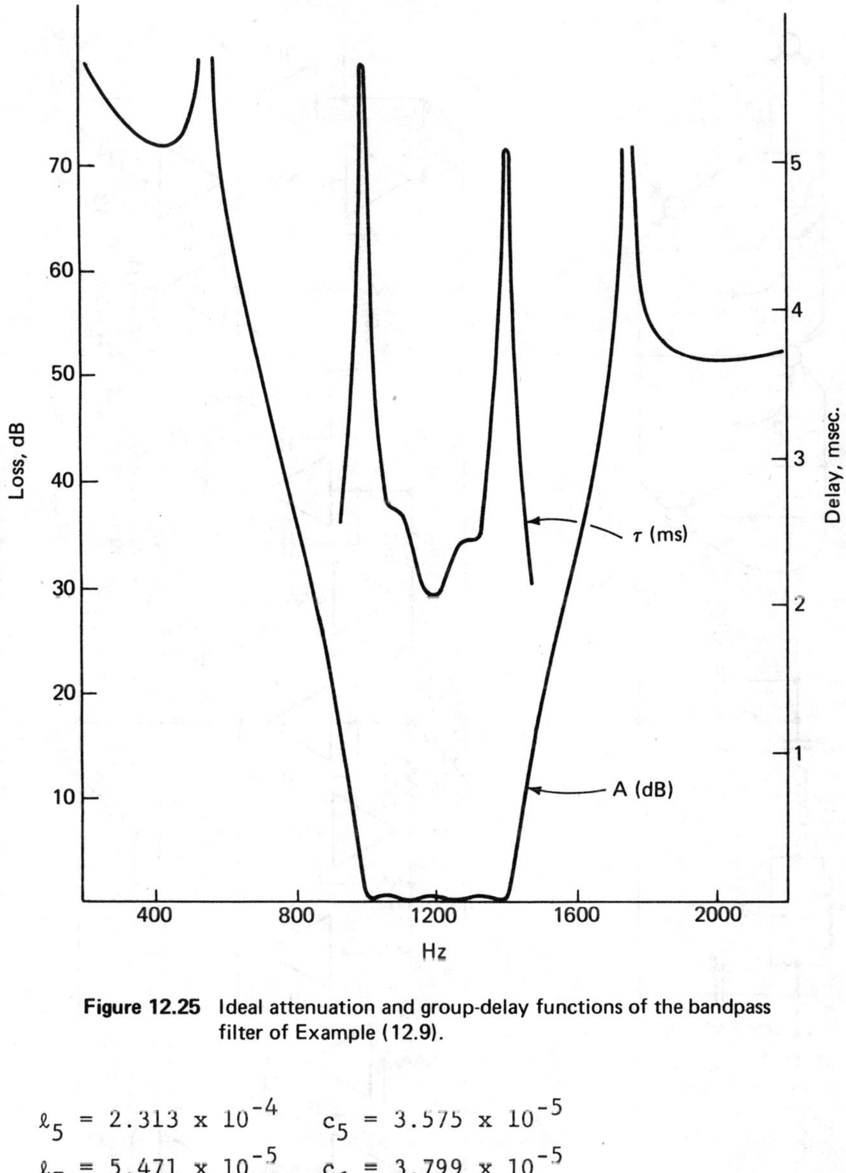

Figure 12.25 Ideal attenuation and group-delay functions of the bandpass filter of Example (12.9).

$$\ell_5 = 2.313 \times 10^{-4} \qquad c_5 = 3.575 \times 10^{-5}$$
$$\ell_7 = 5.471 \times 10^{-5} \qquad c_6 = 3.799 \times 10^{-5}$$
$$c_8 = 3.141 \times 10^{-4}$$

Fig. 12.26c shows the op. amp. circuit implementation of the SFG of Fig. 12.26b. The reader is encouraged to identify the corres-

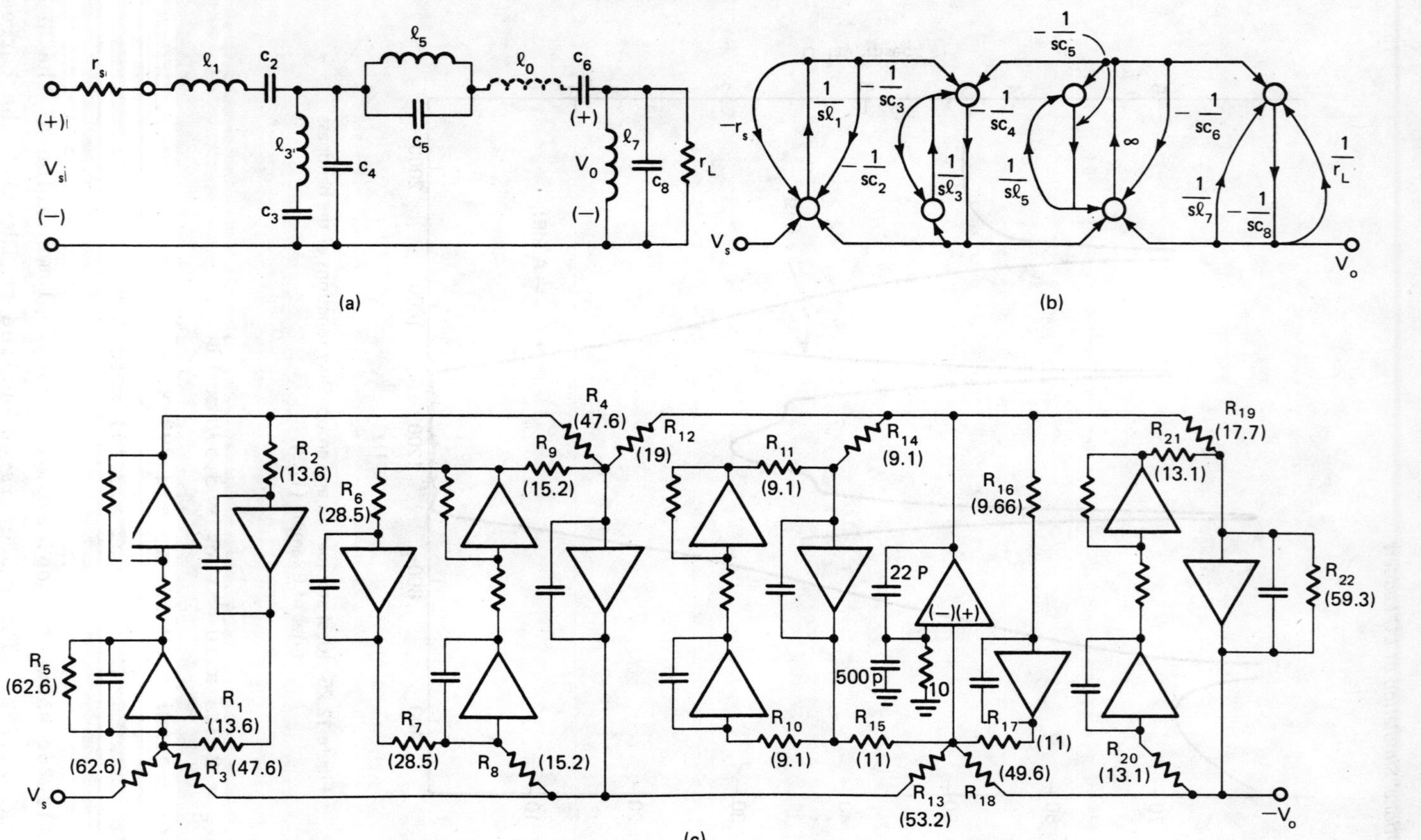

Figure 12.26 Illustrating the design of the eighth order bandpass filter of Example (12.9).

pondence between the circuit and its originating SFG. Note that the Miller-inverter cascade is used for noninverting integrators.

As far as component values are concerned, we have chosen all the integrator capacitors equal to C with C = 10 nF. We identify six coupled two-integrator loops, two termination loops and four reciprocator loops. Equating the loop-gains of every corresponding two-integrator loops and choosing the two time constants of each loop equal results in the following design equations:

$$R_1 = R_2 = \frac{\sqrt{\ell_1 c_2}}{C} = 13.59\ \text{K}\Omega$$

$$R_3 = R_4 = \frac{\sqrt{\ell_1 c_4}}{C} = 47.59\ \text{K}\Omega$$

$$R_6 = R_7 = \frac{\sqrt{\ell_3 c_3}}{C} = 28.45\ \text{K}\Omega$$

$$R_8 = R_9 = \frac{\sqrt{\ell_3 c_4}}{C} = 15.19\ \text{K}\Omega$$

$$R_{10} = R_{11} = \frac{\sqrt{\ell_5 c_5}}{C} = 9.09\ \text{K}\Omega$$

$$R_{20} = R_{21} = \frac{\sqrt{\ell_7 c_8}}{C} = 13.11\ \text{K}\Omega$$

Doing the same for the termination loops obtains

$$R_5 = \frac{\ell_1}{Cr_s} = 62.60\ \text{K}\Omega$$

$$R_{22} = \frac{c_8 r_L}{C} = 59.30\ \text{K}\Omega$$

Now for the loops containing the ℓ_o reciprocator, we assume a fictitious integrator capacitance C_r for the reciprocator and equate loop gains to obtain

$$R_{12}C\,R_{13}\,C_r = c_4 \ell_o$$

$$R_{14}C\,R_{15}\,C_r = c_5 \ell_o$$

$$R_{16}C\,R_{17}\,C_r = c_6 \ell_o$$

$$R_{19} C\, R_{18}\, C_r = c_8 \ell_o$$

These four equations reduce to the set

$$\frac{R_{12}\, R_{13}}{R_{14}\, R_{15}} = \frac{c_4}{c_5}$$

$$\frac{R_{12}\, R_{13}}{R_{16}\, R_{17}} = \frac{c_4}{c_6}$$

$$\frac{R_{12}\, R_{13}}{R_{18}\, R_{19}} = \frac{c_4}{c_8}$$

The component values given in Fig. 12.26c satisfy these equations. Shown also on the circuit schematic, is a stabilizing network for the reciprocator. The need for this network (as well as its design) is explained in the following.

As evident from the circuit diagram, the reciprocator amplifier has a number of negative-feedback paths from its output terminal to its noninverting input terminal. The transfer function of each of these feedback paths is of the form $(- \frac{1}{s\tau})$, where τ is a time constant. The outputs of the integrating feedback networks are summed together by the resistive network composed of R_{15}, R_{13}, R_{18} and R_{17} and fed to the noninverting input. One can therefore calculate the equivalent feedback function which will be of the form $- \frac{1}{s\tau}$. Fig. 12.27b shows a Bode plot of the op. amp. gain (solid line) assuming a single-pole compensation. Also shown is a plot of the inverse of the feedback factor, that is $s\tau$. The two lines intersect at a point where the difference in slope is 12 dB/octave. Now, if these were the only poles of the system it would be stable. However, the reciprocator amplifier has parasitic poles which result in excess phase shift at the intersection frequency. Also, the feedback function will have parasitic poles that will contribute to the phase shift. It is therefore quite possible that the total phase shift at the frequency of intersection will exceed 180°, resulting in oscillations.

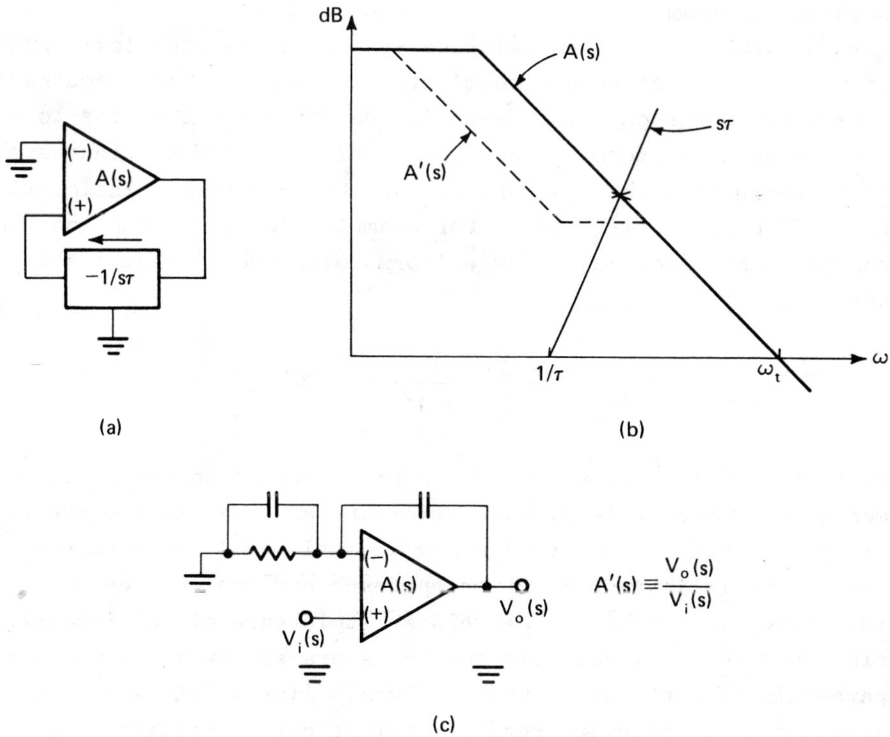

Figure 12.27 Reciprocator stabilization: (a) The reciprocator op. amp. with the feedback paths replaced by an equivalent function $1/s\tau$. (b) Bode plot analysis with the broken line representing the modified op. amp. gain obtained by using the two-capacitor one-resistor network in (c), i.e. $V_0/V_i(s) = A'(s)$.

To guarantee stability of the reciprocator amplifier, we have to modify its open loop response such that at the intersection with the feedback function the slope difference is only 6 dB/octave. This may be accomplished by introducing a zero in the open-loop response as shown by the broken line in Fig. 12.27b. The reader should verify that the two-capacitor one-resistor network connected in the negative feedback path of the reciprocator op. amp. indeed results in the modified response shown.

Sensitivity Considerations

In Chapter 7, we have analytically established the low-sensitivity properties of doubly terminated lossless networks and have presented bounds on the attenuation deviations due to the lossless element and termination variations. These deviations are small throughout the passband and are bounded by a relation with the reflection coefficient. For example, an upper bound on the change in attenuation ΔA for uniform tolerance of elements in this type of filter is given by

$$|\Delta A|_{max} < \frac{8.68|\varepsilon\rho|\omega\tau(\omega)}{1 - |\rho|^2} \text{ dB} \qquad (12.9)$$

where $\varepsilon = \frac{\Delta L}{L} = \frac{\Delta C}{C}$ is the relative tolerance of the reactive elements, ρ is the reflection coefficient, and $\tau(\omega)$ is the group-delay. This formula should apply to active filters designed using the operational simulation approach and implemented using integrators. Here ε is the relative tolerance of the integrator time constants. These time constants are set by RC products and parasitic effects due to the op. amps.' finite gain and bandwidth, all of which are readily estimated (See Section 8.3). Thus, using (12.9) with an SFG active ladder and known components, one can easily estimate $|\Delta A|_{max}$ before proceeding with the actual construction of the active network.

In Chapter 7, we have also obtained results for the attenuation deviation ΔA due to semiuniform parasitic dissipation in the reactive elements of a doubly-terminated lossless ladder. In particular, the attenuation change ΔA due to incremental resistances (Ld_L) in series with inductors and incremental conductances (Cd_C) in parallel with capacitances is given by

$$\Delta A \simeq 4.34 \left[(d_L+d_C)\,\tau(\omega) + (d_L-d_C)\,\text{Im}\left(\frac{\rho_1 + \rho_2}{2\omega}\right)\right] \text{ dB} \qquad (12.10)$$

where Im denotes the imaginary part. Alternatively, the relation in (12.10) can be rewritten in terms of the integrators' Q factors as

$$\Delta A \simeq 4.34 \; [(\frac{1}{Q_L} + \frac{1}{Q_C}) \; \omega\tau(\omega) + \frac{1}{2} \; (\frac{1}{Q_L} - \frac{1}{Q_C}) \; \text{Im} \; (\rho_1 + \rho_2)] \; \text{dB} \tag{12.11}$$

where Q_L denotes the Q-factor of the integrators simulating inuctance elements while Q_C denotes the Q-factor of the integrators simulating the capacitive elements. The values of Q_L and Q_C depend on the circuit configuration chosen for the integrators and are generally proportional to the amplifier open-loop gain at the frequency of interest (See Section 8.3).

It should be noted that in most classes of filter networks the first term on the right hand-side of (12.11) dominates in the passband and the transition band where $\tau(\omega)$ is large, and we can use the approximation,

$$\Delta A \simeq 4.34 \; (\frac{1}{Q_L} + \frac{1}{Q_C}) \; \omega\tau(\omega) \tag{12.12}$$

On the other hand, the second term dominates in the stopband where the reflection coefficients are large. Table 12.1 provides the results of evaluating the deviation ΔA for three different combinations of integrator circuits (See Section 8.3). Case (1) is the usual implementation, such as the one we used in example 12.9. We note the large deviation in the passband attenuation. Case (2), which is obtained by a simple modification of the Miller-inverter integrator, results in reducing the passband distortion term by a factor of two while the stopband deviation term is ideally reduced to almost zero.

A dramatic improvement in performance is obtained in Case (3), which uses the phase-lead integrator [We used this design in a number of examples.]. Thus, using such an integrator in a Leap-Frog or a more general simulation of an LC ladder tremendously extends the useful operating frequency range.

	Capacitance	Inductance	Attenuation deviation, ΔA in dB
1	Miller	Miller-Inverter	$\Delta A = -\frac{4.34}{\lvert A\rvert}[4\ \omega\tau(\omega) + \text{Im}(\rho_1 + \rho_2)]$ In passband, $\Delta A \simeq -\frac{17.36}{\lvert A\rvert}[\omega\tau(\omega)]$
2	Miller	Modified Miller-Inverter	$\Delta A = -\frac{8.68}{\lvert A\rvert}[\omega\tau(\omega)]$
3	Miller	Phase-Lead	$\Delta A = \frac{4.34}{\lvert A\rvert}[\text{Im}(\rho_1 + \rho_2)]$ In passband, $\Delta A \simeq 0$
4	High-Q	Modified High-Q	$\Delta A = -\frac{8.68}{\lvert A\rvert^3}[\omega\tau(\omega)]$

Table 12.1 Deviation in the attenuation function of filters realized as operational simulations of LC ladder filters. Different combinations of integrators (from Fig. 8.11) are used to simulate the capacitance and inductance of the LC ladder. Note that |A| denotes the magnitude of the op. amp. gain.

Example 12.10

To verify the sensitivity results presented above, we discuss their application to the bandpass filter designed in Example 12.9. The circuit was constructed using polystyrene capacitors (TC = -120 ppm) with all time constants set within 0.2 per cent of their nominal values. LM 324 quad op. amps., which have a temperature compensated $f_t \simeq 1$ MHz, were used resulting in four IC packages for the complete filter. Using the data above we find from (12.9) that we should expect a worst case $|\Delta A|_{max} < 0.11$ dB. Also, evaluating ΔA in the first row of Table 12.1 at the upper edge of the passband $\omega = 2\pi \times 1400$ rad. obtains $\Delta A \simeq 0.88$ dB. Thus the passband distortion due to the limited op. amp. bandwidth dominates that due to passive component variations. This could be remedied by using the phase-lead integrators, as outlined above.

In the experimental circuit, all of the integrators were compensated for gain bandwidth limitations by means of small capacitors shunting the integrators' input resistors (See Section 8.3). The resulting measured attenuation is shown in Fig. 12.28. Note that the deviation from ideality in the passband is slightly less than 0.1 dB and that it occurs between the reflection zeros, as predicted by the theory in Chapter 7. Due to the compensation of the amplifiers, all of this passband distortion may be attributed to the passive RC element tolerances and it verifies the theoretical predictions above.

Next, all of the integrator compensation capacitors were removed, and a new passband response measured. This is shown in Fig. 12.29. The worst case deviation occurs at the upper edge of the passband with a value of approximately 0.5 dB, slightly better than predicted.

Other design examples and detailed experimental results can be found in [12] - [14].

Implementation Using Single Amplifier Biquads

The design in the example above provides excellent performance, but many network designers may object to the number of amplifiers required. We would like to point out here that, in many cases where the design tolerances are not too stringent, one can "replace" the three-amplifier biquad subnetworks of the SFG active ladder by simpler and more economical single amplifier biquads (SAB's).

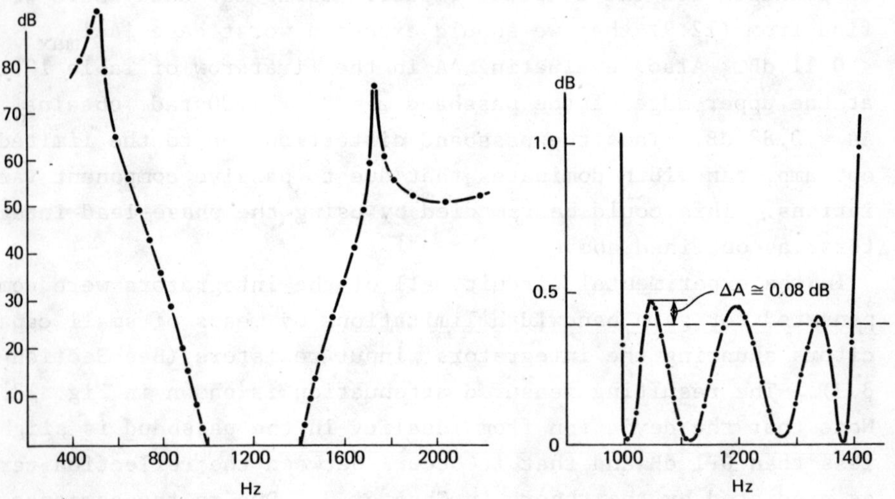

Figure 12.28 Compensated filter response – (Example 12.10). $V_{CC} = \pm 15$ Vdc, Temperature $= +20°$C.

Figs. 12.30a, b, show two SAB's that can be used. Both SAB's have been discussed in Chapter 9. The inverting SAB in Fig. 12.30a is the Friend biquad in its bandpass form (Section 9.7). The noninverting SAB is the positive-feedback biquad of the PF1 type (Section 9.5). Both circuits are capable of producing infinite Q-factors. Note that signal summing is performed by "splitting" the input feedthrough resistor. This technique has been discussed in detail in Chapters 8, 9. We have also used both SAB's in the Leap-Frog circuit of Example 9.3.

The infinite Q-factor in both SAB's is obtained by setting the positive feedback resistive ratio at the appropriate value. This is in contrast to the case of the two-integrator biquad where the infinite Q is independent of passive circuit elements.

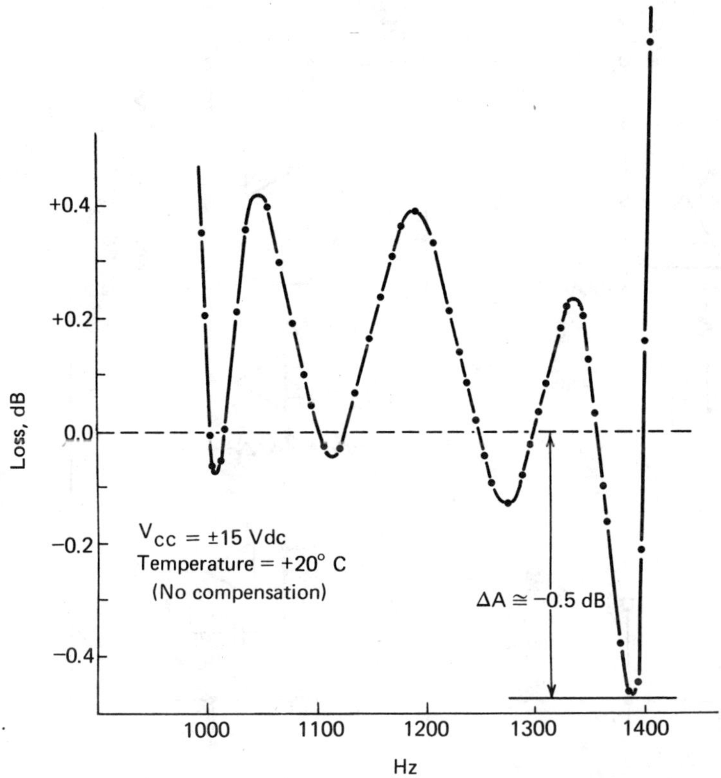

Figure 12.29 Passband response with the amplifiers' compensation removed [Example (12.10)].

Example 12.11

Fig. 12.30c shows the structure of the circuit implementation of the bandpass filter of Example 12.9 where the SAB's of Fig. 12.30a, b, are used to replace the three-amplifier biquads in the original design. A careful comparison of the two circuits indicates how the replacements are accomplished. Note that the cir-

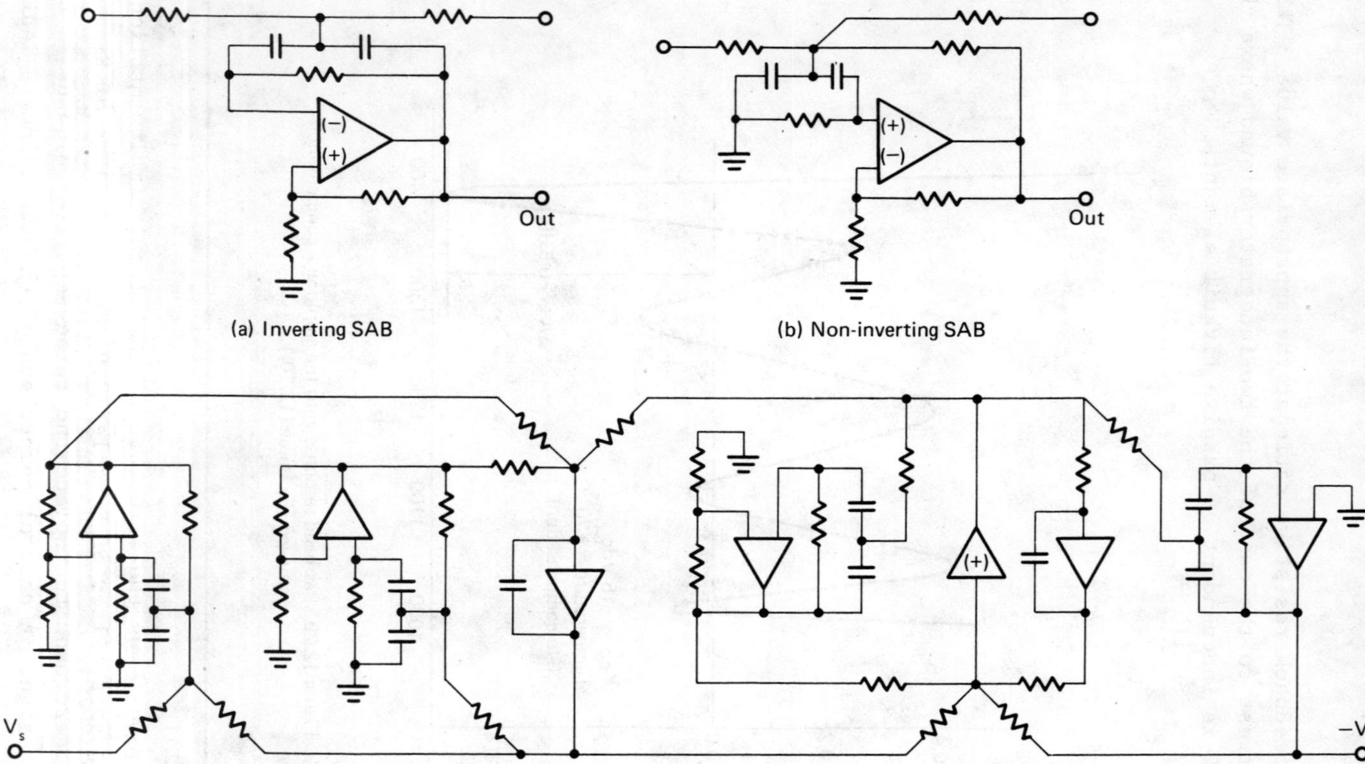

(a) Inverting SAB

(b) Non-inverting SAB

(c) Bandpass filter

Figure 12.30 Illustrating the use of SAB's in the design of SFG filters.

cuit now contains four SAB's, two single integrators, and the reciprocator, for a total of only seven amplifiers versus the fifteen amplifiers for the original design. Of course, the price we pay for this reduction is increased passive element sensitivities, since now passive element variations in the SAB's can produce Q deviations and hence attenuation deviations ΔA as indicated by (12.11). In the original circuit, passive element changes do not produce these Q deviations.

For more details regarding the use of SAB's in the design of SFG filters the reader is referred to [14].

Scaling for Maximum Dynamic Range*

We now present a general method for scaling the component values of the active realization to obtain the widest possible dynamic range for the filter. Recall that in an operational simulation design the output voltages of integrators simulating inductors correspond to the currents through the inductances of the LC prototype. Also, the output voltages of integrators that simulate capacitors correspond to the voltages across the capacitances in the LC prototype. The output voltage of a reciprocator op. amp. corresponds to either the current through the capacitor in the series arm of a capacitive-Π or the voltage across the shunt arm of an inductive T, in the LC prototype.

Let us start with an active realization where the integrators have time constants equal to the elements they are simulating. That is, an inductance ℓ_i in the impedance normalized LC prototype is simulated by an integrator having the transfer function $\frac{1}{s\ell_i}$ while a capacitance c_j is simulated by an integrator having the transfer function $\frac{-1}{sc_j}$. Thus, if the maximum of the absolute value of the voltages and currents in the LC ladder are known, then we also know the maximum of the absolute value of all

*This section is based on [15].

the outputs of the op. amps. in the active filter.

Consider the two-integrator loop corresponding to ℓ_i and c_j of the LC prototype. Let the maximum of the absolute value of the current through ℓ_i be I_{ℓ_i} and the maximum of the absolute value of the voltage across c_j be V_{c_j}. To obtain maximum dynamic range for the active circuit, we have to make the output voltages of the ith integrator (simulating ℓ_i) and the jth integrator (simulating c_j) have equal maxima. This can be achieved by scaling the integrator gains, $\frac{1}{C_iR_i}$ and $\frac{1}{C_jR_j}$, in the ratio of I_{ℓ_i} and V_{c_j}. Note, however, that the loop-gain should be kept constant. Thus, for this loop we have the design equations

$$C_iR_i = \ell_i \left(\frac{I_{\ell_i}}{V_{c_j}}\right)$$

$$C_jR_j = c_j \left(\frac{V_{c_j}}{I_{\ell_i}}\right)$$

This process is repeated for every two-integrator loop. Loops containing reciprocators are treated as follows. Consider an ℓ_o-reciprocator. Let this reciprocator form loops with the simulations of a number of capacitors, say c_m, c_n. In the active realization let the reciprocator have input resistors R_{mr}, R_{nr}, in correspondence with the loops m and n. The design equations then become

$$C_m R_m = c_m \left(\frac{V_{c_m}}{I_{\ell_o}}\right) \qquad R_{mr} = R \left(\frac{I_{\ell_o}}{V_{c_m}}\right)$$

$$C_n R_n = c_n \left(\frac{V_{c_n}}{I_{\ell_o}}\right) \qquad R_{nr} = R \left(\frac{I_{\ell_o}}{V_{c_n}}\right)$$

where R is an arbitrary resistance whose value may be determined from practical considerations. The proof of these equations is left to the reader.

Finally, for the termination loops the design equations are identical to those given for the nonoptimized case. That is, simply equate corresponding time constants.

The voltage and current maximum values may be obtained using a simple ladder analysis program. We should also note that the optimization method given for the all-pole case in Section 12.2 is in a sense a special case of the general method outlined above.

Example 12.12

Redesign the bandpass filter of Example 12.9 scaling the component values for maximum dynamic range.

Solution

The circuit structure is identical to that in Fig. 12.26c and hence will not be redrawn. Here also, we choose all integrator capacitors equal to C = 10 nF. To find the resistance values, we first analyse the LC ladder prototype (Fig. 12.26a) to obtain the maximum absolute values of the ladder voltages and currents. These are found to be

$I_{\ell_1} = 0.7733$ A	$V_{c_2} = 4.099$ v
$I_{\ell_3} = 3.3805$ A	$V_{c_3} = 0.4264$ v
$I_{\ell_5} = 1.580$ A	$V_{c_4} = 1.094$ v
$I_{\ell_7} = 1.974$ A	$V_{c_5} = 3.212$ v
$I_{\ell_o} = 0.7126$ A	$V_{c_6} = 2.970$ v
	$V_{c_8} = 0.6870$ v

Following the design procedure outlined above we obtain:

(A) Reciprocator-Free Loops:

$$R_1 = \frac{\ell_1}{C}\left(\frac{I_{\ell_1}}{V_{c_2}}\right) = 11.81\ K\Omega \qquad R_2 = \frac{c_2}{C}\left(\frac{V_{c_2}}{I_{\ell_1}}\right) = 15.63\ K\Omega$$

$$R_3 = \frac{\ell_1}{C}\left(\frac{I_{\ell_1}}{V_{c_4}}\right) = 44.25\ K\Omega \qquad R_4 = \frac{c_4}{C}\left(\frac{V_{c_4}}{I_{\ell_1}}\right) = 51.18\ K\Omega$$

$$R_7 = \frac{\ell_3}{C}\left(\frac{I_{\ell_3}}{V_{c_3}}\right) = 50.57\ K\Omega \qquad R_6 = \frac{c_3}{C}\left(\frac{V_{c_3}}{I_{\ell_3}}\right) = 16\ K\Omega$$

$$R_9 = \frac{\ell_3}{C}\left(\frac{I_{\ell_3}}{V_{c_4}}\right) = 19.71\ K\Omega \qquad R_8 = \frac{c_4}{C}\left(\frac{V_{c_4}}{I_{\ell_3}}\right) = 11.71\ K\Omega$$

$$R_{10} = \frac{\ell_5}{C}\left(\frac{I_{\ell_5}}{V_{c_5}}\right) = 11.38\ K\Omega \qquad R_{11} = \frac{c_5}{C}\left(\frac{V_{c_5}}{I_{\ell_5}}\right) = 7.27\ K\Omega$$

$$R_{20} = \frac{\ell_7}{C}\left(\frac{I_{\ell_7}}{V_{c_8}}\right) = 15.72 K\Omega \qquad R_{21} = \frac{c_8}{C}\left(\frac{V_{c_8}}{I_{\ell_7}}\right) = 10.93\ K\Omega$$

(B) <u>Termination Loops</u>:

$$R_5 = \frac{\ell_1}{cr_s} = 62.60\ K\Omega \qquad R_{22} = \frac{c_8 r_L}{C} = 59.30\ K\Omega$$

(C) <u>Reciprocator Loops</u>:
Choose R = 60 KΩ

$$R_{12} = \left(\frac{c_4}{C}\right)\left(\frac{V_{c_4}}{I_{\ell_o}}\right) = 55.54\ K\Omega \qquad R_{13} = R\left(\frac{I_{\ell_o}}{V_{c_4}}\right) = 39.08\ K\Omega$$

$$R_{14} = \left(\frac{c_5}{C}\right)\left(\frac{V_{c_5}}{I_{\ell_o}}\right) = 16.11\ K\Omega \qquad R_{15} = R\left(\frac{I_{\ell_o}}{V_{c_5}}\right) = 13.31\ K\Omega$$

$$R_{16} = \left(\frac{c_6}{C}\right)\left(\frac{V_{c_6}}{I_{\ell_o}}\right) = 15.83\ K\Omega \qquad R_{17} = R\left(\frac{I_{\ell_o}}{V_{c_6}}\right) = 14.40\ K\Omega$$

$$R_{19} = \left(\frac{c_8}{C}\right)\left(\frac{V_{c_8}}{I_{\ell_o}}\right) = 30.28\ \text{K}\Omega \qquad R_{18} = R\left(\frac{I_{\ell_o}}{V_{c_8}}\right) = 62.24\ \text{K}\Omega$$

which completes the design.

12.5 REFERENCES AND SOURCES FOR FURTHER READING

(1) O. Wing, "Ladder network analysis by signal flow graph -- Application to analog computer programming," IRE Trans. Circuit Theory, Vol. CT-3, pp. 289-294, Dec. 1956.

(2) F. E. J. Girling and E. F. Good, "Active filters 12. The leap-frog or active ladder synthesis," Wireless World, Vol. 76, pp. 341-345, July 1970.

(3) F. E. J. Girling and E. F. Good, "Active filters 13. Applications of the active ladder synthesis," Wireless World, Vol. 76, pp. 445-450, Sept. 1970.

(4) P. O. Brackett and A. S. Sedra, "Application of signal flow graphs to the synthesis of linear systems: Part I -- A general theory of network stimulation," 1975 IEEE Int. Symp. Circuits and Systems, pp. 242-245, Newton, Mass., 1975.

(5) P. O. Brackett and A. S. Sedra, "Application of signal flow graphs to the synthesis of linear systems: Part II -- Design of active ladder filters," 1975 IEEE Int. Symp. Circuits and Systems, pp. 246-249, Newton, Mass., 1975.

(6) P. O. Brackett and A. S. Sedra, "Direct SFG simulation of LC ladder networks with applications to active filter design," IEEE Trans. Circuits and Systems, Vol. CAS-23, pp. 61-67, Feb. 1976.

(7) S. J. Mason, "Feedback Theory -- Some properties of signal flow graphs," Proc. IRE, Vol. 41, pp. 1144-1156, Sept. 1953.

(8) S. J. Mason, "Feedback theory: further properties of signal flow graphs," Proc. IRE, Vol. 44, pp. 920-926, July 1956.

(9) S. J. Mason and H. Zimmermann, Electronic Circuits, Signals and Systems, Wiley, New York, 1960.

(10) L. P. A. Robichaud, N. Boisvert, and J. Roberts, Signal Flow Graphs and Applications, Prentice-Hall, Englewood Cliffs, N. J. 1962.

(11) A. Fettweis, "Some general properties of signal flow networks," in Network and Signal Theory, Skwirzynski and Scanlon, Eds., Peregrinus, London, England, pp. 48-59, 1973.

(12) P. O. Brackett and A. S. Sedra, "Active compensation for high-frequency effects in op. amp. circuits with applications to active-RC filters," IEEE Trans. Circuits and Systems, Vol. CAS-23, pp. 68-73, Feb. 1976.

(13) P. O. Brackett, "The Operational Simulation of Filter Networks" Ph.D. Thesis, University of Toronto, 1976.

(14) K. Martin, "High-Order Bandpass Active Filters", M.A.Sc. Thesis, University of Toronto, Toronto, 1977.

(15) K. Martin and A. S. Sedra, "Designing Leap-Frog and SFG filters with optimum dynamic range," Proc. IEEE, (Lett.), Vol. 65, No. 7, pp. 1210-1211, Aug. 1977.

PROBLEMS

P12.1 Give a leap-frog realization of a seventh-order Chebyshev low-pass filter. The filter has A_p=0.5dB, $\omega_p=10^3$ rad/sec and DC gain of unity. Choose all integrator capacitors equal and design such that the time constants in each loop are equal.
First give a block diagram, then a circuit using the standard two-integrator-loop and finally a circuit using the phase-lead integrator.

P12.2 Redesign the fourth-order Chebyshev low-pass filter of Example 12.1 to obtain maximum dynamic range.

P12.3 Design a leap-frog circuit, optimized for maximum dynamic range, for a sixth-order maximally-flat all-pole bandpass filter with a center-frequency f_o=10KHz, a 1dB bandwidth of 200Hz and a center-frequency gain of 10. Give two designs: one using the Akerberg-Mossberg biquad and the other using the Friend biquad (with additional inverters as needed).
Optional Part: Which biquadratic section is the most critical in influencing the filter response? If ω_o of this

section changes by +2%, plot the resulting filter response and find the maximum deviation ΔA in the passband attenuation. (Note: This same filter was considered in P6.7, P10.5 and P10.6)

P12.4 Consider a fifth-order Chebyshev low-pass filter with A_p= 1dB and f_p=5.333 KHz. The minimum inductance ladder realization has the element values $r_s=r_\ell=1\Omega$, $c_1=c_5=2.135F$, $c_3=3.001F$, $\ell_2=\ell_4=1.091H$. The group-delay peaks at the passband edge to a value of approximately 400 μsec.

(a) Find a leap-frog design using Miller and Miller-inverter integrators. Assuming that the op amps have f_t=1MHz, use the results of Table 12.1 to evaluate the maximum expected deviation in the passband attenuation (ΔA).

(b) Modify the circuit of (a) using the "Modified Miller-Inverter" integrator given in Fig. 8.11 for inductance simulation. Calculate ΔA.

(c) Modify the circuit of (a) using the "phase-lead" integrator of Fig. 8.11 for inductance simulation. Calculate ΔA.

P12.5 Give the SFG for a seventh-order all-pole low-pass filter.

P12.6 Give the SFG simulation of a sixth-order bandpass filter derived from an all-pole low-pass prototype.

P12.7 Derive the transfer functions $(V_o/V_i)(s)$ of the SFG's in Fig. P12.7

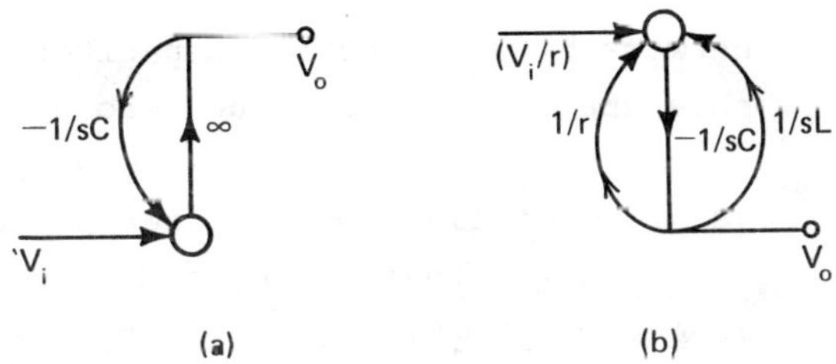

Figure P12.7

P12.8 Use SFG manipulation techniques to convert the SFG obtained in P12.6 to the form shown in Fig. P12.8. Find the transmittances $T_1(s)$, $T_2(s)$ and $T_3(s)$.

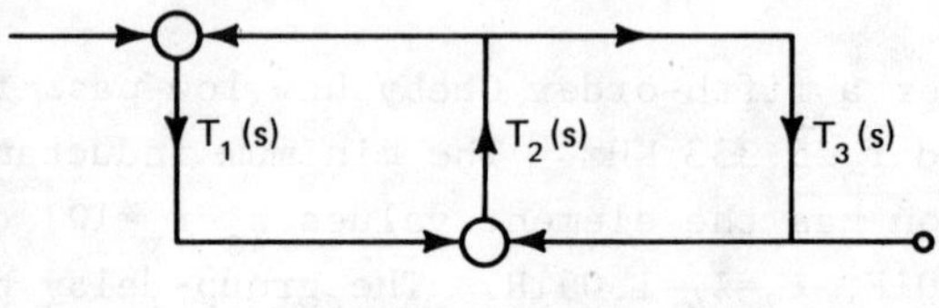

Figure P12.8

P12.9 Find the SFG of a fifth-order elliptic low-pass filter starting from its minimum capacitance LC ladder realization.

P12.10 Find the SFG simulation of the twelfth-order bandpass filter whose LC ladder realization is shown in Fig. P12.10.

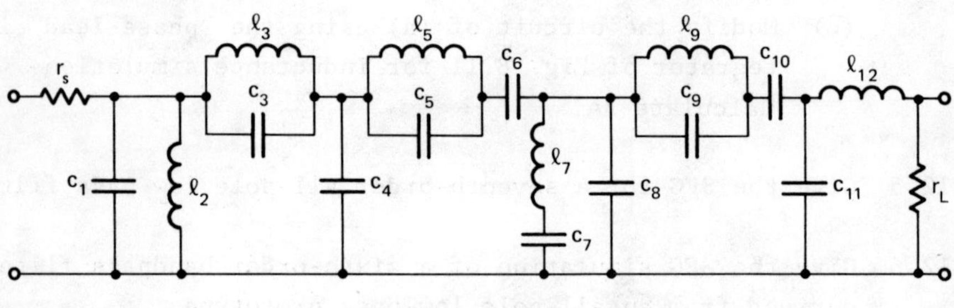

Figure P12.10

P12.11 Consider the reciprocator stability problem illustrated in Fig. 12.27. Denote the components of the stabilizing network by C_a (feedback capacitor), C_b and r. Also denote the frequency of the zero of the modified op amp gain by ω_1 and that of the pole by ω_2. It has been found (experimentally) that the reciprocator will be stable if one arranges that $\omega_2=2\omega_1$ and $\omega_1=0.8\omega_c$, where ω_c is the frequency at which $(s\tau)$ intersects the modified op amp gain function. Show that:

$$C_a = 1/r\ \omega_t$$

$$C_b = \frac{1}{r}\left[\sqrt{\frac{\tau}{0.4\omega_t}} - \frac{1}{\omega_t}\right]$$

P12.12 It is required to obtain a complete SFG design for the fourth-order low-pass filter whose LC ladder realization is shown in Fig. P12.12. Analysis of this ladder results in the current and voltage maxima given by:
V_{c1}=0.6992 V, $I_{\ell 2}$=1.550 A, V_{c2}=1.262 V,
I_t =1.125 A, V_{c3}=0.6900 V, $I_{\ell 4}$=0.8661, A.
Optimize the circuit for maximum dynamic range. Give the reciprocator stabilizing network (use the equations in P12.11 to calculate its values) assuming the op amps to have f_t=1MHz.

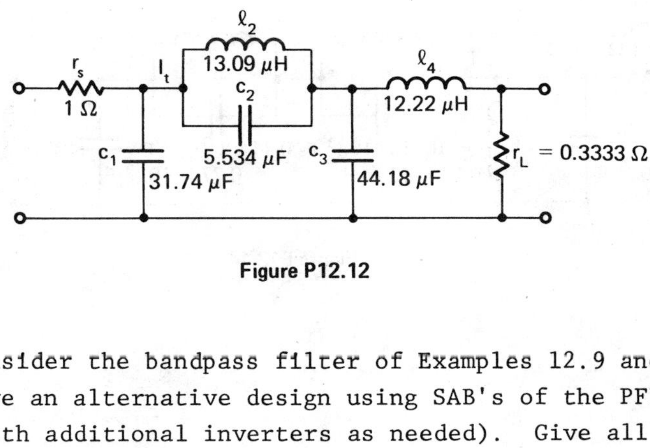

Figure P12.12

P12.13 Consider the bandpass filter of Examples 12.9 and 12.12. Give an alternative design using SAB's of the PF1 type (with additional inverters as needed). Give all component values and optimize the design for maximum dynamic range. [Hint: Use the technique of grounding part of the feed-in component to set each SAB gain to the required value.]

P12.14 Fig. P12.14 shows the impedance normalized LC ladder realization of the sixth-order bandpass filter considered in Example 6.7. Analysis of this network results in the voltage and current maxima:

$$V_{c1}=0.7409\text{ V},\ I_{\ell 1}=7.6671\text{ A},\ I_t=0.6973\text{ A}$$
$$V_{c2}=1.8428\text{ V},\ I_{\ell 2}=2.8064\text{ A},\ V_{c3}=1.8421\text{ V}$$
$$I_{\ell 3}=3.4791\text{ A},\ V_{c4}=0.4999\text{ V},\ I_{\ell 4}=5.0553\text{ A}$$

Give a complete SFG realization for this bandpass filter optimized for maximum dynamic range. Use Miller and phase-lead integrators and include a network for reciprocator stabilization. Assume op amps to have f_t=1MHz. Also, give a realization using SAB's of the ENF type, optimized for maximum dynamic range. (Note: Cascade designs for this filter were considered in P10.7 and P10.8, a PRB design was considered in P10.10 and a component simulation design in P11.10).

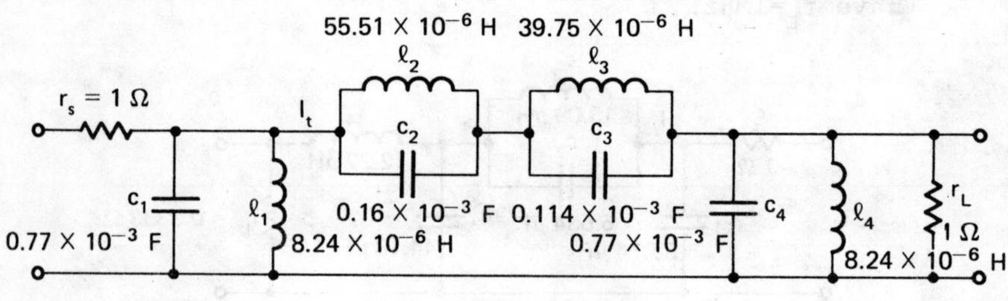

Figure P12.14

INDEX OF IMPORTANT TERMS, ACRONYMS, AND SYMBOLS

INDEX